D0884251

NEUTRON AND X-RAY SPECTROSCOPY

Grenoble Sciences

Grenoble Sciences pursues a triple aim:

- to publish works responding to a clearly defined project, with no curriculum or vogue constraints,
- to guarantee the selected titles' scientific and pedagogical qualities,
- to propose books at an affordable price to the widest scope of readers.

Each project is selected with the help of anonymous referees, followed by a one-year (in average) interaction between the authors and a Readership Committee, whose members' names figure in the front pages of the book. Publication is then confided to the most adequate publishing company by Grenoble Sciences.

(Contact: Tél.: (33) 4 76 51 46 95 – Fax: (33) 4 76 51 45 79
E-mail: Grenoble.Sciences@ujf-grenoble.fr)

Scientific Director of Grenoble Sciences : Jean BORNAREL,
Professor at the Joseph Fourier University, Grenoble 1, France

Grenoble Sciences is supported by the French Ministry of Education and Research and the "Région Rhône-Alpes".

Front cover photo: harmonious composition based on figures extracted from the work:

- The body-centred tetragonal unit cell of the bilayer manganite by CURRAT (after ref. [38])
- Methyl group in lithium acetate dihydrate and in aspirin by JOHNSON (from refs [19] and [49])
- X-PEEM images of permalloy micro-structures and of carbon films by SCHNEIDER (from ref. [11] and taken from ref. [17])
- Interference of outgoing and backscattered photoelectron wave by LENGELER.

Neutron and X-ray Spectroscopy

Edited by

FRANÇOISE HIPPERT
Institut National Physique,
St. Martin d'Hères, France

ERIK GEISSLER
Université J. Fourier de Grenoble,
St. Martin d'Hères, France

JEAN LOUIS HODEAU
Laboratoire de Cristallographie,
CNRS, Grenoble, France

EDDY LELIÈVRE-BERNA
Institut Laue Langevin,
Grenoble, France

and

JEAN-RENÉ REGNARD
Université J. Fourier de Grenoble,
St. Martin d'Hères, France

A C.I.P. Catalogue record for this book is available from the Library of Congress.

ISBN-10 1-4020-3336-2 (HB)
ISBN-13 978-1-4020-3336-0 (HB)
ISBN-10 1-4020-3337-0 (e-book)
ISBN-13 978-1-4020-3337-7 (e-book)

Published by Springer,
P.O. Box 17, 3300 AA Dordrecht, The Netherlands.

www.springer.com

Printed on acid-free paper

Printed in the Netherlands.

TABLE OF CONTENTS

Contributors ... XV

Foreword ... XIX

Introduction ... XXI

X-RAY SPECTROSCOPY

1 - Fundamentals of X-ray absorption and dichroism: the multiplet approach
F. de Groot - J. Vogel ... 3

1. Introduction ... 4
1.1. Interaction of X-rays with matter ... 4
1.2. Basics of XAFS spectroscopy ... 6
1.3. Experimental aspects ... 9

2. Multiplet effects ... 11
2.1. Atomic multiplets ... 14
2.1.1. Term symbols ... 14
2.1.2. Matrix elements ... 18
2.2. Atomic multiplet ground states of $3d^n$ systems ... 20
2.3. j - j coupling ... 20
2.4. X-ray absorption spectra described with atomic multiplets ... 21
2.4.1. Transition metal L_{II-III} edge ... 21
2.4.2. M_{IV-V} edges of rare earths ... 28

3. Crystal field theory ... 31
3.1. The crystal field multiplet Hamiltonian ... 32
3.2. Cubic crystal fields ... 33
3.3. The energies of the $3d^n$ configurations ... 36
3.3.1. Symmetry effects in D_{4h} symmetry ... 40
3.3.2. The effect of the 3d spin-orbit coupling ... 40
3.3.3. The effects on the X-ray absorption calculations ... 42
3.3.4. $3d^0$ systems in octahedral symmetry ... 42
3.3.5. $3d^0$ systems in lower symmetries ... 45
3.3.6. X-ray absorption spectra of $3d^n$ systems ... 46

4. Charge transfer effects ... 47
4.1. Initial state effects ... 47
4.2. Final state effects ... 52
4.3. The X-ray absorption spectrum with charge transfer effects ... 53

5. X-ray linear and circular dichroism ... 59

References ... 64

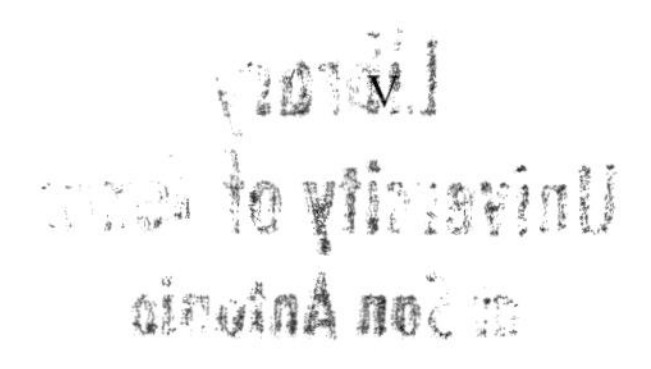

2 - Multiple scattering theory applied to X-ray absorption near-edge structure
P. Sainctavit - D. Cabaret - V. Briois 67

1. **Introduction** 67
2. **MST for real space calculations** 70
   2.1. The absorption cross-section 71
   2.2. Calculation of the cross-section for three different potentials 73
   2.2.1. Free particle 73
   2.2.2. Particle in a spherical potential 75
   2.2.3. Particle in a "muffin-tin" potential 78
3. **Construction of the potential** 83
   3.1. $X-\alpha$ exchange potential 85
   3.2. Dirac-Hara exchange potential 85
   3.3. Complex Hedin-Lundqvist exchange and correlation potential 87
4. **Application of the multiple scattering theory** 89
   4.1. Vanadium K edge cross-section for VO_6 cluster 90
   4.2. Iron K edge cross-section for FeO_6 cluster 93
   4.3. Cross-section calculations in distorted octahedra 94
5. **Spin transition in Fe^{II}(o-phenantroline)$_2$ (NCS)$_2$ followed at the iron *K* edge** 95
6. **Conclusion** 97

Appendix - Expression for the propagators $J_{LL'}^{ij}$ and $H_{LL'}^{ij}$ 97

Acknowledgments 99

References 100

3 - X-ray magnetic circular dichroism
F. Baudelet 103

1. **Introduction** 103
2. **Origins of magnetic circular dichroism** 104
   2.1. Theoretical aspects: role of the spin-orbit coupling 104
   2.2. Light polarization and polarization-dependent selection rules 106
   2.3. Origin of the XMCD signal 108
   2.3.1. Heuristic model: L_{II} - L_{III} edges of transition metals and rare earth elements 108
   2.3.2. Calculation of P_e for the L_{II} - L_{III} edges: $P_e(L_{II})$ and $P_e(L_{III})$ 109
   2.3.3. Calculation of P_e for the K - L_I edges: $P_e(K)$ and $P_e(L_I)$ 112
   2.4. Magnitude of the XMCD signal 113
3. **The sum rules** 114
4. **XMCD signal at the *K* edges (1*s* $\longrightarrow$ *p*)** 117
5. **XMCD and localized magnetism (multiplet approach)** 118
6. **$L_{\text{II-III}}$ edges of rare earths** 121
7. **Magnetic EXAFS** 125
8. **Application to multilayers with GMR** 126
9. **Study of highly correlated [Ce/La/Fe] and [La/Ce/Fe] multilayers** 127

References 128

4 - Extended X-ray absorption fine structure
B. Lengeler ... 131

1. Introduction ... 131

2. X-ray absorption in isolated atoms ... 133

3. X-ray absorption fine structure (XAFS) ... 136
3.1. X-ray absorption fine structure: the basic formula ... 136
3.2. Corrections to the basic EXAFS formula ... 141
3.2.1. Spherical photoelectron wave ... 142
3.2.2. Multiple scattering ... 142
3.3. Set-up for measuring X-ray absorption ... 143
3.4. XAFS data analysis ... 145
3.5. Position and structure of the absorption edge ... 149

4. Some applications of XAS ... 153
4.1. Lattice distortion around impurities in dilute alloys ... 154
4.2. Precipitation in immiscible giant magnetoresistance Ag_{1-x} Ni_x alloys ... 155
4.3. Lattice site location of very light elements in metals by XAFS ... 158
4.4. Valence of iridium in anodically oxidized iridium films ... 160
4.5. Copper-based methanol synthesis catalyst ... 163
4.6. Combined X-ray absorption and X-ray crystallography ... 165

References ... 167

5 - Inelastic X-ray scattering from collective atom dynamics
F. Sette - M. Krisch ... 169

1. Introduction ... 169

2. Scattering kinematics and inelastic X-ray scattering cross-section ... 170

3. Experimental apparatus ... 175

4. The "fast sound" phenomenon in liquid water ... 179

5. Determination of the longitudinal sound velocity in iron to 110 GPa ... 182

6. Conclusions and outlook ... 186

References ... 187

6 - Photoelectron spectroscopy
M. Grioni ... 189

1. Introduction ... 189

2. What is photoemission? ... 190

3. Photoemission in the single-particle limit - Band mapping ... 193

4. Beyond the single-particle approximation ... 201

5. Case studies ... 209
5.1. Fermi liquid lineshape in a normal metal ... 209
5.2. Gap spectroscopy ... 215
5.3. Electronic instabilities in low dimensions ... 220
5.4. Some recent developments ... 229

6. Conclusions ... 235

References ... 236

7 - Anomalous scattering and diffraction anomalous fine structure
J.L. Hodeau - H. Renevier 239

1. Anomalous scattering, absorption and refraction 240

2. Theoretical versus experimental determination of anomalous contribution 241

3. Applications of anomalous dispersion 244
3.1. Structure factor phase solution (MAD method) 245
3.2. Element-selective diffraction (contrast method) 247

4. Diffraction anomalous fine structure data analysis 249
4.1. DAFS and EDAFS formalism 251
4.1.1. Single anomalous site analysis *252*
4.1.2. Multiple anomalous site analysis *253*
4.2. DAFS and EDAFS determination 255
4.3. DAFS and DANES valence determination 258
4.4. Anisotropy of anomalous scattering 260

5. Requirements for anomalous diffraction experiments 262

6. Conclusion 263

References 264

8 - Soft X-ray photoelectron emission microscopy (X-PEEM)
C.M. Schneider 271

1. A "nanoscale" introduction 271

2. Visualizing micro- and nanostructures 271

3. Technical aspects of an Electron Emission Microscope (EEM) 273
3.1. Electron-optical considerations 273
3.2. Transmission and lateral resolution 275

4. Non-magnetic image contrast in X-PEEM 276
4.1. Primary contrast mechanisms 276
4.1.1. Work function contrast 276
4.1.2. Chemical contrast 277
4.2. Secondary contrast mechanisms 280

5. Magnetic contrast in X-PEEM 281
5.1. Magnetic X-ray Circular Dichroism (XMCD) 282
5.1.1. Physics of the magnetic contrast mechanism 282
5.1.2. Contrast enhancement 284
5.1.3. Angular dependence of the image contrast 285
5.1.4. Magnetic domain walls 286
5.1.5. Information depth 288
5.2. X-ray Magnetic Linear Dichroism (XMLD) 290
5.2.1. Properties of the contrast mechanism 290
5.2.2. Imaging domains in antiferromagnets 291

6. Concluding remarks 292

References 293

9 - X-ray intensity fluctuation spectroscopy
M. Sutton 297
1. Introduction....297
2. Mutual coherence functions....301
3. Diffraction by partially coherent sources....304
4. Kinetics of materials....308
5. X-ray intensity fluctuation spectroscopy....310
6. Conclusions....316
References....317

10 - Vibrational spectroscopy at surfaces and interfaces using synchrotron sources and free electron lasers
A. Tadjeddine - P. Dumas 319
1. Introduction....319
2. Infrared synchrotron source and free electron lasers 321
2.1. Infrared synchrotron sources....321
2.1.1. Principles....321
2.1.2. Extraction of the IR beam from the synchrotron ring....324
2.1.3. Fourier transform interferometer....324
2.2. Non linear surface spectroscopy with conventional and free electron lasers....325
2.2.1. Principle of SFG....326
2.2.2. Experimental SFG set-up....328
2.2.3. The CLIO-FEL infrared laser....328
3. Examples of applications in Surface Science 331
3.1. Synchrotron infrared spectroscopy at surfaces....331
3.1.1. Low frequency modes of adsorbed molecules and atoms 333
3.1.2. Vibrational dynamics of low frequency modes 335
3.2. Sum Frequency and Difference Frequency Generation at interfaces....339
3.2.1. Identification of adsorbed intermediate of electrochemical reactions by SFG 340
3.2.2. Vibrational spectroscopy of cyanide at metal-electrolyte interface....342
3.2.3. Vibrational spectroscopy of self-assembled monolayers on metal substrate....346
3.2.4. Vibrational spectroscopy of fullerenes C_{60}, adsorbed on Ag(111), in UHV environment 348
3.2.5. Adsorption of 4-cyanopyridine on Au(111) monitored by SFG 352
4. Conclusion and outlook 356
References....356

NEUTRON SPECTROSCOPY

11 - Inelastic neutron scattering: introduction
R. Scherm - B. Fåk ... 361

1. Interaction of neutrons with matter ... 361

2. Kinematics ... 362
2.1. Energy and momentum conservation ... 362
2.2. Scattering triangle ... 363
2.3. Parabolas ... 364

3. Master equation and $S(Q, \omega)$ ... 364

4. Correlation function ... 366

5. Coherent and incoherent scattering ... 367

6. General properties of $S(Q, \omega)$ ... 369
6.1. Detailed balance ... 369
6.2. Moments ... 370
6.3. Total versus elastic scattering ... 371

7. Magnetic scattering ... 372

8. Response from simple systems ... 373
8.1. Examples ... 373
8.2. Response functions ... 375

9. Instrumentation ... 376
9.1. TAS ... 378
9.2. TOF ... 379

10. How to beat statistics ... 379

References ... 380

12 - Three-axis inelastic neutron scattering
R. Currat ... 383

1. Principle of the technique ... 383

2. The three-axis spectrometer ... 386

3. The TOF versus TAS choice ... 391

4. What determines the TAS count rate? ... 394

5. What determines the size and shape of the resolution function? ... 396

6. Decoupling energy and momentum resolutions: direct space focusing ... 400

7. TAS multiplexing ... 403

8. Phonon studies with TAS ... 408

9. The INS versus IXS choice ... 412

10. Magnetic excitation studies with TAS ... 415

11. Practical aspects ... 420

12. Summary and outlook ... 422

References ... 423

13 - Neutron spin echo spectroscopy
R. Cywinski 427
1. Introduction 427
2. Polarized neutron beams, Larmor precession and spin flippers 428
3. Generalized neutron spin echo 432
4. Polarization dependent scattering processes 436
5. Practicalities: measurement of the spin echo signal 440
6. Applications of neutron spin echo spectroscopy 444
6.1. Soft condensed matter 444
6.2. Glassy dynamics 447
6.3. Spin relaxation in magnetic systems 449
7. Conclusions 453
Bibliography 453
References 454

14 - Time-of-flight inelastic scattering
R. Eccleston 457
1. Introduction 457
2. Classes of TOF spectrometers 458
2.1. Distance-time plots 458
2.2. Kinematic range 460
3. Beamline components 461
3.1. Choppers 461
3.1.1. Fermi choppers 462
3.1.2. Disk choppers 462
3.1.4. T = 0 choppers 462
3.2. Monochromating and analyzing crystals 463
3.3. Filters 463
3.4. Detectors 464
3.5. Neutron guides 464
3.6. Polarizers and polarization analysis 465
4. Direct geometry spectrometers 465
4.1. Chopper spectrometer on a pulsed source 465
4.2. Chopper spectrometers on a steady state source 466
4.3. Multi-chopper TOF spectrometers 467
5. Resolution and spectrometer optimization 469
6. Flux 470
7. Resolution as a function of energy transfer and experimental considerations 471
Single crystal experiments on a chopper spectrometer 472
8. Indirect geometry spectrometers 474
8.1. The resolution of indirect geometry spectrometers 475
8.2. Backscattering spectrometer 475
8.3. Crystal analyzer spectrometers 477

8.4. Deep inelastic neutron scattering 478
8.5. Coherent excitations 479

9. Conclusions 480

Further information 481

References 481

15 - Neutron backscattering spectroscopy
B. Frick 483

1. Introduction 483

2. Reflection from perfect crystals and its energy resolution 485

3. Generic backscattering spectrometer concepts 488
3.1. Neutron optics of the primary spectrometers of reactor-BS instruments 489
3.2. Neutron optics of the primary spectrometer of spallation source-BS instruments ... 493
3.3. Secondary spectrometer 493

4. Total energy resolution of the spectrometers 494

5. How to do spectroscopy? 494
5.1. Spectroscopy on reactor based instruments 494
5.2. Spectroscopy on spallation source-BS instruments 497

6. More details on optical components 498
6.1. BS monochromators and analyzers 498
6.2. How to obtain the best energy resolution in backscattering? 499
6.3. Mosaic crystal deflectors and phase space transformer 500
6.4. Neutron guides 503
6.5. Higher order suppression 503
6.6. Q-resolution 504
6.7. A second time through the sample? 504

7. Examples for backscattering instruments 505
7.1. Reactor instruments 505
7.2. Spallation source instruments 509

8. Data treatment 511

9. Typical measuring methods and examples 513
9.1. Fixed window scans 513
9.2. Spectroscopy 516

References 525

16 - Neutron inelastic scattering and molecular modelling
M.R. Johnson - G.J. Kearley - H.P. Trommsdorff 529

1. Introduction 529

2. Theory framework for tunnelling, vibrations and total energy calculations 532
2.1. Hamiltonians for quantum tunnelling 532
2.2. The dynamical matrix for molecular vibrations 536
2.3. Total energy calculations for determining PES and force constants 538

3. Experimental techniques for measuring rotational tunnelling and molecular vibrations 540

4. Numerical simulations for understanding INS spectra 544
4.1. SPM methyl group tunnelling 544
4.2. Multi-dimensional tunnelling dynamics of methyl groups 546
4.3. Vibrational spectroscopy of molecular crystals 547
5. Discussion 551
References 553
Index 557

Contributors

F. Baudelet
Synchrotron SOLEIL
L'Orme des Merisiers, Saint-Aubin - BP 48, 91192 Gif-sur-Yvette Cedex, France
francois.baudelet@synchrotron-soleil.fr

V. Briois
Synchrotron SOLEIL
L'Orme des Merisiers, Saint-Aubin - BP 48, 91192 Gif-sur-Yvette Cedex, France
valerie.briois@synchrotron-soleil.fr

D. Cabaret
Institut de Minéralogie et de Physique des Milieux Condensés
Université P. et M. Curie, Case 115, 4 place Jussieu, 75252 Paris Cedex 05, France
cabaret@lmcp.jussieu.fr

R. Currat
Institut Laue-Langevin
6 Rue J. Horowitz, BP 156, 38042 Grenoble Cedex 9, France
currat@ill.fr

R Cywinski
School of Physics and Astronomy
University of Leeds, Leeds LS2 9JT, UK
phy6rc@phys-irc.novell.leeds.ac.uk

P. Dumas
Synchrotron SOLEIL
L'Orme des Merisiers, Saint-Aubin - BP 48, 91192 Gif-sur-Yvette Cedex, France
paul.dumas@synchrotron-soleil.fr

R. Eccleston
Sheffield Hallam University, Materials Engineering Research Institute
Howard Street, Sheffield, S1 1WB, UK
r.s.eccleston@shu.ac.uk

B. FÅK
Département de la Recherche Fondamentale sur la Matière Condensée, SPSMS
CEA Grenoble,17 rue des Martyrs, 38054 Grenoble Cedex 9, France
fak@cea.fr

B. FRICK
Institut Laue-Langevin
6 Rue J. Horowitz, BP 156, 38042 Grenoble Cedex 9, France
frick@ill.fr

M. GRIONI
Institut de Physique des Nanostructures
Ecole Polytechnique Fédérale de Lausanne, 1015 Lausanne, Switzerland
marco.grioni@epfl.ch

F. DE GROOT
Department of Inorganic Chemistry and Catalysis
Utrecht University, Sorbonnelaan 16, 3584 CA Utrecht, Netherlands
f.m.f.degroot@chem.uu.nl

J.L. HODEAU
Laboratoire de Cristallographie
CNRS, 25 Avenue des Martyrs, BP 166, 38042 Grenoble Cedex 9, France
hodeau@grenoble.cnrs.fr

M.R. JOHNSON
Institut Laue-Langevin
6 Rue J. Horowitz, BP 156, 38042 Grenoble Cedex 9, France
johnson@ill.fr

G.J. KEARLEY
TU Delft, Interfacultair Reactor Institute
Mekelweg, 15, 2629 JB Delft, Netherlands
kearley@iri.tudelft.nl

M. KRISCH
European Synchrotron Radiation Facility
6 rue Horowitz, BP 220, 38043 Grenoble Cedex 9, France
krisch@esrf.fr

B. LENGELER
RWTH Aachen, II. Physikalisches Institut B
Huyskensweg Turm 28, 52074 Aachen, Germany
lengeler@physik.rwth-aachen.de

H. RENEVIER
Département de la Recherche Fondamentale sur la Matière Condensée, SP2M/NRS
CEA Grenoble, 17 Avenue des Martyrs, 38054, Grenoble Cedex 9, France
Hubert.Renevier@cea.fr

P. SAINCTAVIT
Institut de Minéralogie et de Physique des Milieux Condensés
Université P. et M. Curie, Case 115, 4 place Jussieu, 75252 Paris Cedex 05, France
Philippe.Sainctavit@lmcp.jussieu.fr

R. SCHERM
Physikalisch-Technische Bundesanstalt
Bundesallee 100, 38116 Braunschweig, Germany
Reinhard.Scherm@ptb.de

C.M. SCHNEIDER
Institute of Electronic Properties, Department of Solid State Research (IFF)
Research Center Jülich, 52425 Jülich, Germany
c.m.schneider@fz-juelich.de

F. SETTE
European Synchrotron Radiation Facility
6 rue Horowitz, BP 220, 38043 Grenoble Cedex 9, France
sette@esrf.fr

M. SUTTON
Center for the Physics of Materials
Mac Gill University, Montreal, Québec, Canada, H3A 2T8
mark@physics.mcgill.ca

A. TADJEDDINE
LURE
Université Paris-Sud, Bât. 209D, BP 34, 91898 Orsay Cedex, France
Abderrahmane.Tadjeddine@lure.u-psud.fr

H.P. Trommsdorff
Laboratoire de Spectrométrie Physique
Domaine universitaire, BP 87, 38402 St-Martin-d'Hères, France
hans-peter.trommsdorff@ujf-grenoble.fr

J. Vogel
Laboratoire Louis Néel
CNRS, 25 Avenue des Martyrs, BP 166, 38042 Grenoble, France
vogel@grenoble.cnrs.fr

FOREWORD

This volume is devoted to the use of synchrotron radiation and neutron sources. It is based on a course delivered to young scientists in the HERCULES programme, organised by the Université Joseph Fourier and the Institut National Polytechnique de Grenoble. The success of such a course has been well demonstrated as more than 1000 researchers (about 900 from the European Union) have attended the sessions.

The HERCULES programme is supported by the European Commission, the Ministère de l'éducation nationale, de l'enseignement supérieur et de la recherche, the Centre National de la Recherche Scientifique (CNRS), the Commissariat à l'énergie atomique (CEA), the partner Large Infrastructures (ESRF, ILL, ELETTRA, LLB), the Synchrotron Radiation and Neutron European Integrated Infrastructure Initiatives (I3 and NMI3) and the Région Rhône-Alpes.

F. HIPPERT - E. GEISSLER - J.L. HODEAU - E. LELIÈVRE-BERNA - J.R. REGNARD

This volume is the fifth in a series. Previous volumes are:

Neutrons and Synchrotron Radiation for Condensed Matter Studies: "Theory, instruments and methods", HERCULES Series,
Editors: J. BARUCHEL, J.L. HODEAU, M.S. LEHMANN, J.R. REGNARD,
C. SCHLENKER, Les Editions de Physique - Springer-Verlag, 1993.

Neutrons and Synchrotron Radiation for Condensed Matter Studies: "Applications to solid state physics and chemistry", HERCULES Series,
Editors: J. BARUCHEL, J.L. HODEAU, M.S. LEHMANN, J.R. REGNARD,
C. SCHLENKER, Les Editions de Physique - Springer-Verlag, 1994.

Neutrons and Synchrotron Radiation for Condensed Matter Studies: "Applications to soft condensed matter and biology", HERCULES Series,
Editors: J. BARUCHEL, J.L. HODEAU, M.S. LEHMANN, J.R. REGNARD,
C. SCHLENKER, Les Editions de Physique - Springer-Verlag, 1994.

Neutrons and Synchrotron Radiation for Condensed Matter Studies: "Structure and dynamics of biomolecules", HERCULES Series,
Editors: E. FANCHON, E. GEISSLER, J.L. HODEAU, J.R. REGNARD, P. TIMMINS,
Oxford University Press, 2000.

INTRODUCTION

Large research infrastructures such as synchrotrons for X-ray production and neutron sources have no doubt come to play an indispensable role in modern experimental research into the structure and dynamics of condensed matter and materials. We understand the terms condensed matter and materials to embrace all forms of matter that we encounter in our daily life or that can be produced in the laboratories on earth, including living matter and biological macromolecules. Synchrotron radiation offers intense and collimated photon beams with high peak brightness, short pulse time structure and easy wavelength tunability. Neutron methods provide exquisite sensitivity to magnetic moments, isotope labelling and collective fluctuations. These characteristics are exploited in fundamental research, such as the investigation of strongly correlated electrons or the determination of the structure of proteins; for more applied purposes such as the study of magnetic devices, the analysis of human tissues or the development of pharmaceutical products; and even for very down-to-earth industrial problems concerning e.g., construction or packaging materials.

In spite of their large construction and operating costs (often beyond the resources of a single European country), and of the inconvenience and stress incurred by travelling to distant facilities and the need to complete the experiment in a few frantic days (and nights!), large research infrastructures are ever more popular. This is witnessed on the one hand by the coming on line of new European synchrotron sources and the opening of new beam lines at existing facilities, and on the other, by the start of the new German neutron source in Munich, together with the renewal of instrumentation at European neutron facilities. In order to exploit fully the exciting possibilities offered by synchrotrons and neutron sources to researchers from universities, academic and industrial laboratories, such an expansion must be accompanied by advanced training courses in the continuously evolving experimental methods associated with these facilities. The HERCULES programme, which started in 1989, has played a pivotal role by offering a series of lectures and practical "hands on" sessions on the use of photons and neutrons in science. The first volumes were dedicated to theoretical aspects and applications to particular fields such as condensed matter, biology and chemistry.

The current volume presents recent developments in spectroscopic methods involving neutrons or photons. Experimental developments often arise from concurrent progress in instrumentation and theoretical tools, and methods using photons or neutrons are no exception.

In modern X-ray research, the so called "third generation" synchrotron sources provide undulator beams with unprecedented brilliance, tunable polarization properties and a substantial degree of spatial coherence. The first four chapters of this book, by F. DE GROOT and J. VOGEL, by P. SAINCTAVIT, V. BRIOIS and D. CABARET, by F. BAUDELET and by B. LENGELER, probe the breadth and depth of new possibilities that are offered to absorption-based spectroscopy by these features of modern sources. They also introduce the reader to the theoretical methods needed to understand the meaning and extract the maximum amount of information from the experiments. In particular, the use of polarization-dependent absorption– i.e., dichroism – in the study of magnetic materials, and the analysis of the spectra by sum rules and electronic structure calculations is emphasized.

The three following chapters, by F. SETTE and M. KRISCH, by M. GRIONI and by J.L. HODEAU and H. RENEVIER, discuss the new life of time-honoured techniques that have been undergoing a revolution because of the progress in sources and instrumentation. Inelastic scattering has proven invaluable in the study of the collective excitations of some classes of disordered systems, such as liquids and glasses, thanks to the high brilliance of the sources and the huge resolving power of the spectrometers. Photoemission has evolved into the primary tool for experimental investigations of electronic quasi-particles, band dispersion and Fermi surfaces in solids; and finally, anomalous scattering has been developing into an indispensable tool for the crystallographer, in particular for the characterization of the local environment of a specific atom.

The chapters by C.M. SCHNEIDER and by M. SUTTON illustrate the power of modern microscopic and coherent scattering techniques, respectively. Here the small source size of third generation sources is very important, as it allows a larger number of photons to be brought into a small area (typically micrometric or sub-micrometric). This is promoting the development of new techniques (PEEM microscopy) or the extension of visible laser techniques, such as intensity fluctuation spectroscopy, to the X-ray range.

The chapter by A. TADJEDDINE and P. DUMAS that follows reminds us of the incredible versatility of synchrotrons: here they are shown to be superior sources not just of VUV or X-ray radiation, but also of infrared photons. In the IR energy range the synchrotron brilliance is however surpassed by another accelerator-based source, the Free Electron Laser. Worldwide efforts to extend the spectral range of free electron lasers to ultraviolet and x-ray radiation are ushering in the next revolution, fourth generation light sources.

A feature common to neutron and X-ray experiments is the increasing degree of complexity of data analysis. It has become obvious that a proper understanding of complex material systems, their static or dynamical properties, require modelling using ab initio or empirical methods. The beauty and the simplicity of neutron scattering cross-sections allow direct comparison between experimental data and computer modelling as demonstrated by M.R. JOHNSON, G.J. KEARLEY and H.P. TROMMSDORFF. Even if simulations do not replace experiments, these approaches will develop more widely: modelling gives a microscopic view and can be extended beyond the scope of the experimental measurements.

Neutrons can tell where atoms or magnetic moments are, but more importantly, they can tell what they do. The dynamics of the constituent species of materials determine their characteristics and functions: neutron spectroscopy is a key tool to understand the properties of materials. Both reactor-based facilities and neutron spallation sources have developed new generations of neutron spectroscopy instruments that cover more efficiently large sections in energy transfer (or time windows) over large sections of reciprocal space (or spatial resolution). It is therefore timely to re-address the question of applications of inelastic neutron scattering in general. The review of inelastic processes by R. SCHERM and B. FÅK exposes all the beauty of inelastic neutron scattering and explains the simple connection between neutron data and theoretical quantities. The interplay between neutron spectroscopy measurements and theory is driving the development of many new concepts in science, especially solid state physics. Four different experimental methods are discussed that are related to the two categories of neutron production. Three-axis spectrometers (TAS) presented by R. CURRAT are perfectly suited to reactor-based installations with their high time average neutron fluxes; new developments to enhance data acquisition rates and performance are presented. Neutron TAS methods cover a large area in time and space but complementary techniques (X-ray inelastic scattering and other neutron methods) must be considered, depending on the field of applications.

Neutron time-of-flight techniques (TOF) are used both at steady and at pulsed neutron sources. R. ECCLESTON and B. FRICK introduce the different geometries for spectrometers: space and time domains of application for each class of instruments are presented together with discussion on complementary use and possible overlap. Backscattering spectrometers at reactor and spallation sources are exploited to investigate energy transfers down to the 100 neV range, corresponding to processes with characteristic times shorter than a nanosecond.

Neutron spin echo methods occupy a very particular position because they give access to real time information in the nanosecond to microsecond range. As pointed out by R. CYWINSKI, they very nicely complement TAS and TOF neutron methods, offering gains in energy resolution by taking advantage of Larmor precession of spin-polarized neutrons. A vast domain of effects with slow dynamics up to the microsecond can be studied by these methods.

This collection of chapters represents a useful support for all scientists interested in the study of condensed matter physics and chemistry. Spectroscopic methods using X-rays and neutrons are in fact exploited by a growing scientific community.

Massimo ALTARELLI
Sincrotrone Trieste, Italy

Christian VETTIER
Institut Laue-Langevin, Grenoble, France

X-RAY SPECTROSCOPY

1

FUNDAMENTALS OF X-RAY ABSORPTION AND DICHROISM: THE MULTIPLET APPROACH

F. DE GROOT
Department of Inorganic Chemistry and Catalysis, Utrecht University, Netherlands

J. VOGEL
Laboratoire Louis Néel, CNRS, Grenoble, France

X-ray Absorption Spectroscopy (XAS) using synchrotron radiation is a well-established technique providing information on the electronic, structural and magnetic properties of matter. In X-ray absorption, a photon is absorbed by the atom, giving rise to the transition of an electron from a core state to an empty state above the Fermi level. The absorption cross-section depends on the energy and on the measured element. To excite an electron in a given core level, the photon energy has to be equal or higher than the energy of this core level. This gives rise to the opening of a new absorption channel when the photon energy is scanned from below to above this core-level energy. The energies of the absorption edges therefore correspond to the core-level energies, which are characteristic for each element, making X-ray absorption an element-selective technique.

In general, two regions can be discerned in X-ray absorption spectra: the near-edge and EXAFS (Extended X-ray Absorption Fine Structure) regions. The spectral shape in the near-edge region is determined by electron correlation and density of states (or multiple scattering) effects and gives mainly information about the electronic properties of the absorbing atom. The EXAFS region is dominated by single scattering events of the outgoing electron on the neighbouring atoms, giving information about the local atomic structure around the absorbing site.

Several books and reviews have been published about XANES (X-ray Absorption Near-Edge Spectroscopy) and EXAFS spectroscopy. In this chapter we will treat the theoretical basis of XANES, mainly for systems where electron correlations play an important role (localized electron systems). Multiple scattering effects are treated in the chapter of P. Sainctavit, V. Briois and D. Cabaret, while EXAFS spectroscopy has been described by B. Lengeler.

F. Hippert et al. (eds.), Neutron and X-ray Spectroscopy, 3–66.

This chapter will start with an introduction to the theoretical basis of X-ray absorption spectra. The main absorption edges (1*s*, 2*p* and 3*p* edges of transition metals, 3*d* edges of rare earths) will then be treated in more detail. Emphasis will be given to transitions to relatively localized states, where multiplet effects play an important role.

We will also briefly discuss polarization dependent XAS or X-ray dichroism. Dichroism with circularly polarized X-rays (X-ray Magnetic Circular Dichroism or XMCD) has become a powerful tool for the study of magnetic materials, capable to provide element specific information about spin and orbital magnetic moments.

1. INTRODUCTION

X-ray absorption spectroscopy has become an important tool for the characterization of materials as well as for fundamental studies of atoms, molecules, adsorbates, surfaces, liquids and solids. The particular assets of XAS spectroscopy are its element specificity and the possibility to obtain detailed information without the presence of any long range ordering. Below it will be shown that the X-ray absorption spectrum in many cases is closely related to the empty density of states of a system. As such XAS is able to provide a detailed picture of the local electronic structure of the element studied.

1.1. INTERACTION OF X-RAYS WITH MATTER

In XAS the absorption of X-rays by a sample, as described by Lambert-Beer's law, is measured. The intensity of the transmitted beam at a certain energy $I(\omega)$ is related to the intensity of the beam before the sample $I_0(\omega)$ divided by an exponential containing the length of the sample x times the absorption cross-section $\mu(\omega)$

$$I(\omega) = I_0(\omega)\, e^{-\mu(\omega)x} \qquad \{1\}$$

$$\mu(\omega) = \frac{1}{x}\ln\frac{I_0}{I} \qquad \{2\}$$

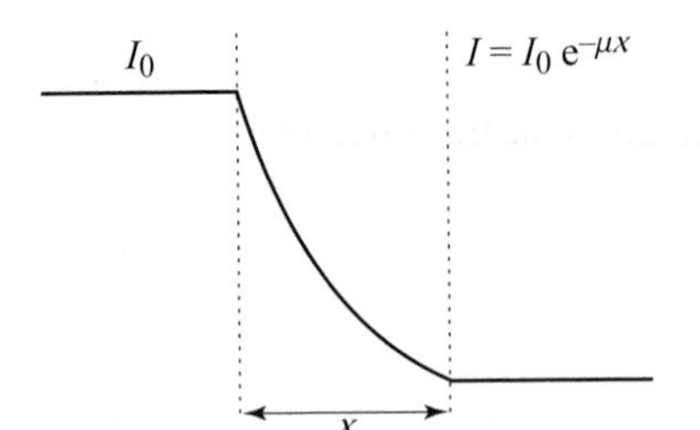

Figure 1 - Transmission through a uniform sample.

How is the absorption cross-section μ determined? As an X-ray passes an electron, its electric field causes oscillations in both direction and strength. In most descriptions of X-ray absorption one uses the vector field $\boldsymbol{A}$ to describe the electromagnetic wave. This vector field is given in the form of a plane wave of electromagnetic radiation

$$\boldsymbol{A} = \hat{\boldsymbol{e}}_q A_0 \cos(kx - \omega t) = \frac{1}{2}\hat{\boldsymbol{e}}_q A_0 \mathrm{e}^{i(kx-\omega t)} \qquad \{3\}$$

The cosine plane wave function contains the wave vector k times the displacement x and the frequency ω times the time t. The cosine function is rewritten to an exponential function $[\cos(kx - \omega t) = \frac{1}{2}(\mathrm{e}^{i(kx-\omega t)} + \mathrm{e}^{-i(kx-\omega t)})]$, for which only the absorbing $e^{-i\omega t}$ term has been retained. The $e^{i\omega t}$ term would induce emissions for atoms where a core-hole is present. $\hat{\boldsymbol{e}}_q$ is a unit vector for a polarization q. The Hamiltonian describing the interaction of X-rays with electrons can be approximated in perturbation theory and its first term can be written as:

$$H_1 = \frac{e}{mc}\boldsymbol{p}\cdot\boldsymbol{A} \qquad \{4\}$$

The interaction Hamiltonian H_1 describes the action of the vector field $\boldsymbol{A}$ on the momentum operator $\boldsymbol{p}$ of an electron. The electric field $\boldsymbol{E}$ is collinear with the vector field, and this term can be understood as the action of an electric field $\boldsymbol{E}$ on the electron moments. The proportionality factor contains the electron charge e, its mass m and the speed of light c.

The Golden Rule states that the transition probability W between a system in its initial state Φ_i and final state Φ_f is given by

$$W_{fi} = \frac{2\pi}{\hbar}|\langle\Phi_f|T|\Phi_i\rangle|^2 \delta_{E_f - E_i - \hbar\omega} \qquad \{5\}$$

The initial and final state wave functions are built from an electron part and a photon part. The photon part of the wave function takes care of the annihilation of a photon in the X-ray absorption process. In the text below, it will not be included explicitly. The delta function takes care of the energy conservation and a transition takes place if the energy of the final state equals the energy of the initial state plus the X-ray energy. The squared matrix element gives the transition rate.

The transition operator T contains all possible transitions and can be separated in parts ($T = T_1 + T_2 + \ldots$), where the transition operator T_1 describes one-photon transitions such as X-ray absorption. It is in first order equal to the first term of the interaction Hamiltonian, i.e. $T_1 = H_1$. The transition rate W is found by calculating the matrix elements of the transition operator T_1. Two-photon phenomena, for example resonant X-ray scattering, are described with the transition operator in second order (T_2). The interaction Hamiltonian is found by inserting the vector field into H_1

$$T_1 \propto \sum_q (\hat{e}_q \cdot \boldsymbol{p}) e^{i\boldsymbol{k}\cdot\boldsymbol{r}} \qquad \{6\}$$

This equation can be rewritten using a Taylor expansion of $e^{i\boldsymbol{k}\cdot\boldsymbol{r}} = 1 + i\boldsymbol{k}\cdot\boldsymbol{r} + \ldots$ Limiting the equation to the first two terms $1 + i\boldsymbol{k}\cdot\boldsymbol{r}$, the transition operator is

$$T_1 \propto \sum_q \left[(\hat{e}_q \cdot \boldsymbol{p}) + i(\hat{e}_q \cdot \boldsymbol{p})(\boldsymbol{k}\cdot\boldsymbol{r}) \right] \qquad \{7\}$$

T_1 contains the electric dipole transition $(\hat{e}_q \cdot \boldsymbol{p})$. The electric quadrupole transition originates from the second term. The value of $\boldsymbol{k}\cdot\boldsymbol{r}$ can be calculated from the edge energy $\hbar\omega_{\text{edge}}$ in eV and the atomic number Z.

$$\boldsymbol{k}\cdot\boldsymbol{r} \approx \sqrt{\hbar\omega_{\text{edge}}} \big/ 80Z \qquad \{8\}$$

In case of the K edges from carbon ($Z = 6$, $\hbar\omega_{\text{edge}} = 284$ eV) to zinc ($Z = 30$, $\hbar\omega_{\text{edge}} = 9659$ eV), the value of $\boldsymbol{k}\cdot\boldsymbol{r}$ lies between 0.03 and 0.04. The transition probability is equal to the matrix element squared, hence the quadrupole transition is smaller by approximately $1.5\ 10^{-3}$ than the dipole transition and can be neglected. This defines the well-known dipole approximation. The transition operator in the dipole approximation is given by

$$T_1 \propto \sum_q (\hat{e}_q \cdot \boldsymbol{p}) \qquad \{9\}$$

Omitting the summation over $\boldsymbol{k}$ and using the commutation law between the position operator $\boldsymbol{r}$ and the atomic Hamiltonian ($p = m / i\hbar [r, \boldsymbol{H}]$) one obtains the familiar form of the X-ray absorption transition operator

$$T_1 = \sum_q (\hat{e}_q \cdot \boldsymbol{r}) \qquad \{10\}$$

1.2. BASICS OF XAFS SPECTROSCOPY

Including this operator into the Fermi golden rule gives

$$W_{fi} \propto \sum_q |\langle \Phi_f | \hat{e}_q \cdot \boldsymbol{r} | \Phi_i \rangle|^2 \delta_{E_f - E_i - \hbar\omega} \qquad \{11\}$$

This equation will form the basis for the rest of this review. This is a rather theoretical and formal description. What happens in practice? If an assembly of atoms is exposed to X-rays it will absorb part of the incoming photons. At a certain energy (depending on the atom) a sharp rise in the absorption will be observed (fig. 2). This sharp rise in absorption is called the absorption edge.

The energy of the absorption edge is determined by the binding energy of a core level. Exactly at the edge the photon energy is equal to the binding energy, or more precisely the edge identifies transitions from the ground state to the lowest empty state. Figure 2 shows the X-ray absorption spectra of manganese and nickel. The $L_{\text{II-III}}$ edges relate to a $2p$ core level and the K edge relates to a $1s$ core-level binding energy.

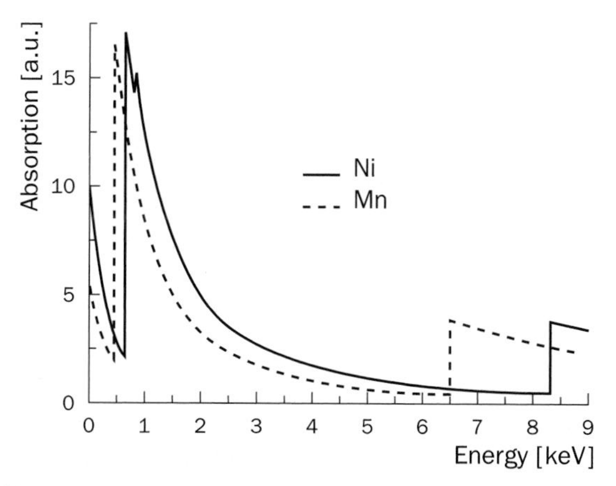

Figure 2 - The X-ray absorption cross-sections of manganese and nickel. Visible are the $L_{\text{II-III}}$ edges at respectively 680 and 830 eV and the K edges at respectively 6500 eV and 8500 eV.

In the case of solids, when other atoms surround the absorbing atom, a typical XAFS spectrum is shown in figure 3.

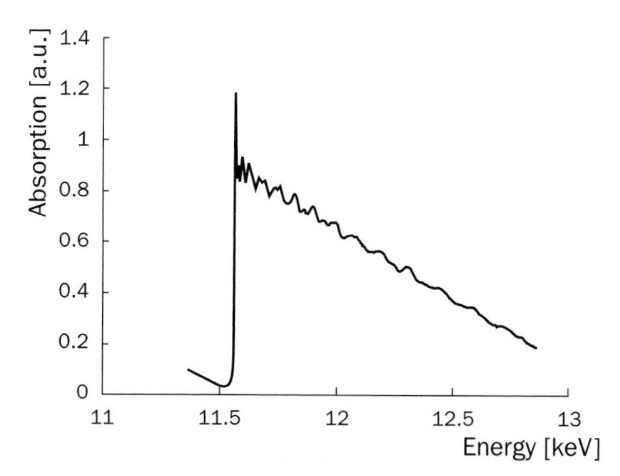

Figure 3 - The L_{III} XAFS spectrum of platinum metal. The edge jump is seen at 11.564 keV and above one observes a decaying background modulated by oscillations.

Instead of a smooth background, oscillations as a function of the energy of the incoming X-rays are visible. These oscillations can be explained assuming that the electron excitation process is a one-electron process. This makes it possible to rewrite the initial state wave function as a core-wave function (c) and the final state wave function as a free electron wave function (ε). One implicitly assumes that all other electrons do not participate in the X-ray induced transition. We will come back to the limitations of this approximation below.

$$|\langle\Phi_f|\hat{e}_q\cdot r|\Phi_i\rangle|^2 = |\langle\Phi_i \underline{c}\,\varepsilon|\hat{e}_q\cdot r|\Phi_i\rangle|^2 = |\langle\varepsilon|\hat{e}_q\cdot r|c\rangle|^2 = M^2 \qquad \{12\}$$

Underligning means that there is a hole in the corresponding level ($\underline{c}$ indicates a core-hole). The squared matrix element M^2 is in many cases a number that is only little varying with energy and one can often assume that it is a constant. The delta function of equation {11} implies that one observes the density of empty states (ρ).

$$I_{XAS} \sim M^2\rho \qquad \{13\}$$

The X-ray absorption selection rules determine that the dipole matrix element M is non-zero if the orbital quantum number of the final state differs by 1 from the one of the initial state ($\Delta L = \pm 1$, i.e. $s \longrightarrow p$, $p \longrightarrow s$ or d, etc.) and the spin is conserved ($\Delta S = 0$). The quadrupole transitions imply final states that differ by 2 from the initial state ($\Delta L = \pm 2$, i.e. $s \longrightarrow d$, $p \longrightarrow f$).

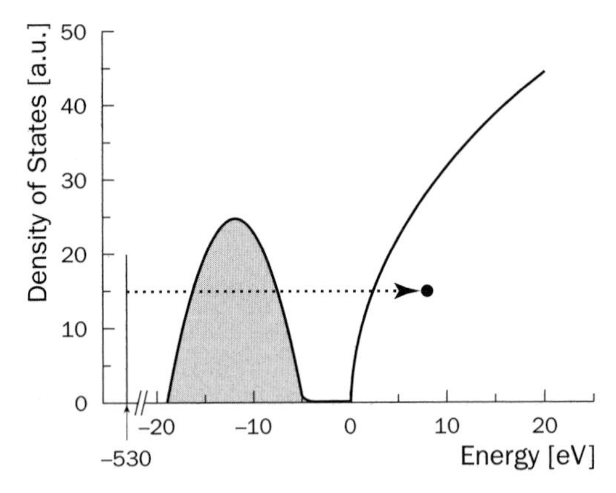

Figure 4 - The schematic density of states of an oxide. The $1s$ core electron of oxygen at 530 eV binding energy is excited to an empty p state giving the information on the oxygen p projected density of states.

In the dipole approximation, the shape of the absorption spectrum should look like the partial density of the ($\Delta L = \pm 1$) empty states projected on the absorbing site, convoluted with a Lorentzian. This Lorentzian broadening is due to the finite lifetime of the core-hole, leading to an uncertainty in its energy according to

Heisenberg's principle. A more accurate approximation can be obtained if the unperturbed density of states is replaced by the density of states in presence of the core-hole. This approximation gives a relatively adequate simulation of the XAS spectral shape when the interaction between the electrons in the final state is relatively weak. This is often the case for $1s \longrightarrow 4p$ transitions (*K* edges) of the $3d$ metals.

1.3. EXPERIMENTAL ASPECTS

An X-ray absorption spectrum originates from the fact that the probability of an electron to be ejected from a core level is dependent on the energy of the incoming beam. For this reason the energy of the X-rays is varied during an experiment, which requires a monochromator to obtain a monochromatic beam.

There are different ways to detect the absorption. The most direct way is *transmission detection*, where the absorption is obtained comparing the incoming photon intensity (I_0) with the intensity transmitted by the sample (I_t). Transmission experiments are standard for hard X-rays. X-rays with energies less than 1 keV have an attenuation length of less than 1 μm, implying that for soft X-rays the sample has to be extremely thin. In addition soft X-rays have a large absorption cross-section with air, hence the experiments have to be performed in vacuum.

For soft X-rays, and sometimes also for hard X-rays, the absorption is usually obtained using the decay of the created core-hole. This decay gives rise to an avalanche of electrons, photons and ions escaping from the surface of the substrate. By measuring any of these decay products, it is possible to measure samples of arbitrary thickness.

An important prerequisite for the use of decay channels is that the channels that are measured are linearly proportional to the absorption cross-section. In general this linear proportionality holds, but there are cases where for example the ratio between radiative a non-radiative decay varies significantly over a relatively short energy range. It turns out that in cases where multiplet effects are important, for example the $2p$ X-ray absorption of $3d$ metals, the fluorescence decay varies by almost a factor of two over an energy range of approximately 15 eV, i.e. from the first to the last peak in the edge region [1]. Because Auger decay dominates over fluorescence (for soft X-rays), this effect is only visible in fluorescence yield detection.

The *fluorescent decay* of the core-hole can be used as the basis for the absorption measurement. The amount of fluorescent decay increases with energy and Auger decay dominates for all core levels below 1 keV. In case of the 3*d* metals, the *K* edges show strong fluorescence and all other edges mainly Auger decay. The photon created in the fluorescent decay has a mean free path of the same order of magnitude as the incoming X-ray, which excludes any surface effect. On the other hand it means that there will be saturation effects if the sample is not dilute [2].

Recently it became possible to use fluorescence detectors with approximately 1 eV resolution to tune to a particular fluorescence channel. This could be denoted as *partial fluorescence yield.* The technique is also known as selective X-ray absorption because one can select for example a particular valence and measure the X-ray absorption spectrum of that valence only [3,4,5]. Other possibilities are the selectivity to the spin orientation and the nature of the neighbouring atoms. Partial fluorescence yield can be considered as a coherent X-ray scattering effect and as such it effectively removes the lifetime broadening of the intermediate state. This effect can be used to measure X-ray absorption spectra with unprecedented spectral resolution.

With the *total electron yield* method one detects all electrons that emerge from the sample surface, independent of their energy. The interaction of electrons with solids is much larger than the interaction of X-rays, which implies that the electrons that escape must originate close to the surface. The probing depth of total electron yield in the soft X-ray range lies in the range of approximately 3 to 10 nm, depending also on the material studied [6,7].

Instead of just counting the escaped electrons, one can detect their respective energies and as such measure the *partial electron yield.* The majority of emitted electrons are true secondary electrons with emission energies of less than 100 eV. However, if one detects an Auger decay channel with high resolution one can perform similar experiments as with high resolution fluorescence detection and do selective X-ray absorption experiments.

Other detection techniques include *ion yield* [8] and *X-ray Raman scattering* [9]. A completely different way to measure X-ray absorption cross-sections is Electron Energy Loss Spectroscopy or EELS. This implies studying the energy that electrons lose in a transmission electron microscope (TEM). TEM-EELS involves the measurement of the energy imparted to a thin specimen by incident electrons with typical energies of 100 keV to 300 keV. Because the interaction of electrons with matter is large, the specimen thickness must be of the order of maximum a few hundred nm in order to obtain a measurable transmission signal. Instead of a TEM microscope, one can also use a dedicated (non-microscopic) electron source with typically lower energy and higher energy resolution [10]. The close correspondence between EELS and XAS spectral shapes lies in an identical transition operator if one uses the dipole approximation in both cases. The enormous advantage of EELS is that it is carried out inside an electron microscope and as such can be applied to obtain spatial resolutions down to a few angström. This combination of high spatial resolution with high energy resolution has given some beautiful applications the last few years. It should be mentioned here that also X-ray absorption can be carried out in a microscopic fashion. Micro-XANES is reaching spatial resolution of some 100 nm and Photo-Emission Electron Microscopy (PEEM) is going towards a few nm resolution [11] (see the chapter by C.M. Schneider in this volume). For more details on the application of TEM-EELS we refer to recent reviews [12,13].

2. *MULTIPLET EFFECTS*

In the case of a transition to states where the interaction between the electrons is strong (strongly correlated or localized states), like the 3*d* states of transition metal ions or the 4*f* states of rare earths, the one-electron approximation does not hold. The interaction between the electrons, as well as the interaction of the electrons with the core-hole after the absorption process, has to be taken into account explicitly. Actually, all systems that contain a partly filled 3*d* shell in the final state of the X-ray absorption process have large interactions between the valence electrons and the 2*p*, 3*s* or 3*p* core-hole. Even the 3*d* metals for which the valence electrons can be described relatively well with mean field methods such as Density Functional Theory, the 2*p*3*d* multiplet interactions remain large and will significantly affect the 2*p* X-ray absorption spectral shapes [14].

Table 1 shows the multiplet interactions between the various possible core-holes and the partly filled valence band. The configurations calculated are respectively s^1d^9, p^5d^9 and d^9d^9 for the final states of Ni^{II}, Pd^{II} and Pt^{II}. The numbers in boldface indicate edges for which multiplet effects will be clearly visible. For multiplet effects to have a significant effect, the value of the Slater-Condon parameters must be larger than or of the same order of magnitude as the spin-orbit coupling separating the two edges. If spin-orbit coupling is very large there still will be an effect from the Slater-Condon parameters but it will be much less pronounced. For example, the 2*p* and 3*p* edges of the 4*d* elements have a large spin-orbit splitting and the multiplet effects are not able to mix states of both sub-edges. However, the effect of the Slater-Condon parameters will still be visible [15,16]. If a multiplet effect will actually be visible in X-ray absorption further depends on the respective lifetime broadenings. Another clear conclusion is that all shallow core levels are strongly affected and the deeper core levels are less affected.

Table 1 - The number in the first line for each element indicates the values of the maximum core valence Slater-Condon parameter for the final states. The second line gives the spin-orbit coupling for each core level of Ni^{II}, Pd^{II} and Pt^{II}. The valence electrons are the $3d$ states for nickel, $4d$ for palladium and $5d$ for platinum. The largest Slater integral is given in the first line and the core level spin-orbit coupling in the second line. Boldface values indicate clearly visible multiplet effects.

	1s	2s	2p	3s	3p	3d	4s	4p	4d	5s	5p	
$\langle ee \rangle$	<0.1	**5**	**8**	**13**	**17**							28 Ni^{II}
$\langle LS \rangle$		**–**	**17**	**–**	**2**							$\underline{c}3d^9$
$\langle ee \rangle$	<0.1	2	2	1	7	**10**	**9**	**13**				46 Pd^{II}
$\langle LS \rangle$		–	160	–	27	**5**	**–**	**5**				$\underline{c}4d^9$
$\langle ee \rangle$	<0.1	2	3	1	5	5	3	10	**10**	**14**	**19**	78 Pt^{II}
$\langle LS \rangle$		–	1710	–	380	90	–	90	**17**	**–**	**12**	$\underline{c}5d^9$

The situation for the $3d$ metals is clear: no visible multiplet effects for the $1s$ core level (K edge) and a significant influence on all other edges. In case of the $4d$ metals, the $3d$, $4s$, $4p$ edges show significant multiplet effects. The most commonly studied Pd edges are however the $1s$ and $2p$ edges that are not ($1s$) and only a little ($2p$) affected. The rare earth systems show large multiplet effects for the $3d$, $4s$, $4p$ and $4d$ core levels, hence all VUV and soft X-ray edges. The often-studied $2p$ core level is much less affected.

Figure 5 shows the comparison of all edges for Ni^{II} with atomic multiplet effects included. A cubic crystal field of 1.0 eV splits the $3d$ states (see section 3). The top three spectra are respectively the $1s$, $2s$ and $3s$ X-ray absorption spectrum calculated as the transition from $1s^2 3d^8 4p^0$ to $1s^1 3d^8 4p^1$. The lifetime broadening has been set to 0.2 eV half-width at half-maximum. Its value is larger in reality. One observes one peak for the 1s spectrum and two peaks for the $2s$ and $3s$ spectra. The reason for the two peaks is the $2s3d$ and $3s3d$ exchange interactions, which are directly related to the Slater-Condon parameters of respectively 5 and 13 eV. The splittings between the parallel and anti-parallel states are respectively ±2.5 and ±6.5 eV, i.e. approximately half the respective Slater-Condon parameter. The experimental $1s$ X-ray absorption spectrum of NiO looks completely different than this single peak, showing essentially an edge jump and transitions from the $1s$ core state to all empty states of p character. The complete spectral shape of K edge X-ray absorption is therefore better described with a multiple scattering formalism.

This single peak reflects just the first white line or leading edge of the spectrum. The 2*s* and 3*s* X-ray absorption spectra are not often measured. The 2*s* spectrum is very broad and therefore adds little information. The 3*s* X-ray absorption spectrum is also not very popular. Instead the $3s^1 3d^n$ final states do play an important role in spectroscopies like 3*s* XPS, 2*p*3*s* resonant X-ray emission, and 2*p*3*s*3*s* resonant Auger [17], as does the charge transfer effect that is discussed in section 4.

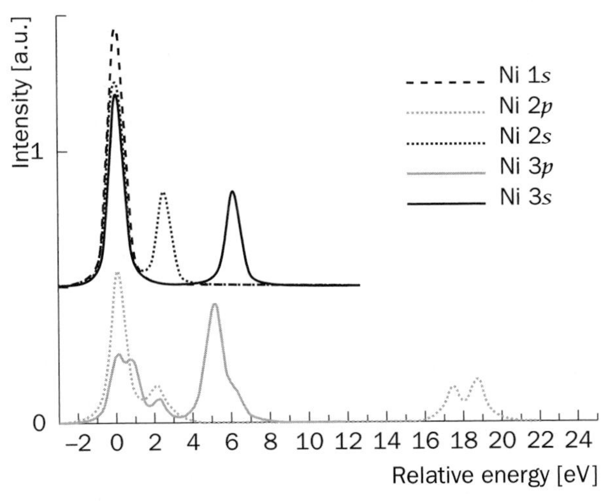

Figure 5 - Calculated X-ray absorption spectra for Ni^{II}. The respective binding energies are 8333, 1008, 870, 110 and 68 eV for the first peak of the 1*s*, 2*s*, 2*p*, 3*s* and 3*p* edges.

The spectra at the bottom in figure 5 are the 2*p* (dashed) and 3*p* (solid) X-ray absorption spectral shapes. These are essentially the well-known 2*p* and 3*p* spectra of NiO and other divalent nickel compounds. The $2p^5 3d^9$ and $3p^5 3d^9$ final states contain one *p* hole and one 3*d* hole that interact very strongly. This gives rise to a multitude of final states. Because the lifetime broadening for 2*p* states is relatively low, these spectral shapes can actually be observed in experiment. This gives 2*p* X-ray absorption, and to a lesser extend 3*p* X-ray absorption, their great potential for the determination of the local electronic structure. In section 5, it will be shown that the magnetic circular dichroism adds the ability to determine the local magnetic moment.

A successful method to analyze these transitions is based on a ligand-field multiplet model. For its description we start with an atomic model, where only the interactions within the absorbing atom are considered, without influence from the surrounding atoms. Solid state effects are then introduced as a perturbation. This can be justified if the intra-atomic interactions are larger than the ones between the atoms.

2.1. ATOMIC MULTIPLETS

In order to show how spectra in strongly correlated electron systems are calculated, we start with the example of a free atom, where there is no influence from the environment. The Schrödinger equation contains the kinetic energy of the N electrons $\left(\sum_N \frac{p_i^2}{2m}\right)$, the electrostatic interaction of the N electrons with the nucleus of charge $+Z$ $\left(\sum_N \frac{-Ze^2}{r_i}\right)$, the electron-electron repulsion $\left(H_{ee} = \sum_{\text{pairs}} \frac{e^2}{r_{ij}}\right)$ and the spin-orbit coupling of each electron $\left(H_{ls} = \sum_N \zeta(r_i) l_i \cdot s_i\right)$. The total Hamiltonian for a free atom, H_{ATOM}, is thus given by

$$H_{ATOM} = \sum_N \frac{p_i^2}{2m} + \sum_N \frac{-Ze^2}{r_i} + \sum_{\text{pairs}} \frac{e^2}{r_{ij}} + \sum_N \zeta(r_i) l_i \cdot s_i \qquad \{14\}$$

The kinetic energy and the interaction with the nucleus are the same for all electrons in a given atomic configuration. They define the average energy of the configuration (H_{av}). The electron-electron repulsion and the spin-orbit coupling define the relative energy of the different terms within this configuration [1]. The main difficulty when solving the Schrödinger equation is that H_{ee} is too large to be treated as a perturbation. A solution to this problem is given by the central field approximation, in which the spherical average of the electron-electron interaction is separated from the non-spherical part. The spherical average $\langle H_{ee} \rangle$ is added to H_{av} to form the average energy of a configuration. In the modified electron-electron Hamiltonian H'_{ee}, the spherical average has been subtracted.

$$H'_{ee} = H_{ee} - \langle H_{ee} \rangle = \sum_{\text{pairs}} \frac{e^2}{r_{ij}} - \left\langle \sum_{pairs} \frac{e^2}{r_{ij}} \right\rangle \qquad \{15\}$$

The two interactions H'_{ee} and H_{ls} determine therefore the energies of the different terms within the atomic configuration.

2.1.1. TERM SYMBOLS

The terms of a configuration are indicated with their orbital moment L, spin moment S and total moment J, with $|L-S| \le J \le L+S$. In the absence of spin-orbit

1. A **configuration** is the assignment of a given number of electrons to a certain set of orbitals, for example $3d^2$ or $2p^5\,3d^7$ (completed filled or empty orbitals are not mentioned). A **term** is an *energy level* of a system. Each configuration, in general, gives rise to a number of energy levels and thus a number of terms.

coupling, all terms with the same L and S have the same energy, giving an energy level which is $(2L + 1)(2S + 1)$-fold degenerate [2]. When spin-orbit coupling is important, the terms are split in energy according to their J value with a degeneracy of $2J + 1$. A term is designed with a so-called term symbol $^{2S+1}X_J$, where X corresponds to a letter according to the value of L. $X = S, P, D, F$... for $L = 0, 1, 2, 3$... The quantity $2S + 1$ is called the spin multiplicity of the term, and the terms are called singlet, doublet, triplet, quartet, etc. according to $S = 0, 1/2, 1, 3/2$...

A single 1s electron has an orbital moment $L = 0$, a spin moment $S = 1/2$ and a total moment $J = 1/2$. There is only one term, with term symbol $^2S_{1/2}$. For one p electron, $L = 1$, $S = 1/2$, and J can be 1/2 or 3/2, corresponding to term symbols $^2P_{1/2}$ and $^2P_{3/2}$. In the case of a transition metal ion, the important configuration for the initial state of the absorption process is $3d^n$. In the final state with a 3s or a 3p core-hole, the configurations are $3s^13d^{n+1}$ and $3p^53d^{n+1}$. The main quantum number has no influence on the coupling scheme, so the same term symbols can be found for 4d and 5d systems, or for 2p and 3p core-holes.

In the case of a $2p^2$ configuration, the first electron has six quantum states available, the second electron only five. This is due to the Pauli exclusion principle that forbids two electrons to have the same quantum numbers n, M_L and M_S. Because the sequence of the two electrons is not important, one divides the number of combinations by two and obtains fifteen possible combinations. These fifteen combinations are indicated in table 2.

Table 2 - The fifteen combinations of states for electrons $\underline{a}$ with $|m_{la}, m_{sa}\rangle$ and $\underline{b}$ with $|m_{lb}, m_{sb}\rangle$ of a $2p^2$ configuration.

$\|m_{la}, m_{sa}\rangle$	$\|m_{lb}, m_{sb}\rangle$	$\|M_L, M_S\rangle$	$\|m_{la}, m_{sa}\rangle$	$\|m_{lb}, m_{sb}\rangle$	$\|M_L, M_S\rangle$
$\|1,+\rangle$	$\|1,-\rangle$	$\|2,0\rangle$	$\|1,-\rangle$	$\|-1,-\rangle$	$\|0,-1\rangle$
$\|1,+\rangle$	$\|0,+\rangle$	$\|1,1\rangle$	$\|0,+\rangle$	$\|0,-\rangle$	$\|0,0\rangle$
$\|1,+\rangle$	$\|0,-\rangle$	$\|1,0\rangle$	$\|0,+\rangle$	$\|-1,+\rangle$	$\|-1,1\rangle$
$\|1,+\rangle$	$\|-1,+\rangle$	$\|0,1\rangle$	$\|0,+\rangle$	$\|-1,-\rangle$	$\|-1,0\rangle$
$\|1,+\rangle$	$\|-1,-\rangle$	$\|0,0\rangle$	$\|0,-\rangle$	$\|-1,+\rangle$	$\|-1,0\rangle$
$\|1,-\rangle$	$\|0,+\rangle$	$\|1,0\rangle$	$\|0,-\rangle$	$\|-1,-\rangle$	$\|-1,-1\rangle$
$\|1,-\rangle$	$\|0,-\rangle$	$\|1,-1\rangle$	$\|-1,+\rangle$	$\|-1,-\rangle$	$\|-2,0\rangle$
$\|1,-\rangle$	$\|-1,+\rangle$	$\|0,0\rangle$			

2. The eigenvalues M_L (M_S) of the operator L_z (S_z), related to the z-component of the orbital (spin) momentum, can take the values $-L \leq M_L \leq L$ ($-S \leq M_S \leq S$).

The presence of $|2,0\rangle$ and $|-2,0\rangle$ states implies that there must be a ^1D term. This term contains five states with $M_S = 0$ and $M_L = -2, -1, 0, 1, 2$, as indicated with D in table 3. In addition, there is a ^3P term, with nine states (P). One state is left, with $M_S = M_L = 0$, giving a ^1S term (S). The $2p^2$ configuration contains therefore the terms ^3P, ^1D and ^1S . It can be checked that total degeneracy adds up to fifteen. Note also that the term symbols of a $2p^2$ configuration form a sub-set of the term symbols of the $2p3p$ configuration, which do not have to obey the Pauli principle. Including spin-orbit coupling, we have the terms $^1D_2$, $^1S_0$ and $^3P_{210}$, a short-hand notation of $^3P_2$ plus $^3P_1$ plus $^3P_0$.

Table 3 - A schematic diagram of the fifteen total symmetry states of a $2p^2$ configuration into the three irreducible representations ^1D, ^3P and ^1S.

	$M_S = 1$	$M_S = 0$	$M_S = -1$
$M_L = 2$	–	D	–
$M_L = 1$	P	$D\ P$	P
$M_L = 0$	P	$D\ P\ S$	P
$M_L = -1$	P	$D\ P$	P
$M_L = -2$	–	D	–

A $3d^1$ configuration has term symbols $^2D_{5/2}$ and $^2D_{3/2}$ with, respectively, six $(2 \times 5/2 + 1)$ and four $(2 \times 3/2 + 1)$ states. The LS term symbols for a $3d^14d^1$ configuration can be found by "multiplying" the term symbols for the configurations $3d^1$ and $4d^1$. This multiplication consists of separately summing L and S of both terms. Multiplication of terms A and B is written as $A \otimes B$. Since both L and S are vectors, the resulting terms have possible values of $|L_A - L_B| \leq L \leq L_A + L_B$ and $|S_A - S_B| \leq S \leq S_A + S_B$. For $^2D \otimes {}^2D$, this gives $L = 0, 1, 2, 3$ or 4 and $S = 0$ or 1. The ten LS term symbols of the $3d^14d^1$ configuration are given in table 4, together with their degeneracy and possible J values. The total degeneracy of the $3d^14d^1$ configuration is 100. In the presence of spin-orbit coupling, a total of eighteen term symbols is found.

Table 4 - Possible term LS term symbols for a $3d^14d^1$ configuration, with their degeneracy and possible J values. The sub-set of LS term symbols for a $3d^2$ configuration is given in boldface.

$3d^14d^1$	$\mathbf{^1S}$	^1P	$\mathbf{^1D}$	^1F	$\mathbf{^1G}$	^3S	$\mathbf{^3P}$	^3D	$\mathbf{^3F}$	^3G	Σ
Degree	**1**	3	**5**	7	**9**	3	**9**	15	**21**	27	{ 100 **45**
J values	**0**	1	**2**	3	**4**	1	**0,1,2**	1,2,3	**2,3,4**	3,4,5	

A $3d^2$ configuration does not have the same degeneracy as the $3d^1 4d^1$ configuration, due to the Pauli exclusion principle. In total there are $10 \times 9/2 = 45$ possible states. Following the same procedure as for the $2p^2$ configuration, one can write out all 45 combinations of a $3d^2$ configuration and sort them by their M_L and M_S quantum numbers. Analysis of the combinations of the allowed M_L and M_S quantum numbers yields the term symbols ^1G, ^3F, ^1D, ^3P and ^1S. This is a sub-set of the term symbols of a $3d^1 4d^1$ configuration. The term symbols can be divided into their J quantum numbers as $^3F_2$, $^3F_3$, $^3F_4$, $^3P_0$, $^3P_1$, $^3P_2$, $^1G_4$, $^1D_2$ and $^1S_0$ as also indicated in table 4.

In the case of a $3d^3$ configuration a similar approach shows that the possible spin states are doublet and quartet. The quartet states have all spins parallel and the Pauli exclusion principle implies that there are two quartet term symbols, respectively ^4F and ^4P. The doublet states have two electrons parallel and for these two electrons the Pauli principle yields the combinations identical to the triplet states of the $3d^2$ configuration. To these two parallel electrons a third electron is added anti-parallel, where this third electron can have any value of its orbital quantum number m_l. Writing out all combinations and separating them into the total orbital moments M_L gives the doublet term symbols ^2H, ^2G, ^2F, ^2D, another ^2D and ^2P. By adding the degeneracies, it can be checked that a $3d^3$ configuration has 120 different states, i.e. $10 \times 9/2 \times 8/3$. The general formula to determine the degeneracy of a $3d^n$ configuration is

$$\binom{10}{n} = \frac{10!}{(10-n)!n!} \qquad \{16\}$$

We can show that the term symbols of a configuration $3d^n$ do also exist in a configuration $3d^{n+2}$, for $n + 2 \leq 5$. Thus the term symbols of $3d^4$ contain all term symbols of $3d^2$, which contains the ^1S term symbol of $3d^0$. Similarly the term symbols of $3d^5$ contain all term symbols of $3d^3$, which contains the ^2D term symbol of $3d^1$. In addition there is a symmetry equivalence of holes and electrons, hence $3d^3$ and $3d^7$ have exactly the same term symbols.

The $2p$ X-ray absorption edge ($2p \longrightarrow 3d$ transition) is often studied for the $3d$ transition metal series, and it provides a wealth of information. Crucial for its understanding are the configurations of the $2p^5 3d^n$ final states. The term symbols of the $2p^5 3d^n$ states are found by multiplying the configurations of $3d^n$ with a ^2P term symbol. The total degeneracy of a $2p^5 3d^n$ state is given as six times the value of equation 16. For example, a $2p^5 3d^5$ configuration has 1512 possible states. Analysis shows that these 1512 states are divided into 205 term symbols, implying in principle 205 possible final states. If all these final states have finite intensity depends on the selection rules.

2.1.2. MATRIX ELEMENTS

Above we have found the number of states of a certain $3d^n$ configuration and their term symbols. The next task is to find the relative energies of the different terms, calculating the matrix elements of these states with the Hamiltonian H_{ATOM}. As discussed in section 2.1, H_{ATOM} consists of the effective electron-electron interaction H'_{ee} and the spin-orbit coupling H_{ls} (cf. eq. {14}). The electron-electron interaction commutes [3] with L^2, S^2, L_z and S_z, which implies that all its off-diagonal elements are zero. The general formulation of the matrix elements of two-electron wave functions is given as

$$\left\langle {}^{2S+1}L_J \left| \frac{e^2}{r_{12}} \right| {}^{2S+1}L_J \right\rangle = \sum_i f_i F^i + \sum_i g_i G^i \qquad \{17\}$$

$F^i(f_i)$ and $G^i(g_i)$ are the Slater-Condon parameters for the *radial (angular)* part of the direct Coulomb repulsion and the Coulomb exchange interaction, respectively. f_i and g_i are non-zero only for certain values of i, depending on the configuration. It is found that the exchange interaction g_i is present only for electrons in different shells. f_0 is always present and the maximum value for i equals two times the lowest value of l. For g_i, i is even if $l_1 + l_2$ is even, and i is odd if $l_1 + l_2$ is odd. The maximum value of i equals $l_1 + l_2$ **[18]**.

A simple example is a $1s2s$ configuration consisting of ^1S and ^3S term symbols. The value of both f_0 and g_0 is 1, and the respective energies are given by

$$\left\langle {}^1S \left| \frac{e^2}{r_{12}} \right| {}^1S \right\rangle = F^0(1s2s) + G^0(1s2s) \qquad \{18\}$$

$$\left\langle {}^3S \left| \frac{e^2}{r_{12}} \right| {}^3S \right\rangle = F^0(1s2s) - G^0(1s2s) \qquad \{19\}$$

This result can be stated as *the singlet and the triplet state are split by the exchange interaction*. This energy difference is $2G^0(1s2s)$. An analogous result is found for a $1s2p$ state for which the singlet and triplet states are split by $(2/3)G^0(1s2p)$. The 2/3 prefactor is determined by the degeneracy of the $2p$ state.

For a $3d^2$ configuration, the electrons come from the same shell hence there are no exchange interactions. There are five term symbols ^1S, ^3P, ^1D, ^3F and ^1G. Their energies are given in table 5. f_0 is equal to the number of permutations

---

3. Two operators ***A*** and ***B*** are said to commute if they have common eigenfunctions, i.e. the eigenfunctions of ***A*** are eigenfunctions of ***B***, and vice versa.

$[N(N-1)/2]$ of n electrons, i.e. equal to 1.0 for two electron configurations. The Slater-Condon parameters F^2 and F^4 have approximately a constant ratio: $F^4 = 0.62\,F^2$. The last column in table 5 gives the approximate energies of the five term symbols.

Table 5 - The relative energies of the term symbols for a $3d^2$ configuration.

	Energy	Relative Energy	Relative Energy
^1S	$F^0 + 2/7\,F^2 + 2/7\,F^4$	$0.46\,F^2$	4.6 eV
^3P	$F^0 + 3/21\,F^2 - 4/21\,F^4$	$0.02\,F^2$	0.2 eV
^1D	$F^0 - 3/49\,F^2 + 4/49\,F^4$	$-0.01\,F^2$	– 0.1 eV
^3F	$F^0 - 8/49\,F^2 - 1/49\,F^4$	$-0.18\,F^2$	– 1.8 eV
^1G	$F^0 + 4/49\,F^2 + 1/441\,F^4$	$0.08\,F^2$	0.8 eV

In case of the $3d$ transition metal ions, F^2 is approximately equal to 10 eV. This gives for the five term symbols the energies as in table 5. The ^3F term symbol has lowest energy and is the ground state of a $3d^2$ system. This is in agreement with Hund's rules, which will be discussed in the next section. The three states ^1D, ^3P and ^1G are close in energy some 1.7 to 2.5 eV above the ground state. The ^1S state has a high energy of 6.4 eV above the ground state, the reason being that two electrons in the same orbit strongly repel each other.

Table 6 gives three related notations that are used to indicate the radial integrals. The Slater-Condon parameters F^k, the normalized Slater-Condon parameters F_k and the Racah parameters A, B and C. The bottom half of table 6 uses the relationship between F^2 and F^4 and it further uses a typical F^2 value of 10 eV and a F^0 value of 8 eV.

Table 6 - A comparison of the Slater-Condon parameters F^k, with the normalized Slater-Condon parameters F_k and the Racah parameters.

Slater-Condon	Normalized	Racah
F^0	$F_0 = F^0$	$A = F_0 - 49\,F_4$
F^2	$F_2 = F^2/49$	$B = F_2 - 5\,F_4$
F^4	$F_4 = F^4/441$	$C = 35\,F_4$
$F^0 = 8.0$	$F_0 = 8.0$	$A = 7.3$
$F^2 = 10.0$	$F_2 = 0.41$	$B = 0.13$
$F^4 = 6.2$	$F_4 = 0.014$	$C = 0.49$

For three and more electrons the situation is considerably more complex. It is not straightforward to write down an anti-symmetrized three-electron wave function. It can be shown that the three-electron wave function can be built from two-electron wave functions with the use of the so-called *coefficients of fractional parentage* [5].

2.2. *ATOMIC MULTIPLET GROUND STATES OF $3d^n$ SYSTEMS*

Based on experimental information Hund formulated three rules to determine the ground state of a $3d^n$ configuration. The three Hund's rules are:

- term symbols with maximum spin S are lowest in energy,
- among these terms, the one with the maximum orbital moment L is lowest,
- in the presence of spin-orbit coupling, the lowest term has $J = |L - S|$ if the shell is less than half full and $J = L + S$ if the shell is more than half full.

A configuration has the lowest energy if the electrons are as far apart as possible. The first Hund's rule "maximum spin" can be understood from the Pauli principle: electrons with parallel spins must be in different orbitals, which on overall implies larger separations, hence lower energies. This is for example evident for a $3d^5$ configuration, where the ^6S state has its five electrons divided over the five spin-up orbitals, which minimizes their repulsion. In case of $3d^2$, the first Hund's rule implies that either the ^3P or the ^3F term symbol must have lowest energy. From the previous section one finds that the ^3F term symbol is lower than the ^3P term symbol, because the ^3F wave function tends to minimize electron repulsion. The effects of spin-orbit coupling are well known in the case of core states. A $2p$ core state has $^2P_{3/2}$ and $^2P_{1/2}$ terms, where $^2P_{3/2}$ has the lowest energy. In the $2p$ XAS or XPS spectrum of nickel, the $^2P_{3/2}$ peak is positioned at approximately 850 eV and the $^2P_{1/2}$ at about 880 eV. Note that the state with the lowest binding energy is related to the lowest energy of the final state configuration. This is in agreement with Hund's third rule: the configuration is $2p^5$, so more than half full, implying that highest J value has lowest energy. The third rule implies that the ground state of a $3d^8$ configuration is $^3F_4$, while it is $^3F_2$ in case of a $3d^2$ configuration.

2.3. *j - j COUPLING*

The splitting scheme that we have used above is known as the Russell-Saunders (RS) coupling scheme: the spins and orbital moments of the individual electrons are coupled to give the total S and L of the configuration, the spin-orbit coupling splits the resulting terms according to their J value. The RS-coupling scheme is valid when the perturbation due to spin-orbit coupling is small compared to the one due to electronic repulsions. This holds for most lighter elements, for which the spin-orbit coupling is small, like the $3d$ transition metals. For heavy elements, spin-orbit coupling becomes more important while electron repulsion starts to decrease (the radial extent of the orbitals becomes larger and the electrons are thus farther apart). This can lead to a breakdown of the RS-coupling scheme.

When the spin-orbit coupling is much larger than the electron repulsions, the orbital moment l and spin moment s of each electron have to be coupled to give the total moment j, and the j values of all the electrons are coupled to give the total J. A configuration is first split into levels according to their j value and the electron repulsions are then treated as a perturbation on these spin-orbit coupling levels. This approach is known as the $\boldsymbol{j}\cdot\boldsymbol{j}$ coupling scheme, and is the direct reverse of the Russell-Saunders scheme. Many heavier elements, like the rare earths, do not conform to either of the two limiting cases, and an *intermediate* coupling scheme has to be applied. This is also the case for the final state after $2p$ absorption in $3d$ transitions metals, since the spin-orbit coupling is strong for the $2p$ hole.

2.4. X-RAY ABSORPTION SPECTRA DESCRIBED WITH ATOMIC MULTIPLETS

2.4.1. TRANSITION METAL $L_{II\text{-}III}$ EDGE

In closed shell systems, the X-ray absorption process excites a $2p$ core electron into the empty $3d$ shell and the transition can be described as $2p^63d^0 \longrightarrow 2p^53d^1$. The ground state has $^1S_0$ symmetry and we find that the term symbols of the final state are $^1P_1$, $^1D_2$, $^1F_3$, $^3P_{012}$, $^3D_{123}$ and $^3F_{234}$. The energies of the final states are affected by the $2p3d$ Slater-Condon parameters, the $2p$ spin-orbit coupling and the $3d$ spin-orbit coupling. The X-ray absorption transition matrix elements to be calculated are

$$I_{XAS} \propto \langle 3d^0 \,|\, \hat{e}_q \cdot \boldsymbol{r} \,|\, 2p^5 3d^1 \rangle^2 \qquad \{20\}$$

The associated term symbols are

$$I_{XAS} \propto \langle [^1S_0] \| [^1P_1] \| [^{1,3}PDF] \rangle^2 \qquad \{21\}$$

Table 7 contains the result of an atomic multiplet calculation for T_i^{IV} ($3d^0$) using the atomic parameters. The twelve states are built from the twelve term symbols according to the matrix given. The irreducible representations, i.e. the states with the same J value block out in the calculation. The $^3P_0$ state has the lowest energy.

This state has zero intensity in an X-ray absorption process (because of the selection rule $|J-1| \leq J' \leq J+1$), but it is possible that another state decays to this $^3P_0$ state *via* Coster-Kronig Auger decay.

Table 7 - The relative energies of an atomic multiplet calculation for Ti^{IV}. The $J = 1$ states (which have finite intensity) are given in boldface.

	$^3P_0$	$^3\mathbf{P}_1$	$^3\mathbf{D}_1$	$^1\mathbf{P}_1$	$^3P_2$	$^3D_2$	$^3F_2$	$^1D_2$	$^3D_3$	$^3F_3$	$^1F_3$	$^3F_4$
−3.281	1.0											
−2.954		**−0.94**	**0.30**	**0.08**								
0.213		**−0.19**	**−0.77**	**0.60**								
5.594		**0.24**	**0.55**	**0.79**								
−2.381					0.81	−0.46	0.01	0.34				
−1.597					−0.03	−0.50	0.56	−0.65				
3.451					0.04	−0.30	−0.82	−0.47				
3.643					−0.57	−0.65	−0.06	0.48				
−2.198									−0.21	0.77	0.59	
−1.369									0.81	−0.19	0.54	
3.777									−0.53	−0.60	0.59	
−2.481												1.0

The symmetry of the dipole transition is given as $^1P_1$, according to the dipole selection rules, which state that $\Delta J = +1,\ 0,\ -1$ but not $J' = J = 0$. Within LS coupling also $\Delta S = 0$ and $\Delta L = 1$. The dipole selection rules reduce the number of final states that can be reached from the ground state. The J value in the ground state is zero, so the J value in the final state must be one, and only the three term symbols $^1P_1$, $^3P_1$ and $^3D_1$ can obtain finite intensity. They are indicated in table 7 in boldface. The problem of calculating the $2p$ absorption spectrum is effectively reduced to solving the three by three energy matrix of the final states with $J = 1$. As discussed above the atomic energy matrix consists of terms related to the two-electron Slater integrals ($H_{ELECTRO} = H'_{ee}$) and the spin-orbit couplings of the $2p$ ($H_{LS\text{-}3d}$) and the $3d$ electrons ($H_{LS\text{-}3d}$). To show the individual effects of these interactions they will be introduced one by one. A series of five calculations will be shown, in which respectively:

a - $H = 0$: All final state interactions are set to zero.

b - $H = H_{LS\text{-}2p}$: Only the $2p$ spin-orbit coupling is included.

c - $H = H_{ELECTRO}$: Only the Slater-Condon parameters are included.

d - $H = H_{ELECTRO} + H_{LS\text{-}2p}$

e - $H = H_{ELECTRO} + H_{LS\text{-}2p} + H_{LS\text{-}3d}$: The $3d$ spin-orbit coupling is added.

a - We start by setting all final state interactions to zero. The results of the $2p$ X-ray absorption spectrum will be given with two 3×3 matrices. The energy levels are given below. They are labelled from top to bottom "^3P", "^3D" and "^1P" states, indicating the approximate term symbol related to the state. The original term symbols ^3P, ^3D and ^1P are given in respectively the first row, second row and bottom row of the eigenvector matrix. The intensity of the states is indicated on the right. With all interactions zero, the complete energy matrix is zero. The states all are the pure *LSJ* states and because of the dipole selection rules all intensity goes to the $^1P_1$ state.

Table 8 - The energy matrix and eigenvectors of the 3×3 matrices of the $2p^5 3d^1$ final states with $J = 1$. The bottom half of the table gives the resulting energies and intensities. All final state interactions are set to zero.

Energy Matrix		Eigenvectors
$\begin{vmatrix} 0 & 0 & 0 \\ 0 & 0 & 0 \\ 0 & 0 & 0 \end{vmatrix}$		$\begin{vmatrix} 1 & 0 & 0 \\ 0 & 1 & 0 \\ 0 & 0 & 1 \end{vmatrix}$
Energy Levels		**Intensities**
0.00	^3P	0.00
0.00	^3D	0.00
0.00	^1P	1.00

b - Inclusion of the $2p$ spin-orbit coupling $H_{LS\text{-}2p}$ of $\zeta p = 3.776$ eV creates non-diagonal elements in the energy matrix. In other words the *LS* character of the individual states is mixed. In case only $2p$ spin-orbit coupling is included the result is rather simple, with the triplet states at $-1/2\ \zeta p$ and the singlet state at $+\ \zeta p$. The eigenvector matrix shows that the three states are mixtures of the three pure states, i.e. the first state is in fact: "^3P" $= 0.5\ ^3P_1 - 0.866\ ^3D_1$. The intensities of the three states are directly given by the square of percentage of $^1P_1$ character. This gives the familiar result that the triplet states, or $2p_{3/2}$ states, have twice the intensity of the singlet, or $2p_{1/2}$, states. This has been indicated also in figure 6.

Table 9 - The energy matrix and eigenvectors of the 3 × 3 matrices of the $2p^5 3d^1$ final states with $J = 1$, after inclusion of the $2p$ spin-orbit coupling.

Energy Matrix				Eigenvectors		
0.944	1.635	2.312		0.5	−0.5	−0.707
1.635	−0.944	1.335		−0.866	−0.288	−0.408
2.312	1.335	0.000		0.0	0.816	−0.577

Energy Levels		Intensities
−1.888	^3P	0.00
−1.888	^3D	0.666
+3.776	^1P	0.333

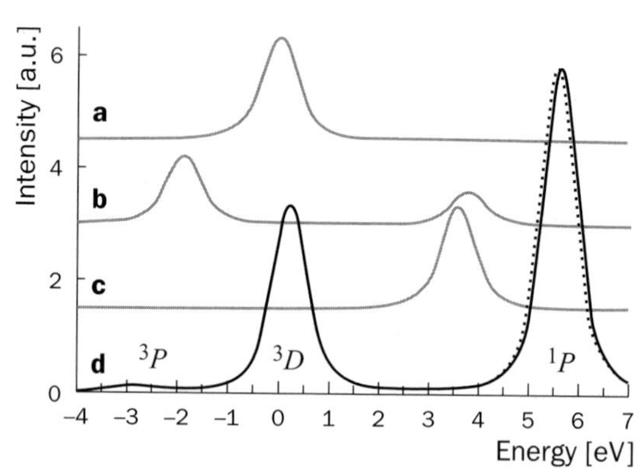

Figure 6 - The effects of the Slater-Condon parameters and the spin-orbit coupling on the atomic multiplet spectrum of a $3d^0$ system: (**a**) no interactions, (**b**) only $2p$ spin-orbit, (**c**) only Slater-Condon, (**d**) both Slater-Condon and $2p$ spin-orbit (solid). The dashed line includes the $3d$ spin-orbit coupling.

c - The Slater-Condon parameters F^2, G^1 and G^3 are respectively 5.042 eV, 3.702 eV and 2.106 eV. Keeping the $2p$ spin-orbit coupling zero, this gives the three states at respectively −1.345, 0.671 and 3.591 eV. Only the $^1P_1$ state has a finite intensity and its energy is shifted to an energy of 3.591 eV above the centre of gravity as indicated in figure 6. The two other states have zero intensity. It can be seen that the pd Slater-Condon parameters are diagonal in the *LS* terms, hence the three states are pure in character.

Table 10 - The energy matrix and eigenvectors of the 3 × 3 matrices of the $2p^53d^1$ final states with J = 1, after inclusion of the $2p3d$ Slater-Condon parameters.

Energy Matrix				Eigenvectors		
−1.345	0	0		1	0	0
0	0.671	0		0	1	0
0	0	3.591		0	0	1

Energy Levels		Intensities
−1.345	^3P	0.00
+0.671	^3D	0.00
+3.591	^1P	1.00

d - In part (**b**) we have seen that the non-diagonal terms of the $2p$ spin-orbit coupling make that all three states are mixtures of the individual term symbols. $2p$ spin-orbit coupling further creates the 2:1 intensity ratio, thereby shifting most of the ^1P character to lower energy. In part (**c**) it was found that the Slater-Condon parameters shift the ^1P state to higher energy and that the triplet states have a considerably lower energy. If one includes both the $2p$ spin-orbit coupling and the pd Slater-Condon parameters, the result will depend on their relative values. In case of the $2p$ core-hole of Ti^{IV}, the Slater-Condon parameters are relatively large and most intensity goes to the $2p_{1/2}$ state. The triplet states are separated by 3 eV and the lowest "^3P" energy state is extremely weak, gaining less than 1% of the total intensity. Figure 6d shows the typical spectral shape with three peaks. In the next section we compare four similar spectra with different ratios of Slater-Condon parameters and core-hole spin-orbit couplings to show the variations in their spectral shapes.

Table 11 - The energy matrix and eigenvectors of the 3 × 3 matrices of the $2p^53d^1$ final states with J = 1, after inclusion of the $2p3d$ Slater-Condon parameters *and* the $2p$ spin-orbit coupling.

Energy Matrix				Eigenvectors		
1.615	1.635	2.312		0.297	−0.776	0.557
1.635	−2.289	1.335		−0.951	−0.185	0.248
2.312	1.335	3.591		0.089	0.603	0.792

Energy Levels		Intensities
−2.925	^3P	0.008
+0.207	^3D	0.364
+5.634	^1P	0.628

e - In figure 6d we also show (dashed line) the calculation including the $3d$ spin-orbit coupling. Because it is only 32 meV, its influence on the spectral shape is negligible in the present case. The energy position of the "^1P" state shifts by 40 meV and its intensity drops by 0.4% of the total intensity. The effects on the intensities and energies of the other two peaks is even smaller. It is noted that the $3d$ spin-orbit coupling can have very significant effects on the spectral shape of $3d$ compounds if the $3d$ shell is partly filled in the ground state.

We now compare a series of X-ray absorption spectra of tetravalent titanium $2p$ and $3p$ edges and the trivalent lanthanum $3d$ and $4d$ edges. The ground states of Ti^{IV} and La^{III} are respectively $3d^0$ and $4f^0$ and they share a ^1S ground state. The transitions at the four edges are respectively

$$Ti^{IV}\ L_{II\text{-}III}\ \text{edge: } 3d^0 \longrightarrow 2p^5 3d^1$$

$$Ti^{IV}\ M_{II\text{-}III}\ \text{edge: } 3d^0 \longrightarrow 3p^5 3d^1$$

$$La^{III}\ M_{IV\text{-}V}\ \text{edge: } 4f^0 \longrightarrow 3d^9 4f^1$$

$$La^{III}\ N_{IV\text{-}V}\ \text{edge: } 4f^0 \longrightarrow 4d^9 4f^1$$

These four calculations are equivalent and all spectra consist of three peaks with $J = 1$. What changes are the values of the atomic Slater-Condon parameters and core-hole spin-orbit coupling. They are given in table 12 for the four situations. The G^1 and G^3 Slater-Condon parameters have an approximately constant ratio with respect to the F^2 value. The important factor for the spectral shape is the ratio of the core spin-orbit coupling and the F^2 value. Finite values of both the core spin-orbit and the Slater-Condon parameters cause the presence of the pre-peak.

Table 12 - The relative intensities, energy, core-hole spin-orbit coupling and F^2 Slater-Condon parameters are compared for four different $^1S_0$ systems.

Edge	Ti $2p$	Ti $3p$	La $3d$	La $4d$
Average Energy (eV)	464.00	37.00	841.00	103.00
Core spin-orbit (eV)	3.78	0.43	6.80	1.12
F^2 Slater-Condon (eV)	5.04	8.91	5.65	10.45
Intensities				
Pre-peak	0.01	10^{-4}	0.01	10^{-3}
$p_{3/2}$ or $d_{5/2}$	0.72	10^{-3}	0.80	0.01
$p_{1/2}$ or $d_{3/2}$	1.26	1.99	1.19	1.99

It can be seen in table 12 that the $3p$ and $4d$ spectra have small core spin-orbit couplings, implying small $p_{3/2}$ ($d_{5/2}$) edges and extremely small pre-peak intensities. The deeper $2p$ and $3d$ core levels have larger core spin-orbit splitting with the result of a $p_{3/2}$ ($d_{5/2}$) edge of almost the same intensity as the $p_{1/2}$ ($d_{3/2}$) edge and a larger pre-peak. Note that none of these systems comes close to the single-particle result of a 2:1 ratio of the p edges or the 3:2 ratio of the d edges. Figure 7 shows the X-ray absorption spectral shapes. They are given on a logarithmic scale to make the pre-edges visible.

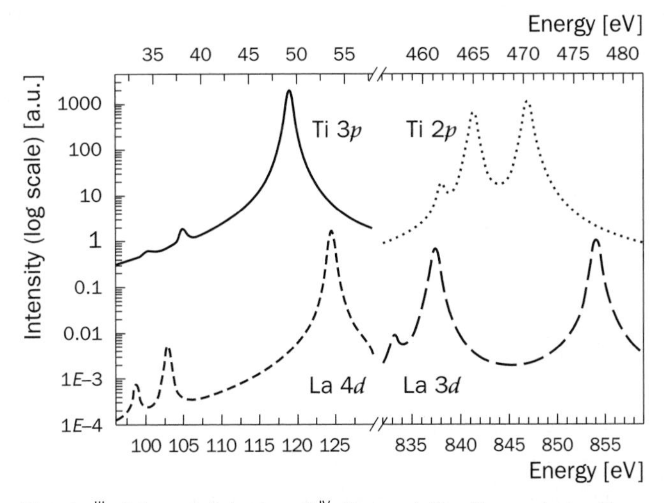

Figure 7 - The La^{III} $4d$ and $3d$ plus Ti^{IV} $3p$ and $2p$ X-ray absorption spectra as calculated for isolated ions. The intensity is given on a logarithmic scale to make the pre-edge peaks visible. The intensities of titanium have been multiplied by 1000.

In table 13 the term symbols of all $3d^n$ systems are given. Together with the dipole selection rules this sets immediately strong limits to the number of final states which can be reached, similar to the case of a $3d^0$ ground state. Consider for example the $3d^3 \longrightarrow 2p^5 3d^4$ transition: the $3d^3$ ground state has $J = 3/2$ and there are respectively 21, 35 and 39 states of $2p^5 3d^4$ with $J' = 1/2$, $J' = 3/2$ and $J' = 5/2$. This implies a total of 95 allowed peaks out of the 180 final state term symbols. From table 13 some special cases can be discriminated: a $3d^9$ system makes a transition to a $2p^5 3d^{10}$ configuration, which has only two term symbols, out of which only the term symbol with $J' = 3/2$ is allowed. In other words, the L_{II} edge has zero intensity. $3d^0$ and $3d^8$ systems have only three respectively four peaks, because of the limited amount of states for the $2p^5 3d^1$ and $2p^5 3d^9$ configurations.

Table 13 - The $2p$ X-ray absorption transitions from the atomic ground state to all allowed final state symmetries, after applying the dipole selection rule: $\Delta J = -1$, 0 or +1.

Transition	Ground	Transitions	Term Symbols
$3d^0 \longrightarrow 2p^5 3d^1$	$^1S_0$	3	12
$3d^1 \longrightarrow 2p^5 3d^2$	$^2D_{3/2}$	29	45
$3d^2 \longrightarrow 2p^5 3d^3$	$^3F_2$	68	110
$3d^3 \longrightarrow 2p^5 3d^4$	$^4F_{3/2}$	95	180
$3d^4 \longrightarrow 2p^5 3d^5$	$^5D_0$	32	205
$3d^5 \longrightarrow 2p^5 3d^6$	$^6S_{5/2}$	110	180
$3d^6 \longrightarrow 2p^5 3d^7$	$^5D_2$	68	110
$3d^7 \longrightarrow 2p^5 3d^8$	$^4F_{9/2}$	16	45
$3d^8 \longrightarrow 2p^5 3d^9$	$^3F_4$	4	12
$3d^9 \longrightarrow 2p^5 3d^{10}$	$^2D_{5/2}$	1	2

Atomic multiplet theory is able to accurately describe the $3d$ and $4d$ X-ray absorption spectra of the rare earths. In case of the $3d$ metal ions, atomic multiplet theory cannot simulate the X-ray absorption spectra accurately because the effects of the neighbours on the $3d$ states are too large. It turns out that it is necessary to include both the symmetry effects and the configuration-interaction effects of the neighbours explicitly.

2.4.2. $M_{IV\text{-}V}$ EDGES OF RARE EARTHS

The rare earths or Lanthanides are elements of the 6th row, with, in the atomic state, electronic configuration $4f^n5d^{0(1)}6s^2$. In the solid state, in most rare earths one $4f$ electron goes to the valence band and the configuration becomes $4f^n(5d6s)^3$. In calculations, usually the rare earth trivalent ions are considered, without the outermost ($5d$ and $6s$) electrons, which do not influence the absorption spectra. The $4f$ electrons are very localized and have little interaction with the environment. They determine the magnetic properties, but do not participate in the chemical bonding. The X-ray absorption edges implying the $4f$ electrons, like the $M_{IV\text{-}V}$ edges ($3d \longrightarrow 4f$ transitions) or $N_{IV\text{-}V}$ edges ($4d \longrightarrow 4f$ transitions) can therefore be described very well within the atomic multiplet theory [19]. The ground states for the different rare earth ions are again given by Hund's rules, and are collected in table 14.

Table 14 - The $4f^n$ atomic ground states of the rare earths.

RE	Ce	Pr	Nd	Sm	Eu	Gd	Tb	Dy	Ho	Er	Tm	Yb	Lu
Conf.	$4f^1$	$4f^2$	$4f^3$	$4f^5$	$4f^6$	$4f^7$	$4f^8$	$4f^9$	$4f^{10}$	$4f^{11}$	$4f^{12}$	$4f^{13}$	$4f^{14}$
Sym.	$^2F_{5/2}$	$^3H_4$	$^4I_{9/2}$	$^6H_{5/2}$	$^7F_0$	$^8S_{7/2}$	$^7F_6$	$^6H_{15/2}$	$^5I_8$	$^4I_{15/2}$	$^3H_6$	$^2F_{7/2}$	^1S

The absorption process can be written as $4f^n \longrightarrow 3d^94f^{n+1}$. The number of possible $3d^94f^{n+1}$ states can be very large, even though in the absorption spectrum only those reachable from the Hund's rule ground state satisfying the $\Delta J = 0, \pm 1$ selection rules will be present. Still, the number of final states that can be reached increases from 3 in lanthanum to 53 in cerium, to 200 in praseodymium and to 1077 in gadolinium. In the end of the series, where the number of $4f$ holes is reduced, it decreases again to 4 in thulium and 1 for ytterbium.

The cases of Tm and Yb involve only configurations with a maximum of respectively one and two holes and can be readily calculated. They will be discussed in detail. The initial state for Yb^{III}, with thirteen $4f$ electrons has $L = 3$ and $S = 1/2$. Two J values are possible $J = 7/2$ and $J = 5/2$ of which the first one, with term symbol $^2F_{7/2}$, is the ground state according to Hund's third rule. The $^2F_{5/2}$ has an energy difference with the ground state that is given by the $4f$ spin-orbit coupling and from atomic calculations one finds an energy difference of 1.3 eV. The final state, after $3d$ (or $4d$) absorption, is given by $3d^94f^{14}$, with term symbols $^2D_{3/2}$ and $^2D_{5/2}$. The energy difference between these terms, corresponding to the $3d$ spin-orbit coupling is 49.0 eV. However, in the X-ray absorption spectrum only the $^2D_{5/2}$ line (corresponding to the M_V edge) is present, since the $^2D_{3/2}$ term can not be reached from the $^2F_{3/2}$ ground state because of the ΔJ selection rules. Crystal field effects, however, can mix some $^2F_{5/2}$ character into the $^2F_{3/2}$ ground state, as will be discussed in section 3.

For Tm, the transition is from the $4f^{12}$ ground state to the $3d^94f^{13}$ final state. The Hund's rule ground state is $^3H_6$. The $3d^94f^{13}$ state after absorption has symmetries that are found after multiplication of a d with a f symmetry state. $^2D \otimes {}^2F$ implies that S is 0 or 1 and L is 1, 2, 3, 4 or 5. This gives five singlet terms ($^1P_1$, $^1D_2$, $^1F_3$, $^1G_4$ and $^1H_5$) and fifteen triplet terms ($^3P_{0,1,2}$, $^3D_{1,2,3}$, $^3F_{2,3,4}$, $^3G_{3,4,5}$ and $^3H_{4,5,6}$), with an overall degeneracy of 10×14 is 140. The $^3H_6$ ground state has $J = 6$ and the final state J must be 5, 6 or 7. There are three states with $J = 5$, respectively $^1H_5$, $^3G_5$ and $^3H_5$ plus one state with $J = 6$, i.e. $^3H_6$. This implies that the $M_{IV\text{-}V}$ edges of Tm exist of four transitions. Figure 8 shows the three peaks at the M_V edge.

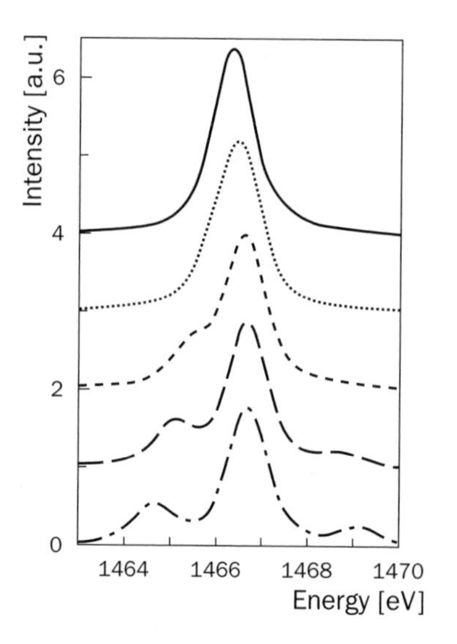

Figure 8 - The three peaks of the M_V edge of Tm^{3+} as a function of the Slater-Condon parameters. The bottom spectrum uses atomic Slater-Condon parameters, i.e. 80% of their Hartree-Fock value. The magnitude of the Slater-Condon parameters is respectively 80%, 60%, 40%, 20% and 0% from bottom to top.

The peak at lowest energy is $^3H_6$ at 1462.38 eV. This is the only term with $J = 6$ and is therefore a pure $^3H_6$ peak also in $\boldsymbol{j \cdot j}$ and thus in intermediate coupling. The other peaks all have $J = 5$, and are mixtures of the three term symbols with $J = 5$ in intermediate coupling. One can determine the exact nature of these three states by solving a 3×3 matrix. The $3d^94f^{13}$ configuration has an average energy of 1482.67 eV and further contains five Slater-Condon parameters, the $4f$ spin-orbit coupling and the $3d$ spin-orbit coupling. The Slater-Condon parameters are respectively $F^2 = 9.09$ eV, $F^4 = 4.31$ eV, $G^1 = 6.68$ eV, $G^3 = 3.92$ eV and $G^5 = 2.71$ eV. The $4f$ spin-orbit coupling has a value of 0.37 eV. The main difference with the initial state is the effect of the large $3d$ spin-orbit coupling of 18.05 eV that is able to strongly mix all states with equal J. This can be seen in table 15.

State I has a wave function $0.455\,|\,^3H_5\rangle - 0.890\,|\,^3G_5\rangle + 0.116\,|\,^1H_5\rangle$. This state is thus approximately 80% pure $^3G_5$ character as also discussed by Pompa and others [20]. The final results of the calculation are the four energies and their respective intensities. After broadening they are given in figure 8.

The $3d \longrightarrow 4f$ transitions have been calculated for all the rare earths [21] using atomic multiplets in intermediate coupling. The electrostatic parameters F^i_f in the initial state and F^i_f and F^i_{fd} in the final state, as well as the exchange parameters G^i_{fd} in the final state, were calculated using an atomic Hartree-Fock program developed by Cowan [22]. These values for scaled down to 80% to account for atomic configuration interaction (CI) effects (see section 4.3).

Table 15 - The energy matrix and eigenvectors of the 3 × 3 matrices of the $3d^94f^{13}$ final states with $J = 5$, after inclusion of the Slater-Condon parameters and the $3d$ and $4f$ spin-orbit couplings. The bottom line includes the energy and intensity of the $J = 6$ final state.

Energy Matrix					Eigenvectors		
1484.706	10.607	−19.163		$^3H_5$	0.455	−0.609	−0.649
10.607	1469.995	−10.088		$^3G_5$	−0.890	−0.302	−0.341
−19.163	−10.088	1486.965		$^1H_5$	0.116	−0.733	0.680

Energy Levels		Intensities
1464.44	I	4.11
1466.89	II	0.52
1510.33	III	0.23
1462.38	$^3H_6$	1.16

The resulting line spectra were then broadened by a Lorentzian to account for the finite lifetime, and an additional Gaussian to reproduce the experimental resolution. For $3d_{3/2}$ transitions, the lineshape is asymmetric due to interactions between the "discrete" $3d_{3/2} \longrightarrow 4f$ transitions and the transitions from $3d_{3/2}$ into the continuum ($6p$, $7p$, $5f$, etc.). This has been taken into account using a Fano lineshape for the M_{IV} edge. The results of the calculations are in very good agreement with the experimental absorption edges.

Some general trends in the spectra are the presence of three distinct groups of lines in the calculated line spectra, giving rise to three peaks in the absorption edge. This is visible especially in the M_{IV} edge for the lighter rare earths and in the M_V edge for the heavier ones. This splitting into three groups is due to the spin-orbit coupling, which, in intermediate coupling, tends to group lines with the same J together. Another trend is the M_V:M_{IV} branching ratio that is almost 1:1 for the beginning of the series, but increases a lot going to the heavier rare earths. The spin-orbit coupling in the $4f$ levels favours $4f_{7/2}$ holes, which are only reachable from the $3d_{5/2}$ level, as was shown above for Yb.

3. *CRYSTAL FIELD THEORY*

Crystal Field Theory is a well-known model used to explain the electronic properties of transition metal systems. It has been developed in the fifties and sixties against the background of explaining optical spectra and EPR data. Books on crystal field theory have been written by Ballhausen [23], Sugano *et al.* [24] and Butler [25]. The starting point of the crystal field model is to approximate the transition metal as an isolated atom surrounded by a distribution of charges, which should mimic the system, molecule or solid, around the transition metal. It turned out that such a simple model was extremely successful to explain a large range of experiments, like optical spectra, EPR spectra, magnetic moments... [24] .

The most important reason of the success of the crystal field model is that the explained properties are strongly determined by symmetry considerations. With its simplicity in concept, the crystal field model could make full use of the results of group theory. Group theory also made possible a close link to atomic multiplet theory. Group theoretically speaking, the only thing crystal field theory does is translate, or branch, the results obtained in atomic symmetry to cubic symmetry and further to any other lower point groups. The mathematical concepts for these branching are well developed. In this section we will focus, using group theory, on the consequences the crystal field effect has on the atomic multiplet states as well as on the spectral shapes. For a discussion of group theory, the reader is referred to books like Weissbluth [18], and Butler [25].

3.1. THE CRYSTAL FIELD MULTIPLET HAMILTONIAN

The crystal field multiplet Hamiltonian consists of the atomic Hamiltonian as outlined in the previous section, to which an electrostatic term is added

$$H_{CF} = H_{ATOM} + H_{FIELD} \qquad \{22\}$$

$$H_{ATOM} = \sum_N \frac{p_i^2}{2m} + \sum_N \frac{-Ze^2}{r_i} + \sum_{\text{pairs}} \frac{e^2}{r_{ij}} + \sum_N \zeta(r_i) l_i \cdot s_i \qquad \{23\}$$

$$H_{FIELD} = -e\phi(r) \qquad \{24\}$$

The electrostatic term consists of the electronic charge e times a potential $\phi(r)$ that describes the surroundings. This potential is written as a series expansion of spherical harmonics Y_{LM}

$$\phi(r) = \sum_{L=0}^{\infty} \sum_{M=-L}^{L} r^L A_{LM} Y_{LM}(\psi,\phi) \qquad \{25\}$$

The electrostatic term due to the crystal field is regarded as a perturbation to the atomic result. This implies that it is necessary to determine the matrix elements of $\phi(r)$ with respect to the atomic $3d$ orbitals $\langle 3d | \phi(r) | 3d \rangle$. One can separate the matrix elements into a spherical part and a radial part, as was done for the atomic Hamiltonian in equation {17}. The radial part of the matrix elements yields the strength of the crystal field interaction. The spherical part of the matrix element can be written completely in Y_{LM} symmetry, where the two 3*d* electrons are written as Y_{2m}. This gives

$$\langle Y_{2m_2} | Y_{LM} | Y_{2m_1} \rangle = (-1)^{m_2} \sqrt{15(2L+1)/4\pi} \begin{pmatrix} 2 & L & 2 \\ -m_2 & M & -m_1 \end{pmatrix} \begin{pmatrix} 2 & L & 2 \\ 0 & 0 & 0 \end{pmatrix} \quad \{26\}$$

The second 3*J* symbol is zero unless *L* is equal to 0, 2 or 4. This limits the crystal field potential for 3*d* electrons to

$$\phi(r) = A_{00}Y_{00} + \sum_{M=-2}^{2} r^2 A_{2M}Y_{2M} + \sum_{M=-4}^{4} r^4 A_{4M}Y_{4M} \quad \{27\}$$

The first term $A_{00}Y_{00}$ is a constant. It will only shift the atomic states and it is not necessary to include this term explicitly if one calculates the spectral shape.

3.2. *CUBIC CRYSTAL FIELDS*

A large range of systems posses a transition metal ion surrounded by six or eight neighbours. The six neighbours are positioned on the three Cartesian axes, or in other words on the six faces of a cube surrounding the transition metal. They form a so-called octahedral field. The eight neighbours are positioned on the eight corners of the cube and form a so-called cubic field. Both these systems belong to the O_h point group. The character table of O_h symmetry is given below. O_h symmetry is a subgroup of the atomic SO_3 group.

The calculation of the X-ray absorption spectral shape in atomic symmetry involved the calculation of the matrices of the initial state, the final state and the transition. The initial state is given by the matrix element $\langle 3d^n | H_{ATOM} | 3d^n \rangle$, which for a particular *J* value in the initial state gives $\Sigma_J \langle J | 0 | J \rangle$. The same applies for the final state matrix element $\langle 2p^5 3d^{n+1} | H_{ATOM} | 2p^5 3d^{n+1} \rangle$, where $\Sigma_{J'} \langle J' | 0 | J' \rangle$ is calculated for the values of J' that fulfil the selection rule. The dipole matrix element $\langle 3d^n | p | 2p^5 3d^{n+1} \rangle$ implies the calculation of all matrices that couple *J* and J': $\Sigma_{J,J'} \langle J | 1 | J' \rangle$. To calculate the X-ray absorption spectrum in a cubic crystal field, these atomic transition matrix elements must be branched to cubic symmetry. This is essentially the only task to fulfil.

Table 16 gives the branching from SO_3 to O_h symmetry. This table can be determined from group theory [25]. An *S* symmetry state in atomic symmetry branches only to a A_1 symmetry state in octahedral symmetry. This is because the symmetry elements of an *s* orbital in O_h symmetry are determined by the character table of A_1 symmetry, i.e. whatever symmetry operation one applies an *s* orbital remains an *s* orbital. This is not the case for the other orbitals. For example, a

p orbital can be described with the characters of the T_1 symmetry state in O_h symmetry. A d orbital or a D symmetry state in SO_3, branches to E plus T_2 symmetry states in octahedral symmetry. This can be related to the character table by adding the characters of E and T_2 symmetry, the properties of d orbitals in O_h symmetry. This is a well-known result: A $3d$ electron is separated into t_{2g} and e_g electrons in octahedral symmetry, where the symmetries include the gerade notation of the complete O_h character table.

Table 16 - Branching rules for the symmetry elements by going from SO_3 symmetry to O_h symmetry.

SO_3		O_h (Butler) [25]	O_h (Mulliken) [24]
S	0	0	A_1
P	1	1	T_1
D	2	$2 + \hat{1}$	$E + T_2$
F	3	$\hat{0} + 1 + \hat{1}$	$A_2 + T_1 + T_2$
G	4	$0 + 1 + 2 + \hat{1}$	$A_1 + E + T_1 + T_2$

One can make the following observations: The dipole transition operator has p symmetry and is branched to T_1 symmetry. Having a single symmetry in O_h symmetry, there will be no dipolar angular dependence in X-ray absorption. The quadrupole transition operator has d symmetry and is split into two operators in O_h symmetry, in other words there will be different quadrupole transitions in different directions. The Hamiltonian is given by the unity representation A_1 of the symmetry under consideration. In O_h symmetry the atomic G symmetry state branches into the A_1 Hamiltonian, which is a confirmation of equation {27} as given above. We can lower the symmetry from octahedral O_h to tetragonal D_{4h} and describe again with a branching table . Table 17 gives the branching table from O_h to D_{4h} symmetry.

Table 17 - Branching rules for the symmetry elements by going from O_h symmetry to D_{4h} symmetry.

O_h (Butler) [25]	O_h (Mulliken) [24]	D_{4h} (Butler) [25]	D_{4h} (Mulliken) [24]
0	A_1	0	A_1
$\hat{0}$	A_2	2	B_1
1	T_1	$1 + \hat{0}$	$E + A_2$
$\hat{1}$	T_2	$1 + \hat{2}$	$E + B_2$
2	E	$0 + 2$	$A_1 + B_1$

An atomic *s* orbital is branched to D_{4h} symmetry according to the branching series $S \longrightarrow A_1 \longrightarrow A_1$. It is still the unity element, and it will always be in all symmetries. An atomic *p* orbital is branched according to $P \longrightarrow T_1 \longrightarrow E + A_2$. A twofold rotation around the *z* axis inverts a *p* orbital, etc. Similarly an atomic *d* orbital is branched according to $D \longrightarrow E + T_2 \longrightarrow A_1 + B_1 + E + B_2$. The dipole transition operator has *p* symmetry and hence is branched to $E + A_2$ symmetry, in other words the dipole operator is described with two operators in two different directions implying an angular dependence in the X-ray absorption intensity. The quadrupole transition operator has *d* symmetry and is split into four operators in D_{4h} symmetry, in other words there will be four different quadrupole transitions in different directions/symmetries. The Hamiltonian is given by the unity representation A_1. As in O_h symmetry, the atomic *G* symmetry state branches into the Hamiltonian in D_{4h} symmetry according to the series $G \longrightarrow A_1 \longrightarrow A_1$. In addition it can be seen that the *E* symmetry state of O_h symmetry branches to the A_1 state in D_{4h} symmetry. The *E* symmetry state in O_h symmetry is found from the *D* and *G* atomic states. This implies that also the series $G \longrightarrow E \longrightarrow A_1$ and $D \longrightarrow E \longrightarrow A_1$ become part of the Hamiltonian in D_{4h} symmetry. This is again a confirmation of equation {27}, where we find that the second term $A_{2M}Y_{2M}$ is part of the Hamiltonian in D_{4h} symmetry. The three branching series in D_{4h} symmetry are in Butlers notation [25] given as $4 \longrightarrow 0 \longrightarrow 0$, $4 \longrightarrow 2 \longrightarrow 0$ and $2 \longrightarrow 2 \longrightarrow 0$ and the radial parameters related to these branches are indicated as X_{400}, X_{420}, and X_{220}. The X_{400} term is important already in O_h symmetry and is closely related to the cubic crystal field term 10 *Dq* as will be discussed below.

The definitions of the crystal field parameters

The X_{400}, X_{420} and X_{220} definition of crystal field operators are used in X-ray absorption, while other definitions like *Dq*, *Ds* and *Dt* are used in optical spectroscopy. In order to compare these definitions, we compare their effects on the set of 3*d* functions. The most straightforward way to specify the strength of the crystal field parameters is to calculate the energy separations of the 3*d* functions. In O_h symmetry there is only one crystal field parameter X_{40}. In tetragonal symmetry (D_{4h}) the crystal field is given by X_{400}, X_{420} and X_{220}. Table 18 gives the action of the X_{400}, X_{420} and X_{220} on the 3*d* orbitals and relates the respective symmetries to the linear combination of *X* parameters, the linear combination of the *Dq*, *Ds* and *Dt* parameters and the specific 3*d* orbital(s) of that particular symmetry.

Table 18 - The energy of the $3d$ orbitals is expressed in X_{400}, X_{420} and X_{220} in the second column and in *Dq, Ds* and *Dt* in the third column.

Γ	Energy expressed in *X* terms	Energy in *D* terms	orbitals
B_1	$30^{-1/2} X_{400} - 42^{-1/2} X_{420} - 2 \times 70^{-1/2} X_{220}$	$6Dq + 2Ds - 1Dt$	$3d_{x2-y2}$
A_1	$30^{-1/2} X_{400} - 42^{-1/2} X_{420} - 2 \times 70^{-1/2} X_{220}$	$6Dq - 2Ds - 6Dt$	$3d_{z2}$
B_2	$-2/3 \times 30^{-1/2} X_{400} + 4/3 \times 42^{-1/2} X_{420} - 2 \times 70^{-1/2} X_{220}$	$-4Dq + 2Ds - 1Dt$	$3d_{xy}$
E	$-2/3 \times 30^{-1/2} X_{400} + 2/3 \times 42^{-1/2} X_{420} + 70^{-1/2} X_{220}$	$-4Dq - 1Ds + 4Dt$	$3d_{xz}$, $3d_{yz}$

3.3. THE ENERGIES OF THE $3d^n$ CONFIGURATIONS

Above, we have given in table 5 the energy levels of a $3d^8$ configuration and in table 13 the ground states of the $3d^n$ configurations in atomic symmetry. The crystal field effect modifies these energy levels by the additional terms in the Hamiltonian as given in equation {22}. We will use the $3d^8$ configuration as an example to show the effects of the O_h and D_{4h} symmetry. Assuming for the moment that the $3d$ spin-orbit coupling is zero, in O_h symmetry the five term symbols in spherical symmetry split into eleven term symbols. Their respective energies can be calculated by adding the effect of the cubic crystal field 10 *Dq* to the atomic energies. The diagrams of the respective energies with respect to the cubic crystal field (normalized to the Racah parameter *B*) are known as the Tanabe-Sugano diagrams.

Figure 9 gives the Tanabe-Sugano diagram for the $3d^8$ configuration. The ground state in O_h symmetry has $^3A_2$ symmetry and its energy is set to zero in figure 9. If the crystal field energy is 0.0 eV one has effectively the atomic multiplet states. From low-energy to high-energy, one can observe respectively the ^3F, ^1D, ^3P, ^1G and ^1S states. Including a finite crystal field strength splits these states, for example the ^3F state is split into $^3A_2 + {}^3T_1 + {}^3T_2$ as indicated in table 19.

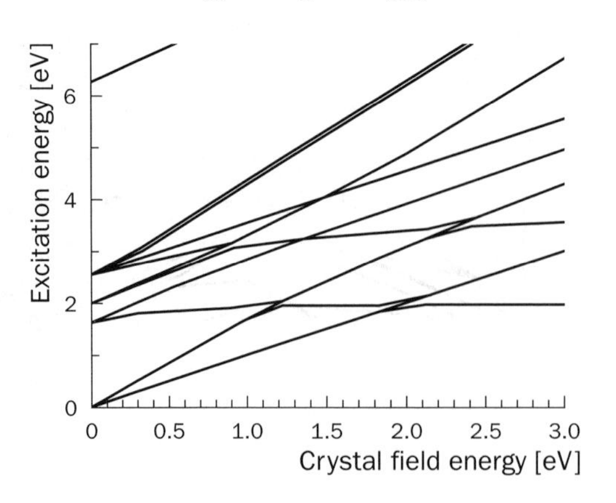

Figure 9 - The Tanabe-Sugano diagram for a $3d^8$ configuration in O_h symmetry.

Table 19 - The five symmetry states of a $3d^8$ configuration in SO_3 symmetry and their respective energies for Ni^{II} are given in columns 1 and 2. Column 3 gives the respective symmetries of these states in O_h symmetry and column 4 in D_{4h} symmetry. In both cases the spin-orbit coupling has not yet been included. The Hund's rule ground state is given in boldface.

	Energy	Symmetries in O_h	Symmetries in D_{4h}
^1S	4.6 eV	$^1A_1$	$^1A_1$
^3P	0.2 eV	$^3T_1$	$^3E + {}^3A_2$
^1D	–0.1 eV	$^1E + {}^1T_2$	$^1A_1 + {}^1B_1 + {}^1E + {}^1B_2$
$\mathbf{^3F}$	–1.8 eV	$\mathbf{^3A_2} + {}^3T_1 + {}^3T_2$	$\mathbf{^3B_1} + {}^3E + {}^3A_2 + {}^3E + {}^3B_2$
^1G	0.8 eV	$^1A_1 + {}^1T_1 + {}^1T_2 + {}^1E$	$^1A_1 + {}^1E + {}^1A_2 + {}^1E + {}^1B_2 + {}^1A_1 + {}^1B_1$

At higher crystal field strengths states start to change their order and they cross. If states actually cross each other or show non-crossing behaviour if their symmetries allow a linear combination of states to be formed. This also depends on the inclusion of the $3d$ spin-orbit coupling.

Figure 10 shows the effect of the reduction of the Slater-Condon parameters. The figure is the same as figure 9 up to a crystal field of 1.5 eV. Then for this crystal field value the Slater-Condon parameters have been reduced from their atomic value, indicated with 80% of their Hartree-Fock value to 0%. The spectrum for 0% has all its Slater-Condon parameters reduced to zero, In other words the $2p3d$ coupling has been turned of and one essentially observes the energies of a $3d^8$ configuration, i.e. of two $3d$ holes. This single particle limit has three configurations, respectively the two holes in $e_g e_g$, $e_g t_{2g}$ and $t_{2g} t_{2g}$ states. The energy difference between $e_g e_g$ and $e_g t_{2g}$ is exactly the crystal field value of 1.5 eV. This figure shows nicely the transition from the single particle picture to the multiplet picture for the $3d^8$ ground state.

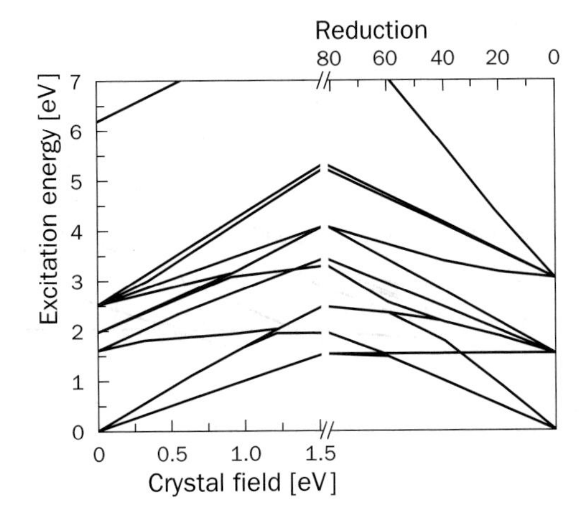

Figure 10 - The Tanabe-Sugano diagram for a $3d^8$ configuration in O_h symmetry, including the effect of reduced Slater-Condon parameters.

The ground state of a $3d^8$ configuration in O_h symmetry always remains $^3A_2$. The reason is clear if one compares these configurations to the single particle description. In a single particle description a $3d^8$ configuration is split by the cubic crystal field into t_{2g} and e_g, following the branching rules from table 17. The eight $3d$ electrons are added one-by-one to these configurations. The t_{2g} configuration has the lowest energy and can contain six $3d$ electrons. The remaining two electrons are placed in the e_g configuration, where both have a parallel alignment according to Hund's rule. The resulting overall configuration is $t_{2g}{}^6 e_{g+}{}^2$, which identifies with the $^3A_2$ configuration.

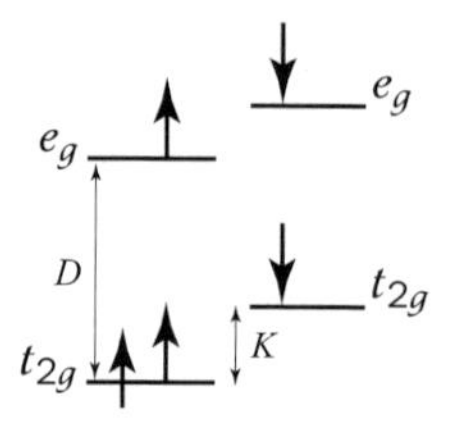

Figure 11 - The splitting of a single $3d$ electron under influence of the cubic crystal field D and the Stoner exchange interaction J. A second electron is indicated with an empty arrow to indicate the energy effects.

Figure 11 shows the splitting of a $3d$ configuration into an e_g and a t_{2g} configuration. Both configurations are further split by the Stoner exchange splitting J. This splitting is given as a linear combination of the Slater-Condon parameters as $J = (F_2 + F_4)/14$. J is an approximation to the effects of the Slater-Condon parameters and in fact a second parameter C, the orbital polarization, can be used in combination with J. It is given as $C = (9F_2 - 5F_4)/98$. Often this orbital polarization is omitted from single particle descriptions. In that case the multiplet configuration $^3A_2$ is not exactly equal to the single particle configuration $t_{2g}{}^6 e_g{}^2$. We assume for the moment that the effect of the orbital polarization will not modify the ground states. Then one can use figure 11 to show that the ground states of $3d^n$ configurations are those as given in table 20.

Table 20 shows that for $3d^4$, $3d^5$, $3d^6$ and $3d^7$ there are two possible ground state configurations in O_h symmetry. A high-spin ground state that originates from Hund's rules plus a low-spin ground state for which first all t_{2g} levels are filled. One can directly related the symmetry of a configuration to the partly filled sub-shell in the single particle model. A single particle configuration with one t_{2g} electron has T_2 symmetry, two t_{2g} electrons imply T_1 symmetry and one e_g electron implies E symmetry. If only the t_{2g} electrons are filled the symmetry is A_2 and if both or none are filled the symmetry is A_1. The nature of the ground state is important, as we will show below that E symmetry states are susceptible to Jahn-Teller distortions and T_1 and T_2 symmetry states are susceptible to the effects of the $3d$ spin-orbit coupling.

Table 20 - The configurations $3d^0$ to $3d^9$ are given in O_h symmetry for all possible high-spin (HS) and low-spin (LS) states. The third column gives the HS term symbols and the last column the LS term symbols. The fourth and fifth columns give the respective occupations of the t_{2g} and e_g orbitals. We use t_{2g+} for a spin-up electron and t_{2g-} for a spin-down

Configuration	SO_3	HS Ground State in O_h		LS Ground State in O_h	
$3d^0$	$^1S_0$	$^1A_1$	–	–	–
$3d^1$	$^2D_{3/2}$	$^2T_2$	$t_{2g+}{}^1$	–	–
$3d^2$	$^3F_2$	$^3T_1$	$t_{2g+}{}^2$	–	–
$3d^3$	$^4F_{3/2}$	$^4A_2$	$t_{2g+}{}^3$	–	–
$3d^4$	$^5D_0$	^5E	$t_{2g+}{}^3\ e_{g+}{}^1$	$t_{2g+}{}^3\ t_{2g-}{}^1$	$^3T_1$
$3d^5$	$^6S_{5/2}$	$^6A_1$	$t_{2g+}{}^3\ e_{g+}{}^2$	$t_{2g+}{}^3\ t_{2g-}{}^2$	$^2T_2$
$3d^6$	$^5D_2$	$^5T_2$	$t_{2g+}{}^3\ e_{g+}{}^2\ t_{2g-}{}^1$	$t_{2g+}{}^3\ t_{2g-}{}^3$	$^1A_1$
$3d^7$	$^4F_{9/2}$	$^4T_1$	$t_{2g+}{}^3\ e_{g+}{}^2\ t_{2g-}{}^2$	$t_{2g+}{}^3\ t_{2g-}{}^3\ e_{g+}{}^1$	^2E
$3d^8$	$^3F_4$	$^3A_2$	$t_{2g+}{}^3\ e_{g+}{}^2\ t_{2g-}{}^3$	–	–
$3d^9$	$^2D_{5/2}$	^2E	$t_{2g+}{}^3\ e_{g+}{}^2\ t_{2g-}{}^3\ e_{g-}{}^1$	–	–

The transition from high-spin top low-spin ground states is determined by the cubic crystal field 10 Dq and the exchange splitting J. Table 21 gives the high-spin and low-spin occupations of the t_{2g} and e_g spin-up and spin-down orbitals t_{2g+}, e_{g+}, t_{2g-} and e_{g-}. The $3d^4$ and $3d^7$ configuration differ by one t_{2g} versus e_g electron hence one time the crystal field splitting D. The $3d^5$ and $3d^6$ configurations differ by $2D$. The exchange interaction J is slightly different for $e_g e_g$ (J_{ee}), $e_g t_{2g}$ (J_{te}) and t_{2g} t_{2g} (J_{tt}) interactions and column 5 of table 21 contains the overall exchange interactions. The last column can be used to estimate the transition point. For this column the exchange splittings were assumed to be equal, yielding the simple rules that for $3d^4$ and $3d^5$ configurations high-spin states are found if the crystal field splitting is less than 3J and for $3d^6$ and $3d^7$ configurations if it is less than $2J$. Because J can be estimated as 0.8 eV, the transition points are approximately 2.4 eV for $3d^4$ and $3d^5$, and 1.6 eV for $3d^6$ and $3d^7$. This means that $3d^6$ and $3d^7$ materials have a tendency to be low-spin compounds. This is particularly true for $3d^6$ compounds because of the additional stabilizing nature of the $3d^6$ $^1A_1$ low-spin ground state.

Table 21 - The high-spin and low-spin distribution of the $3d$ electrons for the configurations $3d^4$ to $3d^7$. The fourth column gives the difference in crystal field energy, the fifth column the difference in exchange energy. For the last column, we have assumed that $J_{te} \sim J_{ee} \sim J_{tt} = J$.

Configuration	High-spin	Low-spin	10 *Dq* (*D*)	Exchange (*J*)	*J*/*D*
$3d^4$	$t_{2g+}^{\ 3}\ e_{g+}^{\ 1}$	$t_{2g+}^{\ 3}\ t_{2g-}^{\ 1}$	$1D$	$3J_{te}$	3
$3d^5$	$t_{2g+}^{\ 3}\ e_{g+}^{\ 2}$	$t_{2g+}^{\ 3}\ t_{2g-}^{\ 2}$	$2D$	$6J_{te} + J_{ee} - J_{tt}$	~3
$3d^6$	$t_{2g+}^{\ 3}\ e_{g+}^{\ 2}\ t_{2g-}^{\ 1}$	$t_{2g+}^{\ 3}\ t_{2g-}^{\ 3}$	$2D$	$6J_{te} + J_{ee} - 3J_{tt}$	~2
$3d^7$	$t_{2g+}^{\ 3}\ e_{g+}^{\ 2}\ t_{2g-}^{\ 2}$	$t_{2g+}^{\ 3}\ t_{2g-}^{\ 3}\ e_{g+}^{\ 1}$	$1D$	$3J_{te} + J_{ee} - 2J_{tt}$	2

3.3.1. *Symmetry effects in D_{4h} symmetry*

In D_{4h} symmetry the t_{2g} and e_g states split further into e_g and b_{2g} respectively a_{1g} and b_{1g}. Depending on the nature of the tetragonal distortion either the e_g or the b_{2g} state has the lowest energy. All configurations from $3d^2$ to $3d^8$ have a low-spin possibility. Only the $3d^2$ configuration with the e_g state as ground state does not posses a low-spin configuration. The $3d^1$ and $3d^9$ configurations contain only one unpaired spin thus they have no possibility to obtain a low-spin ground state. It is important to notice that a $3d^8$ configuration as for example found in Ni^{II} and Cu^{III} can yield a low-spin configuration. This low-spin configuration is found in the trivalent parent compounds of the high T_C superconducting oxides **[26]**. The D_{4h} symmetry ground states are particularly important for those cases where O_h symmetry yields a half filled e_g state, like for $3d^4$ and $3d^9$ plus low-spin $3d^7$. These ground states are unstable in octahedral symmetry and will relax to, for example, a D_{4h} ground state, the well-known Jahn-Teller distortion. This yields the Cu^{II} ions with all states filled except the $^1A_{1g}$ hole.

3.3.2. *The effect of the 3d spin-orbit coupling*

The inclusion of $3d$ spin-orbit coupling will bring one to the multiplication of the spin and orbital moments to a total moment. In this process the familiar nomenclature for the ground states of the $3d^n$ configurations is lost. For example the ground state of Ni^{II} in octahedral symmetry is in total symmetry referred to as T_2 and not as $^3A_2$. In total symmetry also the spin moments are branched to the same symmetry group as the orbital moments, yielding for a $^3A_2$ ground state an overall ground state of $T_1 \otimes A_2 = T_2$. It turns out that in many cases it is better to omit the $3d$ spin-orbit coupling because it is "quenched", for example by solid state effects. This has been found to be the case for CrO_2. A different situation is found for CoO, where the explicit inclusion of the 3*d* spin-orbit coupling is essential for a good description of the 2*p* X-ray absorption spectral shape. This means that 2*p* X-ray absorption is able to determine the different role of the 3*d* spin-orbit coupling in respectively CrO_2 (quenched) and CoO (not quenched).

Figure 12 shows the Tanabe-Sugano diagram for a $3d^7$ configuration in O_h symmetry. Only the excitation energies from 0.0 to 0.4 eV are shown to highlight the high-spin low-spin transition at 2.25 eV and also the important effect of the 3*d* spin-orbit coupling. It can be observed that the atomic multiplet spectrum of Co^{II} has a large number of states at low energy. All these states are part of the $^4F_{9/2}$ configuration that is split by the 3*d* spin-orbit coupling. After applying a cubic crystal field, most of these multiplet states are shifted to higher energies and only four states remain at low energy. These are the four states of the $^4T_1$ Hund's rule ground state. These four states all remain within 0.1 eV from the U_1 ground state. That this description is actually correct has been shown in detail for the 2*p* X-ray absorption spectrum of CoO **[27]**, which has a cubic crystal field of 1.2 eV. At 2.25 eV the high-spin to low-spin transition is evident. A new state is coming from high energy and a *G* symmetry state replaces the U_1 symmetry state at the lowest energy. In fact there is a very interesting complication: due to the 3*d* spin-orbit coupling the *G* symmetry states of the $^4T_1$ and ^2E configurations mix and form linear combinations. Just above the transition point, this linear combination will have a spin state that is neither high-spin nor low-spin and in fact a mixed spin state can be found.

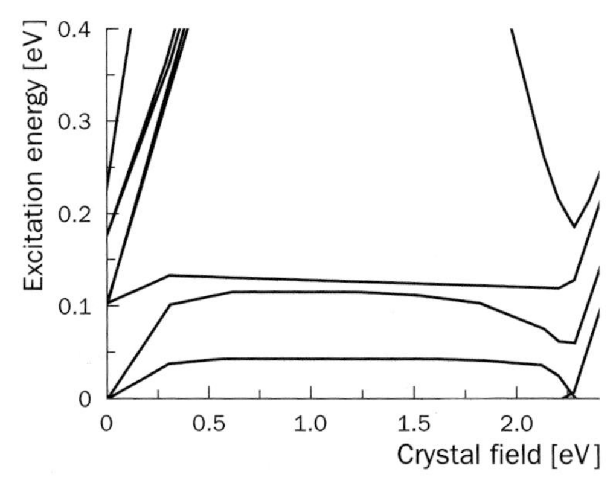

Figure 12 - The Tanabe-Sugano diagram for a $3d^7$ configuration in O_h symmetry.

3.3.3. THE EFFECTS ON THE X-RAY ABSORPTION CALCULATIONS

Table 22 gives all matrix element calculations that have to be carried out for $3d^n \longrightarrow 2p^5 3d^{n+1}$ transitions in SO_3 symmetry for J values up to 4. We will use the transitions $3d^0 \longrightarrow 2p^5 3d^1$ as examples. $3d^0$ contains only $J = 0$ symmetry states. This limits the calculation for the ground state spectrum to only one ground state, one transition and one final state matrix element. We are now going to apply the $SO_3 \longrightarrow O_h$ branching rules to this table.

Table 22 - The matrix elements in SO_3 symmetry needed for the calculation of $2p$ X-ray absorption. Boldface matrix elements apply to a $3d^0$ configuration.

Calculation $3d^n \longrightarrow 2p^5 3d^{n+1}$ in SO_3 symmetry		
Initial State	Transition	Final State
⟨0 | 0 | 0⟩	**⟨0 | 1 | 1⟩**	⟨0 | 0 | 0⟩
⟨1 | 0 | 1⟩	⟨1 | 1 | 0⟩ ⟨1 | 1 | 1⟩ ⟨1 | 1 | 2⟩	**⟨1 | 0 | 1⟩**
⟨2 | 0 | 2⟩	⟨2 | 1 | 1⟩ ⟨2 | 1 | 3⟩	⟨2 | 0 | 2⟩
⟨3 | 0 | 3⟩	⟨3 | 1 | 2⟩ ⟨3 | 1 | 3⟩ ⟨3 | 1 | 4⟩	⟨3 | 0 | 3⟩
⟨4 | 0 | 4⟩	⟨4 | 1 | 3⟩ ⟨4 | 1 | 4⟩	⟨4 | 0 | 4⟩

In octahedral symmetry one has to calculate five matrices for the initial and final states and thirteen transition matrices. Note that this is a general result for all even numbers of $3d$ electrons, as there are only these five symmetries in O_h symmetry. In the $3d^0$ case, the ground state branches to A_1 and only three matrices are needed to generate the spectral shape: $\langle A_1 | A_1 | A_1 \rangle$ for the $3d^0$ ground state, $\langle A_1 | T_1 | T_1 \rangle$ for the dipole transition and $\langle T_1 | A_1 | T_1 \rangle$ for the $2p^5 3d^1$ final state.

3.3.4. $3d^0$ SYSTEMS IN OCTAHEDRAL SYMMETRY

In this section we will focus on the discussion of the crystal field effects on the spectral shape of $3d^0$ systems. The $3d^0$ systems are rather special because they are not affected by ground state effects. Table 23 shows that the $3d^0 \longrightarrow 2p^5 3d^1$ transition can be calculated from a single transition matrix $\langle A_1 | T_1 | T_1 \rangle$ in O_h symmetry. The ground state A_1 matrix is 1×1 and the final state T_1 matrix is 7×7, making the transition matrix 1×7. In other words the spectrum consists of a maximum of seven peaks. Table 7 showed the complete calculation in SO_3 symmetry. The respective degeneracies of the J values in SO_3 symmetry and the degeneracies of the representations in O_h symmetry are collected in table 24. A $2p^5 3d^1$ configuration has twelve representations in SO_3 symmetry that are branched to 25 representations in a cubic field. The overall degeneracy of the $2p^53d^1$ configuration is $6 \times 10 = 60$, implying a possibility of 60 transitions in a system without any symmetry. From these 25 representations, only seven are of interest for the calculation of the X-ray absorption spectral shape, because only these T_1 symmetry states obtain a finite intensity.

Table 23 - The matrix elements in O_h symmetry needed for the calculation of $2p$ X-ray absorption. Boldface matrix elements apply to a $3d^0$ configuration.

Calculation $3d^n \longrightarrow 2p^53d^{n+1}$ in O_h symmetry		
Initial State	Transition	Final State
$\boldsymbol{\langle A_1 \mid A_1 \mid A_1 \rangle}$	$\boldsymbol{\langle A_1 \mid T_1 \mid T_1 \rangle}$	$\langle A_1 \mid A_1 \mid A_1 \rangle$
$\langle T_1 \mid A_1 \mid T_1 \rangle$	$\langle T_1 \mid T_1 \mid A_1 \rangle$ $\langle T_1 \mid T_1 \mid T_1 \rangle$ $\langle T_1 \mid T_1 \mid E \rangle$ $\langle T_1 \mid T_1 \mid T_2 \rangle$	$\boldsymbol{\langle T_1 \mid A_1 \mid T_1 \rangle}$
$\langle E \mid A_1 \mid E \rangle$	$\langle E \mid T_1 \mid T_1 \rangle$ $\langle E \mid T_1 \mid T_2 \rangle$	$\langle E \mid A_1 \mid E \rangle$
$\langle T_2 \mid A_1 \mid T_2 \rangle$	$\langle T_2 \mid T_1 \mid T_1 \rangle$ $\langle T_2 \mid T_1 \mid E \rangle$ $\langle T_2 \mid T_1 \mid T_2 \rangle$ $\langle T_2 \mid T_1 \mid A_2 \rangle$	$\langle T_2 \mid A_1 \mid T_2 \rangle$
$\langle A_2 \mid A_1 \mid A_2 \rangle$	$\langle A_2 \mid T_1 \mid T_2 \rangle$	$\langle A_2 \mid A_1 \mid A_2 \rangle$

Table 24 - The branching of the J values in SO_3 symmetry to the representations in O_h symmetry, using the degeneracies of the $2p^53d^1$ final state in X-ray absorption.

J in SO_3	Degree	Branchings	Γ in O_h	Degree
0	1	A_1	A_1	2
1	3	$3 \times T_1$	A_2	3
2	4	$4 \times E$, $4 \times T_2$	T_1	7
3	3	$3 \times A_2$, $3 \times T_1$,$3 \times T_2$	T_2	8
4	1	A_1, E, T_1, T_2	E	5
Σ	12			25

Table 25 shows the seven T_1 symmetry states calculated with a crystal field splitting of 3.04 eV. Rows one, two and three are related to $J = 1$ final states, where the third row is related to the ${}^1P_1$ state and the intensity of the peak is equal to the square of the values of this row, with the total intensity normalized to 1.0. Rows four, five and six are related to the $J = 3$ states and row seven is related to a $J = 4$ state. Essentially one observes four main peaks, peaks [c, e, f] and [g]. Peak [f] and [g] correspond to the L_{II} edge peaks of respectively t_{2g} and e_g character. They are split by 3.05 eV, about the value of 10 Dq. Peaks [c] and [e] are the L_{III} peaks of t_{2g} and e_g character, also split by 3.05 eV. Peaks [a, b] and [d] are low intensity peaks that originate from the "spin-forbidden transition" in the atomic multiplet calculation.

Table 25 - The T_1 final states of the $2p^5 3d^1$ configuration with 10 Dq = 3.04 eV. The top row gives the energies of the seven final states that are build from seven basis vectors. The third row is related to $^1P_1$ character and is given in boldface.

	460.828	461.641	462.806	464.048	465.859	468.313	471.369
$J = 1$	0.0662	0.0037	0.1550	0.0124	0.4916	0.0404	0.2308
$J = 1$	0.5944	0.0253	0.0007	0.2972	0.0280	0.0078	0.0466
$J = 1$ $^1P_1$	**0.0046** [a]	**0.0091** [b]	**0.1128** [c]	**0.0046** [d]	**0.1845** [e]	**0.2666** [f]	**0.4178** [g]
$J = 3$	0.0161	0.4460	0.0340	0.0980	0.0097	0.2923	0.1039
$J = 3$	0.0020	0.2973	0.2980	0.0791	0.0331	0.2191	0.0714
$J = 3$	0.0044	0.0404	0.3986	0.0116	0.2417	0.1738	0.1294
$J = 4$	0.3124	0.1781	0.0009	0.4972	0.0113	0.0000	0.0001

Figure 13 shows the crystal field multiplet calculations for the $3d^0 \longrightarrow 2p^5 3d^1$ transition in Ti^{IV}. The result of each calculation is a set of seven energies with seven intensities. These seven states are broadened by the lifetime broadening and the experimental resolution. From a detailed comparison to experiment it turns out to be the case that each of the four main lines has to be broadened differently. It is well known that the L_{II} part of the spectrum (i.e. the last two peaks) contains an additional Auger decay that accounts for a significant broadening with respect to the L_{III} part. This effect has been found to be an additional broadening of 0.5 eV half-width at half-maximum (hwhm) [27,28].

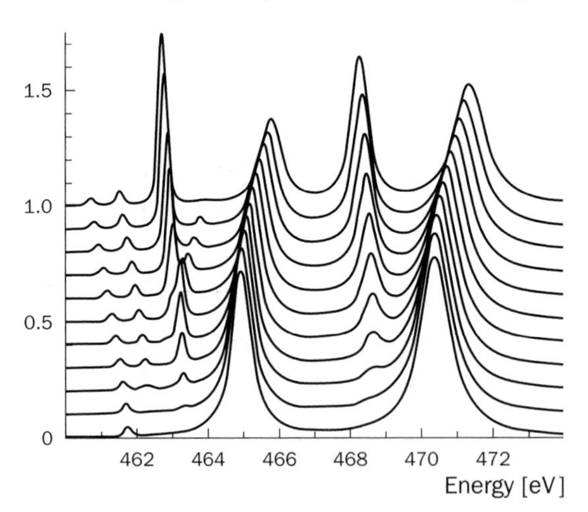

Figure 13 - Crystal field multiplet calculations for the $3d^0 \rightarrow 2p^5 3d^1$ transition in Ti^{4+}. The atomic Slater-Condon parameters and spin-orbit couplings have been used as given in table 12. The bottom spectrum is the atomic multiplet spectrum. Each next spectrum has a value of $10\,Dq$ that has been increased by 0.3 eV. The top spectrum has a crystal field of 3.0 eV.

An additional difference in broadening is found between the t_{2g} and the e_g states. This broadening has been ascribed to differences in the vibrational effects on the t_{2g} respectively the e_g states. Charge transfer multiplet calculations **[29,30]** have indicated that another, more important, cause could be a difference in hybridizational effects. Whatever the origin of the broadening, the comparison with experiment shows that the e_g states must be broadened with an additional 0.4 eV hwhm for the Lorentzian parameter. The experimental resolution has been simulated with a Gaussian broadening of 0.15 eV half-width at half-maximum.

Figure 14 compares the crystal field multiplet calculation of the $3d^0 \longrightarrow 2p^5 3d^1$ transition in Ti^{4+} with the experimental $2p$ X-ray absorption spectrum of $FeTiO_3$. The titanium ions are surrounded by six oxygen atoms in a (slightly) distorted octahedron. The value of $10\,Dq$ has been set to 1.8 eV. The calculation is able to reproduce all peaks that are experimentally visible. In particular the two small pre-peaks can be nicely observed. The similar spectrum of $SrTiO_3$ has an even sharper spectral shape, related to the perfect octahedral surrounding of Ti^{IV} by oxygen **[31,32]**.

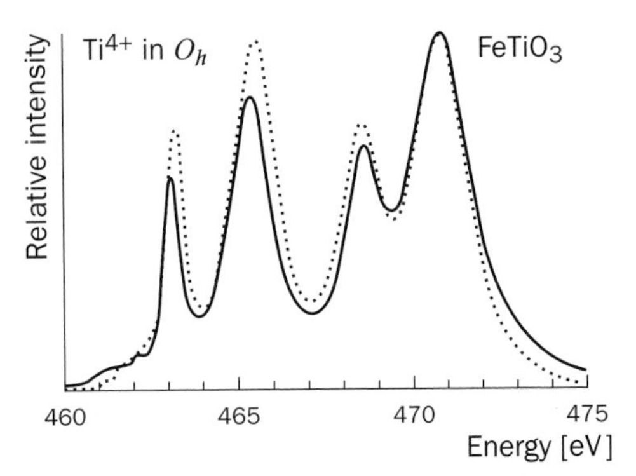

Figure 14 - The $2p$ X-ray absorption spectrum of $FeTiO_3$ (dashed) compared with a crystal field multiplet calculation for Ti^{IV} with a value of 10 Dq of 1.8 eV (solid) (Reprinted with permission from reference **[28]**, © 1990, American Physical Society).

3.3.5. $3d^0$ SYSTEMS IN LOWER SYMMETRIES

If one reduces the symmetry further from O_h to D_{4h} the seven lines in the X-ray absorption spectrum of Ti^{IV} split further. The respective degeneracies of the representations in O_h symmetry and the corresponding symmetries in D_{4h} symmetry are collected in table 26.

A $2p^5 3d^1$ configuration has twelve representations in SO_3 symmetry that are branched to 25 representations in a cubic field. These 25 representations are further branched to 45 representations in D_{4h} symmetry, of the overall degeneracy of 60. From these 45 representations, 22 are of interest for the calculation of the X-ray absorption spectral shape, because they have either E or A_2 symmetry.

Table 26 - The branching of the 25 representations in O_h symmetry to 45 representations in D_{4h} symmetry, using the degeneracies of the $2p^5 3d^1$ final state in X-ray absorption.

Γ in O_h	Degree		Γ in D_{4h}		Degree
A_1	2	A_1	A_1	2 + 5	7
A_2	3	B_1	A_2	7	7
T_1	7	$E + A_2$	B_1	3 + 5	8
T_2	8	$E + B_2$	B_2	8	8
E	5	$A_1 + B_1$	E	7 + 8	15
Σ	25				45

There are now two different final state symmetries possible because the dipole operator is split into two representations. The spectrum of two-dimensional E symmetry relates to the in-plane direction of the tetragon, while the one-dimensional A_2 symmetry relates to the out-of-plane direction. Examples of an angular dependence in D_{4h} and lower symmetries can be found in the study of interfaces, surfaces and adsorbates. A detailed study of the symmetry effects on the calcium $2p$ X-ray absorption spectra at the surface and in the bulk of CaF_2 did clearly show the ability of the multiplet calculations to reproduce the spectral shapes both in the bulk as at the reduced C_{3v} symmetry of the surface [8]. Recently, the group of Anders Nillson performed potassium $2p$ X-ray absorption experiments of potassium adsorbed on Ni(100) as well as the co-adsorption system CO/K/Ni(100) [33].

3.3.6. X-RAY ABSORPTION SPECTRA OF $3d^n$ SYSTEMS

The description of the X-ray absorption spectra of $3d^n$ systems follows the same procedure as for $3d^0$ systems. The matrix elements must be solved for the initial state of the Hamiltonian, the transition operator and the final state Hamiltonian.

A difference between $3d^0$ and $3d^n$ ground states is that the latter are affected by *dd* interactions and crystal field effects. Whether a system is high-spin or low-spin can be determined directly from the shape of the X-ray absorption spectrum. The calculation of the X-ray absorption spectrum has the following parameters to consider:

- The atomic Slater-Condon parameters. For trivalent and tetravalent systems these parameters are sometimes reduced. An effective reduction can also (partly) be achieved by the inclusion of charge transfer effects.
- The inclusion of the cubic crystal field strength 10 *Dq*, optimized to experiment. The value of 10 *Dq* determines the spin state of the $3d^4$ to $3d^7$ systems.
- The inclusion of the atomic 3*d* spin-orbit coupling. Because of an effective quenching of the 3*d* spin-orbit coupling by lower symmetries and/or translational effects, in some cases the 3*d* spin-orbit coupling must be set to zero to achieve a good agreement with experiment. This is for example the case for CrO_2. In contrast, for CoO the 3*d* spin-orbit coupling has to be included to have a good agreement with experimental spectra.
- The inclusion of lower symmetry parameters, for example *Ds* and *Dt*.
- In many systems it is important to extend the crystal field multiplet program with the inclusion of charge transfer effects as will be discussed in section 4.

4. CHARGE TRANSFER EFFECTS

Charge transfer effects are the effects of charge fluctuations in the initial and final states. The atomic multiplet and crystal field multiplet model use a single configuration to describe the ground state and final state. One can combine this configuration with other low-lying configurations similar to the way configuration-interaction works with a combination of Hartree-Fock matrices.

4.1. INITIAL STATE EFFECTS

The charge transfer method is based on the Anderson impurity model and related short-range model Hamiltonians that were applied to core-level spectroscopies. This line of approach has been developed in the eighties by the groups of Kotani and Jo [34], Gunnarsson and Schönhammer [35], Fujimori and co-workers [36] and Sawatzky and co-workers [37,38]. There are variations between the specific methods used, but in this review we sketch only the main line of reasoning behind all these models. For details we refer to the original papers.

The Anderson impurity model describes a localized state, the $3d$ state, which interacts with delocalized electrons in bands. The Anderson impurity model is usually written in second quantization. One starts with the ground state ψ_0 and acts on this state with operators that annihilate ($a^\dagger$) or create (a) a specific electron. For example a 2p to $3d$ X-ray absorption transition is written as $|\psi_0\, a_{2p}^\dagger\, a_{3d}\rangle$. With second quantization one can also indicate the mixing of configurations in the ground state. For example an electron can hop from the $3d$ states to a state in the (empty) conduction band, i.e. $|\psi_0\, a_{3d}^\dagger\, a_{ck}\rangle$, where a_{ck} indicates an electron in the conduction band with reciprocal space vector $\boldsymbol{k}$. Comparison to experiment has shown that the coupling to the occupied valence band is more important than the coupling to the empty conduction band. In other words the dominant hopping is from the valence band to the $3d$ states. If one annihilates an electron in a state and then re-creates it one effectively is counting the occupation of that state, i.e. $a_{3d}^\dagger a_{3d}$ yields n_{3d}. The Anderson impurity Hamiltonian can then be given as

$$H_{AIM} = \varepsilon_{3d}\, a_{3d}^\dagger a_{3d} + U_{dd}\, a_{3d}^\dagger a_{3d} a_{3d}^\dagger a_{3d} + \sum_k \varepsilon_{vk}\, a_{vk}^\dagger a_{vk} + t_{v3d} \sum_k (a_{3d}^\dagger a_{vk} + a_{vk}^\dagger a_{3d}) \quad \{28\}$$

These four terms represent respectively the $3d$ state, the correlation of the $3d$ state, the valence band and the coupling of the $3d$ states with the valence band. One can further extend the Anderson Impurity model to include more than a single impurity, i.e. impurity bands. In addition one can include correlation in the valence band, use larger clusters, etc. In case of multiplet calculations of X-ray absorption these approaches lead in most cases to a too large calculation. There has been much work for the Cu^{II} case, in particular in connection to the high T_c superconductors [39,40] , and there have been calculations concerning the effects of non-local screening on larger clusters for Ni^{II} [41].

Figure 15 sketches the impurity model with a semi-elliptical band of bandwidth w. Instead of a semi-elliptical band one can use the actual band structure that is found from DFT calculations (bottom). Actually, it has been demonstrated that the use of the real band structure instead of an approximate semi-elliptical or square band structure hardly affects the spectral shape [42]. The multiplet model approximates the band usually as a square of bandwidth W, where n number of points of equal intensity are used for the actual calculation. Often one simplifies the calculation further to $n = 1$, i.e. a single state representing the band. In that case the bandwidth is reduced to zero. In order to simplify the notation we will in the following remove the k dependence of the valence band and assume a single state describing the band. It must be remembered however that in all cases one can change back this single state to a real band with bandwidth W.

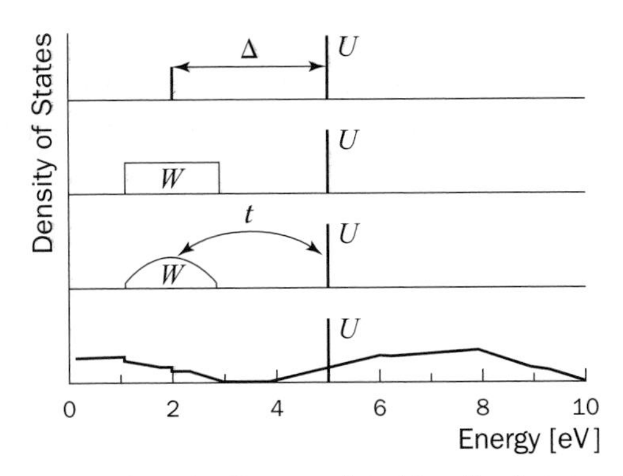

Figure 15 - The interaction of a U_{dd} correlated localized state with delocalized bands. From bottom to top are respectively given: a general DOS, a semi-elliptical valence band, a square valence band and a single valence state.

Removing the k dependence renders the Anderson Impunity Model (AIM) Hamiltonian into

$$H_{AIM} = \varepsilon_{3d}\, a_{3d}^{\dagger} a_{3d} + U_{dd}\, a_{3d}^{\dagger} a_{3d} a_{3d}^{\dagger} a_{3d} + \varepsilon_k\, a_v^{\dagger} a_v + t_{v3d}(a_{3d}^{\dagger} a_v + a_v^{\dagger} a_{3d}) \qquad \{29\}$$

Bringing the multiplet description into this Hamiltonian implies that the single $3d$ state is replaced by all states that are part of the crystal field multiplet Hamiltonian of that particular configuration. This implies that the U_{dd} term is replaced by a summation over four $3d$ wavefunctions $3d_1$, $3d_2$, $3d_3$ and $3d_4$

$$\begin{aligned} H_{AIM} = {} & \varepsilon_{3d}\, a_{3d}^{\dagger} a_{3d} + \varepsilon_k\, a_v^{\dagger} a_v + t_{v3d}(a_{3d}^{\dagger} a_v + a_v^{\dagger} a_{3d}) \\ & + \sum_{\Gamma_1,\Gamma_2,\Gamma_3,\Gamma_4} g_{dd}\, a_{3d1}^{\dagger} a_{3d2}\, a_{3d3}^{\dagger} a_{3d4} + \sum_{\Gamma_1,\Gamma_2} l\cdot s\, a_{3d1}^{\dagger} a_{3d2} + H_{CF} \end{aligned} \qquad \{30\}$$

The term g_{dd} describes all two electron integrals and includes the Hubbard U as well as the effects of the Slater-Condon parameters F^2 and F^4. In addition there is a new term in the Hamiltonian due to the $3d$ spin-orbit coupling. H_{CF} describes the effects of the crystal field potential Φ. This situation can be viewed as a multiplet of localized states interacting with the delocalized density of states, as indicated in figure 16. The energy difference to the centre of gravity of the multiplet is indicated with $\underline{\Delta}$. The effective energy difference to the lowest state of the multiplet is indicated with Δ. One ingredient is still missing from this description, i.e. if the electron is transferred from the valence band to the $3d$ band, the occupation of the $3d$ band changes by one. This $3d^{n+1}$ configuration is again affected by multiplet effects, exactly like the original $3d^n$ configuration. The $3d^{n+1}$ configuration contains a valence band with a hole. Because the model is used mainly for transition metal compounds, the valence band is in general dominated by ligand character, for example the oxygen $2p$ valence band in case of transition metal oxides. Therefore the hole is considered to be on the ligand and is indicated with $\underline{L}$, where $\underline{L}$ indicates a ligand hole, i.e. in an oxide it implies a hole on the oxygen site. The charge transfer effect on the wave function is described as $3d^n + 3d^{n+1}\underline{L}$.

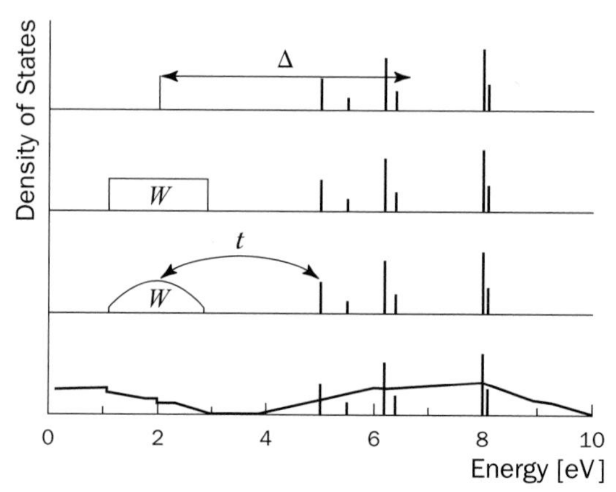

Figure 16 - The interaction of a multiplet of localized states with delocalized bands. From bottom to top are respectively given: a general DOS, a semi-elliptical valence band, a square valence band and a single valence state.

Because the $3d^{n+1}\underline{L}$ state is also affected by multiplet effects, figure 16 is not correct as the $3d^{n+1}\underline{L}$ state is described with a single $3d^{n+1}$ state in combination with $\underline{L}$, a ligand hole that can have either a single energy or be a band. If one includes the effects of the multiplets on the $3d^{n+1}\underline{L}$ in this figure, for a single state one essentially obtains a configuration-interaction between two sets of multiplet states.

Figure 17 gives the crystal field multiplets for the $3d^7$ and $3d^8\underline{L}$ configurations of Co^{II}. The $3d^7$ configurations is centred at 0.0 eV and the lowest energy state is the $^4T_1$ state, where the small splittings due to the $3d$ spin-orbit coupling have been neglected [4]. The lowest state of the $3d^8\underline{L}$ configuration is $^3A_2$, which is the ground state of for example Ni^{II}. The centre of gravity of the $3d^8$ configuration has been set at 2.0 eV, which identifies with a value of $\underline{\Delta}$ of 2.0 eV. The effective charge transfer energy Δ is defined as the energy difference between the lowest states of the $3d^7$ and the $3d^8\underline{L}$ configurations as indicated in figure 17. Because the multiplet splitting is larger for $3d^7$ than for $3d^8\underline{L}$, the effective Δ is larger than $\underline{\Delta}$. The effect of charge transfer is to form a ground state that is a combination of $3d^7$ and $3d^8\underline{L}$. The energies of these states have been calculated on the right half of the figure. If the hopping parameter t is set equal to zero, both configurations do not mix and the states of the mixed configuration are exactly equal to $3d^7$, and at higher energy to $3d^8\underline{L}$. Turning on the hopping parameter, one observes that the energy of the lowest configuration is further lowered. This state will still be the $^4T_1$ configuration, but with increasing hopping, it will have increasing $3d^8\underline{L}$ character. The second-lowest state is split by the hopping and the most bonding combination obtains an energy that comes close to the $^4T_1$ ground state. This excited state is essentially a doublet state and if the energy of this state would cross with the $^4T_1$ state one would observe a charge transfer-induced spin transition. It has been shown that charge transfer effects can lead to new types of ground states, for example in case of a $3d^6$ configuration, crystal field effects lead to a transition of a $S = 2$ high-spin to a $S = 0$ low-spin ground state. Charge transfer effects can also to lead to an $S = 1$ intermediate spin ground state [43].

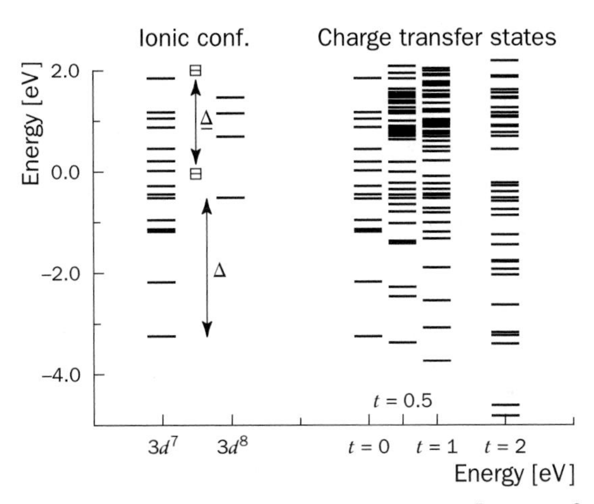

Figure 17 - Left - The crystal field multiplet states of $3d^7$ and $3d^8$ configurations. The multiplet states with energies higher than +2.0 eV are not shown. $\underline{\Delta}$ has been set to +2.0 eV. **Right** - The charge transfer multiplet calculations for the combination of crystal field multiplets as indicated on the left and with the hopping ranging from 0.0 eV to 2.0 eV as indicated below the states.

Figure 17 can be expanded to Tanabe-Sugano diagrams for two configurations $3d^n + 3d^{n+1}\underline{L}$, instead of the usual Tanabe-Sugano diagrams as a function of only one configuration. The energies of such two-configuration Tanabe-Sugano diagrams are affected by the Slater-Condon parameters (often approximated with the B Racah parameter), the cubic crystal field 10 Dq, the charge transfer energy $\underline{\Delta}$ and the hopping strength t. The hopping can be made symmetry dependent and one can add crystal field parameters related to lower symmetries, yielding to an endless series of Tanabe-Sugano diagrams. What is actually important is to determine the possible types of ground states for a particular ion, say Co^{II}. Scanning through the parameter space of F^2, F^4, $10\,Dq$, Ds, Dt, LS_{3d}, t_Γ and Δ one can determine the nature of the ground state. This ground state can then be checked with $2p$ X-ray absorption. After the inclusion of exchange and magnetic fields one has also a means to compare the ground state with techniques like X-ray MCD, optical MCD and EPR.

Comparing figure 15 with figure 17 one observes the transition from a single particle picture to a multiplet configurational picture. One can in principle put more band character into this configurational picture and a first step is to make a transition from a single state to a series of $3d^8\underline{L}$ states, each with its included multiplet but with each a different effective charge transfer energy. One can choose to use a more elaborate cluster model in which the neighbour atoms are actually included in the calculation [39,40,44]. These cluster models are not described further here. In all cases where multiplet effects are important, i.e. with at least two holes in correlated states, these cluster models do in general not lead to significantly different conclusions and ground state descriptions [4].

4.2. *FINAL STATE EFFECTS*

The final state Hamiltonian of X-ray absorption includes the core-hole plus an extra electron in the valence region. One adds the energy and occupation of the $2p$ core-hole to the Hamiltonian as given in equation {30}. The core-hole potential U_{pd} and its higher order terms g_{pd} give rise to equation {31}. This equation describes the overlap of a $2p$ wave function with a $3d$ wave function and is given as a summation over two $2p$ and two $3d$ wavefunctions $2p_1$, $2p_2$, $3d_1$ and $3d_2$

$$H_{2p} = \varepsilon_{2p}\, a_{2p}^{\dagger} a_{2p} + \sum_{\Gamma_1,\Gamma_2,\Gamma_3,\Gamma_4} g_{pd}\, a_{3d1}^{\dagger} a_{2p1} a_{2p2}^{\dagger} a_{3d2} + \sum_{\Gamma_1,\Gamma_2} l\cdot s\, a_{2p1}^{\dagger} a_{2p2} \qquad \{31\}$$

The term g_{pd} describes all two-electron integrals and includes U_{pd} as well as the effects of the Slater-Condon parameters F^2, G^1 and G^3. In addition there is a term in the Hamiltonian due to the $2p$ spin-orbit coupling. There is no crystal field effect on core states.

$$\begin{aligned} H_{AIM} = {} & \varepsilon_{3d}\, a_{3d}^{\dagger} a_{3d} + \varepsilon_k\, a_v^{\dagger} a_v + t_{v3d}\left(a_{3d}^{\dagger} a_v + a_v^{\dagger} a_{3d}\right) \\ & + \sum_{\Gamma_1,\Gamma_2,\Gamma_3,\Gamma_4} g_{dd}\, a_{3d1}^{\dagger} a_{3d2} a_{3d3}^{\dagger} a_{3d4} + \sum_{\Gamma_1,\Gamma_2} l\cdot s\, a_{3d1}^{\dagger} a_{3d2} + H_{CF} \\ & + \varepsilon_{2p}\, a_{2p}^{\dagger} a_{2p} + \sum_{\Gamma_1,\Gamma_2,\Gamma_3,\Gamma_4} g_{pd}\, a_{3d1}^{\dagger} a_{2p1} a_{2p2}^{\dagger} a_{3d2} + \sum_{\Gamma_1,\Gamma_2} l\cdot s\, a_{2p1}^{\dagger} a_{2p2} \end{aligned} \qquad \{32\}$$

The overall Hamiltonian in the final state is given in equation {32}. This equation is solved in the same manner as the initial state Hamiltonian. Using the two configuration description of figure 17, one finds for Co^{II} two final states $2p^5 3d^8$ and $2p^5 3d^9 \underline{L}$. These states mix in a manner similar to the two configurations in the ground state and as such give rise to a final state Tanabe-Sugano diagram. All final state energies are calculated from the mixing of the two configurations. This calculation is only possible if all final state parameters are known. The following rules are used:

- The $2p3d$ Slater-Condon parameters are taken from an atomic calculation. For trivalent ions and higher valences, these atomic values are sometimes reduced.
- The $2p$ and $3d$ spin-orbit coupling are taken from an atomic calculation.
- The crystal field values are assumed to be the same as in the ground state.
- The energies of the configurations, i.e. the charge transfer energy, are given by the values of U_{dd} and U_{pd}. Effectively $\underline{\Delta}_F = \underline{\Delta}_I + U_{dd} - U_{pd}$. Because in general U_{pd} is approximately 1 to 2 eV larger than U_{dd}, one often assumes $\underline{\Delta}_F = \underline{\Delta}_I - 1$ eV or $\underline{\Delta}_F = \underline{\Delta}_F = \underline{\Delta}_I - 2$ eV.
- The hopping parameter t is assumed to be equal in the initial and final states.

Detailed analysis of X-ray absorption and resonant X-ray emission spectra has shown that the crystal field values are smaller by 10 to 20% in the final state [45]. The same observation has been made for the hopping parameters [46]. One can understand these trends from the (slight) compression of the $3d$ wave function in the final state. From the presence of the $2p$ core-hole one would expect a significant compression of the $3d$ wave function, but the effect of the $2p$ core-hole is counteracted by the effect of the extra $3d$ electron in the final state. Because we have seen that U_{dd} is a bit smaller than U_{pd} this counteracting action is not complete and there will be a small compression of the $3d$ wave function. In conclusion it can be said that $\underline{\Delta}$, t and 10 Dq will all be slightly smaller in the final state. Because the reduction of these parameters has counteracting effects on the spectral shape, in most simulations one varies only $\underline{\Delta}$ and keeps t and 10 Dq constant.

4.3. THE X-RAY ABSORPTION SPECTRUM WITH CHARGE TRANSFER EFFECTS

The essence of the charge transfer model is the use of two or more configurations. Ligand field multiplet calculations use one configuration for which it solves the effective atomic Hamiltonian plus the ligand field Hamiltonian, so essentially the following matrices

$$I_{XAS,1} \propto \langle 3d^n | p | 2p^5 3d^{n+1} \rangle^2 \qquad \{33\}$$

$$H_{INIT,1} = \langle 3d^n | \frac{e^2}{r_{12}} + \varsigma_d l_d s_d + H_{LFM} | 3d^n \rangle \qquad \{34\}$$

$$H_{FINAL,1} = \langle 2p^5 3d^{n+1} | \frac{e^2}{r_{12}} + \varsigma_p l_p s_p + \varsigma_d l_d s_d + H_{LFM} | 2p^5 3d^{n+1} \rangle \qquad \{35\}$$

The charge transfer model adds a configuration $3d^{n+1}\underline{L}$ to the $3d^n$ ground state. In case of a transition metal oxide, in a $3d^{n+1}\underline{L}$ configuration an electron has been moved from the oxygen $2p$ valence band to the metal $3d$ band. One can continue with this procedure and add $3d^{n+2}\underline{L}^2$ configuration, etc. In many cases two configurations will be enough to explain the spectral shapes, but in particular for high valence states it can be important to include more configurations [47,48]. As far as X-ray absorption and X-ray emission is concerned, the consequences for the calculations are the replacement of $3d^n$ with $3d^n + 3d^{n+1}\underline{L}$ plus the corresponding changes in the final state. This adds a second initial state, final state and dipole transition

$$I_{XAS,2} \propto \langle 3d^{n+1}\underline{L} | p | 2p^5 3d^{n+2}\underline{L} \rangle^2 \qquad \{36\}$$

$$H_{INIT,2} = \langle 3d^{n+1}\underline{L} | \frac{e^2}{r_{12}} + \varsigma_d l_d s_d + H_{LFM} | 3d^{n+1}\underline{L} \rangle \qquad \{37\}$$

$$H_{FINAL,2} = \langle 2p^5 3d^{n+2}\underline{L} | \frac{e^2}{r_{12}} + \varsigma_p l_p s_p + \varsigma_d l_d s_d + H_{LFM} | 2p^5 3d^{n+2}\underline{L} \rangle \qquad \{38\}$$

The two initial states and two final states are coupled by monopole transitions, i.e. configuration interaction. The mixing parameter t couples both configurations and Δ is the energy difference. The Hamiltonian is abbreviated with t/Δ to describe the monopole interaction

$$H_{MIX\,I1,I2} = \langle 3d^n | t/\Delta | 3d^{n+1}\underline{L} \rangle \qquad \{39\}$$

$$H_{MIX\,F1,F2} = \langle 2p^5 3d^{n+1} | t/\Delta | 2p^5 3d^{n+2}\underline{L} \rangle \qquad \{40\}$$

The X-ray absorption spectrum is calculated by solving the equations 33 to 40. If a $3d^{n+2}\underline{L}^2$ configuration is included its energy is $2\Delta + U_{dd}$, where U_{dd} is the correlation energy between two $3d$ electrons [38]. The formal definition of U_{dd} is the energy difference one obtains when an electron is transferred from one metal site to another, i.e. a transition $3d^n + 3d^n \rightarrow 3d^{n+1} + 3d^{n-1}$. The number of interactions of two $3d^n$ configurations is one more than the number of interactions of $3d^{n+1}$ plus $3d^{n-1}$, implying that this energy difference is equal to the correlation energy between two $3d$ electrons. By analysing the effects of charge transfer it is found that, for systems with a positive value of Δ, the main effects on the X-ray absorption spectral shape are:

- the formation of small satellites,
- the contraction of the multiplet structures.

The formation of small satellites or even the absence of visible satellite structures is a special feature of X-ray absorption spectroscopy. Its origin is the fact that X-ray absorption is a neutral spectroscopy and the local charge of the final state is equal to the charge of the initial state. This implies that there is little screening hence little charge transfer satellites. This effect can be explained by using a two-by-two problem as example. We follow the paper of Hu *et al.* **[26]** to describe the mixing of two configurations that are separated by Δ and mixed by t. This mixing yields a two-by-two determinant

$$H = \begin{vmatrix} 0 & t \\ t & \Delta \end{vmatrix} \qquad \{41\}$$

Solving the determinant yields the two states after mixing: The ground state, or bonding combination, ψ_B has a wave function

$$\psi = \alpha_i \,|3d^n\rangle + \beta_i \,|3d^{n+1}\underline{L}\rangle \qquad \{42\}$$

The energy of the bonding combination is given as

$$E_B = \frac{\Delta}{2} - \frac{1}{2}\sqrt{\Delta^2 + 4t} \qquad \{43\}$$

The parameters α_i and β_i can be defined in Δ and t

$$\alpha_i = \sqrt{\frac{1}{1} + \left(\frac{X-\Delta}{2T}\right)^2}, \quad X = \sqrt{\Delta^2 + 4T^2} \qquad \{44\}$$

$$\beta_i = \sqrt{1-\alpha_i^{\,2}} \qquad \{45\}$$

The anti-bonding combination is given as

$$\Psi_B = \beta_i \,|3d^n\rangle - \alpha_i \,|3d^{n+1}\underline{L}\rangle \qquad \{46\}$$

The energy of the anti-bonding combination is given as

$$E_B = \frac{\Delta}{2} - \frac{1}{2}\sqrt{\Delta^2 + 4t} \qquad \{47\}$$

Figure 18 gives the value of α_i for a series of combinations of Δ and t. It can be observed that, apart from numerical deviations at small charge transfer energies, the value of α_i is essentially proportional to $\sqrt{\Delta / t}$. The dependence on Δ and t is very clear in figure 19. A linear dependence of α_i is observed as a function of t and a square root dependence is found as a function of Δ. This implies for the percentage of $3d^n$ character in the ground state, i.e. α_i^2 that it is proportional to Δ/t^2.

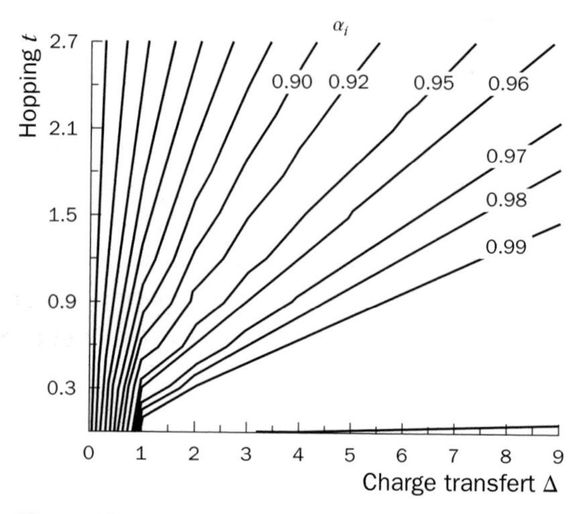

Figure 18 - The initial state value of α as a function of the charge transfer energy Δ and the hopping t.

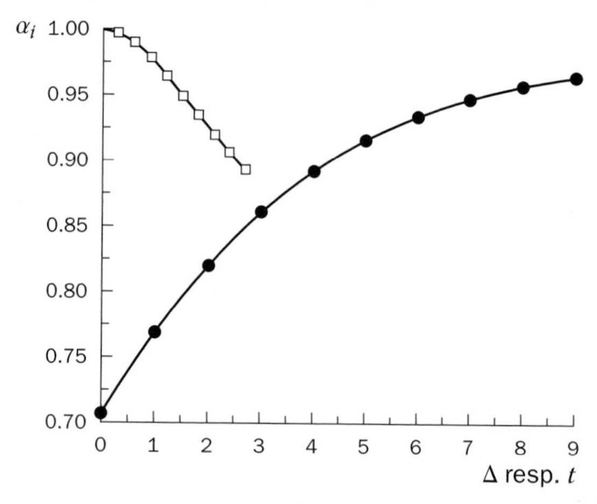

Figure 19 - The initial state value of α as a function of the charge transfer energy Δ for t = 2.7 eV (closed circles) and as a function of the hopping t for Δ = 5 eV (open squares).

We have found that in the final state the value of $\Delta_F \sim \Delta - 1$ eV. This implies that the final state determinant is approximately equal to the initial state determinant

$$H = \begin{vmatrix} 0 & t \\ t_F & \Delta_F \end{vmatrix} \quad \{48\}$$

This yields for α_f and β_f that they are approximately equal to α_i and β_i. The results of the initial and final state equations can be used to calculate the X-ray absorption cross-section. One can make a transition from $3d^n$ to $2p^5 3d^{n+1}$ and from $3d^{n+1}\underline{L}$ to $2p^5 3d^{n+2}\underline{L}$. This implies that the intensity of the main peak is equal to $(\alpha_i\,\alpha_f + \beta_i\,\beta_f)^2$ while the satellite intensity is equal to $(\alpha_i\,\beta_f - \beta_i\,\alpha_f)^2$.

The contraction of the multiplet structure due to charge transfer can also be understood using the two by two matrices. Assume two multiplet states split by an energy δ. They both mix with a charge transfer state that is positioned Δ above the lowest energy multiplet state I. Consequently the charge transfer energy of the second multiplet state II is Δ-δ. Assuming that the hopping terms are the same for these two states, the energy gain of the bonding combination is

$$E_B(I) = \frac{\Delta}{2} - \frac{1}{2}\sqrt{\Delta^2 + 4t} \quad \{49\}$$

$$E_B(II) = \frac{\Delta - \delta}{2} - \frac{1}{2}\sqrt{(\Delta - \delta)^2 + 4t} \quad \{50\}$$

Consider for example a hopping of 1.5 eV. Then the largest energy gain is found for the lowest value of Δ. The higher lying multiplet states have a smaller effective Δ and consequently a larger energy gain. As such their energy comes closer to the lowest energy state and the multiplet appears compressed.

The two-by-two problem in the initial and final state explains the two main effects of charge transfer: a compression of the multiplet structure and the existence of only small satellites. These two phenomena are visible in the figures of Ni^{II} and Co^{II} as given below. In case the charge transfer is negative, the satellite structures are slightly larger because then the final state charge transfer is increased with respect to the initial state and the balance of the initial and final state α's and β's is less good.

Figure 20 shows the effect of the charge transfer energy on divalent nickel. We have used the same hopping t for the initial and final state and reduced the charge transfer energy Δ by one eV. In the top spectrum, $\Delta = 10$ and the spectrum is essentially the ligand field multiplet spectrum of a Ni^{II} ion in its $3d^8$ configuration. The bottom spectrum uses $\Delta = -10$ and now the ground state is almost a pure $3d^9\underline{L}$ configuration. Looking for the trends in figure 20, one finds the increased contraction of the multiplet structure by going to lower values of Δ. This is exactly what is observed in the series NiF_2 to $NiCl_2$ and $NiBr_2$ **[49-52]**. Going from Ni to Cu the atomic parameters change very little, except the $2p$ spin-orbit coupling and the $2p$ binding energy. Therefore the spectra of $3d^n$ systems of different elements are all very similar and the bottom spectrum is also similar to Cu^{II} systems. Therefore one can also use the spectra with negative Δ values for Cu^{III} compounds, such as $La_2Li_{1/2}Cu_{1/2}O_4$ and Cs_2KCuF_6.

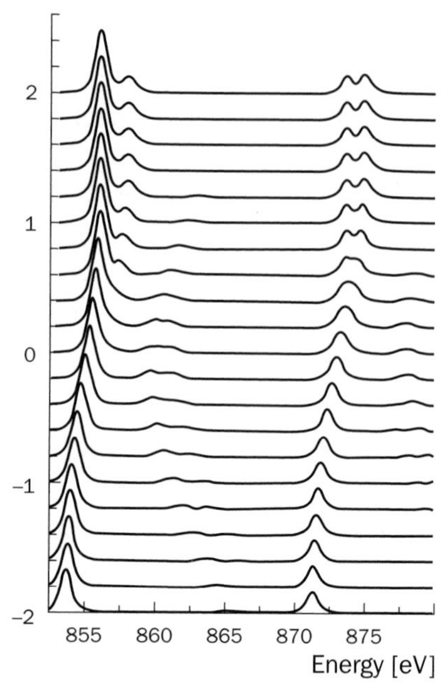

Figure 20 - Series of charge transfer multiplet calculations for the Ni^{II} ground state $3d^8 + 3d^9\underline{L}$. The top spectrum has a charge transfer energy of +10 eV. The bottom spectrum has a charge transfer energy of −10 eV and relates to an almost pure $3d^9$ ground state.

Figure 21 shows the comparison of the $2p$ X-ray absorption spectrum of these two compounds with charge transfer multiplet calculations [53,54]. It can be checked in figure 21 that these calculations look similar to the calculations for Ni^{II} systems with negative values of Δ. For such systems with negative Δ values, it is important to carry out charge transfer multiplet calculations, as no good comparison with crystal field multiplet spectra can be made.

X-ray absorption compared with X-ray photoemission

The crucial difference between X-ray absorption and X-ray photoemission is that XAS is essentially a charge conserving experiment. As discussed this implies small charge transfer effects and small satellites. In XPS and electron escapes from the (local) system and the effects of charge transfer are large. In fact, the spectral shape of photoemission spectra of strongly correlated materials is dominated by charge transfer effects. For details the reader is referred to a number of papers [4,49,50].

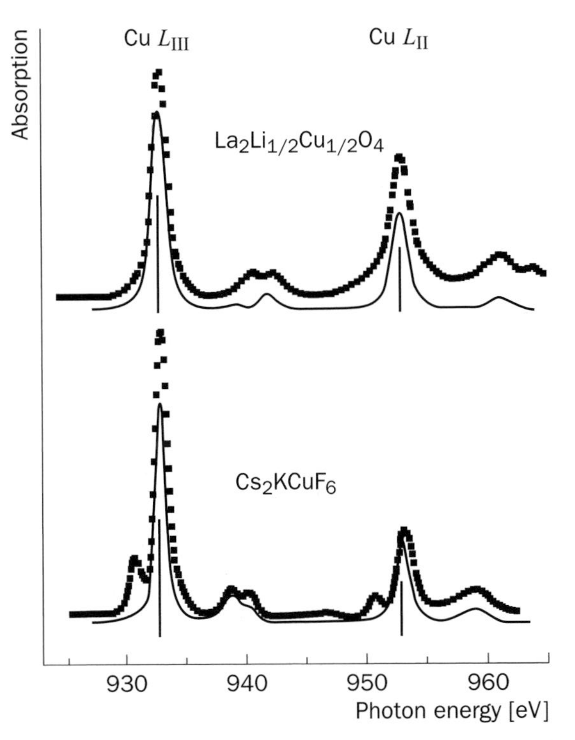

Figure 21 - Results of theoretical simulations of the copper $2p$ X-ray absorption spectra of Cs_2KCuF_6 (bottom) and $La_2Li_{1/2}Cu_{1/2}O_4$ (top), in comparison with the experimental spectra. (Reprinted with permission from ref. **[26]**, © 1998 Elsevier Science).

5. *X-RAY LINEAR AND CIRCULAR DICHROISM*

Dichroism is the property of certain objects showing different colours according to their orientation with respect to the light. It is due to the dependence of the optical response of the object on the relative orientation between the polarization direction of the light and the symmetry axes of the object. With X-rays, in some cases a difference can be observed between the absorption of left and right circularly polarized light (circular dichroism) or for different orientations of the polarization vector of linearly polarized light with respect to a given quantization axis (linear dichroism). Dichroism can only occur when the spherical symmetry of the free atom is broken due to a magnetic or (crystalline) electric field. Magnetic fields can cause both circular and linear dichroism, while a crystal field can only induce linear dichroism.

On the microscopic level, circular dichroism is given by the difference between $\Delta M_J = +1$ and $\Delta M_J = -1$ transitions, while linear dichroism is related to the difference between $\Delta M_J = 0$ and $\Delta M_J = \pm 1$ transitions. (Magnetic) Circular dichroism was first predicted in the $M_{\text{II-III}}$ ($3p$ absorption) edges of transition metals [55], but the first observation of magnetic dichroism was done with linearly polarized light at the $M_{\text{IV-V}}$ edges of rare earths [21,56]. We start with a description of dichroism in these edges, since it can be well understood using the atomic multiplet model.

Another way to write the matrix element for a dipole transition from an initial state with quantum numbers J and M to a final state with J' and M' is given by $|\langle J' M'| \boldsymbol{P}_q | J M \rangle|^2 = \begin{pmatrix} J & 1 & J' \\ -M' & q & M' \end{pmatrix}^2 |\langle J' \| \boldsymbol{P}_q \| J \rangle|^2$. Here, the matrix element is split into an angular (the $3J$ symbol squared) and a radial part using the Wigner-Eckart theorem.

The radial part $|\langle J' \| \boldsymbol{P}_q \| J \rangle|^2$ is also called the *line strength* of the transition. The dependence of the absorption on the polarization of the light is caused by the dependence of the $3J$ symbols on ΔM. The $3J$ symbol is non-zero only if $|J - 1| \leq J' \leq J + 1$ and $q = \Delta M = M - M' = 0, \pm 1$. Transitions with $q = 0$ can be excited only by radiation that has a linearly polarized component along the z axis (i.e. the quantization axis) and $q = \pm 1$ transitions only by components that are respectively left- or right-handed circularly polarized in the plane perpendicular to this axis. The relation between the ΔJ transitions and the polarization is given in table 27.

Table 27 - The relation between ΔJ transitions and the polarization vector $\boldsymbol{q}$.

$\Delta J/q$	-1	0	1
-1	$\frac{J(J-1)-(2J-1)M+M^2}{2J(2J+1)(2J-1)}$	$\frac{J^2-M^2}{J(2J+1)(2J-1)}$	$\frac{J(J-1)+(2J-1)M+M^2}{2J(2J+1)(2J-1)}$
0	$\frac{J(J+1)-M-M^2}{2J(2J+1)(2J-1)}$	$\frac{M^2}{J(2J+1)(J+1)}$	$\frac{J(J+1)+M-M^2}{2J(2J+1)(J+1)}$
1	$\frac{(J+1)(J+2)+(2J+3)M+M^2}{2(2J+3)(2J+1)(2J-1)}$	$\frac{(J+1)^2-M^2}{(2J+3)(2J+1)(J+1)}$	$\frac{(J+1)(J+2)-(2J+3)M+M^2}{2(2J+3)(2J+1)(2J-1)}$

In spherical symmetry, the Hund's rule ground state is $(2J + 1)$ fold degenerate and all M_J levels are equally populated. It can be shown that in that case the polarization of the light does not have any influence on the absorption spectrum [57]. A magnetic or crystalline electric field splits up the ground state in levels with different values of M_J and at low enough temperature levels with different M_J will be unequally occupied. The simplest example is for an Yb^{3+} ion in an applied magnetic field. The magnetic field will cause a Zeeman splitting of the levels of the $^2F_{7/2}$ ground state in fifteen levels with $M_J = -7/2, -5/2 \ldots 7/2$ and energy $\mu_0 gHM_J$. At zero kelvin, only the lowest lying level, with $M_J = -7/2$, will be occupied. In the final state, only the $^2D_{5/2}$ state is available, with $-5/2 \leq M_J \leq 5/2$.

As a result, at zero kelvin only $\Delta M = +1$ transitions can take place, so only left circularly polarized light or linearly polarized light, incident along the quantization (in this case, the magnetization direction) axis. Right circularly polarized light, or linearly polarized light with its polarization vector parallel to the magnetization axis will not be absorbed. At higher temperatures, also higher M_J levels will be occupied, according to a Boltzmann distribution, and the other transitions (first $\Delta M = 0$ for $M_J = -5/2$, than also $\Delta M = -1$ for $M_J = -3/2$), will become possible. However, some polarization dependence of the absorption spectra will persist until $kT >> \mu_0 gH$. X-ray dichroism can thus be used to measure magnetic ordering, with the element selectivity inherent to X-ray absorption spectroscopy. It can be shown that circular dichroism (difference between $\Delta M = -1$ and $\Delta M = -1$ transitions) is proportional to $\langle M_J \rangle$, the average M_J value of the ion, which corresponds to its magnetic moment ($|\boldsymbol{M}| = \langle M \rangle \mu_0 g \alpha_J$). Linear dichroism is related to $\langle M_J^2 \rangle$, the average of the square of M_J. It is therefore sensitive to magnetic order, but it can not give the sign, or direction, of the magnetic moments. In contrast to (magnetic) circular dichroism, linear dichroism is also sensitive to crystalline electric fields with a less than cubic symmetry. This often limits the interest of linear dichroism for magnetic studies of rare earths, since in many magnetic compounds containing rare earths the rare earth ion is in a low symmetry crystal field environment.

As discussed above, in many rare earth $M_{IV\text{-}V}$ edges three peaks are visible corresponding to the different ΔJ transitions. The cross-section $\sigma^q{}_J$ for a transition from a single ground-state level $|JM^k{}_J\rangle$ of the free ion to all final states allowed by the selection rules and for a photon polarization q is given by $\sigma^q{}_J = \Sigma_{J'} |\, 3J \text{ symbol} \,|^2 \; \sigma_{J'}$ This means that for any M_J level and any $q = \Delta M$, the transition can be written as a linear combination of the three ΔJ components. As an example, in figure 22. we give the calculated ΔJ components for the M_V edge of Dy (ground state $^6H_{15/2}$). In the same figure, we give the different ΔM transitions for the lowest lying level in a magnetic field, $M_J = -15/2$. As can be seen, the ΔM transitions closely correspond to ΔJ transitions. For the other M_J levels, every ΔM contains the three ΔJ components, with different weights. The occupation of the different M_J levels, which depends on the splitting as well as the temperature, determines therefore the polarization dependence of the spectrum.

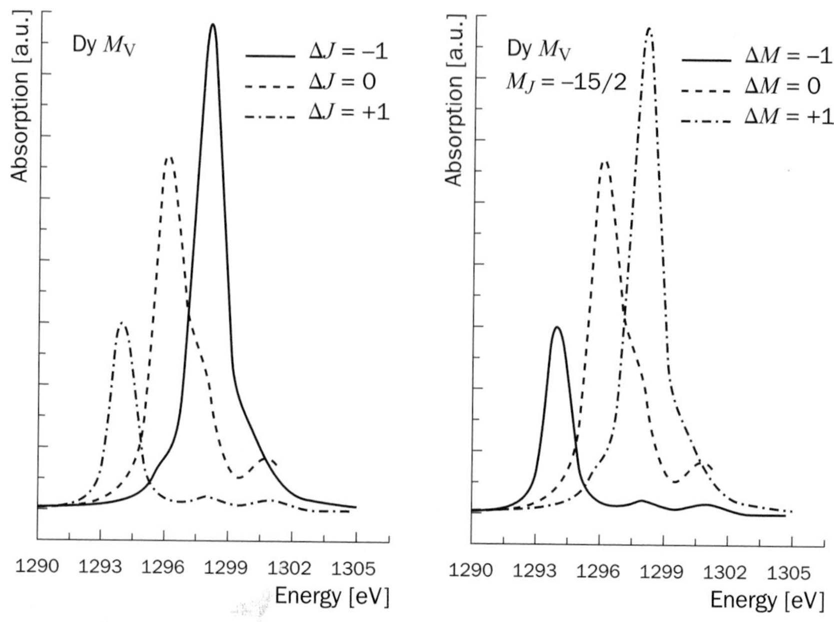

Figure 22 - a - Calculated ΔJ components for the M_V edge of Dy (ground state $^6H_{15/2}$)
b - Different ΔM transitions for the lowest lying level in a magnetic field, $M_J = -15/2$.

As an example, we give in figure 23 the spectra for left and right circularly polarized light, as well as their difference (the circular dichroism) as a function of temperature for a Dy^{3+} ion in a magnetic field. The spectra and dichroism are given for different values of the reduced temperature $T_R = kT/\mu_B gH$, where μ_B is the Bohr magneton, g is the Landé factor and H is the magnetic field strength. $\mu_B gH$ defines the splitting between the different Zeeman levels and kT their occupation. This shows that in general the spectra depend both on the (internal) magnetic field acting on the ion and on the temperature. The size of the dichroism gives a direct measure of the average magnetic moment in the direction of propagation of the X-rays. The right panel of figure 23 shows that the size of the dichroism depends on the temperature and magnetic field, but that the shape of the curves does not change. It can be shown that this is generally the case for both circular and linear dichroism in rare earths (see for instance ref. **[58]**), as long as the energy between the Hund's rule ground state and the first excited state is large.

In a crystal field, the M_J levels are split according to their absolute value $|M_J|$, i.e. every energy level is a mixture of different M_J levels, where M_J and $-M_J$ have the same weight. Since the spectra for M_J, ΔM are the same as for $-M_J$, $-\Delta M$ a crystal field cannot introduce circular dichroism. However, a crystal field with a symmetry lower than cubic will give rise to linear dichroism, like ferromagnetic but also antiferromagnetic ordering.

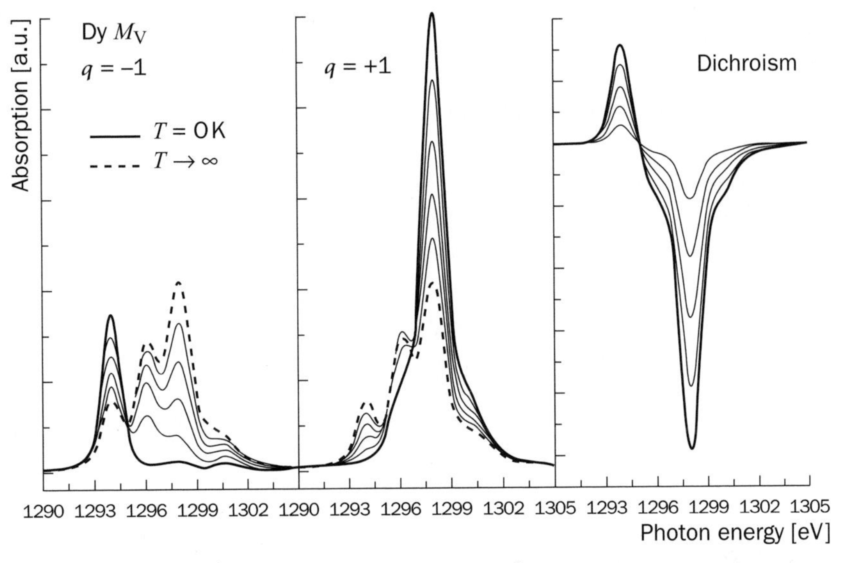

Figure 23 - Calculated M_V edge absorption for a Dy^{3+} ion in a magnetic field for right (first panel) and left (second panel) circularly polarized light. The difference (or XMCD) curves are given on the right. The spectra are for different values of the reduced temperature T_R (see text), with T_R = 0, 2, 4, 7, 15 and $T_R \to \infty$.

For metallic transition-metal atoms, the *d* electrons are delocalized (or itinerant) and the atomic approach for the absorption and dichroism is not valid. This can be partly overcome using a configuration interaction approach, where the ground state of the absorbing atom is written as the superposition of atomic states of different *d* count, with different weights. In most cases, band-structure effects play an important role in the shape of spectra of metallic transition metals, and a one-electron approach is generally used to treat the excitations [4,59]. The use of XMCD for magnetic studies became very important after the development of sum rules, which allow the spin and orbital magnetic moment of the measured element to be separately determined from the measured dichroism spectra [60,61]. X-ray dichroism is has recently also been used for magnetic imaging, where domains in both ferromagnetic (circular dichroism) and antiferromagnetic (linear dichroism) materials can be observed with very good spatial resolution [62-64] (see also chapter by C.M. Schneider in this volume). In complex magnetic structures (multilayers, spin-valves) it can be used to perform layer-resolved measurements of magnetization dynamics with sub-nanosecond time resolution [65]. In the chapter of F. Baudelet more details about XMCD in absorption will be given.

REFERENCES

[1] F.M.F. DE GROOT, M.A. ARRIO, P. SAINCTAVIT, C. CARTIER & C.T. CHEN - *Solid State Comm.* **92**, 991 (1994)

[2] S.L.M. SCHROEDER, G.D. MOGGRIDGE, R.M. LAMBERT & T. RAYMENT - *In "Spec. for Surface Science"*, 1 (1998)

[3] K. HAMALAINEN, C.C. KAO, J.B. HASTINGS, D.P. SIDDONS, L.E. BERMAN, V. STOJANOFF & S.P. CRAMER - *Phys. Rev. B* **46**, 14274 (1992)

[4] F.M.F. DE GROOT - *J. Elec. Spec.* **67**, 529 (1994)

[5] F.M.F. DE GROOT - *Chem. Rev.* **101**, 1779 (2001)

[6] M. ABBATE, J.B. GOEDKOOP, F.M.F. DE GROOT, M. GRIONI, J.C. FUGGLE, S. HOFMANN, H. PETERSEN & M. SACCHI - *Surf. Int. Anal.* **18**, 65 (1992)

[7] J. VOGEL & M. SACCHI - *J. Elec. Spec.* **67**, 181 (1994)

[8] F.J. HIMPSEL, U.O. KARLSSON, A.B. MCLEAN, L.J. TERMINELLO, F.M.F. DE GROOT, M. ABBATE, J.C. FUGGLE, J.A. YARMOFF, B.T. THOLE & G.A. SAWATZKY - *Phys. Rev. B* **43**, 6899 (1991)

[9] U. BERGMANN, O.C. MULLINS & S.P. CRAMER - *Analytical Chemistry* **72**, 2609 (2000)

[10] J. FINK, M. KNUPFER, S. ATZKERN & M.S. GOLDEN - *J. Elec. Spec.* **117-118**, 287 (2001)

[11] F. NOLTING, A. SCHOLL, J. STOHR, J.W. SEO, J. FOMPEYRINE, H. SIEGWART, J.P. LOCQUET, S. ANDERS, J. LUNING, E.E. FULLERTON, J.M.F. JONEY, M.R. SCHEINFEIN & H.A. PADMORE - *Nature* **405**, 767 (2000)

[12] V.J. KEAST, A.J. SCOTT, R. BRYDSON, D.B. WILLIAMS & J. BRULEY - *J. Micros.* **203**, 135 (2001)

[13] C. COLLIEX, M. TENCE, E. LEFEVRE, C. MORY, H. GU, D. BOUCHET & C. JEANGUILLAUME - *Mikrochimica Acta* **114**, 71 (1994)

[14] J. FINK, T. MÜLLER-HEINZERLING, B. SCHEERER, W. SPEIER, F.U. HILLEBRECHT, J.C. FUGGLE, J. ZAANEN & G.A. SAWATSKY - *Phys. Rev. B* **32**, 4889 (1985)

[15] F.M.F. DE GROOT, Z.W. HU, M.F. LOPEZ, G. KAINDL, F. GUILLOT & M. TRONC - *J. Chem. Phys.* **101**, 6570 (1994)

[16] F.M.F. DE GROOT - *Physica B* **209**, 15 (1995)

[17] M. FINAZZI, N.B. BROOKES & F.M.F. DE GROOT - *Phys. Rev. B* **59**, 9933 (1999)

[18] M. WEISSBLUTH - *Atoms and Molecules*. Plenum Press, New-York, (1978)

[19] B.T. THOLE, G. VAN DER LAAN, J.C. FUGGLE, G.A. SAWATZKY, R.C. KARNATAK & J.M. ESTEVA - *Phys. Rev. B* **32**, 5107 (1985)

[20] M. POMPA, A.M. FLANK, P. LAGARDE, J. RIFE, I. STEKHIN, M. NAKAZAWA, H. OGASAWARA & A. KOTANI - *J. Physique IV* **7**(C2), 159 (1997)

[21] B.T. THOLE, G. VAN DER LAAN & G.A. SAWATZKY - *Phys. Rev. Lett.* **55**, 2086 (1985)

[22] R.D. COWAN - *The Theory of Atomic Structure and Spectra*, U. of California Press, Berkeley (1981)

[23] C.J. BALLHAUSEN - *Introduction to ligand Field Theory*, McGraw-Hill, New-York (1962)

[24] S. SUGANO, Y. TANABE & H. KAMIMURA - *Multiplets of Transition Metal Ions*, Academic Press, New-York (1970)

[25] P.H. BUTLER - *Point Group Symmetry Applications*, Plenum Press, New-York (1981)

[26] Z. HU, C. MAZUMDAR, G. KAINDL, F.M.F. DE GROOT, S.A. WARDA & D. REINEN - *Chem. Phys. Lett.* **297**, 321 (1998).

[27] F.M.F. DE GROOT - *X-ray absorption of transition-metal oxides*, Ph.D. Thesis (1991)

[28] F.M.F. DE GROOT, J.C. FUGGLE, B.T. THOLE & G.A. SAWATZKY - *Phys. Rev. B* **41**, 928 (1990)

[29] K. OKADA, T. UOZUMI & A. KOTANI - *Jap. J. Appl. Phys.* **32**(S32-2), 113 (1993)

[30] T. UOZIMI, K. OKADA & A. KOTANI - *J. Phys. Soc. Japan* **62**, 2595 (1993)

[31] F.M.F. DE GROOT, M.O. FIGUEIREDO, M.J. BASTO, M. ABBATE, H. PETERSEN & J.C. FUGGLE - *Phys. Chem. Min.* **19**, 140 (1992)

[32] M.O. FIGUEIREDO, A.C. DOS SANTOS, M.F. MELO, F.M.F. DE GROOT & M. ABBATE - *Adv. Sci. Tech. 3B*, 733 (1995)

[33] J. HASSELSTROM, A. FOHLISCH, R. DENECKE, A. NILSSON & F.M.F. DE GROOT - *Phys. Rev. B* **62**, 11192 (2000)

[34] T. JO & A. KOTANI - *J. Phys. Soc. Japan* **57**, 2288 (1988)

[35] O. GUNNARSSON & K. SCHÖNHAMMER - *Phys. Rev. B* **28**, 4315 (1983)

[36] A. FUJIMORI & F. MINAMI - *Phys. Rev. B* **30**, 957 (1984)

[37] G.A. SAWATZKY & J.W. ALLEN - *Phys. Rev. Lett.* **53**, 2339 (1984)

[38] J. ZAANEN, G.A. SAWATZKY & J.W. ALLEN - *Phys. Rev. B* **55**, 418 (1985)

[39] K. OKADA & A. KOTANI - *Physica B* **237**, 383 (1997)

[40] K. OKADA & A. KOTANI - *J. Phys. Soc. Japan* **66**, 341 (1997)

[41] M.A. VAN VEENENDAAL & G.A. SAWATZKY - *Phys. Rev. Lett.* **70**, 2459 (1993)

[42] F. JOLLET, V. ORTIZ & J.P. CROCOMBETTE - *J. Elec. Spec.* **86**, 83 (1997)

[43] R.H. POTZE, G.A. SAWATZKY & M. ABBATE - *Phys. Rev. B* **51**, 11501 (1995)

[44] M.A. VAN VEENENDAAL & G.A. SAWATZKY - *Phys. Rev. B* **49**, 3473 (1994)

[45] S.P. CRAMER, F.M.F. DE GROOT, Y.J. MA, C.T. CHEN, F. SETTE, C.A. KIPKE, D.M. EICHHORN, M.K. CHAN, W.H. ARMSTRONG & E. LIBBY - *J. Am. Chem. Soc.* **113**, 7937 (1991)

[46] S.M. BUTORIN - *J. Elec. Spec.* **110**, 213 (2000)

[47] P. MAHADEVAN & D.D. SARMA - *Phys. Rev. B* **61**, 7402 (2000)

[48] C. DE NADAI, A. DEMOURGUES, J. GRANNEC & F.M.F. DE GROOT - *Phys. Rev. B* **63**, 125123 (2001)

[49] G. VAN DER LAAN, J. ZAANEN, G.A. SAWATZKY, R.C. KARNATAK & J.M. ESTEVA - *Phys. Rev. B* **33**, 4253 (1986)

[50] K. OKADA & A. KOTANI - *J. Phys. Soc. Japan* **61**, 449 (1992)

[51] K. OKADA & A. KOTANI - *J. Phys. Soc. Japan* **61**, 4619 (1992)

[52] K. OKADA, A. KOTANI & B.T. THOLE - *J. Elec. Spec.* **58**, 325 (1992)

[53] Z. HU, G. KAINDL, S.A. WARDA, D. REINEN, F.M.F. DE GROOT & B.G. MULLER - *Chem. Phys.* **232**, 63 (1998).

[54] S.A. WARDA, W. MASSA, D. REINEN, Z. HU, G. KAINDL & F.M.F. DE GROOT - *J. of Solid State Chem.* **146**, 79 (1999).

[55] J.L. ERSKINE & E.A. STERN - *Phys. Rev. B* **12**, 5016 (1975)

[56] G. VAN DER LAAN, B.T. THOLE, G.A. SAWATZKY, J.B. GOEDKOOP, J.C. FUGGLE, J.M. ESTEVA, R.C. KARNATAK, J.P. REMEIKA & H.A. DABKOWSKA - *Phys. Rev. B* **34**, 6529 (1986)

[57] M. SACCHI & J. VOGEL - *In "Magnetism and Synchrotron Radiation"*, E. BEAUREPAIRE, F. SCHEURER, G. KRILL & J.P. KAPPLER eds, *Lecture Notes in Physics* **565**, 87, Springer, Berlin (2001)

[58] J.PH. SCHILLÉ, J.P. KAPPLER, PH. SAINCTAVIT, C. CARTIER DIT MOULIN, G. BROUDER & G. KRILL - *Phys. Rev. B* **48**, 9491 (1993)

[59] J. STÖHR & Y. WU - *In "New directions in research with third-generation soft X-ray synchrotron radiation sources"* A.S. SCHLACHTER & F.J. WUILLEUMIER eds, Kluwer, Dordrecht (1994)

[60] B.T. THOLE, P. CARRA, F. SETTE & G. VANDERLAAN - *Phys. Rev. Lett.* **68**, 1943 (1992)

[61] P. CARRA, B.T. THOLE, M. ALTARELLI & X.D. WANG - *Phys. Rev. Lett.* **70**, 694 (1993)

[62] G. SCHONHENSE - *J. Phys.: Cond. Matter* **11**, 9517 (1999)

[63] H. OHLDAG, T.J. REGAN, J. STOHR, A. SCHOLL, F. NOLTING, J. LUNING, C. STAMM, S. ANDERS & R.L. WHITE - *Phys. Rev. Lett.* **8724**, 7201 (2001)

[64] H. OHLDAG, A. SCHOLL, F. NOLTING, S. ANDERS, F.U. HILLEBRECHT & J. STOHR - *Phys. Rev. Lett.* **86**, 2878 (2001)

[65] M. BONFIM, G. GHIRINGHELLI, F. MONTAIGNE, S. PIZZINI, N.B. BROOKES, F. PETROFF, J. VOGEL, J. CAMARERO & A. FONTAINE - *Phys. Rev. Lett.* **86**, 3646 (2001)

2

MULTIPLE SCATTERING THEORY APPLIED TO X-RAY ABSORPTION NEAR-EDGE STRUCTURE

P. SAINCTAVIT - D. CABARET
Laboratoire de Minéralogie Cristallographie, Université Pierre et Marie Curie, Paris, France

V. BRIOIS
Synchrotron SOLEIL, Gif-sur-Yvette, France

1. INTRODUCTION

Multiple scattering theory (MST) is one of the theories that are used to calculate X-ray absorption spectra. This theory stems from the work by Korringa (1947) [1] who first developed the multiple scattering equations in 1947. The set of equations can be either solved in a real space or reciprocal space calculations. As far as X-ray absorption is concerned, MST has almost always been applied to real space calculations. It allowed to compute the discrete energy levels of a collection of atoms, a cluster, below the zero energy. It was also applied to compute the energy levels above the zero energy belonging to the continuum. The method was fully developed by K.H. Johnson (1973) [2] and applied by Dill and Dehmer [3] to the determination of photoionization differential cross-sections with an application to the nitrogen K edge in N_2.

In its usual form, the MST is used to compute the final state wave function of a photoelectron in an inner shell absorption process. Photoabsorption for the inner shell is governed by the matrix element between an initial state $|i\rangle$ and a final state $|f\rangle$ coupled by an interaction Hamiltonian describing the interaction between matter and X-rays. The initial state is strongly localized on the absorbing atom so that photoabsorption probes the value of the final state wavefunction $|f\rangle$ on the absorbing atom. The energy of the final state wavefunction is the kinetic energy of the photoelectron, so that $|f\rangle$ is well above the Fermi energy and belongs to the continuum of the Hamiltonian describing the absorbing sample. The value of the

F. Hippert et al. (eds.), Neutron and X-ray Spectroscopy, 67–101.

wavefunction $|f\rangle$ on the absorbing atom depends on the nature and the position of the atoms neighbouring the absorbing atom. This point is not obvious and asks for explanation. For very large kinetic energy of the photoelectron, the potential of the cluster can be neglected in a first approximation and the solutions of the Schrödinger equation are plane waves or linear combinations of plane waves. In any case, these are strongly delocalized solutions. In a second step, if the potential of the cluster is introduced as a perturbation one then expects that the plane wave functions will be modified by the potentials of all the atoms in the cluster, even for those that are far from the absorbing atom. In summary, MST is developed to compute the value of the final state wave function on the absorbing atom and this value depends on the potentials of all the atoms surrounding the absorbing atom.

In order to fix the terminology, we recall that the K edge corresponds to transitions from a $1s$ inner shell to empty levels. Similarly L_{I}, M_{I} and N_{I} edges are respectively transition from $2s$, $3s$ or $4s$ inner shell levels. $L_{\mathrm{II\text{-}III}}$ are transitions from the $2p$ level and similar denominations for $M_{\mathrm{II\text{-}III}}$ ($3p$ inner shell) and $N_{\mathrm{II\text{-}III}}$ ($4p$ inner shell). The two labels "II-III" are there to describe the splitting of the experimental edges due to the spin-orbit coupling acting on the hole of the inner shell. L_{II} corresponds to transitions where the orbital momentum of the p hole, $l = 1$, and spin momentum, $s = 1/2$, are coupled to give $j = l - s = 1/2$ while L_{III} edge is for the coupling $j = l + s = 3/2$. $M_{\mathrm{IV\text{-}V}}$ edges are spin-orbit split transitions originating from the $3d$ level with M_{IV} corresponding to a coupling $j = l - s = 3/2$ and M_{V} to a coupling $j = l + s = 5/2$ (and similar definitions for $N_{\mathrm{IV\text{-}V}}$ edges that concern $4d$ inner shell levels).

In what follows we shall be considering that matter and X-rays interaction is given in the electric dipole approximation. This is the first term in the development of the complete photon field. It is usually enough for K edges of most elements although higher order terms can be necessary for elements heavier than silver **[4]** or in very special cases where electric dipole cross-section is zero (for instance X-ray natural circular dichroism in chiral systems). At $L_{\mathrm{II\text{-}III}}$ edges, the electric dipole Hamiltonian is enough for most cases. A rule of thumb to size the influence of the electric quadrupole term is to compare the wavelength λ of the X-rays with the spatial extension of the inner shell: When λ is smaller than the spatial extension of the inner shell the electric quadrupole term cannot be neglected. In the electric dipole approximation, the selection rules state that the variation of orbital momentum between the initial state and the final state is either +1 or –1. K, L_{I} and M_{I} edges probe the empty levels with p symmetry seen from the absorbing atom. $L_{\mathrm{II\text{-}III}}$, $M_{\mathrm{II\text{-}III}}$, and $N_{\mathrm{II\text{-}III}}$ probe the empty levels with d and s symmetry. $M_{\mathrm{IV\text{-}V}}$ and $N_{\mathrm{IV\text{-}V}}$ edges probe empty levels with f and p symmetry.

In this volume, de Groot and Vogel present the Ligand Field Multiplet Theory (LFMT), which is also applied to the calculation of X-ray absorption cross-sections. We shall here detail the experiments for which LFMT has to be applied and for which MST has to be applied. The two theories originate from very different areas in physics. LFMT is developed in the framework of atomic physics and ingredients necessary to apply it to solid state compounds have been introduced in a second step. It is well adapted to the description of transitions where the final state wavefunction is strongly localized and where Coulomb interactions between the electrons in this shell are very strong. This is the case close to the edge (the first 10 to 20 eV) for L_{II-III} edges of $3d$ transition elements in oxides where the dominant final state wavefunction has a $3d$ character and also the case for the M_{IV-V} edges of lanthanides where the final state has essentially a $4f$ character. In LFMT Coulomb repulsion and spin-orbit coupling are treated in a Hartree-Fock model for a given atomic configuration. The influence of the ligands on the open shells is treated by an effective electrostatic field. It is also possible to simulate the covalency of the chemical bond by configuration interactions. For all these terms building the Hamiltonian of the sample, the angular dependence of the Hamiltonian is treated exactly by making extensive use of the symmetry and the radial integrals remain the only parameters. These parameters such as Slater integrals, spin-orbit coupling constant, crystal field parameters are "free" parameters (constrained by other types of spectroscopy, such as optical spectroscopy, EPR) that are extracted from the experimental data by a fitting procedure. LFMT is a parameterized, non *ab initio* method but draws its strength from the determination of the parameters with a precise physical meaning. MST is well adapted to transitions where the final state wave function is a rather delocalized wave function. This is the case for all edges far from the edge and also for edges where the final state wave function is so strongly hybridized with the neighbours that it has lost all its atomic character. This is the case for almost all K edges of elements heavier than lithium and also for L_{II-III} edges of elements heavier than cadmium. There are cases where it is difficult to sort out which method is adapted to the analysis of the spectra. For instance in the case of L_{II-III} edges of nickel, LFMT is well adapted to the interpretation of Ni edge in nickel oxides (NiO, $NiCr_2O_4$) or nickel cyanides ($CsNiCr(CN)_6$, 2 H_2O). With configuration interaction and the extra parameters it brings, LFMT can also be applied to the case of metallic nickel, although the large number of parameters and the small number of spectral features partly impairs its predictive potential [5]. Due to the strong hybridization of the $3d$ shell in metallic nickel, MST can be applied to compute nickel L_{II-III} edges. Though MST yields the general trends, it fails to reproduce precisely the branching ratios of isotropic and XMCD signals [6]. The L_{II-III} edges of metallic nickel are a case in between where neither of the two theories is excellent and where further theoretical developments are still needed.

There are at least two ways to introduce the MST: The scattered wave method and the Green function method. In the scattered wave method, the final state wave function is computed as being the sum of two parts: an outgoing wave plus a scattered wave. This method is rather pedestrian and makes contact with the usual theoretical presentations of X-ray absorption. This method was the heart of the course given by Rino Natoli in the 1980s [7], put in notes by Christian Brouder [8] and reproduced in the Ph.D. Thesis of Claire Levelut [9]. This is the method that we shall describe in the following. The Green function method is more straightforward but partially relies on the knowledge of the solution of a linear differential equation. It has been briefly developed in various papers and also in the Mittelwihr course on multiple scattering by Christian Brouder [10]. An account of this approach can also be found in the Ph.D. Thesis by Delphine Cabaret [11].

2. *MST FOR REAL SPACE CALCULATIONS*

We present the method of the scattered wave function. The calculation of the absorption cross-section is made in the framework of time dependent perturbation theory where all the initial state and final state wavefunctions are written in the harmonic representation [12]. Starting from the expression for the cross-section, the essential difficulty is the computation of the wavefunction for the photoelectron in the final state. We consider the simple case of an electromagnetic wave that interacts with an electron considered as a quantum particle with no spin, with charge q and mass m. For a deep understanding of all the ingredients present in the calculation, the wave function of the photoelectron is calculated for three different environments:

- no potential; this is the case of a free particle
- a spherical potential with a finite range
- a "muffin-tin" potential (that shall be defined later)

In this approach the electromagnetic field is not quantized and it should be remembered that a full quantum treatment leads to the same results for the absorption [13]. In all cases we restrict ourselves to the Schrödinger equation: spin-orbit coupling is neglected and the spin-up and spin-down channels are supposed to have the same cross-section. Moreover, one considers that the problem can be restricted to a single-electron computation.

2.1. THE ABSORPTION CROSS-SECTION

The expression for the cross-section follows from a straightforward application of the results given in the books by Cohen-Tannoudji and collaborators [12,14]. The absorption cross-section between a core state $|\phi_i\rangle$ normalized with $\langle\phi_i|\phi_j\rangle = \delta_{ij}$ and continuum states $|b\rangle$ normalized by $\langle b|b'\rangle = n(b)\,\delta(b-b')$ is given by the Fermi Golden rule

$$\sigma(\hbar\omega) \equiv \sigma(E) = 4\pi^2\alpha\hbar\omega\sum_{\beta}\rho(E,\beta)\left|\langle b(E,\beta)|\varepsilon\cdot r|\phi_j\rangle\right|^2 \qquad \{1\}$$

$E = \hbar\omega + E_i$ ($E_i < 0$ is the core level binding energy), $\rho(E,\beta)$ is called the density of states and $\rho(E,\beta) = \frac{|J[b(E,\beta)]|}{n[b(E,\beta)]}$ where $n(b)$ is defined by the normalization relation and J is the Jacobian that relates the variables describing the continuum states to the energy: $\rho(E,\beta)$ gives the number of states per energy. The sum in β is made over the energy allowed final states $|b\rangle$; alternatively if $\rho(E,\beta)$ is multiplied by $\delta(E-E_i-\hbar\omega)$ to account for energy conservation the sum can be made over all possible final i states.[1] α is the fine structure constant and ε is the X-ray polarization vector. In the following we only consider linear polarization for simplicity.

We want to write the functions $|b\rangle$ in the harmonic representation, that is in the basis built with the spherical functions $|kL\rangle$ defined by

$$\langle r|kL\rangle = j_l(kr)Y_L(\hat{r})$$

with the compound index $L = (l,m)$, the spherical Bessel function $j_l(kr)$ and the spherical harmonic function $Y_L(\hat{r})$. The properties of the $|kL\rangle$ functions are the following

1. $|kL\rangle$ are solutions of the Schrödinger equation for free particles

$$-\frac{\hbar^2}{2m}\Delta\phi(r) = E\phi(r)$$

2. The normalization of the $|kL\rangle$ functions is given by

$$\langle kL|k'L'\rangle = \frac{\pi}{2k^2}\delta(k-k')\delta_{L,L'}$$

3. Then the norm is $n(k,L) = \pi/(2k^2)$.

---

1. When one is dealing with transitions between non degenerate bound states, the cross section is $\sigma(\hbar\omega) = 4\pi^2\alpha\hbar\omega\Sigma_{\phi_f}|\langle\phi_f|\varepsilon.r|\phi_j\rangle|^2\delta(E_{\phi_f}-E_{\phi_j}-\hbar\omega)$. This expression is not well suited to deal with transitions towards states of the continuum, for which an infinite number of eigenstates exist at a given energy E_{ϕ_f} and in such a case equation 1 is to be prefered.

When transforming the expressions of the $|kL\rangle$ functions expressed with the variables (*k*,*L*) as functions of the variables (*E*,*L*), the Jacobian *J*(*E*,*L*) is given by $|J(E,L)| = dk/dE = \frac{m}{\hbar^2 k}$ and then the density of states $\rho(E,L)$ is

$$\rho(E,L) = \frac{|J(E,L)|}{n(k,L)} = \frac{m}{\hbar^2 k}\frac{2k^2}{\pi} = \frac{k}{\pi}\frac{2m}{\hbar^2}$$

In the harmonic representation the X-ray absorption cross-section per time unit is given by

$$\sigma(E) = 4\pi\alpha\hbar\omega k\frac{2m}{\hbar^2}\sum_L \left|\langle kL|\boldsymbol{\varepsilon}\cdot\boldsymbol{r}|\phi_j\rangle\right|^2 \qquad \{2\}$$

Relation {2} is the basic equation to which we shall refer in the following. In the literature the $|kL\rangle$ functions can be normalized in different ways. For instance with $|\widetilde{kL}\rangle = \sqrt{\frac{k_{u.a.}}{\pi}}\,|kL\rangle$ if *E* is in rydberg and *k* in atomic units then $\frac{\hbar^2}{2m} = 1$ and the cross-section is

$$\sigma(E) = 4\pi\alpha\hbar\omega k\sum_L \left|\langle\widetilde{kL}|\boldsymbol{\varepsilon}\cdot\boldsymbol{r}|\phi_j\rangle\right|^2 \qquad \{3\}$$

This normalization is useful since it makes contact with the current expression for the cross-section (cf. eq. {1}).

The main interest of the calculation of the cross-section in the harmonic representation is the existence of the theorem of "asymptotic completeness" for differential equations such as the Schrödinger equation.

Theorem of asymptotic completeness:

If $\{|\Psi_{Lk}\rangle\}_{Lk}$ is a set of eigenfunctions of the Hamiltonian $H = H_0 + V$ (where $H_0 = \Delta + k^2$) with *V* a time independent potential and if $\lim_{V\to 0}|\Psi_{Lk}\rangle = |kL\rangle$. Then the set of functions $\{|\Psi_{Lk}\rangle\}_{Lk}$ is a complete basis set with normalization conditions $\langle\Psi_{Lk}|\Psi_{L'k'}\rangle = \frac{\pi}{2k^2}\delta(k-k')\delta_{L,L'}$ and $1 = \sum_{kL}|\Psi_{Lk}\rangle\langle\Psi_{Lk}|\frac{2k^2}{\pi}$.

In the $\{|\Psi_{Lk}\rangle\}_{Lk}$ basis the cross-section is

$$\sigma(E) = 4\pi\alpha\hbar\omega\frac{2m}{\hbar^2}\sum_L \left|\langle\Psi_{Lk'}|\boldsymbol{\varepsilon}\cdot\boldsymbol{r}|\phi_j\rangle\right|^2 \qquad \{4\}$$

The target of multiple scattering theory is to compute the complete basis set $\{|\Psi_{Lk}\rangle\}_{Lk}$, eigenstates of *H*, that have the correct asymptotic behaviour when $V \to 0$.

2.2. CALCULATION OF THE CROSS-SECTION FOR THREE DIFFERENT POTENTIALS

We restrict ourselves to the case of K or L_I edges. The core level $|\phi_i\rangle$ is a $1s$ or $2s$ state of the absorbing atom. $|\phi_i\rangle$ is simply obtained by numerical integration as solution of the Schrödinger equation for an atomic potential. The only problem resides in the calculation of the final state $|b(E,\beta)\rangle$. To exemplify the way the functions $|b(E,\beta)\rangle$ are determined, we first calculate them, as solutions of the Schrödinger equation, for a null potential, then for a spherical potential and then for a "muffin-tin" potential.

2.2.1. FREE PARTICLE

The essence of this section can be found in textbooks [12]. For a free particle ($H_0 = \frac{-\hbar^2}{2m}\Delta$), the cross-section has been given in equation {2} and we shall develop this result. The equation to solve is

$$-\frac{\hbar^2}{2m}\Delta\Psi_{Lk}(\boldsymbol{r}) = E\Psi_{Lk}(\boldsymbol{r}) \qquad \{5\}$$

With $k^2 = \frac{2m}{\hbar^2}E$, this simplifies to

$$[\Delta + k^2]\Psi_{Lk}(\boldsymbol{r}) = 0 \qquad \{6\}$$

As a starting point, we notice that the spherical functions $\langle \boldsymbol{r} \,|\, kL\rangle = j_l(kr)Y_L(\hat{r})$ are solutions of equation {6} and are regular throughout space. From equation {6} one can write the radial Schrödinger equation that is satisfied by spherical Bessel functions. Let us decompose $\Psi_{Lk}(\boldsymbol{r})$ into its radial part $\phi(r)$ and its angular part $Y_L(\hat{r})$: $\Psi_{Lk}(\boldsymbol{r}) = \phi(r)Y_L(\hat{r})$. With the expression for the Laplacian in spherical coordinates

$$\Delta = \frac{1}{r^2}\frac{\partial}{\partial r}\left(r^2\frac{\partial}{\partial r}\right) - \frac{L^2}{r^2}$$

Equation {6} reduces to

$$\left[\frac{\partial^2}{\partial r^2} + \frac{2}{r}\frac{\partial}{\partial r} + k^2 - \frac{l(l+1)}{r^2}\right]\phi(r) = 0$$

With the notation $kr = \rho$ and $\varphi(\rho) = \phi(r)$, this transforms to

$$\left[\frac{\partial^2}{\partial \rho^2} + \frac{2}{\rho}\frac{\partial}{\partial \rho} + 1 - \frac{l(l+1)}{\rho^2}\right]\varphi(\rho) = 0 \qquad \{7\}$$

The solutions of equation {7} are the well known spherical Bessel functions j_l et n_l. The function j_l is the real solution, regular at the origin ($r = 0$), while the function n_l, that is also called Neumann function, is the real solution, irregular at the origin. For completeness, one defines the Hankel functions by: $h_l^+(r) = j_l(r) + i\, n_l(r)$ and $h_l^-(r) = j_l(r) - i\, n_l(r)$. We have gathered some useful properties of the Bessel and Neumann functions.

$$\lim_{\rho\to 0} j_l(\rho) = \lim_{\rho\to 0} \frac{\rho^l}{(2l+1)!!}$$

$(2l + 1)!!$ means odd factorial, i.e.: $(2l + 1)!! = (2l + 1).(2l - 1).(2l - 3)\ldots 3.1$.

$$\lim_{\rho\to\infty} j_l(\rho) = \lim_{\rho\to\infty} \frac{1}{\rho}\cos(\rho - l\frac{\pi}{2})$$

$$\lim_{\rho\to 0} n_l(\rho) = \lim_{\rho\to 0} -\frac{(2l-1)!!}{\rho^{l+1}}$$

$$\lim_{\rho\to\infty} n_l(\rho) = \lim_{\rho\to\infty} -\frac{1}{\rho}\sin(\rho - l\frac{\pi}{2})$$

$$\lim_{\rho\to 0} h_l^{\pm}(\rho) = \pm i \lim_{\rho\to 0} n_l(\rho) = \mp i \frac{(2l-1)!!}{\rho^{l+1}}$$

$$\lim_{\rho\to\infty} h_l^{+}(\rho) = i^{-l-1} \lim_{\rho\to\infty} \frac{1}{\rho} e^{i\rho}$$

$$\lim_{\rho\to\infty} h_l^{-}(\rho) = i^{l+1} \lim_{\rho\to\infty} \frac{1}{\rho} e^{-i\rho}$$

One is looking for the solutions of equation {7} that are regular everywhere and in particular at the origin. This excludes the Neumann functions. Then the set of $j_l(kr)Y_L(\hat{r})$ functions builds a complete basis set of the regular eigenstates of H_0. The cross-section is then

$$\sigma_{free}(E) = 4\pi\alpha\hbar\omega k \frac{2m}{\hbar^2} \sum_L \left| \langle kL \,|\boldsymbol{\varepsilon}\cdot \boldsymbol{r}\,|\, \phi_i\rangle \right|^2 \qquad \{8\}$$

The particular case of a particle in a constant potential $V \neq 0$ can be solved in a very similar way. Starting with $\left[\frac{-\hbar^2}{2m}\Delta + V\right]\Psi_{Lk}(r) = E\,\Psi_{Lk}(r)$ and k given by $k^2 = \frac{2m}{\hbar^2}(E - V)$ instead of $\frac{2m}{\hbar^2}E$, one finds that the cross-section for a particle in a constant potential V is $\sigma_V(E) = \sigma_{free}(E - V)$. A non-zero potential is then equivalent to an energy shift of the spectrum.

2.2.2. PARTICLE IN A SPHERICAL POTENTIAL

We consider a particle in the spherical potential $U(r)$ that has a finite range. We mean that $U(r)$ is zero beyond a certain value p. We shall solve the Schrödinger equation by separating the space into two different regions: zone I for $r > \rho$ and zone II for $r \leq \rho$.

We look for the solutions of

$$\left[-\frac{\hbar^2}{2m}\Delta + U(r)\right]\Psi(r) = E\Psi(r) \qquad \{9\}$$

This case is also essentially treated in the literature, although without reference to X-ray absorption. See references **[12]** and **[15]** for more information. With the transformations $V(r) = \frac{2m}{\hbar^2}U(r)$ and $k^2 = \frac{2m}{\hbar^2}E$, equation {9} then writes

$$[\Delta + k^2 - V(r)]\Psi(r) = 0 \qquad \{10\}$$

We are looking for the functions $\left\{|Y_{Lk}\rangle\right\}_{Lk}$ that are solutions of equation {10} and such that $\lim_{V\to 0}|\Psi_{Lk}\rangle = |kL\rangle$. If the potential $V(r)$ is discontinuous, the second derivative of Ψ_{Lk} cannot be continuous. So that the strongest requirement concerning Ψ_{Lk} is that it be continuous and with a continuous derivative Ψ'_{Lk} : Ψ_{Lk} needs to be a C^1 function and differentiable at second order almost everywhere. One is then looking for a function of the type $\Psi_{Lk}(r) = \phi_{lk}(r)Y_L(\hat{r})$ that has to be C^1 on the whole of space R^3.

In region I ($r > \rho$): $V(r) = 0$ equation {10} reduces to equation {6}

$$[\Delta + k^2]\Psi_{Lk}(r) = 0$$

with $\Psi_{Lk}(r) = \phi_{lk}(r)Y_L(\hat{r})$. The radial equation is:

$$\left[\frac{\partial^2}{\partial r^2} + \frac{2}{r}\frac{\partial}{\partial r} + k^2 - \frac{l(l+1)}{r^2}\right]\phi_{lk}(r) = 0$$

The solutions are given by linear combinations of Bessel and Neumann (or equivalently Hankel) functions. We set for the solution $\Psi_{Lk}(r) = \phi_{lk}(r)Y_L(\hat{r})$

$$\phi_{lk}(r) = j_l(kr) + it_l h_l^+(kr) \qquad \{11\}$$

Contrary to the free particle case, the irregular solutions (Neumann or Hankel functions) are sound physical solutions since they are only irregular at the origin, which is a point that is by definition excluded from region I.

In region II ($r \leq \rho$): since the potential is spherically symmetric, one is looking for a solution of the form $\Psi_{Lk}(\mathbf{r}) = \phi_{lk}(r)Y_L(\hat{r})$. The radial equation satisfied by $\phi_{lk}(r)$ is

$$\left[\frac{\partial^2}{\partial r^2} + \frac{2}{r}\frac{\partial}{\partial r} + k^2 - V(r) - \frac{l(l+1)}{r^2}\right]\phi_{lk}(r) = 0 \qquad \{12\}$$

The shape of potential $V(r)$ is generally unknown and then it is impossible to find an analytical solution for $\phi_{lk}(r)$. By numerical integration, one determines a solution $R_l(r)$ that is regular at the origin [16,17]. All the regular solutions of equation {12} are proportional. This is easily demonstrated by computing the Wronskian of two different solutions (the Wronskian of two functions f and g is $W[f,g] \equiv f\,g' - f'g$). The solution $\phi_{lk}(r)$ inside the sphere with radius ρ is then proportional to the solution $R_l(r)$.

$$\phi_{lk}(r) = C_l R_l(r) \qquad \{13\}$$

One is looking for a C^1 solution $\Psi_{Lk}(\mathbf{r})$. With the preceding choices in equations {11} and {13}, one only has to ensure that $\Psi_{Lk}(\mathbf{r})$ is C^1 at the limit between the two regions, for $r = \rho$. The angular part of $\Psi_{Lk}(\mathbf{r})$ is obviously continuous at $r = \rho$. The problem concerns the radial part of $\phi_{lk}(r)$ and its derivative. A way to write that ϕ_{lk} and ϕ'_{lk} are continuous at $r = \rho$ is to say that ϕ_{lk} and its logarithmic derivative $\frac{1}{\phi_{lk}}\frac{d}{dr}\phi_{lk}$ are continuous. The continuity of the logarithmic derivative is easy to handle since the C_l coefficient disappears from the ratio

$$\frac{1}{C_l R_l(r)}\frac{d}{dr}\left[C_l R_l(r)\right] = \frac{1}{R_l(r)}\frac{d}{dr}\left[R_l(r)\right]$$

The continuity for $\frac{1}{\phi_{lk}}\frac{d}{dr}\phi_{lk}$ is written

$$\frac{1}{R_l(r)}\frac{d}{dr}\left[R_l(r)\right] = \frac{1}{j_l(kr) + it_l h_l^+(kr)}\left[\frac{d}{dr}j_l(kr) + it_l\frac{d}{dr}h_l^+(kr)\right]$$

and

$$it_l\left[h_l^+(kr)\frac{d}{dr}R_l(r) - R_l(r)\frac{d}{dr}h_l^+(kr)\right] = -\left[j_l(kr)\frac{d}{dr}R_l(r) - R_l(r)\frac{d}{dr}j_l(kr)\right] \qquad \{14\}$$

One recognizes in the right and left hand side two Wronskian expressions between j_l and R_l (r.h.s.) and h_l^+ and R_l (l.h.s.). Equation {14} must be true for $r = \rho$

$$t_l = i\frac{W[j_l(k\rho), R_l(\rho)]}{W[h_l^+(k\rho), R_l(\rho)]} \qquad \{15\}$$

The continuity of $\phi_{lk}(r)$ at $r = \rho$ gives the relation between t_l and C_l:

$$C_l = \frac{j_l(k\rho) + it_l h_l^+(k\rho)}{R_l(\rho)} \quad \{16\}$$

Some properties of t_l:

- $$t_l = i\frac{W[j_l(k\rho), R_l(\rho)]}{W[h_l^+(k\rho), R_l(\rho)]} = i\frac{\Re\left(W[h_l^+(k\rho), R_l(\rho)]\right)}{W[h_l^+(k\rho), R_l(\rho)]}$$

with $W[h_l^+(k\rho), R_l(\rho)] = \beta e^{i\alpha}$ then $t_l = i\frac{\cos(\alpha)}{e^{i\alpha}} = \cos(\alpha)e^{i(-\alpha+\pi/2)}$. If one sets $\delta_l = \pi/2 - \alpha$, then there exists a quantity δ_l, called the phase shift, such that

$$t_l = \sin(\delta_l)e^{i\delta_l} \quad \{17\}$$

The well known result, concerning the phase shifts, is that

$$\lim_{r/\rho\to\infty} \phi_{lk}(r) = \lim_{r\to\infty} \frac{e^{i\delta_l}}{kr}\cos(kr - l\pi/2 + \delta_l) \quad \{18\}$$

By a comparison of the asymptotic expressions for $\phi_{lk}(r)$ and $j_l(kr)$, one sees that δ_l is the phase difference between the solution $\phi_{lk}(r)$ for potential $V(r)$ and the solution $j_l(kr)$ with no potential. This property is often considered as being the definition of the phase shifts.

- One notices also that $\text{cotg}(\delta_l) = i + t_l = \frac{W[n_l(k\rho), R_l(\rho)]}{W[j_l(k\rho), R_l(\rho)]}$ and then $C_l = t_l \frac{W[n_l(k\rho), R_l(\rho)]}{W[j_l(k\rho), R_l(\rho)]}$. Since $W[n_l(k\rho), j_l(k\rho)] = -\frac{1}{k\rho^2}$ then

$$C_l = t_l \frac{1}{k\rho^2 W[j_l(k\rho), R_l(\rho)]} \quad \{19\}$$

Finally the solutions of the problem for $H = H_0 + V(r)$ with $V(r) = 0$ if $r > \rho$ are:

$$r \le \rho \qquad \Psi_{Lk}(r) = t_l \frac{-1}{k\rho^2 W[j_l(k\rho), R_l(\rho)]} R_l(r)Y_L(\hat{r}) = t_l \underline{R_l}(r)Y_L(\hat{r})$$

$$r > \rho \qquad \Psi_{Lk}(r) = [j_l(kr) + it_l h_l^+(kr)]Y_L(\hat{r})$$

where $\underline{R_l}(r)$ is a "renormalized" $\underline{R_l}$ function and $t_l = i\frac{W[j_l(k\rho), \underline{R_l}(\rho)]}{W[h_l^+(k\rho), \underline{R_l}(\rho)]}$.

After some fastidious calculations, it can be shown that when l is large, the phase shift δ_l is proportional to the discontinuity of the potential in ρ **[15,18]**. If $V \to 0$, then $\delta_l \to 0$. The same is true for $t_l = \sin(\delta_l)\, e^{-i\delta_l}$. Then for $r > \rho$, the solutions $|\Psi_{Lk}\rangle$ tend towards $|kL\rangle$. The $|\Psi_{Lk}\rangle$ have then the same normalization conditions as $|kL\rangle$ for $r \ge \rho$ and by continuity for $r \le \rho$.

The X-ray absorption cross-section then becomes

$$\sigma(E) = 4\pi\alpha\hbar\omega k \frac{2m}{\hbar^2} \sum_L \left| \langle \Psi_{Lk} | \boldsymbol{\varepsilon} \cdot \boldsymbol{r} | \phi_i \rangle \right|^2$$

with
$$M_L = \int_{sphere} d\boldsymbol{r}\, \underline{R_l(r) Y_L(\hat{r})} (\boldsymbol{\varepsilon} \cdot \boldsymbol{r}) \phi_i(\boldsymbol{r})$$

The integration can be reduced to the volume of the sphere $r \leq \rho$ since for a core state $\phi_i(\boldsymbol{r}) \approx 0$ if $r > \rho$. The cross-section for the absorption of a photon with energy $\hbar\omega$ by a core level $\phi_i(\boldsymbol{r})$ in a spherical potential is

$$\sigma(E) = 4\pi\alpha\hbar\omega k \frac{2m}{\hbar^2} \sum_L | t_l |^2 | M_L |^2 \qquad \{20\}$$

It should be noticed that the electric dipole selection rules are contained in M_L and not in t_l. From the definitions of t_l, C.R. Natoli was the first to make contact with the optical theorem

$$\sigma(E) = 4\pi\alpha\hbar\omega k \frac{2m}{\hbar^2} \sum_L | M_L |^2 \Im | t_l |$$

2.2.3. *PARTICLE IN A "MUFFIN-TIN" POTENTIAL*

The method was first developed by Korringa and the same approach was rederived 7 years later by Kohn and Rostoker for band structures calculations [19]. It has been applied by Johnson to the computation of bound states [2,20,21]. The method has been generalized by Dill and Dehmer to be applied to the continuum states [3]. During the 80's, many versions have been published mainly under the impulsion of Natoli [7,22]. An approximation to this theory has then been implemented by Rehr and co-workers through an efficient and fast algorithm [23].

The absorbing atom belongs to a cluster of atoms that is represented in the so-called "muffin-tin" approximation. To determine the $\Psi_{Lk}(\boldsymbol{r})$ functions we want to solve the Schrödinger equation for such a potential. In the "muffin-tin" approximation, the potential is divided into three regions (fig. 1).

ZONE I: This is the outer region. It is situated outside the "Outer Sphere", that is the smallest sphere containing all the atomic spheres. In general, the potential in this region is spherical and tends towards zero at infinity.

ZONE II: This is the interstitial region. It is inside the outer sphere and outside the atomic spheres. The interstitial potential is constant in zone II.

ZONE III: The atomic spheres. The atomic potentials are spherical potentials delimited by finite atomic spheres. There is only one atom in each atomic sphere and it is at the centre of the sphere. There is no overlap between the atomic spheres.

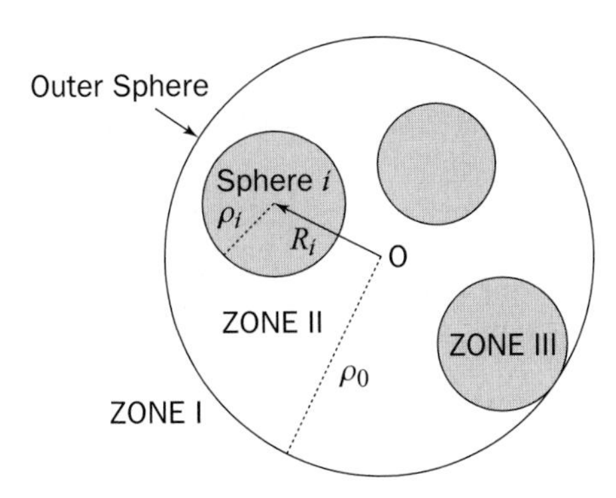

Figure 1 - Definition of various zones in a "muffin-tin" potential.

For simplification, in this course, one assumes that the potential is null in both zone I and zone II (One remembers that a shift of potential in zone II is for a free particle equivalent to a shift of the cross-section). A discussion of the importance of such an assertion in actual calculations has been given by Delphine Cabaret in her Ph.D. Thesis [11]. One looks for a complete set of solutions $\{|\Psi_{Lk}\rangle\}_{Lk}$ such that $\lim_{V\to 0} |\Psi_{Lk}\rangle = |kL\rangle$.

In zones I and II, the problem can be solved simultaneously since the potential is null. One sets that the general solution of the Schrödinger equation can be written as

$$\Psi_{Lk}(\boldsymbol{r}) = j_l(kr)Y_L(\hat{r}) - i\sum_{jL'} B^j_{L'}(L)h^+_{l'}(kr_j)Y_{L'}(\hat{r_j}) \qquad \{21\}$$

where $\boldsymbol{r}_j = \boldsymbol{r} - \boldsymbol{R}_j$ and $\boldsymbol{R}_j$ is the position of atom j relative to the absorbing central atom.

One first needs to prove that $\Psi_{Lk}(\boldsymbol{r})$ as defined in equation $\{21\}$ is a solution of the Schrödinger equation in zones I and II. Then we shall show that the choice of Ψ_{Lk} in equation $\{21\}$ gives enough coefficients $B^j_{L'}$ to solve the continuity problem at the boundary of the atomic spheres. In zones I and II one needs to solve $[\Delta + k^2]\Psi_{Lk}(\boldsymbol{r}) = 0$. The part $j_l(kr)Y_L(\hat{r})$ of $\Psi_{Lk}(\boldsymbol{r})$ is a solution for a free particle. The second part in equation $\{21\}$ is a sum of terms $h^+_{l'}(kr_j)Y_{L'}(\hat{r_j})$. These terms are solutions for the free particle problem centred in j since the Laplacian is invariant by translation. This is easily shown by writing the Laplacian in Cartesian coordinates, i.e. $[\Delta_{\boldsymbol{R}_j} + k^2] = [\Delta_0 + k^2]$. The function $h^+_{l'}(kr_j)Y_{L'}(\hat{r_j})$ is a solution of $[\Delta_{\boldsymbol{R}_j} + k^2]\Phi(\boldsymbol{r}) = 0$ and then it is also a solution of $[\Delta_0 + k^2]\Phi(\boldsymbol{r}) = 0$.

If the sum of spherical harmonics is extended up to a given l_{max} in the sphere j, the number of coefficients $B^j_{L'}(L)$ for a given L is given by

$$q = \sum_{l=0}^{l_{max}} \sum_{m=-l}^{m=l} 1 = \sum_{l=0}^{l_{max}} (2l+1) = (l_{max}+1)^2$$

For N atoms in the cluster the total amount of coefficients $B^j_{L'}(L)$ is $N(l_{max}+1)^2$. Inside sphere i, one sets the same type of solution as for the case of a spherical potential.

$$\Psi_{Lk}(\boldsymbol{r}) = \sum_{L'} C^i_{L'}(L) R^i_{l'}(r_i) Y_{L'}(\hat{r_i}) \qquad \{22\}$$

where $R^i_{l'}(r_i)$ is the regular solution of the radial Schrödinger equation (eq. {11}) for l'. The number of coefficients $C^i_{L'}(L)$ is $(l_{max}+1)^2$ per sphere, hence $N(l_{max}+1)^2$ for all the spheres. We want to calculate the coefficient $B^j_{L'}(L)$ and $C^i_{L'}(L)$. We need to write the continuity conditions for $\Psi_{Lk}(\boldsymbol{r})$ and its logarithmic derivative at the border of the spheres. Since the solution Ψ_{Lk} inside sphere i (eq. {22}) is written as a function of r_i, we need to rewrite the solutions Ψ_{Lk} in zone I and II (eq. {21}) as functions of r_i. In order to do that we shall use the translation operators $H^{ij}_{LL'}$ and $J^{i0}_{L'L}$ for the spherical functions. The two operators $H^{ij}_{LL'}$ and $J^{i0}_{L'L}$ depend on the coordinates of atoms i and j and are independent of the potential. These translations read

$$-ih^+_{l''}(kr_j)Y_{L''}(\hat{r_j}) = \sum_{L'} j_{l'}(kr_i)Y_{L'}(\hat{r_i})H^{ij}_{L'L''} \qquad \{23\}$$

$$j_l(kr)Y_L(\hat{r}) = \sum_{L'} j_{l'}(kr_i)Y_{L'}(\hat{r_i})J^{i0}_{L'L} \qquad \{24\}$$

Equation {23} is true if the distance between sites i and j is larger than r_i (that is $r_i < \| \boldsymbol{R}_i - \boldsymbol{R}_j \|$). On the other hand, equation {24} is true without restriction. Equation {24} states that a spherical wave centred on 0, with orbital momentum L, is seen from site i as a superposition of spherical waves whose angular momentum can take all possible L' values. Demonstration of these equations with definitions for H and J can be found in reference **[24]**, p. 88. A flavour of the demonstration for $j^{i0}_{L'L}$ is given in the appendix. With equations {23} and {24} one rewrites equation {21}

$$\begin{aligned}\Psi_{Lk}(\boldsymbol{r}) &= j_l(kr)Y_L(\hat{r}) - i\sum_{jL'} B^j_{L'}(L)h^+_{l'}(kr_j)Y_{L'}(\hat{r_j}) \\ &= \sum_{L'} j_{l'}(kr_i)Y_{L'}(\hat{r_i})J^{i0}_{L'L} - i\sum_{L'} B^i_{L'}(L)h^+_{l'}(kr_i)Y_{L'}(\hat{r_i}) \\ &\qquad + \sum_{L'',j\neq i} B^j_{L''}(L)\sum_{L'} j_{l'}(kr_i)Y_{L'}(\hat{r_i})H^{ij}_{L'L''}) \end{aligned} \qquad \{25\}$$

hence

$$\Psi_{Lk}(\boldsymbol{r})=\sum_{L'}\left[j_{l'}(kr_i)J^{i\,0}_{L'L}-iB^i_{L'}(L)h^+_{l'}(kr_i)+\sum_{L'',j\neq i}B^j_{L''}(L)j_{l'}(kr_i)H^{i\,j}_{L'L''}\right]Y_{L'}(\hat{r_i}) \quad \{26\}$$

We have expressed the function Ψ_{Lk} in the interstitial region as a function of $\boldsymbol{r}_i$. If we write the continuity conditions for the logarithmic derivative of Ψ_{Lk} at the border of sphere $i(r_i = \rho_i)$ find the following equations for any L' and centre i

$$B^i_{L'}(L) = -t^i_{l'}J^{i\,0}_{L'L}-t^i_{l'}\sum_{j\neq i}\sum_{L''}B^j_{L''}(L)iH^{i\,j}_{L'L''} \quad \{27\}$$

with t^i_l defined for each site in a similar way as in equation {15}

$$t^i_l = i\frac{W[j_l(k\rho_i),R^i_l(\rho_i)]}{W[h^+_l(k\rho_i),R^i_l(\rho_i)]} \quad \{28\}$$

Equation {27} can be written as a vectorial equation with the following definitions

$\boldsymbol{B}(L)\equiv[B^i_{L'}(L)]$ is a column vector with $N(l_{\max}+1)^2$ terms

$\boldsymbol{J}(L)\equiv[J^{i\,0}_{L'L}]$ is a column vector with $N(l_{\max}+1)^2$ terms

$H\equiv[H^{i\,j}_{L'L''}]$ is a $N(l_{\max}+1)^2\times N(l_{\max}+1)^2$ matrix

$T_a\equiv\frac{-1}{k}[t^i_l\delta_{L,L'}\delta_{i,j}]$ is a $N(l_{\max}+1)^2\times N(l_{\max}+1)^2$ matrix

With these definitions, equation {27} becomes

$$\boldsymbol{B}(L) = \left[1-kT_aH\right]^{-1}kT_a\boldsymbol{J}(L) \quad \{29\}$$

The matrix H and vector $\boldsymbol{J}(L)$ depend on the atomic coordinates in the cluster and are independent of the potential, while the diagonal T_a matrix depends only on the potential. If the absorbing atom is in position n (usually it is $n = 0$), one gets:

Inside zones I and II, the solution is given by

$$\Psi_{Lk}(\boldsymbol{r})=\sum_{L'}\left[j_{l'}(kr_n)J^{n\,0}_{L'L}-iB^n_{L'}(L)h^+_{l'}(kr_n)+\sum_{L'',j\neq n}B^j_{L''}(L)j_{l'}(kr_n)H^{n\,j}_{L'L''}\right]Y_{L'}(\hat{r_n})\{30\}$$

From equation {27}, one has

$$\sum_{j\neq n}\sum_{L''}B^j_{L''}(L)H^{n\,j}_{L'L''} = \frac{-1}{t^n_{l'}}B^n_{L'}(L)-J^{n\,0}_{L'L}$$

Equations {27} and {30} then gives

$$\Psi_{Lk}(\boldsymbol{r}) = \sum_{L'}\left[\frac{-1}{t^n_{l'}}B^n_{L'}(L)j_{l'}(kr_n)-iB^n_{L'}(L)h^+_{l'}(kr_n)\right]Y_{L'}(\hat{r_n}) \quad \{31\}$$

Inside sphere n, the solution is:

$$\Psi_{Lk}(r) = \sum_{L'} C_{L'}^{n}(L) \underline{R_{l'}^{n}}(r_n) Y_{L'}(\hat{r_n}) \qquad \{32\}$$

with $\underline{R_{l'}^{n}}(r_n)$ a renormalized $R_{l'}^{n}(r_n)$ function such that

$$\underline{R_{l'}^{n}}(r_n) = \frac{-R_{l'}^{n}(r_n)}{k(\rho_n)^2 W\left[j_{l'}(k\rho_n), R_{l'}^{n}(\rho_n)\right]}$$

With this normalization condition one has

$$\underline{R_{l'}^{n}}(\rho_n) = \frac{1}{t_{l'}^{n}} j_{l'}(k\rho_n) + i h_{l'}^{+}(k\rho_n) \qquad \{33\}$$

hence

$$\Psi_{Lk}(\rho_n) = \sum_{L'} \left[\frac{1}{t_{l'}^{n}} C_{L'}^{n}(L) j_{l'}(k\rho_n) + i C_{L'}^{n}(L) h_{l'}^{+}(k\rho_n) \right] Y_{L'}(\hat{\rho}_n) \qquad \{34\}$$

From equations {31} and {34}, it follows that

$$C_{L'}^{n}(L) = -B_{L'}^{n}(L) \qquad \{35\}$$

And the general form of the solution is

$$\Psi_{Lk}(r) = -\sum_{L'} B_{L'}^{n}(L) \underline{R_{l'}^{n}}(r_n) Y_{L'}(\hat{r_n}) \qquad \{36\}$$

with the coefficients $B_{L'}^{n}(L)$ given by

$$\boldsymbol{B}(L) = \left[1 - kT_a H\right]^{-1} kT_a \boldsymbol{J}(L)$$

If $V \to 0$ then $\delta_l \to 0$ [18] and then $t_l^i = \sin(\delta_l^i)\, \mathrm{e}^{i\delta_l^i} \to 0$. The T_a matrix then tends towards 0. It follows that $\boldsymbol{B}(L)$ is 0 so that $\lim_{V \to 0} |\Psi_{Lk}\rangle = |kL\rangle$. The cross-section can then be written

$$\sigma(E) = 4\pi\alpha\hbar\omega k \frac{2m}{\hbar^2} \sum_{L} \left| \langle \Psi_{Lk} | \boldsymbol{\varepsilon} \cdot \boldsymbol{r} | \phi_i \rangle \right|^2 \qquad \{37\}$$

If instead of using complex spherical harmonics, one uses only real spherical harmonics (this is the case for several multiple scattering codes), then for a real potential and linear polarization (i.e. real polarization vector) the $M_{L'}$ terms are real. One then gets the following formula where the $B_{L'}^{n}(L)$ are nevertheless complex

$$\langle \Psi_{Lk} | \boldsymbol{\varepsilon} \cdot \boldsymbol{r} | \phi_i \rangle = -\sum_{L'} [B_{L'}^{n}(L)]^{*} M_{L'} \qquad \{38\}$$

with the notation

$$M_{L'} = \int_{sphere\ n} \underline{R_{l'}^{n}}(r_n) Y_{L'}^{*}(\hat{r_n})(\boldsymbol{\varepsilon} \cdot \boldsymbol{r}) \phi_i(r_n) dr_n \qquad \{39\}$$

and the cross-section is

$$\sigma(E) = 4\pi\alpha\hbar\omega k \frac{2m}{\hbar^2} \sum_{L'L''} M_{L'} M_{L''} \sum_{L} \left[B^n_{L''}(L)\right]^* B^n_{L'}(L) \qquad \{40\}$$

The whole work in a multiple scattering calculation consists in the construction of the H matrix and the $\boldsymbol{J}$ vector, the computation of the t_l^i coefficients and then the inversion of the matrix $[1 - kT_a H]$ for each energy point in order to obtain the $B_{L'}(L)$ coefficients. The electric dipole selection rules are expressed in the $M_{L'}$ and $M_{L''}$ coefficients and select the various l' and l'' on which the sum $\sum_{L'L''}$ is performed in equation {40}.

3. CONSTRUCTION OF THE POTENTIAL

When the potential is known, the calculation of the cross-section for a specific edge is straightforward. One of the important points in the MST is indeed the building of the potential. We shall review here some of the aspects of this question., Due to electron-electron repulsion the potential should be a multi-electron potential. When one wants to reduce the potential to a single-electron one, one usually needs to express it as a "local" potential, i.e. a potential $V(\boldsymbol{r})$ (that only depends on the electronic density and eventually on its various derivatives at the same position $\boldsymbol{r}$. To represent the repulsive Coulomb potential between the electrons, one considers a mean field potential to which an exchange and correlation part is added.

In a schematic presentation a "muffin-tin" potential is built as follows. One first selects a cluster, i.e., a collection of atoms surrounding the absorbing atom. Usually the cluster is built with one hundred to a few hundreds atoms. For each atom in the cluster, one computes an atomic self consistent field (SCF) potential. Around each atom, one defines a "muffin-tin" radius. This is a free parameter, that has no real physical meaning and whose determination is somewhat arbitrary. It is usually chosen so as to reduce the volume of the interstitial region and also minimize the discontinuities between the potentials of adjacent atomic spheres and also between the potential of the atomic spheres and the interstitial potential. To stick to "orthodox" MST, one should not let the "muffin-tin" spheres overlap each other. In most calculations, a small overlap (around 10%) is allowed in order to reduce the size of the interstitial region, where the description of the potential is the worst. Inside each atomic sphere, one superimposes the SCF atomic charge densities with the tails of the SCF charge densities of the neighbouring atoms. For an illustration, one can check in figure 2 the influence of a 10% overlap on the vanadium *K* edge in the cluster $V(H_2O)_6$.

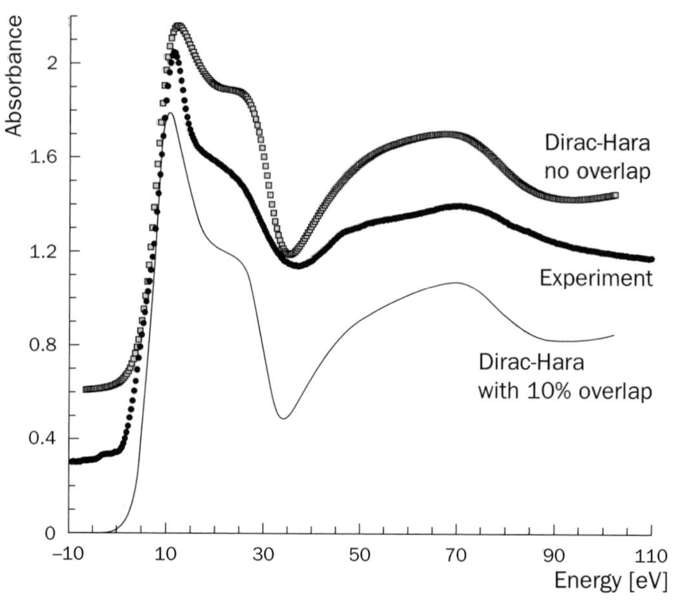

Figure 2 - Vanadium *K* edge in the cluster $V(H_2O)_6$ with and without overlapping atomic spheres.

A 10% overlap means that the sum of the "muffin-tin" radii of two adjacent atomic spheres is less than 10% larger than the distance between the centres of the two adjacent spheres. The differences between the calculation with and without overlapping spheres are not large. In the example of figure 2, the calculation with 10% overlap is only slightly better than without. Then inside each atomic sphere, the charge density is spherically averaged. In the interstitial region, the charge density is set constant and is an average of the contributions from all the atoms of the cluster. A mean potential is calculated by solving the Poisson equation for the charge density in the whole cluster. An exchange and correlation potential, V_{xc}, is added to the mean potential. V_{xc} is also a function of the local charge density.

In most cases, the potential in the interstitial region is set to be the zero potential at infinity, so that all eigenstates with energy larger than the interstitial potential shall belong to the continuum. In some cases, a SCF procedure on the cluster potential is developed to obtain an SCF "muffin-tin" potential. The main drawback of such a potential (either SCF or not) resides in its "muffin-tin" shape. Various routes have been followed to cure this problem: pseudo-potentials, full potential linearized augmented plane waves, potential defined on a grid for finite difference methods,

Rino Natoli's non "muffin-tin" potentials. These methods can strongly improve the agreement between calculated spectra and experiments but they also introduce severe complications and are not covered in this chapter.

We shall review in the following the various exchange and correlation potentials that are commonly used to perform X-ray absorption calculations in the "muffin-tin" approximation.

3.1. $X-\alpha$ EXCHANGE POTENTIAL

This potential takes into account only the exchange term that results from the Pauli exclusion principle. One can show that the average exchange energy per electron can be written for a given atom [25]

$$U_{X-\alpha}\,[\text{ryd}] = -\frac{3\alpha}{\pi}k_F\,[\text{u.a.}] \qquad \{41\}$$

where $k_F = [3\pi^2\rho(r)]^{1/3}$ is the local Fermi momentum defined at position r, where the electronic density is $\rho(r)$. The α coefficients depend on the considered chemical element and have been tabulated by Schwartz [26] for atoms with $Z \leq 41$. They vary between 0.77298 for helium and 0.70383 for niobium. The $X-\alpha$ exchange potential does not take into account the dependence of the exchange potential on the kinetic energy of the photoelectron and it does not include any inelastic losses to which the photoelectron is subject. In figure 3, one sees that the $X-\alpha$ calculation gives quite good agreement with experiment as long as one does not consider the too highly pronounced features of the calculation.

3.2. DIRAC-HARA EXCHANGE POTENTIAL

The exchange potential first proposed by Dirac and formulated by Hara is the exchange potential experienced by an electron above the Fermi level in a Fermi liquid [27]. One obtains for an electron with kinetic energy $\frac{\hbar^2k^2}{2m}$:

$$U_{D.H.}(k) = -\frac{3\alpha}{\pi}k_F F\left(\frac{p}{k_F}\right)$$

where the local photoelectron momentum, p, is defined by $\sqrt{k^2+k_F^2}$ and F is given by

$$F(x) \equiv 1+\frac{1-x^2}{2x}\ln\frac{|1+x|}{|1-x|}$$

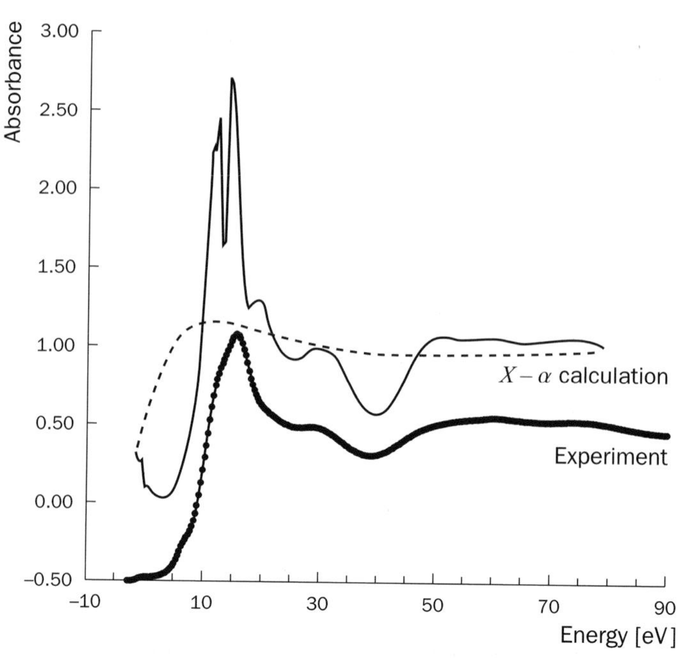

Figure 3 - Comparison of the experimental and $X-\alpha$ theoretical spectra at the iron *K* edge of $Fe(phen)_2(NCS)_2$ (see section 5). The smooth dashed line is the theoretical atomic cross-section.

One notes that the knowledge of p requires a knowledge of the effective single-electron potential U that requires the knowledge U^{ex}. One can notice that with $\alpha = 2/3$ one makes contact with the expression for the $X-\alpha$ potential for electrons at the Fermi level. The Dirac-Hara exchange potential depends on the energy of the photoelectron, though no inelastic scattering is present in it. The spectra calculated with $X-\alpha$ and Dirac-Hara exchange potentials present very sharp features because there is no broadening due to the conjugated effect of the finite lifetime of the core-hole and also to the inelastic scattering of the photoelectron by the other electrons of the absorbing compound.

A comparison with experiment is usually easier when broadening has been introduced in the calculations. This can be done in an *ad hoc* way by convoluting the $X-\alpha$ spectra by a Lorentzian function (intrinsic and extrinsic losses) or a Gaussian function (instrumental broadening). This can also be done in a more *ab initio* way by computing the self-energy of the photoelectron.

3.3. *COMPLEX HEDIN-LUNDQVIST EXCHANGE AND CORRELATION POTENTIAL*

We shall not fully develop the formalism of the complex Hedin-Lundqvist potential and we shall concentrate on its specific importance for the simulation of X-ray absorption cross-sections. The potential felt by the photoelectron in the single-electron approach contains a mean part, an exchange part and what is left receives the name of correlation potential. Expressions for the exchange and correlation parts have been developed by Hedin and Lundqvist [28] as a by-product of the calculation of the self-energy of the photoelectron. The self-energy contains all the interactions of the photoelectron with the other electrons of the medium. It is a complex function where the real part is the sum of the exchange and correlation energy and the imaginary part is a damping term representing the inelastic interactions of the photoelectron with the medium. The medium is simulated by a dielectric function, usually defined in the plasmon pole approximation and the interaction of the photoelectron with the medium is expressed as a series of the screened interaction. Only the first term is computed (this is the so called GW approximation). When the photoelectron kinetic energy is larger than the plasmon energy of the valence electrons (that is the plasmon energy associated with the electronic density in the interstitial region), the photoelectron is undergoing severe inelastic losses. This corresponds to a sharp increase of the imaginary part of the photoelectron self-energy. Following a simple model derived from the WKB procedure, one can relate the imaginary part of the photoelectron self-energy ($\Im\Sigma$) in the atomic spheres or in the interstitial region to an estimation of the photoelectron mean free path $\lambda_{el}(E)$, as a function of the kinetic energy, E, of the photoelectron.

$$\lambda_{el}(E) = \frac{\hbar^2 k}{2m}\frac{1}{\Im\Sigma}$$

One sees (fig. 4) that below the plasmon energy (here it is ≈ 9 eV, $\lambda_{el}(E)$ is infinite.

If one considers also the finite life time of the core-hole, one can calculate an effective mean free path for a specific edge. It is given by

$$\lambda_{eff} = \frac{\hbar}{\Gamma_{tot}(E)}\sqrt{\frac{2E}{m}}$$

with $$\Gamma_{tot}(E) = \Gamma_h + \Gamma_{el}(+\Gamma_{exp})$$

where Γ_h is the core-hole width and Γ_{el} is the damping function related to the photoelectron mean free path $\lambda_{el}(E), \Gamma_{el} = \frac{\hbar}{\lambda_{el}(E)}\sqrt{\frac{2E}{m}}$. An eventual Γ_{exp} term can be added to mimic instrumental broadening, although this should experimentally be reduced at maximum.

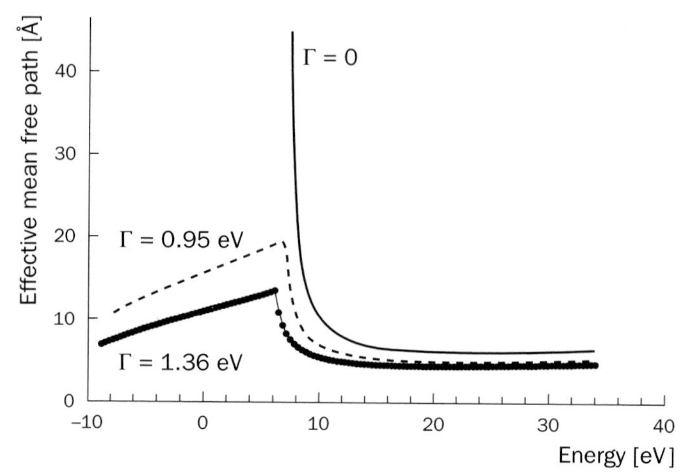

Figure 4 - Mean free path $\lambda(E)$ of the photoelectron calculated from the imaginary part of the complex Hedin-Lundqvist potential for a bcc cluster of iron atoms and various core-hole lifetimes Γ.

In figure 4 we have plotted some mean free path curves. One sees that below the plasmon energy, the finite value of λ_{eff} depends only on the constant part $\Gamma_h + \Gamma_{\text{exp}}$ and the effective mean free path then varies as the square root the photoelectron kinetic energy. Above the plasmon energy, the contribution of λ_{el} to λ_{eff} is dominating and $\Gamma_h + \Gamma_{\text{exp}}$ can be neglected.

The Hedin-Lundqvist potential allows us to calculate spectra that can be directly compared to the experimental ones. A more pedestrian way to compute broadened spectra is to compute cross-sections for a real potential and to convolute the spectra with a broadening function (mixture of Lorentzian and Gaussian functions). This has the disadvantage of introducing parameters in an *ab initio* calculation.

Figure 5 is a comparison of various exchange and correlation potentials at the vanadium K edge in $V(H_2O)_6$. For the Hedin-Lundqvist calculation a core-hole width of $\Gamma_h = 1$ eV has been used. Both $X-\alpha$ and Dirac-Hara calculations have been convoluted with a Lorentzian function of width $\Gamma_h = 1$ eV to account for core-hole lifetime. From the comparison with the experimental spectrum, one finds that the Dirac-Hara calculation seems to be the best. This is often the case when one is only considering the region very close to the edge (less than 70 eV). When the region farther from the edge is computed, the Hedin-Lundqvist potential usually yields better results.

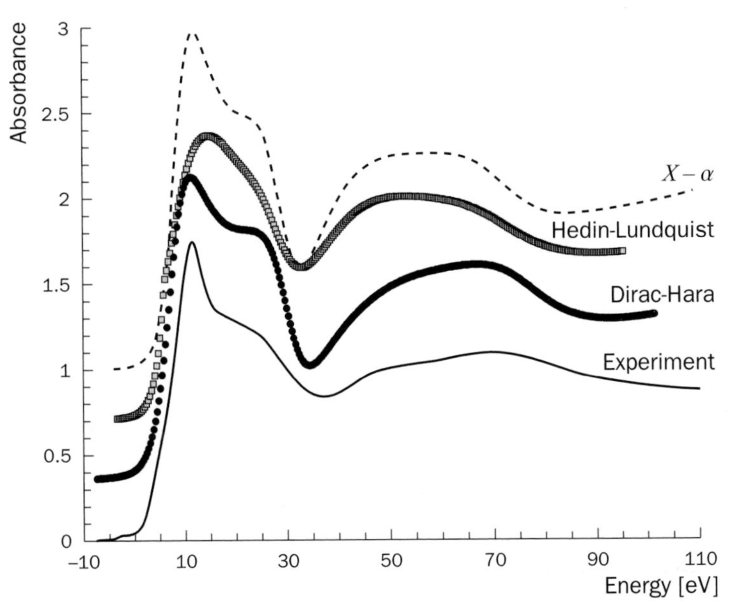

Figure 5 - Vanadium *K* edge in VO_6 for various types of exchange and correlation energies.

4. *APPLICATION OF THE MULTIPLE SCATTERING THEORY*

The goal of this section is twofold:

1. to show the practice of Multiple Scattering calculations to reproduce XANES structures and dichroism effects.
2. to show the influence of the main parameters (potentials, cluster size, structure) on the resonance (intensity, number, position).

The following section is an illustration of the theoretical part of section 2.

We would like to discuss the influence of the two main parameters (potential and structure) on the results of a MST calculation. Calculations of the metal *K* edge XANES spectra of three $M(H_2O)_6^{2+}$ complexes with M = V, Fe and Cu recorded in solution [29] will be used for this task (fig. 6). The metallic ions are in sixfold coordination for the three complexes. The vanadium and iron complexes are in O_h symmetry with a mean oxygen-metal distance equal to 2.16 Å. The copper complex is in D_{4h} symmetry with two oxygen-metal distances 1.96 Å and 2.18 Å.

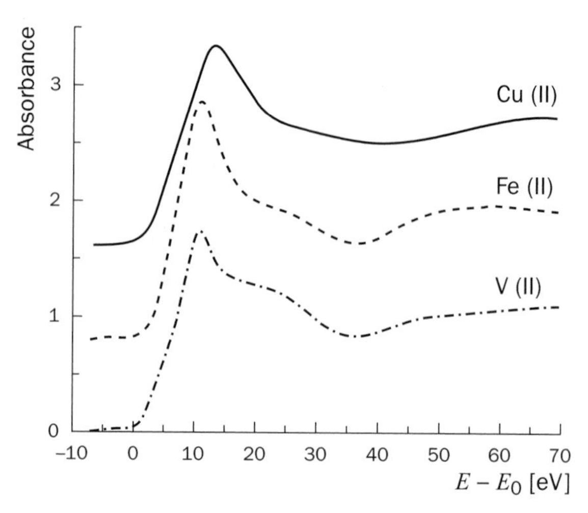

Figure 6 - Metal *K* edge XANES spectra of $M(H_2O)_6^{2+}$ complexes with M = V, Fe and Cu as a function of the relative energy to the absorption edge (E_0 = 5465 eV for V, 7112 eV for Fe and 8979 eV for Cu).

4.1. *VANADIUM K EDGE CROSS-SECTION FOR VO_6 CLUSTER*

The VO_6 cluster considered for this calculation is made of 7 atoms (we do not consider the hydrogen atoms due to the well known weak scattering power of this element). The oxygen atoms are located along the x, y and z axes as displayed in figure 7. The symmetry is O_h.

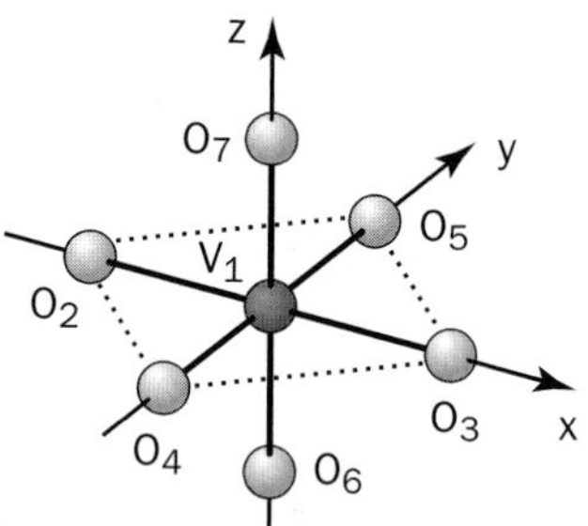

Figure 7 - VO_6 cluster.

The electronic configuration for vanadium in the ground state is simply "23 protons + $1s^2\,2s^2\,2p^6\,3s^2\,3p^6\,4s^2\,3d^3$". One usually starts the calculation with SCF atomic orbitals for neutral atoms. Since the $1s$ wave function is only slightly potential dependent, the precise determination of the initial state potential has no great importance. For the final state, the determination of the potential is on the contrary more important. In the final state a core-hole is present on the vanadium $1s$ shell so that the configuration is "23 protons + $1s^2\,2s^2\,2p^6\,3s^2\,3p^6\,4s^2\,3d^3$". The potential to consider is the fully relaxed potential with a core-hole. A good approximation to this state is the configuration "24 protons + $1s^2\,2s^2\,2p^6\,3s^2\,3p^6\,4s^2\,3d^3$" where an extra proton and an extra electron have been added on the nucleus and the $1s$ shell. This configuration corresponds to the chromium ground state (with no core-hole). Substituting the vanadium orbitals with a core-hole by the chromium ones with no core-hole is called the $Z + 1$ approximation. Doing so is intended to mimic the relaxation of the vanadium orbitals under the attraction of the core-hole. The final state configuration (either "23 protons + $1s^2\,2s^2\,2p^6\,3s^2\,3p^6\,4s^2\,3d^3$" or "24 protons + $1s^2\,2s^2\,2p^6\,3s^2\,3p^6\,4s^2\,3d^3$") is positively charged, so that dielectric response of the absorbing compound tends to screen the positive charge. To mimic the screening effect, one can add an extra electron or a fraction of it on the outer shells.

In figure 8, one can notice the influence of the type of potential. When only relaxation is considered ($Z + 1$ configuration without extra electron on the outer shell) the potential is too attractive and this yields too sharp a rising edge just above the zero energy. The screened and relaxed potential yields calculation in better agreement with the experiment. The importance of the screening is difficult to foresee.

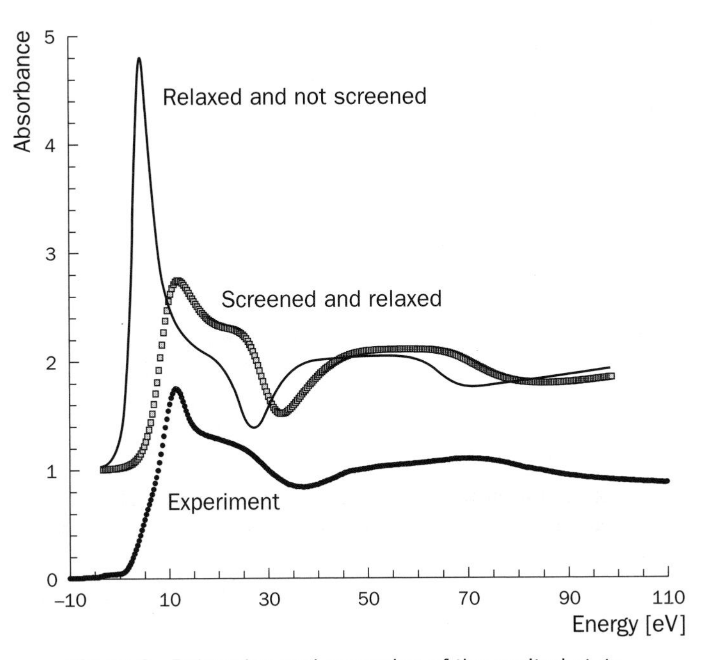

Figure 8 - Relaxation and screening of the excited state.

For instance at the Al K edge it has been shown that it depends on the aluminium site symmetry and coordination [30]. On can also perform a calculation by considering that the final state is equal to the ground state with no core-hole.

In figure 9 the final state is either equal to the ground state or to the excited state in the relaxed and screened scheme developed just above. One finds that the difference between the two configurations is not large. When the screening is efficient the core-hole is almost not felt by the photoelectron close to the edge.

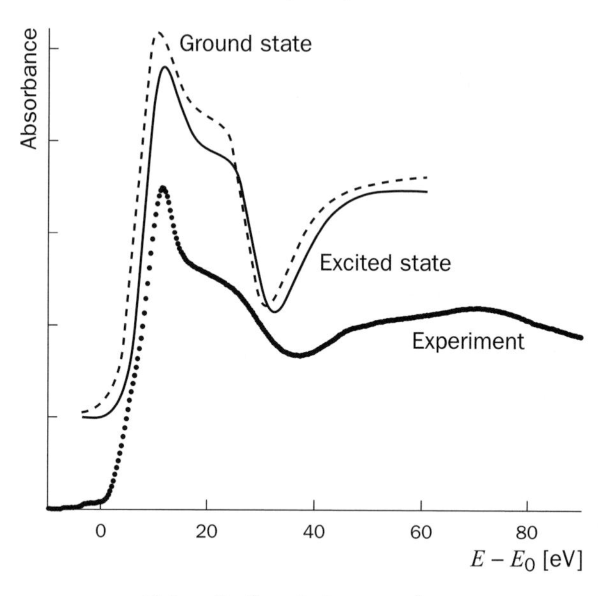

Figure 9 - Core-hole screening.

To check the importance of the cluster geometry, we now consider that the VO_6 cluster has suffered an isostatic compression, which has rescaled the metal-oxygen distance by a factor of compression of 3%. The spectra for the two geometries are plotted in figure 10. The potential prescriptions (screened and relaxed) are the same for the two calculations. One notices that the spectral features are shifted to higher energies under the compression. This is in agreement with the famous Natoli rule that states that $Ed^2 = Const$ where E is the energy of a spectral feature relative to the zero energy and d is a length characteristic of the absorbing atom environment. The formula $Ed^2 = Const$ derives from the fact that the radial coordinate essentially appears in the Multiple Scattering equations through the spherical Bessel functions whose argument is kr. Evidently $kr = Const$ is similar to $Ed^2 = Const$.

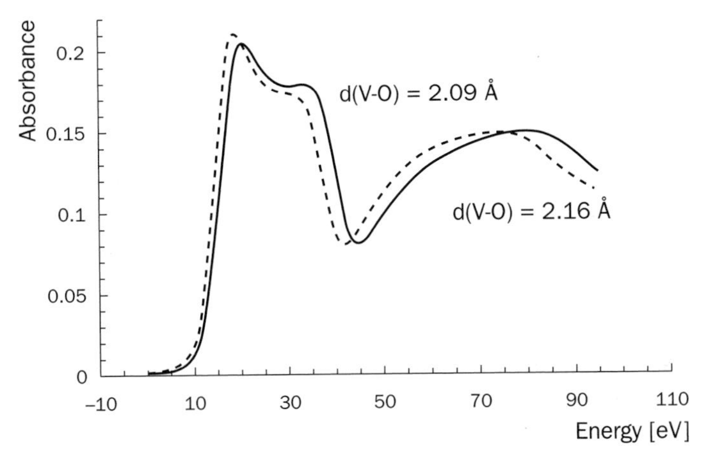

Figure 10 - Vanadium *K* edge in VO_6 for two V-O distances.

4.2. *IRON K EDGE CROSS-SECTION FOR FeO_6 CLUSTER*

The comparison between iron and vanadium *K* edges can be seen in the insert of figure 11. From the comparison between the two spectra, one observes an inversion of resonances A and B: at iron *K* edge, A is larger and B smaller than at vanadium *K* edge.

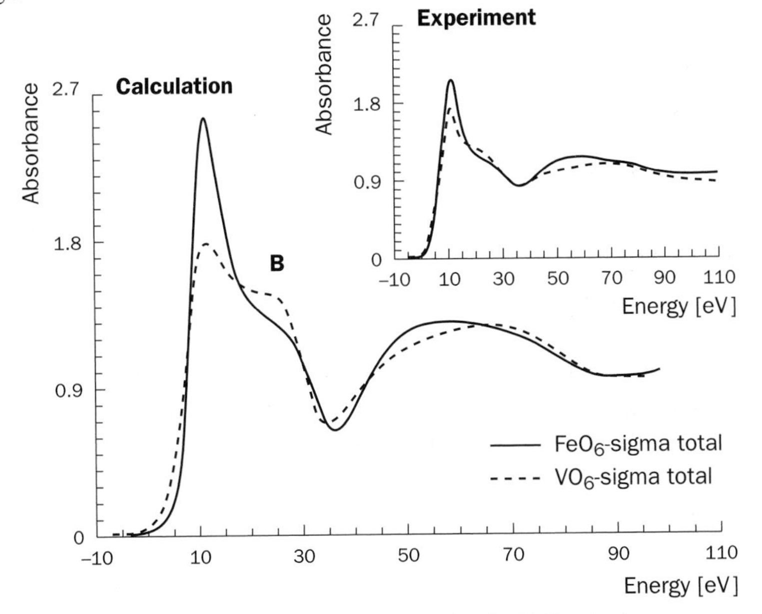

Figure 11 - Iron *K* edge cross-section for FeO_6 cluster.

From figure 11, one also sees that the calculation can nicely reproduce these experimental trends. Since both complexes have the same structural parameters in the cluster, the H free propagator matrix and the $\boldsymbol{J}$ vector are identical for both complexes. Then the differences between the two spectra only originate from the T_a matrices of the two clusters. Again since the oxygen neighbours are identical for the two clusters, the only difference is in the t_l coefficients for the central absorbing atom. A careful check of the t_l coefficients shows that the scattering power of iron is larger than that of vanadium very close to the edge.

4.3. CROSS-SECTION CALCULATIONS IN DISTORTED OCTAHEDRA

We focus on the effect of symmetry lowering around the metal atom and on its influence on the XANES features. The CuO_6 cluster is in D_{4h} symmetry. Due to Jahn-Teller distortion the Cu-O distances are spread into two groups, one with four oxygen atoms and one with two oxygen atoms. We consider a complex with axial ligands (along z) at longer distances from the absorbing atom than the ligands in the (x,y) plane (fig. 12). Unlike in O_h symmetry, the electric dipole term is not isotropic, so we can examine the X-ray Natural Linear Dichroism (XNLD). The electric dipole operator can be decomposed into two irreducible representations (irreps) of D_{4h}. When $\boldsymbol{\varepsilon} \| C_4$ axis, $\boldsymbol{\varepsilon} \cdot \boldsymbol{r} = z$ is a basis for the irrep A_{2u} of D_{4h}. When $\boldsymbol{\varepsilon} \perp C_4$ axis, $\boldsymbol{\varepsilon} \cdot \boldsymbol{r}$ can be equal to x or y, and (x,y) is a basis for the irrep E_u of D_{4h}.

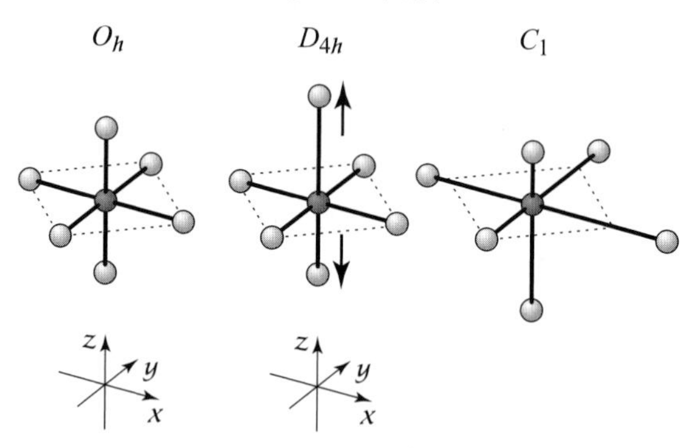

Figure 12 - Various symmetries around metallic ions.

Since the initial state is a $1s$ level that belongs to the A_{1g} irrep of D_{4h}, the matrix element $\langle \Psi_{Lk} | \boldsymbol{\varepsilon} \cdot \boldsymbol{r} | \phi_i \rangle$ is different from zero if and only if Ψ_{Lk} belongs to either A_{2u} or E_u (or a linear combination of them). Since the allowed final states are spread on two irreps, this tends to broaden features and promote shoulders absent on the spectrum of the fully symmetric compound. This can be verified in figure 13, where the intensity of the main rising edge is slightly decreased with the lowering of symmetry and where a shoulder is also present below the maximum. In figure 13 a calculation with Cu sitting in C_1 configuration has also been represented. One sees that the influence of the symmetry can be quite large and is difficult to foresee without calculation.

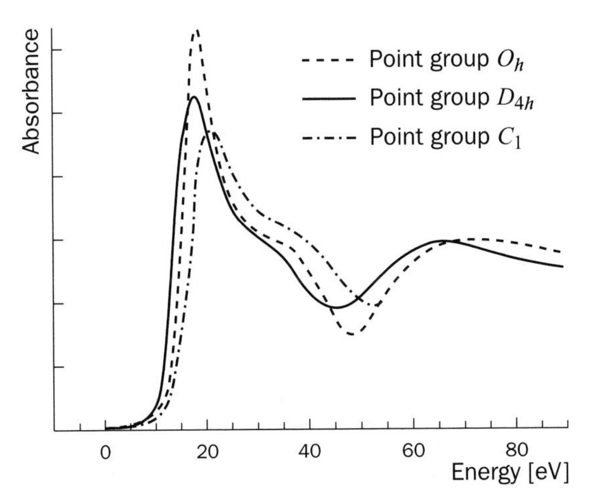

Figure 13 - Cross-sections at Cu *K* edge for various symmetries.

5. *SPIN TRANSITION IN* Fe^{II}*(O-PHENANTROLINE)*$_2$ *(NCS)*$_2$ *FOLLOWED AT THE IRON K EDGE*

Iron ion in Fe^{II}(o-phenantroline)$_2$ (NCS)$_2$ presents an abrupt and complete spin transition from $S = 2$ high-spin (HS) state to $S = 0$ low-spin (LS) state. The iron *K* edges are very dependent on the spin state as can be seen from the experimental spectra in figures 14 and 15. EXAFS analysis at the iron *K* edge has shown that during the spin conversion one observed also a large contraction of the coordination shell of iron ($\Delta R = 0.24$ Å). The question is to determine the origin of the modifications of the iron *K* edges during spin conversion. They may be due to the iron spin state or more simply to the bond distance contraction. To solve this puzzle, one can perform simulation at the iron *K* edge in the MST. The spin is not an ingredient in the building of the potential for most MST calculations. Then if MST calculations are able to reproduce the experimental modifications of *K* edges during spin conversion, that means that they are due only to bond distance contraction. This is what was done by Briois *et al.* [31].

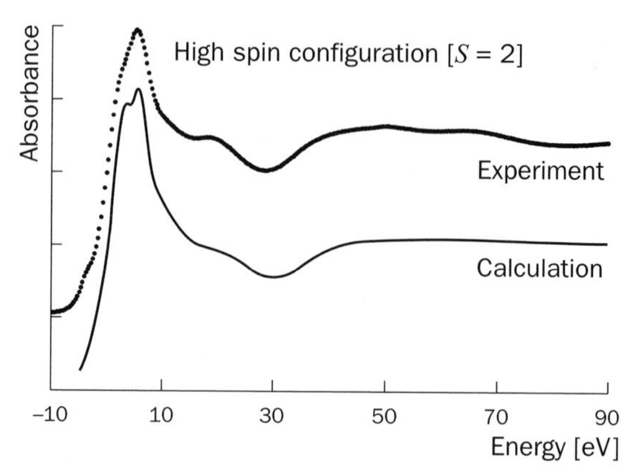

Figure 14 - Iron K edge in Fe^{II}(o-phenantroline)$_2$(NCS)$_2$ at $T = 300$ K. Iron spin $S = 2$.

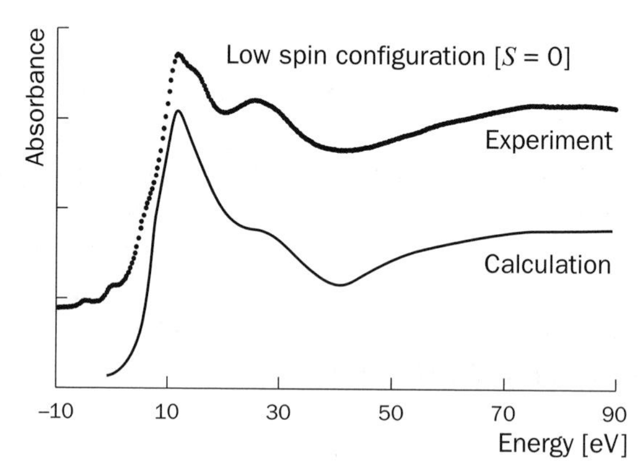

Figure 15 - Iron K edge in Fe^{II} (o-phenantroline)$_2$ (NCS)$_2$ at $T = 77$ K. Iron spin $S = 0$.

Both calculations for the HS and LS states nicely reproduce the experimental spectra. The cluster is made of all the atoms belonging to the molecule Fe^{II} (o-phenantroline)$_2$ (NCS)$_2$. The potential is built from the relaxed and screened prescriptions for the final state, with $X-\alpha$ exchange potential. The calculated spectra have been convoluted with a Lorentzian function, the width of which is equal to $\Gamma_h = 1.25$ eV. The agreement between experiments and calculations clearly shows that the observed modifications on the iron K edges are simply due to the crystallographic modification experienced by the molecule during the spin conversion. A parallel study at the iron $L_{\text{II-III}}$ edges developed in the framework of the Ligand Field Multiplet method showed that the modifications of the iron $L_{\text{II-III}}$ edges are, in this case, due to the spin conversion itself.

6. CONCLUSION

We have presented the Multiple Scattering Theory applied to the calculation of X-ray absorption spectra. We have specially outlined that in this technique the contributions from the potential and from the structure could be separated. The essence of the theory resides in solving equations of continuity that have a compact expression in the case of "muffin-tin" potentials. The construction of the potential has been developed with some emphasis concerning the exchange and correlation potential of common use for X-ray absorption calculations. The application of the Multiple Scattering Theory has been exemplified on a series of *K* edges for divalent vanadium, iron and copper in solution. The influence of various parameters has been illustrated: formulations of the final state, influence of the bond distance, resonance of the phase shifts and local symmetry around the absorbing atom. A more complicated case has also been illustrated where the influence of spin conversion and coordination sphere contraction at iron *K* edges could be separated through calculations in the MST framework.

The MST is still a very active field and many theoretical developments took place during the last seven years around Rino Natoli (X-ray Natural Circular Dichroism), John Rehr (the various FEFF codes), Dave Foulis (non "muffin-tin" code), Adriano Filipponi and Andrea di Cicco (coupling Molecular Dynamics and XANES calculations), Yves Joly (comparative approach of MST and Finite Differences Method), Christian Brouder (semi-relativistic approach) or Hubert Ebert (XMCD at *K* and $L_{\text{II-III}}$ edges).

APPENDIX - EXPRESSION FOR THE PROPAGATORS $J_{LL'}^{ij}$ AND $H_{LL'}^{ij}$

1. In the momentum representation the closure relation is

$$\int d^3p \,|\, p\rangle \frac{1}{2\pi^3} \langle p \,|=1 \qquad \{42\}$$

$$\langle p \,|\, q\rangle = (2\pi)^3 \delta(p-q) \qquad \{43\}$$

2. In the harmonic representation, it is

$$\sum_{kL} |\, kL\rangle \frac{2k^2}{\pi} \langle kL \,| = 1 \qquad \{44\}$$

with $\sum_{kL} = \sum_{L} \int dk$.

3. The translation operator P of vector $\boldsymbol{R}$ applied to the ket $|\boldsymbol{r}\rangle$ is defined by

$$P(\boldsymbol{R})|\boldsymbol{r}\rangle = |\boldsymbol{r}+\boldsymbol{R}\rangle \qquad \{45\}$$

$$\langle k|P(\boldsymbol{R})|\boldsymbol{r}\rangle = \langle k|\boldsymbol{R}+\boldsymbol{r}\rangle = \mathrm{e}^{-ik(\boldsymbol{R}+\boldsymbol{r})} = \mathrm{e}^{-i\boldsymbol{k}\cdot\boldsymbol{R}}\langle \boldsymbol{k}|\boldsymbol{r}\rangle$$

on noticing that the adjoint operator of $P(\boldsymbol{R})$ is $P(-\boldsymbol{R})$ one obtains that

$$P(\boldsymbol{R})|\boldsymbol{k}\rangle = \mathrm{e}^{-i\boldsymbol{k}\cdot\boldsymbol{R}}|\boldsymbol{k}\rangle \qquad \{46\}$$

4. Bauer's identity reads

$$\langle \boldsymbol{p}|\boldsymbol{r}\rangle = \mathrm{e}^{-i\boldsymbol{p}\cdot\boldsymbol{r}} = 4\pi\sum_L i^{-l} j_l(pr) Y_L(\hat{p}) Y_L(\hat{r}) \qquad \{47\}$$

By definition $\langle \boldsymbol{r}|kL\rangle = j_l(kr)Y_L(\hat{r})$, then $\langle \boldsymbol{p}|\boldsymbol{r}\rangle = 4\pi\sum_L i^{-l}\langle \boldsymbol{r}|pL\rangle Y_L(\hat{p})$. For real spherical harmonics, equation {47} is equivalent to

$$\langle \boldsymbol{p}|\boldsymbol{r}\rangle = 4\pi\sum_L i^{-l}\langle pL|\boldsymbol{r}\rangle Y_L(\hat{p}) \qquad \{48\}$$

$$\langle \boldsymbol{p}|\boldsymbol{r}\rangle = \frac{2}{\pi}\sum_{kL} k^2\langle \boldsymbol{p}|kL\rangle\langle kL|\hat{r}\rangle \qquad \{49\}$$

One integrates over $\boldsymbol{r}$ the right hand side of equation {49}

$$\begin{aligned}
\int d^3r\langle \boldsymbol{r}|kL\rangle\langle \boldsymbol{p}|\boldsymbol{r}\rangle &= \int d^3r\langle \boldsymbol{r}|kL\rangle 4\pi\sum_{L'} i^{-l'}\langle pL'|\boldsymbol{r}\rangle Y_{L'}(\hat{p}) \\
&= 4\pi\sum_{L'} i^{-l'}\langle pL'|kL\rangle Y_{L'}(\hat{p}) \\
&= 4\pi\sum_{L'} i^{-l'}\delta(p-k)\delta_{L,L'}\frac{\pi}{2k^2}Y_{L'}(\hat{p}) \\
&= \frac{2\pi^2}{k^2} i^{-l} Y_L(\hat{p})\delta(p-k)
\end{aligned}$$

and

$$\begin{aligned}
\int d^3r\langle \boldsymbol{r}|kL\rangle\langle \boldsymbol{p}|\boldsymbol{r}\rangle &= \int d^3r\langle \boldsymbol{r}|kL\rangle\frac{2}{\pi}\sum_{k'L'} k'^2\langle \boldsymbol{p}|k'L\rangle\langle k'L'|\boldsymbol{r}\rangle \\
&= \frac{2}{\pi}\sum_{k'L'} k'^2\langle \boldsymbol{p}|k'L'\rangle\langle k'L'|kL\rangle \\
&= \frac{2}{\pi}\sum_{k'L'} k'^2\langle \boldsymbol{p}|k'L'\rangle\frac{\pi}{2k^2}\delta(k-k')\delta_{L,L'} \\
&= \langle \boldsymbol{p}|kL\rangle
\end{aligned}$$

and then

$$\langle \boldsymbol{p}|kL\rangle = \frac{2\pi^2}{k^2} i^{-l} Y_L(\hat{p})\delta(p-k)$$

From the definition in equation {42}, we can write $P(\boldsymbol{R}).f(r) = f(r+\boldsymbol{R})$ in the harmonic representation

$$P(\boldsymbol{R})|kL\rangle = \frac{1}{8\pi^3}\int d^3p P(\boldsymbol{R})|\boldsymbol{p}\rangle\langle \boldsymbol{p}|kL\rangle$$

With the respective use of equations {46}, {44}, {50} and {47}, one obtains

$$
\begin{aligned}
P(\boldsymbol{R})|kL\rangle &= \frac{1}{8\pi^3}\int d^3p\,\mathrm{e}^{-i\boldsymbol{p}\cdot\boldsymbol{R}}\,|\boldsymbol{p}\rangle\langle\boldsymbol{p}|kL\rangle \\
&= \frac{1}{8\pi^3}\sum_{k'L'}\int d^3p\,\mathrm{e}^{-i\boldsymbol{p}\cdot\boldsymbol{R}}\langle k'L'|\boldsymbol{p}\rangle\langle\boldsymbol{p}|kL\rangle|k'L'\rangle\frac{2k'^2}{\pi}
\end{aligned}
$$

$$
=\frac{1}{8\pi^3}\frac{2}{\pi}\sum_{k'L'}\int d^3p\,\mathrm{e}^{-i\boldsymbol{p}\cdot\boldsymbol{R}}k'^2\frac{2\pi^2}{k'^2}i^{l'}Y_{L'}(\hat{p})\delta(p-k')\frac{2\pi^2}{k^2}i^{-l}Y_L(\hat{p})\delta(p-k)|k'L'\rangle
$$

$$
\begin{aligned}
P(\boldsymbol{R})|kL\rangle &= \sum_{L'}\int d^3p\,\mathrm{e}^{-i\boldsymbol{p}\cdot\boldsymbol{R}}i^{l'}Y_{L'}(\hat{p})\frac{1}{k^2}i^{-l}Y_L(\hat{p})\delta(p-k)|pL'\rangle \\
&= \sum_{L'}\int d^3p\,4\pi\sum_{L''}i^{-l''}j_{l''}(pR)Y_{L''}(\hat{p})Y_L(\hat{R})i^{l'-l}Y_{L'}(\hat{p})\frac{1}{k^2}Y_L(\hat{p})\delta(p-k)|pL'\rangle
\end{aligned}
$$

If one remembers that $\frac{1}{k^2}d^3p=\frac{1}{k^2}dp\,dS=\frac{p^2}{k^2}dp\,d\Omega$ then

$$
\begin{aligned}
P(\boldsymbol{R})|kL\rangle &= 4\pi\sum_{L'L''}i^{l'-l-l''}j_{l''}(kR)Y_{L''}(\hat{R})\int d\Omega\,Y_{L''}(\hat{p})Y_{L'}(\hat{p})Y_L(\hat{p})|kL'\rangle \\
&= \sum_{L'}P_{LL'}(k\boldsymbol{R})|kL'\rangle
\end{aligned}
$$

from which

$$
P_{LL'}(k\boldsymbol{R}) = 4\pi\sum_{L''}i^{l'-l-l''}j_{l''}(kR)Y_{L''}(\hat{R})\int d\Omega\,Y_{L''}(\hat{p})Y_{L'}(\hat{p})Y_L(\hat{p})
$$

by definition and then

$$
J_{LL'}^{ij} \equiv P_{L'L}(k\boldsymbol{R}_{ji})
$$

with $\boldsymbol{R}_{ji}=\boldsymbol{R}_j-\boldsymbol{R}_i$ that is

$$
J_{LL'}^{ij} = 4\pi\sum_{L''}i^{l+l''-l'}j_{l''}(kR_{ij})Y_{L''}(\hat{R}_{ij})\int d\Omega\,Y_{L''}(\hat{p})Y_{L'}(\hat{p})Y_L(\hat{p})
$$

a similar expression can be obtained for $H_{LL'}^{ij}$

$$
H_{LL'}^{ij} = -4\pi i\sum_{L''}i^{l+l''-l'}h_{l''}^{+}(kR_{ij})Y_{L''}(\hat{R}_{ij})\int d\Omega\,Y_{L''}(\hat{p})Y_{L'}(\hat{p})Y_L(\hat{p})
$$

ACKNOWLEDGMENTS

We are glad to acknowledge the help of Christian Brouder during the redaction of this chapter.

REFERENCES

[1] J. KORRINGA - *Physica* **13**, 392 (1947)

[2] K.H. JONHSON - *Advances in Quantum Chemistry* **7**, 143 (1973)

[3] D. DILL & J.L DEHMER - *J. Chem. Phys.* **61**, 692 (1974);
J.L DEHMER & D. DILL - *Phys. Rev. Lett.* **35**, 213 (1975)

[4] J.E. MULLER & J.W. WILKINS - *Phys. Rev. B* **29**, 4331 (1984)

[5] G. VAN DER LAAN & B.T. THOLE - *J. Phys.: Condens. Matter* **4**, 4181 (1992)

[6] H. EBERT - *In "Spin-Orbit-Influenced spectroscopies of magnetic solids", Lecture notes in physics* **466**, 159, Springer Verlag (1995)

[7] C.R. NATOLI, D.K. MISEMER, S. DONIACH & F.W. KUTZLER -
Phys. Rev. A **22**, 1104 (1980)

[8] C. BROUDER - *J. Phys.: Condens. Matter* **2**, 701 (1990)

[9] C. LEVELUT - *Organisation structurale locale dans les composés CdS_xSe_{1-x} massifs ou inclus dans une matrice vitreuse*, Ph.D. Thesis, U. Paris 6 (1991);
C. LEVELUT, P. SAINCTAVIT, A. RAMOS & J. PETIAU -
J. Phys.: Condens. Matter **7**, 2353 (1995)

[10] C. BROUDER - *In "Magnetism and Synchrotron Radiation"*, E. BEAUREPAIRE, B. CARRIÈRE & J.P. KAPPLER eds, Les Editions de physique, Les Ulis (1997)

[11] D. CABARET - *Théorie de la diffusion multiple comme modèle de l'absorption des rayons X : application aux seuils K de l'aluminium et du magnésium dans les géomatériaux*, Ph.D. Thesis, U. Paris 6 (1997);
D. CABARET, C. BROUDER & P. SAINCTAVIT - *In "Proceedings of the first International Alloy conference (Athens, 1996)"*,
A. GONIS, A. MEIKE & P. TURCHI eds, Plenum Press, New York (1997)

[12] C. COHEN, B. DIU & F. LALOË - *Mécanique quantique*, Hermann, Paris (1973);
C. COHEN-TANNOUDJI, B. DIU & F. LALOË - Quantum *Mechanics*, Wiley - Interscience, New York (1996)

[13] R. LOUDON - *The Quantum Theory of Light*, Clarendon Press, Oxford (1973)

[14] C. COHEN, J. DUPOND-ROC & G. GRYNBERG - *Processus d'interaction entre photons et atomes*, Editions du CNRS, Paris (1988);
C. COHEN-TANNOUDJI, J. DUPOND-ROC & G. GRYNBERG - *Atom-photon interactions: Basic processes and applications*, Wiley - Interscience, New York (1998)

[15] A. MESSIAH - *Quantum Mechanics*, vol. 1 & 2, North-Holland, Amsterdam, (1961)

[16] F. SCHEID - *Theory and problems of numerical analysis*,
Schaum's Outline Series, Mc Graw-Hill Book Company, New York, 224 (1968)

[17] G.W. PRATT - *Phys. Rev.* **88**, 1217 (1952)

[18] C. BROUDER - *Proceedings of the MRS conference*, Boston (1991)

[19] W. KOHN & N. ROSTOKER - *Phys. Rev. B* **94**, 1111 (1954)

[20] K.H. JOHNSON - *J. Chem. Phys.* **45**, 3085 (1966)

[21] K.H. JOHNSON & F.C. SMITH JR - *Phys. Rev. B* **5**, 831 (1972)

[22] F.W. KUTZLER, C.R. NATOLI, D.K. MISEMER, S. DONIACH & K.O. HODGSON - *J. Chem. Phys.* **73**, 3274 (1980)

[23] J.J. REHR, R.C. ALBERS & S.I. ZABINSKY - *Phys. Rev. Lett.* **69**, 3397 (1992)

[24] P. LLOYD & P.V. SMITH - *Adv. Phys.* **21**, 69 (1972)

[25] J.C. SLATER (1972) - *Advances in Quantum Chemistry* **6**, 1 (1972)

[26] K. SCHWARTZ (1972) - *Phys. Rev. B* **5**, 2466 (1972)

[27] S. HARA (1967) - *J. Phys. Soc. Japan* **22**, 710 (1967)

[28] L. HEDIN & S. LUNDQUIST - *In "Solid State Physics"* **23**, 1, F. SEITZ, D. TURNBULL & H. EHRENREICH eds, Academic Press, New York (1969)

[29] V. BRIOIS, P. LAGARDE, C. BROUDER, P. SAINCTAVIT & M. VERDAGUER - *Physica B* **208-209**, 51 (1995)

[30] D. CABARET, P. SAINCTAVIT, P. ILDEFONSE, G. CALAS & A.M. FLANK - *Phys. Chem. Min.* **23**, 226 (1996)

[31] V. BRIOIS, C. CARTIER DIT MOULIN, P. SAINCTAVIT, C. BROUDER & A.M. FLANK - *J. Am. Chem. Soc.* **117**, 1019 (1995)

3

X-RAY MAGNETIC CIRCULAR DICHROISM

F. BAUDELET
Synchrotron SOLEIL, Gif-sur-Yvette, France

X-ray Magnetic Circular Dichroism (XMCD) is an element-specific as well as a symmetry selective probe of microscopic magnetic properties. Using different X-ray energy domains, it is possible to extract information on the spin and orbital polarization of the conduction band or localized orbitals, which are responsible for itinerant or localized magnetism respectively. In the soft X-ray range, atomic calculations reproduce quantitatively the XMCD signal, providing a powerful element-selective magnetometry. In the hard X-ray domain, the theoretical description based on band calculations is more difficult, but magnetic information is currently extracted from XMCD signals.

1. INTRODUCTION

X-ray Magnetic Circular Dichroism is a technique that uses the polarization properties of X-rays to probe microscopic magnetism. Intense linearly or circularly polarized X-rays produced in synchrotron sources have enabled new probes of the magnetic structure of materials to be developed. Magnetic resonant scattering and magnetic X-ray dichroism use linearly or circularly polarized photons [1] with energy near an absorption edge. X-ray Magnetic Circular Dichroism is the difference between left and right circularly polarized X-ray absorption cross-section of a ferromagnetic or ferrimagnetic material.

The interest of using circular polarized light comes from the fact that the magnetic absorption cross-section is directly proportional to $\langle \boldsymbol{M} \rangle$, the mean value of the macroscopic magnetic moment. Two models generally describe magnetism in metals and alloys:

F. Hippert et al. (eds.), Neutron and X-ray Spectroscopy, 103–130.

- The localized model where the electrons responsible for the magnetic moment are quasi-atomic. This model will describe the magnetism of the 4f orbitals in rare earth compounds and d orbitals in some transition metal compounds. These orbitals are probed by "soft X-ray" absorption (< 3 keV).
- The itinerant model where the spin-polarized electrons are considered to be delocalized. In this case the exchange and correlation interactions yield magnetism because of the different filling of the spin-up and spin-down conduction bands. These conduction bands are probed by absorption of "hard X-rays". In that range (2 $\longrightarrow$ 20 keV) the final states (or the projection of the final states) of X-ray absorption spectroscopy are of p symmetry for $3d$ transition metals and of $5d$ symmetry for $4f$ rare earths.

$3d$ bands of $3d$ transition metal are an intermediate case. They are usually described by rigid band models with two sub-bands of opposite spin direction, split by exchange energy. $3d$ states are of course delocalized states. Nevertheless the $L_{\text{II-III}}$ absorption edges of $3d$ transition metals ($2p \longrightarrow 3d$ transition) are usually described in the literature with good accuracy by atomic models (see section 5).

In the hard and soft energy ranges circular magnetic X-ray dichroism is a selective probe of the element for both the itinerant and localized models of magnetism. The selectivity concerns both the nature of the probed element and the symmetry of the probed band.

In this lecture we will focus on the origin of the XMCD effect in core-hole spectroscopy. Readers not familiar with core-hole spectroscopy can, for example, read the related chapter in reference **[2]**.

2. *ORIGINS OF MAGNETIC CIRCULAR DICHROISM*

2.1. *THEORETICAL ASPECTS*
ROLE OF THE SPIN-ORBIT COUPLING

The origin of magnetic dichroism can be found in the photon-matter interaction process. In X-ray absorption it is not directly due to the interaction between the transverse field $\boldsymbol{B}$ of the photon and the spin of the electrons but rather to the interaction between the spin and the orbit, which couples the spin and real space.

Quantum field theory is not mandatory to describe correctly the X-ray absorption process. A semi-classical model may be used, where the atom is quantified and the electromagnetic field of the photons is described by Maxwell's equations.

The interaction Hamiltonian is written as

$$H_{\text{int}} = -\left[\sum_i \frac{q}{m} p_i \cdot A(r_i) - \sum_i \frac{q^2}{2m} A(r_i)^2\right] - \left[\sum_i g_i \frac{q}{m} S \cdot B(r_i)\right] \qquad \{1\}$$

where the electrons are characterized by their mass m_e, charge q, momentum $\boldsymbol{p}$ and spin $\boldsymbol{S}$. The photons are described by an electromagnetic field ($\boldsymbol{\varepsilon}$ and $\boldsymbol{B}$), where $\boldsymbol{\varepsilon}$ is derived from the vector potential $\boldsymbol{A}$.

This interaction Hamiltonian consists of three main terms, the third of which contains the spin dependence. Nevertheless this term does not act on the spatial part of the wave function, so in a first approximation it cannot couple the initial and the final state of absorption in the X-ray range. This spin part of the dipolar magnetic contribution to XMCD is negligible because the principal quantum number of the initial state is different from that of the final state. Thus, in a first order approximation the wave functions of the initial and final states remain perpendicular during the absorption process ("sudden" approximation) and the $\boldsymbol{S} \cdot \boldsymbol{B}$ matrix element is zero. Of course this term will be preponderant for intra-band transitions induced by visible light absorption.

Only the first term is relevant to X-ray absorption and its "spin sensitivity", which must be found within the canonical dipolar approximation.

The probability transition from an initial state $|i\rangle$ to a final state $|f\rangle$ is given by the Fermi Golden Rule, which comes from a time-dependent perturbation calculation for harmonic potential

$$\Gamma_{i,f} = \frac{2\pi}{\hbar} \sum_{i,f} \left|\langle f | \boldsymbol{r} \cdot \boldsymbol{\varepsilon} | i \rangle\right|^2 \delta(E_f - E_i - \hbar\omega) \qquad \{2\}$$

where $\boldsymbol{\varepsilon}$ is the X-ray electric field.

This expression, which involves all the initial and final states of the absorption process, may be simplified by using the one-electron approximation. In this approximation only one electron changes its quantum number during the absorption process, the other electrons remaining as spectators.

The above term does not directly depend on the spin, which means that spin is conserved in the absorption process: $\Delta m_s = 0$. Spin-dependent absorption will be induced in our case by the spin-orbit coupling either in the final state of the excited photoelectron or in the initial state of the core level. The spin-orbit coupling of the core level is larger than in the valence state because the amplitude of the spin-orbit coupling is proportional to the potential experienced by the electron (or by the hole).

The challenge in the theoretical development is to describe the final state $|f\rangle$ correctly. This is quite easy for the case of transitions to a localized state within the multiplet framework, and represents the most important application of XMCD. More challenging are transitions toward delocalized states (itinerant magnetism). The former is extensively described in the chapter by J. Vogel and F. de Groot in this volume. The greatest impact of XMCD in the field of thin film and interfacial magnetism is given special attention in the chapter by C.M. Schneider. Here, we limit ourselves to the underlying basis of the XMCD effect and comments on the theoretical approach of XMCD for the itinerant magnetism case.

2.2. LIGHT POLARIZATION AND POLARIZATION-DEPENDENT SELECTION RULES

The interaction Hamiltonian depends on the polarization $\boldsymbol{\varepsilon}$ of the incident electric field.

- **Linear polarization:** The electromagnetic field vector has a constant direction in a plane perpendicular to the propagation vector.
- **Circular polarization:** The electromagnetic field vector rotates around the propagation vector direction (fig. 1).

The selection rules are polarization-dependent, i.e., the difference between the transition probability for left and right circularly polarized light gives circular magnetic dichroism:

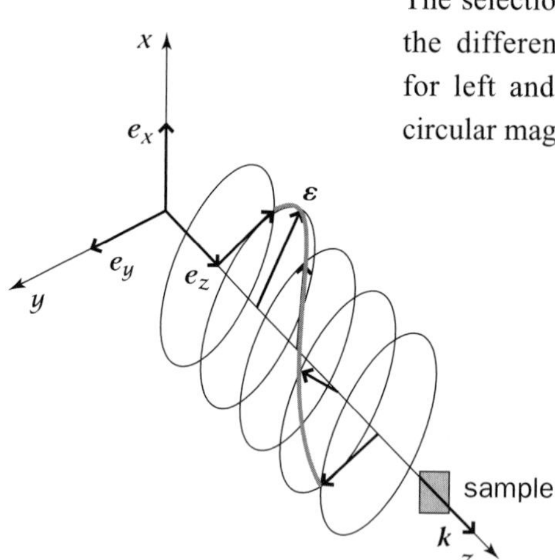

Figure 1 - Circular polarization: the electromagnetic field vector turns around the direction of the propagation vector. In the figure the X-rays are left circularly polarized, the electromagnetic field turns to the left for an observer placed at the sample position.

For a left circularly polarized beam propagating along Oz, the expression for the electric field is

$$\boldsymbol{\varepsilon}^+ = \frac{e_x + ie_y}{\sqrt{2}}$$

and for a right circularly polarized beam

$$\boldsymbol{\varepsilon}^- = \frac{\boldsymbol{e}_x - i\boldsymbol{e}_y}{\sqrt{2}}$$

The position vector is expressed as

$$\boldsymbol{r} = x\boldsymbol{e}_x + y\boldsymbol{e}_y + z\boldsymbol{e}_z$$

The expressions for the polarization-dependent dipolar operators are

$$\boldsymbol{r}\cdot\boldsymbol{\varepsilon}^+ = (x+iy)$$
$$\boldsymbol{r}\cdot\boldsymbol{\varepsilon}^- = (x-iy)$$

Using spherical coordinates

$$x = r\sin\theta\cos\varphi = -\sqrt{\frac{2\pi}{3}}r(Y_1^1 - Y_1^{-1})$$

$$y = r\sin\theta\sin\varphi = i\sqrt{\frac{2\pi}{3}}r(Y_1^1 + Y_1^{-1})$$

$$z = r\cos\theta = \sqrt{\frac{4\pi}{3}}rY_1^0$$

The expression for the polarization-dependent dipolar operators can be written as a combination of spherical harmonics

$$\boldsymbol{r}\cdot\boldsymbol{\varepsilon}^+ = (x+iy) = -r\sqrt{\frac{4\pi}{3}}Y_1^1$$

$$\boldsymbol{r}\cdot\boldsymbol{\varepsilon}^- = (x-iy) = r\sqrt{\frac{4\pi}{3}}Y_1^{-1}$$

The transition matrix element, which is then a combination of spherical harmonics, is non-zero only if

$$\Delta l = l_f - l_i = \pm 1$$

and

$$\Delta m_l = m_{lf} - m_{li} = +1 \qquad \text{for left circular polarization}$$

$$\Delta m_l = m_{lf} - m_{li} = -1 \qquad \text{for right circular polarization}$$

where l is the orbital angular momentum and m_l its projection along the z direction.

We have always $\Delta m_s = 0$.

Finally the two absorption cross-sections for left (σ_+) and right (σ_-) circular polarization are

$$\sigma_\pm = 4\pi^2\,\hbar\omega\,\alpha\sum_{fi}\left|\langle i\,|\mp\sqrt{\frac{4\pi}{3}}rY_1^{-1}\,|\,f\rangle\right|^2\delta(E_f - E_i - \hbar\omega) \qquad \{3\}$$

which give the asymmetry R measured by the XMCD

$$R = \frac{\sigma_+ - \sigma_-}{\sigma_+ + \sigma_-}$$

The conditions necessary for a non-zero value for R are:

- spin-orbit coupling in the final state of the excitation, either of the core-hole or of the photo-electron,
- macroscopic magnetization (ferro- or ferrimagnetic compounds) with the magnetization direction non perpendicular to the propagation direction of the incident photons,
- a circularly polarized light beam.

These three conditions will be illustrated in section 2.3.

2.3. *ORIGIN OF THE XMCD SIGNAL*

2.3.1. *HEURISTIC MODEL: L_{II} AND L_{III} EDGES OF TRANSITION METALS AND RARE EARTH ELEMENTS*

The necessary conditions to obtain an XMCD signal are presented here in a very simple model for the case of L_{II} and L_{III} edges of transition metals and rare earth elements. In an L edge the transition of a photoelectron is from a p level ($l = 1$) to a d state ($l = 2$) where we neglect the spin-orbit coupling (fig. 2).

For an L_{II} edge, left circularly polarized light gives a larger probability of creating photoelectrons of majority spin ($\alpha^+{}_\downarrow = 75\%$), i.e., of creating photoelectrons with spin (↓). The probability of creating a photoelectron of minority spin (↑) is then $\alpha^+{}_\uparrow = 25\%$ (see section 2.3.2). For right circular polarized light the larger probability is for photo-electrons with spin ↑ (75%). $\alpha^+{}_\downarrow$ and $\alpha^+{}_\uparrow$ are the photo-electron excitation rate with majority and minority spin respectively at an L_{II} edge with left circularly polarized light.

The XMCD signal is the difference between two cross-sections with right and left circular polarized light, respectively

$$\sigma_- = K(\alpha^-{}_\uparrow \rho\uparrow + \alpha^-{}_\downarrow \rho\downarrow) \quad \text{and} \quad \sigma_+ = K(\alpha^+{}_\uparrow \rho\uparrow + \alpha^+{}_\downarrow \rho\downarrow)$$

$\rho\uparrow$ and $\rho\downarrow$ are respectively the empty state densities of spin-up and spin-down and are different only for compounds that exhibit a magnetic moment. The XMCD signal is zero for non-magnetic or antiferromagnetic compounds.

Finally the experimental signal is proportional to

$$XMCD = \frac{\sigma_+ - \sigma_-}{\sigma_+ + \sigma_-} = \frac{\alpha' - \alpha}{\alpha' + \alpha}\frac{\rho\uparrow - \rho\downarrow}{\rho\uparrow + \rho\downarrow} = \frac{1}{2}P_e \cdot P_c \cdot \frac{\Delta\rho}{\rho}$$

with $\alpha^+{}_\uparrow = \alpha^-{}_\downarrow = \alpha'$ and $\alpha^+{}_\downarrow = \alpha^-{}_\uparrow = \alpha$ (see remarks, section 2.3.2)

$$P_e = \frac{\alpha' - \alpha}{\alpha' + \alpha} \quad \text{and} \quad \frac{\Delta\rho}{\rho} = \frac{\rho\uparrow - \rho\downarrow}{\rho\uparrow + \rho\downarrow}$$

where the degree of circular polarization of the X-ray photons is accounted for by the factor P_c.

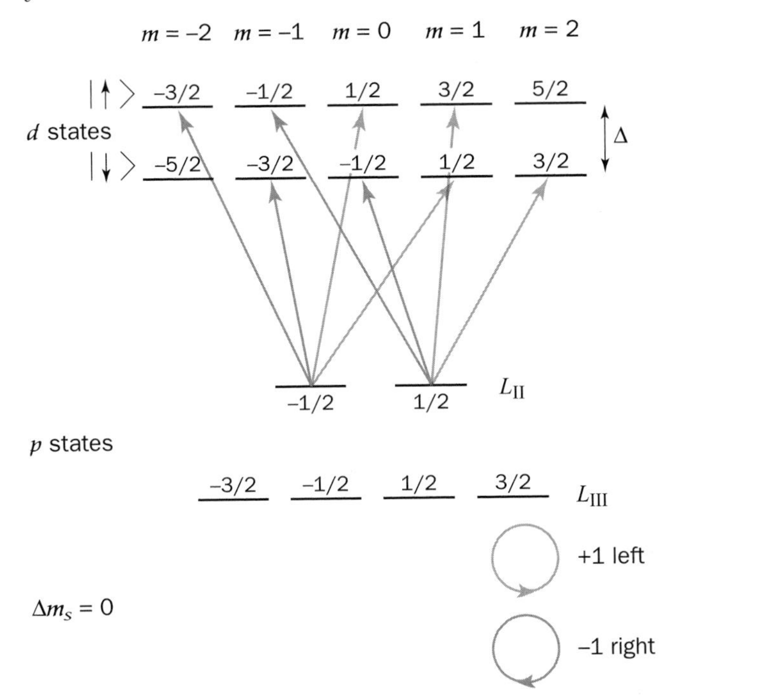

Figure 2 - For an L_{II} edge, Dipole transitions from a $J = 1/2$ level for photons of helicity +1 (left polarization, $\Delta m_l = +1$), for linearly polarized photons ($\Delta m_l = 0$) and for photons of helicity −1 (right polarization, $\Delta m_l = -1$). The p level is split by spin-orbit coupling interaction. The d level is split by exchange interaction Δ.

2.3.2. CALCULATION OF P_E FOR THE L_{II} AND L_{III} EDGES: $P_E(L_{II})$ AND $P_E(L_{III})$

Due to spin-orbit coupling the $2p$ band is split into two sub-bands of kinetic moment $j = 1/2$ and $j = 3/2$. This is a final state effect. The two sublevels can be separated by energy as large as 1000 eV in the case of 5d transition metals or in rare earths.

During the transition process, the overlap between the radial wave function of the core $2p$ state and the excited d state is assumed to be constant and is written

$$\mathrm{R}(j) = \int R^*(l=2;j)R(l=1)r^3\,\mathrm{d}r$$

We denote $\{|l, s; m_l, m_s\rangle\}$ the eigenstate basis of l^2, l_z, S^2, S_z with $m_s = +1/2$ (respectively $-1/2$) represented by $\uparrow$ (respectively $\downarrow$). The decomposition of the eigenstates of J^2 and J_z over this basis is

For $j = 3/2$ (L_{III} edge)

$$\left|\frac{3}{2};\frac{3}{2}\right\rangle = \left|1,\frac{1}{2};1,\uparrow\right\rangle$$

$$\left|\frac{3}{2};\frac{1}{2}\right\rangle = \sqrt{\frac{1}{3}}\left|1,\frac{1}{2};1,\downarrow\right\rangle + \sqrt{\frac{2}{3}}\left|1,\frac{1}{2};0,\uparrow\right\rangle$$

$$\left|\frac{3}{2};-\frac{1}{2}\right\rangle = \sqrt{\frac{2}{3}}\left|1,\frac{1}{2};0,\downarrow\right\rangle + \sqrt{\frac{1}{3}}\left|1,\frac{1}{2};-1,\uparrow\right\rangle$$

$$\left|\frac{3}{2};-\frac{3}{2}\right\rangle = \left|1,\frac{1}{2};-1,\downarrow\right\rangle$$

and for $j = 1/2$ (L_{II} edge)

$$\left|\frac{1}{2};\frac{1}{2}\right\rangle = \sqrt{\frac{2}{3}}\left|1,\frac{1}{2};1,\downarrow\right\rangle - \sqrt{\frac{1}{3}}\left|1,\frac{1}{2};0,\uparrow\right\rangle$$

$$\left|\frac{1}{2};-\frac{1}{2}\right\rangle = \sqrt{\frac{1}{3}}\left|1,\frac{1}{2};0,\downarrow\right\rangle - \sqrt{\frac{2}{3}}\left|1,\frac{1}{2};-1,\uparrow\right\rangle$$

The dipolar electric operator has the spherical harmonic expression $rY_l^{m_l}$, see section 2.2.

We can write the basis function $|l, s; m_l, m_s\rangle = |s; m_s\rangle|l; m_l\rangle$ because the spherical harmonics act only on the angular part of the wave function.

We derived the expression of the probability to excite a spin-up $\uparrow$ (resp. spin-down $\downarrow$) photo-electron with a left circular polarization for the L_{II} and L_{III} edges

$$\alpha^{+}{\uparrow}(L_{\mathrm{III}}) = \frac{4\pi}{3}\left[\left|\left\langle 2;2\middle|Y_1^1\middle|1;1\right\rangle\right|^2 + \frac{2}{3}\left|\left\langle 2;1\middle|Y_1^1\middle|1;0\right\rangle\right|^2 + \frac{1}{3}\left|\left\langle 2;0\middle|Y_1^1\middle|1;-1\right\rangle\right|^2\right]\mathrm{R}\left(\frac{3}{2};\uparrow\right)^2$$

$$\alpha^{+}{\downarrow}(L_{\mathrm{III}}) = \frac{4\pi}{3}\left[\frac{1}{3}\left|\left\langle 2;2\middle|Y_1^1\middle|1;1\right\rangle\right|^2 + \frac{2}{3}\left|\left\langle 2;1\middle|Y_1^1\middle|1;0\right\rangle\right|^2 + \left|\left\langle 2;0\middle|Y_1^1\middle|1;-1\right\rangle\right|^2\right]\mathrm{R}\left(\frac{3}{2};\downarrow\right)^2$$

$$\alpha^{+}{\uparrow}(L_{\mathrm{II}}) = \frac{4\pi}{3}\left[\frac{1}{3}\left|\left\langle 2;1\middle|Y_1^1\middle|1;0\right\rangle\right|^2 + \frac{2}{3}\left|\left\langle 2;0\middle|Y_1^1\middle|1;-1\right\rangle\right|^2\right]\mathrm{R}\left(\frac{1}{2};\uparrow\right)^2$$

$$\alpha^{+}{\downarrow}(L_{\mathrm{II}}) = \frac{4\pi}{3}\left[\frac{2}{3}\left|\left\langle 2;2\middle|Y_1^1\middle|1;1\right\rangle\right|^2 + \frac{1}{3}\left|\left\langle 2;1\middle|Y_1^1\middle|1;0\right\rangle\right|^2\right]\mathrm{R}\left(\frac{1}{2};\downarrow\right)^2$$

Using the matrix element values

$$\sqrt{\frac{4\pi}{3}}\left\langle 2;2\middle|Y_1^1\middle|1;1\right\rangle=\sqrt{\frac{2}{5}}\qquad \sqrt{\frac{4\pi}{3}}\left\langle 2;1\middle|Y_1^1\middle|1;0\right\rangle=\sqrt{\frac{1}{5}}\qquad \sqrt{\frac{4\pi}{3}}\left\langle 2;0\middle|Y_1^1\middle|1;-1\right\rangle=\sqrt{\frac{1}{15}}$$

we obtain

$$\alpha^{+}{\uparrow}(L_{\mathrm{III}})=\frac{5}{9}\mathrm{R}_{\mathrm{III}}\qquad \alpha^{+}{\uparrow}(L_{\mathrm{II}})=\frac{1}{9}\mathrm{R}_{\mathrm{II}}$$
$$\alpha^{+}{\downarrow}(L_{\mathrm{III}})=\frac{3}{9}\mathrm{R}_{\mathrm{III}}\qquad \alpha^{+}{\downarrow}(L_{\mathrm{II}})=\frac{3}{9}\mathrm{R}_{\mathrm{II}}$$

assuming $\mathrm{R}\left(\frac{1}{2};\uparrow\right)^2=\mathrm{R}\left(\frac{1}{2};\downarrow\right)^2=\mathrm{R}_{\mathrm{II}}$ and $\mathrm{R}\left(\frac{3}{2};\uparrow\right)^2=\mathrm{R}\left(\frac{3}{2};\downarrow\right)^2=\mathrm{R}_{\mathrm{III}}$.

The probabilities to excite a spin-up ↑ (resp. spin-down ↓) photo-electron with left circular polarization for the L_{II} and L_{III} edges are

$$\frac{\alpha^{+}{\uparrow}}{\alpha^{+}{\uparrow}+\alpha^{+}{\downarrow}}(L_{\mathrm{III}})=62.5\%\qquad \frac{\alpha^{+}{\uparrow}}{\alpha^{+}{\uparrow}+\alpha^{+}{\downarrow}}(L_{\mathrm{II}})=25\%$$
$$\frac{\alpha^{+}{\downarrow}}{\alpha^{+}{\uparrow}+\alpha^{+}{\downarrow}}(L_{\mathrm{III}})=37.5\%\qquad \frac{\alpha^{+}{\downarrow}}{\alpha^{+}{\uparrow}+\alpha^{+}{\downarrow}}(L_{\mathrm{II}})=75\%$$

and the expressions for P_e ($L_{\text{II-III}}$) are

$$P_e(L_{\mathrm{III}})=\frac{\alpha^{+}{\uparrow}-\alpha^{+}{\downarrow}}{\alpha^{+}{\uparrow}+\alpha^{+}{\downarrow}}=0.25$$
$$P_e(L_{\mathrm{II}})=\frac{\alpha^{+}{\uparrow}-\alpha^{+}{\downarrow}}{\alpha^{+}{\uparrow}+\alpha^{+}{\downarrow}}=-0.5$$

Remarks

- It is left to the reader to prove the following equalities

$$\alpha^{+}{\uparrow}(L_{\mathrm{III}})=\frac{5}{9}\mathrm{R}_{\mathrm{III}}=\alpha^{-}{\downarrow}(L_{\mathrm{III}})\qquad \alpha^{+}{\uparrow}(L_{\mathrm{II}})=\frac{1}{9}\mathrm{R}_{\mathrm{II}}=\alpha^{-}{\downarrow}(L_{\mathrm{II}})$$
$$\alpha^{+}{\downarrow}(L_{\mathrm{III}})=\frac{3}{9}\mathrm{R}_{\mathrm{III}}=\alpha^{-}{\uparrow}(L_{\mathrm{III}})\qquad \alpha^{+}{\downarrow}(L_{\mathrm{II}})=\frac{3}{9}\mathrm{R}_{\mathrm{II}}=\alpha^{-}{\uparrow}(L_{\mathrm{II}})$$

  This shows that it is equivalent to reverse the direction of the circular polarization of the light or to reverse the direction of the external applied magnetic field (i.e., the spin-up and spin-down population). Changing the direction of the external magnetic field (which is applied to align the magnetic domains along the z axis) is equivalent to exchanging the majority and minority spin populations. Experimentally therefore, we can measure the XMCD signal by using only photons of left (or right) helicity and by reversing the direction of the applied magnetic field.

- We immediately verify the equality

$$\alpha^{+}{\uparrow}(L_{\mathrm{III}})+\alpha^{+}{\uparrow}(L_{\mathrm{II}})\;=\;\alpha^{+}{\downarrow}(L_{\mathrm{III}})+\alpha^{+}{\downarrow}(L_{\mathrm{II}})$$

  There is no XMCD effect in the absence of spin-orbit coupling, i.e., if there is no energy separation between the $2p_{1/2}$ and $2p_{3/2}$ levels.

2.3.3. CALCULATION OF P_E FOR THE K AND L_I EDGES: P_E (K) AND P_E (L_I)

At the K and L_I edges, there is no spin-orbit coupling in the initial level (s) of the transition. The XMCD effect comes from the spin-orbit coupling in the final (p) level of the transition.

The overlap between the radial wave function of the core s state and the excited p state is

$$\mathrm{R}(j) = \int R^*(l=1;j)R(l=0)r^3\,\mathrm{d}r$$

We denote by $\mathrm{R}(3/2)$ and $\mathrm{R}(1/2)$ the matrix elements corresponding to transition towards $j = 3/2$ and $j = 1/2$ respectively.

Spin-orbit coupling is absent in the initial s state. In this case the angular parts of the dipole matrix element are the same for both circular polarizations. Spin polarization results from the difference between $\mathrm{R}(3/2)$ and $\mathrm{R}(1/2)$ [3].

For left circular polarization the matrix element between the two spin states m_s and m_s' is

$$\sqrt{\frac{4\pi}{3}}\left\langle l',\frac{1}{2};m'_l,m'_s\middle|rY_1^1\middle|0,\frac{1}{2};0,m_s\right\rangle =$$

$$\frac{1}{\sqrt{3}}\sum_{j'}\left\langle l;\frac{1}{2};m'_l;m'_s\middle|l;\frac{1}{2};j';m'\right\rangle \mathrm{R}(j')\left\langle 1;\frac{1}{2};j';m'\middle|l;\frac{1}{2};l;m_s\right\rangle$$

The probability of exciting a spin-up ↑ (resp. spin-down ↓) photo-electron with left circular polarization is

$$\alpha^+{}_\uparrow = \sum_{m_s}\left|\left\langle m'_l,\uparrow\middle|Y_1^1\middle|0,m_s\right\rangle\right|^2 \qquad \alpha^+{}_\downarrow = \sum_{m_s}\left|\left\langle m'_l,\downarrow\middle|Y_1^1\middle|0,m_s\right\rangle\right|^2$$

Replacing the Clebsch-Gordan coefficients by their value

$$\alpha^+{}_\uparrow = \frac{1}{3}\left\{\mathrm{R}^2\left(\frac{3}{2}\right)+\frac{2}{9}\left[\mathrm{R}\left(\frac{3}{2}\right)-\mathrm{R}\left(\frac{1}{2}\right)\right]^2\right\}$$

$$\alpha^+{}_\downarrow = \frac{1}{3}*\frac{1}{9}\left[\mathrm{R}\left(\frac{3}{2}\right)+2*\mathrm{R}\left(\frac{1}{2}\right)\right]^2$$

$$P_e = \frac{\alpha^+{}_\uparrow-\alpha^+{}_\downarrow}{\alpha^+{}_\uparrow+\alpha^+{}_\downarrow} = \frac{2\left[\mathrm{R}\left(\frac{3}{2}\right)-\mathrm{R}\left(\frac{1}{2}\right)\right]*\left[2*\mathrm{R}\left(\frac{3}{2}\right)+\mathrm{R}\left(\frac{1}{2}\right)\right]+\left[\mathrm{R}\left(\frac{3}{2}\right)-\mathrm{R}\left(\frac{1}{2}\right)\right]^2}{2\left[\mathrm{R}\left(\frac{3}{2}\right)-\mathrm{R}\left(\frac{1}{2}\right)\right]^2+\left[2*\mathrm{R}\left(\frac{3}{2}\right)+\mathrm{R}\left(\frac{1}{2}\right)\right]^2}$$

With no spin-orbit in the p state $\mathrm{R}(3/2) = \mathrm{R}(1/2)$ and $P_e = 0$.

The range order of P_e is about 1% for the K and L_I edges.

Remarks

- This simple model gives an idea of the origin of the XMCD signal. It is not appropriate as a quantitative description of XMCD. More realistic models, which depend of the X-ray energy range, are described below. The present formalism is a one-electron model. The relaxation induced by Coulomb interactions by the created core-hole is stronger in metals than in insulators, where many-body effects are less sensitive. Nevertheless this relaxation is roughly the same for the two upper bands of majority and minority spins. We also have to be very careful with the interpretation of the XMCD signal at the $L_{II\text{-}III}$ edges of rare earths, where many-body effects are induced in the probed $5d$ bands by the presence of the localized open $4f$ bands.
- In the hard X-ray range the transitions accessible in transition metals (K edges) and in rare earths ($L_{II\text{-}III}$ edges) do not probe directly the bands responsible for magnetism, respectively $3d$ and $4f$, but rather the $4p$ and $5d$ bands. The XMCD signal is strongly influenced by the neighbourhood of the probed atom.
- In the soft X-ray range, XMCD probes directly the $3d$ bands of transition metal or $4f$ bands of rare earths, which carry the main part of the magnetic moment. XMCD signals can be modelled by multiplet calculations and quantitative values are extracted. See chapter by F. de Groot and J. Vogel and reference **[4]**.

2.4. MAGNITUDE OF THE XMCD SIGNAL

The polarization rate P_e can be estimated approximately by atomic calculations. It is about 1% for K edge and larger for L_{II} and L_{III} edges. The difference between $L_{II\text{-}III}$ edges and the K edge is due to the presence of a strong spin-orbit coupling in the initial state of the $L_{II\text{-}III}$ edges ($2p$ bands), which is not present in the initial state of the K edge ($1s$ band).

Realistic values for P_e are listed in the following table

Edges	K	L_I	L_{II}	L_{III}
P_e	+0.01	–0.15	–0.50	+0.25

The absolute sign of the XMCD signal is convention-dependent. These conventions applied to every edge enable us to determine the sign of the coupling between the majority spin of the 3d and 4f band, for example in a transition metal / rare earth binary compound.

XMCD of Gd Fe_2 illustrates this, where the K edge of iron and the L_{III} edge of gadolinium are shown in figure 3. The gadolinium moment is greater than that of iron. From the sign of the XMCD we can conclude that the Gd moments are aligned parallel to the external applied magnetic field, whereas those of iron are antiparallel, which proves the antiferromagnetic coupling between the gadolinium and iron moments.

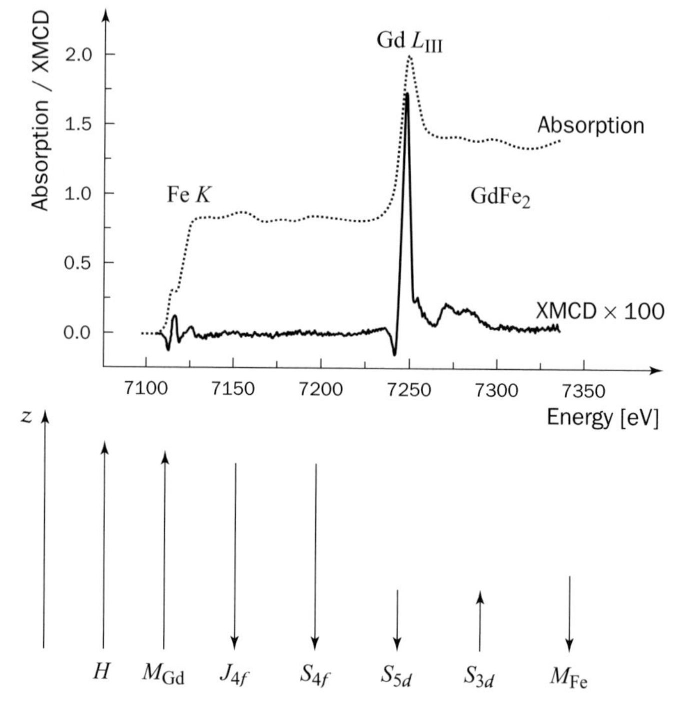

Figure 3 - From the sign of XMCD we know the direction of the magnetic moment in the applied magnetic field ***H***. Here we see antiparallel coupling between the iron and the gadolinium moment (D11 LURE).

It means also that Gd behaves like a heavy rare earth. We see the large difference in amplitude between a K edge signal of a $3d$ transition metal (no spin-orbit coupling in the $1s$ initial state) and the L_{III} edge of a rare earth (strong spin-orbit coupling in the $2p$ initial state).

3. *THE SUM RULES*

The first sum in X-ray absorption spectroscopy, developed by B.T. Thole and G. Van der Laan [5] give the branching ratio R of $M_{IV\text{-}V}$ and $L_{II\text{-}III}$ edges. $R = I(M_V)/[I(M_V)+I(M_{IV})]$ or $I(L_{III})/[I(L_{III})+I(L_{II})]$. $R = R^0 + A\langle i|Z|i\rangle$ where Z

is the angular part of the spin-orbit operator in the initial state $|i\rangle$. A is a constant depending on the number of holes n_h in the open probed band. $A = -1/3n_h$ for the L edges and $A = -4/15n_h$ for the M edges. R^0 is equal to the statistical value for negligible core-valence electrostatic interaction and depends on the Slater core-valence integrals in other cases. I is the total intensity of the absorption edge and is proportional to the integral of the absorption coefficient μ.

From the XMCD experimental signal amplitudes, the authors of reference **[5]** and **[6]** give two magneto-optical sum rules, which allow to calculate the value of the angular momentum $\langle L_z \rangle$ and spin momentum $\langle S_z \rangle$ to be determined.

The orbital part of the magnetic moment m_l results from the interplay of several effects, the Coulomb interaction and the spin-orbit correlation, hybridization and crystal field effects. Thole *et al.* **[7]** give a quite simple expression to extract the value of the orbital angular momentum from XMCD measurements.

For a transition between a core level (of orbital quantum number c) to a valence level (of orbital quantum number l) with n electrons in the ground state, neglecting the $l = c - 1$ channel, the orbital sum rule is

$$\frac{\int_{J^-+J^+}(\mu^+ - \mu^-)\mathrm{d}\omega}{\int_{J^-+J^+}(\mu^+ - \mu^- + \mu^0)\mathrm{d}\omega} = -\frac{\langle 0|L_z|0\rangle}{l\, n_h} \qquad \{5\}$$

n_h is the number of holes in the l shell: $n_h = 4l + 2 - n$.

$\langle L_z \rangle$ is the mean value of the z component of the orbital momentum operator L. z is along the quantization axis in the wave vector direction of the photons. The integral is over J^+ and J^- where $J^{\pm} = c \pm 1/2$. When the orbital moment is zero, the integral is zero, which in the L edges cases corresponds to a ratio of -2 between the normalized $L_{II\text{-}III}$ XMCD signals. The ratio is -1 before normalization.

This sum rule is applied successfully at the L edges of transition metals and M edges of rare earths. It fails at the L edges of rare earths because the approximations included are no longer valid in that case, especially the energy and spin independence of the radial matrix elements of the transition (see section 6).

The same authors give a second sum rule, called "spin sum rule", considering again only the $l = c + 1$ channel

$$\frac{\int_{J^+}(\mu^+ - \mu^-)\mathrm{d}\omega - \frac{c+1}{c}\int_{J^-}(\mu^+ - \mu^-)\mathrm{d}\omega}{\int_{J^-+J^+}(\mu^+ - \mu^- + \mu^0)\mathrm{d}\omega} = \frac{-2}{3n_h}\left(\langle 0|S_z|0\rangle + \frac{2l+3}{l}\langle 0|T_z|0\rangle\right) \qquad \{6\}$$

The conditions of validity are the same as for the orbital sum rule with, in addition, the requirement of a large gap between the J^+ and J^- edges to avoid overlapping. T_z is the dipole magnetic operator; it describes any correlation between the spin and the electron distribution around the absorbing atom. This term is at the origin of many restrictions to the application of the second sum rule. Many authors [9-11] emphasize that the $\langle T_z \rangle$ can be important and is not always negligible compared to $\langle S_z \rangle$. The value of $\langle T_z \rangle$ is temperature and crystal symmetry dependent. In the case of a pure J state, $\langle T_z \rangle$ can be calculated; this is the case for $M_{\text{IV-V}}$ edges of rare earths. It is questionable, however, if such a rule derived from a single ion model can be applied to complex real materials such as transition metals, where the valence and conduction electrons are described by their band structure [12].

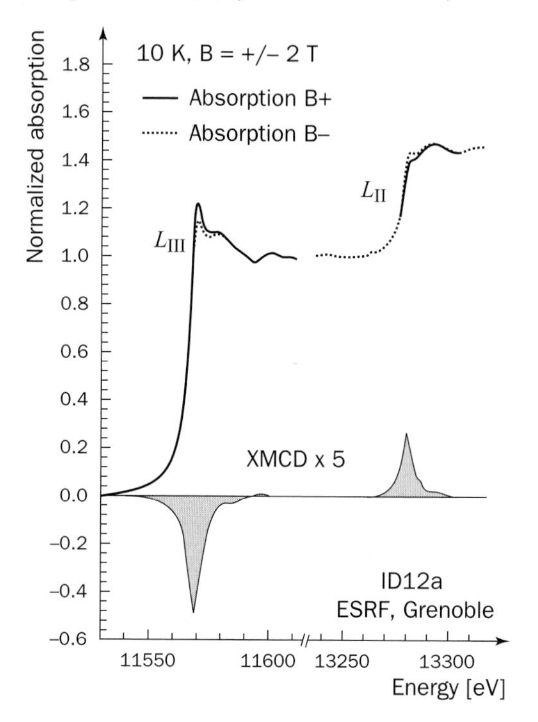

Figure 4
$PtFe_3$ XMCD $L_{\text{II-III}}$ edges. The Pt orbital and spin magnetic moment extracted from the sum rules are respectively: $m_l^{5d} = 0.055\ \mu_B$ and $m_s^{5d} = 0.25\ \mu_B$. The Pt orbital magnetic moment is 16% of the total magnetic moment [8].

High precision photo absorption measurement of XMCD at the $L_{\text{II-III}}$ edges of iron and nickel demonstrate the applicability of individual orbital and spin sum rules [13]. One of the nicest sum rule applications was made by Odin [8], who shows quenching of the platinum orbital moment under pressure in $PtFe_3$ invar compounds (fig. 4). The platinum magnetic moment is found to be $\langle M \rangle = 0.305\ \mu_B$ with $\langle L_z \rangle = 0.055\ \mu_B$. This orbital part of the orbital magnetic moment of platinum is quenched around 4 GPa.

Application of sum rules is one of the most useful tools provided by XMCD spectroscopy [14,15].

4. XMCD SIGNAL AT THE K EDGES (1s ⟶ p)

The *K* edge case is delicate because the XMCD signal arises from many contributions and mainly to spin-orbit coupling in the final state (the *p* probed state) of the photoelectron. The probability transitions α and α' are not different, as in the case of the $L_{\text{II-III}}$ edges (section 2.3), if we neglect the spin-orbit coupling in the final state. The spin-orbit interaction in the final state (*p* band) is dependent on the orbital kinetic moment $\langle L_z \rangle$, which is generally considered to be zero because of crystal field quenching in 3*d* metals. In fact, this orbital moment is weak but not zero and gives a contribution to the magnetic moment. In pure 3*d* transition metals its values are: 0.09 μ_B for iron, 0.15 μ_B for cobalt and 0.07 μ_B for nickel. The presence of the 3*d* open band with a large orbital moment plays an important role in the XMCD signal formation.

It is therefore quite difficult to extract absolute quantitative information from XMCD at the *K* edges. The comparison between XMCD signals in different compounds can give information about the spin orientation in the p band, about the variation of the magnetic moment of an element with temperature, or with concentration [16]. Information is found more directly by simply detecting the presence of spin polarization of an element for one symmetry band. For example it as been shown that the 4*p* valence band of copper is spin polarized in Cu/Fe and Cu/Co multilayers [16] (see section 8).

A very interesting development of the qualitative interpretation of XMCD using multiple scattering Green function formalism is given by Brouder [17]. This complex model describes scattering of the excited photoelectron by the potential of the neighbouring atoms. The XMCD signal involves many contributions (see expression 62 of reference [18]). The first is purely atomic: it is the XMCD signal expected from an isolated atom. Another contribution comes from the spin polarization of the probed *p* band of the absorbing atom, induced by the neighbouring atoms. Finally a very significant part of the signal comes from the spin-orbit interaction of the excited photoelectron with the *p* and *d* band of the neighbours. This last part gives a great sensitivity of XMCD at the *K* edge to the spin polarization of the d band of the transition metal. This had been suspected since the early comparison between measurements on 3*d* transition metals (1987). Comparing the XMCD at the *K* edges of nickel, cobalt and iron (fig. 5), we see the presence of a positive peak for iron that does not exist for nickel and cobalt. This positive peak has been correlated to the weak ferromagnetic nature of iron, i.e., to the presence of a majority empty state in the *d* band. From multiple scattering calculations [17] it is clearly seen that the positive peak is due to the spin-orbit coupling of the photoelectron with the $3d$ band of the neighbouring iron atoms. From XMCD measurements at the K edge of $3d$ transition metals we can thus detect the weak or strong nature of the ferromagnetism. This has been applied in CoFe alloys and multilayers [16]. We mention that another theoretical approach by J. Igarashi and K. Hirai using the tight binding approximation leads to the same conclusion, that the spin-orbit interaction of the $3d$ states of neighbours with the photoelectron gives rise to a dominant contribution to the XMCD [18].

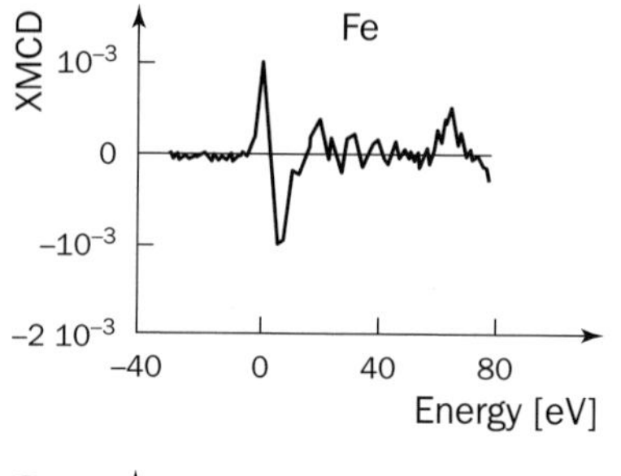

Figure 5 - K edge XMCD signal of metallic Fe, Ni and Co. The positive peak of the Fe signal is due to spin-orbit interaction of the photo-electron with the d bands of the neighbouring atoms (D11 LURE).

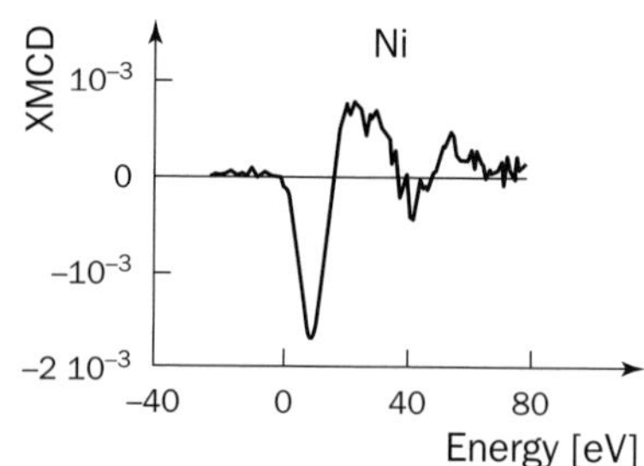

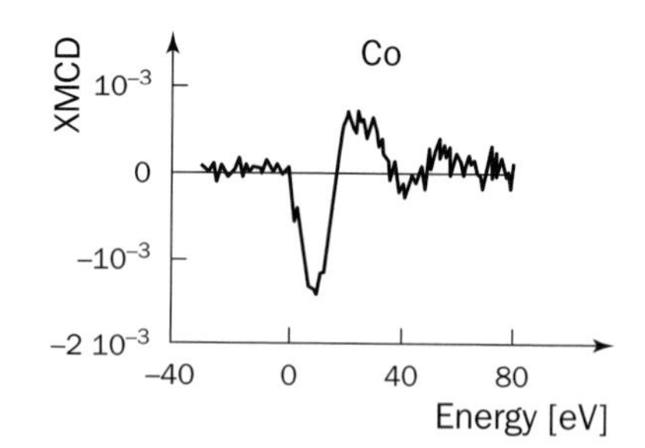

G.Y. Guo [19] within the framework of a spin- and orbital-polarized relativistic multiple scattering theory came to the same conclusion that the K edge XMCD (and $4p$ orbital moment) is mainly due to the spin-orbit coupling of the $3d$ states through $4p$ - $3d$ hybridization.

5. XMCD AND LOCALIZED MAGNETISM (MULTIPLET APPROACH)

XMCD at the $M_{IV\text{-}V}$ edges in rare earths and at the $L_{II\text{-}III}$ edges in transition metals, which give transitions to $4f$ and $3d$ states respectively, is easier to calculate because the final states are more localized. Absorption edges involving a localized final state are described in terms of local models [20,21].

The description involves calculation of the discrete energy levels of the initial and final N particle wave functions (multiplet). The spectrum is the superposition of all selection rule-allowed transitions from the ground state to the levels of the final state multiplet (fig. 6).

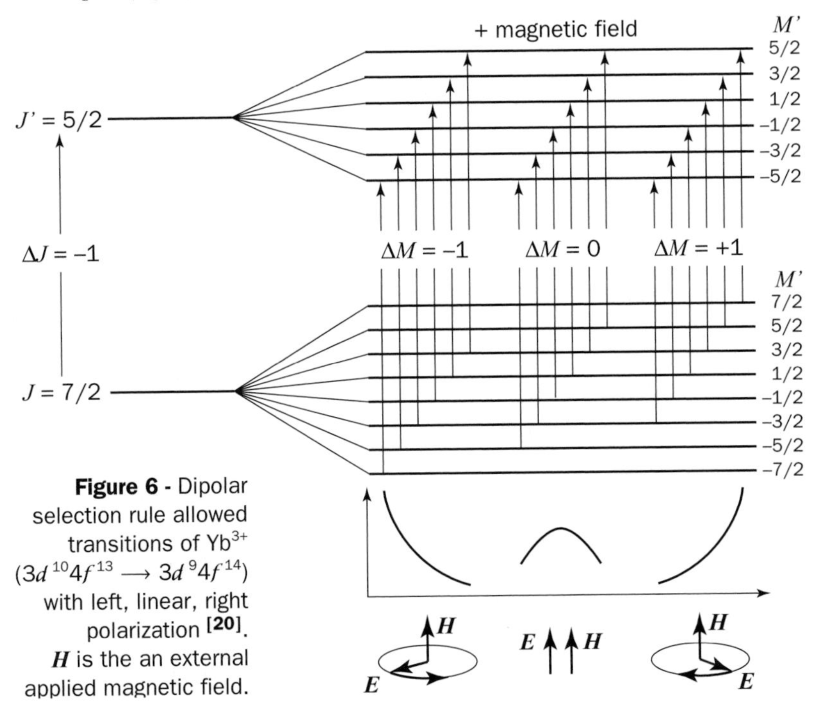

Figure 6 - Dipolar selection rule allowed transitions of Yb^{3+} ($3d^{10}4f^{13} \longrightarrow 3d^{9}4f^{14}$) with left, linear, right polarization **[20]**. ***H*** is the an external applied magnetic field.

The presence of a core-hole is taken into account naturally in this atomic-like picture. The dichroic signal is given by the selection rules applied to the multiplet final state created by symmetry breaking due to an external magnetic field. The splitting due to the Zeeman effect (10 meV) is not directly visible because it is two orders of magnitude smaller than the experimental linewidths.

Nevertheless, occupation of the Zeeman levels is temperature dependent. The Boltzmann average occupation of the final state confers temperature dependence on the absorption strength for each allowed transition. The temperature dependent terms of the absorption cross-section can be expressed **[22]** as a combination of $\langle \boldsymbol{M} \rangle$ and $\langle \boldsymbol{M}^2 \rangle$ where $\langle \boldsymbol{M} \rangle$ is related to the magnetic moment $\boldsymbol{M}$ by the expression $\langle \boldsymbol{M} \rangle = \boldsymbol{M}/\mu_B g_j$. It is easily demonstrated that the circular magnetic dichroism signal is proportional to $\langle \boldsymbol{M} \rangle$ and also that the linear polarization dependent magnetic absorption is proportional to $\langle \boldsymbol{M}^2 \rangle$ (fig. 7). See the chapter by F. de Groot and J. Vogel in this volume and reference **[4]**.

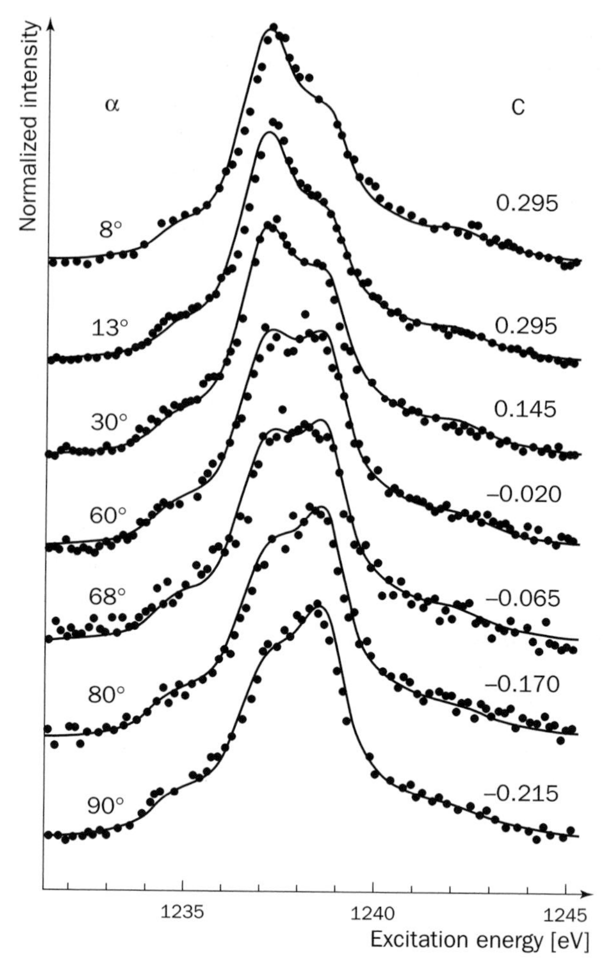

Figure 7 - The first experimental proof of magnetic X-ray dichroism is a linear dichroic signal recorded at LURE in 1986 on TbIG (terbium iron garnet) at the M_V edge of Tb. α is the angle between the [111] magnetization direction and the polarization vector of the X-ray photons. The solid lines are fits using the C parameter, which is proportional to the mean square of the magnetic moment [21,22].

XMCD at the $M_{IV\text{-}V}$ edges in rare earths and at the $L_{II\text{-}III}$ edges in transition metals is currently used to perform element-selective measurements of spin and orbital magnetic moments.

In this atomic approach, we do not take into account bonding with other neighbouring atoms. This covalency induces a decrease or an increase of the charge density of the probed atom. This can be included by the theoretical approach of Jo and Sawatzky [23], which includes configuration mixing in multiplets to explain the experimental results of Sette *et al.* (XMCD at the *L* edges of Ni) [24].

They find in particular that the orbital moment is non zero, as indicated by the "quenching picture", and that 10% of the total moment (0.6 μ_B) is due to orbital moment (0.07 μ_B). Another important result is the interpretation of the presence of a satellite very close to the main excitation. This satellite is 6 eV above the peak in the absorption spectra and 4 eV in the difference spectra (the XMCD signal). This surprising result led the authors to develop atomic calculations with a mixing configuration in the fundamental state, i.e., 15-20% $|3d^{10}\rangle$ + 60-70% $|3\,d^9\rangle$ + 15-20% $|3\,d^8\rangle$. This last contribution is responsible for the difference in energy of the satellite position between the absorption spectra and the XMCD spectra.

A very interesting study has been made by M.A. Arrio [25] on the high T_c molecular based magnet $Cs^I[Ni^{II}Cr^{III}(CN)_6].2H_2O$ at the $L_{II\text{-}III}$ edges of nickel. Ni^{II} is in high spin configuration and is ferromagnetically coupled to the surrounding Cr^{III}. A complete calculation taking into account the multiplet coupling effect and the covalent hybridization allows the total magnetic moment carried by Ni^{II} to be determined. The parameters to be taken into account are the spin-orbit coupling in the 2*p* and 3*d* states of nickel, the crystal field splitting, the Zeeman effect, the breaking of cubic symmetry due to anisotropy in a powder in a magnetic field, and a configuration mixing of 90% $3d^8$ and 10% $3d^9\boldsymbol{L}$. $3d^9\boldsymbol{L}$ means that an electron is transferred from the ligands to the Ni atom. All these ingredients are necessary to fit correctly the experimental data (fig. 8).

6. $L_{II\text{-}III}$ EDGES OF RARE EARTHS

For the $L_{II\ III}$ edges of rare earths, the probed 5*d* band is strongly influenced by the presence of the localized 4*f* band. Conversely to the case of $L_{II\text{-}III}$ edges of 5*d* transition metals, the XMCD signal in rare earths is usually of opposite sign and composed of many structures [26]. Influence of the 4*f* band, through exchange interaction between 4*f* and 5*d* electrons, results in an orbital polarization of the 5*d* state [27]. This means that each 5*d* state with different quantum number m_l do not have the same energy and the same occupation number. This would explain the multi-peak shape of the XMCD signals having the wrong sign (fig. 9).

Another idea to obtain the correct sign was proposed by X. Wang *et al.* [28]. They take into account the influence of the 4*f* band on the radial part of the 5*d* wave function.

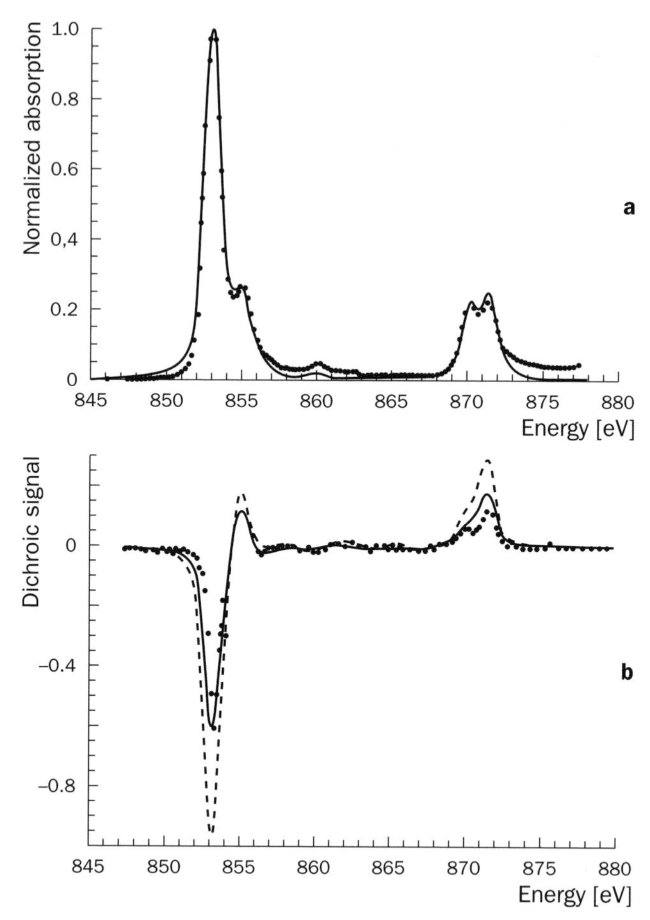

Figure 8 (from ref. **[25]**) - **a** - Experimental (dots) and theoretical (line) absorption profile of $Cs^{I}[Ni^{II}Cr^{III}(CN)_6].2H_2O$ at the $L_{II,III}$ Ni edges. **b** - Experimental (dots) and theoretical (dashed line) XMCD profile. The theoretical profile is normalized to the experimental circular polarization rate.

Because the exchange interaction of the 5*d* majority spin band with the 4*f* moment is larger than for the 5*d* minority spin band, the 5*d* majority spin band is pulled toward the core of the absorbing atom and the overlap with the core wave function is larger than for the 5*d* minority spin band. This gives a larger absorption cross-section for majority spin electrons than for the minority ones. On the other hand, the cross-section is expected to be proportional to the empty density of state, which is larger for the minority state (as described in section 2.3) Therefore, taking the radial part of the wave function into account can give a signal opposite to that previously expected.

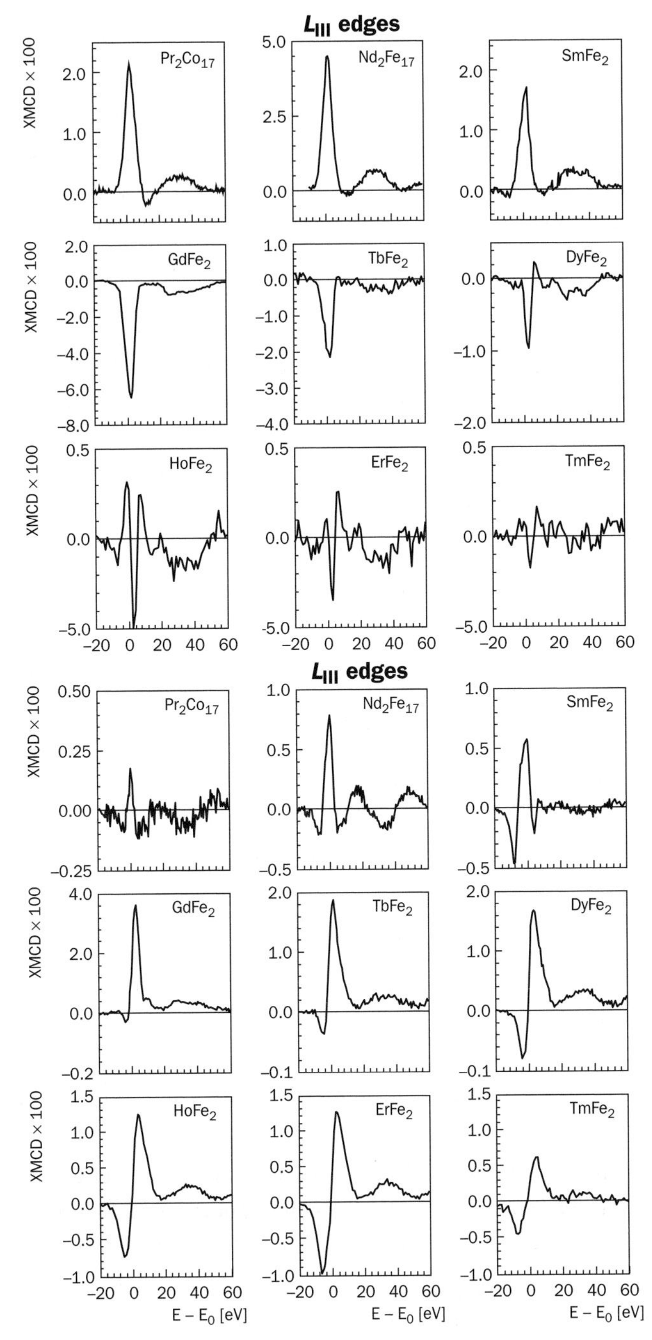

Figure 9 - Dichroic signal at the $L_{II\text{-}III}$ edges of various rare earth compounds (D11 LURE).

These two models were followed by recent theoretical developments based on atomic calculations.

The first begins with the observation that most rare earth $L_{II\text{-}III}$ edge XMCD signals look like a derivative of the total absorption cross-section (fig. 10a), because of the exchange splitting of the 5*d* band. The absorption cross-section resulting from transitions predominantly toward states of spin-up and -down are not at the same energy. In a first approximation the XMCD signal has a derivative shape resulting from the difference of these two absorption cross-sections [29,30]. The radial part of the wave function is then introduced, giving an enhancement of the majority cross-section and an asymmetric derivative XMCD signal. This derivative model has been applied recently by Neumann [31] to various compounds of rare earths. Good agreement is found for the L_{II} edge, but the derivative model fails for the L_{III} edge.

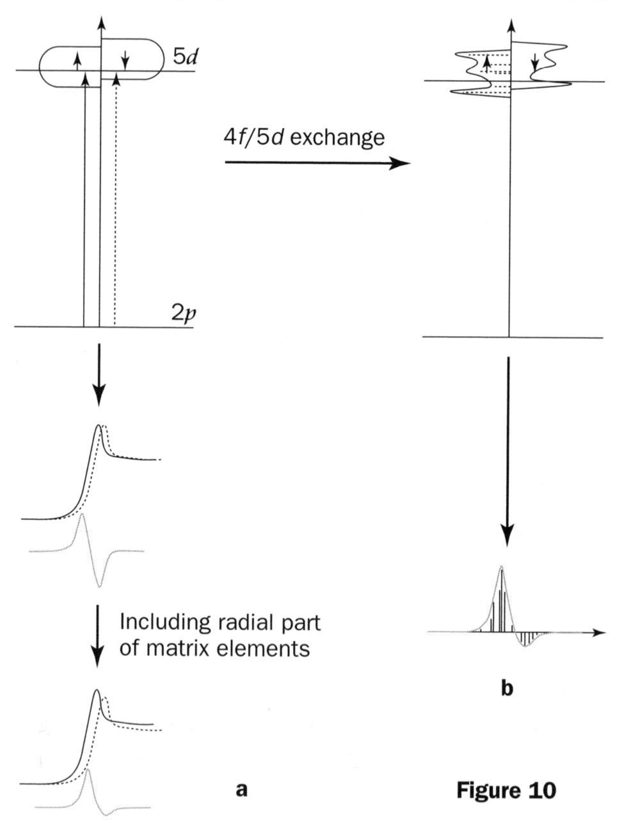

Figure 10

a - The XMCD signal at the $L_{II\text{-}III}$ edges of rare earths seems to be a derivative shape of absorption. Different radial wave functions of 5*d* majority and minority spin cause the asymmetry. **b** - Real signals can sometimes be fitted by including 5*d*/4*f* exchange, giving orbital polarization of the 5*d* band.

In the second model (fig. 10b) the orbital polarization of the $5d$ band induced by $4f/5d$ exchange was used by A. Kotani [32,33] who did not rely on derivative based arguments. Kotani presents an atomic model where corrections due to the radial part of the wave function are taken into account. Even if the model remains phenomenological, it is probably one of the most straightforward to describe the L edges of rare earths.

A model based on atomic calculations neglects many solid state effects and can probably not describe correctly transitions where a delocalized $5d$ band is involved. In rare earth insulators the $5d$ band is less delocalized than in metal compounds and attempts to apply atomic models are justified. Nevertheless much effort seems to be required before a full description of experimental results can be given. In any case a model based on band calculations [34,35] is probably better to fit experimental data but the presence of two open shells with orbital polarization make it difficult to achieve and a huge amount of theoretical work will be required before XMCD at the L edges of rare earths can be used for magnetometry.

The **presence of quadrupolar transitions** ($2p$ $4f$) in XMCD at the L edges ($2p$ $5d$) of rare earths was pointed out by Carra and Altarelli [36]. Neglected in classical absorption spectroscopy, this contribution can be very important in the rare earth $L_{\text{II-III}}$ edges because of the strong spin polarization of the $4f$ band. These authors showed that the quadrupolar contribution can be observed through its angular dependence θ, which is different from that of the dipolar transition (θ is the angle between the magnetic moment and the X-ray wave vector). After many unsuccessful experiments, some experimental evidence of quadrupolar effects in rare earth $L_{\text{II-III}}$ edges has been obtained [37,38].

7. MAGNETIC EXAFS

Readers not familiar with EXAFS may refer to the chapter by B. Lengeler in this volume and to reference [39]. Theoretical work on magnetic EXAFS by C. Brouder [40] shows that magnetic EXAFS can be interpreted within the multiple scattering approach. XMCD structure seems to be due to all neighbours and the rigid band picture can be used at high energy. This rigid band model allows a pseudo exchange energy value ΔE to be extracted from XMCD EXAFS oscillations. This is different from the exchange energy value at the Fermi level. The relation between both values remains to be understood. Some magnetic EXAFS studies have been published [41]. The possibility of detecting only the spin-polarized neighbours by Fourier transform of magnetic EXAFS oscillations is still a matter of controversy.

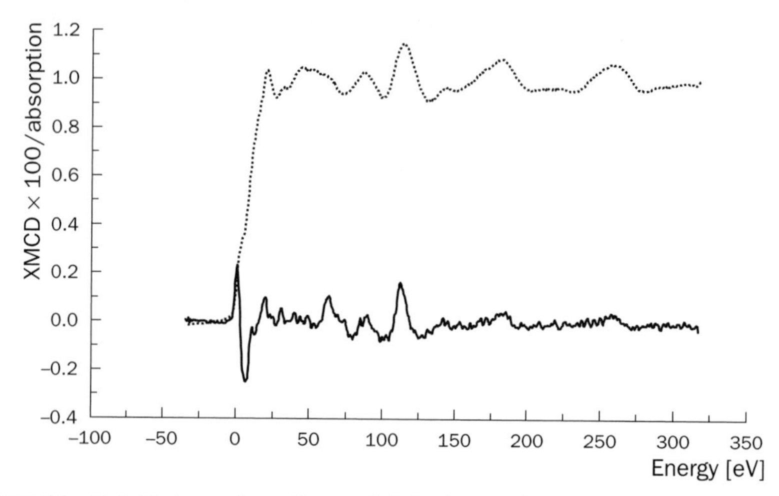

Figure 11 - Metallic iron: absorption and dichroism at the *K* edge of Fe (D11 LURE).

8. *APPLICATION TO MULTILAYERS WITH GMR*

Multilayers are intensively studied because of their exceptional magnetic properties and technological applications in magnetic recording. One of the most promising properties is giant magnetoresistance GMR [42-44], which is the basis of very sensitive magnetic recording heads.

The most famous GMR systems are multilayers of Fe/Cu or Co/Cu which are known to exhibit oscillations of the magnetic coupling between magnetic layers as a function of the copper spacer thickness. Models based on a partial confinement of the d band within the spacer layer [45] or based on the RKKY theory are able to predict the periods and the amplitudes of the coupling oscillations. XMCD was used to probe the polarization of the spacer layer. The *K* edge measurements of S. Pizzini [16] show a large magnetic polarization induced on the sp band of copper. In contrast, measurements by M.G. Samant [46] at the *L* edge show a weak 3*d* moment induced in copper.

K edge measurements can be seen in figure 12, which compares the Cu *K* edge XMCD signal with the corresponding Co or Fe *K* edges. The most striking result is the amplitude of copper, which is only three times smaller than Co or Fe with the same sign (fig. 12). There is spin polarization of the 4*p* band of copper with a magnetic moment of about 0.02 μ_B parallel to the Co-4*p* magnetic moment. The measurements at the *L* edges [46] demonstrate the presence of an induced Cu-3*d* magnetic moment. This moment is very weak, 0.01 μ_B for 13 Å of Cu and 0.05 μ_B for 4 Å Cu, compared to the 1.7 μ_B 3*d* magnetic moment of bulk cobalt. In contrast, the 4*p* induced Cu moment is of the same order as that of 4*p* Co, even for a 20 Å thick copper layer, as can be seen simply by comparing the XMCD amplitudes in figure 12. The polarization of Cu is induced by the hybridization of the Cu-4*p* band with the Co-3*d* band at the Cu/Co interface. In conclusion these results agree with the formation of quantum well states because the 4*p* magnetization of copper is not restricted to the interface as in the case of 3*d* states.

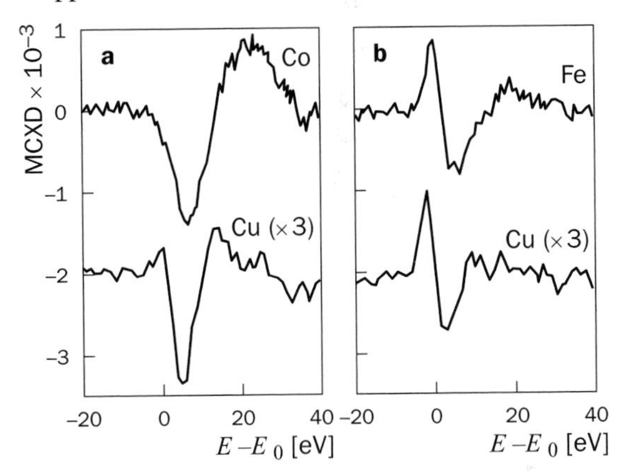

Figure 12 - Cu, Co and Fe XMCD *K* edges of $Co_{12Å}Cu_{8Å}$ (**a**) and $Fe_{12Å}Cu_{8Å}$ (**b**) multilayers (from ref. **[16]**).

9. *STUDY OF HIGHLY CORRELATED [CE/LA/FE] AND [LA/CE/FE] MULTILAYERS*

In these layers, Ce adopts an α phase-like electronic structure, which is usually non magnetic. Strong 3*d*-5*d* and 3*d*-4*f* hybridization occurs at the interfaces. Fe induces a 5*d* and 4*f* magnetic moment of Ce and a 5*d* magnetic moment of La. Measurements at the $M_{\text{IV-V}}$ and $L_{\text{II-III}}$ of Ce in Fe/La/Ce/La/Fe multilayers show that a magnetic polarization of the 5*d* states in the α Ce phase can exist without magnetic polarization of the 4*f* states. By contrast, in Fe/Ce/La/Ce/Fe, Fe induces magnetic polarization in both the 4*f* and the 5*d* states of Ce. 4*f* polarization requires direct contact between Ce and Fe. 5*d* polarization is then of more complex origin because it does not require such direct overlap with the $3d$ band of iron. It seems to be an intrinsic property of the α Ce phase, which may be close to a magnetic instability. The energy difference between the usual paramagnetic state of α Ce and the ferromagnetic state seen in these multilayers is very small. We imagine that subtle electronic or strain effects in multilayer environments can induce long-range magnetic order [47].

REFERENCES

[1] G. SCHÜTZ, W. WAGNER, W. WILHELM, P. KIENLE, R. ZELLER, R. FRAHM & G. MATERLIK *Phys. Rev. Lett.* **58**, 737 (1987)

[2] D.C. KONINGBERGER & R. PRINS - *X-ray Absorption: Principles, Applications, Techniques of EXAFS, SEXAFS and XANES,* A Wiley interscience publication, John Wiley & Sons, New York (1987)

[3] U. FANO - *Phys. Rev.* **178**, 131 (1969)

[4] M. SACCHI & J. VOGEL - *Magnetism and synchrotron radiation, Lecture Notes in Physics* **565**, 87, Springer, Berlin (2001)

[5] B.T. THOLE & G. VAN DER LAAN - *Phys. Rev. A* **38**, 1943 (1988)

[6] P. CARRA, B. T. THOLE, M. ALTARELLI & X. WANG - *Phys. Rev. Lett.* **70**, 694 (1993)

[7] B.T. THOLE, P. CARRA, F. SETTE & G. VAN DER LAAN - *Phys. Rev. Lett.* **68**, 1943 (1992)

[8] S. ODIN, F. BAUDELET, E. DARTYGE, J.P. ITIÉ, A. POLIAN, J.C. CHERVIN, J.P. KAPPLER, A. FONTAINE & S. PIZZINI - *Phil. Mag. B* **80**, 155 (2000)

[9] M. ALTARELLI & P. SAINCTAVIT - *In "Magnetism and Synchrotron Radiation",* E. BEAUREPAIRE, B. CARRIÈRE & J.P. KAPPLER eds, Les Editions de physique, Les Ulis, 65 (1997)

[10] J. VOGEL, A. FONTAINE, V. CROS, F. PETROFF, J.P. KAPPLER, G. KRILL, A. ROGALEV & J. GOULON - *Phys. Rev. B* **55**, 3663 (1997)

[11] W. GRANGE, M. MARET, J.P. KAPPLER, J. VOGEL, A. FONTAINE, F. PETROFF, G. KRILL, A. ROGALEV, J. GOULON, M. FINAZZI & N.B. BROOKES - *Phys. Rev. B* **58**, 6298 (1998)

[12] R. WU, D. WANG & A.J. FREEMAN - *J. Magn. Magn. Mater.* **132**,103 (1994)

[13] C.T. CHEN, Y.U. IDZERDA, H.J. LIN, N.V. SMITH, G. MEIGS, E. CHABAN, G.H. HO, E. PELLEGRIN & F. SETTE - *Phys. Rev. Lett.* **75**, 152 (1995)

[14] D. WELLER, J. STÖHR, R. NAKAJIMA, A. CARL, M.G. SAMANT, C. CHAPPERT, R. MÉGY, P. BEAUVILLAIN, P. VEILLET & G.A. HELD - *Phys. Rev. Lett.* **75**, 3752 (1995)

[15] Y. WU, J. STÖHR, B.D. HERMSMEIER, M.G. SAMANT & D. WELLER - *Phys. Rev. Lett.* **69**, 2307 (1992)

[16] S. PIZZINI, A. FONTAINE, E. DARTYGE, C. GIORGETTI, F. BAUDELET, J.P. KAPPLER, P. BOHER & F. GIRON - *Phys. Rev. B* **50**, 3779 (1994)
S. PIZZINI, A. FONTAINE, C. GIORGETTI, E. DARTYGE, J.F. BOBO, M. PIECUCH & F. BAUDELET - *Phys. Rev. Lett.* **74**, 1470 (1995)

[17] C. BROUDER & M. HIKAM - *Phys. Rev. B* **43**(5), 3809 (1991)

[18] J. IGARACHI & K. HIRAI - *Phys. Rev. B* **50**, 17820 (1994)
J. IGARACHI & K. HIRAI - *Phys. Rev. B* **53**, 6442 (1996)

[19] G.Y. GUO - *Phys. Rev. B* **55**, 11619 (1997)

[20] J.B. GOEDKOOP, B.T. THOLE, G. VAN DER LAAN, G.A. SAWATZKY, F.M.F. DE GROOT & J.C. FUGGLE - *Phys. Rev. B* **37**, 2086 (1988)

[21] G. VAN DER LAAN, B.T. THOLE, G.A. SAWATZKY, J.B. GOEDKOOP, J.C. FUGGLE, J.M. ESTEVA, R. KARNATAK, J.P. REMEIKA & H.A. DABKOWSKA - *Phys. Rev. B* **34**, 6529 (1986)

[22] B.T. THOLE, G. VAN DER LAAN, G. SAWATZKY - *Phys. Rev. Lett.* **55**, 2086 (1985)

[23] T. JO & G. SAWATZKY - *Phys. Rev. B* **43**, 8771 (1991)

[24] F. SETTE, C.T. CHEN, Y. MA, S. MODESTI & N.V. SMITH - *Phys. Rev. B* **42**, 7262 (1990)

[25] M.A. ARRIO, PH. SAINCTAVIT, CH. CARTIER DIT MOULIN, CH. BROUDER, F.M.F. DE GROOT, T. MALLAH & M. VERDAGUER - *J. Phys. Chem.* **100**, 4679 (1996)

[26] E. DARTYGE, F. BAUDELET, CH. GIORGETTI & S. ODIN - *Journal of Alloys and Compounds* **275-277**, 526 (1998)

[27] T. JO & S. IMADA - *J. Phys. Soc. Japan* **62**, 3721 (1994)

[28] X. WANG, T.C. LEUNG, B.N. HARMON & P. CARRA - *Phys. Rev. B* **47**, 9087 (1993)

[29] M. VAN VEENENDAAL, J.B. GOEDKOOP & B.T. THOLE - *Phys. Rev. Lett.* **78**, 1162 (1997)

[30] M. VAN VEENENDAAL & R. BENOIST - *Phys. Rev. B* **58**, 3741 (1998)

[31] C. NEUMANN, B.W. HOOGENBOOM, A. ROGALEV & J.B. GOEDKOOP - *Solid State Comm.* **110**, 375 (1999)

[32] H. MATSUYAMA, I. HARADA & A. KOTANI - *J. Phys. Soc. Japan* **66**, 337 (1997)

[33] H. MATSUYAMA, K. FUKUI, K. OKADA, I. HARADA & A. KOTANI - *J. Elec. Spec.* **92**, 31 (1998)

[34] CH. BROUDER, M. ALOUANI & K. H. BENNEMANN - *Phys. Rev. B* **54**, 7334 (1996)

[35] H. EBERT - *In "Spin-Orbit Influenced Spectroscopies of Magnetic Solids"* H. EBERT & G. SCHÜTZ eds, *Lecture Notes in Physics Berlin* **466**, Springer, 159 (1996)

[36] P. CARRA & M. ALTARELLI - *Phys. Rev. Lett.* **64**, 1286 (1990)

[37] J.C. LANG, G. SRAJER, C. DETLEFS, A.I. GOLDMAN, H. KÖNIG, X. WANG, B.N. HARMON & R.W. MCCALLUM - *Phys. Rev. Lett.* **74**, 4935 (1995);
CH. GIORGETTI, E. DARTYGE, C. BROUDER, F. BAUDELET, C. MEYER, S. PIZZINI, A. FONTAINE & R.M. GALÉRA - *Phys. Rev. Lett.* **75**, 3186 (1995);
J. CHABOY, L.M. GARCÍA, F. BARTOLOMÉ, A. MARCELLI, G. CIBIN, H. MARUYAMA, S. PIZZINI, A. ROGALEV, J.B. GOEDKOOP & J. GOULON - *Phys. Rev. B* **57**, 8424 (1998)

[38] CH. GIORGETTI, E. DARTYGE, F. BAUDELET & C. BROUDER - *Appl. Phys. A* **73**, 703 (2001)

[39] A. FONTAINE - *In "Neutron and Synchrotron Radiation for Condensed Matter Studies", HERCULES vol. 1,* J. BARUCHEL, J.L. HODEAU, M.S LEHMANN, J.R. REGNARD & C. SCHLENKER eds, Les Editions de Physique - Springer-Verlag, 323 (1993)

[40] CH. BROUDER, M. ALOUANI, CH. GIORGETTI, E. DARTYGE & F. BAUDELET - *In "Spin-Orbit Influenced Spectroscopies of Magnetic Solids"* H. EBERT & G. SCHÜTZ eds, *Lecture Notes in Physics* **466**, Springer, Berlin, 259 (1996)

[41] E. DARTYGE, F. BAUDELET, C. BROUDER, A. FONTAINE, C. GIORGETTI, J.P. KAPPLER, G. KRILL, M.F. LOPEZ & S. PIZZINI - *Physica B* **208-209**, 751 (1994)
M. KNÜLLE, D. AHLERS & G. SCHÜTZ - *Solid State Comm.* **94**, 267 (1995)

[42] S.S.P. PARKIN, R. BHADRA & K.P. ROCHE - *Phys. Rev. Lett.* **66**, 2152 (1991)

[43] F. PETROFF, A. BARTHÉLEMY, D. H. MOSCA, D.K. LOTTIS, A. FERT, P.A. SCHROEDER, W.P. PRATT JR, R. LOLOEE & S. LEQUIEN - *Phys. Rev. B* **44**, 5355 (1991)

[44] P. BRUNO & C. CHAPPERT - *Phys. Rev. B* **46**, 261 (1992)

[45] D.M. EDWARDS & J. MATHON - *J. Magn. Magn. Mater.* **93**, 85 (1991)

[46] M.G. SAMANT, J. STÖHR, S.S.P. PARKIN, G.A. HELD, B.D. HERMSMEIER, F. HERMAN, M. VAN SCHILFGAARDE, L.C. DUDA, D.C. MANCINI, N. WASSDAHL & R. NAKAJIMA - *Phys. Rev. Lett.* **72**, 1112 (1994)

[47] M. AREND, M. FINAZZI, O. SCHUTTE, M. MÜNZENBERG, A.M. DIAS, F. BAUDELET, CH. GIORGETTI, E. DARTYGE, P. SCHAAF, J.P. KAPPLER, G. KRILL & W. FELSCH - *Phys. Rev. B* **57**, 2174 (1998)

4

EXTENDED X-RAY ABSORPTION FINE STRUCTURE

B. LENGELER
II. Physikalisches Institut B, RWTH, Aachen, Germany

1. INTRODUCTION

X-ray diffraction has contributed to a large extent to the understanding of the atomic arrangement in condensed matter. The structure of many thousands of inorganic and organic compounds has been determined in this way. However, for applying that technique it is necessary that the atoms are arranged in a lattice with long range order and that the sample is at least a polycrystal with crystallites of the order of 10 nm or more in size. In that case the position and the intensity of the Bragg peaks give the lattice and the position of all atoms in the unit cell. On the other hand, there are cases where the long-range order is not the relevant feature for understanding the mode of operation of a chemical or physical reaction in condensed matter. Take, for instance, the influence of an active centre in a catalyst. This is, in general, a particular atomic species that mediates a chemical reaction between 2 or more partners. This reaction happens on a length scale given by the range of chemical bonds, which is of the order of a few tenths of nanometers. For that purpose a technique which probes the local geometric and chemical structure around a given atomic species is particular useful. X-ray absorption spectroscopy (XAS) is such a technique. It is the purpose of this lecture to give a first introduction to the field of XAS. Special topics like magnetic dichroism are treated in different lectures. A few textbooks on XAS are quoted in the references [1-5].

When X-rays pass through matter they are attenuated. The intensities $I_1(E)$ and $I_2(E)$ in front and behind an absorber of thickness d are related by the Lambert-Beer law

$$I_2(E) = I_1(E)\mathrm{e}^{-\mu(E)d} \qquad \{1\}$$

F. Hippert et al. (eds.), Neutron and X-ray Spectroscopy, 131–168.

The linear absorption coefficient $\mu(E)$ depends on the photon energy $E = \hbar\omega$ and on the absorbing material. Figure 1 shows the absorption coefficient of X-rays for atomic copper and for metallic copper. The basic photoabsorption process is an atomic process, characterized by absorption edges and a strong increase of absorption at the edges. Energy can only be removed from the incoming beam by absorption of quanta $\hbar\omega$. This is done by exciting an electron from an occupied core level to an unoccupied state, as required by the Pauli exclusion principle. The excited states are discrete states or free states above the ionization level. At an absorption edge, photons become able to excite strongly bound electrons into the continuum. A new absorption channel is opened and μ increases drastically with photon energy. At a K edge $1s$ electrons are excited and at L_I L_{II} and L_{III} edges $2s$, $2p^{1/2}$ and $2p^{3/2}$ electrons are removed from the atom.

In condensed matter, as e.g. in metallic copper, there is an additional fine structure at and above the edges, as shown in figure 1. This fine structure is due to backscattering of the photoelectron by other atoms which surround the absorbing atom. It contains the structural and electronic information we are interested in. Before we will discuss this point we will consider in somewhat more detail the basic absorption effect observed in isolated atoms.

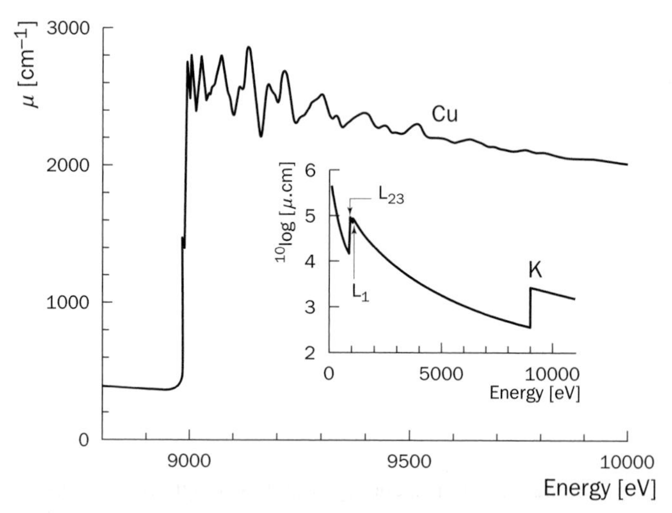

Figure 1 - Absorption coefficient μ in the vicinity of the K-absorption edge in metallic Cu and in atomic Cu (insert).

2. X-RAY ABSORPTION IN ISOLATED ATOMS

We consider a plane electromagnetic wave with frequency ω, polarization $\hat{\boldsymbol{\varepsilon}}$ along the x-direction and propagating along the z-direction

$$\boldsymbol{E} = E_0 \hat{\boldsymbol{\varepsilon}} \cdot \cos(\boldsymbol{q} \cdot \boldsymbol{r} - \omega t), \quad \hat{\boldsymbol{\varepsilon}} = (1,0,0) \tag{2}$$

impinging on individual atoms with atomic number Z. The energy density u of the incoming wave is

$$u = E_0^2 / 4\pi \tag{3}$$

Let N_{at} be the number density of atoms in the volume V. N_{at} can be expressed by the mass density ρ, Avogadro's number N_{a} and the atomic mass A as

$$N_{\text{at}} = \rho N_{\text{a}} / A \tag{4}$$

Energy is removed from the beam by photoabsorption in quanta $\hbar\omega$. The energy absorbed per second is

$$\frac{du}{dt} = -N_{\text{at}} \hbar\omega R_{if} \tag{5}$$

R_{if} is the transition rate for photoabsorption. According to Fermi's golden rule [6] it is given by

$$R_{if} = \frac{2\pi}{\hbar} \sum_f |\langle f | H_1 | i \rangle|^2 \delta(\varepsilon_f - \varepsilon_i - \hbar\omega) \tag{6}$$

Since absorption is strongest for tightly bound inner electrons, we assume absorption by a hydrogen-like $1s$ electron in an atom with atomic number Z. In the dipole approximation the perturbation H_1 that mediates the absorption can be written as **[6]**

$$H_1 = e_0 E_0 (\hat{\boldsymbol{\varepsilon}} \cdot \boldsymbol{r}) \mathrm{e}^{-i\omega t} \tag{7}$$

where $\boldsymbol{r}$ is the radius vector of the $1s$ electron relative to the nucleus of the absorbing atom.

The dipole approximation assumes a constant electric field over the dimension of $1s$ orbit. Let us see if this assumption is justified in the case of a Cu $1s$ state. At the Cu K edge (ε_i = 8,982 eV binding energy) the wavelength hc/E of X-rays is 1.38 Å. The spatial extension of the hydrogen-like $1s$ state is $a_0/Z = 0.02$ Å ($a_0 = 0.529$ Å being the Bohr radius). Hence $2\pi a_0/(\lambda Z) = 0.083$ is much smaller than 1, thus justifying the dipole approximation. The transition rate now reads

$$R_{if} = \frac{2\pi}{\hbar} e_0^2 E_0^2 \sum_f |\langle f | \hat{\boldsymbol{\varepsilon}} \cdot \boldsymbol{r} | i \rangle|^2 \delta(\varepsilon_f - \varepsilon_i - \hbar\omega) \tag{8}$$

The sum over the final states $|f\rangle$ will be specified by the experimental conditions. We assume, at first, a photon energy $E = \hbar\omega$ much larger than the binding energy ε_i. Then the final electron states are described by plane waves

$$\langle \boldsymbol{r} | f \rangle = V^{-1/2} e^{i\boldsymbol{k}\cdot\boldsymbol{r}} \qquad \{9\text{ a}\}$$

where $\hbar\boldsymbol{k}$ is the photoelectron momentum, related to its kinetic energy by

$$E_{\text{kin}} = E - \varepsilon_i = \hbar^2 k^2 / 2m_0 = 13.6\,\text{eV}(k\,a_0)^2 \qquad \{9\text{ b}\}$$

The hydrogen-like $1s$ initial state of the electron is

$$\langle \boldsymbol{r} | i \rangle = \sqrt{Z^3 / \pi a_0^3}\; e^{-Zr/a_0} \qquad \{10\}$$

Inserting equations {9} and {10} into equation {8} gives

$$|\langle f | \hat{\boldsymbol{\varepsilon}} \cdot \boldsymbol{r} | i \rangle|^2 = 2^{10} \frac{\pi}{V} \frac{1}{k^{10}} \frac{Z^5}{a_0^5} \sin^2\theta \cos^2\phi \qquad \{11\}$$

Since we consider the absorption as independent from the direction of ejection of the photoelectron we must integrate over k, θ and ϕ. Assuming $\varepsilon_i \ll E \equiv \hbar\omega$ and using

$$\frac{V}{(2\pi)^3} \int dk\, k^2 \sin\theta\, d\theta\, d\phi \frac{1}{k^{10}} \sin^2\theta \cos^2\phi\, \delta\left(\frac{\hbar^2 k^2}{2m_0} - E\right)$$

$$= \frac{V}{(2\pi)^3} \frac{2\pi}{3} \left(\frac{\hbar^2}{2m_0}\right)^{7/2} \frac{1}{E^{9/2}} \qquad \{12\}$$

we finally obtain for the transition rate in all possible final states

$$R_{\text{iall}} = \frac{8\pi}{3} \frac{4\text{Ry}}{\hbar} \left(n_q \frac{a_0^3}{V}\right) Z^5 \left(\frac{4\text{Ry}}{E}\right)^{7/2} \qquad \{13\}$$

Here n_q is the average number of photons per volume in the beam

$$E_0^2 / 4\pi = n_q \hbar\omega / V \qquad \{14\}$$

and 1 Ry $= e_0^2/2a_0$ is the Rydberg unit of energy in atomic physics.

In general, the photoabsorption is described by the linear absorption coefficient μ_0. It is related to the absorption rate by equations {5} and {13}

with
$$\frac{du}{dt} = \frac{du}{dz} \frac{dz}{dt} = c \frac{du}{dz} \qquad \{15\}$$

we get
$$\frac{du}{dz} = -\frac{16\pi}{3} \frac{1}{a_0} (N_{at}\, a_0^3)\, \alpha\, Z^5 \left(\frac{4\text{Ry}}{E}\right)^{7/2} u \qquad \{16\}$$

This differential equation has the solution

$$u(z) = u(0)e^{-\mu_0 z} \qquad \{17\}$$

with the linear absorption coefficient μ_0.

$$\mu_0 = \frac{16\pi}{3}\frac{1}{a_0} N_{at} a_0^3 \alpha Z^5 \left(\frac{4\mathrm{Ry}}{E}\right)^{7/2} \qquad \{18\}$$

$$\frac{\mu_0}{\rho} = \frac{16\pi}{3}\frac{N_a}{A} a_0^2 \alpha Z^5 \left(\frac{4\mathrm{Ry}}{E}\right)^{7/2} \qquad \{19\}$$

Here, $\alpha = e_0^2/\hbar c_0$ is the fine structure constant. This behaviour is expected far above the absorption edges. The strong decrease of μ/ρ with the photon energy is visible in figure 1. Two points are noteworthy about equations {13} and {19}. First, the photoabsorption process is not an instantaneous process. According to equation {13} it takes in the averages the time $1/R_{\mathrm{iall}}$ for a photon to be absorbed by an atom. Below, we will need this result in order to understand the extended fine structure in the absorption process. Second, equation {19} shows the strong dependence of the mass absorption coefficient μ/ρ on the atomic number Z and on the photon energy. However equation {19} does not fit the experimental data very well, because it relies on a crude model. The screening of the nuclear charge by the other electrons was neglected. We therefore expect a lower power than 5 for the Z dependence of μ/ρ. The Coulomb attraction of the photoelectron by the ionized absorber atom was also neglected. A widely accepted approximation for μ_0/ρ between absorption edges is

$$\frac{\mu_0}{\rho} = \mathrm{const}\, N_\mathrm{a} \frac{Z}{A}\left(\frac{Z}{E}\right)^3 \qquad \{20\}$$

Since Z/A is 0.45 ± 0.05 for all elements (except for hydrogen) the mass absorption coefficient varies roughly like $(Z/E)^3$. This strong variation of μ_0/ρ with Z/E is a striking feature of X-ray absorption in matter. For comparison with experimental data, we consider μ_0/ρ above the K edge of copper in the energy range from 9 to 70 keV. The data collected by Veigele **[7]** show an energy dependence as $E^{-2.78}$. At 20 keV μ_0/ρ varies as $Z^{2.64}$ for the elements Ca to Nb and at 30 keV as $Z^{2.73}$ for the elements Ca to Cd. The experimental exponents are in reasonable agreement with equation {20}.

The strong variation of μ_0/ρ on Z and E does not only show up between the edges. The height of the absorption edges is another noteworthy feature in this regard. As an example, let us consider again copper at the K edge. μ_0/ρ increases from 40.7 to 297 cm^2/g **[7]**, although only 2 more electrons (out of a total of 29) contribute to the edge jump. This overwhelming contribution of the 2 inner $1s$ electrons to the absorption coefficient and the strong dependence of μ_0/ρ on Z/E both rely on the same effect: a photon cannot be absorbed by a free electron, as follows from energy and momentum conservation. Indeed, for a free electron with momenta $\boldsymbol{p}_1$ and $\boldsymbol{p}_2$ before and after absorption and a photon with momentum $\hbar\boldsymbol{q}$ and energy $\hbar\omega$, momentum and energy conservation read

$$\begin{aligned} \hbar\boldsymbol{q} + \boldsymbol{p}_1 &= \boldsymbol{p}_2 \\ \hbar\omega + \sqrt{m_0^2c^4 + p_1^2c^2} &= \sqrt{m_0^2c^4 + p_2^2c^2} \end{aligned} \qquad \{21\}$$

Inserting p_2 from the first equation into the second equation and squaring twice in order to eliminate the roots gives

$$p_1^2\sin^2\theta + m_0^2c^2 = 0 \qquad \{22\}$$

where θ is the angle between $\boldsymbol{q}$ and $\boldsymbol{p}_1$. Equation {22} cannot be fulfilled. In other words, a free electron can scatter a photon, but it cannot absorb a photon without emitting another photon. However, it can absorb a photon when it is bound in an atom, which guarantees the balance of momentum. The absorption is favoured by a strong binding of the electron and by a low kinetic energy of the ejected photoelectron. This explains the large contribution of the inner electrons to photoabsorption and the observed variation of μ_0/ρ on a high power of Z/E.

3. X-RAY ABSORPTION FINE STRUCTURE (XAFS)

As shown in figure 1, the X-ray absorption in metallic copper shows a fine structure above the K edge. This fine structure is characteristic for condensed matter and contains information on the local structure around the absorbing atoms and on some electronic properties of the absorber. We will discuss these points separately.

3.1. X-RAY ABSORPTION FINE STRUCTURE: THE BASIC FORMULA

We first consider the fine structure above the edge. This structure is observed typically over an energy range of about a thousand eV. It is due to the backscattering of the photoelectron by the surrounding atoms, when the absorber is incorporated in condensed matter (fig. 2). In that case, the final state $|f\rangle$ in the transition rate contains besides the outgoing wave, an additional scattering contribution [1-5].

$$|f\rangle = |f_0\rangle + |f_{sc}\rangle = (1+B)|f_0\rangle \qquad \{23\}$$

$$B = i\,3\cos^2(\theta_j)F_j(k)\,e^{i\beta_j}\,\frac{e^{i2kr_j}}{kr_j^2}\,e^{2i\delta_1} \tag{24}$$

$|f_0\rangle$ is an outgoing spherical wave and the only term for an isolated absorbing atom. The second term $|f_{sc}\rangle$ is the backscattered wave emanating from the scatterer B at position $\boldsymbol{r}_j$ (fig. 2). Partial wave scattering theory [6] states that the scattered wave has the same symmetry as the incoming wave and that it is modified in the following way: It contains a phase factor $e^{2i\delta_1}$ which takes care of the phase shift suffered by the photoelectron within the potential of the absorbing atom. In addition, the factor $kF_j(k)e^{i\beta_j}$ is the complex amplitude for backscattering by the neighbouring atom B. The term e^{i2kr_j} is the geometrical phase shift suffered by the photoelectron with wave number k on is trajectory to the neighbour and back again. Finally, the term $3\cos^2\theta_j$ takes care of the angular dependence of the photoelectron emission in the absorption process. The photoelectron is emitted preferentially along the direction of $\boldsymbol{E}$. θ_j is the angle between the direction of the $\boldsymbol{E}$ field and the line joining absorber and backscatterer. Inserting equations {23} and {24} into equation {8} gives

$$\mu = \mu_0\,|1+B|^2 = \mu_0(1+B+B^*) = \mu_0(1+\chi_j) \tag{25}$$

$$\chi_j(k) = 3\cos^2\theta_j\;F_j(k)\frac{1}{kr_j^2}\sin\left[2kr_j+\Phi_j(k)\right] \tag{26}$$

$$\Phi_j(k) = 2\delta_1(k)+\beta_j(k)+\pi = \Phi_a(k)+\Phi_b(k) \tag{27}$$

where Φ_a and Φ_b are the phases for absorber and backscatterer.

$\chi(k)$ is the normalized oscillatory part in the linear absorption coefficient,

$$\chi = (\mu-\mu_0)/\mu_0 \tag{28}$$

In deriving equation {26} it has been assumed that the spherical photoelectron wave can be approximated by a plane wave. The term $|B|^2$ in equation {25} has been neglected because backscattering is a weak effect, in other words $\mu-\mu_0 << \mu_0$.

The expression equation {26} for the extended X-ray absorption fine structure (EXAFS) $\chi(k)$ has a simple interpretation. The photoelectron emitted in the absorption process propagates as a spherical wave from the absorber atom. If other atoms are located in the vicinity of the absorber, the photoelectron is scattered by the neighbours. The incoming and the scattered waves may interfere, as indicated in figure 2. For constructive interference, the probability to find the photoelectron outside of the absorber is increased compared to the case of an isolated absorber. Hence, $\mu > \mu_0$ in this case. For destructive interference $\mu < \mu_0$.

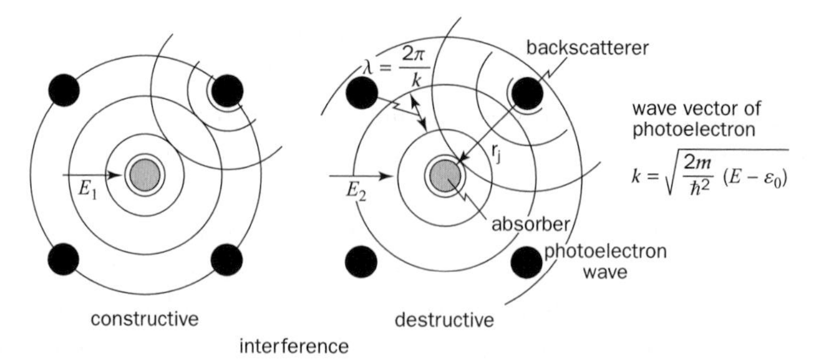

Figure 2 - Interference of outgoing and backscattered photoelectron wave responsible for XAFS oscillation. $E = \hbar\omega$ is the incoming photon energy.

Therefore, χ is expected to vary periodically with k. One period is completed when the ratio of double distance $2r_j$ between absorber and backscatterer and photoelectron wavelength has increased by one. This is the meaning of the expression $\sin 2kr_j$ in equation {26}. The photoelectron suffers energy-dependent phase shifts in the absorber and in the backscatterer. They are summarized in the scattering phase shift $\Phi_j(k)$ in equation {26}. $F_j(k)$ describes the scattering strength of the backscatterer located in r_j. Up to now, we have considered only one backscatterer in the vicinity of the absorber atom. When the absorber is embedded in a crystal a number of modifications have to be applied to equation {26}. When the absorber is surrounded by shells of atoms, the expression equation {26} has to be summed over neighbouring shells.

$$\chi(k) = \sum_j \chi_j(k) \qquad \{29\}$$

In a polycrystalline sample with no preferred orientation equation {29} has to be averaged over the angle θ_j giving

$$3\sum_j \cos^2\theta_j = \sum_j N_j \qquad \{30\}$$

where N_j is the coordination number in the jth shell. In general, the atoms in condensed matter will thermally vibrate around their equilibrium position. Defects may generate static displacements from the ideal lattice position. These displacements reduce the EXAFS amplitude. If the atoms in the coordination shell j are distributed according to a Gaussian with standard deviation σ_j around the average distance r_j, then the EXAFS amplitude is damped by a Debye-Waller factor $e^{-2\sigma_j^2 k^2}$ with

$$\sigma_j^2 = \left\langle [\boldsymbol{r}_j \cdot (\boldsymbol{u}_j - \boldsymbol{u}_0)]^2 \right\rangle \qquad \{31\}$$

the mean-square average of the difference of the displacement $\boldsymbol{u}_j$ of the backscatterer relative to the displacement $\boldsymbol{u}_0$ of the absorber. Note that in X-ray diffraction it is the mean square displacement $\left\langle (\boldsymbol{r}_j \cdot \boldsymbol{u}_j)^2 \right\rangle$ that enters into the Debye-Waller factor.

The question now arises, how many shells will contribute to the EXAFS spectrum. In that regard, it is important to realize that photoabsorption is not an instantaneous effect, as mentioned already above. According to Fermi's golden rule it takes a time $\tau_0 = 1/R_{if}$ for the transition to occur. Let us estimate the average absorption time τ_0 by a classical argument. Since emission and absorption are inverse processes it also takes in the average a time τ_0 for the emission of a photon to occur. We consider an atomic dipole $p(t)$ oscillating with an amplitude a_0/Z and an angular frequency ω_K with $\hbar\omega_K = e_0^2 Z^2/2\, a_0$. This corresponds to an electron bound in the ground state with radius a_0/Z to a nucleus of charge Ze. The power emitted by this classical oscillator is given by

$$\mathrm{P} = \frac{2}{3c^3}\overline{(\ddot{p})^2} = \frac{1}{3c^3}\left(\frac{e_0 a_0}{Z}\right)^2 \omega_K^4 \qquad \{32\}$$

In the average it takes a time ($\alpha = e_0^2/\hbar c$)

$$\tau_0 = \frac{\hbar\omega_K}{P} = \frac{12\hbar}{13.6\mathrm{eV}}\frac{1}{\alpha^3 Z^4} \qquad \{33\}$$

for the emission of a photon to take place. For a Cu 1s electron, τ_0 has a value of 2.10^{-15} s. If no other broadening processes are effective, then the photoelectron has to be described by a wave train of coherence length

$$l_l = \hbar k \tau_0 / m = \sqrt{\frac{E_{kin}}{13.6\ \mathrm{eV}}}\frac{24}{\alpha^3 Z^4} a_0 \qquad \{34\}$$

The present estimation attributes a longitudinal coherence length l_l of 125 Å to a photoelectron emitted from a copper atom with a kinetic energy of 100 eV. In reality, XAFS oscillations are rarely visible for shells further apart from the absorber than 10 Å. Inelastic scattering processes of the photoelectron with other electrons reduces substantially the coherence length. This effect is taken care of by the additional damping factor $D_j\,(k)$ inserted in equation {26}.

$$D_j(k) = \mathrm{e}^{-2r_j/\Lambda} \qquad \{35\}$$

where $\Lambda(k)$ is a mean free path for the photoelectron.

The main merit of equation {34} is not the absolute value of l_l, but the point that there is a longitudinal coherence length of the photoelectron at all. If the absorption process were instantaneous it would not be understandable that the backscattered photoelectron can alter the photoabsorption process.

Adding all factors, the final EXAFS formula for K-shell absorption in polycrystals now reads

$$\chi(k) = \sum_j \frac{N_j F_j(k)}{k r_j^2} e^{-2k^2\sigma_j^2} e^{-2r_j/\Lambda} \sin(2k r_j + \Phi_j) \qquad \{36\}$$

A similar expression is obtained for absorption above L_{III} edges, when $2p^{3/2}$ electrons are excited into the continuum. The main excitation channel for the photoelectrons is from occupied $2p$ states into empty d states. There is also a small probability for excitations from $2p$ into empty s states. Since it is of the order of 2% compared to the p-d transition its contribution is generally neglected [1,2].

The concept of data analysis is now clear. We are interested in the interatomic distances r_j, their spread σ_j and in the coordination numbers N_j for the first shells $j = 1, 2, 3\ldots$ This information can be extracted from the oscillatory part χ of the absorption coefficient, provided the scattering amplitudes $F_j(k)$ and the phases $\Phi_j(k)$ are known.

The phases and amplitudes have been calculated for all elements in the periodic table [9,10]. Some typical calculated values of the backscattering amplitudes $F(k)$ and of the phases $\Phi_a(k)$ and $\Phi_b(k)$ for absorber and backscatterer are shown in the figures 3-5. Figure 3 shows the variation of $F(k)$ for the elements C, Si, Ge, Sn and Pb [9]. Even if these data are not the newest available in the literature they nevertheless show the trend in the behaviour of the amplitudes and phases with the photoelectron energy and with the atomic number Z of absorber and backscatterer.

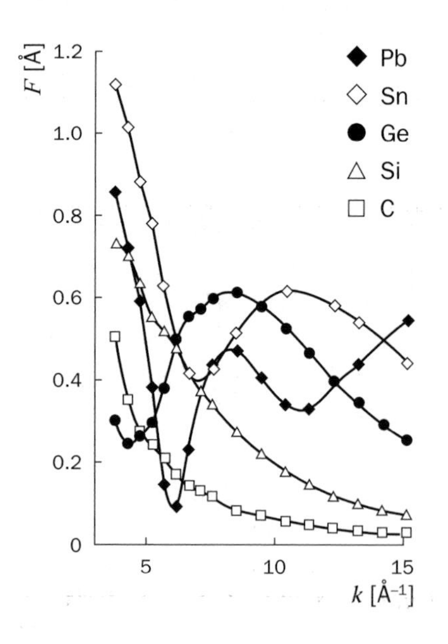

Figure 3 - Backscattering amplitude $F(k)$ for C, Si, Ge, Sn and Pb with Ramsauer-Townsend resonances in the heavy elements [9].

The backscattering strength decreases with increasing photoelectron energy and increases with Z, i.e. with the number of electrons in the backscatterer. However, for the heavy elements one or more dips are superposed to this monotonic behaviour, which result from Ramsauer-Townsend resonances. At certain electron energies the backscatterers are almost transparent. Lead Pb has two resonances between 3 and 15 A^{-1}. The resonances show also up in the phase of the backscatterers (fig. 4). The phases of the absorber, on the other hand, always decrease monotonically with increasing k (fig. 5). In the meantime more sophisticated theories and computer codes for XAFS data analysis have been developed, as will be outlined in the next section.

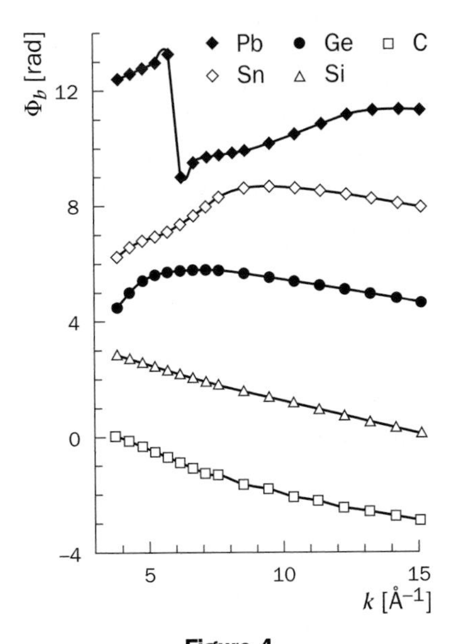

Figure 4
Phase shift $\Phi_b(k)$ suffered by the photoelectron when back-scattered by C, Si, Ge, Sn or Pb [9].

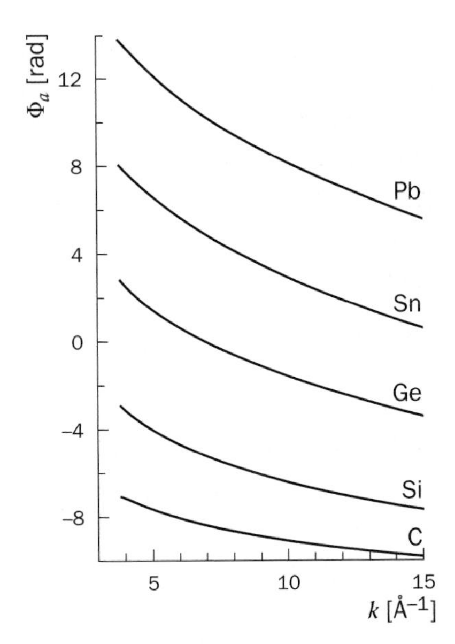

Figure 5
Phase shift $\Phi_a(k)$ suffered by the photoelectron in the absorber atoms C, Si, Ge, Sn or Pb [9].

3.2. CORRECTIONS TO THE BASIC EXAFS FORMULA

A number of assumptions for deriving the standard EXAFS formula {36} have been treated in the previous section. But there are two more simplifying assumptions whose influence will now be assessed. In the last ten years improved expressions for $\chi(k)$ have been developed which take these effects into account.

3.2.1. SPHERICAL PHOTOELECTRON WAVE

In deriving equation {36} it has been assumed that the spherical photoelectron wave impinging on the neighbouring atoms can be replaced by a plane wave. The approximation is justified for more distant neighbours and high values of k, but not for the nearest neighbours and low k values. In this latter regime the effective size of the backscatterer is similar to the interatomic distance, whereas for high k values the photoelectron penetrates deeper into the backscatterer before scattering occurs. Hence, at high k the reduced effective atomic size makes the small-atom approximation more satisfactory. The X-ray absorption for spherical photoelectron waves has been treated in a number of papers. In the end, it has turned out that the EXAFS formula equation {36} can we kept, but with the complex backscattering amplitude $F(k)\ e^{i\Phi_b(k)}$ replaced by an effective scattering amplitude, which depends on k and r_j. Very efficient computer codes have been developed which take, among other things, the curved wave of the photoelectron into account. Some commonly used cotes include EXCURVE [11], FEFF [12,13] and GNXAS [14]. These and others are described in the International XAFS Society catalogue (http://ixs.iit.edu/ISX/catalog/XAFS_Programs).

3.2.2. MULTIPLE SCATTERING

The main contribution to the interference of the photoelectron is its reflection by a neighbouring atom directly back to the absorber. However, a neighbouring atom may reflect the wave towards another neighbour, which then reflects the wave to the original absorber. Besides these double reflections more scatterers may be involved in the backscattering. These multiple scattering processes are neglected in the simple EXAFS formula equation {36}. They may play a non-negligible role and are particularly important in a shadowing configuration for three atoms in a row, as formed, for instance, by an absorber and its first and fourth neighbour in a face-centred cubic lattice (three atoms along the diagonal of a cubic face). The intermediate atom (first neighbour) between the absorber and the backscatterer (fourth neighbour) works as a lens for the photoelectron wave. This results in a strong influence on the modulus and the phase of the backscattered amplitude, which may exceed in magnitude the simple backscattering. The XAFS computer codes mentioned above [11-15] take multiple scattering into account in addition to treating the photoelectron as a spherical wave. These codes have substantially improved the XAFS data analysis eliminating the need for tabulated phase shifts [9,10].

3.3. SET-UP FOR MEASURING X-RAY ABSORPTION

Figure 6 shows a typical set-up for measuring X-ray absorption. Since the spectroscopy needs tunability of the X-rays over about 1 keV the most common source are bending magnets at storage rings. A silicon double crystal monochromator selects a narrow band of 1 to a few eV from the white radiation. For X-rays from 2 to 10 keV a Si(111) cut is adequate. The range between 10 and 20 keV is covered best by a Si(311) cut and above about 20 keV a Si(511) cut is appropriate. It is necessary to reject the harmonics from the beam. This is done by detuning the monochromator crystals in such a way that the broader fundamental reflection transmits the beam whereas the narrower higher harmonics are sufficiently shifted in angle in order to block that energy. This task is alleviated by Si planes $(2n+1, 1, 1)$ since the reflections $(4n+2, 2, 2)$ are forbidden in silicon. For very low photon energies this type of rejection is insufficient and totally reflecting mirrors are added with a glancing angle that transmits the fundamental but rejects the higher harmonics. This is often combined with mirrors that focus the beam at the sample site on a spot of about 1 mm^2. The intensities I_1 and I_2 in front of and behind the sample are measured by ionization chambers or by scattering foils. The sample environment depends strongly on the problem to be studied and on the energy range to be covered in the experiment.

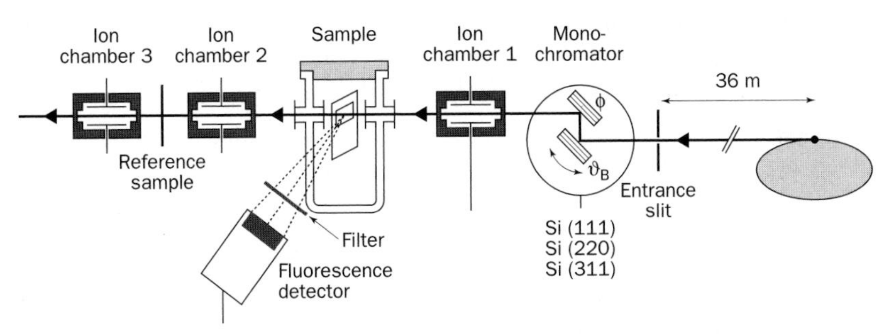

Figure 6 - Typical set-up for measuring hard X-ray absorption in transmission and/or fluorescence. Sometimes the ionization chambers are replaced by scattering foils and photodiodes.

If ever possible XAFS spectra should be measured at low temperature in order to reduce the Debye-Waller damping and to cover a k-range of the data set as large as possible. Liquid nitrogen temperature is often adequate and easy to control technically. Above about 7 keV absorption in air is low enough to let the beam pass through air when bridging the gap between different devices in the set-up. The relatively high penetration depth of X-rays above 7 keV allows investigation of samples in realistic environments like gases or liquids as required in catalytic studies. Below 2 keV the absorption in air and in windows is so strong that the beam has to be confined to an UHV environment.

Equation {1} is the basis for the measurement of χ by a transmission experiment. This is not the only technique for determining $\chi(k)$. In dilute systems and at low photon energies it may be favourable to measure $\chi(k)$ by a fluorescence technique [8]. When a core-hole has been created by photoabsorption the atom is left in an excited state. The core-hole will be filled by an electron from an upper shell. This process is accompanied by the emission of a fluorescence photon or by the emission of Auger electrons. Fluorescence dominates for deep core-holes and Auger processes are favoured at low photon energies. The radiation probability (fluorescence yield ε_f) is monotonically increasing with Z. For iron ($Z = 26$) $\varepsilon_f = 0.30$ and for molybdenum ($Z = 42$) $\varepsilon_f = 0.75$. When the incident radiation of energy E and the fluorescent radiation of energy E_f make an angle of 45° with the sample normal, the fluorescence counting rate is given by [8]

$$I_f = I_0 \frac{\varepsilon_f (\Omega/4\pi)\, \mu_x(E)}{\mu_t(E)+\mu_t(E_f)} \left(1 - e^{[-\mu_t(E)-\mu_t(E_f)]d}\right) \qquad \{37\}$$

Ω is the solid angle subtended by the detector. μ_x is the absorption coefficient for the species of interest and μ_t is the total absorption coefficient (signal plus background). For a thin, concentrated sample

$$\frac{I_f}{I_0} = \varepsilon_f (\Omega/4\pi)\, \mu_x(E) \quad \text{for} \quad \left[\mu_t(E) - \mu_t(E_f)\right] d << 1 \qquad \{38\}$$

For a thick, dilute sample

$$\frac{I_f}{I_0} = \varepsilon_f (\Omega/4\pi)\, \mu_x(E) \left[\mu_t(E) + \mu_t(E_f)\right]^{-1} \quad \text{for} \quad \left[\mu_t(E) - \mu_t(E_f)\right] d >> 1 \qquad \{39\}$$

In both cases I_f/I_0 is basically linearly dependent on the quantity $\mu_x(E)$ of interest. However, for a thick, concentrated sample the EXAFS oscillations may be strongly suppressed by self-absorption. The effectively illuminated sample volume and hence the total number of fluorescent atoms varies with energy, being larger for a minimum of μ and smaller for a maximum of μ. This effect has to be taken care of in any meaningful data analysis.

It is advisable to measure the fluorescence signal by a detector whose energy resolution separates the fluorescence signal from the elastically scattered incoming intensity. Multi-element Ge detectors are used very often for that purpose. When the incoming beam is linearly polarized the detector should be located in the direction of the ***E*** vector. This is the direction of minimum scattered intensity. It helps to improve the signal to background ratio.

Since the Auger electrons originate within about 3 nm of the surface, X-ray absorption spectroscopy in Auger mode is strongly surface sensitive. Therefore it is a common technique used for surface EXAFS (SEXAFS), when the structure of adsorbates on a surface is of interest. See, e.g., the chapter 10 by J. Stöhr in reference **[2]**.

A final word should be added about sample preparation. Homogeneous samples are a prerequisite for a good data set. Pinholes, large grains giving rise to strong Bragg reflections and a not optimized sample thickness are the source of systematic errors which affect in particular the determination of the coordination number. No data analysis can recover what has been lost by a careless sample preparation.

3.4. XAFS DATA ANALYSIS

In this section, we will discuss at some length an EXAFS data analysis. The experimentally determined intensities I_1 and I_2 are related by

$$\ln(I_1/I_2) = \mu d + f(\mu_1 d_1, \mu_2 d_2, \ldots) \qquad \{40\}$$

The correction term f describes a smooth background due to the energy dependence of the detection in the chambers 1 and 2. It can be determined experimentally by measuring the absorption with and without sample. Subtraction of the curves is the first step in data analysis (fig. 7). Then, the zero of kinetic energy, ε_0, has to be chosen. ε_0 is a poorly defined quantity. In general it is not the Fermi level. A common procedure is to choose ε_0 as the energy at the first inflection point in the edge and to treat it as a free parameter in the subsequent data analysis. Note that the background decreases with energy. According to Heitler **[16]** it varies in good approximation like

$$\mu_0 d = J(1-\beta k^2) \quad \text{with } \beta = 8\hbar^2 / 6m\varepsilon_0 = 0.746\text{Å}^2(13.6\text{ eV}/\varepsilon_0) \qquad \{41\}$$

where J is the height of the jump at the edge. Hence

$$\chi = \frac{(\mu - \mu_0)d}{J(1-\beta k^2)} \qquad \{42\}$$

The quantity $(\mu - \mu_0)d$ is extracted from the data by a set of cubic splines. As a rule of thumb the number of splines should be chosen as the length of the data set divided by 3 or 4 Å^{-1}.

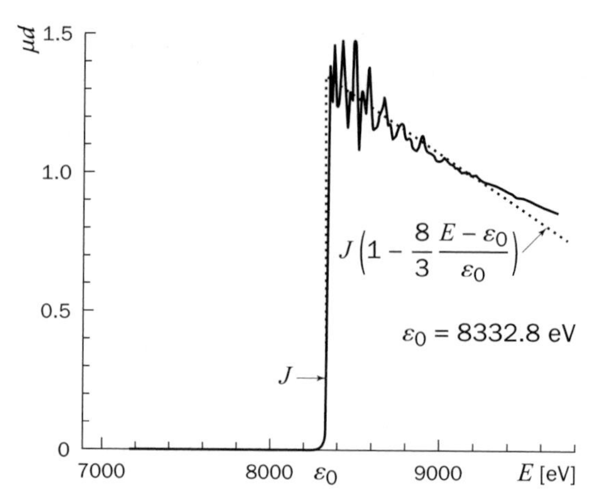

Figure 7 - X-ray absorption spectrum of metallic Ni at 77 K with background removed, showing the jump J at the edge and $J(1-\beta k^2)$ according to equation {41} with $8(E-\varepsilon_0)/3\,\varepsilon_0 = \beta k^2$.

This allows for a good separation of the background, without the splines following the XAFS oscillations. Finally the energy axis is converted into a k-axis according to equation {9b}, and the data are multiplied by k^n with n typically 1,2 or 3. Large values of n suppress the low k values, where multiple scattering plays a dominant role. For low Z backscatterers n should be small, about 1, since $F(k)$ decreases rapidly. In order to separate the different shells in the subsequent Fourier transformation it is advisable to have a signal of almost equal amplitude in the whole k range. Figure 8 shows the XAFS χk^2 for metallic nickel measured at 77 K.

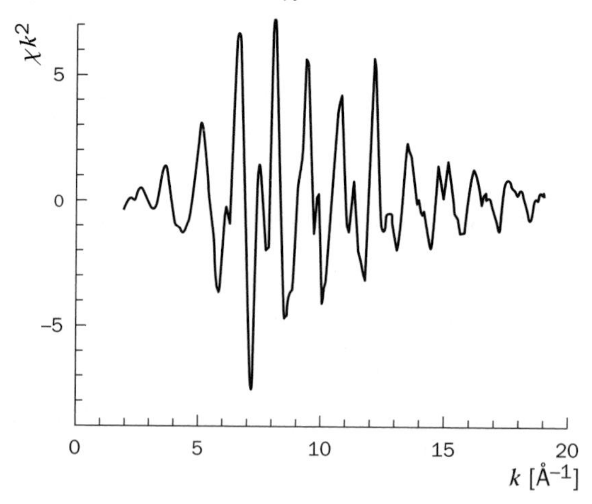

Figure 8 - XAFS χk^2 for metallic Ni measured at 77 K.

The next step in the data analysis is a Fourier transformation of χk^n, as illustrated in figure 9. Since the data set χk^3 is effectively longer than the set χk^1 (in the sense that the signal χk is basically vanishing for large values of k), the different shells in the Fourier transformation are separated much better for n = 3 than for n = 1. The contribution of the nearest neighbour shell may be separated by a window function. A common window increases from 0 to 1 like a cosine (fig. 9).

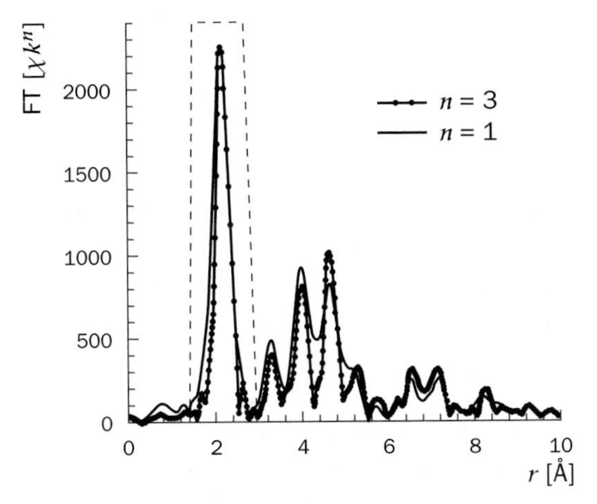

Figure 9 - Fourier transformation FT(χk^n) with n = 3 and n = 1 of the XAFS for metallic Ni measured at 77 K, with the window function used for separating the first neighbour contribution.

The envelope and the phase of the back-transformed first neighbour contribution are shown in the figures 10 and 11.

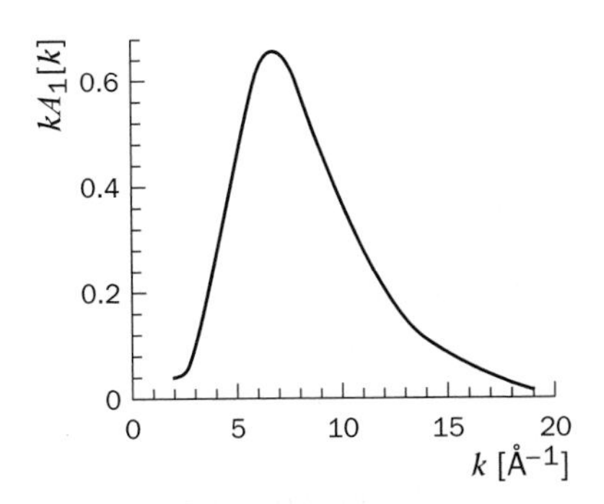

Figure 10 - Envelope $kA_1(k)$ from the first neighbour shell in metallic Ni measured at 77 K. A_1 is the prefactor in $\chi_1 = A_1 \sin(2kr_1 + \Phi_1)$.

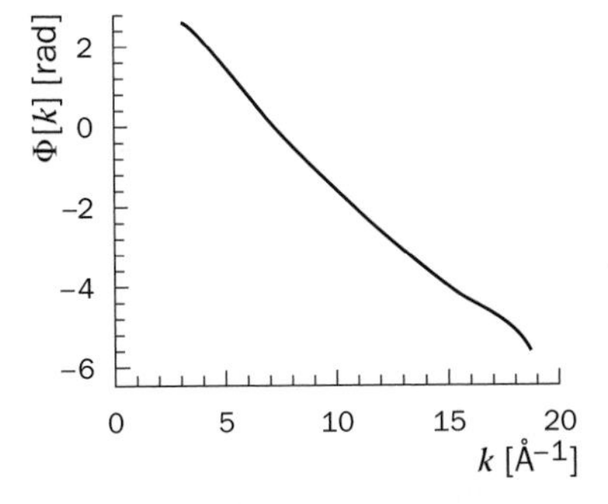

Figure 11 - Phase $\Phi_1(k)$ from the first neighbour shell in metallic Ni measured at 77 K.

Besides the use of sophisticated computer codes for data analysis there are cases where model compounds can be very favourable. This simple procedure works very well when the model compound is close in structure and chemical state to the system to be analyzed, thus reducing substantially systematic errors in the data analysis. If, for instance, the lattice distortion around a Mn atom in a dilute **Ni**Mn alloy should be investigated by XAFS then the ordered intermetallic compound $MnNi_3$ is an excellent model. Here and in the following the matrix of the dilute alloy is indicated by bold letters. In $MnNi_3$ every Mn atom is surrounded by 12 Ni neighbours in a well-known distance. The phases and amplitudes extracted from the Mn-XAFS in $MnNi_3$ are transferred to the alloy **Ni**Mn. The transferability of the amplitudes and phases is well established, in particular, for heavy backscatterers [15]. This procedure will be used in an example of section 4. Here, we will illustrate the procedure by analyzing the data of copper measured at room temperature with the Cu data measured at 77 K. The XAFS $\chi(k)$ may be written as

$$\chi = \frac{N}{N_M}\frac{r_M^2}{r^2}\,e^{-2(\sigma^2-\sigma_M^2)k^2}A_M(k)\sin(2kr+\Phi_M) \qquad \{43\}$$

The model distance r_M and coordination number N_M are known. The model amplitudes A_M and phases Φ_M are extracted from the measured model data. By means of a non-linear least square fit the unknown distance r_1, coordination number N_1 and Debye-Waller damping $\sigma_1^2 - \sigma_M^2$ is extracted from $\chi(k)$. As a fourth fitting parameter per shell we use a shift $\delta\varepsilon_0$ in ε_0. Figure 12 shows the fit of the first shell of Cu at room temperature with the Cu data measured at 77 K. Table 1 gives the result of the fit.

Table 1 - First shell of Cu at room temperature fitted with Cu data measured at 77 K.

	r_1	N_1	$\delta\sigma_1^2$	$\delta\varepsilon_0$
Fit	2.546 Å	12.06	0.00442 Å^2	−1.48 eV
to be expected	2.556 Å	12		

In the present case the model is almost ideal and the errors in r_1 and N_1 are small. Typical values are 0.01 to 0.02 Å for the first neighbour distance and 5 to 10% for N_1 and $\delta\sigma_1^2$. The quantities N_1 and $\delta\sigma_1^2$ are strongly correlated. This is one of the major reasons for the large errors in these quantities. On the other hand, r_1 and $\delta\varepsilon_0$ are not strongly correlated, so that the error in the poorly defined zero of energy ε_0 has only little influence on the accuracy of r_1. This behaviour is due to the fact that, according to the phase factor 2 kr_1, in equation {36}, r_1 is mainly determined by the large values of k and these are only weakly dependent on ε_0 and hence, on its error $\delta\varepsilon_0$.

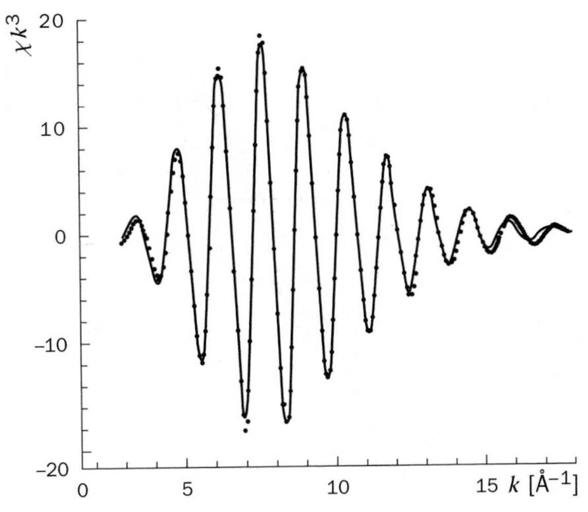

Figure 12 - Fit of the first shell Cu data at room temperature with the first shell Cu data measured at 77 K.

It is not always possible to separate the different shells by a window function. For instance, in a bcc metal the first and second neighbour are located at $a\sqrt{3}/2$ and at a. The difference in only 13% and the corresponding shells merge into one. In such a case a two-shell fit has to be used. Now the question arises how many free parameter may be determined in a significant way by the fit. The answer is given by the ratio of the window width Δr_F and the resolution in Fourier space $(k_{max} - k_{min})^{-1}$

$$N_{free} \leq \frac{2}{\pi} \Delta r_F (k_{max} - k_{min}) \qquad \{44\}$$

For the data in figure 12, $\Delta r_F = 0.8$ Å and $k_{max} - k_{min} = 15$ Å^{-1}, so that $N_{free} \leq 8$. Since not every good-looking fit is a significant fit, equation {44} is an important criterion in data analysis.

3.5. *POSITION AND STRUCTURE OF THE ABSORPTION EDGE*

Figure 13 shows the K edge of Cu in three Cu oxides (Cu_2O, CuO and $KCuO_2$), in metallic copper and in a high T_c superconductor $YBa_2Cu_3O_{6.97}$. There are distinct differences in the form and in the position of the edges. It turns out that the position of the edge is shifted to higher energies with increasing formal valence of the absorber. Note that $\tilde{\mu}$ is the absorption coefficient normalized to an edge jump of 1 (for above relative to far below the edge).

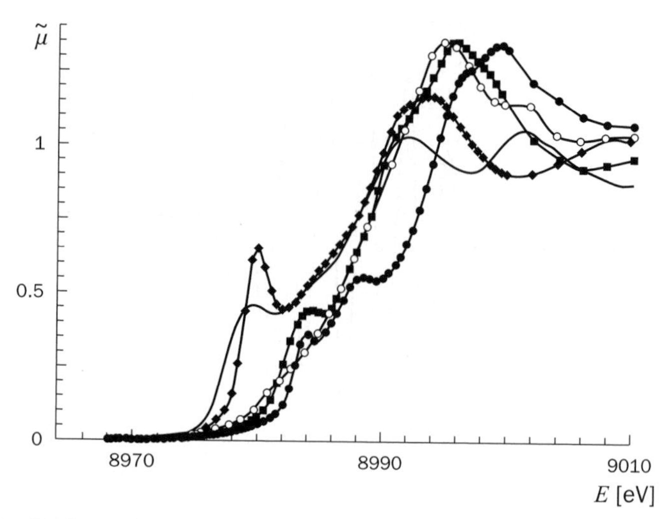

Figure 13 - *K* edges of Cu in copper metal (–) and in formally monovalent Cu_2O (◆), divalent CuO (■), trivalent $KCuO_2$ (●) and in $YBa_2Cu_3O_{6.97}$(○).

This is confirmed by the figures 14 and 15 showing the *K* edge of selenium in metallic selenium, in formally tetravalent Li_2SeO_3 and in hexavalent Li_2SeO_4 and the *K* edge for arsenic in metallic arsenic, formally trivalent As_2O_3 and pentavalent As_2O_5 and in $KAsF_6$. The edge shift is most pronounced in the most electronegative ligands like oxygen and fluorine. Fluorine being more electronegative than oxygen, the shift is larger in $KAsF_6$ than in As_2O_5. A qualitative argument for the shift of the absorption edges is as follows. With increasing formal valence and increasing electronegativity of the ligands more electronic charge is removed from the cell of the metallic ion.

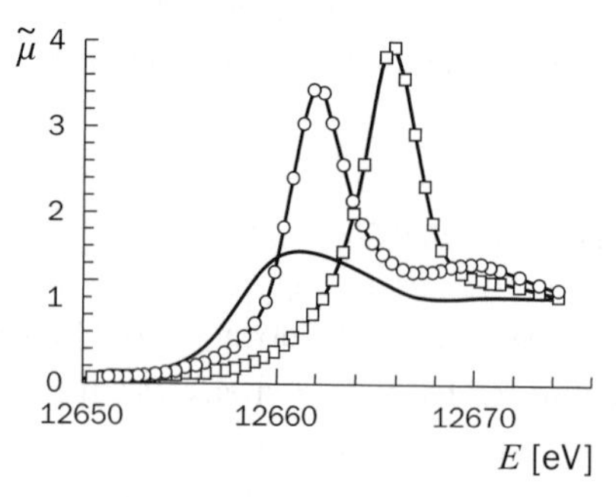

Figure 14 – *K* edges of Se in metallic Se (—), formally tetravalent Li_2SeO_3 (○) and hexavalent Li_2SeO_4 (◇)

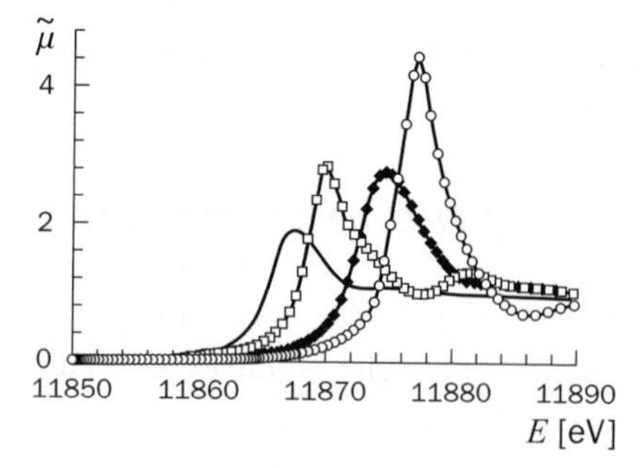

Figure 15 - *K* edges of As in metallic As (—), formally trivalent As_2O_3 (□), pentavalent As_2O_5 (◆) and $KAsF_6$ (○)

Hence its nucleus is less well screened, resulting in an increased binding energy of the deepest core levels. The shift can be as large as 13 eV. The simple, well established relation between formal valency and edge position for metallic ions surrounded by electronegative ligands can be used to determine the unknown valence of an atom in a compound. Figure 16 shows an example. Co is trivalent in the compound $YBa_2Cu_{2.5}Co_{0.5}O_{7.4}$ (which is no longer a superconductor) as is obvious by comparison with divalent CoO and trivalent $LiCoO_2$. More examples will be treated below.

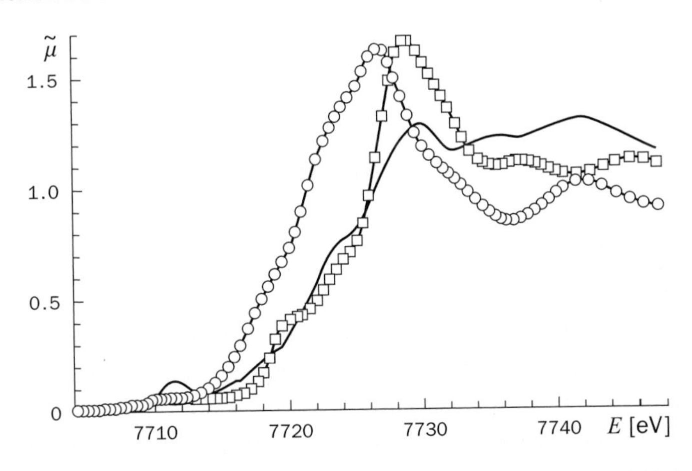

Figure 16 - By comparison with divalent CoO(○) and trivalent $LiCoO_2$(□). Co in $YBa_2Cu_{2.5}Co_{0.5}O_{7.4}$ (—) is found to be trivalent.

We consider now the form of the absorption edges. It reflects the empty electronic density of states and is strongly dependent on the type of coordination like linear, tetrahedral or octahedral coordination. In figure 17 is shown an energy diagram for the absorption process. Only those photons can be absorbed by, say, a 1*s* electron whose energy is larger than their binding energy. Due to the conservation of angular momentum only those final states are allowed which have *p* character about the absorber atom. Before absorption the 1*s* electron has no angular momentum, whereas the photon carries (in the dipole approximation) the momentum $\hbar$. After absorption the photon has disappeared, so that the electron has to carry the angular momentum $\hbar$. Therefore, the difference Δl in electron angular momentum has to be

$$|\Delta l| = 1 \qquad \{45\}$$

When the photons are absorbed by *p* electrons (e.g., L_{II} and L_{III} edges) only electron transitions into empty *s* and *d* states are possible. However, calculations

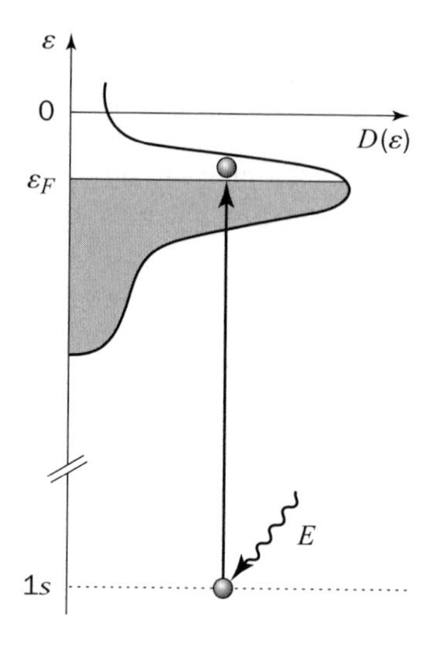

have shown that the cross-section for transition into empty d states is about 50 times larger than that into empty s states [9]. According to reference **[17]** the near-edge structure $\mu(E)$ can be written as a product

$$\mu(E) = M(E)\,\rho_l(E) \qquad \{46\}$$

Figure 17 - Energy diagram for the absorption of a photon with energy E by a $1s$ electron.

The matrix element $M(E)$ gives the probability for a transition from an occupied, say, $1s$ state into an empty p state. The projected (l-dependent) density of states $\rho_l(E)$ gives the probability to find a final state of appropriate symmetry. Band structure calculations have shown that $M(E)$ is only weakly energy dependent. Thus the structure observed in the absorption edges must reflect the structure in the projected densities of states $\rho_l(E)$. There are a number of calculations of $\rho_l(E)$ for elements and for stoichiometric compounds that confirm this conclusion [18-20]. An example shows that this is also true for dilute alloys [20]. Since the X-ray absorption is specific to individual species, it is possible to probe the projected densities of states locally around the impurity in a dilute alloy. Figure 18 shows the Mn K edge in a dilute NiMn 2 at. % alloy. A self-consistent KKR band structure calculation with exchange and correlation in the local density approximation has been performed for a Mn atom in a Ni matrix [20]. The p density of states in the Mn cell shows strong variations in the energy range considered (up to 40 eV above the Fermi level). The matrix element $M(E)$ is only weakly energy dependent. The calculated quantity $M\rho_l$ has to be convoluted with a Lorentzian that takes into account the finite lifetime of the core-hole, the finite lifetime of the end states and the energy resolution of the spectrometer. It is obvious that this broadening smears out many details of the calculation (fig. 19). The agreement between calculated and measured data is satisfactory. The main results are the following:

- The structure in the absorption edges reflects the structure in the density of states.
- The measurement of absorption edges is a simple and stringent test of the quality of band structure calculations.

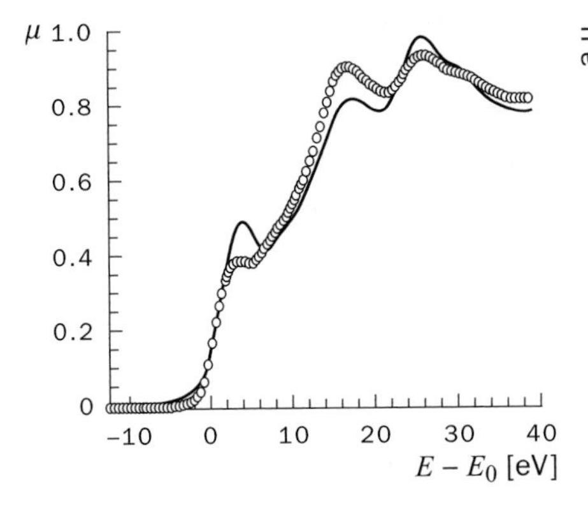

Figure 18 - Mn *K* edge in **Ni**Mn 2 at. % (○) and result of a band structure calculations (—)

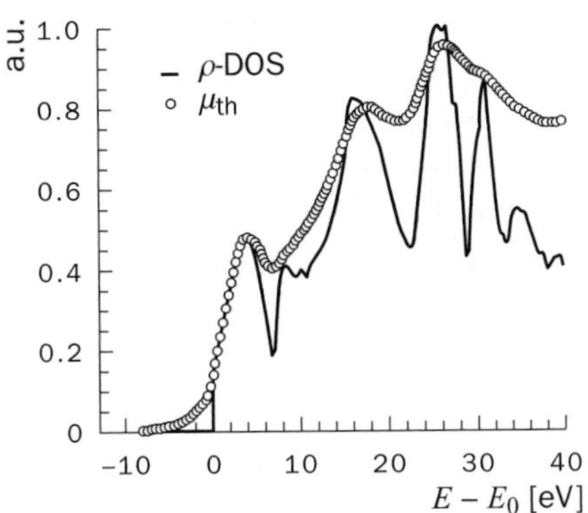

Figure 19 - Effect of energy smearing on the calculated (—) local empty DOS $\rho_1(E)$ for the *K* edge of Mn in **Ni**Mn 2 at. %

Considerable effort has been invested in simulating the near-edge structure (XANES) in the X-ray absorption for systems lacking long-range order so that standard band structure calculations cannot be applied. However, improved scattering potentials and consideration of many-body effects are needed before XANES can deploy its full power for the investigation of the chemical and electronic structure of absorbing species [35,36] and chapter by P. Sainctavit, V. Briois and D. Cabaret in this volume.

4. SOME APPLICATIONS OF XAS

XAS has found wide application in many areas of materials science, structural chemistry, catalysis and biology. Experience has shown that it is not trivial to obtain a unique answer to a structural problem by XAS alone and that it is highly advisable to combine the technique with other structural tools like X-ray and neutron diffraction, electron microscopy and others. Whenever the absorbing species is contained in the sample in different configurations with different local environments, then it is difficult to separate the contribution from the different configurations. Scattering technique and nuclear probes may help in some cases to disentangle the contributions. We will now give a subjective choice of examples, which should illustrate the possibilities, and the drawbacks of XAS.

4.1. LATTICE DISTORTION AROUND IMPURITIES IN DILUTE ALLOYS [21]

Technically important properties of metals and semiconductors are largely influenced by structural and chemical defects. Among those are impurity atoms which expand or compress the host lattice and change the average lattice parameter, measured e.g. by X-ray diffraction. However, the local distortion near the impurity, which is responsible for a number of changes in the mechanical and electronic properties of the alloy, cannot be deduced from lattice parameter measurement. It is accessible to XAFS. As an example, we consider an alloy with 2 at. % of Mn homogeneously dissolved in Ni. The absorption and the XAFS χk^2 were measured in transmission.

Figure 20 shows the Fourier transformation and the window for extracting the first shell. The data were analyzed with $MnNi_3$ as model compound.

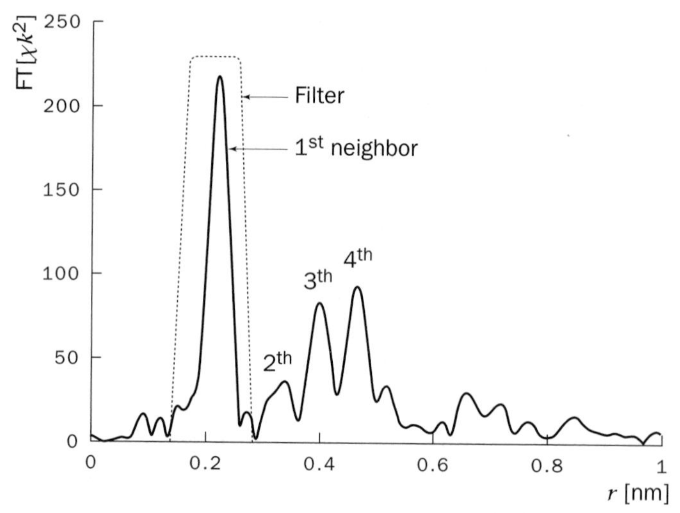

Figure 20 - Fourier transform of χk^2 at the Mn *K* edge in **Ni**Mn 2 at. % measured at 77 K. The window used to separate the first shell contribution around Mn impurities is shown.

The fit of the data (fig. 21) reveals that the substitutional Mn atom shifts its 12 nearest Ni neighbours by + (0.023 ±0.004) Å. A systematic analysis has been performed for impurities in Al, Ni, Cu, Pd, Ag, V, Fe and Nb. The shift of neighbour atoms is caused by defect-induced changes of the charge density, illustrating the connection between electronic and atomic structure.

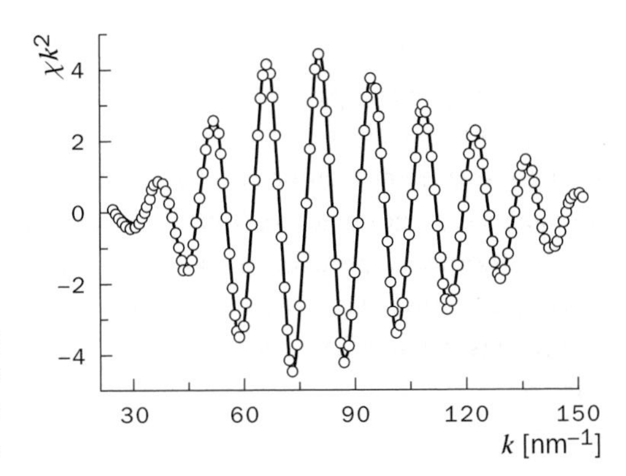

Figure 21 - Fit of the first shell contribution from figure 20 with phases and amplitudes transferred from $MnNi_3$.

Figure 22 shows a comparison of measured and calculated lattice distortions of 3*d* row elements in Cu [22]. There are two main effects that determine the lattice distortion. The first is the valence difference between impurity and host and the corresponding change of charge density in the impurity cell. It gives rise to a parabolic dependence of Δr_1 with atomic number *Z*. The second one is a magneto-elastic contribution. For Cr, Mn and Fe in copper, the majority and minority bands are split giving rise to strong magnetic moments and low density of states at the Fermi level. However, a low DOS at the Fermi level reduces the binding and hence induces a lattice expansion.

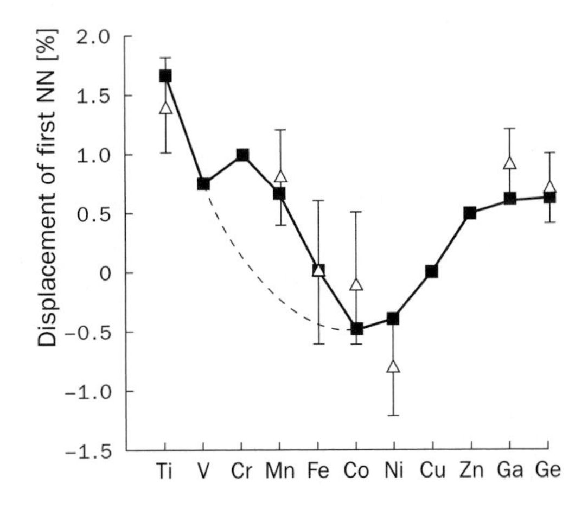

Figure 22 - Comparison of calculated (■) and measured (△) lattice distortions around 3*d* row impurities in copper (NN: nearest neighbours) [21,22].

4.2. *PRECIPITATION IN IMMISCIBLE GIANT MAGNETORESISTANCE $Ag_{1-x}Ni_x$ ALLOYS*

Magnetic thin films are technically of great importance because they can show the giant magnetoresistance (GMR) effect which forms the basis of sensors for reading magnetic bits in data storage discs [23]. However, the deposition of magnetic and non-magnetic layers is a delicate and costly process. Therefore, the observation of the GMR effect in alloys produced by simple co-deposition of two immiscible constituents, a magnetic transition metal, like Ni, and a simple metal, like Ag, and subsequent annealing has found great interest as a possible alternative to the expensive layer deposition. In that regard it is necessary to know which types of precipitants form upon annealing from the 2 immiscible components Ag and Ni. X-ray absorption spectroscopy has been used for that purpose [24,25]. The authors have used alloys $Ag_{1-x}Ni_x$ with x = 0.35 and 0.37. Figure 23 shows the Fourier transforms for the EXAFS at the Ni *K* edge with the annealing temperature as parameter.

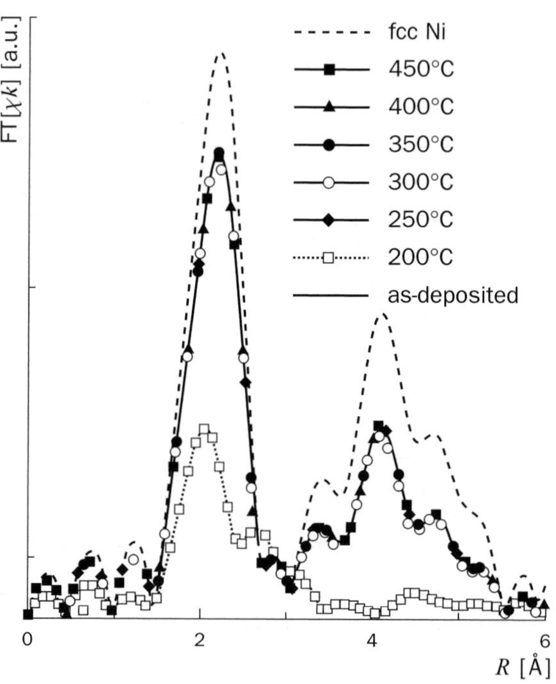

Figure 23 - Fourier transforms of the EXAFS χk at the Ni *K* edge in $Ni_{0.37}\,Ag_{0.63}$ and in pure nickel after annealing at the indicated temperatures.

The EXAFS results can be summarized as follows:

- The main effect of annealing on the local structure occurs between 200 and 250°C.
- A Ni atom has Ni neighbour at 2.48 Å (as in pure nickel) and Ag neighbour at 2.75 Å. The number of Ni-Ni neighbours increases from 6.7 in the as-deposited alloy to 8.6 at 250°C.
- The alloy is a heterogeneous mixture of almost pure nickel precipitates in an Ag matrix with some dissolved nickel.

The size of the Ni precipitates increases from 1.1 nm in the as-deposited alloy, to 2.5 nm at 250°C and about 5 nm at 400°C. As in many complex systems it is not possible to solve the structural problem in an unambiguous way only on the basis of X-ray absorption data. Therefore, the authors have performed additional investigations:

- X-ray diffraction: from the position, the intensity and the width of the Bragg peaks the authors have deduced the crystallographic structure, the size and the orientation of the different phases.
- Small angle X-ray scattering gives the size and the volume fraction of the nickel precipitates.

The different pieces of evidence from all 3 techniques have led the authors to the following model for the $Ag_{0.65}Ni_{0.35}$ alloy:

- The alloy is composed of 2 phases: pure nickel and a supersaturated solution of silver with nickel. The nickel concentration in the solid solution decreases from 7-8 at. % in the as-deposited alloy to almost pure Ag after annealing at 400°C.
- The size of the Ni precipitates grows from about 1 nm in the as-deposited state to 3-5 nm after annealing at 400°C. The size of the Ag rich phase stays at 4-5 nm.
- Both types of crystallites, pure Ni and the Ag rich phase, are preferentially oriented with the dense (111) planes parallel to the surface.
- Although there are significant changes in the structure occurring between room temperature and 250°C, almost no changes were observed in the magneto-resistance in that annealing range. On the other hand, in the temperature range from 250°C to 400°C little changes are seen in the structure, but the magnetoresistance changes substantially [25].

Two conclusions are drawn from this example:

- None of the 3 structural techniques alone (X-ray absorption, diffraction and small angle scattering) is able to give an unambiguous structural model. This can at best be achieved by a combination of all 3 techniques.
- It is certainly necessary to know the geometric structure of the alloy in order to understand the electronic and magnetic transport properties in an alloy. However, the knowledge of the geometric structure is not always sufficient for explaining other relevant properties, as the transport behaviour in the present case.

4.3. LATTICE SITE LOCATION OF VERY LIGHT ELEMENTS IN METALS BY XAFS

The backscattering strength of hydrogen with only one electron is too weak to be observable in XFAS. Therefore, hydrogen shows up in XAS mainly by lattice distortion. This effect is not specific enough to determine the lattice site occupied by the hydrogen. Nevertheless, XAFS can be used in many cases to determine the lattice site of hydrogen and other light atoms [26]. When the hydrogen is located on the line joining the absorber and the backscatterer, then it changes the phases and the amplitude of the photoelectron on its way to the backscatterer and back, i.e. the hydrogen works as a lens for the photoelectron. This effect mentioned already above in connection with multiple scattering is demonstrated in figure 24 for Ni and $NiH_{0.85}$. In both systems Ni forms a f.c.c. lattice, in which the hydrogen occupies octahedral sites. Those are the sites between an absorber and its second neighbour. The hydrogen acting as a lens increases the second-neighbour amplitude by 50%. The reason for the hydrogen acting as a lens but not as a mirror for the photoelectron is found in the angular dependence of the scattering strength. The forward amplitude is about 10 times larger than the backward amplitude.

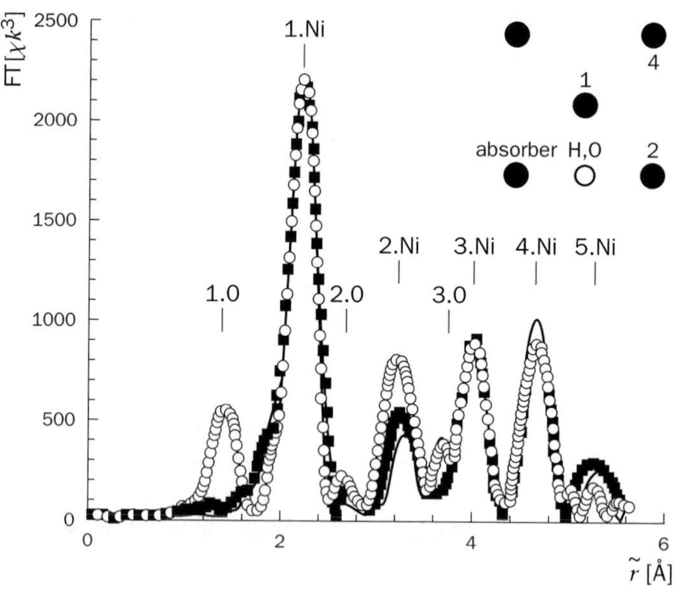

Figure 24 - Fourier transforms of the EXAFS χk^3 of Ni (—), $NiH_{0.85}$ (■) and NiO (○). The abscissae for the hydride and the oxide have been rescaled in order to eliminate the influence of the lattice expansion. The insert shows the lattice position of H and O, which shadow a second Ni neighbour.

The lens effect is not limited to hydrogen. But hydrogen shows only the lens effect. When the hydrogen is replaced by a heavier element like oxygen in NiO, which also has the NaCl structure, then the second-neighbour contribution is enhanced (fig. 24). The oxygen shows up as a backscatterer as well.

This effect will now be used in order to study the interaction of hydrogen with yttrium impurity atoms in palladium [27]. Yttrium in Pd generates a strong lattice expansion. Y pushes the first shell of 12 Pd neighbours outwards by 0.06 Å. When hydrogen is dissolved in Pd, the question arises whether the hydrogen prefers the vicinity of the Y atoms where the lattice has already been expanded, i.e. whether or not the Y traps the hydrogen. Figure 25 shows the Fourier transform of the EXAFS at the Y *K* edge in **Pd**Y 2 at. % and in **Pd**Y 2% H 80%. Pd having a f.c.c. lattice, it is again the second-neighbour shell that is of interest here. It is obvious that in the hydrogen-loaded sample the second shell is not enhanced by the hydrogen. This leads to the conclusion that Y does not trap the hydrogen. In view of the large hydrogen concentration one must even conclude that the Y expels the hydrogen from its neighbourhood. If the hydrogen were distributed statistically, 4.8 out of 6 nearest octahedral sites would be occupied with hydrogen for an alloy with 80% hydrogen. This is not the case, as shown in figure 25. In other words, the hydrogen avoids the vicinity of the Y atoms. This conclusion is supported by an analysis of the lattice distortion around Y.

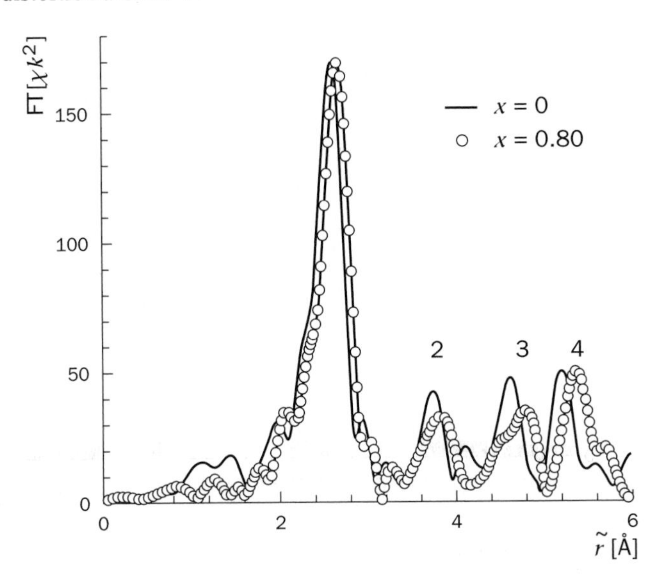

Figure 25 - Fourier transforms of the EXAFS χk^2 at the Y *K* edge in **Pd**Y 2 at. % (—) and in **Pd**$YH_{0.80}$ (○).

Figure 26 shows the interatomic distances r_1 and r_3 of the first and third shells around Y in Pd. Y expands the Pd lattice. The hydrogen dissolved in pure Pd also expands the lattice (full line in fig. 26). The addition of hydrogen to the dilute alloy obviously expands the lattice locally less than what corresponds to the addition of both effects. If the hydrogen were trapped by Y, the points at 73 and 80% H should be above the dashed curves. That they are below this line implies that the hydrogen is expelled from the vicinity of Y. It also implies that the high hydrogen concentration outside the Y vicinity acts as an external pressure shifting the Pd atoms in the nearest shells towards the Y. The same observation was made for a great number of other oversized impurities in Pd [27]. The lens effect has turned out to be very helpful in the lattice site location of other light elements like boron in palladium [28].

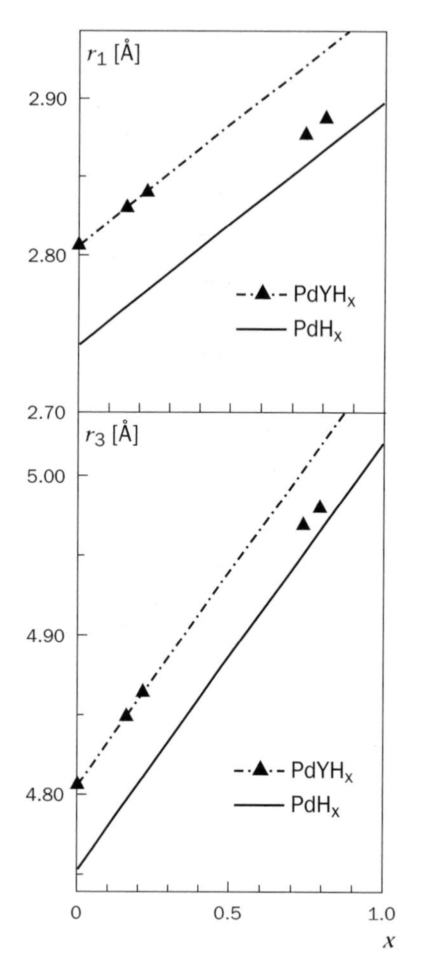

Figure 26 - Interatomic distances r_1 and r_3 of the first and third neighbours in Pd and PdY 2 at. % as a function of the hydrogen concentration x. In the concentrated alloy r_1 and r_3 are smaller than the value expected from the sum of the local lattice expansion of the impurity and the overall lattice expansion due to the hydrogen (broken line). This is indicative of a repulsive interaction between the hydrogen and Y.

4.4. *VALENCE OF IRIDIUM IN ANODICALLY OXIDIZED IRIDIUM FILMS*

Iridium oxide films have attracted some attention due to their high electrocatalytic activity for chlorine and oxygen evolution. In addition, they are chemically stable and may be used as pH sensors. Finally, they are electrochromic with a colour change from transparent to black. XAS at the Ir L_I (13,424 eV) and Ir L_{III} (11,212 eV) edges is well suited to study *in situ* the reactions occurring in iridium during a cycle in the voltammogram [29]. Using Ir oxalate and Ir dioxide as models for tri- and tetravalent Ir, the change in valence has been determined as a function of the potential (fig. 27).

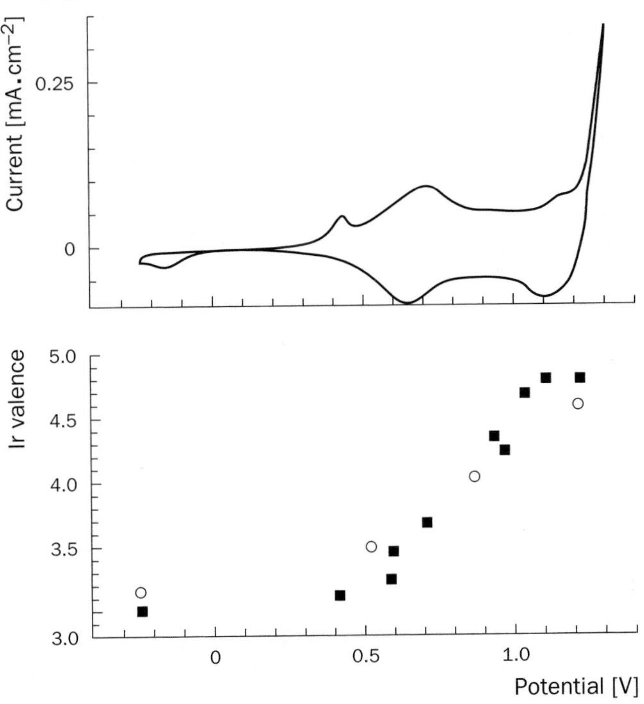

Figure 27 - Ir valence of anodically oxidized iridium films versus potential in the chemical cell. The squares were deduced from the L_{III} edge, the circles from the L_I edge.

The valence changes from 3 to 4.8 between –0.2 and +1.2 volt. At the transition from transparent to black (+0.7 volt) the valence exceeds the value 3.8. At that transition the nearest neighbour shell of oxygen ligands starts to shrink and at the same time there occurs a strong increase in static disorder (fig. 28). This is attributed to a change in coordination from $Ir(OH)_3$ in the transparent phase to $IrO(OH)_2$ and $IrO_2(OH)$ in the black phase as a consequence of charge exchange during a voltammogram cycle as has been proposed by Kötz *et al.* [30]. Figure 29 shows the changes in the structure during a cycle. Static disorder in the first shell is due to the difference in Ir-O and Ir-OH distances, as not uncommon in oxyhydroxides [31].

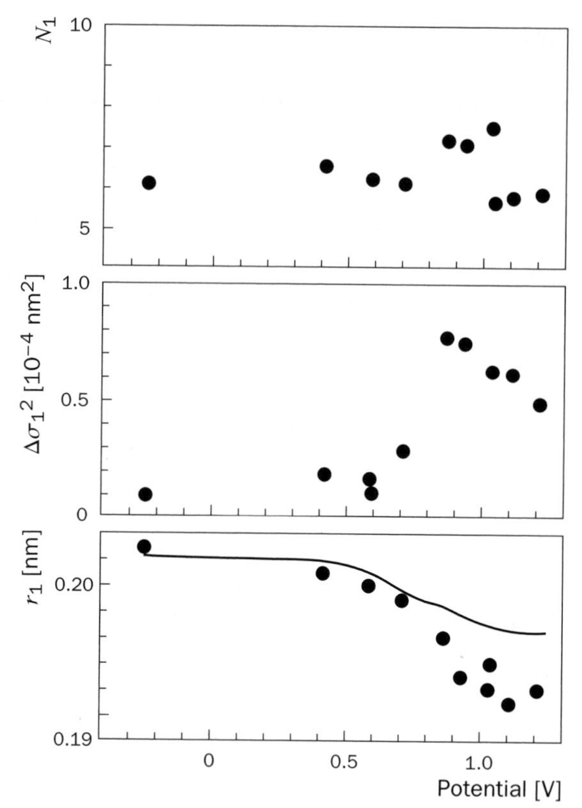

Figure 28 - Interatomic distance r_1, coordination number N_1 and Debye-Waller factor $\Delta\sigma_1^2$ relative to IrO_2 for the first shell around Ir in an oxidized Ir film versus voltage in the electrochemical cell. IrO_2 was used as a model for the backscattering amplitude and the scattering phases.

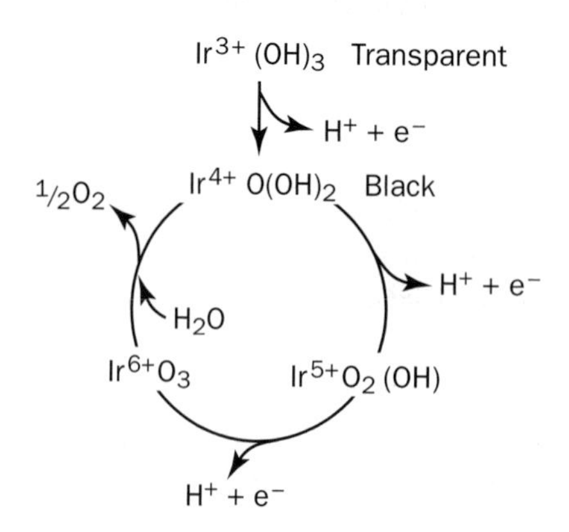

Hexavalent IrO_3 is not visible in XAS because it reacts immediately with water under oxygen evolution. As a summary, XAS is in agreement with the Kötz model developed on the basis of UPS data.

Figure 29 - Changes in the structure of iridium oxide films during the cycle of a voltammogram (after Kötz *et al.* [30]).

4.5. COPPER-BASED METHANOL SYNTHESIS CATALYST

Cu/ZnO catalysts are used on an industrial basis for the synthesis of methanol according to the reaction

$$CO_2 + 3\,H_2 \longrightarrow CH_3OH + H_2O$$

Numerous investigations have been done in order to elucidate the nature of the catalytically active centre. The most efficient approach has again turned out to be a combination of different techniques: X-ray absorption spectroscopy, X-ray diffraction and on-line catalytic measurements by mass spectrometry [32]. A small capillary microreactor offers ideal conditions for catalysis and ensures that the structural information by the different *in situ* techniques is obtained at the same location, which is also the location with well defined catalytic conditions. Cu/ZnO catalysts with 4.5 wt% loading were used in the analysis. Mass spectrometry data are shown in figure 30 during the reduction phase of the Cu catalyst in Ar with 0.5% CO, 5% CO_2, and 4% H_2 at different temperatures between 120°C and 220°C. The increase of water ($m/e = 18$) and the decrease of hydrogen ($m/e = 2$) in mass spectrometry show that the reduction of the catalysts starts around 180°C.

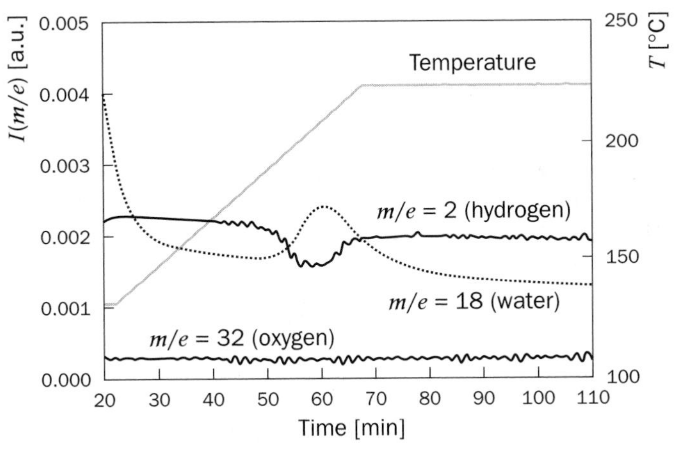

Figure 30 - Mass spectrometer data taken during the reduction of a CuO/ZnO catalyst in a reduction gas (0.5% CO, 5% CO_2, 4% H_2 in argon) when temperature is increased from 120°C to 220°C at a rate of 0.5°C/min [32].

This is supported by the EXAFS analysis (fig. 31). The Cu-O peak at 1.5 (uncorrected for phase shift) disappears at 220°C at the expense of a Cu-Cu peak at 2.2 Å showing up simultaneously. The near-edge structure changes from that for CuO to metallic Cu (insert in fig. 31).

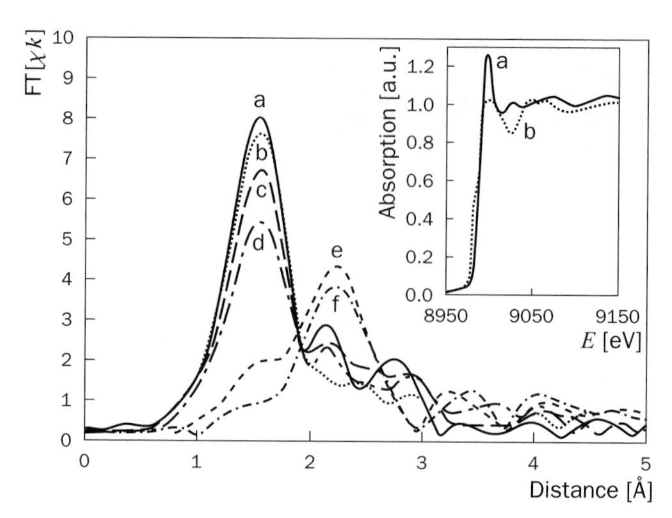

Figure 31 - Fourier transforms of the EXAFS χk at the Cu K edge of CuO/ZnO from figure 30. **a** - 120°C, **b** - 130°C, **c** - 180°C, **d** - 190°C, **e** - 210°C and **f** - 220°C. The insert shows the Cu edge at 120°C (**a**) and 220°C (**b**) [32].

Hence, both on-line mass spectrometry and XAS show the formation of metallic Cu particles at about 200°C. In X-ray diffraction only the ZnO phase is observed after reduction at 220°C. The lack of CuO lines before and of metallic Cu after reduction is due to the very small copper particles at a loading of 4.5 wt %. However, the use of a gas with higher reduction potential (more CO relative to H_2) leads to larger Cu crystallites, which became visible in X-ray diffraction (fig. 32). A more detailed analysis of the EXAFS data for the reduced Cu particles gives a Cu-Cu distance of 2.54, very close to the value 2.55 for bulk copper. The first shell coordination number was 8.5 when anharmonic effects in the small particles are taken into account [33]. For a small spherical particle a substantial fraction of atom is located at the surface with a coordination number substantially below 12. To an average coordination number of 8.5 corresponds a particle size of 10-15 Å. As explained in reference [32], there is strong indication that the small copper particles reversibly undergo strong morphological changes depending on the reaction conditions, which have a strong influence on their catalytic activity.

In summary, this study stands for many examples of XAS on catalysts. The high degree of dispersion and the missing long range order make XAS a very valuable tool for the study of active centres in catalysts when the technique is done *in situ* and when it is supported by other techniques like diffraction, mass spectrometry and others.

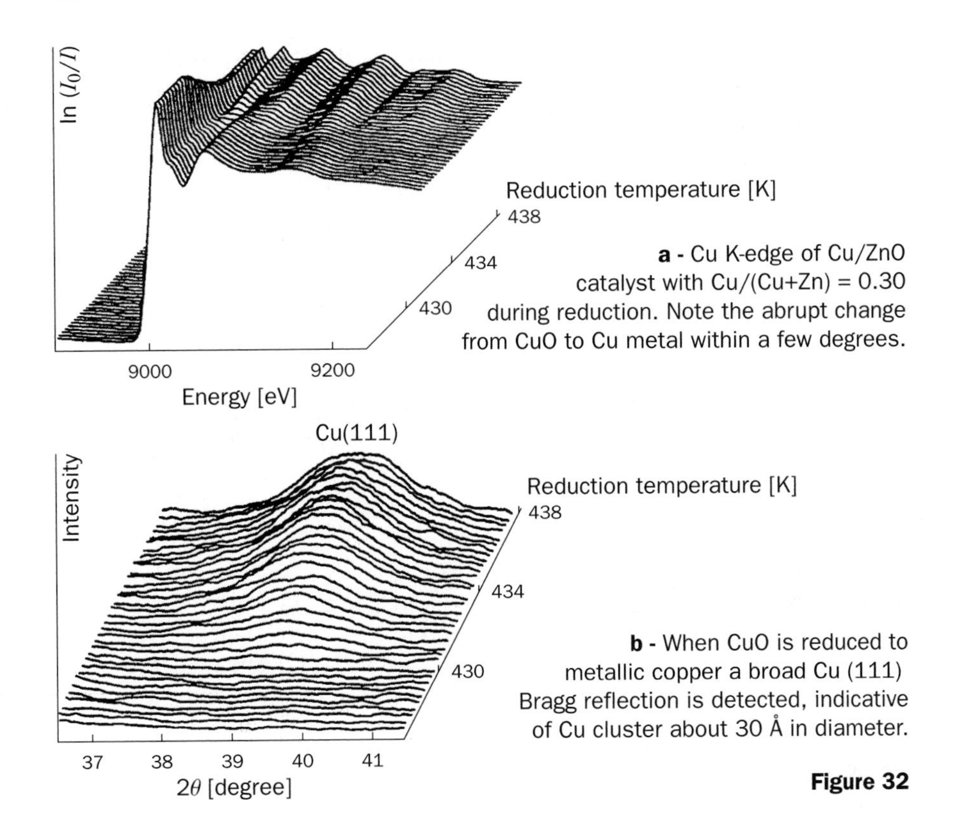

a - Cu K-edge of Cu/ZnO catalyst with Cu/(Cu+Zn) = 0.30 during reduction. Note the abrupt change from CuO to Cu metal within a few degrees.

b - When CuO is reduced to metallic copper a broad Cu (111) Bragg reflection is detected, indicative of Cu cluster about 30 Å in diameter.

Figure 32

4.6. COMBINED X-RAY ABSORPTION AND X-RAY CRYSTALLOGRAPHY FOR STRUCTURE-FUNCTION STUDIES OF METALLOPROTEINS [34]

Protein crystallography (PX) is a common tool for investigating the structure-function relation in biomolecules. However, it is difficult to reach atomic resolution (1.2 Å and below) in PX for large macromolecules. This is needed for an unambiguous structure determination around the catalytically relevant metal centre. Here, high quality XAFS data have turned out as a very valuable additional tool. This will be illustrated for the case of copper nitrite reductase [34]. This protein plays an essential role in the cycling of nitrogen in the biosphere, catalyzing the reaction from NO_2^- to NO, which finally ends up as N_2 gas. X-ray crystallography data with a resolution of 3.0, 1.9 and 1.04 Å were obtained for native oxidized copper nitrite reductase. These data were used to simulate EXAFS around the Cu *K* edge by means of the EXCURVE code mentioned in section 3. Figure 33 shows a comparison of the measured and calculated EXAFS data for the 3 crystallographic models (corresponding to resolution 3.0, 1.9 and 1.04 Å) and the corresponding Fourier transforms. It is obvious that the structure models deduced from 3.0 Å and 1.9 Å data are not able to fit the experimental EXAFS data, whereas there is excellent agreement with the crystallographic model based on atomic resolution data (1.04 Å).

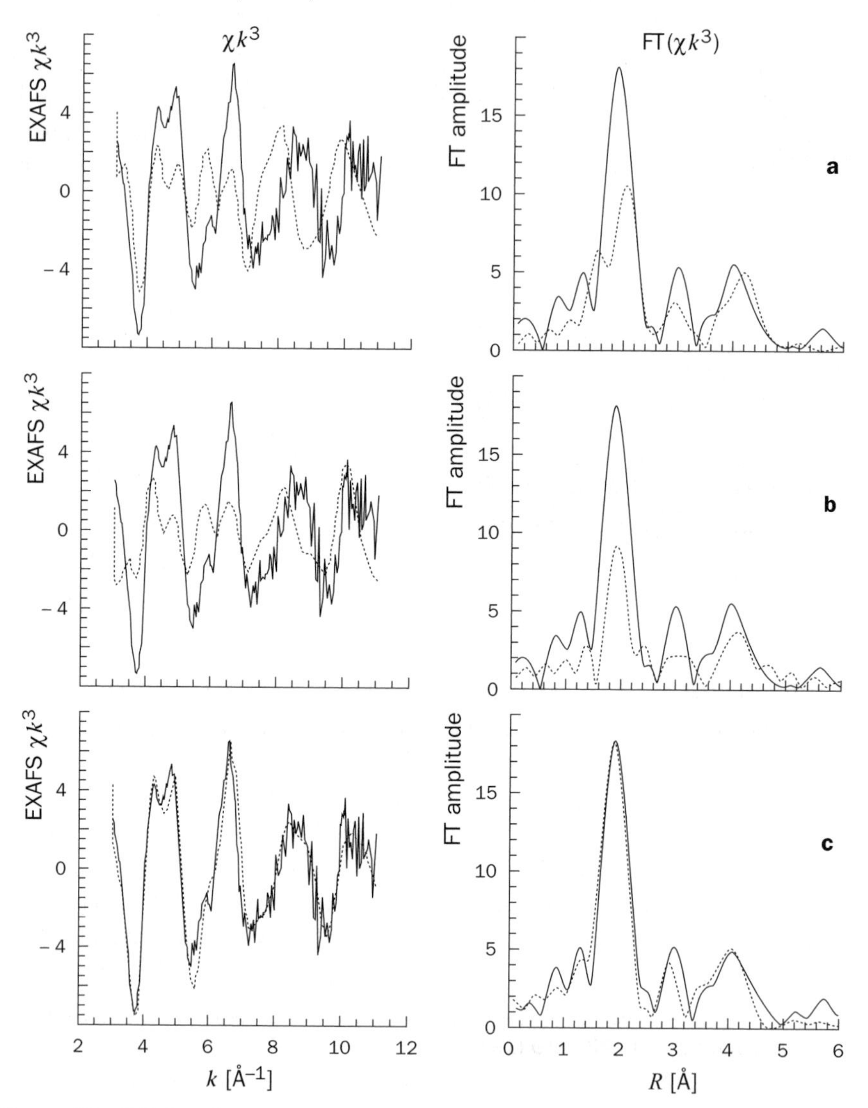

Figure 33 - EXAFS χk^3 and its Fourier transforms FT of the type 2 Cu site in oxidized copper nitrite reductase (full lines) and three simulations (dashed lines) calculated using crystallographic data at 3.0 Å (**a**), 1.9 Å (**b**) and 1.04 Å (**c**) resolution.

In the oxidized state the Cu is surrounded by a water molecule and 3 other ligands, called His94, His129 and His300. The correct Cu-OH_2 distance is larger by 0.25 Å compared to the inadequate models and also the Cu-N distances to the histidine vary by 0.10 to 0.14 Å from model to model. The message from this example is the following: XAFS excellently complements PX. Crystallography probes the long-range order whereas XAFS probes the local structure around specific atomic species. When the PX data do not have atomic resolution (a quality difficult to achieve for large biomolecules) then XAFS can provide the local resolution needed to determine the biologically relevant part of the structure. In addition, it is highly advisable in PX to measure the near-edge structure of the relevant metals in the protein in order to know the oxidation state of the protein whose structure is to be determined.

REFERENCES

[1] E.A. STERN & S.M. HEALD - *In "Handbook on Synchrotron Radiation"*, E.E. KOCH ed., Vol. 1b, 955 (1983)

[2] D.C. KONINGSBERGER & R. PRINS - *X-Ray Absorption*, J. Wiley (1986)

[3] A. FONTAINE - *In "Neutron and Synchrotron Radiation for Condensed Matter Studies", HERCULES vol. 1*, J. BARUCHEL, J.L. HODEAU, M.S. LEHMANN, J.R. REGNARD & C. SCHLENKER eds, Les Editions de Physique - Springer-Verlag, 323 (1993)

[4] D.C. KONINGSBERGER - *In "Neutron and Synchrotron Radiation for Condensed Matter Studies", HERCULES vol. 2*, J. BARUCHEL, J.L. HODEAU, M.S. LEHMANN, J.R. REGNARD & C. SCHLENKER eds, Les Editions de Physique - Springer-Verlag, 213 (1994)

[5] J. ALS-NIELSEN & D. MCMORROW - *Elements of Modern X-ray Physics*, chapter 6, J. Wiley (2001)

[6] R. SHANKAR - *Principles of Quantum Mechanics*, Plenum Press (1994)

[7] W.J. VEIGELE - *In "Handbook of Spectroscopy"* J.W. ROBINSON ed., Vol. I, CRC Press (1974)

[8] J. JAKLEVIC, J.A. KIRBY, M.P. KLEIN, A.S. ROBERTSON, G.S. BROWN & P. EISENBERGER *Solid State Comm.* **23**, 679 (1977)

[9] B.K. TEO & P.A. LEE - *J. Amer. Chem. Soc.* **101**, 2815 (1979)

[10] A.G. MCKALE, G.S. KNAPP & S.K. CHAN - *Phys. Rev. B* **33**, 841 (1986)

[11] N. BINSTED, J.W. CAMPBELL, S.J. GURMAN & P.C. STEPHENSON - EXCURVE, SERC Daresbury Laboratory, Worringten, UK (1991)

[12] J.J. REHR, J. MUSTRE DE LEON, S.I. ZABINSKI & R.C. ALBERS - *J. Am. Chem. Soc.* **113**, 5136 (1991)

[13] J.J. REHR & R.C. ALBERS - *Rev. Modern Physics* **72**, 621 (2000)

[14] A. FILIPPONI, A. DI CICCO & C.R. NATOLI - *Phys. Rev. B* **52**, 15122 (1995)

[15] B. LENGELER - *J. Phys.* **47**, C8, 75 (1986).

[16] W. HEITLER - *The Quantum Theory of Radiation*, 204, Oxford University Press (1984)

[17] J.E. MÜLLER, O. JEPSEN & J.W. WILKINS - *Solid State Comm.* **42**, 365 (1982)

[18] B. LENGELER & R. ZELLER - *Solid State Comm.* **51**, 889 (1984)

[19] G. MATERLIK, J.E. MÜLLER & J.W. WILKINS - *Phys. Rev. Lett.* **50**, L67 (1983)

[20] R. ZELLER, G. STEGEMANN & B. LENGELER - *J. Phys.* **47**, C8, 1101 (1986)

[21] U. SCHEUER & B. LENGELER - *Phys. Rev. B* **44**, 9883 (1991)

[22] N. PAPANIKOLAOU, R. ZELLER, P.H. DEDERICHS & N. STEFANOU - *Phys. Rev. B* **55**, 4157 (1997)

[23] see for instance J. MATHON - *In "Spin Electronics"*, M. ZIESE & M.J. THORNTON eds, Springer Lecture Notes in Physics (2001)

[24] O. PROUX, J.R. REGNARD, I. MANZINI, C. REVENANT, B. RODMACQ & J. MIMAULT - *Eur. Phys. J.: AP* **9**, 115 (2000)

[25] O. PROUX, J. MIMAULT, J.R. REGNARD, C. REVENANT-BRIZARD, B. MEVEL & B. DIENY - *J. Phys.: Cond. Matter* **12**, 3939 (2000)

[26] B. LENGELER - *Phys. Rev. Lett.* **53**, 74 (1984)

[27] B. LENGELER - *J. Phys.* **47**, C8, 1015 (1986)

[28] B. LENGELER - *Solid State Comm.* **55**, 679 (1985)

[29] M. HÜPPAUFF & B. LENGELER - *J. Electrochem. Soc.* **140**, 598 (1993)

[30] R. KÖTZ, C. BARBERO & O. HAAS - *J. Electroanalyt. Chem.* **296**, 37 (1990)

[31] A.F. WELLS - *Structural Inorganic Chemistry*, Clarendon Press (1975)

[32] J.D. GRUNWALDT, A.M. MOLENBROEK, NY TOPSOE, H. TOPSOE & B.S. CLAUSEN - *J. Catalysis* **194**, 452 (2000)

[33] B.S. CLAUSEN & J.K. NORSKOV - *Top. Catal.* **10**, 221 (2000)

[34] S.S. HASNAIN & R.W. STRANGE - *J. of Synchrotron Radiation* **10**, 9 (2003)

[35] A. ANKUDINOV, B. RAVEL, J.J. REHR & S. CONRADSON - *Phys. Rev. B* **58**, 7565 (1998)

[36] M. BENFATTO, S. DELLA LONGA & C.R. NATOLI - *J. of Synchrotron Radiation* **10**, 51 (2003)

5

INELASTIC X-RAY SCATTERING FROM COLLECTIVE ATOM DYNAMICS

F. SETTE - M. KRISCH
European Synchrotron Radiation Facility, Grenoble, France

1. INTRODUCTION

The study of atomic dynamics in condensed matter at momentum transfer, $\boldsymbol{Q}$, and energy, E, characteristic of collective motions is, traditionally, the domain of neutron spectroscopy. The experimental observable is the dynamic structure factor $S(\boldsymbol{Q},E)$, which is the space and time Fourier transform of the density-density correlation function. Neutrons as probing particle are particularly suitable, since

- the neutron-nucleus scattering cross-section is sufficiently weak to allow for a large penetration depth,
- the energy of neutrons with wavelengths of the order of inter-particle distances is about 100 meV, and therefore comparable to the energy of collective excitations associated with density fluctuations such as phonons, and
- the momentum of the neutron may be used to probe the whole dispersion scheme out to several $Å^{-1}$.

This is in contrast to inelastic light scattering techniques such as Brillouin and Raman scattering, which can only determine acoustic and optic modes, respectively, at very small momentum transfers.

While it has been pointed out in several text books [1,2] that X-rays can in principle be utilized as well to determine the $S(\boldsymbol{Q},E)$, it was stressed that this would represent a formidable experimental challenge, mainly owing to the fact that an X-ray instrument would have to provide an extremely high energy resolution. This is understood considering that photons with a wavelength of $\lambda = 0.1$ nm have an energy of about 12 keV. Therefore, the study of phonon excitations in condensed matter, which are in the meV region, requires a relative energy resolution of at least

F. Hippert et al. (eds.), Neutron and X-ray Spectroscopy, 169–188.

$\Delta E/E \approx 10^{-7}$. On the other hand, there are situations where the use of photons has important advantages over neutrons. A specific case is based on the general consideration that it is not possible to study acoustic excitations propagating with the speed of sound v_S using a probe particle with a speed v smaller than v_S. This limitation is not particularly relevant in neutron spectroscopy studies of crystalline samples. Here, the translation invariance allows one to study the acoustic excitations in high order Brillouin zones, thus overcoming the above-mentioned kinematic limit on phonon branches with steep dispersions. On the contrary, the situation is very different for topologically disordered systems such as liquids, glasses and gases. In these systems, in fact, the absence of periodicity imposes that the acoustic excitations must be measured at small momentum transfers. Thermal neutrons have a velocity in the range of 1000 m/s, and only in disordered materials with a speed of sound smaller than this value (mainly fluids of heavy atoms and low density gases) the acoustic dynamics can be effectively investigated [3]. Another advantage of the inelastic X-ray technique arises from the fact that very small beam sizes of the order of a few tens of micrometers can presently be obtained at third generation synchrotron sources. This opens the possibility to study systems available only in small quantities down to a few 10^{-6} mm^3 and/or their investigation in extreme thermodynamic conditions, such as very high pressure. These differences with respect to inelastic neutron scattering motivated the development of the very high resolution inelastic X-ray scattering (IXS) technique, and following the pioneering experiments in 1986 [4,5], the IXS technique rapidly evolved. To date there are four instruments operational at the ESRF (2), APS (1) and Spring-8 (1), and several more under construction.

The aim of the present article is to give the reader an introduction to the IXS technique, where the similarities and differences with respect to coherent inelastic neutron scattering (INS) shall be highlighted, therefore providing a natural link to other chapters of volume *V*. The present capabilities of the IXS technique are illustrated by discussing two representative experiments in some detail.

2. *SCATTERING KINEMATICS AND INELASTIC X-RAY SCATTERING CROSS-SECTION*

The inelastic scattering process is depicted schematically in figure 1. The momentum and energy conservation impose that

$$\boldsymbol{Q} = \boldsymbol{k}_i - \boldsymbol{k}_f \qquad \{1a\}$$

$$E = E_i - E_f \qquad \{1b\}$$

$$Q^2 = k_i^2 + k_f^2 - 2k_i k_f \cos(\theta) \qquad \{1c\}$$

where θ is the scattering angle between the incident and scattered photons. The relation between momentum and energy in the case of photons is given by $E(k) = \hbar ck$. Considering that the energy losses or gains associated to phonon-like excitations are always much smaller than the energy of the incident photon ($E \ll E_i$), one obtains

$$(Q / k_i) = 2\sin(\theta / 2) \qquad \{2\}$$

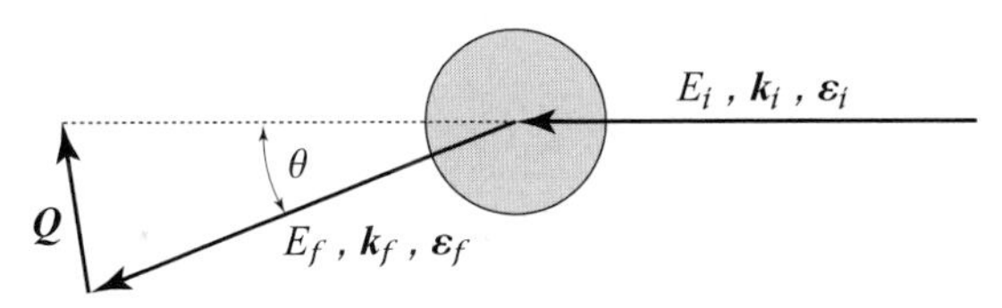

Figure 1 - Schematics of the inelastic scattering process. The incident and scattered photons are characterized by their energy E, wave vector $\boldsymbol{k}$ and polarization vector $\boldsymbol{\varepsilon}$. See text for details.

The ratio between the exchanged momentum and the incident photon momentum is completely determined by the scattering angle. Therefore, for IXS, there are no limitations in the energy transfer at a given momentum transfer for phonon-like excitations, in strong contrast to INS where a strong coupling between energy and momentum transfer exists.

The Hamiltonian, describing the electron-photon interaction in a scattering process, is composed, in the weak relativistic limit, of four terms [6]. Neglecting resonance phenomena close to X-ray absorption thresholds and the much weaker magnetic couplings, only the term arising from the Thomson interaction Hamiltonian has to be retained

$$H_{X-Th} = \frac{1}{2} r_0 \sum_j \boldsymbol{A}^2(\boldsymbol{r}_j, t) \qquad \{3\}$$

where $r_0 = e^2/m_e c^2$ is the classical electron radius and $\boldsymbol{A}(\boldsymbol{r}_j,t)$ is the electromagnetic field vector potential in the $\boldsymbol{r}_j$, coordinate of the jth electron. The sum extends over all the electrons in the system. The double differential cross-section is proportional to the number of incident probe particles scattered within an energy range ΔE and momentum variation into a solid angle $\Delta\Omega$. In the process, where a photon of energy E_i, wavevector $\boldsymbol{k}_i$, and polarization $\boldsymbol{\varepsilon}_i$, is scattered into a final state of energy E_f, wavevector $\boldsymbol{k}_f$, and polarization $\boldsymbol{\varepsilon}_f$, and the electron system goes from the initial state $|I\rangle$ to the final state $|F\rangle$, we obtain

$$\frac{\partial^2 \sigma}{\partial \Omega \partial E} = r_0^2 (\boldsymbol{\varepsilon}_i \cdot \boldsymbol{\varepsilon}_f)^2 \frac{k_i}{k_f} \sum_{I,F} P_I \left| \langle F | \sum_j e^{iQr_j} | I \rangle \right|^2 \delta(E - E_f - E_i) \qquad \{4\}$$

The sum over the initial and final states is the thermodynamic average, and P_I corresponds to the thermal population of the initial state. From this expression, which implicitly contains the *correlation function of the electron density*, one arrives to the *correlation function of the atomic density*, on the basis of the following considerations:

- We assume the validity of the adiabatic approximation. This allows one to separate the system quantum state $|S\rangle$ into the product of an electronic part, $|S_e\rangle$, which depends only parametrically from the nuclear coordinates, and a nuclear part, $|S_n\rangle$: $|S\rangle = |S_e\rangle|S_n\rangle$. This approximation is particularly good for exchanged energies that are small with respect to the excitations energies of electrons in bound core states: this is indeed the case in basically any atomic species when considering phonon energies. In this approximation, one neglects the portion of the total electron density contributing to the delocalized bonding states in the valence band region.
- We limit ourselves to consider the case in which the electronic part of the total wave function is not changed by the scattering process, and therefore the difference between the initial state $|I\rangle = |I_e\rangle|I_n\rangle$ and the final state $|F\rangle = |I_e\rangle|F_n\rangle$ is due only to excitations associated to atomic density fluctuations. Using these two hypotheses we then obtain

$$\frac{\partial^2 \sigma}{\partial \Omega \partial E} = r_0^2 (\boldsymbol{\varepsilon}_i \cdot \boldsymbol{\varepsilon}_f)^2 \frac{k_i}{k_f} \left\{ \sum_{I_n, F_n} P_{I_n} \left| \langle F_n | \sum_k f_k(\boldsymbol{Q}) \, e^{i\boldsymbol{Q}\boldsymbol{R}_k} | I_n \rangle \right|^2 \delta(E - E_f - E_i) \right\} \{5\}$$

where $f_k(Q)$ is the atomic form factor of the atom k and $\boldsymbol{R}_k$ its position vector. The expression in the curly brackets contains the dynamic structure factor $S(\boldsymbol{Q},E)$. Assuming that all the scattering units in the system are equal, this expression can be simplified by the factorization of the form factor of these scattering units

$$S(\boldsymbol{Q},E) = |f(\boldsymbol{Q})|^2 \frac{1}{2\pi} \frac{1}{N} \int dt \, e^{iEt/\hbar} \sum_{jl} \langle e^{i\boldsymbol{Q}\boldsymbol{R}_l(t)} e^{-i\boldsymbol{Q}\boldsymbol{R}_k(0)} \rangle \qquad \{6\}$$

where N is the number of atoms in the system. The brackets indicate the thermal average. In this presentation, $S(\boldsymbol{Q},E)$ is the space and time Fourier transform of the pair correlation function $G_p(\boldsymbol{R},t)$, which is identified with the probability to find two different particles at positions $\boldsymbol{R}_k(0)$ and $\boldsymbol{R}_l(t)$, separated by the distance $\boldsymbol{R}$ and at different times t. The double differential cross section can then be written as

$$\frac{\partial^2 \sigma}{\partial \Omega \partial E} = r_0^2 (\boldsymbol{\varepsilon}_i \cdot \boldsymbol{\varepsilon}_f)^2 \frac{k_i}{k_f} |f(\boldsymbol{Q})|^2 S(\boldsymbol{Q},E) = \left(\frac{\partial \sigma}{\partial \Omega}\right)_{Th} |f(\boldsymbol{Q})|^2 S(\boldsymbol{Q},E) \quad \{7\}$$

For this specific case, the coupling characteristics of the photons to the system, the Thomson scattering cross section, is separated from the dynamical properties of the system, and the atomic form factor $f(\boldsymbol{Q})$ appears only as a multiplicative factor. The corresponding INS cross section is obtained by replacing the Thomson cross-section and the atomic form factor by the coherent neutron scattering length b of the element under study. For this specific case, INS and IXS both probe the $S(\boldsymbol{Q},E)$ in the same fashion.

As already mention above, the cross-section of equation {7} is valid for a system composed of a single atomic species. This derivation, however, can be easily generalized to molecular or crystalline systems by substituting the atomic form factor with either the molecular form factor, or the elementary cell form factor, respectively. This is in close analogy to INS, although the different atomic species contribute differently to the $S(\boldsymbol{Q},E)$ for neutrons and X-rays. This arises from the fact that both the Q dependence and the atomic number, Z, dependence of the X-ray form factor and of the neutron scattering length are substantially different. The situation becomes more involved if the system is multi-component and disordered. In this case the factorization of the form factor is still possible only assuming some distribution among the different atoms. In the limit case that such distribution is completely random, an incoherent contribution appears in the X-ray scattering cross-section [3].

The functional form of the $S(\boldsymbol{Q},E)$ depends on whether the system is crystalline or amorphous. In the first case, the formalism for phonons is well established. The inelastic structure factor, symmetry considerations, and phonon selection rules are identical to the ones for INS. For a detailed description the interested reader is referred to the chapter dedicated to INS and standard textbooks [7,8]. For amorphous systems one has to extend the macroscopic hydrodynamic theory in order to properly take into account that the wavelength of the probing particle approaches interatomic distances and to include eventual relaxation processes. This is subject of the so-called generalized hydrodynamic theory [3]. One of the most common methods to derive this generalization is to introduce the concept of memory functions.

To conclude this paragraph, further similarities and differences between the IXS and INS cross section are summarized below:

- X-rays couple to the electrons of the system with a cross-section proportional to the square of the classical electron radius, $r_0 = 2.82 \times 10^{-13}$ cm, i.e. with a strength comparable to the neutron-nucleus scattering cross-section *b*.

- The IXS cross-section is proportional to $f_j(\mathbf{Q})^2$. In the limit $Q \to 0$, the form factor is equal to the number of electrons in the scattering atom, Z; for increasing values of Q, the form factor decays with decay constants of the order of the inverse of the atomic wavefunction dimensions of the electrons in the atom.

- The total absorption cross-section of X-rays above 10 keV energy is limited in almost all cases ($Z > 4$) by the photoelectric absorption process, and not by the Thomson scattering process. The photoelectric absorption, whose cross-section is roughly proportional to Z^4, determines therefore the actual sample size along the scattering path. Consequently the Thomson scattering channel is not very efficient for system with high Z in spite of the Z^2 dependence of its cross-section. The flux of scattered photons in an energy interval ΔE and in a solid angle $\Delta\Omega$ can be written as

$$N = N_0 \frac{\partial \sigma}{\partial \Omega \partial E} \Delta\Omega \Delta E \, \rho L e^{-\mu L} \qquad \{8\}$$

where N_0 is the flux of the incident photons, ρ is the number of scattering units per unit volume, L is the sample length and μ is the total absorption coefficient. The behaviour of the signal intensity as a function of atomic number in monatomic systems with an optimal sample length $L = 1/\mu$ and $f(Q) = Z$ is reported in figure 2. We show the quantity σ_c/σ_t, with $\sigma_c = (r_o Z)^2$ and $\sigma_t = \mu/\rho$ which provides a measure of the efficiency of the method for a given photon energy. The step between $Z = 44$ and $Z = 45$ signifies the energy of the K-shell absorption edge with respect to the incident photon energy (in this case 22 keV). If the photon energy is smaller than the K-shell binding energy, more sample can be probed, whereas in the other case an additional absorption channel opens up, and consequently the scattering volume is reduced. From the above, it is obvious that low momentum transfer studies of disordered materials become increasingly difficult for large Z materials. For single crystal materials this limitation can be overcome by working in higher Brillouin zones, where the Q^2-increase of the inelastic cross section partly compensates for the decrease of the $f(Q)$.

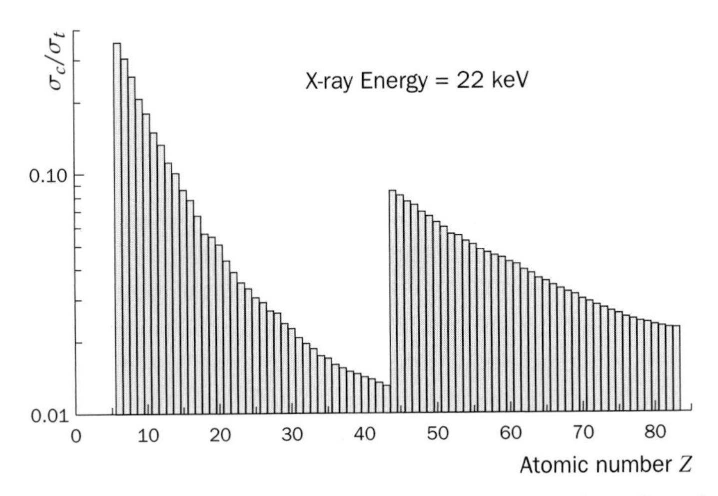

Figure 2 - Efficiency of the IXS method. Ratio between the total number of photons scattered by the Thomson process and those lost through all the other processes (predominantly through photoelectric absorption) in a sample of optimum length $1/\mu$ as a function of the atomic number Z for an incident photon energy of 22 keV

3. EXPERIMENTAL APPARATUS

The energy resolution requirements of an IXS instrument are not only very demanding for the incident X-rays, but as well for the energy analysis of the scattered photons. An effective method to obtain X-rays with high resolving power is based on Bragg reflection from perfect crystals. It can be shown that the resolving power ($E/\Delta E$) is given by [9]

$$\left(\frac{\Delta E}{E}\right) = \frac{d_{hkl}}{\pi \Lambda_{ext}} \qquad \{9\}$$

where d_{hkl} denotes the lattice spacing, associated with the (*hkl*) reflection order, and Λ_{ext} the primary extinction length, a quantity deduced within the framework of the *Dynamical Theory of X-ray Diffraction* [9]. Λ_{ext} increases with increasing reflection order. In order to reach a high resolving power, it is therefore necessary to use high order Bragg reflections, and to have highly perfect crystals. A perfect crystal can be defined as a periodic lattice without defects and/or distortions in the reflecting volume capable to induce relative variations of the distance between the diffracting planes, $\Delta d/d$, larger than the desired relative energy resolution: within this volume $\Delta d/d \ll (\Delta E/E)_h \approx 10^{-8}$. This stringent requirement practically limits the choice of the material to silicon.

Geometrical conditions, besides the energy resolution issues, are another important aspect to use efficiently a high order Bragg reflection. From the differentiation of Bragg's law, one obtains a contribution to the relative energy resolution due to the angular divergence $\Delta\theta$ of the beam impinging on the crystal, $\Delta E/E = \cot(\theta_B)\,\Delta\theta$. In order to reach the intrinsic energy resolution of the considered reflection, this angular contribution should be comparable or smaller than the intrinsic energy resolution. In typical Bragg reflection geometries $\cot(\theta_B) \approx 1$: consequently for high order reflections with $(\Delta E/E)_h \approx 10^{-8}$, the required angular divergence should be in the nrad range, i.e., values much smaller than the collimation of X-ray beams available even at the new third generation synchrotron radiation sources. This geometrical configuration would induce a dramatic reduction of the number of photons Bragg-reflected from the monochromator and analyzer crystals within the desired spectral bandwidth. Starting from the pioneer work of Bottom [10] and Maier-Leibnitz [11], an elegant solution to this problem has been found by introducing the *extreme backscattering* geometry, i.e. the use of Bragg angles very close to 90°. This provides very small values of $\cot(\theta_B)$ ($\theta_B \approx 89.98°$, $\cot(\theta_B) \approx 10^{-4}$). In such a way the acceptable values of $\Delta\theta$ are increased to values well above ~ 20 μrad, and therefore they become comparable to the typical divergence of synchrotron radiation from an undulator source.

Although the problems connected to the energy resolution are conceptually the same for the monochromator and for the analyzer, the required angular acceptances are very different. The monochromator can be realised with a *flat* perfect crystal. In the case of the analyzer crystal, however, the optimal angular acceptance is dictated by the desired momentum resolution. Considering typical values of ΔQ in the range of $0.2 \div 0.5$ nm^{-1} the corresponding angular acceptance of the analyzer crystal must be ~ 10 mrad or higher, which is again an angular range well above acceptable values, i.e. also larger than the deviation of the Bragg angle from 90°. The only way to obtain such large angular acceptance is by the use of a focussing system, which, nevertheless, has to preserve the crystal perfection properties, necessary to obtain the energy resolution. This automatically excludes considering elastically bent crystals. A solution consists in laying a large number of undistorted perfect *flat* crystals on a *spherical* surface, with the aim to use a 1:1 pseudo-Rowland circle geometry with aberrations kept such that the desired energy resolution is not degraded. The spectrometers of the ESRF IXS beam lines are equipped with such "*perfect* silicon crystal with *spherical* shape", consisting of a spherical substrate of $R = 6{,}500$ mm and 100 mm diameter, on which approximately 12,000 silicon perfect crystals of surface size 0.6×0.6 mm^2 and thickness 3 mm are bonded [12].

In figure 3 we report the layout of the main optical elements of the ID16 IXS beam line at the ESRF. The source consists of linear undulators, providing X-rays, linearly polarized in the horizontal plane. Typically, emission from the 3rd, 5th or 7th harmonics is utilized. This X-ray beam has an angular divergence of approximately $40 \times 15\ \mu$rad (full-width at half-maximum (fwhm) horizontal × vertical), a spectral bandwidth $\Delta E/E \sim 10^{-2}$, and an integrated power within this divergence of the order of 200 W. This beam is first pre-monchromated to $\Delta E/E \approx 2 \times 10^{-4}$ using a silicon (111) double crystal device kept in vacuum and at the cryogenic temperature of ~ 110 K.

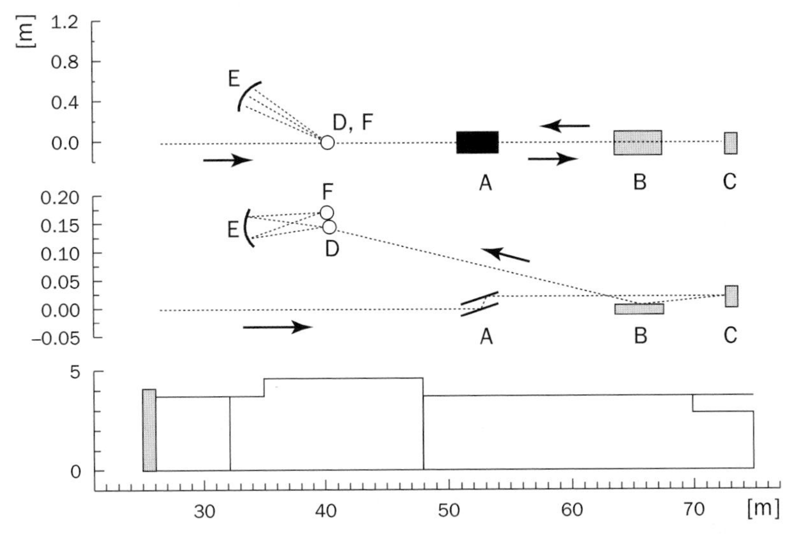

Figure 3 - Schematic layout of the key optical elements of the ID16 IXS beam line at the ESRF. A - High heat-load silicon (1,1,1) premonochromator; B - Toroidal mirror; C - High energy resolution monochromator; D - Scattering sample; E - High energy resolution analyzer; F - Detector. The x-axis denotes the distance from the undulator X-ray source. Top: top view; middle: side view; bottom: schematic lead hutch structure.

The photons from the pre-monochromator reach the high energy resolution backscattering monochromator, consisting of a flat symmetrically cut silicon crystal oriented along the (111) direction, and operating at a Bragg angle of 89.98°. This highly monochromated beam impinges on a toroidal mirror, focussing at the sample position to a beam size of 250 (horizontal) × 150 (vertical) μm^2 fwhm. The scattered X-rays are energy-analyzed by the "*perfect* silicon crystal with *spherical* shape" described above, operating at the same reflection order as the monochromator, and in a 1:1 Rowland circle geometry. The refocussed, energy-analyzed X-rays are detected by an inclined silicon diode detector with an equivalent thickness of 2.5 mm. There are in fact five independent analyzer systems with a fixed angular offset between them, mounted on a 7 m long arm that can rotate around a vertical axis through the scattering sample. This rotation allows one to choose the scattering angle θ_S for the five analyzers, and therefore the corresponding exchanged momentum, $Q = 2\,k_i\,\sin(\theta_S/2)$. The arm operates between 0° and 15° (up to 55° in the case of ID28). Further components of the spectrometer are an entrance pinhole, slits in front of the analyzer crystal to set the desired momentum resolution, and an exit pinhole in front of the detector.

Differently from traditional triple axis spectrometry, however, as a consequence of the backscattering geometry, the energy difference between analyzer and monochromator cannot be varied modifying the Bragg angle on one of the two crystals. The energy scans are performed by keeping the Bragg angle constant, and by changing the relative temperature between the two crystals with the effect of varying their relative lattice parameter, and therefore the value of the reflected energies. Specifically, the analyzer is kept at constant temperature while the monochromator temperature, and therefore E_i, is varied. Considering that $\Delta E/E = \Delta d/d = \alpha\,\Delta T$, with $\alpha = 2.58 \times 10^{-6}\,\mathrm{K}^{-1}$ in silicon at room temperature, in order to obtain an energy step of about one tenth of the energy resolution, i.e. $\Delta E/E \sim 10^{-9}$, it is necessary to control the monochromator crystal temperature with a precision of about 0.5 mK, while keeping the analyzer temperatures constant within typically a mK over hours. This difficult task has been reached with a carefully designed temperature bath, controlled with an active feedback system [13]. The key performances of the instrument are summarized in Table 1.

Table 1 – Key parameters of the ID16 and ID18 instruments. Measured flux of the X-ray beam from the high energy resolution monochromator with a storage ring current of 200 mA, incident photon flux and energy resolution ΔE_M, and total energy resolution ΔE_{tot}, for the indicated silicon reflections

Reflection	Energy (keV)	Flux (photon/s)	ΔE_M (meV)	ΔE_{tot} (meV)
7 7 7	13.840	9.0×10^{10}	5.0	7.0
8 8 8	15.816	6.0×10^{10}	3.8	5.4
9 9 9	17.793	1.8×10^{10}	1.8	3.0
11 11 11	21.748	4.4×10^{9}	0.8	1.5
12 12 12	23.725	3.3×10^{9}	0.7	–
13 13 13	25.702	9.1×10^{8}	0.5	0.9

4. *THE "FAST SOUND" PHENOMENON IN LIQUID WATER*

An example that serves to highlight advantages, limitations and characteristics of state of the art inelastic X-ray scattering experiments on disordered systems is the determination of the $S(\boldsymbol{Q},E)$ of liquid water. For a more extended account on the topic, the interested reader should refer to the review article of Ruocco and Sette, and references cited therein [14]. This work was motivated by previous molecular dynamics (MD) simulations and coherent inelastic neutron experiments. MD simulations [15] revealed that in the Q region between 3 and 6 nm^{-1} there exist two distinct features in the $S(\boldsymbol{Q},E)$. Under the hypothesis that the two excitations are associated with propagating modes, the corresponding sound velocities, v, result to be ~1500 m/s and ~3000 m/s. The first value is comparable to the sound velocity as determined by ultrasound and Brillouin light scattering techniques. The other mode was interpreted as a band of excitations associated to the hydrogen bond network propagating with a much higher sound velocity, and the expression *fast sound* was coined. A first INS experiment [16] detected the weakly dispersing lower energy excitation, but could not reach the excitations branch associated to the fast sound due to the kinematics limitations. In a second experiment [17] the kinematics region could be extended up to the expected dispersion law of the fast sound thanks to the use of higher energy neutrons, and the existence of a collective excitation propagating with $v \sim 3300$ m/s in the $Q = 3.5$ to 6 nm^{-1} region was demonstrated. The low energy band, observed in the earlier INS experiment [16], however, was not observed because of the limited energy resolution ($\Delta E = 4.5$ meV). A series of IXS experiments were performed in order to cover the Q-E region of the two branches in a single experiment – at that time not attempted using INS –, and possibly witness the simultaneous existence of two sound speeds (as revealed by the MD simulations), and more generally, gain further insight into the microscopic dynamics of this important liquid.

Representative IXS spectra are shown in figure 4, recorded with a total energy resolution of 1.5 meV, using the silicon (11,11,11) reflection order at an incident photon energy of 21,747 eV. The spectra are characterized by a strong (quasi-)elastic line and weak inelastic features. Positive (negative) energies correspond to creation (annihilation) of a phonon, and their intensity ratio is governed by the detailed balance factor. In order to extract significant spectroscopic parameters from the data such as an average excitation frequency and the energy width of these excitations, the data were fitted to a model function. This was constructed by the convolution of the experimentally determined resolution function with a theoretical model representing the quasi-elastic central line with a Lorentzian and the inelastic signal with a *Damped Harmonic Oscillator* (DHO) **[18]**. This model can be theoretically derived within the generalized classical hydrodynamics as the high frequency limit of the viscoelastic model **[3]**. Using this fitting procedure, one obtains the spectroscopic parameters $\Omega(Q)$, $\Gamma(Q)$ and $I(Q)$ entering the DHO, which represent the energy of the excitation, its width and its intensity, respectively. While in the IXS spectra in the Q-range 1-2.5 nm^{-1} (left panel of figure 4) only one inelastic feature can be identified, the spectra for $Q > 4$ nm^{-1} clearly reveal a second feature, at energies below the first excitation (see insets in the right panel of figure 4).

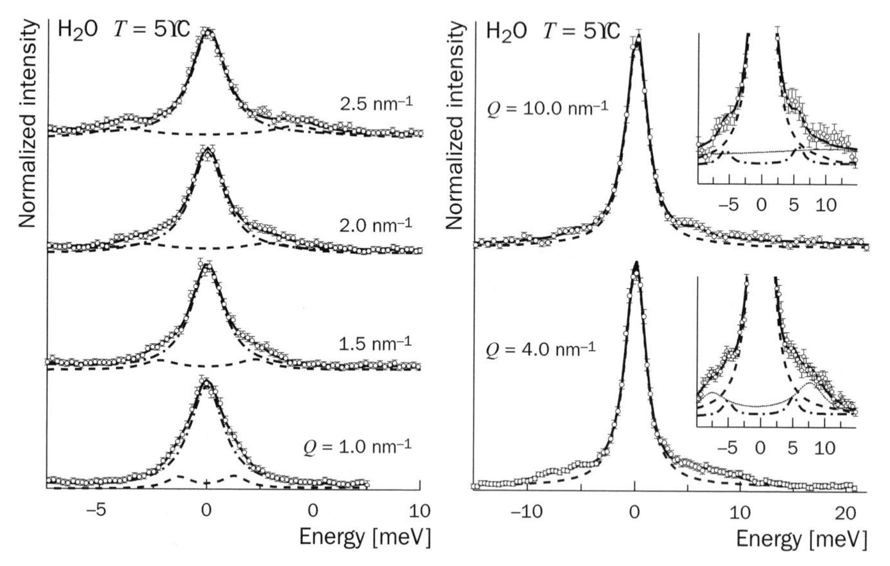

Figure 4 - IXS spectra from water at T = 5°C (o) at the indicated Q values, shown together with the total fit and its individual components. The data are normalized to their central peak maximum intensity. Typical count rates on the maximum were 1 to 3 counts per second, and the integration time per point was 300 to 400 s. The insets in the right panel show a zoom of the low energy region, where a second excitation at about 4 meV is visible.

The values of $\Omega(Q)$ of these excitations are reported in figure 5, together with the corresponding quantities derived from the neutron experiments, also using the DHO model. At the lowest Q of 1 nm^{-1} the observed excitation energy corresponds to a sound velocity of 1500 m/s, a value very close to the sound velocity in the low frequency hydrodynamic limit. In the Q-range 1.5 to 4 nm^{-1}, $\Omega(Q)$ increases in such a fashion that at 4 nm^{-1} a sound velocity of 3,200 ± 100 m/s is reached, a value which is approximately maintained up to the highest recorded spectrum at

$Q = 14$ nm^{-1}. In the common Q-range a very good agreement between the X-ray and neutron data can be observed for the "*fast sound*" branch, in spite of the fact that the neutron data were taken on D_2O and the X-ray ones on H_2O. Moreover, the energy position of the second, low lying excitation, appearing in the spectra with $Q > 4$ nm^{-1}, are also in good agreement with the first INS experiment.

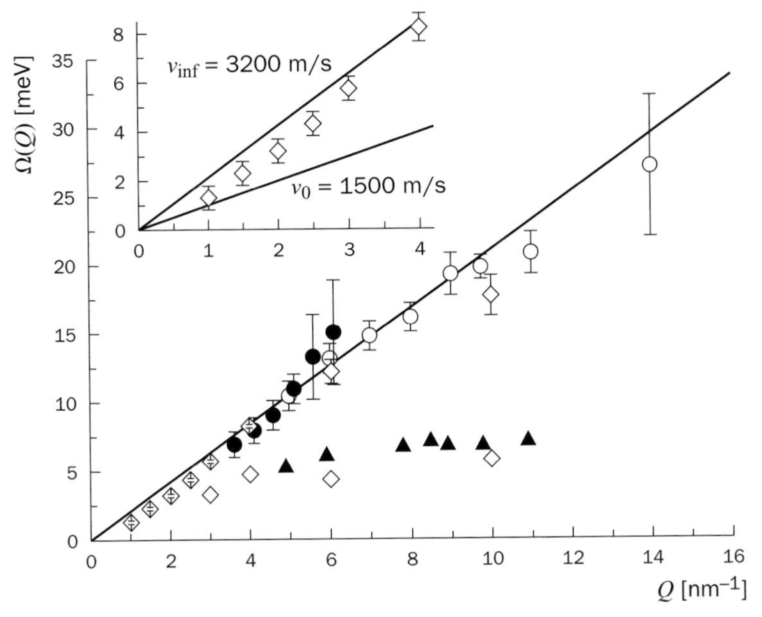

Figure 5 - Excitation energies $\Omega(Q)$ obtained using the DHO model for the group of IXS and INS data. Open circles: IXS, total energy resolution $\Delta E_T = 5$ meV; open diamonds: IXS, total energy resolution $\Delta E_T = 1.5$ meV; full circles: INS, Teixeira *et al.* [17]; full triangles: INS, Bosi *et al.* [16]. The full line corresponds to a sound speed of 3200 m/s, and is a result of a linear fit to the IXS experimental points for $Q \geq 4$ nm^{-1}. The insert shows the IXS dispersion at low Q values on an expanded scale, chosen to emphasize the change with Q of the velocity of sound from the high Q value of $v_{inf} = 3200$ m/s towards the low Q value of $v_0 = 1500$ m/s.

The results of the IXS experiments can be summarized as follows:

- In the Q region between 4 and 10 nm^{-1} there exists an excitation in liquid water that propagates with a velocity of sound more than double the hydrodynamic sound velocity, the so-called fast sound.
- The energy of the excitation at a given Q is basically identical in H_2O and D_2O, without relevant shifts to be related to the mass difference between H and D: it is therefore possible to conclude that this excitation involves the centre of mass of the whole molecule, and not only the lighter hydrogen atoms.

- The simultaneous existence of two modes, as predicted by molecular dynamics calculations, has been firmly demonstrated for $Q > 4$ nm^{-1}.
- The fast sound branch arises from the evolution of the ordinary sound mode at high Q values. The large increase in sound velocity takes place in the range between $Q = 1$ and 4 nm^{-1} at 5°C.

Further studies of this sound dispersion from the ordinary sound, v_0, to the "*fast sound*", v_{inf}, as a function of temperature, made it possible to relate the origin of this phenomenon to a structural relaxation process, involving the making and the breaking of hydrogen bonds. Within this framework, it is then possible to recognize a hydrodynamic "*normal*" regime and a "*solid-like*" regime, where in the latter, the sound velocity is close to the one of crystalline ice I_h, and the second, low energy excitation can be identified as a reminiscence of a transverse dynamics.

5. *DETERMINATION OF THE LONGITUDINAL SOUND VELOCITY IN IRON TO 110 GPA*

The determination of the pressure-dependent compression (v_p) and shear (v_s) velocities of iron are of fundamental importance for comparison with global seismic models of the Earth, which provide acoustic wave velocity profiles. Various experimental techniques, such as X-ray diffraction (XRD) lattice strain measurements, nuclear resonant inelastic X-ray scattering (NRIXS) and Raman spectroscopy have been used to determine the elastic properties of iron, and derive from them the relevant sound velocities. The inconsistency found among the various experiments and techniques might be partly attributed to the fact that none of these techniques provide a direct measurement of acoustic wave velocities. This limitation can be overcome by IXS and INS, where the sound velocity can be directly derived from the dispersion of the acoustic phonon energy. High-pressure INS experiments are, however, limited typically to 10 GPa. This limitation arises from the fact that a sample volume in the order of one mm^3 is needed in order to obtain a sufficient inelastic signal, which in turn limits the maximum pressure achievable, if a reasonably compact high-pressure press is to be used. In contrast to this, as already mentioned above, X-rays are routinely focussed down to a few tens of micrometers. Moreover, the optimum sample thickness ($t = 1/\mu$) for medium Z materials is in the same range. These dimensions are well adapted to typical sample volumes that can be accommodated in diamond anvil high-pressure cells (DAC) with which pressures beyond 100 GPa can be reached.

The experiment [19] was performed on a polycrystalline sample of iron since it is impossible to preserve a single crystal while crossing the phase boundary from the low-pressure body-centred cubic structure (b.c.c. or α-phase) to the h.c.p. structure (ε-phase) between 12 and 15 GPa. The reported results thus correspond to an orientational average of the sound velocity. Data were recorded with an energy resolution of 5.5 meV, utilizing the silicon (8,8,8) reflection order, at 3 pressures in the α-phase (ambient pressure, 0.2 GPa and 7 GPa) and at six pressures in the ε-phase.

Typical IXS spectra and their corresponding fits are shown as a function of energy transfer in figure 6. The knowledge of the phonon dispersion curves of iron [20] and diamond [21] at ambient pressure allows an unambiguous assignment of the features. The strong inelastic signal at high energy transfer (i.e. high acoustic wave velocity) at the lowest recorded Q value corresponds to the transverse acoustic (TA) phonon branch of the diamond anvils, whereas the remaining peak can be attributed to the longitudinal acoustic (LA) phonon of iron. These inelastic contributions shift towards higher energies with increasing Q values, so that the inelastic contribution from diamond is rapidly out of the energy transfer window at Q values larger than 4 nm^{-1} due to its higher speed of sound with respect to iron. At 8.31 and 10.46 nm^{-1} an additional feature is visible between the elastic line and the LA phonon of iron. Wave velocities derived from the energy position of these excitations strongly suggest that they correspond to the transverse acoustic (TA) phonon of iron. Since the TA phonons could only be detected at 2 or 3 momentum transfers at two pressures, a further analysis, deriving the pressure dependence of the shear velocities has not been attempted. In contrast to this, the LA phonon branch is clearly observed over the whole momentum and pressure range explored.

The longitudinal sound velocity at each pressure was determined by fitting the dispersion curve with a sine function

$$E\,[\mathrm{meV}] = 4.192\times10^{-4}\cdot v_p\,[\mathrm{m.s^{-1}}]\cdot Q_{\max}\,[\mathrm{nm^{-1}}]\sin\left(\frac{\pi}{2}\frac{Q[\mathrm{nm^{-1}}]}{Q_{\max}[\mathrm{nm^{-1}}]}\right) \qquad \{10\}$$

from which the average sound velocity v_p as well as the position of the edge of the first Brillouin zone, $Q_{\max}$, can be derived. The numerical factor accounts for the conversion of the sound velocity in m s^{-1}. Data recorded at four to five momentum transfers have been used in each dispersion curve to constrain longitudinal v_p velocities within an estimated error of about 3%, with exception of the 110 GPa data point, for which only two momentum transfer data could be used. Except for the highest pressure point, the parameter $Q_{\max}$ was left free during the fitting procedure, and yielded excellent agreement with those calculated from the equation of state [22].

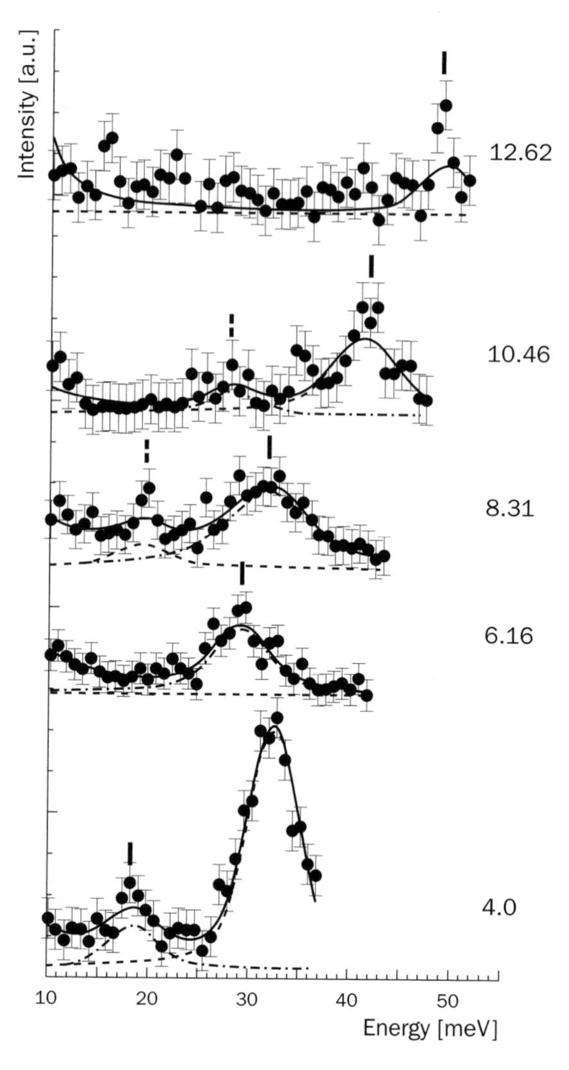

Figure 6 - IXS spectra of hcp iron at 28 GPa and room temperature at the indicated Q values in nm^{-1}. The energy position of the iron longitudinal (transverse) acoustic phonon is indicated by solid (broken) ticks. The integration time for each point was of the order of 600 to 700 s, obtained by a summation of four to six IXS scans. Data are shown together with their best fits (solid line) and their individual components (dashed line).

Results for the high-pressure ε-phase are displayed in figure 7. Here, the sound velocity v_p is plotted as a function of the specific mass. This type of representation shows the validity of a Birch law [23], namely a linear dependence between v_p and density. The data indeed follow this law, thus providing a convenient relationship for the extrapolation to higher pressures. In the range of overlap, the IXS data are in excellent agreement with Hugoniot shock measurements [24]. As a matter of fact, in shock experiments pressure and temperature are not independent parameters (the higher the pressure, the higher the temperature); therefore the good agreement with the IXS data provides an additional proof for the validity of a Birch law in h.c.p. iron. Seismic data [25] reported in figure 7 clearly fall out of such an extrapolation, suggesting that the Earth's inner core is slightly lighter than pure iron, as proposed in earlier work. The comparison of the present results with other more indirect experimental techniques are in fair agreement with XRD measurements [26] and NRIXS data [27] below 9,500 kg $\times$ m^{-3}, corresponding to 50 GPa, but display an increasing discrepancy above these values, where both XRD and NRIXS show a higher sound velocity. Their extrapolation to core pressures suggests that the Earth's inner core is slightly heavier than pure iron, unfortunately leading to a conclusion opposite to the one inferred from the IXS and shock wave results. In order to finally settle this central geophysical issue, further studies on iron need to be extended to iron compounds and alloys which are suspected to be potential candidates for the composition of the Earth's inner core.

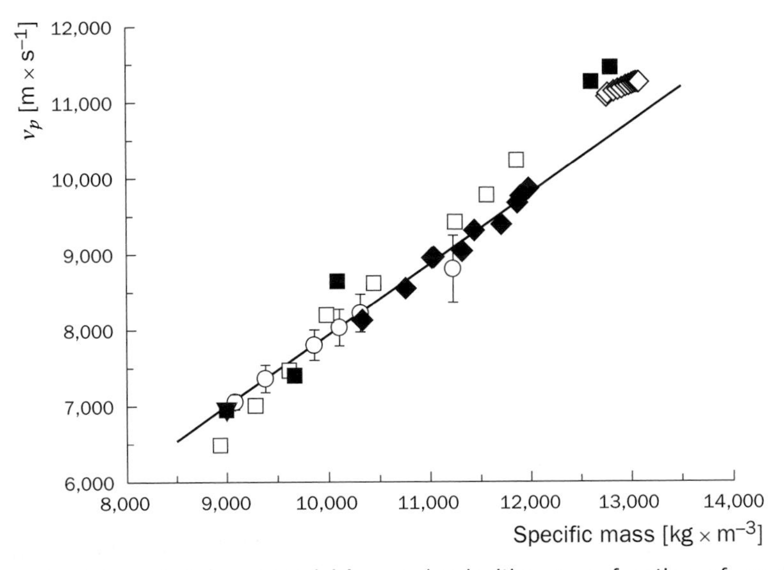

Figure 7 - Orientational averaged LA sound velocities as a function of specific mass. Open circles: IXS [19], full diamonds: shock wave Hugoniot [24]; full triangle: ultrasound [25]; open squares: NRIXS [27]; full squares: XRD [25]; open diamonds: preliminary reference earth model (PREM) [26]. The solid line is the result of a linear fit to the IXS data.

6. CONCLUSIONS AND OUTLOOK

Within the framework of the present article it was not possible to give a concise overview of all the various systems studied by IXS, and the interested reader is referred to a recent review article by Burkel [28]. The two examples here were specifically chosen to emphasize two of the most important differences with respect to inelastic neutron spectroscopy:

1. the absence of kinematic limitations, and therefore the capability to access a previously unexploited *Q-E* region, and
2. the possibility to study materials only available in small quantities and/or under very high pressures.

The first point obviously applies to the whole class of disordered systems, and a large amount of studies have been devoted to systems such as liquid metals and semiconductors, many glass forming liquids and glasses [29], as well as quantum liquids [28,30] and polymers [31].

The second point applies to novel materials where high quality samples cannot be synthesized in sufficiently large quantities to be studied by INS. Recent examples comprise GaN, a potentially interesting material for opto-electronic applications [32], and the determination of optical phonon dispersion in electron-doped high temperature superconductors [33]. As discussed in the second example, the possibility of studying the pressure evolution of the phonon dispersion enables to address important questions in many fields of research. For polycrystalline samples average dispersions and sound velocities can be determined, whereas high-pressure experiments performed on single crystals give access to the complete dispersion scheme, as has been shown in the case of argon [34]. This allows, for example, the precise determination of elastic constants and evidence phonon anomalies associated with the onset of structural or electronic phase transitions.

Inelastic X-ray scattering with high energy resolving power is still a relatively novel technique, and judging from the increasing number of instruments operational or under development, it should open new and important opportunities in the investigation of the atomic dynamics in condensed matter.

REFERENCES

[1] N. ASHCROFT & D. MERMIN - *Solid State Physics* (1976)

[2] W. COCHRAN - *In "Phonons"* R.W.H. STEVENSON ed., Oliver and Boyd, London (1966)

[3] U. BALUCANI & M. ZOPPI - *Dynamics of the Liquid State*, Oxford Science Publ., Oxford (1994)

[4] B. DORNER, E. BURKEL & J. PEISL - *Nucl. Instr. Meth. A* **246**, 450 (1986)

[5] B. DORNER, E. BURKEL, T. ILLINI & J. PEISL - *Z. Phys.: Condens. Matter* **69**, 179 (1987)

[6] W. SCHÜLKE - *In "Handbook on Synchrotron Radiation"*, Vol. 3, G. BROWN & D.E. MONCTON eds, Elsevier Science Publ. (1991)

[7] S.W. LOVESEY - *Theory of Neutron Scattering from Condensed Matter*, Clarendon Press, Oxford (1984).

[8] B. DORNER - *In "Coherent Inelastic Neutron Scattering in Lattice Dynamics"*, Springer, Berlin (1982)

[9] H. ZACHARIASEN - *Theory of X-ray Diffraction in Crystals*, Dover, New York (1944)

[10] V.E. BOTTOM - *A. Acad. Brasileira Ciencias* **37**, 407 (1965)

[11] H. MAIER-LEIBNITZ - *Nucleonik* **8**, 61 (1966)

[12] C. MASCIOVECCHIO, U. BERGMANN, M. KRISCH, G. RUOCCO, F. SETTE & R. VERBENI - *Nucl. Instr. Meth. B* **111**, 181 (1996);
C. MASCIOVECCHIO, U. BERGMANN, M. KRISCH, G. RUOCCO, F. SETTE & R. VERBENI - *Nucl. Instr. Meth. B* **117**, 339 (1996)

[13] R. VERBENI, F. SETTE, M. KRISCH, U. BERGMANN, B. GORGES, C. HALCOUSSIS, K. MARTEL, C. MASCIOVECCHIO, J.F. RIBOIS, G. RUOCCO & H. SINN - *J. Synchrotron Radiation* **3**, 62 (1996)

[14] G. RUOCCO & F. SETTE - *J. Phys.: Condens. Matter* **11**, R259 (1999)

[15] A. RAHMAN & F.H. STILLINGER - *Phys. Rev. A* **10**, 368 (1974)

[16] P. BOSI, F. DUPRÉ, F. MENZINGER, F. SACCHETTI & M.C. SPINELLI - *Lett. Nuovo Cimento* **21**, 436 (1978)

[17] J. TEIXEIRA, M.C. BELLISSENT-FUNEL, S.H. CHEN & B. DORNER - *Phys. Rev. Lett.* **54**, 2681 (1985)

[18] B. FÅK & B. DORNER - Institut Laue Langevin (Grenoble, France), Technical report N° 92FA008S (1992)

[19] G. FIQUET, J. BADRO, F. GUYOT, H. REQUARDT & M. KRISCH - *Science* **291**, 468 (2001)

[20] V.J. MINKIEWICZ, G. SHIRANE & R. NATHANS - *Phys. Rev.* **162**, 528 (1967)

[21] J.L. WARREN, J.L. YARNELL, G. DOLLING & R.A. COWLEY - *Phys. Rev.* **158**, 805 (1967)

[22] H.K. MAO, Y. WU, L.C. CHEN, J. SHU & A.P. JEPHCOAT - *J. Geophys. Res.* **95**, B13, 21737 (1990)

[23] F. BIRCH - *in "Solids Under Pressure"*, W. PAUL & D.M. WARSCHAUER eds, McGraw-Hill, New York, 1963;
F. BIRCH - *Geophys. J. R. Astron. Soc.* **4**, 295 (1961)

[24] J.M. BROWN & R.G. MCQUEEN - *J. Geophys. Res.* **91**, B7, 7485 (1986)

[25] H.K. MAO, J. SHU, G. SHEN, R.J. HEMLEY, B. LI & A.K. SINGH - *Nature* **396**, 741 (1998); correction *Nature* **399**, 280 (1999)

[26] A.M. DZIEWONSKI & D.L. ANDERSON - *Phys. Earth Planet. Int.* **25**, 297 (1981)

[27] H.K. MAO, J. XU, V.V. STRUZHKIN, J. SHU, R.J. HEMLEY, W. STURHAHN, M.Y. HU, E.E. ALP, L. VOCALO, D. ALFÉ, G.D. PRICE, M.J. GILLEM, M. SCHWOERER-BOHNING, D. HAUSERMANN, P. ENG, G. SHEN, H. GIEFERS, R. LUBBERS & G. WORTMANN - *Science* **292**, 914 (2001)

[28] E. BURKEL - *Rep. Prog. Phys.* **63**, 171 (2000)

[29] F. SETTE, M.H. KRISCH, G. MASCIOVECCHIO, G. RUOCCO & G. MONACO - *Science* **280**, 1550 (1998)

[30] A. CUNSOLO, G. PRATESI, G. RUOCCO, M. SAMPOLI, F. SETTE, R. VERBENI, F. BAROCCHI, M. KRISCH, C. MASCIOVECCHIO & M. NARDONE - *Phys. Rev. Lett.* **80**, 3515 (1998)

[31] A. MERMET, A. CUNSOLO, E. DUVAL, M. KRISCH, C. MASCIOVECCHIO, S. PERGHEM, G. RUOCCO, F. SETTE, R. VERBENI & G. VILIANI - *Phys. Rev. Lett.* **80**, 4205 (1998)

[32] T. RUF, J. SERRANO, M. CARDONA, P. PAVONE, M. PABST, M. KRISCH, M. D'ASTUTO, T. SUSKI, I. GRZEGORY & M. LESZCZYNSKI - *Phys. Rev. Lett.* **86**, 906 (2001)

[33] M. D'ASTUTO, P.K. MANQ, P. GIURA, A. SHUKLA, P. GHIGNA, A. MIRONE, M. BRADEN, M. GRAVEN, M. KRISCH & F. SETTE - *Phys. Rev. Lett.* **88**, 167002 (2002)

[34] F. OCCELLI, M. KRISCH, P. LOUBEYRE, F. SETTE, R. LE TOULLEC, C. MASCIOVECCHIO & J.P. RUEFF - *Phys. Rev. B* **63**, 224306 (2001)

6

PHOTOELECTRON SPECTROSCOPY

M. GRIONI
Institut de Physique des Nanostructures,
Ecole Polytechnique Fédérale de Lausanne, Switzerland

1. INTRODUCTION

Most properties of materials reflect, directly or indirectly, the nature of their electronic states. Electrons interact with the ions – and therefore feel the translation symmetry of the lattice – but also, namely in interesting materials, with other electrons. This leads to complex correlated states. Not surprisingly, the most straightforward way to investigate electronic properties is to remove the electrons from the solid, and to measure them far from the interacting system. This is the general idea of a photoemission experiment. Perhaps more surprisingly, such a simple measurement contains crucial information on the interacting system. The purpose of this chapter is to illustrate this simple and powerful idea.

The impressive results obtained over the past four decades, and the huge number of experiments performed every year, demonstrate that photoemission has reached the venerable status of a standard probe of the electronic properties of solids. And yet, new and exciting developments, often associated with synchrotron radiation, are reshaping the practice and scope of the technique, bridging the gap with conventional "thermodynamic" probes of the electronic states. The forefront of research has moved from studies of the band structure on the eV scale, to the investigation of elementary excitations, electronic instabilities, and new exotic properties of correlated systems, at the meV range and with high momentum resolution. In this perspective, this chapter gives only a very brief account of traditional aspects of photoelectron spectroscopy, while it addresses, with examples from the recent literature, the spectral properties of correlated electron systems. For a broader view and a much more detailed description of the technique the reader is referred to some excellent reviews [1,2].

F. Hippert et al. (eds.), Neutron and X-ray Spectroscopy, 189–237.

This chapter is organized as follows: section 2 gives a brief qualitative description and an intuitive interpretation of a photoemission experiment. Section 3 presents a discussion of a photoemission spectrum based on the simple and widespread 3-step model. Correlation effects and their manifestation in the spectral properties are introduced in section 4. These ideas are illustrated by selected case studies in section 5.

2. *WHAT IS PHOTOEMISSION?*

Photoelectron spectroscopy (PES) is a photon in-electron out experiment (figure 1). The interaction of a monochromatic beam of UV or soft X-ray photons with a sample generates photoelectrons with a broad distribution of emission angles and kinetic energies. The target may be indifferently a solid, a liquid or a gas, but in the following we will implicitly consider the more common case of a solid. The emitted electrons are then collected over a broad (angle-integrated PES, or simply PES) or a narrow (angle resolved PES or ARPES) acceptance angle. The subsequent measurement of the distribution of kinetic energies, typically performed by an electrostatic analyzer, yields a *spectrum* or *energy distribution curve* (EDC). The EDC represents the number of photoelectrons measured as a function of kinetic energy within the energy and angular acceptance windows of the analyzer. The measured intensity depends in a non-trivial way on various parameters that can be controlled, at least in principle, independently. They include the energy, polarization and incidence angle of the photon beam, the orientation and temperature (and magnetization etc.) of the sample, the collection angle and the angular acceptance window of the analyzer. A measurement of the spin may be added to the energy analysis, but the specific topic of spin-resolved PES and ARPES, which is of obvious interest in the study of magnetic systems, will not be covered here.

In this chapter we will address the crucial question: "*What kind of information is contained in a photoelectron spectrum?*" at various levels of sophistication. From the outset our physical intuition suggests that a PES spectrum must reflect the energy distribution of electrons inside the solid. When a photon is absorbed, its energy is transferred to an electron, which "jumps" from its initial state at energy ε_i to an excited state. If the photon is sufficiently energetic, the final state lies above the vacuum level and the photoelectron can escape from the solid. Energy conservation then determines the kinetic energy ε_K of the photoelectron

$$\varepsilon_K = \varepsilon_i + \hbar\Omega - \phi \tag{1}$$

Here $\hbar\Omega$ is the photon energy and ϕ is the work function of the sample, i.e. the energy difference between the Fermi level (E_F) and the vacuum level, and energies are measured with respect to E_F. Because of the energy conservation relation {1}, structures in the energy distribution of the photoelectrons should reproduce the structures in the occupied density of states (DOS). PES is obviously blind to the unoccupied states, which can be probed by X-ray absorption (XAS) described in another chapter of this book, or by Inverse photoemission (IPES) [3]. IPES is an electron in-photon out experiment (figure 1), which may be considered as the "mirror" (time-reverse) experiment of PES. A combined PES/IPES measurement yields a complete view – modulated by matrix elements – of the DOS. We shall see later on (section 4) that a deeper connection between the two techniques is revealed by the unifying concept of spectral function.

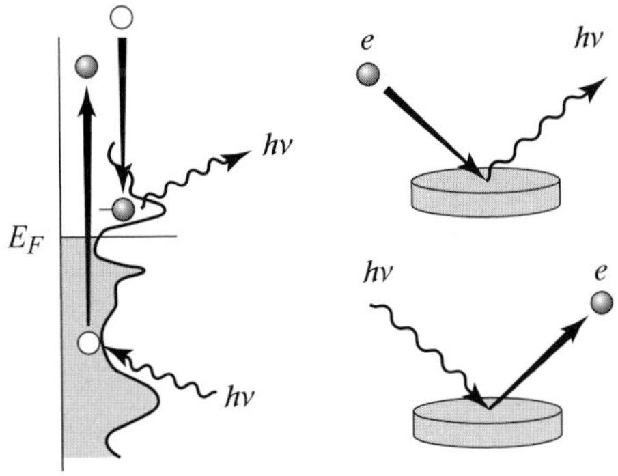

Figure 1 - Schematic view of the Photoemission (bottom) and Inverse Photoemission (top) processes.

The PES spectrum from a polycrystalline silver sample shown in figure 2 confirms this intuitive picture. It is dominated by an intense feature 4-8 eV below the Fermi level (the $E = 0$ reference) from the 10 electrons of the completely filled Ag 4*d* band. The weak emission between the 4*d* band and the Fermi level is due to the single electron in the broad and partially occupied 5*s* band. The PES spectrum is cut at E_F, but the IPES data shows that the conduction band is continuous beyond the high-energy limit of the PES spectrum. The Fermi level position can be accurately determined from the inflection point of the "metallic Fermi edge" of the spectrum. This sharp temperature-dependent feature reflects the rapid change of occupation of the electronic states across E_F, expressed by the Fermi-Dirac distribution, and its width is limited by the experimental resolution. Since the sample and the spectrometer are in electrical contact during the measurement, their Fermi levels are aligned, and the kinetic energy of a photoelectron emitted from the Fermi surface is

$$\varepsilon_K = \hbar\Omega - \phi_A \qquad \{2\}$$

Notice that, for a given photon energy, the Fermi level position depends exclusively on the analyzer's work function ϕ_A. It is a constant reference valid for all materials, including insulators, which obviously do not exhibit a metallic Fermi edge.

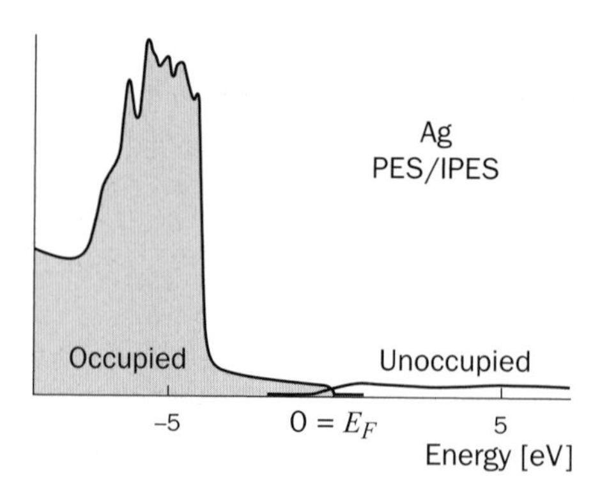

Figure 2 - PES and IPES spectra of polycrystalline silver, plotted with a common Fermi level (at E = 0). Together they probe the density of states (DOS).

The information on the occupied (unoccupied) DOS, which is immediately available from a PES (IPES) spectrum like that of figure 2, is extremely valuable, and cannot be obtained in such a direct way by any other technique. In an optical absorption experiment, for instance, occupied and unoccupied states are deeply intertwined and, for a given photon energy, only an average over many coupled states is measured. The drawback – but also a peculiar and interesting aspect – of PES is its large surface sensitivity. Photoelectrons can travel only a very short distance (~ 5-20 Å for kinetic energies in the 5-2000 eV range) inside a solid without suffering inelastic collisions with other electrons, phonons, or impurities. For most materials the energy dependence of the electron mean free path is expected or believed to follow the universal "U-shaped" curve of figure 3. Actually, this curve may not be so universal, and much work, especially on the high-energy side, is necessary – and partly in progress – to put it on firmer experimental grounds. Nonetheless, the existence of a minimum in the range of interest for PES and ARPES seems well established. Scattered electrons rapidly lose memory of their initial state, and their contribution to the spectrum is a shapeless *inelastic background* rising away from the Fermi level and peaking at very low kinetic energy.

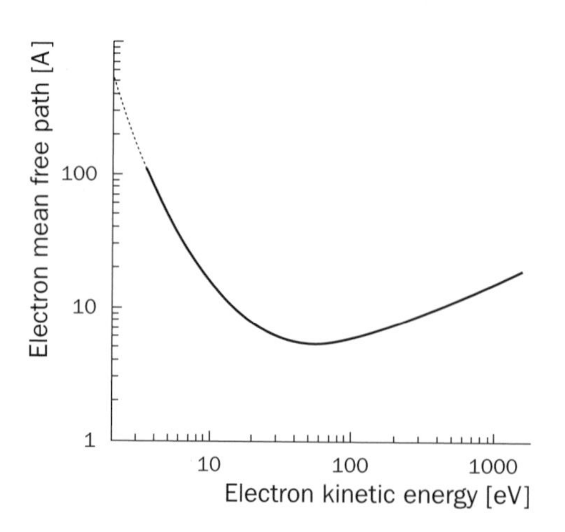

Figure 3 - "Universal curve" of the photoelectron elastic mean free path as a function of kinetic energy.

The *intrinsic* spectrum originates in the very first atomic layers of the sample. Therefore it is extremely important to prepare an atomically clean surface, and to limit the amount of subsequent contamination by keeping the pressure in the experimental apparatus at the lowest possible value. This explains why PES measurements of solids are usually performed in ultra-high (10^{-9}-10^{-11} torr) vacuum.

In conventional PES or ARPES experiments, UV photons are produced in rare gas discharges that generate sharp and intense emission lines [4]. He lamps are by far the most common laboratory sources. They produce a very intense line at 21.2 eV (He I) and a 10 to 20 times weaker satellite at 40.8 eV (He II). In the best lamps, the energy width of these lines is of the order of 1 meV, and their spectral brilliance is comparable to that of state-of-the-art SR beam lines. The higher energies necessary to probe core levels are produced by X-ray sources including standard (UHV-compatible) X-ray tubes or more sophisticated rotating-anode sources. When monochromated, the latter provide X-rays (usually at the Al Kα emission line at 1486 eV) of reasonable intensity with an energy resolution $\Delta E \sim 0.3$-0.5 eV.

In the laboratory the available excitation energies are limited to few UV and X-ray lines. The situation is radically different at a storage ring, where modern beam lines provide very brilliant and continuously tunable sources of photons, from the UV to the hard X-ray region. The properties of SR and the performances of SR beam lines are described in other Chapters of this Series, and will not be discussed here. PES and ARPES practitioners often exhibit an irrational affection for 21 and 40 eV photons even at synchrotrons. Nonetheless, even conservative users recognize the remarkable advantages (intensity, brilliance, polarization, time structure, tunability) offered by SR. The unique properties of SR open entirely new possibilities to photoelectron spectroscopy, some of which will be described in the next sections.

3. *PHOTOEMISSION IN THE SINGLE-PARTICLE LIMIT BAND MAPPING*

The intuitive description of photoemission sketched in the previous section provides a simple and qualitative interpretation of a PES experiment. In order to appreciate the full power of the technique, we need to consider more carefully the physical processes leading to the formation of the spectrum. In this section we will derive an expression for the PES and ARPES signal, in the simplifying hypothesis of non-interacting electrons. These results are the basis of band mapping, an important and widespread application of photoelectron spectroscopy. Correlation effects will be discussed in the next section.

The process at the heart of photoemission is the interaction of a photon with a many-electron system. We will assume that the system is initially in its ground state $|N,0\rangle$, with energy $E_0(N)$. We now consider transitions under the influence of the monochromatic e.m. field, represented by the perturbing Hamiltonian H_{int}, to the final states $|N,s\rangle$ of the N-electron system, where one of the particles is the photoelectron. The $|N,s\rangle$ are eigenstates of the unperturbed Hamiltonian $H_0(N)$, and "s" stands for a set of quantum numbers that completely identify the eigenstate. The transition probability from the ground state to a final state where the photoelectron is detected with a kinetic energy ε_K is given by the Fermi Golden Rule expression

$$p(\varepsilon_K) = \left(\frac{2\pi}{\hbar}\right) \sum_s \left| \langle N,s | H_{\text{int}} | N,0 \rangle \right|^2 \delta(E_s{}^N - E_0{}^N - \hbar\Omega) \qquad \{3\}$$

where $\hbar\Omega$ is the photon energy, and the Dirac δ ensures that the total energy is conserved in the transition. For the (limited) intensities encountered even at synchrotrons, only one-photon processes are important and one can retain in the interaction Hamiltonian only terms linear in the vector potential $\boldsymbol{A}$

$$H_{\text{int}} = \frac{e}{2m} \sum_i (\boldsymbol{p}_i \cdot \boldsymbol{A} + \boldsymbol{A} \cdot \boldsymbol{p}_i) \qquad \{4\}$$

Here $\boldsymbol{A} = \boldsymbol{A}_0\, e^{i\boldsymbol{q}\cdot\boldsymbol{r}}$, $\boldsymbol{p} = -i\hbar\nabla$ is the electron momentum operator, and the sum is over all electrons. In a homogeneous medium, and at least in the UV range, where photon wavelengths are considerably larger than characteristic interatomic distances, $\boldsymbol{A}$ can be considered constant: $\boldsymbol{A} \sim \boldsymbol{A}_0$ (dipole approximation), and the interaction term reduces to

$$H_{\text{int}} = \frac{e}{2m} \boldsymbol{A}_0 \cdot \sum_i \boldsymbol{p}_i \qquad \{4'\}$$

Notice that this approximation may not satisfactory at the surface of a solid, where the vector potential has a strong spatial dependence. Alternative forms of H_{int}, and the validity of the dipole approximation are extensively discussed in another Chapter of this series, and are omitted here.

In the limit of non-interacting electrons, the N-electron wavefunction can be written as the product of N independent single-particle wavefunctions $|j\rangle$. Exploiting the orthogonality of the $(N-1)$-particle states, the transition probability {3} becomes

$$\begin{aligned} p(\varepsilon_\kappa) &= \left(\frac{2\pi}{\hbar}\right)\sum_{k,j}\left|\langle\kappa|H_{\text{int}}|k\rangle\right|^2\left|\langle(N-1)^*,j|(N-1)^*,k\rangle\right|^2\delta(\varepsilon_\kappa-\varepsilon_j-\hbar\Omega) \\ &= \left(\frac{2\pi}{\hbar}\right)\sum_{k}\left|\langle\kappa|H_{\text{int}}|k\rangle\right|^2\delta(\varepsilon_\kappa-\varepsilon_k-\hbar\Omega) \\ &= \left(\frac{2\pi}{\hbar}\right)\sum_{k}\left|M_{k\kappa}\right|^2\delta(\varepsilon_\kappa-\varepsilon_k-\hbar\Omega) \end{aligned} \qquad \{5\}$$

This expression has a simple interpretation. The photon is absorbed by an individual electron, which makes a transition from an initial state $|k\rangle$ inside the solid, to a photoelectron state $|\kappa\rangle$ in vacuum. The remaining $(N-1)$ electrons are just "spectators", unaffected by the transition. The photoelectrons spectrum consists of "spikes" at energies $\varepsilon_\kappa = \varepsilon_k + \hbar\Omega$, with intensities determined by the single-particle dipole matrix element $M_{k\kappa}$.

In angle-resolved PES experiments, not only the energy ε_κ of the photoelectron is measured, but also its momentum $\boldsymbol{\kappa}$. We must therefore add to equation {5} a condition for momentum conservation. The momentum carried by the photon can be neglected with respect to the electron momentum in the UV or soft X-ray range ($\hbar k_{\text{ph}} = 5\ 10^7\ \text{cm}^{-1}$, and $\hbar k_{\text{el}} = 1.6\ 10^9\ \text{cm}^{-1}$ at 1 keV) and the condition takes the form $\delta(\boldsymbol{k}-\boldsymbol{\kappa})$. The probability of measuring a photoelectron with energy ε_κ and momentum $\boldsymbol{\kappa}$ is

$$p(\varepsilon_\kappa,\boldsymbol{\kappa}) = \left(\frac{2\pi}{\hbar}\right)\sum_{k}\left|M_{k\kappa}\right|^2\delta(\varepsilon_\kappa-\varepsilon_j-\hbar\Omega)\delta(\boldsymbol{k}-\boldsymbol{\kappa}) \qquad \{6\}$$

The simultaneous conservation of energy and momentum represents a very strict condition, involving the matching of the dispersion relations for the photoelectron in the vacuum, and the electrons in the system. The condition can be fulfilled only for special values of k. In general, for a given value of ε_k and an arbitrary emission angle – and therefore for arbitrary $\boldsymbol{\kappa}$ – no photoelectron will be observed.

Equation {6} could be used as the starting point to calculate transition probabilities from an initial state inside the solid (a Bloch state) to a free photoelectron state. This more rigorous *one-step* approach requires a careful definition and matching of the initial and final wavefunctions over the whole space (inside and outside the solid). It can be implemented on a computer, but it may be difficult to grasp the physics behind the calculated output. It is common practice to resort instead to the approximate but physically transparent *three-step model* [5], which breaks the photoemission process into three logically separated steps:

1. absorption of the photon,
2. propagation of the hot electron to the surface, and
3. escape into the vacuum.

They are described below.

Step 1: absorption - The first step describes the photon absorption and the transition from an initial one-particle Bloch state $|k\rangle$ to an excited final state $|f\rangle$ of energy $\varepsilon^* = \varepsilon_\kappa + \hbar\Omega$ ***inside*** the solid. It is interesting to note that the simultaneous energy and momentum conservation forbids the absorption of a photon in the in the free-electron limit. As shown in figure 4, *vertical* momentum-conserving transitions are incompatible with the free-electron parabolic dispersion (things are different at a surface, which acts as a "momentum reservoir"). Vertical transitions are possible only with the help of the lattice, which can provide (discrete quantities of) momentum to the electron. The conserved quantity is not *momentum*, but *crystal momentum*. The corresponding condition is $\boldsymbol{\kappa} = \boldsymbol{k} + \boldsymbol{G}$, and transitions are vertical only if all vectors are reduced to the first Brillouin zone (BZ). This is schematically illustrated in figure 4b, which shows one-dimensional nearly free-electron bands in the repeated zone scheme, and a free-electron parabola representing the limit of vanishing lattice potential. We consider the absorption of a photon of energy $\hbar\Omega$ by an electron initially in a state A within the lowest band, and we assume that the photon energy is sufficient to bring the electron into the second band. The electron can absorb the photon only if the energy difference between the initial state A, and the final state A' on the second band folded in the 1st BZ is equal to $\hbar\Omega$. The vertical transition $A \longrightarrow A'$ in the first BZ actually involves one reciprocal lattice vectors $\boldsymbol{G}$. The absorption process can be seen as the result of the scattering of the electron from A to B, followed by the vertical transition in the second zone. One should however be careful, and remember that points A and B are equivalent, and represent the same Bloch state. Transitions involving a larger number of reciprocal lattice vectors (e.g. *via C*) are also possible, but the dominant contribution is the one for which the final state lies on the "free electron" parabola.

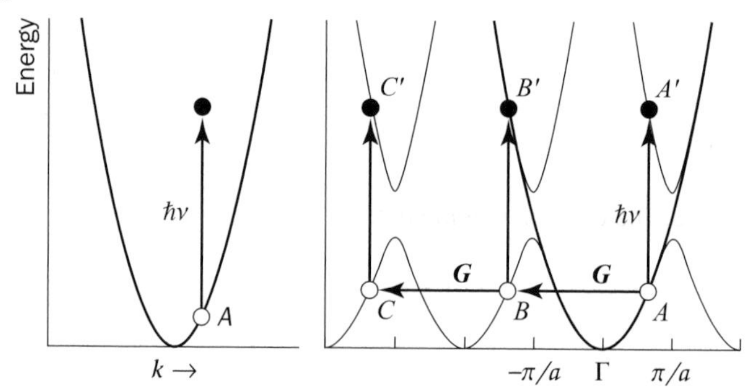

Figure 4 - An impossible "vertical transition" for a free-electron (left), and transitions involving reciprocal lattice vectors in a periodic potential.

Step 2-3: Transport to the surface and escape into the vacuum - The "hot" electrons generated in the absorption step must now propagate through the solid towards the surface, and then escape into the vacuum. While propagating they can be scattered out of their Bloch state by other electrons, phonons etc. These are the processes that determine the electron escape depth (figure 3). The scattered electrons represent a large portion of the photoelectron current and, with certain assumptions, it is possible to calculate their spectrum [5]. Here we are only interested in the intrinsic signal, which contains information on the band structure, and we will therefore neglect inelastic processes.

The "hot" electrons can escape from the solid only if several conditions are satisfied. First of all, the electron must be travelling towards the surface, and its total energy must be larger than that of the vacuum level. This however is not a sufficient condition to overcome the potential barrier at the surface. Figure 5 shows the constant energy curves for the "hot" electron in the solid (wavevector $\boldsymbol{k}$) and in the vacuum (wavevector $\boldsymbol{\kappa}$), as a function of the components of the wavevectors parallel ($k_{//}$) and perpendicular ($k_{\perp}$) to the surface. The photoelectron kinetic energy is $\varepsilon_{\kappa} = \varepsilon_k - eV_0$, where V_0 is the potential step at the surface. In the vacuum the photoelectron is free, and the surface of constant energy $\varepsilon_{\kappa} = \hbar^2\kappa^2/2\mathrm{m}$ is a sphere. For the "hot" electron in the solid, we also use a free electron approximation. The periodicity of the potential along the surface, both in vacuum and in the solid, requires that $\kappa_{//} = k_{//}$ (*modulo* a surface vector $\boldsymbol{G}_{//}$) while the only condition on $k_{\perp}$ is imposed by energy conservation. Therefore the electron is *refracted* at the surface, i.e. its direction is changed. Notice that, for a given value of ε_k, there is a limit value (k_L) of $k_{//}$ beyond which matching is not possible. Electrons approaching the surface at an angle larger than the corresponding limit angle are totally reflected by the surface.

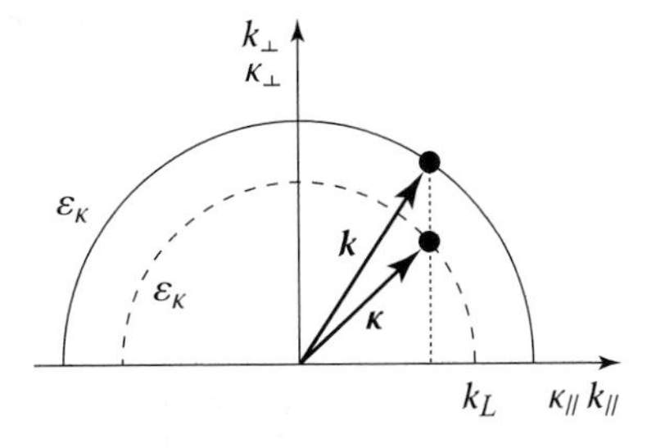

Figure 5 - Escape conditions for a "hot" electron at the sample surface. Only the component of the electron momentum along the surface is conserved. The electron is *refracted*.

Thanks to the free-electron relation a measurement of the kinetic energy of the photoelectron immediately yields the photoelectron momentum κ and, since $\kappa_{//} = k_{//}$, also the parallel component of the electron's wavevector in the solid

$$k_{//} = \frac{\sqrt{2m\varepsilon_{\kappa}}}{\hbar}\sin\theta \sim 0.512\varepsilon_{\kappa}^{1/2}\sin\theta \qquad \{7\}$$

where θ is the emission angle with respect to the surface normal, and energies are in eV. A complete determination of the electron's wavevector is more difficult because $k_\perp$ is not conserved and cannot be determined unless the dispersion relation for the hot electron is known. The choice of a free-electron dispersion (as in figure 4) is satisfactory for large enough energies, for which the wavefunctions should not be too much distorted by the ionic potential. When available, improved calculated final state bands can of course be used. Notice also that $k_\perp$ is not precisely known, so that final state bands are "blurred". This uncertainty reflects the partial localization of the photoelectron near the surface (the mean free path of the hot electron, in the 3-step model). The corresponding partial relaxation of momentum conservation sets a fundamental limit on the accuracy of any band structure determination.

Band mapping is easier in two-dimensional (2D) systems, e.g. at surfaces and in layered materials, where only $k_{//}$ is a good quantum number. We shall see later that these systems are particularly suitable for ARPES for yet another reason: only in 2D the spectral signatures of the hole (the interesting part!) and of the "hot" photoelectron can be decoupled. Full 3D band mapping requires collecting ARPES data from different crystal surfaces, and the initial states are identified by a triangulation procedure. An account of these rather specialized techniques is given in ref. **[2]**. The band states of real 3D materials (e.g. Cu or Ag) have been obtained with considerable precision using these techniques. However the recent popularity of ARPES is at least in part due to the quasi-2D nature of the superconducting cuprates, which justifies the use of the simpler approach.

In an ARPES experiment the kinetic energy of the photoelectrons is measured in a narrow angular window, as a function of the emission direction (polar and azimuthal). In practice this is done either keeping the sample in a fixed position relative to the photon beam, and moving the analyzer in the half space above the surface, or having the analyzer fixed and rotating the sample around one or two axes. The second option is technically simpler but it has the disadvantage of changing the angle between the polarization vector of the light and the sample. This introduces matrix elements, the effects of which may severely modify the measured intensity distribution (but, of course, not the energies).

Band mapping is time consuming, and ARPES spectra are usually measured only along high symmetry directions, where a comparison with band structure calculations is possible. Typically the polar angle is varied at fixed azimuthal angle, in a high-symmetry plane, thus varying the parallel momentum according to equation {7}. The outcome of such an experiment on the conduction band of metallic sample is schematically illustrated in figure 6. The energy of the ARPES peak changes with the emission angle (and $k_{//}$), reproducing the band dispersion. Eventually, at the Fermi wavevector $\boldsymbol{k}_F$, the peak reaches the Fermi energy. Beyond this angle (wavevector) the peak disappears, and the band cannot be followed any longer. Different strategies to sample the momentum space, e.g. in circles at fixed polar angles, are of course possible.

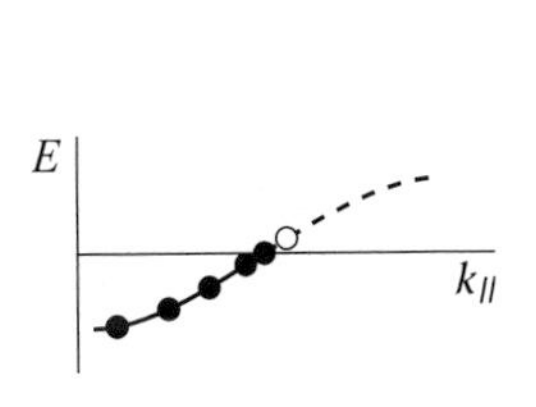

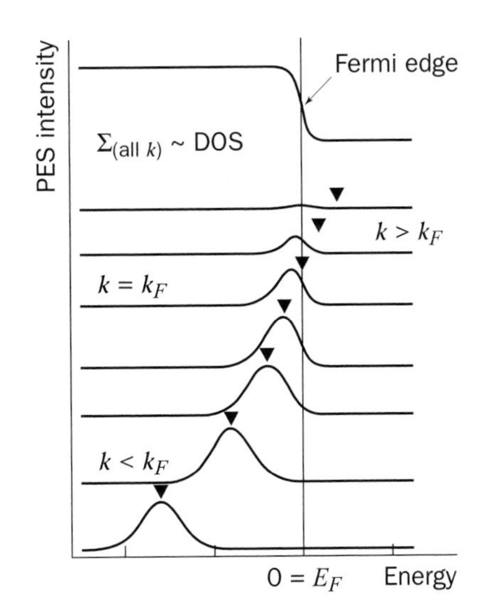

Figure 6
An ARPES experiment on a metal. The energy of the ARPES peak changes with the emission angle (with parallel momentum), reproducing the dispersion of the underlying band.

More recently a new generation of electron analyzers, equipped with two-dimensional detectors, allow spectra from different emission angles to be collected in parallel within a defined angular window, typically ~ 12° × 1°, or larger. This parallel detection mode has largely improved the efficiency of band mapping, and made possible a dense sampling of the band structure, with various "paving" options of the Brillouin zone. The results of an angular scan can be displayed in two equivalent ways, as in figure 7. In the most common representation (left) a series of spectra measured at different emission angles, are stacked on top of each other, in a "real life" version of figure 6. The intensity and angular dependence of the spectra can be mapped into a two-dimensional "ARPES intensity map" (right) of the measured intensity $I(k,E)$. Each column of the map corresponds to a spectrum of the stack. This representation is quite appealing because it is directly comparable with the calculated band structure. Subtle two-dimensional correlations are also more easily captured by the eye from the $I(k,E)$ map than from the raw spectra.

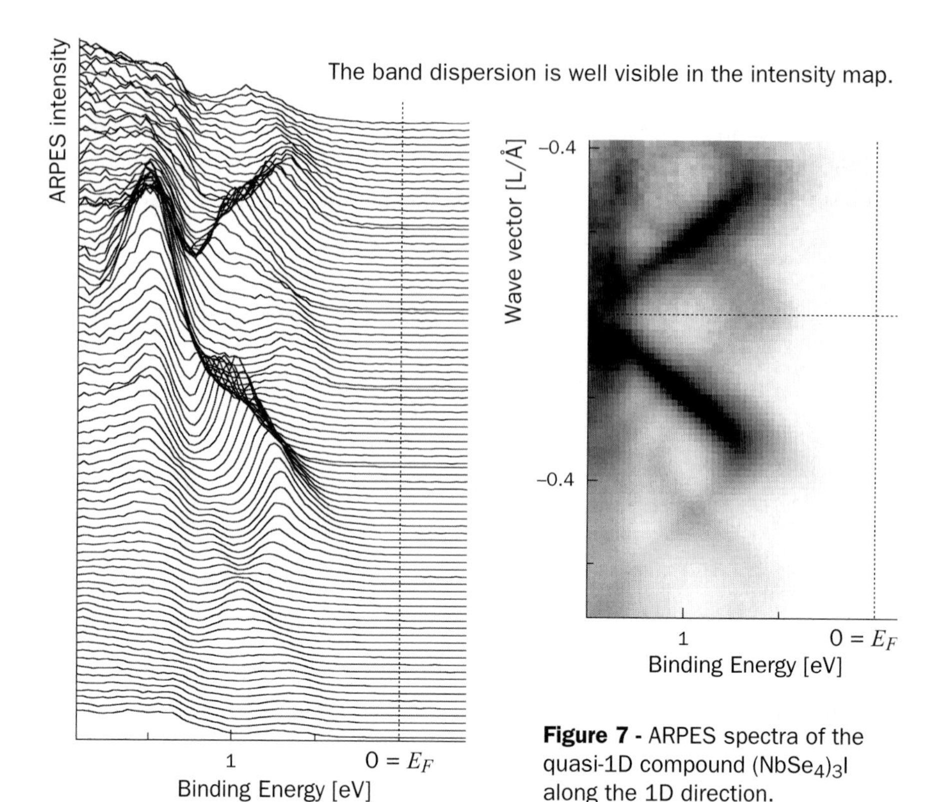

Figure 7 - ARPES spectra of the quasi-1D compound $(NbSe_4)_3I$ along the 1D direction.

Data like those of figure 7 contain, for one particular direction of momentum space, information on the whole band structure. Sometimes this is more than is actually needed. Often the most crucial information is the momentum distribution of electron states at the Fermi level, i.e. the Fermi surface (FS), but measuring the complete band dispersion is a rather inefficient way to map the FS. As an interesting alternative, the energy window of the analyzer can be set exactly at E_F, and the complete angular distribution of the photoelectrons can be mapped at this particular energy. The resulting intensity map is a stereographic projection on the FS, which is then transformed into a projection along the two surface components of the wavevector, as in figure 8 [6]. In the particular case of a two-dimensional system the map directly represents the FS. The capability of ARPES of producing Fermi surface maps has been thoroughly exploited especially in studies of the high-T_c cuprates. ARPES is certainly not the only probe of the FS of a metal. However, it has two interesting advantages over more conventional magneto-transport techniques:

1. ARPES potentially yields the detailed shape of the FS, and not just the areas of particular orbits,
2. FS mapping by ARPES does not require very low temperatures or extremely pure samples.

Note that neither of these conditions could be fulfilled in the case of the SC cuprates.

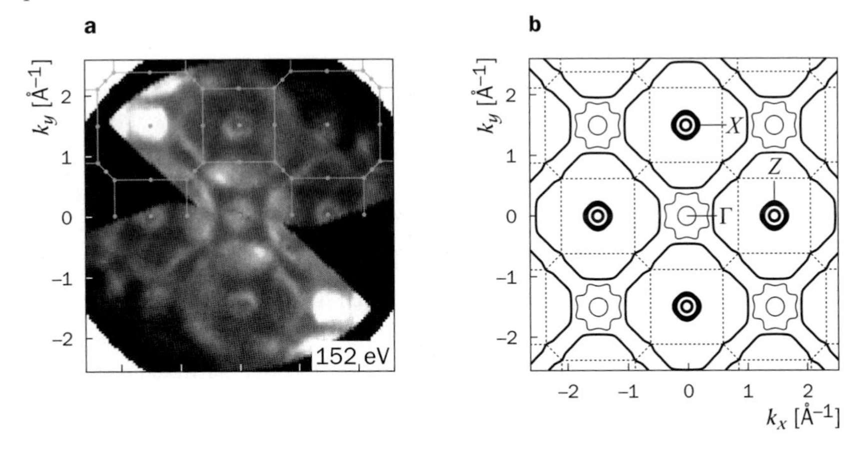

Figure 8 - ARPES intensity map of $LaRu_2Si_2$ measured at $E = E_F$, compared with a theoretical Fermi surface from ref. **[6]**.

4. BEYOND THE SINGLE-PARTICLE APPROXIMATION

The independent-particle hypothesis, which was the basis of our discussion in the previous Section, is certainly questionable when interactions among the electrons, or between the electrons and other excitations, are large. These effects are of course always present to some extent, but for weakly interacting systems they introduce only quantitative perturbations to the predictions of the simple model. In other cases, the nature of the ground state and of the excitations depends qualitatively on interactions, and the spectral properties are correspondingly modified. New structures appear, which contains extremely useful information on the interacting system. In order to understand and exploit the spectral features of interacting systems, we need to develop a more general framework.

The starting point is again equation {3}, which describes in general terms the transition from the ground state to the final state. We will not assume that the N-particle electronic wavefunctions is a product of independent one-particle wavefunctions, but only that the photoelectron is instantaneously decoupled from remaining the $(N-1)$-electron system. Semi-quantitative arguments on the response time of the spectator electrons can be used to justify this *sudden approximation*. From a practical point of view, its validity is proven *a posteriori* by a comparing the experimental results with the theoretical predictions. With this assumption, and after some juggling with operators, we obtain the following expression

$$\begin{aligned} p(\varepsilon_\kappa, \boldsymbol{k}) &= \left(\frac{2\pi}{\hbar}\right) \sum_s \left| \langle N,s | H_{\text{int}} | N,0 \rangle \right|^2 \delta(E_s{}^N - E_0{}^N - \hbar\Omega) \\ &= \left(\frac{2\pi}{\hbar}\right) |M_{k\kappa}|^2 \sum_s \left| \langle (N-1),s | a_k | N,0 \rangle \right|^2 \delta(\varepsilon_\kappa - \varepsilon_s - \hbar\Omega) \end{aligned} \quad \{8\}$$

$|(N-1), s\rangle$ is an eigenstate of the $(N-1)$-electron Hamiltonian, the operator a_k destroys one electron with wavevector $\boldsymbol{k}$, and $\varepsilon_s = E_0^N - E_s^{N-1} = (E_0^N - E_0^{N-1}) - (E_s^{N-1} - E_0^{N-1}) = \varepsilon_0 - \Delta\varepsilon_s$. This is a very important expression, which illustrates the physics of the PES process in the sudden limit. The photon absorption instantaneously creates an $(N-1)$-electron state $a_k|N,0\rangle$, which is not an eigenstate of $H_0(N-1)$, the Hamiltonian of the $(N-1)$ particle system. The spectrum is the projection of this "frozen state" onto the "fully relaxed" eigenstates $|(N-1), s\rangle$. The "fundamental" peak (0), at kinetic energy $\varepsilon_{\kappa 0} = E_0^N - E_s^{N-1} + \hbar\Omega$, corresponds to a transition, which leaves the $(N-1)$ system in its ground state $|(N-1),0\rangle$. The spectrum also exhibits a number of extra peaks corresponding to the excited eigenstates $|(N-1), s\rangle$, with kinetic energies reduced by the corresponding excitation energies: $\varepsilon_{\kappa s} = E_0^N - E_s^{N-1} + \hbar\Omega = \varepsilon_{\kappa 0} - \Delta\varepsilon_s$ as schematically illustrated by figure 9. It is hard to overstate the importance of this point. No matter how complex the physical system, the correct approach to interpret a PES spectrum is to consider all the transitions allowed by symmetry from the ground state to the final states.

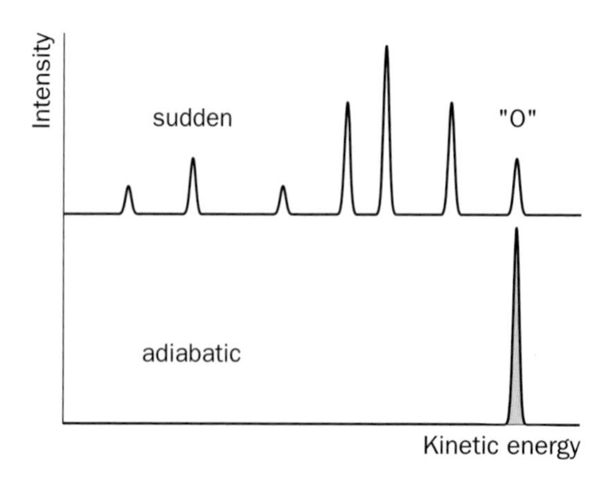

Figure 9 - Artist's view of an ARPES spectrum in the opposite sudden and adiabatic limits. The "0" transition corresponds to the ground state of the $(N-1)$-electron system.

Notice that the energy position of the peaks does not depend on the validity of the sudden approximation. Even if the sudden approximation is not satisfied, the photoelectron is eventually completely decoupled from the remaining $(N-1)$ electrons. Due to energy conservation, a measurement of the photoelectron kinetic energy is equivalent to measuring the energy of the $(N-1)$-electron system, and the only possible results are the eigenvalues of $H_0(N-1)$. The relative intensities of the various peaks on the other hand depend on how fast the decoupling takes place. In the adiabatic limit, where the photoelectron is removed on a time scale which is large with respect to the relaxation time, the $(N-1)$-electron system directly evolves into its ground state, and all the spectral intensity goes into the "0" peak. Any intermediate situation is in principle possible.

THE SPECTRAL FUNCTION - A THEORETICAL INTERLUDE

Equation (8) is all we need to interpret a PES spectrum. However it is instructive to write the expression for the photocurrent in a different way, which makes use of two fundamental theoretical concepts: the single particle Green's function $G(\boldsymbol{k},\omega)$ and the spectral density function $A(\boldsymbol{k},\omega)$. The interested reader will find exhaustive descriptions of these quantities in references **[7-9]**. Here we just recall that the one-electron Green's function – or propagator – $G(x,x';t,t')$ describes the following chain of events: at time t' an electron $(t>t')$ or hole $(t<t')$ is added to the ground state of the many-body system at x'; it interacts with the other electrons between time t' and t, and it is then removed from the system at time t and coordinates x. $G(x,x';t,t')$ expresses the probability amplitude that, after the addition, interaction, and removal of the extra particle, the N-electron system remains in its ground state. The probability will in general be smaller than 1 in interacting systems. One should notice that the Green's function deals with the addition or removal of a real electron (or hole), and it is therefore deeply related to the photoemission and inverse photoemission process.

The spectral density function $A(\boldsymbol{k},\omega)$, or spectral function, in short, appears in the spectral representation of $G(\boldsymbol{k},\omega)$, the position and time Fourier transform of $G(x,x';t,t')$

$$G(\boldsymbol{k},\omega) = \int \frac{A(\boldsymbol{k},\omega)}{\omega-\omega^* + i\eta\,\mathrm{sgn}(\omega^*)}\mathrm{d}\omega^* \qquad \{9\}$$

where ω is the energy measured from the chemical potential. After a skilful manipulation equation {8} one arrives at the following important result

$$p(\varepsilon_\kappa,\boldsymbol{k}) = \left(\frac{2\pi}{\hbar}\right)\left|M_{\boldsymbol{k}\kappa}\right|^2 A(\boldsymbol{k},\omega) = \left(\frac{2\pi}{\hbar}\right)\left|M_{\boldsymbol{k}\kappa}\right|^2 \frac{1}{\pi}\left|\mathrm{Im}G(\boldsymbol{k},\omega)\right| \qquad \{10\}$$

which gives the ARPES ($\omega < 0$) and IPES ($\omega > 0$) spectra in terms of $A(\boldsymbol{k},\omega)$, and demonstrates the deep connection between the two techniques. Conversely, a combined PES and IPES experiment yields, apart from dipole matrix elements, the complete spectral function and, through {9}, the propagator. This justifies the interest for electron spectroscopy, since $G(\boldsymbol{k},\omega)$ is a fundamental theoretical tool for the description of interacting systems. Notice that the above expressions are valid for $T = 0$. At finite temperature the spectrum depends explicitly on temperature through a Fermi-Dirac factor. There is also an implicit dependence through the T-dependence of G. The Green's function – and therefore the spectral function – exhibits a normal T-dependence (thermal excitations, electron-phonon scattering), but an extraordinary dependence may occur at a phase transition. Examples of both kinds of T-dependences are illustrated below.

The deep connection between PES/IPES spectra and the one-particle Green's function is not unexpected if we consider that $G(\boldsymbol{k},\omega)$ is precisely defined in terms of one electron removal and addition. It is also a direct consequence of the sudden character of the perturbation, and therefore of the large excitation energy. This directly contradicts the common criticism according to which in high-energy spectroscopy the perturbation is so strong that all "fine details" are lost. We see that quite the opposite is true: since the excitation energy is large, and the photoelectron is rapidly removed, the experiment reveals the one-particle spectral function. The spectrum in principle contains structures at all relevant energy scales of the problem, down to the low-energy excitations, which shape the low temperature physical properties of a material. Therefore spectra measured with keV photons can probe meV excitations. Retrieving all this information from the experimental spectrum is a technical problem, not a fundamental one.

The expressions for the propagator and the spectral function are particularly simple in the limit of non-interacting particles

$$G_0(\boldsymbol{k},\omega) = \frac{1}{\omega - \varepsilon_k - i\eta} \qquad \{11\}$$

$$A_0(\boldsymbol{k},\omega) = \delta(\omega - \varepsilon_k) \qquad \{12\}$$

The Green's function of a system of non-interacting fermions has poles – with unitary weight – at the single-particle energies ε_k (η is an infinitesimal quantity). The spectral function (that is the PES/BIS spectrum) simply consists of delta functions at wavevectors $\boldsymbol{k}$, and corresponding energies ε_k, according to the simple picture of figure 6.

Many-body effects modify the properties of a real system, and produce changes in the one-particle Green's function. Simple electrons and holes are not stable excitations, but decay with time. The simplest approach to the many-body problem is to add to the free-particle Green's function an exponential decay with a phenomenological *lifetime* $\tau = (1/\Gamma)$. After a Fourier transform, this yields

$$G_\Gamma(\boldsymbol{k},\omega) = \frac{1}{\omega - \varepsilon_k + i\Gamma} \qquad \{13\}$$

which now has complex poles, and **[9]**

$$A_\Gamma(\boldsymbol{k},\omega) = \frac{1}{\pi}\frac{\Gamma}{(\omega - \varepsilon_k)^2 + \Gamma^2} \qquad \{14\}$$

This is an intuitive results: because of the finite lifetime of the excitations, the peaks in the spectrum acquire a finite width $\Delta E = 2\Gamma$, and a Lorentzian lineshape. The energy position of the peaks, on the other hand, remains unchanged.

Although conceptually useful, the Green's function {13} is an oversimplification, and a more sophisticated approach is required to construct a Green's function with the "good" analytical properties. It is customary to lump the effects of interactions in the self-energy operator $\Sigma(\boldsymbol{k},\omega)$, which yields the simple expression, analogous to the independent particle case

$$G(\boldsymbol{k},\omega) = \frac{1}{\omega - \varepsilon_k - \Sigma(\boldsymbol{k},\omega)} \qquad \{15\}$$

Close to the Fermi surface then the poles of $G(\boldsymbol{k},\omega)$ are defined self-consistently by the Dyson equation

$$\omega_s = E_k = \varepsilon_k + \mathrm{Re}\,\Sigma(\boldsymbol{k}, E_k) \qquad \{16\}$$

In itself this is no real progress, because the difficulty has only been shifted to the determination of Σ. Luckily, the powerful machinery of perturbation theory can be deployed, and solutions can be found for important cases [8].

We are interested in the structure and properties of the spectral function

$$A(\boldsymbol{k},\omega) = \frac{1}{\pi}\frac{|\mathrm{Im}\,\Sigma(\boldsymbol{k},\omega)|}{|\omega - \varepsilon_k - \mathrm{Re}\,\Sigma(\boldsymbol{k},\omega)|^2 + |\mathrm{Im}\,\Sigma(\boldsymbol{k},\omega)|^2} \qquad \{17\}$$

Unlike the case of exponentially decaying excitations, the spectral function does not have a symmetric Lorentzian line shape. The peak width $\Delta E = 2\,\mathrm{Im}\,\Sigma(\boldsymbol{k},\omega)$ is energy dependent, and the peak position is shifted by the $\mathrm{Re}\,\Sigma$ term. The lifetime of excitations is infinite at the Fermi surface, and decreases away from it because an electron (or hole) can be scattered away from its initial state, with the creation of electron-hole excitations across the Fermi surface. For a normal metal (a *Fermi liquid*, see below) the self-energy has the asymptotic behaviour $\mathrm{Im}\,\Sigma(\boldsymbol{k},\omega) \sim \omega^2$, and $\mathrm{Re}\,\Sigma(\boldsymbol{k},\omega) \sim -\omega$ near the Fermi surface.

Expanding the denominator, one can rewrite equation {17} as

$$A(\boldsymbol{k},\omega) = Z_k \frac{1}{\pi} \frac{|\mathrm{Im}\,\Sigma(\boldsymbol{k},\omega)|}{|\omega - E_k|^2 + |\mathrm{Im}\,\Sigma(\boldsymbol{k},\omega)|^2} + A_{\mathrm{inc}} \qquad \{18\}$$

where $Z_k^{-1} = (1 - \partial \mathrm{Re}\Sigma / \partial \omega)$, and A_{inc} is the *incoherent* part of the spectral function. This expression is worth a few words of comment. Near the Fermi surface, for a given $\boldsymbol{k}$, the spectral function has an almost Lorentzian line shape centred at $E_k \approx \varepsilon_k$. Therefore, we can identify a one-to-one correspondence between the electron and hole states of the independent particle problem, and the excitations of the many body system. The lifetime of these *quasi-particles* (QPs) $\tau = (2\ \mathrm{Im}\ \Sigma)^{-1} = (1/\Gamma)$ is long enough to grant them the status of individual excitations. This correspondence between the one-particle states and the quasi-particles is a cornerstone of Landau's Fermi liquid (FL) theory. A QP may be pictured as a bare particle, surrounded or *dressed* by a cloud of virtual excitations. In particular, when $\boldsymbol{k} = \boldsymbol{k}_F$, the spectral function is a delta function. This is directly related to another characteristic property of Fermi liquid, the persistence of a sharp Fermi surface in the presence of interactions.

A major difference between the independent particle {12} and the many-body {18} expressions is the smaller integrated weight ($Z_k < 1$) in the latter. This tells us that electrons and holes are not proper excitations of the system. The QP weight Z_k is the weight of the corresponding state $|k\rangle$ of the non-interacting system (the "bare particle"), while $(1 - Z_k)$ is the weight of the correlation cloud, i.e. the weight of other excitations that contribute to the QP state. The spectral weight missing in the QP peak is redistributed, possibly over a much wider energy range, to A_{inc} which reflects different degrees of freedom of the system, like charge-transfer excitations or collective excitations like plasmons or phonons.

The spectral function satisfies an important sum-rule

$$n_k = \int_{-\infty}^{0} A(k < k_F, \omega)\, \mathrm{d}\omega < 1 \qquad \{19\}$$

where n_k represents the average occupation of the one-particle state $|k\rangle$. Since $n_k < 1$ there is some spectral weight above the chemical potential even for $\boldsymbol{k} < \boldsymbol{k}_F$. The factor $Z(k_F) < 1$ that renormalizes the QP weight at the Fermi level, is equal to the step of the n_k function at the Fermi surface. It is also the factor that renormalizes the density of states (and the related physical properties: χ, γ, m^*) of the interacting system. Qualitatively, strong correlations yield small Z values, and a correspondingly small QP spectral weight. This is summarized in figure 10, which compares the spectral function of a hypothetical non-interacting metal and of a

Fermi liquid. In the non-interacting case the sharp peaks – their width is resolution limited – cross the chemical potential following the band dispersion, as in figure 6. The momentum distribution function exhibits a step $n(k) = 1$ at k_F. In the interacting system the peaks acquire a finite, energy dependent width, and lose spectral weight, which is moved to the *incoherent* sidebands. The dispersion is renormalized, and $n(k) = Z < 1$.

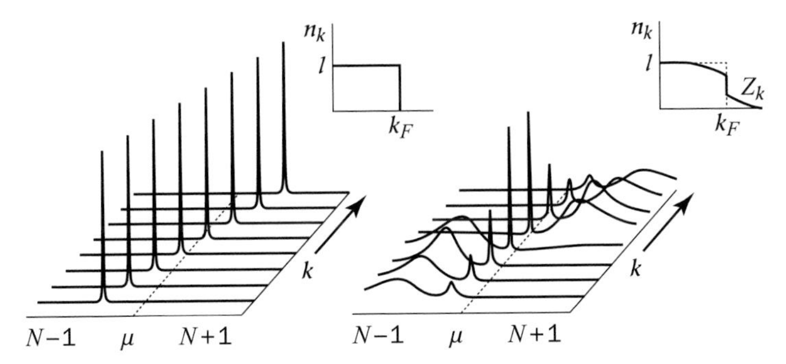

Figure 10 - Artist's view of the *k*-dependent ARPES spectral function is the free-electron limit (left) and in an interacting system (right) [adapted from M. Meinders, PhD Thesis, Groningen (1994)].

The predicted consequences of interactions are experimentally observed as in the classic example of figure 11 [10]. Sodium is a nearly free-electron system, where one might expect negligible correlations effects, but the measured dispersion indicates a 30% renormalization of the electronic mass with respect to local density (LDA) calculations. Using the language of the previous paragraphs, this is a direct manifestation of a nonzero Re Σ term. For stronger correlations the QP renormalization is more dramatic, and the spectrum can be drastically modified.

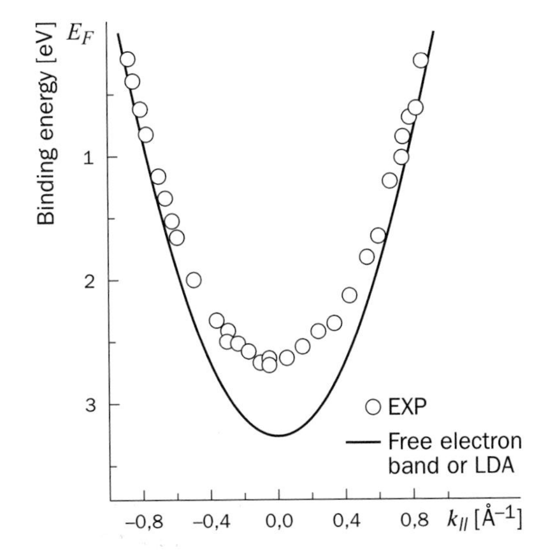

Figure 11 - Theoretical (line) and experimental (o) dispersion of the conduction band of sodium from ref. **[10]**. The band narrowing is a typical consequence of electronic correlations.

A typical example is that of the "heavy fermion" materials, where masses 10^3 times larger than the bare electron mass are not uncommon. In this case the QP band is compressed in a very narrow energy region, and its weight is vanishingly small.

Further correlation effects are revealed by an analysis of the QP line shape, which can be performed in two complementary ways, as schematically illustrated in figure 12. The experimental ARPES intensity map $I(k,E)$, which reproduces the band dispersion, can be "sliced" either at constant momentum (k^*), or at constant energy (E^*). The constant momentum cut yields the usual EDC $I(k^*,E)$. The QP displays the characteristic FL line shape (eq. {17}), and the energy width yields the QP scattering rate $\Gamma = (1/\tau)$. The constant energy cut yields a Momentum Distribution Curve (MDC), i.e. the momentum distribution of the spectral weight for a fixed energy $I(k^*,E)$. The information contained in the MDC is complementary to that of an EDC, but the line shape is simpler because the electron self-energy Σ depends only weakly on k. If this dependence is neglected altogether, equation {17} shows that the MDC $A(k,E^*)$ is a Lorentzian, and its intrinsic linewidth is inversely proportional to the QP coherence length: $\Delta k = (1/l)$. Notice that l and τ are related to each other through the band's group velocity: $l = v_{QP}{}^*\tau$.

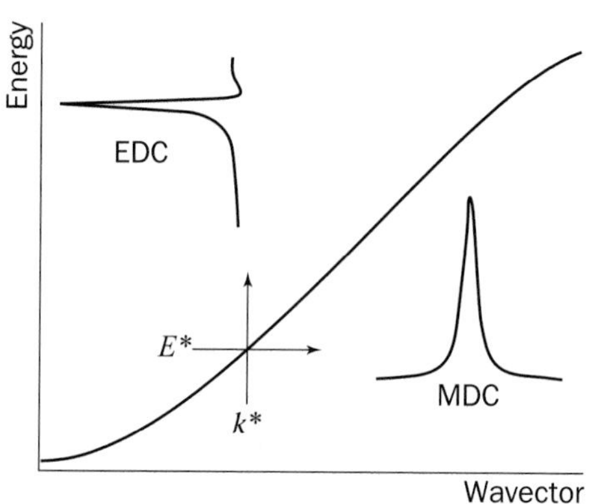

Figure 12 - Cuts of the ARPES intensity map produce Energy distribution curves (EDCs, at constant k) or Momentum distribution curves (MDCs, at constant energy)

Extracting the intrinsic information on the QP lifetime and coherence length contained in the spectral line shape faces two kinds of obstacles. The experimental energy and momentum broadening must be small enough not to mask the intrinsic contributions, but this is not a sufficient condition. There is a further complication, which we have neglected so far. We assumed (eq. {10}) that the ARPES spectrum reflects the one-particle excitation in the many body system, i.e. the hole spectral function. In reality, the spectrum is the product of the spectral functions of the hole and of the excited "hot electron". The electron – at energy $\varepsilon^* = \varepsilon_k + h\nu$ – also interacts with the many-body system, and this limits its lifetime and mean free path. The width of the QP spectrum contains contributions from both scattering rates Γ_h and Γ_e. It is possible to show [11] that the electron's contribution is weighed by the ratio of the hole and electron group velocities perpendicular to the surface. At normal emission the total spectral width is given by

$$\Gamma = \frac{\Gamma_h + \Gamma_e \left| v_{h\perp} / v_{e\perp} \right|}{\left| 1 - (v_{h\perp} / v_{e\perp}) \right|} \qquad \{20\}$$

In the limit of a zero perpendicular hole velocity, i.e. in the absence of perpendicular band dispersion, the linewidth simply reflects the hole lifetime. Once again, we verify that ARPES is especially well suited for the study of 2D systems, like surface states and, to a good approximation, layered quasi-2D materials like the cuprates.

Besides "natural" 2D systems, artificial systems can be tailored to investigate the spectral properties of electrons in metals. When thin and highly perfect overlayers are grown epitaxially over a single crystal, it is possible to obtain a two-dimensional electron gas, which is confined in the perpendicular direction by the discontinuities of the potential at the overlayer-vacuum and overlayer-substrate interfaces. In this realization of the classical "particle in a box" problem, electron states are allowed only when standing wave conditions are satisfied along the surface normal. Since the perpendicular dispersion is zero, such a configuration is ideal for a measurement of the hole spectral properties. Figure 13 [12] shows normal emission spectra from Ag layers of various thickness grown epitaxially over a Fe(100) substrate. The sharp peaks represent allowed states with $k_{//} = 0$, and it is clear that the number of states increases with the layer thickness, as expected from the particle in the box solution. The sharpness of the peaks proves the quality of the film, namely the high reflectivity at both interfaces (an analogy can be made with the "finesse" of a Fabry-Perot resonator). Studying the parallel dispersion of these quantum well states, and their temperature-dependent linewidth, the scattering rate of carriers at the Fermi surface of Ag was evaluated. It was also possible to determine the Fermi wavevector with very high accuracy, improving by 1% the accepted value obtained by traditional de Haas-Van Alphen measurements.

5. *CASE STUDIES*

5.1. *FERMI LIQUID LINESHAPE IN A NORMAL METAL*

The close relation between the ARPES spectrum and the hole spectral function $A(k,\omega)$ can be exploited to test the theoretical predictions of the Fermi liquid theory for a normal metal. Such experiments have a fundamental interest, because the combined energy and momentum resolution of ARPES can probe unique aspects of the theory. They also have a more practical goal. There is a large current interest in materials, like the superconducting cuprates or the colossal magnetoresistance manganites, for which the theoretical framework is much less developed. ARPES is one of the leading tools of investigation, and it is extremely important to dispose of a reliable Fermi liquid reference, against which the "unusual" or novel aspects of the spectral properties can be compared.

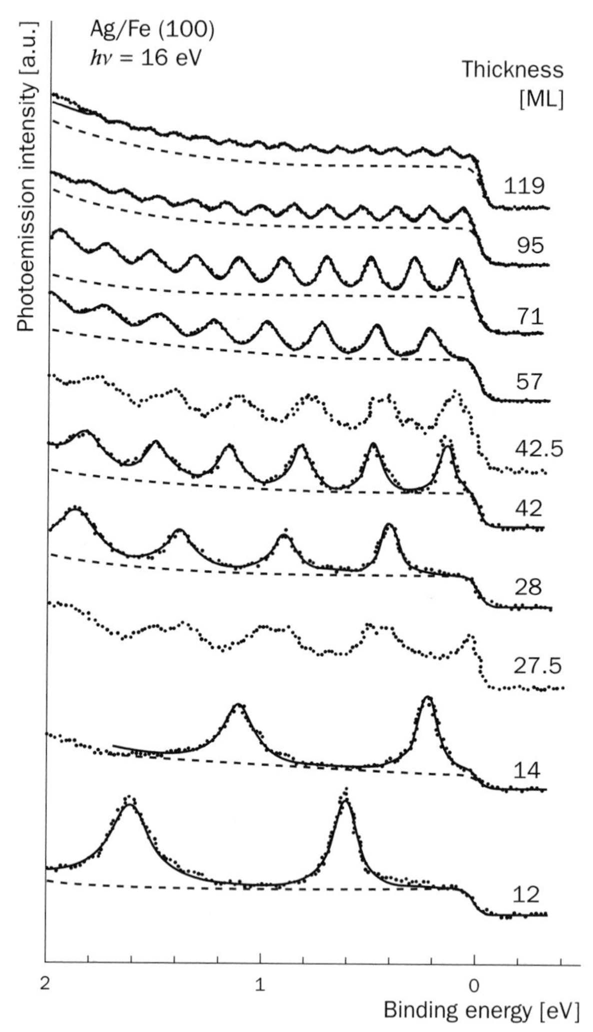

Figure 13 - Normal emission spectra from quantum well states in epitaxial Ag(100) films of varying thickness from ref. **[12]**.

We have seen in the previous section that quasi-2D materials are more appropriate than 3D metals like Cu, or Al, for an accurate ARPES determination of the hole spectral function. The layered TM chalcogenides MX_2 (X = S, Se, Te) exhibit quasi-2D properties due to the arrangement of the metal atoms in planes, sandwiched between layers of chalcogen atoms, and weakly interacting with each other. Specifically, *1T*-$TiTe_2$ (1 T indicates a polytype where the Ti cations are octahedrally coordinated to the Te anions) is a normal metal with moderately weak correlations. The ARPES spectra of figure 14 [13] illustrate the dispersion of the Ti 3d conduction band of $TiTe_2$, along the high-symmetry ΓM direction of the hexagonal Brillouin zone. The band, which lies above E_F at Γ (0°), forms an electron pocket around the *M* point (~ 27°), and the figure clearly shows a QP feature moving below E_F for $\theta > 16°$. The very weak background and the absence of other interfering features in this part of the *BZ*, make it a very favourable case for a line shape analysis. Near the Fermi level crossing (at 16°) the spectra are reproduced by a Fermi liquid line shape, with the asymptotic form of the Fermi liquid self-energy $\mathrm{Im}\,\Sigma = \beta E^2$, plus an energy-independent broadening.

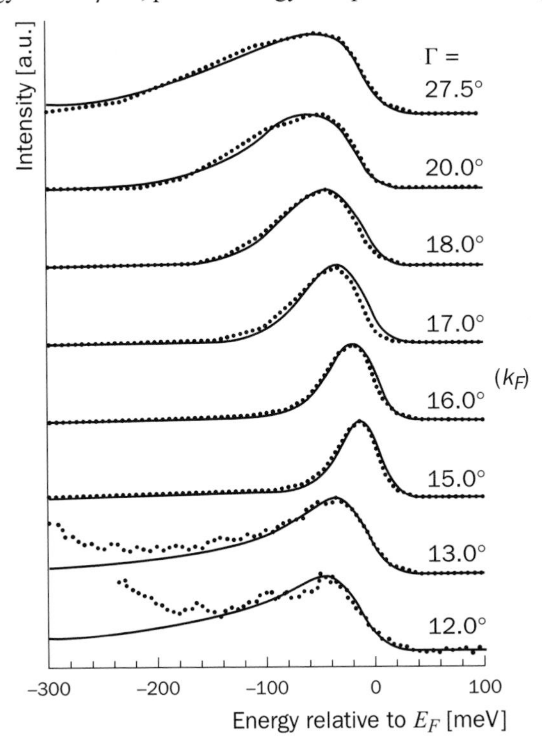

Figure 14 - ARPES Fermi liquid line shapes in the 2D metal $TiTe_2$ from ref. **[13]**. Notice the typical FL tail beyond k_F.

This simple form describes the increasing peak width (remember that $\Delta E = 2\ \mathrm{Im}\,\Sigma$) and the decreasing lifetime when the QP moves away from the Fermi surface. With a more elaborate expression for Σ, it is also possible to reproduce the strongly asymmetric line shape even far from the Fermi surface (e.g. at 20° and 27.5°). One noteworthy aspect of these data is the weak structure (arbitrarily enhanced in the figure) seen at $k < k_F$ ($\theta < 16°$), representing the tail of the QP peak, which lies above the Fermi energy. This tail is a distinctive many-body signature, and is reproduced by the FL expression. The observation of this feature confirms a fundamental theoretical prediction: the ground state of the interacting system contains contributions from one-particle states lying outside the Fermi surface.

The relation $\mathrm{Im}\,\Sigma = \beta E^2$ implies an infinite QP lifetime and an infinitely sharp QP spectral peak, at the Fermi surface. A quantitative analysis of the spectrum at $k = k_F$ in figure 14 reveals a residual broadening which cannot be attributed to the experimental resolution. It shows that the previous description is incomplete in (at least) two respects:

1. it only considers scattering between QPs, and
2. it is only valid at $T = 0$.

QPs can be scattered also by impurities and phonons (and other collective excitations), and all these processes limit their lifetime. While impurity scattering is temperature independent, phonon scattering is strongly T-dependent, and represents the most important broadening term near E_F at finite T. Notice that the Fermi liquid self-energy also exhibits a T^2 dependence, which however smaller can usually be neglected.

The phonon term can be evaluated from the temperature dependence of the ARPES spectra. Figure 15 [14] shows spectra of $TiTe_2$ measured at the Fermi surface at temperatures between 13 K and 237 K. With increasing temperature the linewidth increases and the peak position, which is resolution limited at $T = 13$ K, moves to $E = 0$. Part of these changes is due to the temperature dependence of the Fermi function, which sharply cuts the QP spectrum at the Fermi, level at the lowest temperature. This effect, like the effect of the finite experimental resolution, can be taken into account in a fit of the spectra with a temperature-dependent self-energy $\Sigma(T)$ based on an appropriate theory of electron-phonon scattering [7]. The fit then yields the intrinsic temperature-dependent linewidth Γ(inset). The linear slope is proportional to the electron-phonon coupling parameter λ, and offers a direct determination of this important quantity. For $TiTe_2$ one obtains $\lambda = 0.22$, and the moderate coupling explains the absence of instabilities (superconductivity, charge-density waves) in this compound. At low temperature Γ saturates to a finite value, while both the electron-electron and electron-phonon contributions give a null contribution in the $T = 0$ limit. This residual term includes impurity scattering, which also determines the residual value of the electrical resistivity. Independent considerations, however, indicate that the largest contribution is not due to QP scattering mechanisms, but rather to the hot electron lifetime, according to equation {20}. The small but finite coupling between metallic planes, and the associated dispersion (and finite perpendicular group velocity) here set a limit to the width of the observed QP features.

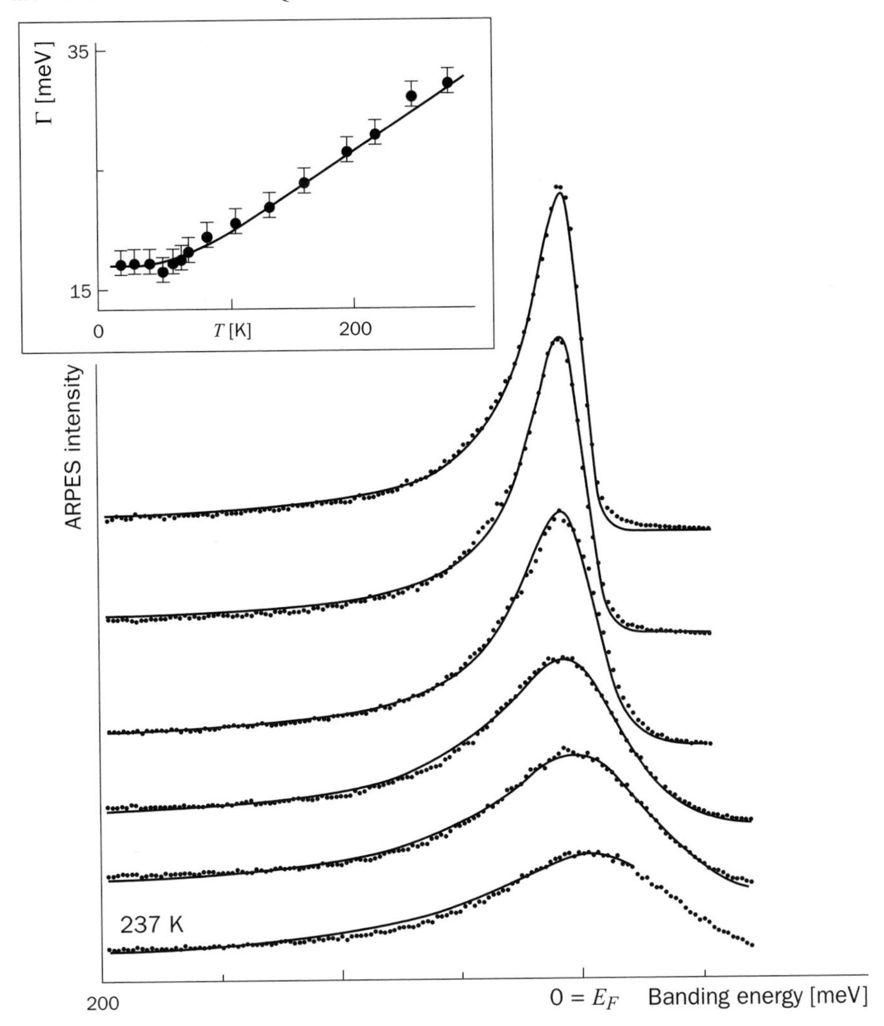

Figure 15 - QP spectra of $TiTe_2$ measured at the Fermi surface at various temperatures adapted from ref. **[14]**. The temperature-dependent linewidth reflects the increasing electron-phonon scattering at high T.

The effect of electron-phonon coupling on the spectral properties of the QPs is not limited to the Fermi surface. The QP energy and line shape are affected over an energy range determined by the width of the phonon spectrum. This is illustrated in figure 16 **[15]** showing the dispersion of a surface state at the Be(0001) surface. The 2D nature of these electronic states again favours the analysis of the ARPES spectra. The QP peak sharpens approaching the Fermi level. Furthermore, a peculiar double structure develops when the binding energy is comparable with the largest phonon energy in the system (~ 70 meV).

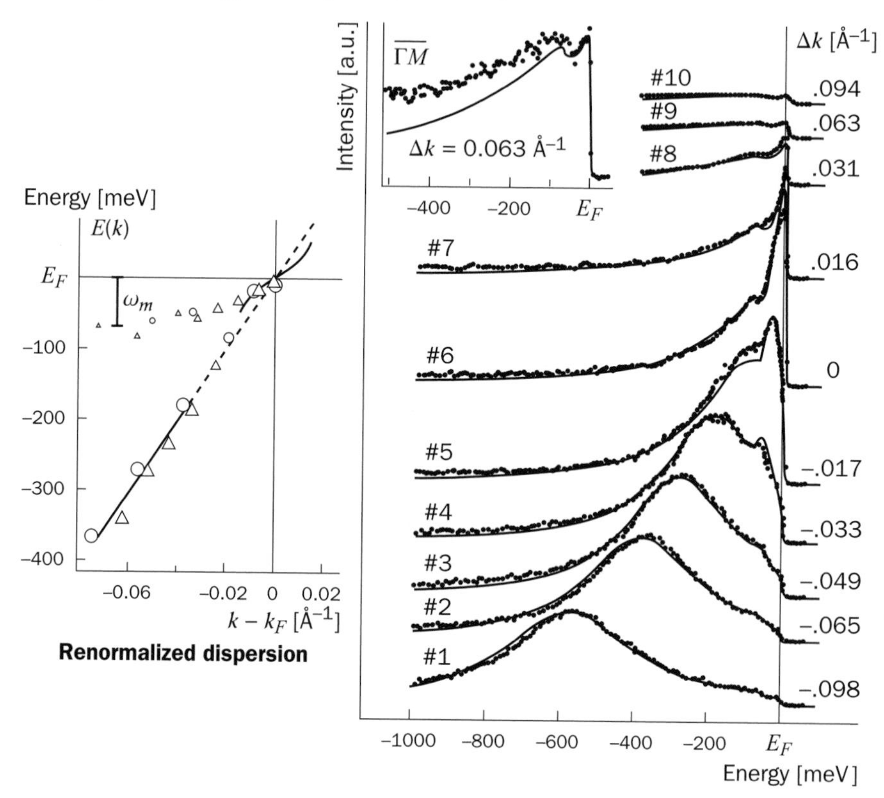

Figure 16 - Experimental and theoretical ARPES spectra of a surface state on Be(0001), showing the effect of a strong electron-phonon interaction from ref. **[15]**.

The experimental binding energies are plotted in figure 16 (left). Besides the low energy feature, which has no counterpart in the band structure, we observe a flattened dispersion near k_F, corresponding to a smaller Fermi velocity and a larger QP mass, both being renormalized by the electron-phonon interaction. All spectral features are well reproduced by theory. Unlike the previous example, the data of figure 16 were collected at only one temperature, but the coupling parameter λ can be evaluated from the difference between the measured QP energies (E_k) and the extrapolated linear dispersion ε_k, because $E_k - \varepsilon_k = \mathrm{Re}\Sigma \sim \lambda E$. This comparison yields the rather strong coupling value $\lambda = 1.18$. The Be surface state offers a "clean" example of electron-phonon coupling in a metal. It is interesting to note that similar unusual ARPES line shapes have been recently, and perhaps unexpectedly, observed also in the spectra of several superconducting cuprates [16]. A similar analysis, and a comparison with the "clean" reference, then established that electron-phonon interactions might play a relevant role in those materials.

5.2. *GAP SPECTROSCOPY*

Interactions in a Fermi liquid may be strong enough to destabilize the "normal" metallic ground state, and force a transition to a new phase characterized by a different symmetry. The phase transition may deeply modify the electronic structure, producing characteristic signatures in various response functions. This is the case when a gap opens at the Fermi level in the density of states. The momentum selectivity of ARPES offers a unique view of the electronic states at the transition. The expected spectral changes are illustrated by a comparison of the cartoons of figure 6 and figure 17. In the non-metallic phase the band feature does not cross the Fermi level, and its closest approach is at the gap edge.

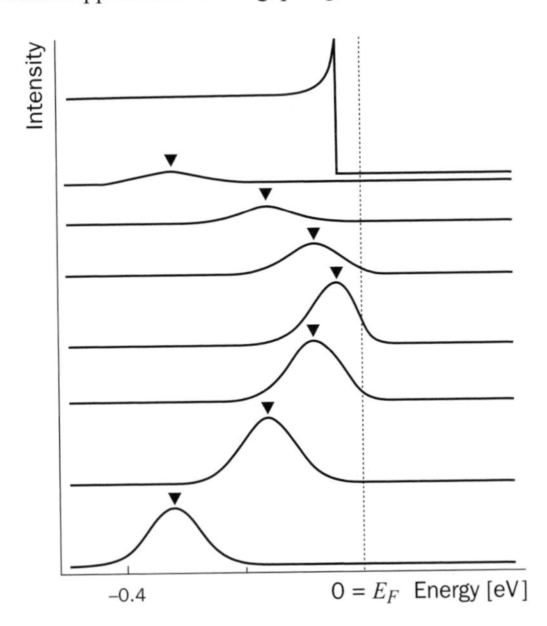

Figure 17 - Expected dispersion and DOS (top) in a superconducting (SC) or charge-density wave (CDW) state, within a mean-field scenario. The topmost curve is the momentum-integrated spectrum.

We may assume that a new periodicity, corresponding to the Fermi wavevector of the metallic phase, appears in the ordered phase. This is verified both for a BCS superconductor, and for a charge (or spin) density wave (see below) [17,18]. The ARPES peak will feel the new periodicity, and disperse back away from E_F beyond that point, with a reduced intensity, as a shadow band. The figure also shows (for $T = 0$) the corresponding momentum-integrated spectrum. Here, the opening of an energy gap removes spectral weight at the Fermi level, and piles it up in the sharp feature (a divergence in the BCS picture) at the gap edge.

The small size of the energy gap and the low transition temperature in conventional superconductors have for a long time prevented experimental verification of these spectral changes. Initial attempts suggested unusual temperature dependence across T_c, but the effect was masked by insufficient energy resolution [19]. A real breakthrough came with the discovery of high-T_c superconductivity in the cuprates, where the superconducting gap size is larger than the experimental resolution of state-of-the-art spectrometers. The ARPES literature on this subject is extremely rich, and it is impossible to give here even a coarse summary of the many interesting results obtained over the last decade [20]. When the spectrum of a superconductor is measured at the Fermi surface above and below T_c, large changes are observed, as in the case of Bi-2212 ($Bi_2Sr_2CaCu_2O_{8-\delta}$, optimally doped; $T_c = 90$ K) shown in figure 18 [21].

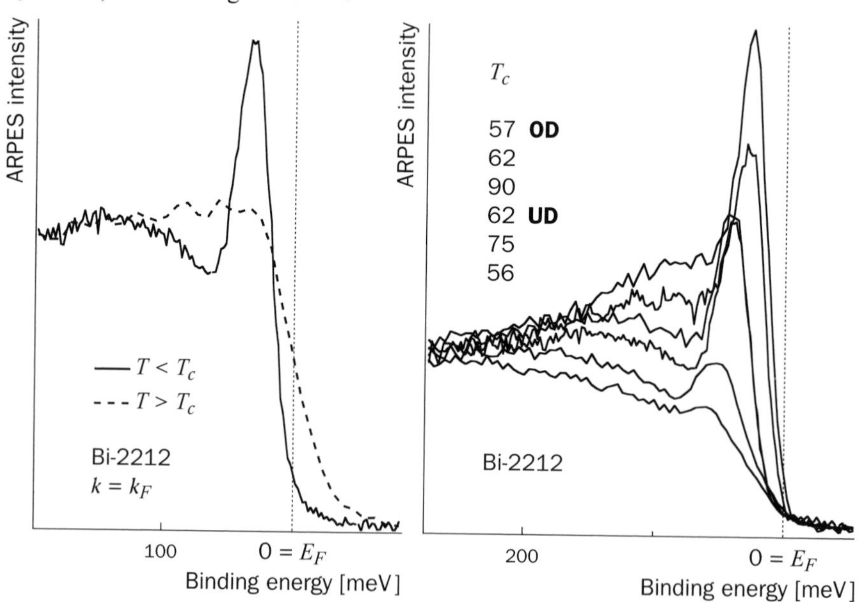

Figure 18 - Gap opening in the SC cuprate Bi-2212 (left). Evolution of the SC gap as a function of doping in Bi-2212 (right) from ref. **[21]**.

The high-temperature spectrum, measured at 100 K near $k = (\pi/2,0)$ on the Fermi surface, shows the high-energy part of a broad QP peak cut by the Fermi function. In the SC phase the peak sharpens considerably, and recedes from E_F. Assuming as a first approximation the validity of a BCS scenario, the binding energy of the peak yields the temperature-dependent gap $\Delta(T)$. Such an analysis showed that the reduced gap Δ_0/k_BT_c in the cuprates is considerably larger than the usual weak coupling BCS value of 3.5. The SC gap size depends strongly on the "doping level", i.e. on the carrier (hole) density in the CuO planes, and it is not simply proportional to the critical temperature. The ARPES data of figure 18 illustrate this point. The spectra also show that the gap *increases* while T_c decreases in the "underdoped" (UD) part of the phase diagram, a rather counterintuitive observation that confirms the unconventional nature of superconductivity in these materials. The intensity of the QP peak is however strongly reduced, following the decrease of the SC superfluid density. Notice also that in a rather strongly overdoped (OD) samples ($T_c = 57$ K) the reduced ratio Δ_0/k_BT_c approaches the BCS value.

In conventional ("*s* wave") BCS superconductors the gap is a constant, independent of the location on the Fermi surface. By contrast, in the cuprates the gap exhibits strong momentum dependence. Unlike other types of spectroscopy (tunnelling, optics) ARPES can directly probe this angular dependence by measurements over the whole Fermi surface. Within the experimental accuracy it is found that the gap has 4 "nodes", equally spaced by 90°, at 45° from the direction of the Cu-O bonds in the CuO plane, as shown for one quadrant in figure 19 [22]. Near optimum doping the size of the gap follows the expression

$$\Delta = \Delta_0 \,|\cos(ak_x) - \cos(bk_y)\,|,$$

where *a* and *b* are the lattice constants of the Cu-O plane, consistent with "*d* wave" symmetry.

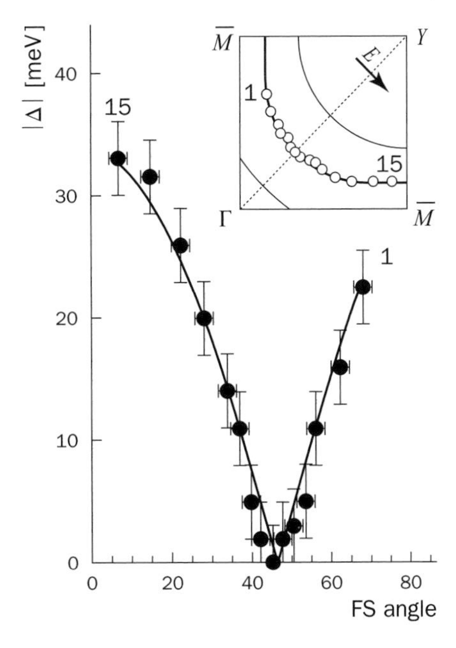

Figure 19 - Angular dependence of the SC gap in Bi-2212 extracted from ARPES data from ref. **[22]**. The dependence is consistent with a "*d* wave" scenario.

Note that ARPES is sensitive to the amplitude, but not to the phase of the gap, and the measured angular dependence would also be compatible with "asymmetric *s* wave" symmetry. Phase-sensitive measurements (tunnelling), which favour the "*d* wave" scenario, are necessary to distinguish between the two possibilities. ARPES data like those of figures 18 and 19 have undoubtedly much contributed to the present understanding of high-T_c superconductors. Unfortunately, sample surface quality problems have so far limited the scope of the ARPES studies, and most of the available data concern the Bi-2212 (and Bi-2201) compounds. These materials are easily cleaved within the Bi-O double layer, and the rather inert cleavage surface shields quite effectively the interesting Cu-O planes from contamination and other surface-related problems. ARPES studies on YBCO ($YbBa_2Cu_3O_{7-\delta}$) have been far less successful precisely because of surface problems like oxygen loss or reconstruction. However, recent promising results on LSCO ($La_{2-x}Sr_xCuO_4$) single crystals, and on thin films suggest that stoichiometric superconducting surfaces may be stabilized and investigated.

Steady improvements of the experimental conditions achieved over the last decade, have finally brought even conventional "low-temperature" superconductors within the reach of ARPES studies. The well-understood predictions of BCS theory can therefore be directly tested for the first time. Figure 20 [23] illustrates the results of a PES experiment on V_3Si, an A15 superconductor with $T_c = 18.6$ K.

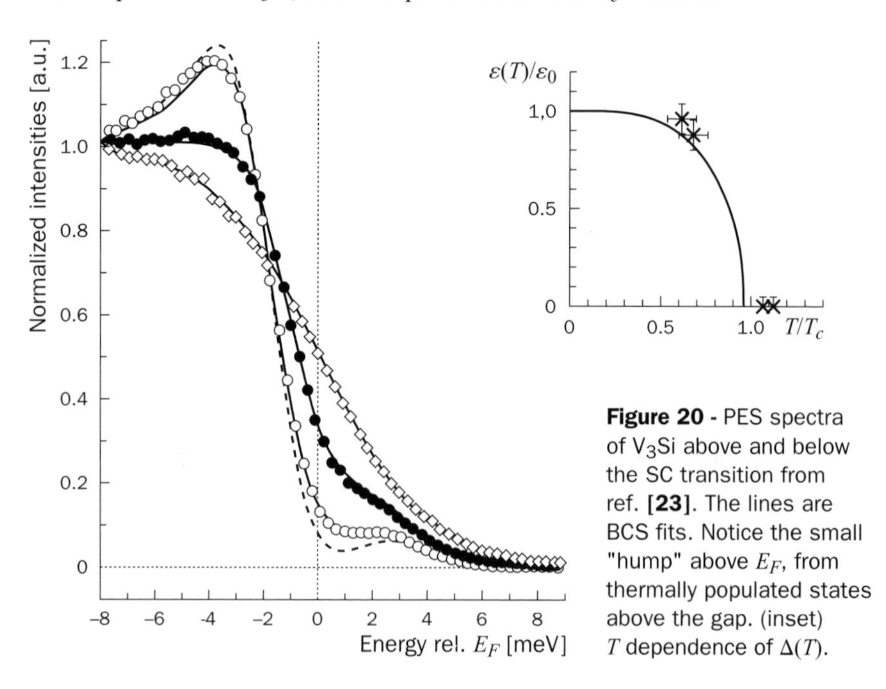

Figure 20 - PES spectra of V_3Si above and below the SC transition from ref. **[23]**. The lines are BCS fits. Notice the small "hump" above E_F, from thermally populated states above the gap. (inset) T dependence of $\Delta(T)$.

Below T_c, the spectrum changes as expected: intensity is lost at E_F and a gap opens. The line shape can be reproduced quite accurately by a BCS function, with the slightly reduced near-surface $T_c' = 17.1$ eV, and the temperature dependence of the SC gap follows quite accurately the expected BCS dependence. The divergence of the BCS function at the gap edge is of course smeared by the finite experimental resolution ($\Delta E \sim 3$ meV). A closer look at the spectrum reveals a second, weaker, but remarkable feature at positive (~ 3 meV) energy, which corresponds to the second of the two symmetrically located [*at* $\pm\Delta(T)$] divergences of the BCS DOS function. At the low but finite measurement temperature (12 K) this part of the DOS is partially populated. This provides a very convincing verification of the BCS scenario, and of the sensitivity of photoemission to low-energy electronic excitations.

The need to take into account, and possibly remove, the effect of the Fermi distribution to directly access the spectral function, is common to all high-resolution photoemission studies. In principle, this can be done without further assumptions, by *directly* dividing the raw spectra by the Fermi function. In practice, this is difficult because:

1. the raw spectrum is broadened by the finite experimental resolution,
2. the position of the Fermi level and the sample temperature must be known with high accuracy,
3. both the spectrum and the Fermi function drop to values close to zero within a few k_BT of E_F.

If all these difficulties are overcome, small changes in the spectral function, which would be hardly visible in the raw spectra, can be visualized. Figure 21 [24] shows the temperature evolution of the DOS in a polycrystalline sample of the "Kondo insulator" CeRhAs. Each spectrum was divided by the Fermi function, with a temperature suitably increased to take into account the additional experimental broadening. The resulting curves represent the DOS up to an energy of few k_BT above E_F. The dip around E_F, and up to an energy $\Delta_0 = 90\text{-}100$ meV, corresponds to the formation of a deep and asymmetric pseudo-gap, due to the interaction of the strongly correlated 4*f* states with the extended conduction bands. The size Δ_0 of the pseudo-gap is temperature-independent, and of the order of the characteristic "Kondo temperature" of this material. A second, narrow dip develops at the lowest temperature within the wide pseudo-gap, and the width of this structure (~ 10 meV) is compatible with the low-temperature transport properties. A similar manipulation of the spectra of an Au film yields a temperature-invariant DOS, confirming the validity of the analysis.

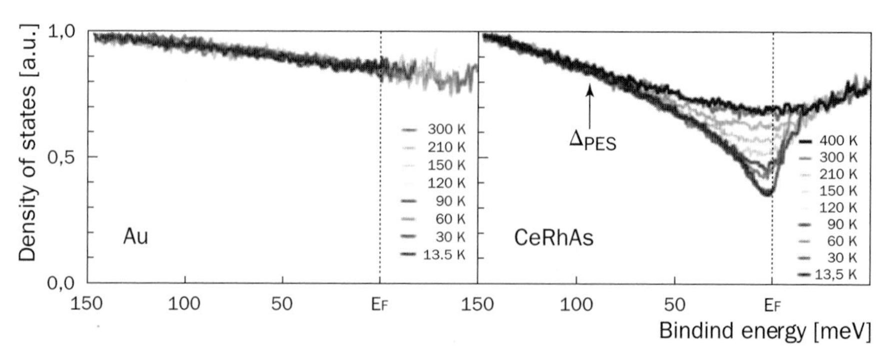

Figure 21 - Temperature-dependent spectra of a gold reference and of the "Kondo insulator" CeRhAs illustrate the opening of a pseudo-gap at E_F in the latter from ref. [**24**]. The spectra were divided by the Fermi function.

The strong energy-dependent perturbation of the Fermi function can also be eliminated by a proper symmetrization of the spectra, which consists in adding to a spectrum its mirror image with respect to the Fermi level. This approach, which is often used in the high-T_c literature, does not require the extreme data quality and low noise level of the previous method, but relies on an assumption. It is easy to verify that, if the spectral function is symmetric with respect to the Fermi level, i.e. if $A(k,\omega) = A(k,-\omega)$, the "symmetrized" spectrum $I^*(k,\omega) = I(k,\omega) + I(k,-\omega)$ is just proportional to $A(k,-\omega)$, and the Fermi function has been removed. This assumption can only be validated for $k = k_F$, and one may expect that it still holds in the proximity of the Fermi surface. Figure 22 [25] illustrates the procedure for PES measurements of an epitaxial C_{60} monolayer grown on the Ag(100) surface. The temperature-dependent spectral changes near E_F are more evident in the symmetrized spectra, which show the opening of a gap already near room temperature, suggesting the intriguing possibility of high-temperature interface superconductivity.

5.3. ELECTRONIC INSTABILITIES IN LOW DIMENSIONS

In the previous paragraph (figures 18 and 21) we have encountered examples of phase transitions in low-dimensional systems. Quite generally, when the dimensions of an electronic system are reduced below 3, the normal metallic state becomes more vulnerable to electronic instabilities [18]. The most typical example of these electronic instabilities is the Peierls transition in 1D, leading to a "charge-density wave" (CDW) ground state. In the CDW phase the electronic density – and the lattice spacing – are modulated in space with a wavevector $Q_{CDW} = 2\,k_F$. This modulation removes the 1D Fermi surface (2 points, *at* $\pm k_F$) and the system is driven to an insulating state.

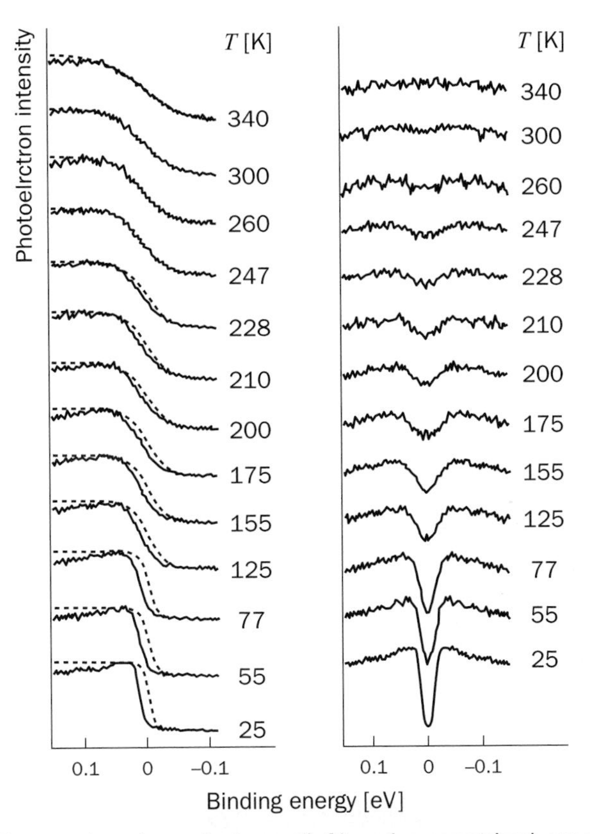

Figure 22 - Temperature-dependent raw (left) and symmetrized spectra of a C_{60} monolayer on the Ag(100) surface, showing the opening of a gap at E_F from ref. **[25]**.

Perfect nesting of the FS, i.e. the possibility of translating the entire FS onto itself by a translation $\Delta k = Q_{\mathrm{CDW}} = 2k_F$, is at the origin of the instability. The degeneracy of states at the Fermi surface connected by Q_{CDW} causes the divergence of the generalized charge susceptibility $\chi(2k_F)$ and, *via* electron-phonon coupling, the spontaneous charge and lattice modulation. Perfect nesting cannot be achieved in 2D or 3D, except for pathological situations, like a square FS. Nevertheless, partial nesting conditions over a good portion of the FS are often realized in quasi-2D materials, like the layered transition metal compounds. The ensuing enhancement of $\chi(2k_F)$ may be sufficient to drive the system into a CDW phase. Unlike the singular 1D situation, partial nesting removes only part – the nested portion – of the FS, and the transition is not a metal to insulator, but a metal-to-metal one.

Both 1D and 2D instabilities, and the corresponding changes in electronic states, have been studied by ARPES. The "geometrically" simpler 1D case presents some subtleties, which mask the simple scenario just outlined, and we will consider it later. We first review examples of the more straightforward 2D transitions. Here the issue is analogous to that of mapping the FS shape and the gap in the SC cuprates, and the techniques employed are the same. Figure 23 [26] shows the result of a FS mapping experiment in *2H*-$NbSe_2$, a 2D CDW system with a critical temperature $T_{CDW} = 33$ K. It is a layered compound, structurally similar to $TiTe_2$, which we considered earlier (figures 14 and 15), and the prefix *2H* refers to the local trigonal prismatic coordination of the Nb ions. The ARPES intensity map at energy $E = 0$ shows a hexagonal FS sheet centred at Γ, and additional sheets around the (inequivalent) K and K' points. Apart from the strong threefold symmetry, which can be explained as a matrix element effect, the experimental FS is compatible with band structure calculations. The predicted fine structure within each sheet, and a smaller "pancake-like" FS sheet at Γ, are not seen in these experimental data (but they are in another data set collected with improved conditions, see below). The most interesting aspect of these data is the direct observation of possible nesting. In fact, the distance between the parallel sides of the central hexagon coincides with the CDW wavevector, which is independently measured in X-ray or neutron scattering experiments. This part of the FS is therefore well nested for three equivalent translations at 120° from each other, consistent with the actual threefold degenerate CDW.

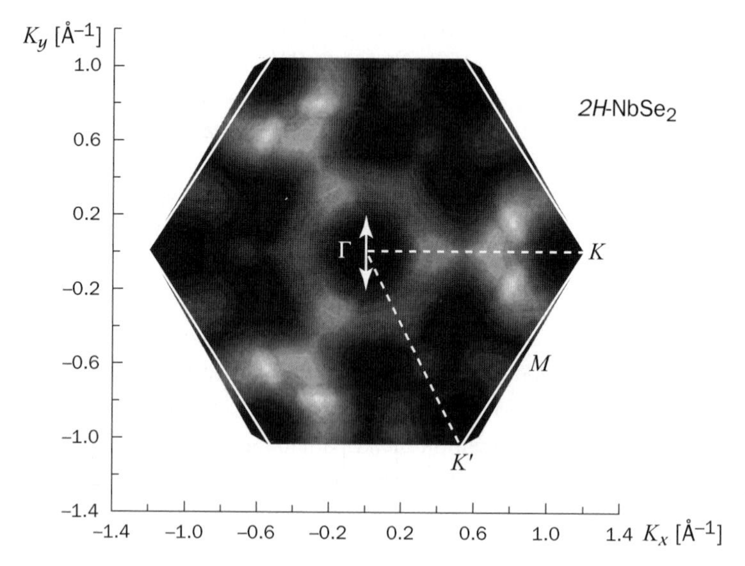

Figure 23 - ARPES Fermi surface(FS) map of the 2D CDW material *2H*-$NbSe_2$ from ref. **[26]**. The arrow shows a possible nesting vector.

Notice however that other parts of the FS are poorly nested by these translations. Clearly, nesting is only one element concurring to the formation of the complex ordered state, which could involve a more extensive rearrangement of the bands, and a strong modulation of bond distances which are not included in the simple weak-coupling scenario. A better understanding of the origin of the CDW demands, similarly to the cuprates, a more thorough mapping of the bands and of the CDW gap.

The selective removal of parts of the FS in the CDW phase can be studied by high resolution ARPES. Figure 24 [27] illustrates the case of *2H*-$TaSe_2$, a similar compound with a CDW transition at $T = 122$ K, and a Fermi surface similar to that of figure 23. The intensity map shows nearly parabolic band dispersion along the ΓK direction, in the gap between the hexagonal FS sheets centred at the Γ and K points. The temperature dependence of the spectra at the two distinct Fermi level crossings is quite different. At k_{F1}, on the inner hexagon, the only temperature dependence between 130 K (above T_c) and 75 K (below T_c) is a sharpening of the QP peak, from the narrower Fermi function, while on the FS sheet centred around k_{F2} there is a clear shift at low temperature. From these data it is clear that the CDW gap opens only at the second location, notwithstanding the good nesting conditions at the central hexagon.

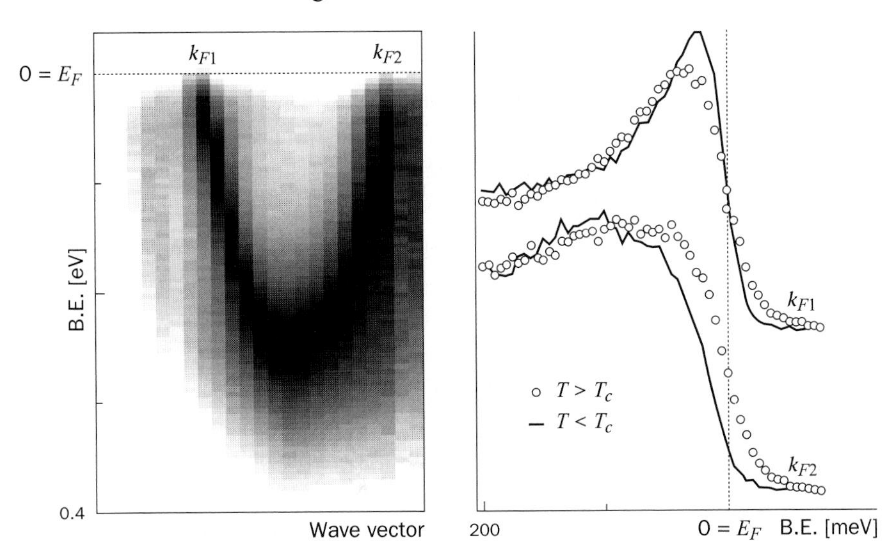

Figure 24 - ARPES intensity map of the 2D CDW system *2H*-$TaSe_2$ along the ΓK direction (left). Spectra measured at the 2 FS crossings, above and below T_c. A gap opens only at k_{F2} (right) from ref. **[27]**.

Sometimes two different instabilities compete (or cooperate) in the same systems, giving rise to a more complex and interesting behaviour. We show here two examples of coupled instabilities in 2D systems. The first is that of $2H$-$NbSe_2$ (fig. 23) where the CDW ($T_c = 33$ K) coexists with superconductivity below $T_{sc} = 7.2$ K. The strong electron-phonon interaction in this material is at the origin of both ordered phases, but the electronic states involved in the two instabilities are not necessarily the same. Namely, different portions of the complex FS may be differently affected by the transitions. This issue is ideally addressed by exploiting the momentum selectivity of ARPES. The experiment is quite demanding because of the low SC transition temperature and the correspondingly small gap, which require very low temperatures and an extremely high-energy resolution. Figure 25 [28] shows state-of-the-art ARPES data above (10 K) and below (5.3 K) T_{sc}. The first couple of spectra was measured on the inner "pancake-like" FS sheet (inset), which could not be resolved in the map of figure 23. The edge of the 5.3 K spectrum is slightly sharper, but not shifted within the experimental accuracy. Therefore the Se $4p$ states, which form this portion of the FS, are ungapped by the SC transition, at least at this temperature. By contrast, the spectra measured on the hexagonal FS sheet, generated by Nb $4d$ states, present clear signatures of the transition, namely a $\Delta = 1.0 \pm 0.1$ meV energy gap. Assuming a BCS temperature dependence, one obtains the reduced gap value $2\Delta(0)/k_B T_c = 3.6 - 3.9$, corresponding to moderately strong coupling for the Nb $4d$ electrons.

$1T$-TaS_2, another layered chalcogenide material, presents a different and unique example of coupled instabilities in 2D. The main instability is a CDW with a transition (at $T = 180$ K) from a quasi-commensurate (QC) to a commensurate (C) phase. In the QC phase CDW domains with a $(\sqrt{13} \times \sqrt{13})$ periodicity and finite size, are separated by unmodulated boundaries – or discommensurations – and form a regular hexagonal array. In the C phase the CDW is coherent over the whole crystal. Surprisingly, at the QC-C transition the electrical resisitivity increases by more than one order of magnitude, and it has been suggested that electronic correlations suddenly trigger a metal-insulator Mott transition. This scenario is confirmed by the ARPES data of figure 26 [29]. The intensity maps show the dispersion of the Ta $5d$ valence band along the high-symmetry ΓM direction. In the metallic QC phase the band is broad and crosses the Fermi level, but in the C phase the CDW superlattice potential splits it into 3 sub-bands. From simple electron counting, the topmost sub-band should be half filled, and therefore straddle the Fermi level. It is instead separated from E_F by a 0.2 eV gap, and the small dispersion suggests localization of the corresponding electrons.

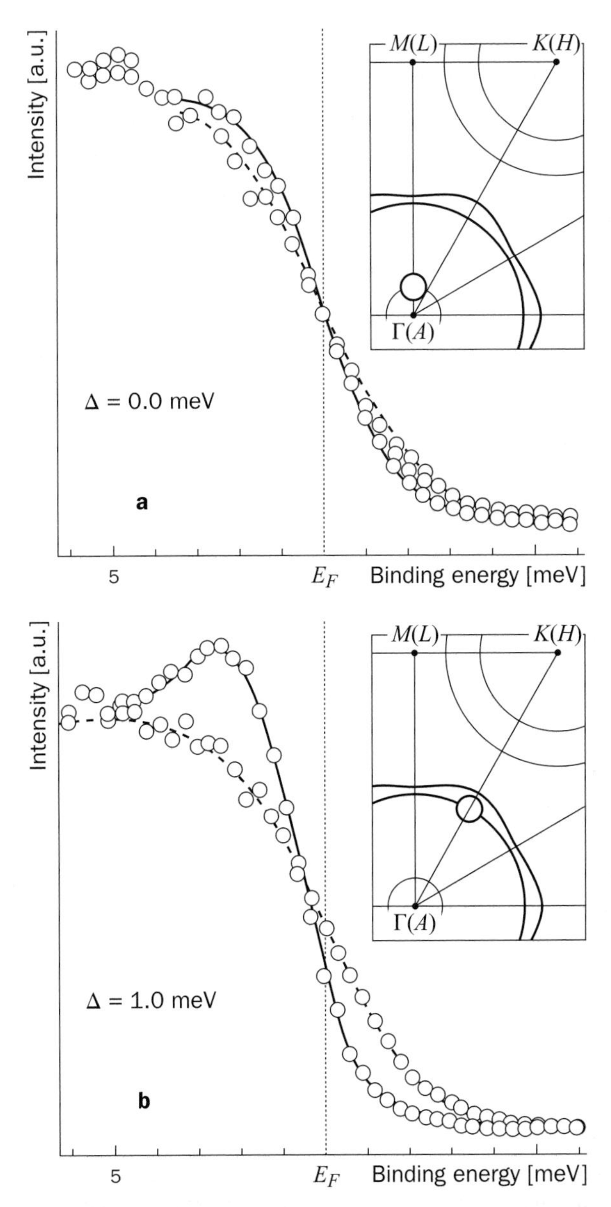

Figure 25 - Very high-resolution spectra of *2H*-$NbSe_2$, measured on two different FS sheets across the SC transition (T_{sc} = 7.2 K) from ref. **[28]**.

The opening of a gap in the C phase is also clearly illustrated by normal emission ARPES spectra, where intensity is lost at E_F and a new sharp feature appears at finite binding energy. Strong electronic correlations split the half filled topmost band into a filled lower Hubbard band (LHB) seen by ARPES, and an empty upper Hubbard band (UHB) above E_F, separated by a correlation gap. The transition is therefore a many-body effect, triggered by the CDW, which splits the valence band. Clearly, the momentum-dependent information from ARPES provides an invaluable support to the coupled CDW-Mott scenario.

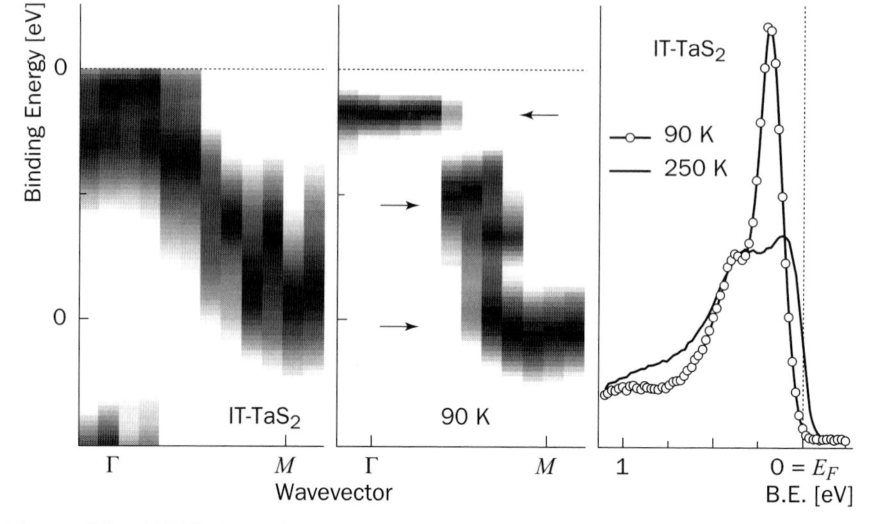

Figure 26 - ARPES intensity map of *1T*-TaS_2 in the metallic (left) and insulating (centre) phase. The opening of a Mott-Hubbard gap is also evident from the spectra (right) from ref. **[29]**.

The Peierls instability is strongest in 1D, and one might expect to observe there the clearest spectral signatures of the ordered CDW state. In reality, the whole mean field scenario of the metal-non-metal transition (figure 17) is questionable in 1D, and may be completely masked by stronger effects. Experimentally, one finds that metallic 1D systems do not exhibit the typical spectral signatures of a metal – Fermi level crossings in ARPES and a Fermi step in momentum integrated spectra – illustrated in figure 6. Bands with strong 1D character, i.e. no perpendicular dispersion, and with the symmetry properties expected from the underlying lattice, are observed by ARPES, e.g. in the organic conductor TTF-TCNQ (figure 27) [30]. The ARPES map shows two bands, from the segregated TTF and TCNQ chains, generally consistent with the calculated band structure (inset). However, the spectrum measured at the Fermi wavevector does not exhibit a sharp QP peak at the Fermi level, but rather a broad feature and a pseudo-gap. The leading edge is much broader than the resolution-limited Fermi edge of a metallic silver film measured at the same temperature (solid line).

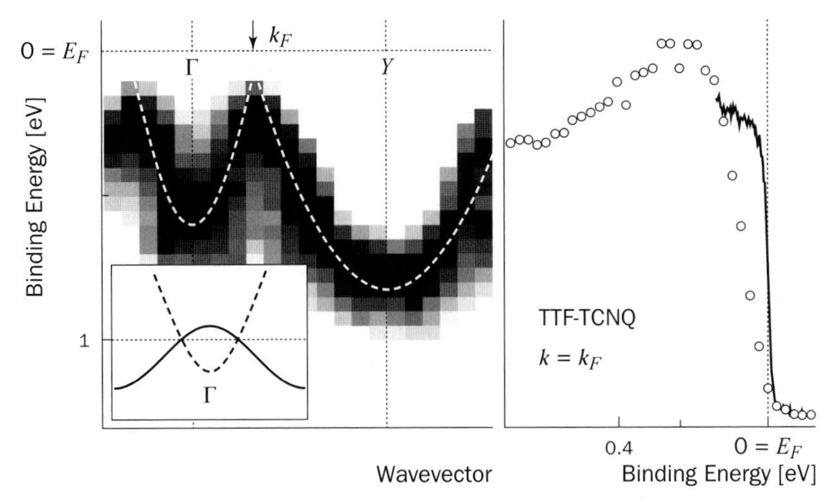

Figure 27 - ARPES intensity map for the organic 1D compound TTF-TCNQ (left). The spectrum at $k = k_F$ exhibits a deep pseudo-gap (right) from ref. **[30]**.

The origin of the peculiar behaviour, which is observed time and again in 1D materials, is still unclear, and different scenarios are possible. There is the intriguing possibility that the QP picture, our guideline throughout this chapter and the cornerstone of the physics of interacting Fermions, may break down in these systems. This is precisely what is predicted by theory strictly in 1D, where QPs do not exist as long-lived independent excitations [31]. An electron, or hole, added to a 1D system immediately breaks down into collective spin and charge modes, which propagate with different velocities, and therefore separate. This *spin-charge separation* has dramatic consequences on the spectral function of such a *Luttinger liquid*, shown in figure 28 [32]. For a given value of the momentum, the characteristic QP peak of the Fermi liquid (figure 10) is replaced by two separate divergences representing the *spinon* and *holon* excitations. The details of the line shape depend on the non-universal exponent ("α") of the Luttinger model, which characterizes the strength of the interaction. Large α values (of order 1), corresponding to strong and long-range interactions, yield line shapes that are similar to the experimental ones, and exhibit a "pseudo-gap" feature. Distinct spinon and holon peaks have not yet been convincingly identified in experimental ARPES spectra, although there are some indications of spin-charge separation in organic and inorganic 1D conductors [33]. An indirect spectral evidence of spin charge separation is found in the peculiar dispersion measured in 1D Mott insulators like $SrCuO_2$, which, however, are not LLs, due to the charge gap. (figure 29) [34]. The 1D dispersion is "fuzzy" and considerably broader than the dispersion of a related 2D oxide. These differences can be reproduced by model

1D calculations, and are the result of the strikingly different dynamics of a hole in 2D and 1D. However, the issue is not solved, and other interpretations of the peculiar 1D line shapes are possible. For instance, they may be due to strong (but not "pathological" as in the 1D Luttinger liquid) interactions. Such strong interactions, e.g. with the lattice, could considerably reduce the coherent QP weight (Z_k in eq. {18}) and move large amounts of spectral weight to the incoherent part of the spectral function at lower kinetic energy, as discussed above.

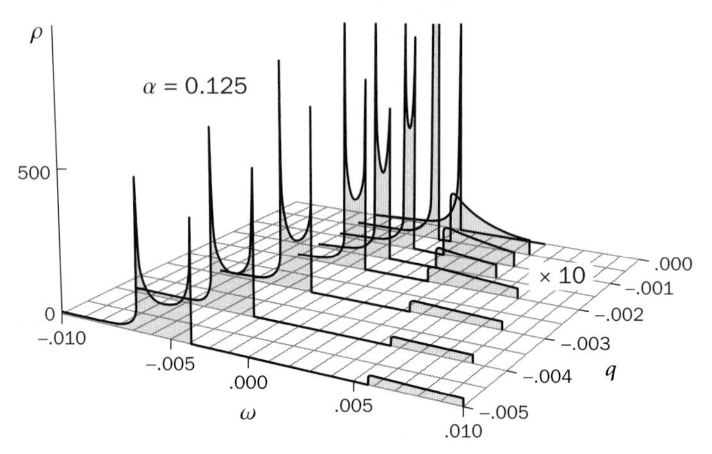

Figure 28 - Calculated spectral function of the Luttinger Liquid, showing separate spinon and holon excitations from ref. **[32]**.

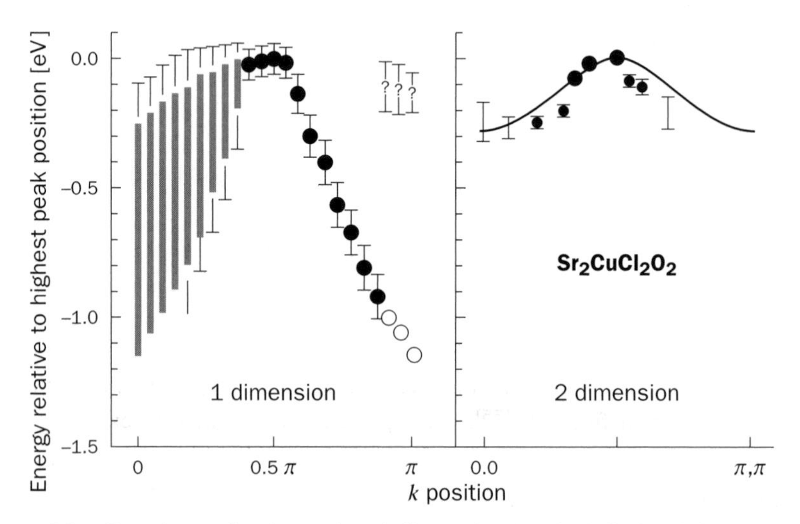

Figure 29 - Experimental valence band dispersion in 1D $SrCuO_2$ (left), and in a 2D cuprate (right) from ref. **[34]**. The broader 1D dispersion is compatible with spin-charge separation in 1D (for the π points marked with ?, see comments in ref. **[34]**).

5.4. SOME RECENT DEVELOPMENTS

So far we have implicitly considered ARPES experiments with low-energy photons. Energies in the range 10-40 eV lead to optimum energy and momentum resolution, and are commonly used for FS mapping and gap spectroscopy. These experiments are often performed with conventional sources, even if the higher brilliance, the polarization control, and the tunability of synchrotron radiation present obvious advantages. There are however good reasons to move to higher energies, where synchrotrons are the only viable option. Experiments like core-level spectroscopy and resonant photoemission have become traditional SR techniques during the last two decades, but they are receiving a new impulse from the improved experimental conditions. Moreover, totally new experiments, which were simply unfeasible, can now exploit the exceptional performances of third generation SR sources.

Resonant PES (RESPES) is a typical SR-based experiment. It is really a very efficient way of making use of the energy dependence of matrix elements near an absorption edge. To be specific, let us consider a measurement of the Ce $4f$ electrons in a cerium compound, a problem of considerable interest in the context of intermediate-valence, or heavy fermion phenomena. When the photon energy is tuned at the Ce $N_{4,5}$ ($4d \longrightarrow 4f$; $h\nu \sim 120$ eV) absorption edge, $4f$ electrons can be emitted in two processes:

1. the normal PES process,
2. an "indirect" process where a $4d$ core electron is excited to an empty $4f$ level, after which one $4f$ electron fills the $4d$ hole, and another $4f$ electron is ejected in a Coster-Kronig process: $3d^{10}4f^{N} \longrightarrow 3d^{9}4f^{N+1} \longrightarrow 3d^{10}4f^{N-1}$.

Our description is oversimplified because the absorption and further emission should really be treated as a single coherent process, but what is most important here is the fact that the normal and the indirect process have the same initial and final states, and therefore interfere. As a consequence the $4f$ photo-ionization cross-section exhibits a strong resonance, and draws intensity from the sharp $4d$ absorption cross-section. The $4f$ signal is rapidly enhanced and in a few eV it goes from a minor contribution to the main feature of the PES spectrum, as in the case of $CePd_3$ shown in figure 30 [35]. As expected, the $4f$ intensity exhibits a Fano profile (inset). Notice that because of the crucial role played by the localized Ce $4d$ core level, and of dipole selection rules, the intensity enhancement only concerns the Ce $4f$ electrons. Therefore RESPES selects the site (the chemical species) and the orbital character of the enhanced states. Off- and on-resonance measurements thus provide the best possible contrast for the states of interest.

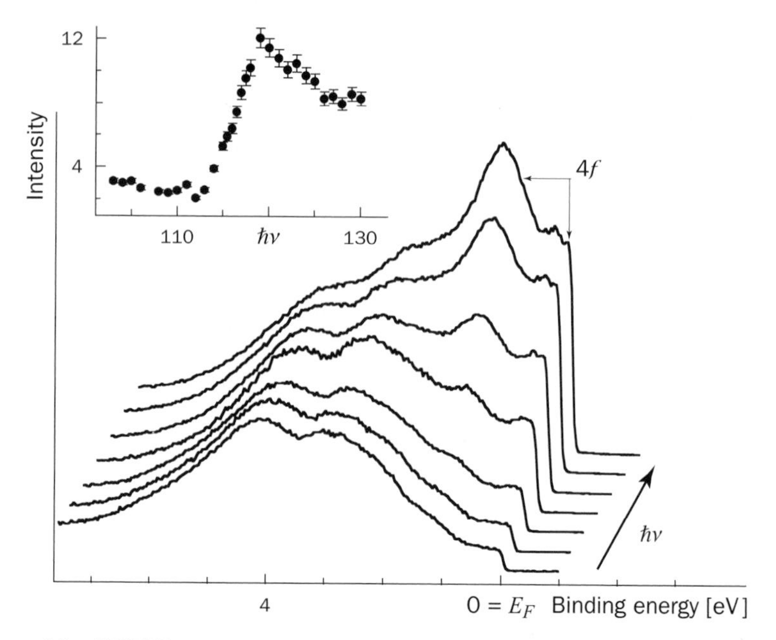

Figure 30 - RESPES spectra of the Kondo system $CePd_3$ measured across the Ce 4*d* absorption edge. Photon energy dependence of the 4*f* signal (inset) from ref. **[35]**.

The large and selective intensity enhancement of RESPES has been recently exploited in sophisticated experiments which probe the nature of the ground state, namely the spin state, in strongly correlated systems like the cuprate high-T_c superconductors. Theory predicts that the lowest energy local configuration of a doped Cu site is a 2-hole hybrid Cu-O state with singlet character, rather than a triplet as suggested by the atomic Hund's rule. This prediction cannot be directly verified by PES or ARPES, nor by more elaborate spectroscopic techniques, like spin-resolved PES, and magnetic X-ray dichroism (XMCD). In fact, these materials do not exhibit a net magnetization, and the spin signal from different sites averages to zero. This difficulty is overcome using circularly polarized light at the L_3 ($2p_{3/2} \longrightarrow 3d$) absorption edge. The underlying idea is to simulate hole doping in the superconducting materials by a photoemission process in a model Cu^{2+} (d^9, 1 hole) material like CuO or an "undoped" site in SC Bi-2212 **[36,37]**. The lowest binding energy final state of PES (the first ionization state) is the first state that would be occupied when a hole is added by conventional chemical doping. RESPES actually only probes the (small) d^8 part of the hybrid 2-hole state, but the resonant intensity enhancement is sufficient to make a further spin analysis – which has an efficiency of only few percent – possible. From the selection rules for the *local* $d^9 \longrightarrow 2p^5d^{10} \longrightarrow d^8$ resonant process excited by circularly polarized light the transition rates to triplet and singlet d^8 final states can be calculated. They do not depend on the orientation of the local moments – the hole spin sets the local reference on each site – and an overall singlet-triplet asymmetry can be measured. Figure 31 shows the decomposition of the spectrum in singlet and triplet components, near the Fermi level. Clearly the singlet part is predominant at E_F, and this substantiates the theoretical predictions.

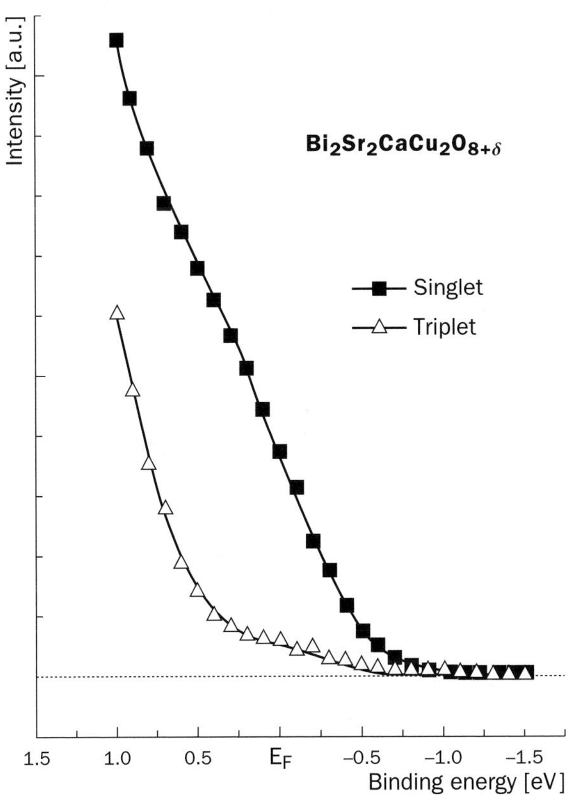

Figure 31 - Decomposition into triplet and singlet components of the spectrum of the SC compound Bi-2212 from ref. **[37]**. Spin-resolved data were taken at the Cu 2*p*-3*d* resonance.

The interest for RESPES has been considerably revamped by the commissioning of instruments (beam lines and spectrometers), which can perform experiments with an energy resolution approaching that of standard low-energy measurements. Figure 32 **[38]** compares high resolution RESPES data of the heavy fermion $CeCu_2Si_2$ collected at the Ce $4d \longrightarrow 4f$ (120 eV) and at the $3d \longrightarrow 4f$ (at 880 eV) resonances. The resonant signal of the Ce $4f$ signal is also possible at the $3d$ absorption edge, since the $3d \longrightarrow 4f$ transition is also dipole-allowed. Both spectra reflect almost entirely the Ce 4*f* states, but the line shapes are extremely different, due to the different surface sensitivities. At a photon energy of 120 eV the escape depth of the photoelectrons is just ~ 5 Å, close to the minimum of the escape depth curve (figure 3). The spectrum probes the 4*f* state near the surface. The 880 eV spectrum is considerably less (λ ~ 15 Å) surface sensitive, and probes Ce atoms in a more bulk-like configuration. The different line shape can be understood along these lines and within the theoretical framework of the Anderson model **[39]**. At the surface the hybridization between Ce 4*f* and conduction electrons, which is at the origin of the heavy Fermion behaviour, is weak. This justifies the strong "ionization peak" (f^0 final state) around 2 eV. The more bulk-sensitive spectrum is dominated by a shallow feature, the tail of the many-body Kondo resonance, which reflects the stronger hybridization.

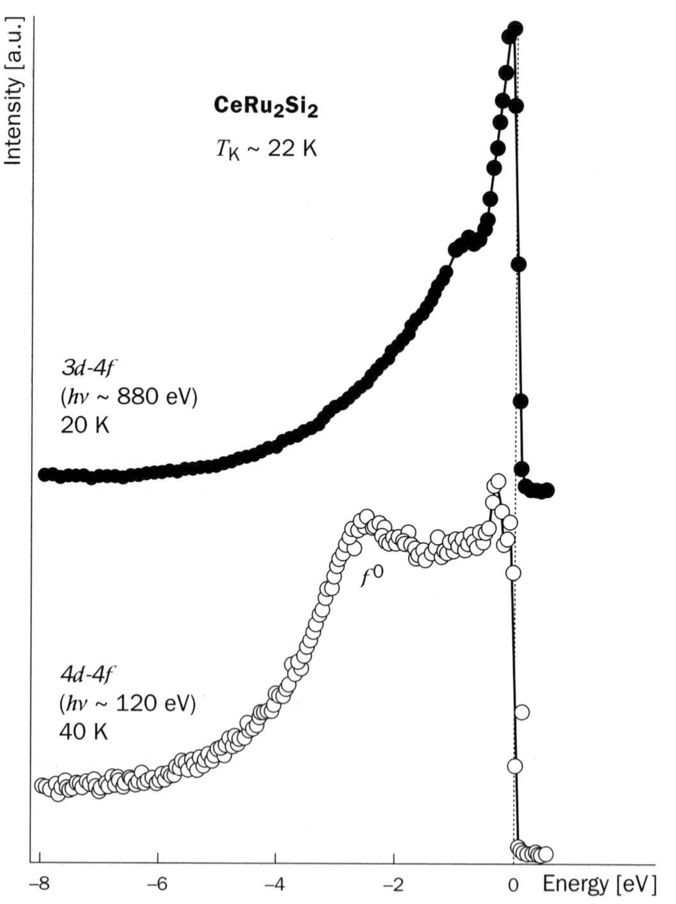

Figure 32 - RESPES spectra of the Kondo system $CeRu_2Si_2$ measured at the Ce 3*d* and Ce 4*d* edges, with different surface sensitivity from ref. **[38]**.

The previous example demonstrates that the probing depth of PES can be effectively adjusted by an appropriate choice of photon energy. There is an obvious interest in pushing towards much higher energies, perhaps to 6-8 keV, to take advantage of the (expected) increase in the escape depth. It should be realized that experiments in such extreme conditions are very challenging. On one hand, cross sections rapidly decrease with photon energy. On the other hand the energy and momentum analysis of the photoelectron becomes more and more difficult. Both factors conjure to reduce dramatically the measured signal. Such experiments therefore require the high photon fluxes only available from undulator beam lines at third generation sources like the ESRF. A recent result illustrates the potential rewards of this development. Figure 33 **[40]** shows spectra of the 3*d* core levels of metallic Sm, measured at photon energies between 1486 (a standard Al Kα X-ray source) and 6 keV. The line shape is complex. The spin-orbit-split ($j = 5/2$ and $j = 3/2$) manifolds exhibit structure, which reflects the accessible final states of the Sm^{3+} ($3d^9 4f^5$) configuration. Moreover, at lower binding energy are visible the signatures of an Sm^{2+} ($3d^9 4f^6$) configuration. The divalent configuration is energetically favoured at the surface, and the spectrum reflects the inhomogeneous mixed valence within the probing depth of the PES measurement. The figure shows that the Sm^{2+} contribution is largest at the lowest photon energy, for which the kinetic energy of the 3*d* electrons is $\varepsilon_k \sim 400$ eV, and much smaller at 6 keV.

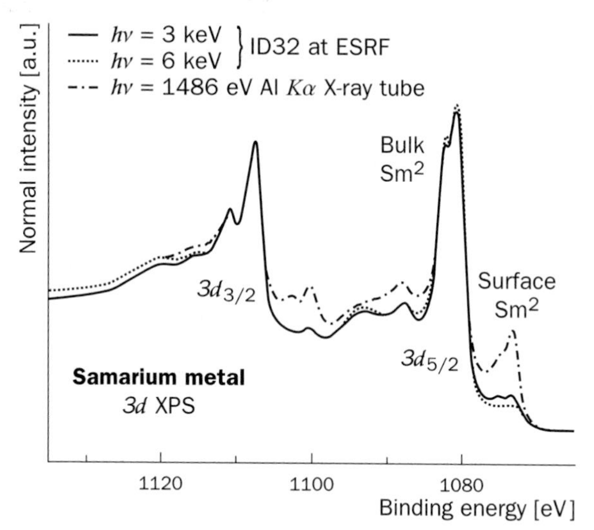

Figure 33 - Spectra of the 3*d* core levels of metallic samarium, at 3 different photon energies from ref. **[40]**. The Sm^{2+} surface component is drastically reduced in the bulk-sensitive high-energy spectrum.

The ability of separately probing the surface and bulk environments is relevant for many materials, including lanthanides and transition metal compounds of fundamental and technological interest. Obviously only the bulk sensitive measurement can be correlated in a meaningful way with other bulk measurements (transport, thermodynamic, magnetic), while the surface sensitive information may be crucial to understand and tailor catalytic or interface properties. This shows the interest of improving the control over the probing depth of PES. The ultimate frontier may be represented by a bulk-sensitive experiment operating at very high energy (8-10 keV) with moderately high ($\Delta E \sim$ 20-40 meV) resolution, a goal that could be reached in the next few years.

The trend towards higher energy is not limited to momentum-integrated PES, but also concerns ARPES. Increased bulk sensitivity is a factor, but not the only one. High-energy band mapping is appealing because at high energy the mapping "trajectories" in k-space become simpler, and the interpretation of results from 3D materials is more straightforward. When high-energy photoelectrons are measured near normal emission, equation (7) shows that a very small opening angle is sufficient to cover the whole BZ along the surface (at 500 eV, for instance, 5° corresponds to 1 Å^{-1}). For such small angular changes $k_{\perp}$ is essentially constant, and band mapping of 3D solids can be performed taking constant $k_{\perp}$ cuts of the band structure, as a function of photon energy.

Figure 34 [41] illustrates the results of a band mapping experiment on the Cu(100) surface, performed at beam line ID08 of the ESRF with photons of 580 eV. The intensity map shows the band dispersion along the (010) direction, centred on normal emission. The narrow d bands are clustered in the energy range 2-6 eV below E_F. Above the 3d manifold the parabolic dispersion of the weaker 4s band, and its crossings of the Fermi level, are well visible. Similar results are quite promising, even if high-energy ARPES remains a difficult experiment. Besides the need for a special undulator source, the high kinetic energy and compressed angular scale put stringent requirements on the performance of the electron analyzer, and on the planarity of the sample surface. Moreover, fundamental potential obstacles, namely the increasing weight of phonon-assisted "indirect" transitions, which may broaden the spectral features in momentum space, are not well understood and must be quantitatively investigated.

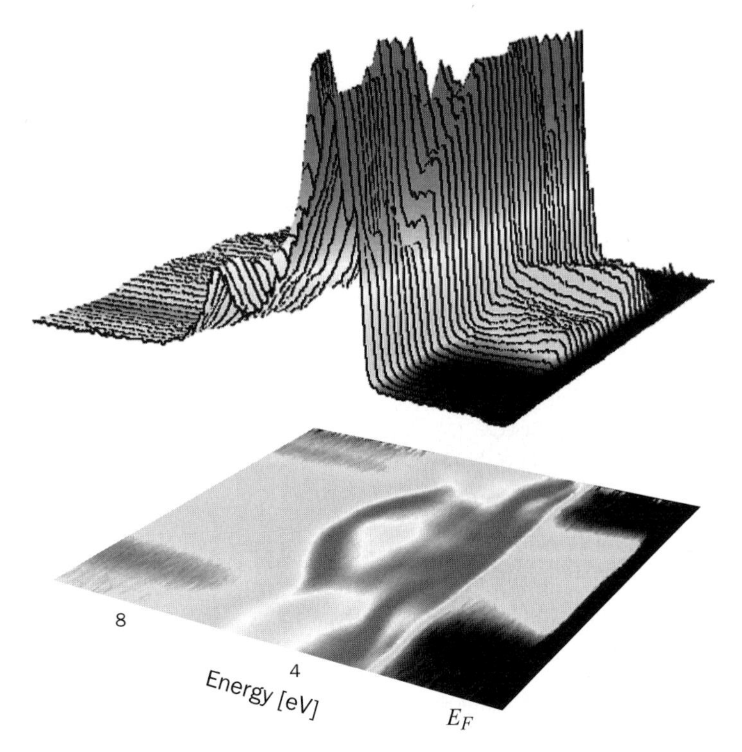

Figure 34 - Band mapping at the Cu(100) surface with high-energy photons ($h\nu$ = 580 eV) from ref. **[41]**.

6. *CONCLUSIONS*

Photoelectron spectroscopy is probably the best practical realization of the theorist's dream of probing the properties of a single particle in an interacting many-body system. Remarkably, although photoemission is certainly a "high-energy" spectroscopy, the large excitation energy does not destroy even subtle signatures of correlations. On the contrary, the close connection with the one-particle Green's function ensures that the spectrum carries momentum-resolved information at all the relevant energy scales, down to the low-lying excitations that shape the thermodynamic properties. It was the goal of this chapter to illustrate this simple but important idea, which is validated by recent experimental results both on model systems and on exciting new materials. This guideline, and the continuous improvements in experimental conditions, namely exploiting the unique properties of SR sources, holds promise for exciting future developments.

REFERENCES

[1] S.D. KEVAN ed. - *Angle-resolved Photoemission. Theory and current application* - Elsevier, Amsterdam (1992)

[2] S. HÜFNER - *Photoelectron Spectroscopy*, Springer Verlag, Berlin (1995)

[3] J.C. FUGGLE & J.E. INGLESFIELD eds. - *Unccupied Electronic States*, Springer-Verlag (1992)

[4] J.A.R. SAMSON - *Techniques of Vacuum Ultraviolet Spectroscopy*, J. Wiley and Sons, New York (1967)

[5] C.N. BERGLUND & W.E. SPICER - *Phys. Rev.* **136**, A1044 (1964)

[6] J.D. DENLINGER, G.H. GWEON, J.W. ALLEN, C.G. OLSON, M.B. MAPLE, J.L. SARRAO, P.E. ARMSTRONG, Z. FISK & H. YAMAGAMI - *J. Elec. Spec.* **117-118**, 347 (2001)

[7] G.D. MAHAN - *Many-Particle Physics*, Plenum, New York (1981)

[8] R.D MATTUCK - *A Guide to Feynman Diagrams in the Many-body Problem*, 2nd ed., Mc Graw-Hill, New York (1976)

[9] L. HEDIN & S. LUNDQVIST - *Solid State Physics* **23**, 1 (1969)

[10] I.W. LYO & E.W. PLUMMER - *Phys. Rev. Lett.* **60**, 1558 (1988)

[11] N.V. SMITH, P. THIRY & Y. PETROFF - *Phys. Rev. B* **47**, 15476 (1993)

[12] J.J. PAGGEL, T. MILLER & T.C. CHIANG - *Science* **283**, 1709 (1999)

[13] S. HÜFNER, R. CLAESSEN, F. REINERT, TH. STRAUB, V.N. STROCOV & P. STEINER - *J. Elec. Spec.* **100**, 191 (1999)

[14] L. PERFETTI, C. ROJAS, A. REGINELLI, L. GAVIOLI, H. BERGER, G. MARGARITONDO, M. GRIONI, R. GAAL, L. FORRO & F. RULLIER-ALBENQUE - *Phys. Rev. B* **64**, 115102 (2001)

[15] M. HENGSBERGER, L. FRÉSARD, D. PURDIE, P. SEGOVIA & Y. BAER - *Phys. Rev. B* **60**, 10796 (1999)

[16] A. LANZARA, P.V. BOGDANOV, X.J. ZHOU, S.A. KELLAR, D.L. FENG, E.D. LU, T. YOSHIDA, H. EISAKI, A. FUJIMORI, K. KISHIO, J.I. SHIMOYAMA, T. NODA, S. UCHIDA, Z. HUSSAIN & Z.X. SHEN - *Nature* **412**, 510 (2001)

[17] J.R. SCHRIEFFER - *Theory of Superconductivity*, W.A. Benjamin, New York (1964)

[18] G. GRUNER - *Density Waves in Solids*, Addison-Wesley, Reading, Massachusets (1994)

[19] M. GRIONI, D. MALTERRE, B.DARDEL, J.M. IMER, Y. BAER, J. MULLER, J.L. JORDA & Y. PETROFF - *Phys. Rev. B* **43**, 1216 (1991)

[20] A. DAMASCELLI, Z.X. SHEN & Z. HUSSAIN - *Rev. Mod. Phys.* **75**, 473 (2003).

[21] I. VOBORNIK, H. BERGER, L. FORRO, M. GRIONI, G. MARGARITONDO & F. RULLIER-ALBENQUE - *Phys. Rev. B* **61**, 11248 (2000)

[22] H. DING, M.R. NORMAN, J.C. CAMPUZANO, M. RANDERIA, A.F. BELLMAN, T. YOKOYA, T. TAKAHASHI, T. MOCHIKU & K. KADOWAKI - *Phys. Rev. B* **54**, R9678 (1996)

[23] F. REINERT, G. NICOLAY, B. ELTNER, D. EHM, S. SCHMIDT, S. HÜFNER, U. PROBST & E. BUCHER - *Phys. Rev. Lett.* **85**, 3930 (2000)

[24] H. KUMIGASHIRA, T. SATO, T. YOKOYA, T. TAKAHASHI, S. YOSHII & M. KASAYA - *Phys. Rev. Lett.* **82**, 1943 (1999)

[25] C. CEPEK, I. VOBORNIK, A. GOLDONI, E. MAGNANO, G. SELVAGGI, J. KRÖGER, G. PANACCIONE, G. ROSSI & M. SANCROTTI - *Phys. Rev. Lett.* **86**, 3100 (2001)

[26] W.C. TONJES, V.A. GREANYA, R. LIU, C.G. OLSON & P. MOLINIÉ - *Phys. Rev. B* **63**, 235101 (2001)

[27] M. GRIONI, I. VOBORNIK, F. ZWICK & G. MARGARITONDO - *J. Elec. Spec.* **100**, 313 (1999)

[28] T. YOKOYA, T. KISS, A. CHAINANI, S. SHIN, M. NOHARA & H. TAKAGI - *Science* **294**, 2518 (2001)

[29] F. ZWICK, D. JÉROME, G. MARGARITONDO, M. ONELLION, J. VOIT & M. GRIONI - *Phys. Rev. Lett.* **81**, 2974 (1998)

[30] F. ZWICK, H. BERGER, I. VOBORNIK, G. MARGARITONDO, L. FORRO, C. BEELI, M. ONELLION, G. PANACCIONE, A. TALEB & M. GRIONI - *Phys. Rev. Lett.* **81**, 1058 (1998)

[31] F.D.M. HALDANE - *J. Phys. C* **14**, 2585 (1981)

[32] J. VOIT - *Rep. Prog. Phys.* **58**, 977 (1995)

[33] M. GRIONI & J. VOIT - *In "Electron Spectroscopies Applied to Low-Dimensional Materials"*, H.P. HUGHES & H.I. STARNBERG eds., Kluwer, Dordrecht, 209 (2000)

[34] C. KIM, A.Y. MATSUURA, Z.X. SHEN, N. MOTOYAMA, H. EISAKI, S. UCHIDA, T. TOHYAMA & S. MAEKAWA - *Phys. Rev. Lett.* **77**, 4054 (1996)

[35] M. ZACCHIGNA, J. ALMEIDA, I. VOBORNIK, G. MARGARITONDO, D. MALTERRE, B. MALAMAN & M. GRIONI - *Eur. Phys. J. B* **2**, 463 (1998)

[36] L.H. TJENG, B. SINKOVIC, N.B. BROOKES, J.B. GOEDKOOP, R. HESPER, E. PELLEGRIN, F.M.F. DE GROOT, S. ALTIERI, S.L. HULBERT, E. SHEKEL & G.A. SAWATZKY - *Phys. Rev. Lett.* **78**, 1126 (1997)

[37] N.B. BROOKES, G. GHIRINGHELLI, O. TJERNBERG, L.H. TJENG, T. MIZOKAWA, TW. LI & A.A. MENOVSKY - *Phys. Rev. Lett.* **87**, 237003 (2001)

[38] A. SEKIYAMA, T. IWASAKI, K. MATSUDA, Y. SAITOH, Y. ONUKI & S. SUGA - *Nature* **403**, 396 (2000)

[39] D. MALTERRE, M. GRIONI & Y. BAER - *Adv. Phys.* **45**, 299 (1996)

[40] C. DALLERA, L. DUÒ, G. PANACCIONE, G. PAOLICELLI, L. BRAICOVICH & A. PALENZONA - *ESRF Highlights*, 77 (2001).

[41] O. TJENBERG, C. DALLERA & N. BROOKES - private communication.

7

ANOMALOUS SCATTERING AND DIFFRACTION ANOMALOUS FINE STRUCTURE

J.L. HODEAU
Laboratoire de Cristallographie, CNRS, Grenoble, France

H. RENEVIER
Département de recherche fondamentale sur la matière condensée, CEA, Grenoble, France

Anomalous scattering and Diffraction Anomalous Fine Structure methods described herein are based on the variation with energy of the contribution of the atomic scattering factor to the structure factor of Bragg reflections. This *anomalous scattering* variation is important near the edge of a given atom *"A"* and induces a variation in the contribution of this specific atom to diffracted intensities. This property is used to extract directly the phase of the structure factor for complex crystallographic structures as in bio-crystallography in the so-called *MAD* analysis. It is also used for *contrast* analysis to perform element-selective diffraction experiments.

On the other hand, the *anomalous* (or *resonant*) *scattering* contribution which varies as a function of energy can be extracted from the diffracted intensities for a given atomic site or compound, so that diffraction and spectroscopy techniques can be combined to perform a site-selective spectroscopy by means of *Diffraction Anomalous Fine Structure*. Furthermore, this spectroscopy method can make use of *Anisotropy of Anomalous Scattering* to extract information on site symmetry or distortion.

It should be noted that this spectroscopic effect on diffraction intensity also yields an enhancement of the X-ray magnetic interaction by means of X-ray Resonant Magnetic Scattering. The latter effect, which is used to probe both the electronic structure and the magnetic properties, is not discussed in this chapter. In this contribution, after recalling the resonant scattering process and the properties of the resonant factors, we present a few examples of these applications and finally we focus on the Diffraction Anomalous Fine Structure method.

F. Hippert et al. (eds.), Neutron and X-ray Spectroscopy, 239–269.

1. ANOMALOUS SCATTERING, ABSORPTION AND REFRACTION

X-ray anomalous or resonant diffraction is related to the close relationship between scattering, absorption and refraction. It refers to the modification of the scattering intensity due to absorption processes. In this way, the long range structural information contained in diffraction peaks is combined with the chemical and local structure selectivity of X-ray absorption spectroscopy. Based on the anomalous scattering process, the Diffraction Anomalous Fine Structure (DAFS) method uses the variation of the diffracted intensities with photon energy and combines X-ray diffraction and X-ray absorption fine structure (XAFS) in the same experiment. Consequently, DAFS can provide site-selective and chemically selective structural information.

Elastic scattering from electrons or a nucleus of an atom *"A"* is proportional to the square of the modulus of the scattering length (b_A or f_A) of the atomic scatterer. For neutrons, b_A varies from nucleus to nucleus in a non-systematic way and can be negative. For X-rays scattered from electrons in an atom, f_A is defined as the ratio of the amplitude of the wave scattered by the atom to the amplitude of the wave scattered by a free electron. This term f_A is complex ($f_A = f_{0A} + f'_A + i\,.f''_A$) and varies with the scattering vector $\boldsymbol{Q}$ and the photon energy E [$|\boldsymbol{Q}| = (4\pi/\lambda)\sin\theta$, $E = \hbar\omega$, $\lambda = 2\pi c/\omega$)]. The Thomson scattering factor $f_{0A}(\boldsymbol{Q})$ is frequency or energy independent and is the Fourier transform of the electronic density in the atom, $\rho_A(r)$. This real form factor f_{0A} decreases with increasing $\boldsymbol{Q}$ and, in the forward scattering limit, tends to the atomic number Z_A in electron units (e.u.). A fraction of the X-rays is absorbed by the photoelectric effect and therefore the atomic scattering factor contains a complex quantity [$f'_A(\omega,\boldsymbol{Q}) + i\,.f'_A(\omega,\boldsymbol{Q})$], named anomalous or resonant contribution. When the photon frequency ω is close to an atomic resonance, this photon can interact with the corresponding bound inner electron and be absorbed. This complex term depends on the chemical nature of the scatterer and becomes substantial near absorption edges. Anomalous effects are more important near absorption edges occurring in the soft X-ray energy range.

Basics on the origin of X-ray anomalous scattering are reported by James (1965). In the forward scattering limit, the refraction index $n(\omega)$ is related to the scattering factor $f_A(\omega,0)$ of the N_A atoms *"A"* **[1-3]** by

$$n = 1 - \delta = 1 - (2\pi e^2/m\omega^2)[\,\Sigma_A N_A f_A(\omega,0)] \qquad \{1\}$$

Far from any atomic resonance, $f_A(\omega,0)$ may be taken as equal to the atomic number Z_A. Close to an absorption edge, $f_A(\omega,0) = Z_A + f'_A(\omega,0) + i\,.f''_A(\omega,0)$ and $n(\omega)$ vary with the energy as

$$n = 1 - \alpha - i\beta \quad \text{with} \quad \begin{aligned} \alpha &= \Sigma_A(2\pi\, N_A\, e^2/m\omega^2)[Z_A + f'_A(\omega,0)] \\ \beta &= \Sigma_A(2\pi\, N_A\, e^2/m\omega^2)[f''_A(\omega,0)] \end{aligned} \qquad \{2\}$$

Thus the refraction index is a complex expression and the wave of frequency ω propagating in the medium with index $n(\omega)$ can be expressed by

$$\begin{aligned} E &= E_0\, e^{i\omega(t - nr/c)} = E_0\, e^{i\omega(t - (1-\alpha)r/c)}\, e^{-\beta\omega r/c} \\ I &= E\,E^* = E_0^2\, e^{-2\beta\omega\, r/c} = E_0^2\, e^{-\mu r} \end{aligned} \qquad \{3\}$$

Its decrease corresponds to an absorptive refraction term β, proportional to the absorption μ

$$\beta = \mu c/2\omega \ , \ \mu = \Sigma_A \mu_A(\omega) = \Sigma_A(4\pi N_A\, e^2/m\omega\, c)[f''_A(\omega,0)] \qquad \{4\}$$

Thus, scattering, absorption and refraction are closely related and this allows different possibilities to measure or to use anomalous dispersion terms

- the imaginary term $f''_A(E)$ is proportional to $E\,\mu_A(E)$,
- the real term $f'_A(E)$ can be obtained from the real part of the refractive index $n(E)$,
- the causality relationship between an applied field and its effect leads to the Kramers-Kronig relation between $f'_A(\omega)$ and $f''_A(\omega)$ [2,4]

$$f'_A(\omega) = +2/\pi \int_0^{\infty} \omega' f''_A(\omega') / (\omega^2 - \omega'^2)\, d\omega' \qquad \{5\}$$

and

$$f''_A(\omega) = -2\omega/\pi \int_0^{\infty} f'_A(\omega') / (\omega^2 - \omega'^2)\, d\omega' \qquad \{6\}$$

2. *THEORETICAL VERSUS EXPERIMENTAL DETERMINATION OF ANOMALOUS CONTRIBUTION*

Simple theoretical calculations of anomalous dispersion are used to determine the smooth variation of the $f'_A(E)$ and $f''_A(E)$ factors and give reliable values far from the absorption edges, where damping and binding effects can be neglected [5,6]. They give the step-like shape at the edge and the variation between edges but do not give near-edge features sensitive to chemical effects (fig. 1). Thus an experimental determination of $f'_A(E)$ or $f''_A(E)$ must be undertaken near edges on the sample itself, to take into account all sample-specific features.

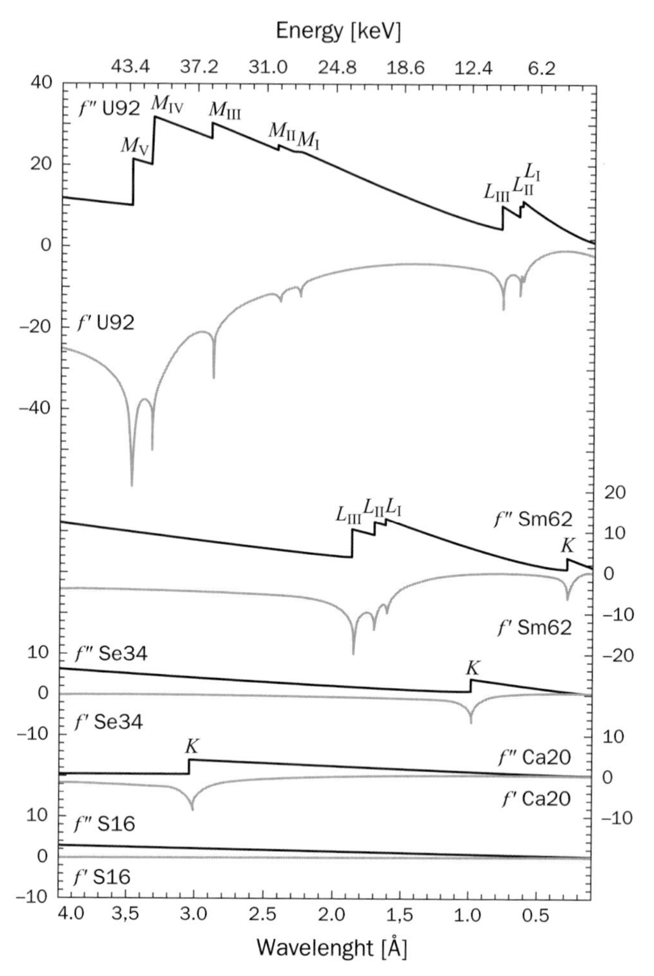

Figure 1 - Variations (in electron units) obtained from theoretical calculations for some anomalous atomic scattering factors f', f'' with λ or E near absorption edges (from ref. **[7]**).

This can be done in different ways:

- **1** - measurement of $f'_A(E) + i f''_A(E)$ from Bragg reflections [8-10], since the structure factor is related to the atomic scattering factor,
- **2** - determination of $f'_A(E)$ from reflectivity measurement performed by refraction of X-rays through a prism [11] or by total external reflection on a surface [12,13],
- **3** - determination of $f'_A(E)$ from X-ray interferometry measurements [14],

- **4** - determination of $f'_A(E)$ from the small angular shift $(2d\,(n^2 - \cos^2\Theta)^{1/2} = k\lambda)$ of the diffracting angle Θ relative to the Bragg angle $(2d \sin\Theta_B = k\lambda)$ with n the refractive index **[15,16]**,
- **5** - finally, direct determination of $f''_A(E)$ from an X-ray absorption measurement, as $\mu(E)$ is proportional to $(\rho/E).(\Sigma_A n_A f''_A(E)/\Sigma_A n_A M_A)$ with ρ the density, M_A the molar mass of the element "A" and n_A the number of these different atoms in the unit cell.

Methods (**2**) (**3**) & (**4**) are based on the variation of the refractive index and of the real $f'_A(E)$ contribution, $f''_A(E)$ being deduced by means of the Kramers-Kronig relation. On the other hand, method (**5**) gives direct access to $f''_A(E)$ and the dispersive term $f'_A(E)$ is obtained through the Kramers-Kronig relation **[17-19]** (fig. 2). The last method (**5**) is the most often used since, even with non-standard samples, absorption spectra are relatively easy to obtain with high signal-to-noise ratio. EXAFS oscillations and chemical shifts can be measured on the sample itself. Nevertheless, to use the Kramers-Kronig relation, $f''_A(E)$ must be known over a large energy range and a theoretical determination far from the edge is generally used.

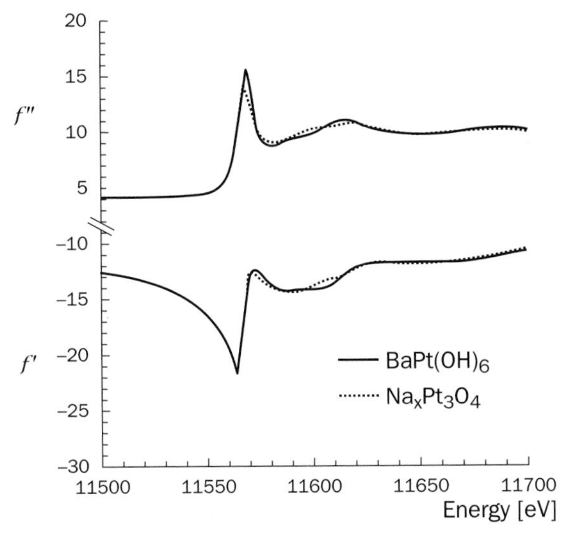

Figure 2 - f'_{Pt} calculated using a Kramers-Kronig transform from the observed f''_{Pt} spectrum extracted from absorption on platinum oxides and theoretical values calculated far from the Pt L_{III} edge (from ref. **[19]**).

All these methods give equivalent results, see for example the comparison between $f'_A(E)$ terms, obtained on a Cu compound by X-ray absorption measurements and the Kramers-Kronig dispersion relation [20], diffraction refinements of Bragg reflections from single crystal [8], and X-ray interferometry measurements near the Cu K edge [11,21]. It is also possible, on the same sample and during the same diffraction experiment, to extract joint measurements of $f'_A(E)$ from the 2Θ diffraction angle shift and $f''_A(E)$ from the diffracted reflection width – related to the absorption penetration length – and their corresponding Kramers-Kronig transformation [15]. The good agreement obtained in such a comparison on multilayers shows that, although the determination of $f''_A(E)$ from absorption data is the easiest, direct $f'_A(E)$ determinations are reliable at low energies.

3. *APPLICATIONS OF ANOMALOUS DISPERSION*

Anomalous scattering applications are based on the variation of a given atomic scattering factor contribution to the structure factor of Bragg reflections

$$\mathrm{F}(\boldsymbol{Q}, E) = \mathrm{F}(hkl) = \Sigma_{\mathrm{cell}} \left(f_A\, e^{2\pi i (hx_A + ky_A + lz_A)}\, e^{-DW_A}\right) \qquad \{7\}$$

with $f_A = f_{0A}(\boldsymbol{Q}) + f\ _A(E) + i f\ _A(E)$.

Because of diffuseness of outer electron shells, the normal term f_{0A} decreases rapidly with $\sin\Theta/\lambda$. On the contrary, the f'_A and f''_A terms originate mainly from inner electrons which are concentrated near the atomic nucleus and these anomalous terms do not decrease with the scattering angle Θ. As the optical theorem states that the total absorption cross-section is related to the imaginary anomalous scattering term, $f''_A(E)$ is proportional to $E\mu_A(E)$ and exhibits energy variations that are directly related to those of the absorption coefficient. As seen in the few examples shown in figure 1, the maximum amplitude of this variation is relatively weak, of the order of 10% of the amplitude of the Thomson scattering f_{0A}, but it increases for absorption edges lying in the soft X-ray energy range. Such variations of $f'_A(E)$, $f''_A(E)$ near the edge of one specific atom "A" produce a variation of the contribution of this atom to diffracted intensities. Its non-oscillating contribution – called herein the "smooth" anomalous contribution – can be used for contrast measurement using its chemical sensitivity. The oscillating contribution of $f'_A(E)$ and $f''_A(E)$ can also be extracted from diffracted intensities. Thus diffraction and spectroscopy information can be combined. With appropriate data analysis, properties of anomalous dispersion are currently used for:

- solution of the structure factor phase (*MAD method*),
- element selective diffraction (*Contrast method*),
- selective site spectroscopy (*DAFS method*). This last method is based on the chemical sensitivity and the anisotropy of anomalous scattering.

3.1. Structure factor phase solution (MAD method)

This method, which was first described more than forty years ago [22,23], uses the variation of the anomalous factors with X-ray photon energy to extract reflection phase information from its different magnitudes at different wavelengths/energies. It was described by Fourme in an earlier volume of HERCULES series [7]. Here, we only briefly outline its basic concepts. The complex structure factor can be represented geometrically as in figure 3. In the absence of anomalous dispersion factors, the Friedel pairs F^+, F^- have the same modulus (fig. 3a). The phase change, induced by the anomalous absorptive $f''(E)$ term, which has a $\pi/2$ phase shift with respect to f_0 and $f'(E)$, gives rise to a change in their magnitude (fig. 3b). Such a difference between Friedel pairs can be used for the absolute determination of molecules and is one of basic principles of *MAD method* (Multi-wavelength Anomalous Dispersion).

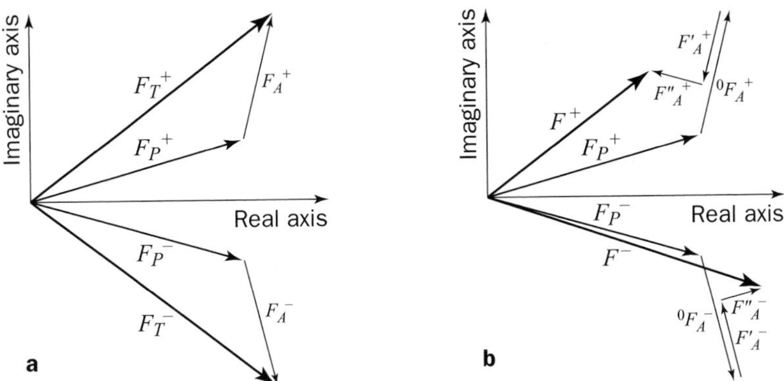

Figure 3 - Effect of anomalous dispersion on diffracted intensities $F(hkl)$ and $F(-h-k-l)$ (geometrical representation) (from ref. [7]).
$F^+ = F(hkl)$; $F^- = F(-h-k-l)$; $F_T = |F_T|\, e^{i\Phi_T}$; $F_P = |F_P|\, e^{i\Phi_P}$; $F_A = |F_A|\, e^{i\Phi_A}$;
$F = F_P + F_A + F'_A + i\,F''_A = F_T + F'_A + i\,F''_A$; F_P = scattering f_0 and smooth anomalous terms of normal atoms; F_A = scattering f_0 of anomalous atoms;
$F'_A + i\,F''_A$ = contribution of anomalous scattering factors (f', f'') of anomalous atoms; F_T = non anomalous scattering f_0 of all atoms; $F_T = F_P + F_A$ (including those of the anomalous scatter).

The *MAD method* uses the variation of both $f'(E)$ and $f''(E)$. Karle developed an algebraic analysis that separates energy-independent contributions F_T from energy-dependent ones $(F'_A + i\,F''_A)$ [24-26]

$$\begin{aligned} F(H) &= F(hkl) \\ &= (\Sigma_m f_{0m}\, e^{2\pi i\,(Hr_m)} e^{-DW_m} + \Sigma_j f_{0j}\, e^{2\pi i\,(Hr_j)} e^{-DW_j}) \\ &\quad + \Sigma_j (f_j + i f_j)\, e^{2\pi i\,(Hr_j)} e^{-DW_j} \end{aligned} \qquad \{8\}$$

$$F(H) = [\Sigma_m (F_P)_m + \Sigma_j (F_A)_j] + \Sigma_j (F'_A + i F''_A)_j = F_T + F'_A + i F''_A \qquad \{9\}$$

with $F^+ = F(h\,k\,l),\ F^- = F(-h-k-l),$

$$F_T = |F_T|\, e^{i\Phi_T},\ F_A = |F_A|\, e^{i\Phi_A},\ \delta\Phi = \Phi_T - \Phi_A$$

In the case of only one kind of anomalous scatterer "A", for $I(h\,k\,l)$ and $I(-h-k-l)$, this leads to

$$I(H)^+ \alpha |F_T|^2 + [(f_{A}^{2} + f_{A}^{2})/f_{0A}^{2}]\,|F_A|^2 + 2(f_{A}/f_{0A})\,|F_T|\,|F_A|\cos(\delta\Phi) + 2(f_{A}/f_{0A})\,|F_T|\,|F_A|\sin(\delta\Phi) \qquad \{10\}$$

$$I(H)^- \alpha |F_T|^2 + [(f_{A}^{2} + f_{A}^{2})/f_{0A}^{2}]\,|F_A|^2 + 2(f_{A}/f_{0A})\,|F_T|\,|F_A|\cos(\delta\Phi) - 2(f_{A}/f_{0A})\,|F_T|\,|F_A|\sin(\delta\Phi) \qquad \{11\}$$

These equations contain only three energy-independent parameters $|F_T|$, $|F_A|$ and $\delta\Phi = \Phi_T - \Phi_A$. The term f''_A shows up in the *Bijvoet (or anomalous) differences* $(I^+ - I^-)$. Using several energy data, *dispersive differences* $(I_{E_i} - I_{E_j})$ are sensitive to f'_A. For *MAD* experiments f'_A and f''_A are assumed to be known at each energy, thus $|F_T|$, $|F_A|$ and $\delta\Phi$ can be determined. Using dispersive and Bijvoet differences as Fourier coefficients, we can calculate dispersive and Bijvoet Patterson maps, which give the positions and the phase Φ_A of the anomalous scatterers. This last information gives access to the phase Φ_T of the structure factor and allows the electron density map calculation **[26-29]**. As example in a [Fe-Ni-Se] hydrogenase where the cysteine residues bound to the Ni were replaced by a seleno-cysteine, difference Fourier synthesis, computed with data collected at energies close to both the Fe K edge and the Se K edge, clearly shows the presence of the three [4Fe-4S] clusters and the Fe-Ni active site, but an unexpected mononuclear Fe site also appears (fig. 4) **[30]**.

Figure 4 - Anomalous difference Fourier synthesis computed for the [Fe-Ni-Se] hydrogenase from *Desulfomicrobium baculatum*; the high density regions (represented by black dots) correspond to the iron contribution: the three [4Fe-4S] clusters, the bi-nuclear Fe-Ni active centre and an additional Fe atom (from ESRF Highlights (1997/98) and ref. **[30]**).

3.2. ELEMENT-SELECTIVE DIFFRACTION (CONTRAST METHOD)

For X-ray diffraction, the ability to distinguish elements is related to the electron number ratio for both atoms, and thus the discrimination between elements with close atomic numbers can be difficult. Neutron diffraction can be used to separate neighbouring elements in the periodic table, since for neutrons the nuclear scattering length depends on the isotope and nuclear spin states and is not related to the atomic number (and nearly zero or negative scattering also could be possible). Nevertheless, for small samples only X-ray analyses can be used, and around absorption edges the anomalous atomic scattering factor variation ranges from 5 to 20 electron units (e.u.) per atom and provides an additional contrast [31-34]. This was illustrated on the high Tc superconductor $Yba_2Cu_{3-x}M_xO_{7-\delta}$ where the substitution of Cu by M = Fe, Co, Ni and Zn cations has a drastic effect on the superconducting properties (fig. 5) [35]. Such a chemical contrast was also applied to extract site occupancies of several neighbouring elements in the periodic table in various compounds such as spinel oxides [36], Zn-exchanged Na-zeolite-Y [37] or high-pressure experiments to increase contrast between In and Sb [38].

For complex structural studies, resonant diffraction can also be used to improve the data redundancy; it allows data collection of different diffraction patterns at the edges of several elements. This was used to solve substituted samples in doped high Tc superconductors or in ternary alloys [39-43]. In such a problem, complementarity of neutron data and resonant X-ray for contrast studies can properly be used, as was the case for the structure determination of the non-stoichiometric $La(Ni_{1-x}Cu_x)_{5+\delta}$ compound [44]. A comparative analysis of the contrast efficiency of resonant X-ray scattering relative to neutron diffraction was also reported by Warner *et al.* (1991) [45]. In their two refinements they obtain transition metal occupancies in agreement at the level of two estimated standard deviations (1-2%) for the two types of radiation and confirm the equivalence of both techniques. It should be noted that for neutron data, scattering lengths can present large variations between elements, and in particular the contrast can be tremendously improved for a compound with an atom having a negative neutron scattering length.

On single crystals, the use of anomalous scattering for contrast studies was also reported by Wulf (1990) [46] who differentiates experimentally lead and bismuth in the mineral structure of the galenobismuthite $PbBi_2S_4$. For this analysis, the author uses the dispersive difference map which depends mainly on the variation of the real term f' and shows up only at the atomic positions of the resonant atoms.

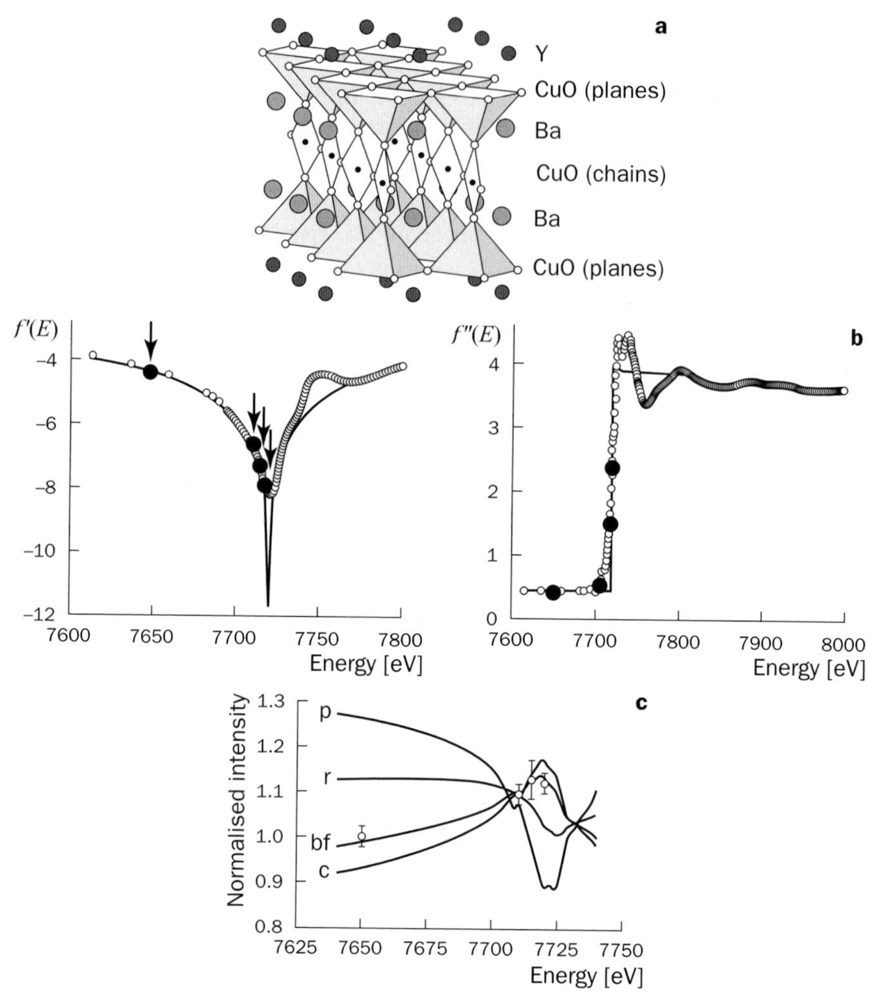

Figure 5 - a - Representation of the $Yba_2(Cu_{3-x}M_x)O_7$ structure, with the two Cu sites (planes & chains) where M cations can substitute. **b** - Variation *versus* the energy of the real dispersive f' and absorptive f'' contributions of the M = Co scattering factor; experimental measurements are collected at four energies (arrows & dots). **c** - Intensity variation of (004) reflection for different localization of Co in substituted $Yba_2(Cu_{3-x}M_x)O_7$ oxides; Experimental measurements at four energies for M = Co; x = 0.5 (dots) and calculated curves for (p) a plane "CuO" site occupancy, (c) a chain "CuO" site occupancy, (r) a random occupancy and (bf) the best experimental fit (from ref. **[35]**).

We can also use the atomic selectivity of resonant diffraction for the analysis of modulated structures: the energy dependence of the intensity of satellite reflections gives specific information about displacements of a given atom **[47,48]**. An example is the study of quasi-one-dimensional metals that exhibit Peierls transitions: the materials change from a metallic to a semi-conducting or isolating state and the transition is generally associated with displacive modulation of the metallic atoms. By working at the absorption edge of such atoms in $(TaSe_4)_2I$, anomalous scattering differentiates their scattering and reveals tetramerization of the tantalum atoms [49].

The contrast effect of resonant scattering is also used for other analyses like the study of diffuse scattering to extract short-range pair correlation of atoms in crystalline solid solutions such as binary or ternary alloys [50,51]. It is used on artificial samples like multilayers where the number of reflections may be rather low owing to poor epitaxy: for the same sample, resonant scattering increases the number of data sets with different atomic contributions and allows better density profile analysis of a specific element in super-lattices [52,53]. For small angle scattering experiments, in compositional and topological in-homogeneity studies, anomalous dispersion helps to separate the contributions and was used to study the solute partitioning during alloy decomposition [54]. We must add that differential anomalous scattering and partial structure factor analyses are efficient for studying amorphous-like and/or complex materials such as glasses, liquids, nanoparticles and catalysts [34,55-60].

4. *DIFFRACTION ANOMALOUS FINE STRUCTURE DATA ANALYSIS*

At energies close to the absorption edge of the atom *"A"*, the three $\mu_A f''_A f'_A$ spectra exhibit features that are sensitive to its chemical state, its short-range environment and its local symmetry:

- For a given chemical state, some absorbing atoms exhibit a sharp resonance peak related to transitions from a deep core level to energetically dense final states [61]. This resonance is generally broader for low energy absorption edges, its minimum line-width varying from less than 1 eV at low energy to a few tens of eV at high energy [62,63].
- An energy shift of the edge position occurs for different valence states [64] (fig. 6) and, at energies close to the edge (0-50 eV), the X-ray Absorption Near Edge Structure (*XANES*) of the same element may be very sensitive to the geometrical environment of the absorbing atom [65,66].

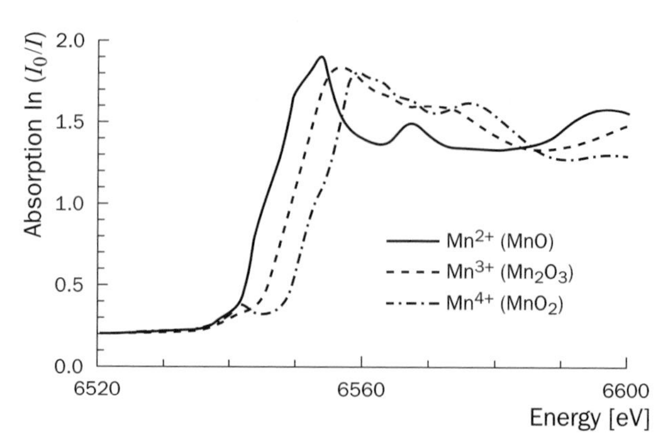

Figure 6 - Energy shifts of the *K* edge position in manganese oxides due to different valence states; the site symmetry and the valence state of the resonant atom affect the edge region (*XANES*, pre-peaks, energy shift of the absorption edges) (from ref. **[67]**).

- Owing to this local environment sensitivity, for low local site symmetry of the absorbing atom, $\mu_A(E)$, $f''_A(E)$ and $f'_A(E)$ are polarization dependent **[68]**. Thus, the resonant scattering part, $f'_A(E)+i\,f''_A(E)$, of the atomic form factor should be represented by a complex *tensor* rather than by a complex scalar number **[69-71]**.
- Photons are sensitive to magnetization densities **[72]** and, although for many years X-ray magnetic scattering was considered a curiosity, spectacular enhancements of this interaction occur near absorption edges due to resonant phenomena which exhibit a polarization dependence closely related to that occurring for charge scattering **[73,74]**.
- Extended oscillations, due to interference effects of photoelectrons with neighbours, appear at energies above the edge (50-1000 eV) in absorption spectra (*EXAFS*) **[75]**; they also affect resonant scattering factors and diffracted intensities.

All these features enable *anomalous diffraction* to be exploited for spectroscopic studies using Diffraction Anomalous Fine Structure (DAFS). The *DAFS method* contains the chemical and short-range order sensitivity of absorption as well as the long-range order sensitivity of diffraction. It offers new possibilities for obtaining chemical information on mixed valence oxides, phase selective spectroscopy of mixed compounds, selective absorption spectra on multilayers, etc.

As stated above, DAFS is a probe based on the principle that absorption-like information can be obtained from scattering measurements by measuring Bragg peak intensities as a function of energy through an absorption edge. Although the first work on fine structure in X-ray diffraction was reported long ago [76-78], almost no further contribution to these studies was published during the sixties and seventies. Only in the eighties, with the development of synchrotron radiation sources, a few reports showed the possibilities of DAFS measurements [9,79,80] and, in 1992, Stragier *et al.* [81] presented a demonstration of this method on a copper single crystal. Several groups then applied this method to different types of samples (thin films, multilayers, powders, single crystals, nanostructures) [19,82-89].

The Bragg intensity can be expressed by

$$I(\boldsymbol{Q},E) = S.D(E)\, Abs(\boldsymbol{Q},E)\, L(\boldsymbol{Q},E).P(\boldsymbol{Q})\, |F(\boldsymbol{Q},E)|^2 \qquad \{12\}$$

where S is a scale factor; D takes into account the detector efficiency; Abs is the correction for the bulk absorption of the sample together with the geometrical effects; L and P are the Lorentz and polarization corrections and $F(\boldsymbol{Q},E)$ is the structure factor. The atomic scattering factor $f_{Aj}(\boldsymbol{Q},E)$ above the edge of an atom "A" on site j may be split into *smooth* and *oscillatory* parts [81]

$$f_{Aj}(\boldsymbol{Q},E) = f_{0A}(\boldsymbol{Q},E) + f'_{0A}(E) + i\, f''_{0A}(E) + \Delta f''_{0A}(E)\, [\chi'_{Aj}(E) + i\, \chi''_{Aj}(E)] \qquad \{13\}$$

where f_{0A} is the Thomson scattering, f'_{0A} and f''_{0A} are the *bare* or *smooth* atom anomalous corrections to f_{Aj} and $\Delta f''_{0A}$ is the imaginary part of resonant contribution due to the excited core electron. χ'_{Aj} and χ''_{Aj} are the *oscillatory* parts of the anomalous contribution, which is also called extended-DAFS. This energy dependent modulation of the peak intensity contains local structural information similar to that of extended-XAFS (fig. 7) [81,83].

4.1. DAFS AND EDAFS FORMALISM

The formalism for analysing the DAFS spectra and EDAFS oscillations is in most cases very similar to that of EXAFS. In relation to XAFS, the advantage of DAFS for studying complex systems with different phases or different atomic sites is to give structural information about a specific atom by choosing corresponding Bragg peaks (spatial- or site-selective Bragg peaks). Spatial selectivity is easy to obtain by selecting some Bragg peaks corresponding to the different phases. The site selectivity can be more complex and, depending on the crystallographic site selectivity of the diffracted reflections, two different kinds of situations are encountered when analysing the EDAFS spectra.

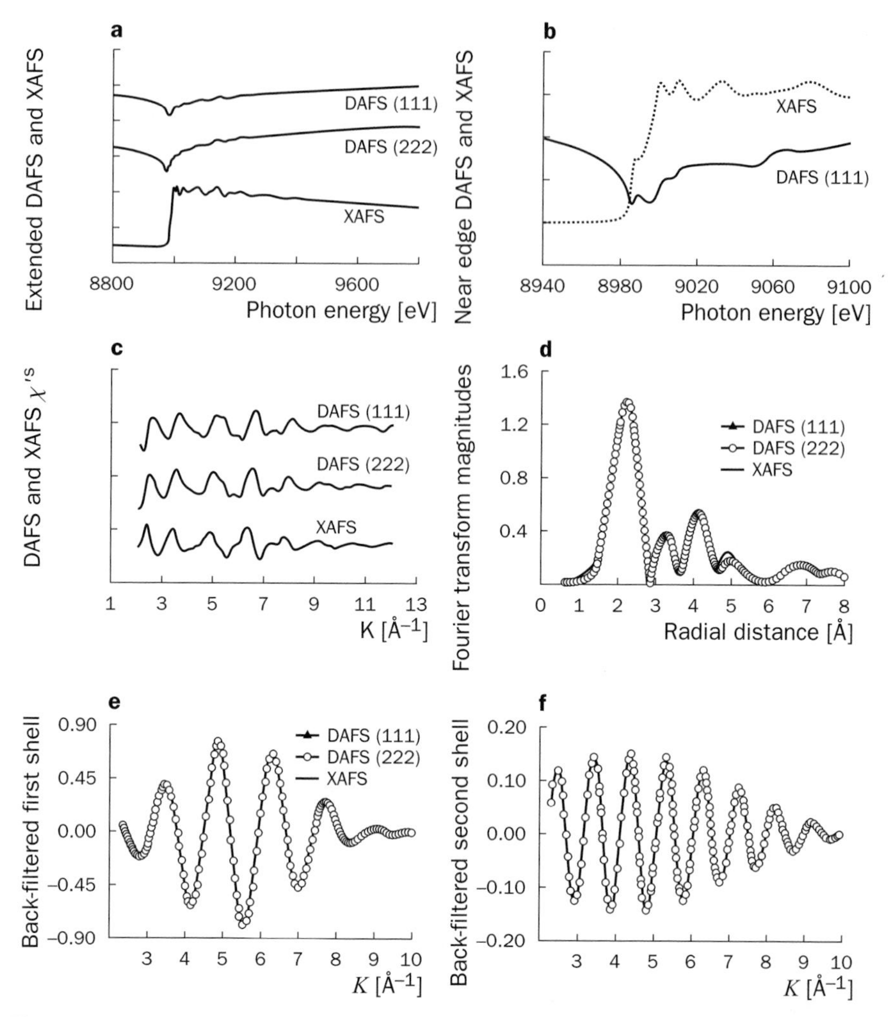

Figure 7 - Comparison of the XAFS and DAFS spectra (**a** - **b**) of the EXAFS and the EDAFS spectra (**c**) obtained on a Cu single crystal at the Cu *K* edge. Same Fourier transforms (**d** - **e** - **f**) are obtained from the different spectra (from ref. **[83]**).

4.1.1. SINGLE ANOMALOUS SITE ANALYSIS

The easiest situation is the *single anomalous site analysis*, which corresponds to only one anomalous crystallographic site contribution to a given diffracted intensity: the DAFS spectrum is obtained with a site-selective Bragg reflection. The XAFS-like structural information may be obtained without knowing the crystallographic structure and an iterative Kramers-Kronig method may be used to extract successively the experimental χ'_A and χ''_A spectra. Once χ''_A is obtained, the EXAFS oscillations are analyzed with a standard EXAFS data analysis package.

This first procedure was initiated successfully by Stragier *et al.* (1992) [81] and is the most convenient one. This direct procedure of site- or phase-selective χ'_A and χ''_A determination was also used for Fe/Ir multilayers to separate information from the Ir buffer and from the multilayer [90]. It can be used for samples whose symmetry space group gives rise to extinction rules and where, consequently, reflections exist for which only anomalous scatterers located in one site could contribute, such as in ferrite or Co_3O_4 [82].

4.1.2. MULTIPLE ANOMALOUS SITE ANALYSIS

We must use a *multiple anomalous site analysis* when we have several anomalous site contributions to the diffracted intensities, with different weights. In that case knowledge of the crystallographic structure is generally needed to determine the different weights of each site on reflections and to extract the XAFS-like information about each anomalous site.

For the general case of non-centro-symmetric structure with N_A anomalous atoms, the structure factor can be written

$$F(\boldsymbol{Q},E) = F_T(\boldsymbol{Q},E)\, e^{i\varphi_T(\boldsymbol{Q})} + \Sigma_j^{N_A} |\alpha_{Aj}(\boldsymbol{Q})|\, e^{\varphi_{Aj}(\boldsymbol{Q})}\, [f'_{Aj}(E) + i\, f''_{Aj}(E)] \quad \{14\}$$

with $\quad f'_{Aj} = f'_{0A} + \Delta f''_{0A}\chi'_{Aj} \qquad f''_{Aj} = f''_{0A} + \Delta f''_{0A}\chi''_{Aj}$

where the $F_T(\boldsymbol{Q},E)$ term includes the net contribution of all non anomalous atoms *and* the Thomson scattering of all anomalous atoms. Thus it corresponds to a complex structure factor of phase φ_T. Each anomalous factor is weighted by a structural contribution, $\alpha_j(\boldsymbol{Q}) = c_{Aj}\, e^{-DW_{Aj}} e^{\varphi_{Aj}(\boldsymbol{Q})}$, where c_{Aj} is the occupation factor of atom "A" on site j, $\varphi_{Aj} = \boldsymbol{Q}\cdot\boldsymbol{r}_i$ contains the corresponding atomic positions, and $e^{-DW_{Aj}}$ the Debye-Waller factor. All these structural terms contribute to the *smooth* variation of the diffracted intensity and correspond to a complex *smooth* structure factor $F_{\text{smooth}}(\boldsymbol{Q},E)$ with a phase $\varphi_{\text{smooth}}(\boldsymbol{Q},E)$. It can be calculated without taking into account the complex fine structure χ'_{Aj} and χ''_{Aj}. Assuming that the *bare* or *smooth* atom anomalous corrections (f'_{0A}, f''_{0A}) are identical for all anomalous atoms, we can write [91]

$$|F_{\text{smooth}}(\boldsymbol{Q},E)|^2 = |F_T|^2\, [(\cos(\varphi_T - \varphi_A) + \beta f'_{0A})^2 + (\sin(\varphi_T - \varphi_A) + \beta f''_{0A})^2] \quad \{15\}$$

where $\quad \beta = |\alpha_A|/|F_T|$ and $|\alpha_A(\boldsymbol{Q})|\, e^{i\varphi_{Aj}(\boldsymbol{Q})} = \Sigma_j^{N_A} |\alpha_{Aj}(\boldsymbol{Q})|\, e^{i\varphi_{Aj}(\boldsymbol{Q})}$

This expression shows that the *smooth* variation with energy of the diffracted intensity near an absorption edge gives access to the *phase difference* $\Delta\Phi = \varphi_T - \varphi_A$ and the relative *anomalous contribution* β of the corresponding reflection. Therefore it gives important and precise information on the crystallographic structure itself and determines the shape of the DAFS spectrum. This explains why diffracted intensities can be modulated in different ways by the $f'(E)$ and $f''(E)$ variations versus energy due to their different $\Delta\Phi$ and β parameters, (fig. 8) **[19]**.

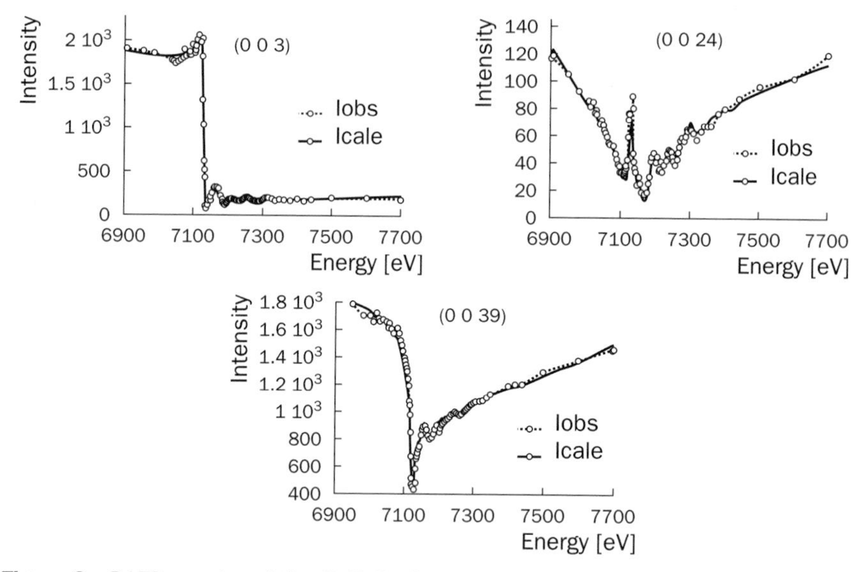

Figure 8 - DAFS spectra of the $BaZnFe_6O_{11}$ hexaferrite (0 0 3), (0 0 24) and (0 0 39) reflections and intensity fits using bulk $\mu(E)$, refined individual $f'_{Fe}(E)$ and $f''_{Fe}(E)$ curves for tetrahedral and octahedral sites; their *smooth* shape versus energy is totally different due to their different $\Delta\Phi$ and β parameters which give information on the Fe/Zn occupancies of the tetrahedral sites (from ref. **[19]**).

In the case of a centro-symmetric structure with several anomalous sites, the site dependent oscillatory contribution χ'_{Aj} and χ''_{Aj} can be extracted from the f'_{Aj} and f''_{Aj} terms, using an iterative Kramers-Kronig procedure between real and imaginary parts, and taking into account the structural weights of the different sites **[83]**. For a non-centro-symmetric structure with several anomalous sites, such a simple expression cannot be found due to crossed terms $f''_{Aj}\, f''_{Ak}$ $(j \neq k)$ in the expression $|F(\boldsymbol{Q},E)|^2$. However we can often neglect these crossed order terms, since they are generally small, and this approximation allows the first order EDAFS oscillations for a given reflection to be extracted directly. The experimental intensity spectrum is thus normalized and its oscillating contribution $\chi_Q(k)$ can be expressed as a function of the different site contributions χ'_{Aj} and χ''_{Aj} by

$$\chi_Q(k) = (|F_{smooth}|/2|\alpha_A|\Delta f''_{0A})\,[(I_{exp} - I_{smooth-exp})/I_{smooth-exp}]$$
$$= \Sigma_j^{N_A} W_{Aj}\,[(\cos(\varphi_{smooth} - \varphi_{Aj})\chi'_{Aj} + (\sin(\varphi_{smooth} - \varphi_{Aj})\chi''_{Aj}] \qquad \{16\}$$

where I_{exp} is the experimental intensity corrected for the absorption and for the fluorescence background, $I_{smooth-exp}$ the *smooth* diffracted intensity. The phase difference $(\varphi_{smooth} - \varphi_{Aj})$ and the weight contribution of each anomalous site W_{Aj} contain the crystallographic structure information and can be calculated from the crystallographic structure. The normalization factor $S_D = (|F_{smooth}|/2|\alpha_A|\Delta f''_{0A})$ can be either calculated from the structure or obtained directly by fitting the intensity spectrum with the reflection dependant parameters $\Delta\Phi$ and β in the equation

$$S_D = [(\cos\Delta\Phi + \beta f'_{0A})^2 + (\sin\Delta\Phi + \beta f''_{0A})^2]^{1/2}/(2\beta\Delta f''_{0A}) \qquad \{17\}$$

Finally, as the imaginary contribution χ''_{Aj} is equivalent to an EXAFS signal and the real contribution χ'_{Aj} is related to χ''_{Aj} via the Kramers-Kronig relation, the oscillating EDAFS contribution $\chi_Q(k)$ can be expressed in a formalism very similar to that of EXAFS, using a weighted contribution of the different sites j with a structural phase shift $\Delta\Phi_j$ that is additional to that of photoelectron scattering. For the simple case of one anomalous site this structural phase difference $\Delta\Phi_j$ can be extracted directly from the fit of the intensity *smooth* spectrum [91]. Such an EXAFS-like representation has the advantage of avoiding the use the iterative Kramers-Kronig transformation and, after a specific extraction of the oscillating term $\chi_Q(k)$, allows an EDAFS analysis with classical EXAFS software packages.

4.2. DAFS AND EDAFS DETERMINATION

Thus, several procedures can be used for extracting the site-selective structural information. The procedure using the refinement of individual f'_{Aj} and f''_{Aj} at each energy has been used for single crystal analysis of Platinum oxide [19] and hexaferrite $BaZnFe_6O_{11}$ [92]. The structure of this latter compound contains four octahedral sites occupied by Fe cations and two tetrahedral sites occupied by Fe and Zn atoms. Using the data refinement versus energy of eleven (*00l*) reflections, the fits of the *smooth* spectra (fig. 8) give the Fe/Zn occupancies and the refined f'_{Fe}/f''_{Fe} spectra of both octahedral and tetrahedral sites reveal their different local environment symmetry (fig. 9).

Using the DAFS formalism described previously, another procedure for extracting the site-selective structural information is to deduce the individual χ'_{Aj} and χ''_{Aj} from a linear system of equations at each energy and use the Kramers-Kronig transforms as an additional constraint to relate χ'_{Aj} and χ''_{Aj} [83,91].

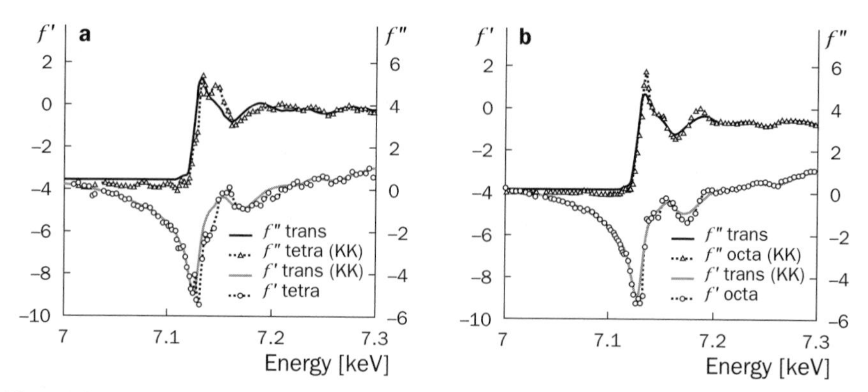

Figure 9 - $f'_{\text{tetra}}(E)$, $f''_{\text{tetra}}(E)$ and $f'_{\text{octa}}(E)$, $f''_{\text{octa}}(E)$ spectra for the tetrahedral and octahedral sites of hexaferrite, derived from DAFS refinements and the Kramers-Kronig relation; the average $f''_{\text{KK}}(E)$ spectrum is obtained from transmission XAFS experiments (from ref. **[92]**).

This procedure is basically the same as refining the individual f'_{Aj} and f''_{Aj} as a function of energy, except that the second order terms are neglected. Such a linear system of equations giving site-selective χ'_{Aj} and χ''_{Aj} terms has been used for different studies of epitaxial thin films or multilayers. Most of these experiments have been undertaken to separate the different atomic local structures in the different layers, which is not possible using only absorption spectroscopy since the contributions of all equivalent atoms in the different parts of the material are averaged. In the example shown in figure 10, EDAFS spectra were used in Fe/Ir multilayers to differentiate first neighbour distances for Fe atoms located in the middle and in the interface of the Fe layer [85]. As the number of DAFS spectra is often not much larger than the number of unknowns, to get reliable results it is important to choose reflections for which the contributions to the diffracted intensities of each anomalous site are rather different. Thus, in the example displayed in figure 10, satellite reflections having inter-reticular *d* spacing characteristic of Fe neighbours, such as the weak (0 0 36) one, were important for stabilizing the refinement.

Diffracted intensity is proportional to the product of the squared modulus of the structure factor and the bulk absorption of the sample, and EDAFS experiments are performed at absorption edges. Thus the absorption process (sample shape, absorption length…) must be perfectly controlled and corrected for. Consequently, accurate EDAFS spectra were obtained mainly on epitaxial thin films or multilayers since, in contrast to small single crystals or powders, we can simplify the absorption correction.

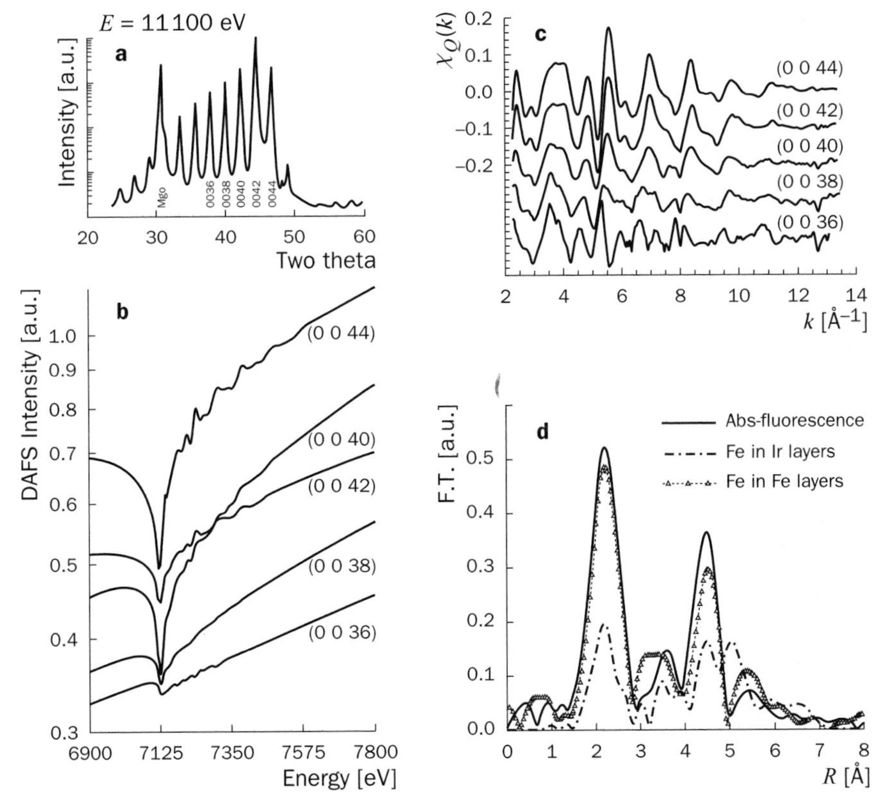

Figure 10 - **a** - **b** - Raw DAFS spectra obtained from Ir/Fe multilayer Bragg reflections at the Fe *K* edge. **c** - EDAFS oscillation spectra for (0 0 36) to (0 0 44) reflections. **d** - Fourier transforms for the Fe atoms in Ir-rich layers and in the Fe-rich layers and that extracted from fluorescence EXAFS data (from ref. **[85]**).

The data quality of these spectra gives a precise first neighbour environment, obtained either by an EXAFS-like single shell analysis using experimental model phases and amplitudes or by a multi-shell analysis (see e.g. ref. **[85]**, **[88]**, **[91]**, **[93]**). The multi-shell analysis, using theoretical phases and amplitudes generated by the GNXAS or FEFF and FEFFIT programs **[94-96]** has also been used to extract the second and third nearest neighbour shell. We show in figure 11 the overall best fit and the individual contributions of the different neighbours obtained in a single GaAsP epilayer **[91]**. EDAFS experiment can also be used to extract information on nano-systems and such an experiment has been performed on self-assembled InAs/InP quantum wires **[89]**. As reported by Ravel *et al.* (1999) **[97]**, EDAFS spectroscopy could be used to separate $\chi(E)$ oscillations of different X-ray absorption edges when two or more absorption edges are too close in energy in a given material, such as for the Ti *K* and the Ba *L* edges in $BaTiO_3$.

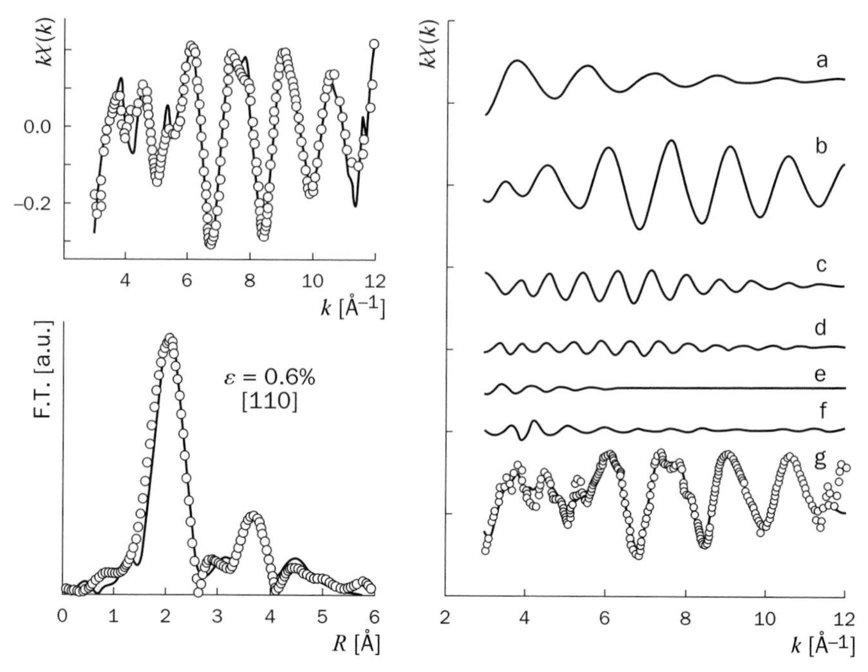

Figure 11 - (006) EDAFS oscillation spectrum, Fourier transform and individual contributions of the different scattering paths to the EDAFS spectrum (g) of a single Ga-As-P epilayer ; Ga-As (a), Ga-P (b), Ga-Ga II-shell (c), Ga-As III-shell (d), Ga-P III-shell (e), Ga-As-Ga MS (f) (from ref. **[91]**).

4.3. DAFS AND DANES VALENCE DETERMINATION

For crystals with inequivalent sites, Diffraction Anomalous Near Edge Structure experiments (DANES) can assign near edge features to these different sites. Using the DAFS formalism described in § 4.1, site selective DANES spectra can detect any energy shift of the edge position, giving information on the valence state of the resonant atom with the same sensitivity as XANES experiments. Such experiments gave very good results on thin films. For instance, the site selectivity of DANES was nicely demonstrated by Sorensen *et al.* (1994) **[84]** on a thin film of the high Tc superconductor $YBa_2Cu_3O_{6.6}$: the individual f''_A curves of the two copper sites were extracted by using a co-refinement of different reflections. From XANES data, the assignment of near edge features characteristic of each Cu site (chains and planes) was difficult, whereas the contribution from the two Cu sites is well defined in Bragg reflections: DANES measurements provide an easier and clean assignment of these features (fig. 12).

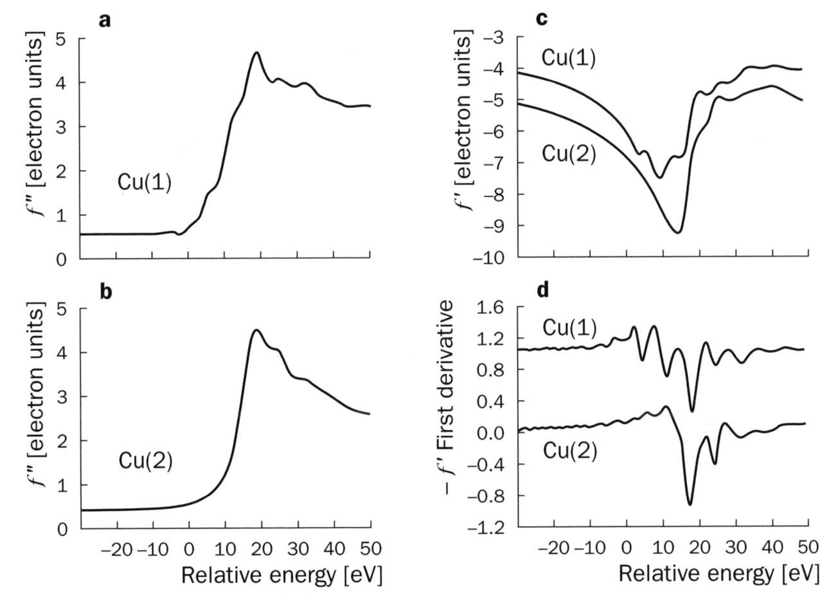

Figure 12 - Site separated $f''_{\mathrm{Cu}}(E)$ spectra of the Cu(1) (chains) and Cu(2) (planes) atoms, extracted from a DAFS analysis on $YBa_2Cu_3O_{6.6}$ thin film (from ref. **[83]**).

Another demonstration of the DANES site-selective efficiency for charge order studies is provided by a study of α-NaV_2O_5 crystal. This compound exhibits a structural and a magnetic phase transition below 35 K and there was some controversy over the assignment of a possible vanadium charge ordering related to the presence of magnetic V^{4+}-V^{4+} dimers. Anomalous X-ray diffraction studies of superstructure reflections reveal the zig-zag charge order in the vanadium ladders present in this structure. Using the anomalous X-ray diffraction study of the (15/2 ½ ¼) reflection, a single layer zig-zag model was first proposed **[98]**, then studies of 20 different superstructure reflections showed the coexistence of different zig-zag models corresponding to different stacking sequences along the *c* direction (fig. 13) **[99]**.

Such analyses can be useful in solid state chemistry, but samples are often available only in powder form and there are some intrinsic limits of DANES method when applied to powders (low count rate, low f'' sensitivity, sampling problems and fluorescence scattering) **[33,82,100,101]**. Nevertheless measurements performed on a highly absorbing powder $La_4Mn_5Si_4O_{22}$ have revealed three valence states of manganese distributed over three different sites **[67]**. However, in such powders the statistics of DANES experiments is poorer than for single crystals and the extended-DAFS spectra cannot be extracted.

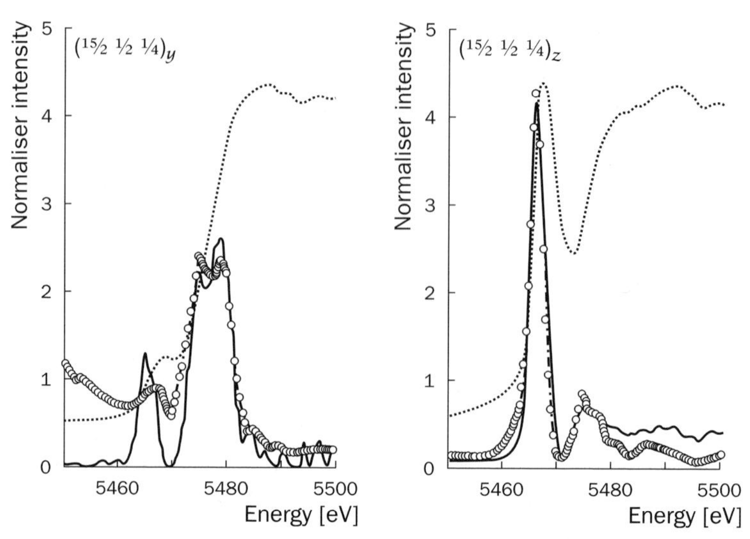

Figure 13 - Experimental (dots) and calculated (line) intensities of (15½ ½ ¼) superstructure reflection, and absorption spectra (dot-dash line), taken at the vanadium *K* edge on a α-NaV_2O_5 crystal for polarization along y and along z. The intensity comes from vanadium charge ordering, the difference between both polarization states comes from the anisotropy of the vanadium site (from ref. **[99]**).

4.4. ANISOTROPY OF ANOMALOUS SCATTERING

As synchrotron radiation sources provide polarized X-rays, DANES spectra near absorption edges, such as XANES spectra, are orientation dependent on the local chemical environment of the anomalous scatterer with respect to the beam polarization direction. This occurs when the resonant atom exhibits coordination with a low symmetry [69,102]. Due to the anisotropy of anomalous scattering, this dichroism can also be observed on more symmetric crystals such as in the $P2_13$ cubic $NaUO_2(C_2H_3O_2)_3$ compound, in which each uranyl ion lies on a threefold axis [69]. Thus, the anomalous scattering factor must be replaced by a *tensor* to take into account polarization of both incident and diffracted beams. This effect is a sophistication of resonant effects, but it is also a new means of solving problems [71]. It explains the observation of reflections that are forbidden by a screw axis or a glide plane and why their intensity changes with the azimuthal angle [103-105]. As point defects and atomic thermal vibrations decrease the symmetry of the local atomic environment, they can also contribute to additional anisotropy of the resonant scattering and to forbidden reflections near absorption edges [106,107].

This *tensor* properties of DANES spectra are sensitive to the empty orbital symmetries and occupations, thus DANES is also a new probe for studying site-specific orbital maps [108]. Nevertheless, as phenomena like orbital ordering are generally associated with lattice distortion, we cannot claim that DANES is a simple and direct probe for studying orbital ordering [109]. In $LaMnO_3$, for instance, the DANES or resonant spectrum of the forbidden reflection (0 0 3) is mainly due to the closely associated Jahn-Teller distortion rather than to orbital ordering itself [110]. In another example, resonant diffraction was used to investigate electronic fluctuations of the octahedral iron atoms in magnetite, which is considered to be the prototype of charge localization [111,112]. Near edge DAFS spectra of (0 0 2) and (0 0 6) forbidden reflections were measured, above and below the Verwey transition, as a function of energy, for different crystal azimuthal angles (fig. 14). Their appearance implies that all octahedral sites display the same slight 3-fold distortion and that charge localization, if it exists, does not affect the (0 0 1) resonant spectra.

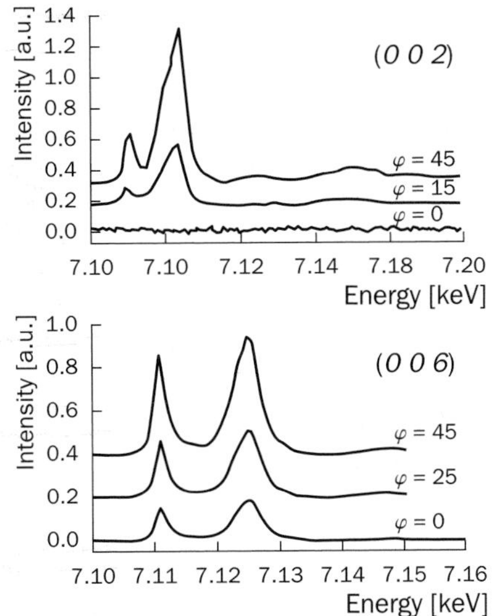

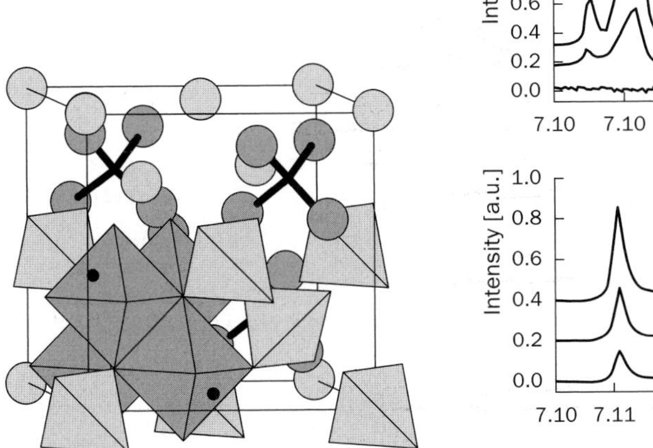

Figure 14 - (*0 0 2*) and (*0 0 6*) forbidden reflections measured at the Fe *K* edge in magnetite Fe_3O_4 at 300 K. The energy dependence of the integrated intensity is proportional for different azimuthal angles. This intensity variation is due to a *3*-fold distortion (through the black dot) of the Fe octahedral sites (from ref. **[112]**).

It must be recalled here that for any phase transition analysis using valence selectivity and/or the anisotropy property of anomalous scattering, it is essential to select *several* superstructure or satellite reflections sensitive to different sites, atoms and aspects of the concerned transition. This was nicely shown in α-NaV_2O_5 (fig. 13) [99] and also in $LaPrMnO_2$ to separate the charge contribution from the distortion-/orbital-ordering contribution [113].

5. REQUIREMENTS FOR ANOMALOUS DIFFRACTION EXPERIMENTS

As f' and f'' variations are not large, very accurate data are needed for anomalous scattering and DAFS experiments. We recall here some specific experimental requirements.

- For DAFS experiments, *accurate energy calibration* is essential. An external check is needed and the best way is to measure directly the absorption of the sample itself in fluorescence mode. In this case, the absorption reference used is such that the sample and energy resolution of the absorption spectrum and of the DAFS is the same. Such a calibration is very important since at the edge, the "step-like" feature of different Bragg reflections can exhibit shifts of a few eV due to real chemical shifts (fig. 6) and/or to effects of different structural parameters $|F_T|$, $|F_A|$ and $\delta\Phi$. The energy stability is also crucial to get reliable values of f' and f''. Due to their high source stability and high energy resolution, third generation synchrotron radiation sources are well suited for anomalous experiments [114].
- As anomalous measurements are performed at absorption edges, another important step of the analysis is a precise absorption correction since the bulk absorption could be important for samples having large anomalous scattering content and introduce intensity oscillations after the edge [115]. These oscillations are caused by the average absorption of all anomalous atoms present on all sites of the structure. Thus, for the absorption correction, μ must be measured on the sample itself. Strong reflections are very sensitive to the bulk absorption effects and to extinction [19,116]; in contrast, weak reflections are more sensitive to variations of f'_A and f''_A and exhibit much more pronounced DAFS oscillations. As absorption corrections can be sensitive to thickness variations of a few μm, an accurate empirical absorption correction procedure can be developed for highly absorbing single crystals [19].
- For the use of the anisotropy of the anomalous scattering, we need high energy resolution. The maximum variation of the scattering amplitude with respect to the polarization direction can increase from 1 to 8 electron units by improving the energy resolution from 10 to 1 eV (fig. 15) [117]. Furthermore the analysis of this anisotropic contribution requires a polarization selective detector using an analyzer crystal.

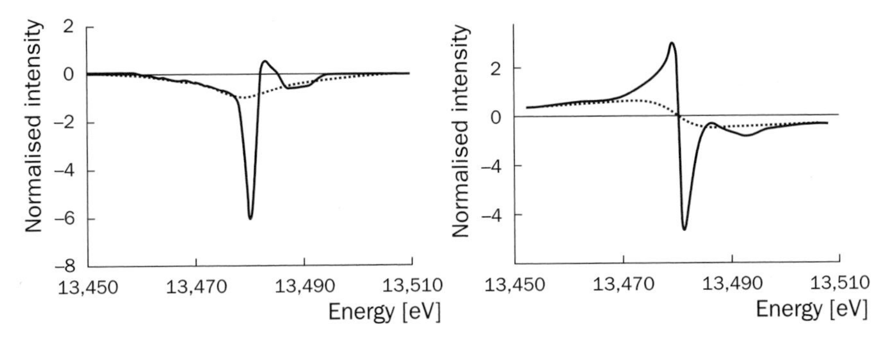

Figure 15 - Effect of energy resolution on the polarization anisotropy of dispersive terms in sodium bromate represented here by the difference patterns $f_\sigma'' - f_\pi''$ and $f_\sigma' - f_\pi'$ at the Br K edge, broken curves are convoluted with a Lorentzian broadening of ~ 10-11 eV (from ref. **[117]**).

6. *CONCLUSION*

Anomalous or resonant diffraction can be used to obtain much information about a given element in a crystal. However this information requires sophisticated tools (synchrotron rings), accurate experiments and complex analysis. This is a price that still needs to be paid to go beyond the anomalous complexity and/or correction and to fully extract and use this rich "*anomalous*" information. However, several applications are nowadays available for the MAD method to determine experimentally the phase and for the contrast method that uses the chemical sensitivity of resonant diffraction. This is also the case for the DAFS method: the fine structure of the diffracted intensity as a function of the energy can now be analyzed in very complex structures or systems.

ACKNOWLEDGEMENTS

We would like to thank all our present and past collaborators. We are particularly pleased to acknowledge P. Wolfers, J. Vacinova, B. Ravel, G. Subias, M.G. Proietti, H. Palancher, E. Lorenzo, Y. Joly, S. Grenier, J. Garcia, V. Favre-Nicolin, S. Bos, and J.F. Berar for helpful discussions and collaborations.

REFERENCES

[1] R.W. JAMES - *The optical principles of the diffraction of X-rays*, Cornell Univ. Press, Ithaca, NY, USA (1965) Ox. Box Press, Woodbridge, CT, USA (1982)

[2] J.J. SAKURAI - *Advanced Quantum Mechanics*, Addison-Wesley, Reading, Mass., USA (1967, 1984)

[3] J. ALS-NIELSEN - *In "Neutron and Synchrotron Radiation for Condensed Matter Studies", HERCULES vol. 1*, J. BARUCHEL, J.L. HODEAU, M.S. LEHMANN, J.R. REGNARD & C. SCHLENKER eds, Les Editions de Physique - Springer-Verlag, 3 (1993)

[4] J. TOLL - *J. Phys. Rev.* **104**, 1760 (1956)

[5] D.T. CROMER & D. LIBERMAN - *J. Chem. Phys.* **53**, 1891 (1970)

[6] S. SASAKI - *Anomalous scattering factors for synchrotron radiation users calculated using the Cromer and Liberman method*, National Laboratory for High Energy Physics, KEK, Tsukuba, Japan (1984)

[7] R. FOURME, W. SHEPARD, M. SCHILTZ, M. RAMIN & R. KAHN - *In "Structure and Dynamics of Biomolecules", HERCULES vol. IV*, E. FANCHON, E. GEISSLER, J.L. HODEAU, J.R. REGNARD & P. TIMMINS eds, Oxford University Press, Oxford (2000)

[8] A. FREUND - *In "Anomalous scattering"*, S. RAMASESHAN & S.C. ABRAHAMS eds, Munksgaard Copenhagen, 69 (1975)

[9] D.H. TEMPLETON, L.K. TEMPLETON, J.C. PHILLIPS & K.O. HODGSON - *Acta Cryst. A* **36**, 436 (1980)

[10] P. SUORTTI, J.B. HASTING & D.E. COX - *Acta Cryst. A* **41**, 413 (1985)

[11] B. LENGELER - *In "Resonant anomalous X-ray scattering"*, G. MATERLIK, C.J. SPARKS & K. FISCHER eds, Elsevier Science B.V., Amsterdam, 35 (1994)

[12] T.S. FUKAMACHI, S. HOSOYA, T. KAWAMURA, S. HUNTER & Y. NAKANO - *Jap. J. Appl. Phys.* **17**, 326 (1978)

[13] F. STANGLMEIER, B. LENGELER, W. WEBER, H. GÖBEL & M. SCHUSTER - *Acta cryst. A* **48**, 626 (1992)

[14] R. BEGUM, M. HART, K.R. LEA & D.P. SIDDONS - *Acta Cryst. A* **42,** 456 (1986)

[15] L. SÈVE, J.M. TONNERRE & D. RAOUX - *J. Appl. Cryst.* **31,** 700 (1998)

[16] U. STAUB, O. ZAHARKO, H. GRIMMER, M. HORISBERGER & F. D'ACAPITO - *Eur. Phys. Lett.* **56**, 241 (2001)

[17] P. DREIER, P. RABE, W. MALZFELDT & W. NIEMAN - *J. Phys. C* **17**, 3123 (1984)

[18] J.P. QUINTANA, B.D. BUTLER & D.R. HAEFFNER - *J. Appl. Cryst.* **24**, 184 (1991)

[19] J. VACINOVA, J.L. HODEAU, P. WOLFERS, J.P. LAURIAT & E. ELKAIM - *J. Synchrotron Radiation* **2**, 236 (1995)

[20] J.J. HOYT, D. DE FONTAINE & W.K. WARBURTON - *J. Appl. Cryst.* **17**, 344 (1984)

[21] U. BONSE, I. HARTMANN-LOTSCH & H. LOTSCH - *Nucl. Instr. Meth.* **208**, 603 (1983)

[22] Y. OKAYA & R. PEPINSKY - *Phys. Rev.* **103**, 1645 (1956)

[23] C.M. MITCHELL - *Acta Cryst.* **10**, 475 (1957)

[24] J. KARLE - *Int. J. Quant. Chem.* **7**, 357 (1980)

[25] J. KARLE - *Physics Today*, June issue, 22 (1989)

[26] W.A. HENDRICKSON - *Science* **254**, 51 (1991)

[26] R. KAHN, R. FOURME, R. BOSSHARD, M. CHIADMI, J.L. RISLER, O. DIDEBERG & J.P. WERY - *FEBS Lett.* **179**, 133 (1985)

[27] J.M. GUSS, E.A. MERRITT, R.P. PHIZACKERLEY, B. HEDMAN, M. MURATA, K.O. HODGSON & H.C. FREEMAN - *Science* **241**, 806 (1988)

[28] M.R. PETERSON, S.J. HARROP, S.M. MCSWEENEY, G.A. LEONARD, A.W. THOMPSON, W.N. HUNTER & J.R. HELLIWELL - *J. Synchrotron Radiation* **3**, 24 (1996)

[29] W. SHEPARD, W.B.T. CRUSE, R. FOURME, E. DE LA FORTELLE & T. PRANGÉ - *Structure* **6**, 849 (1998)

[30] A. VOLBEDA, E. GARCIN, C. PIRAS, A.L. DE LACEY, V.M. FERNANDEZ, E.C. HATCHIKIAN, M.C. FREY & J.C. FONTECILLA-CAMPS - *J. Am. Chem. Soc.* **118**, 12989 (1996)

[31] J.P. ATTFIELD - *Mater. Sci. Forum* **228-231**, 201 (1996)

[32] G. HEGER - *In "Neutron and Synchrotron Radiation for Condensed Matter Studies", HERCULES vol. 2,* J. BARUCHEL, J.L. HODEAU, M.S. LEHMANN, J.R. REGNARD & C. SCHLENKER eds, Les Editions de Physique - Springer-Verlag, Berlin, 23 (1994)

[33] D. COX & A.P. WILKINSON - *In "Resonant anomalous X-ray scattering",* G. MATERLIK, C.J. SPARKS & K. FISCHER eds, Elsevier Science B.V., Amsterdam, 195 (1994)

[34] Y. WASEDA - *Anomalous X-ray scattering for materials characterisation,* Springer Tracts in Modern Physics Vol. 179, Springer Verlag, Berlin (2002)

[35] R.S. HOWLAND, T.H. GEBALLE, S.S. LADERMAN, A. FISHER-COLBRIE, M. SCOTT, J.M. TARASCON & P. BARBOUX - *Phys. Rev. B* **39**, 9017 (1989)

[36] J. LORIMIER - *Problématique des valences mixtes dans les ferrites nano-métriques : possibilités offertes par la diffraction résonnante des rayons X,* PhD Thesis, Dijon University (2000)

[37] A.P. WILKINSON, A.K. CHEETHAM, S.C. TANG & W.J. REPPART - *Chem. Comm.*, 1485 (1992)

[38] R.J. NELMES, P.D. HATTON, M.I. MCMAHON, R.O. PILTZ, J. CRAIN, R.J. CERNIK & G. BUSHNELL-WYE - *Rev. Sci. Instr.* **63**, 1039 (1992)

[39] M.A.G. ARANDA, D.C. SINCLAIR, J.P. ATTFIELD & A.P. MCKENZIE - *Phys. Rev. B* **51**, 12747 (1995)

[40] G.H. KWEI, R.B. VON DREELE, S.W. CHEONG, Z. FISK & J.D. THOMPSON - *Phys. Rev. B* **41**, 1889 (1990)

[41] P. COPPENS, P. LEE, Y. GAO & H.S. SHEU - *J. Phys. Chem. Solids* **52**, 1267 (1991)

[42] M.D. MARCOS, M.A.G. ARANDA, D.C. SINCLAIR & J.P. ATTFIELD - *Physica C* **235-240**, 967 (1994)

[43] J.M. JOUBERT, R. CERNY, M. LATROCHE, A. PERCHERON-GUEGUAN & K. YVON - *J. Appl. Cryst.* **31**, 327 (1998)

[44] M. LATROCHE, J.M. JOUBERT, A. PERCHERON-GUÉGAN & P.H.L. NOTTEN - *J. of Solid State Chem.* **146**, 313 (1999)

[45] J.K. WARNER, A.P. WILKINSON, A.K. CHEETHAM & D.E. COX - *J. Phys. Chem. Solids* **52**, 1251 (1991)

[46] R. WULF - *Acta. Cryst. A* **46**, 681 (1990)

[47] A.H. MOUDDEN, D. DURAND, M. BESSIERE & S. LEFEBVRE - *Phys. Rev. B* **37**, 7655 (1988)

[48] Y. SOEJIMA, K. YAMASAKI & K.F. FISCHER - *Acta Cryst. B* **53**, 415 (1997)

[49] V. FAVRE-NICOLIN, S. BOS, E. LORENZO, J.L. HODEAU, J.F. BERAR, P. MONCEAU, R. CURRAT, F. LEVY & H. BERGER - *Phys. Rev. Lett.* **87**, 015502 (2001)

[50] S. HASHIMOTO, H. IWASAKI, K. OHSHIMA, J. HARADA, M. SAKATA & H. TERAUCHI - *J. Phys. Soc. Japan* **54**, 3796 (1985)

[51] G.E. ICE & C.J. SPARKS - *In "Resonant anomalous X-ray scattering"*, G. MATERLIK, C.J. SPARKS & K. FISCHER eds, Elsevier Science B.V., Amsterdam, 265 (1994)

[52] M. DE SANTIS, A. DE ANDRES, D. RAOUX, M. MAURER, M. PIECUCH & M. RAVET - *Phys. Rev. B* **46**, 15465 (1992)

[53] A. DÉCHELETTE, M.C. SAINT-LAGER, J.M. TONNERRE, G. PATRAT, D. RAOUX, H. FISCHER, S. ANDRIEU & M. PIECUCH - *Phys. Rev. B* **60**, 6623 (1999)

[54] J.P. SIMON, O. LYON & D. DE FONTAINE - *J. Appl. Cryst.* **18**, 230 (1985)

[55] P.H. FUOSS, P. EISENBERGER, W.K. WARBURTON & A. BIENENSTOCK - *Phys Rev. Lett.* **46**, 1537 (1981)

[56] E. MATSUBARA & Y. WASEDA - *In "Resonant anomalous X-ray scattering"*, G. MATERLIK, C.J. SPARKS & K. FISCHER eds, Elsevier Science B.V., Amsterdam, 345 (1994)

[57] D. RAOUX - *In "Resonant anomalous X-ray scattering"*, G. MATERLIK, C.J. SPARKS & K. FISCHER eds, Elsevier Science B.V., Amsterdam, 323 (1994)

[58] C. MENEGHINI, A.F. GUALTIERI & C. SILIGARDI - *J. Appl. Cryst.* **32**, 1090 (1999)

[59] R. REVEL, D. BAZIN, E. ELKAIM, Y. KIHN & H. DEXPERT - *J. Phys. Chem. B* **104**, 9828 (2000)

[60] D. BAZIN, L. GUCZI & J. LYNCH - *J. Appl. Catalysis A* **226**, 87 (2002)

[61] M. BROWN, R.E. PEIERLS & E.A. STERN - *Phys. Rev. B* **15**, 738 (1977)

[62] U. ARP, A. MATERLINK, G. MEYER & M. RICHTER - *In "X-ray Absorption Fine Structure"*, S.S. HASNAIN ed., 44 (1991)

[63] P. COPPENS - *Synchrotron Radiation Crystallography*, Academic Press, London (1992)

[64] J. WONG, F.W. LYTLE, R.P. MESSMER & D.H. MAYLOTTE - *Phys. Rev. B* **30**, 5596 (1984)

[65] C. BROUDER - *J. Phys.: Condens. Matter* **2**, 701 (1990) and references therein.

[66] M. BENFATTO, C.R. NATOLI & E. PACE eds - *Theory and computation for synchrotron radiation spectroscopy,* A.I.P., New York (2000)

[67] S. BOS - *Etudes DANES : Méthodologie et application à des échantillons absorbants,* PhD Thesis, Grenoble University (1999)

[68] D.H. TEMPLETON & L.K. TEMPLETON - *Acta Cryst. A* **36**, 237 (1980)

[69] D.H. TEMPLETON & L.K. TEMPLETON - *Acta Cryst. A* **38**, 62 (1982)

[70] D.H. TEMPLETON - *In "Resonant anomalous X-ray scattering"*, G. MATERLIK, C.J. SPARKS & K. FISCHER eds, Elsevier Science B.V., Amsterdam, 1 (1994)

[71] V.E. DMITRIENKO - *Acta Cryst. A* **39**, 29 (1983); *Acta Cryst. A* **40**, 89 (1984)

[72] F. DE BERGEVIN & M. BRUNEL - *Phys. Lett. A* **39**, 141 (1972)

[73] M. BLUME - *In "Resonant anomalous X-ray scattering"*, G. MATERLIK, C.J. SPARKS & K. FISCHER eds, Elsevier Science B.V., Amsterdam, 495 (1994)

[74] E. BEAUREPAIRE, B. CARRIÈRE & J.P. KAPPLER eds - *Magnetism and synchrotron radiation*, Les Editions de Physique, les Ulis (1997)

[75] D.E. SAYERS, E.A. STERN & F.W. LYTLE - *Phys. Rev. Lett.* **27**, 1024 (1971)

[76] Y. CAUCHOIS - *C.R. Acad. Sci.* **242**, 100 (1956)

[77] Y. CAUCHOIS & C. BONNELLE - *C.R. Acad. Sci.* **242**, 1596 (1956)

[78] Y. HENO - *C.R. Acad. Sci.* **242**, 1599 (1956)

[79] T.S. FUKAMACHI, S. HOSOYA, T. KAWAMURA & J.B. HASTINGS - *J. Appl. Cryst.* **10**, 321 (1977)

[80] I. ARCON, A. KODRE, D. GLAVIC & M. HRIBAR - *J. Physique C* **9**, 1105 (1987)

[81] H. STRAGIER, J.O. CROSS, J.J. REHR, L.B. SORENSEN, C.E. BOULIN & J.C. WOICIK - *Phys. Rev. Lett.* **21**, 3064 (1992)

[82] I.J. PICKERING, M. SANSONE, J. MARCH & G.N. GEORGE - *J. Am. Chem. Soc.* **115**, 6302 (1993)

[83] L.B. SORENSEN, J.O. CROSS, M. NEWVILLE, B. RAVEL, J.J. REHR, H. STRAGIER, C.E. BOULIN & J.C. WOICIK - *In "Resonant anomalous X-ray scattering"*, G. MATERLIK, C.J. SPARKS & K. FISCHER eds, Elsevier Science B.V., Amsterdam, 389 (1994)

[84] J.O. CROSS , M. NEWVILLE, L.B. SORENSEN, H.J. STRAGIER, C.E. BOULDIN & J.C. WOICIK - *J. Physique IV* **7**(**C2**), 745 (1997)

[85] H. RENEVIER, J.L. HODEAU, P. WOLFERS,S. ANDRIEU, J. WEIGELT & R. FRAHM - *Phys. Rev. Lett.* **78**, 2775 (1997)

[86] M.A. QING, J.F. LEE & D.E. SAYERS - *Physica B* **208-209**, 663 (1995)

[87] J.L. HODEAU & J. VACÍNOVÁ - *Synchrotron Radiation News* **9**, 15 (1996)

[88] J.C. WOICIK, J.O. CROSS, C.E. BOULDIN, B. RAVEL, J.G. PELLEGRINO, B. STEINER, S.G. BOMPADRE, L.B. SORENSEN, K.E. MIYANO & J.P. KIRKLAND - *Phys. Rev. B* **58**, R4215 (1998)

[89] S. GRENIER, M.G. PROIETTI, H. RENEVIER, L. GONZALEZ, J.M. GARCIA & J. GARCIA - *Euro. Phys. Lett.* **57**, 499 (2002)

[90] H. RENEVIER, J. WEIGELT, S. ANDRIEU, R. FRAHM & D. RAOUX - *Physica B* **208-209**, 217 (1995)

[91] M.G. PROIETTI, H. RENEVIER, J.L. HODEAU, J. GARCIA, J.F. BERAR & P. WOLFERS - *Phys. Rev. B* **59**, 5479 (1999)

[92] J.L. HODEAU, J. VACÍNOVÁ, Y. GARREAU, A. FONTAINE, E. ELKAÏM, J.P. LAURIAT, M. HAGELSTEIN, J. MULLER & A. COLLOMB - *Nucl. Instr. Meth. B* **97**, 115 (1995)

[93] D.C. MEYER, K. RICHTER, A. SEIDEL, J. WEIGELT, R. FRAHM & P.J. PAUFLER - *J. Synchrotron Radiation* **5**, 1275 (1998)

[94] A. FILIPPONI, A. DI CICCO & C.R. NATOLI - *Phys. Rev. B* **52**, 15122 (1995)

[95] S.I. ZABINSKI, J.J. REHR, A. ANKUDINOV, R.C. ALBERS & M.J. ELLER - *Phys. Rev. B* **52**, 2995 (1995)

[96] E.A. STERN, M. NEWVILLE, B. RAVEL, Y. YACOBY & D. HASKEL - *Physica B* **208-209**, 117 (1995)

[97] B. RAVEL, C.E. BOULDIN, H. RENEVIER, J.L. HODEAU & J.F. BERAR - *Phys. Rev. B* **60**, 778 (1999)

[98] H. NAKAO, K. OHWADA, N. TAKESUE, Y. FUJII, M. ISOBE, Y. UEDA, M.VON ZIMMERMANN, J.P. HILL, D. GIBBS, J.C. WOICIK, I. KOYAMA & Y. MURAKAMI - *Phys. Rev. Lett.* **85**, 4349 (2000)

[99] S. GRENIER, A. TOADER, J.E. LORENZO, Y. JOLY, B. GRENIER, S. RAVY, L.P. REGNAULT, H. RENEVIER, J.Y. HENRY, J. JEGOUDEZ & A. REVCOLEVSKI - *Phys. Rev. B* **65**, 180101 (2002)

[100] A.P. WILKINSON, A.K. CHEETHAM & D.E. COX - *Acta Cryst. B* **47**, 155 (1991)

[101] J. VACINOVA, J.L. HODEAU, P. BORDET, M. ANNE, D. COX, A. FITCH, P. PATTISON, W. SCHWEGGLE, H. GRAAFSMA & A. KVICK - *Mater. Sci. Forum* **228-231**, 241 (1996)

[102] D.H. TEMPLETON & L.K. TEMPLETON - *Acta Cryst. A* **41**, 365 (1985)

[103] D.H. TEMPLETON & L.K. TEMPLETON - *Acta Cryst. A* **42**, 478 (1986)

[104] D.H. TEMPLETON & L.K. TEMPLETON - *Acta Cryst. A* **43**, 573 (1987)

[105] A. KIRFEL - *In "Resonant anomalous X-ray scattering"*, G. MATERLINK, C.J. SPARKS & K. FISCHER eds, Elsevier Science B.V., Amsterdam, 231 (1994)

[106] V.E. DMITRIENKO & E.N. OVCHINNIKOVA - *Acta Cryst. A* **56**, 340 (2000)

[107] E.N. OVCHINNIKOVA & V.E. DMITRIENKO - *Acta Cryst. A* **56**, 2 (2000)

[108] K.D. FINKELSTEIN, M. HAMRICK & Q. SHEN - *In "Resonant anomalous X-ray scattering"*, G. MATERLINK, C.J. SPARKS & K. FISCHER eds, Elsevier Science B.V., Amsterdam, 91 (1994)

[109] Y. MURAKAMI, J.P. HILL, D. GIBBS, M. BLUME, I. KOYAMA, M. TANAKA, H. KAWATA, T. ARIMA, Y. TOKURA, K. HIROTA & Y. ENDOH - *Phys. Rev. Lett.* **81**, 582 (1998)

[110] M. BENFATTO, Y. JOLY & C.R. NATOLI - *Phys. Rev. Lett.* **83**, 636 (1999)

[111] K. HAGIWARA, M. KANAZAWA, K. HORIE, J. KOKUBUN & K. ISHIDA - *J. Phys. Soc. Japan* **68**, 88 (1999)

[112] J. GARCIA, G. SUBIAS, M.G. PROIETTI, H. RENEVIER, Y. JOLY, J.L. HODEAU, J. BLASCO, M.C. SANCHEZ & J.F. BERAR - *Phys. Rev. Lett.* **85**, 578 (2000)

[113] M. VON ZIMMERMANN, J.P. HILL, D. GIBBS, M. BLUME, D. CASA, B. KEIMER, Y. MURAKAMI, Y. TOMIOKA & Y. TOKURA - *Phys. Rev. Lett.* **83**, 4872 (1999)

[114] H. RENEVIER, S. GRENIER, S. ARNAUD, J.F. BERAR, B. CAILLOT, J.L. HODEAU, A. LETOUBLON, M.G. PROIETTI & B. RAVEL - *J. Synchrotron Radiation* **10**, 435 (2003)

[115] J.L. HODEAU, V. FAVRE-NICOLIN, S. BOS, H. RENEVIER, E. LORENZO & J.F. BERAR - *Chem. Rev.* **101**, 1843 (2001)

[116] D.C. MEYER, K. RICHTER, A. SEIDEL, J. WEIGELT, R. FRAHM & J.P. PAUFLER - *J. Synchrotron Radiation* **10**, 144 (2003)

[117] D.H. TEMPLETON & L.K. TEMPLETON - *Acta Cryst. A* **45**, 39 (1989)

8

SOFT X-RAY PHOTOELECTRON EMISSION MICROSCOPY (X-PEEM)

C.M. SCHNEIDER
Institute of Electronic Properties, Research Center Jülich, Germany

1. A "NANOSCALE" INTRODUCTION

Surface and nanoscale aspects are becoming more and more important in modern technology. The ongoing trend for smaller and yet more powerful devices in micro-electronics and data storage technology pushes the relevant lateral dimensions far into the sub-micrometer regime. In microelectronics, for example, the smallest lateral dimension of elements in a Random Access Memory (RAM) cell are currently reaching down to about 100 nm and the 65 nm technology node is projected to come within reach in the year 2007 [1]. The bit size in commercially available magnetic data storage has decreased to about 100×500 nm [2], and yet higher storage densities resulting in smaller bit sizes are demonstrated in various research labs throughout the world. At the same time, the relevant vertical dimensions have dropped to the nanometer regime. In order to observe single electron tunnelling phenomena in nonmagnetic or spin-dependent transport effects in magnetic systems, extremely thin films of 1 - 2 nm thickness must be prepared. Paired with these technological developments is a strong scientific activity in the fields of surface physics, surface chemistry, and materials science, which also concerns the creation of a wide variety of nanoscale physical systems.

2. VISUALIZING MICRO- AND NANOSTRUCTURES

It is obvious that this situation asks for high-resolution imaging techniques, in order to visualize the system itself and to investigate its underlying physical and

F. Hippert et al. (eds.), Neutron and X-ray Spectroscopy, 271–295.

chemical properties on a small lateral scale. In addition, these techniques must have a certain surface sensitivity, if studies of surface-related effects or thin film systems are concerned. With respect to the experimental realization, two principal imaging approaches may be distinguished. In *scanning probe techniques* a finely focussed electron beam (Scanning Electron Microscopy) [4] or a tip with nanometer tip radius (Scanning Tunnelling Microscopy) [5] is scanned across the sample surface and the information is collected sequentially on a point-by-point basis. These scanning techniques are well established in surface physics and have been constantly improved in lateral resolution. They have been specialized with respect to the contrast mechanisms to meet various needs. In some cases, even studies of dynamic processes have been carried out. Also dedicated variations of these techniques for the investigation of magnetic surfaces have been developed. These are, for example, Scanning Electron Microscopy with Spin Polarization Analysis (SEMPA) [3], Magnetic Force Microscopy (MFM) [6], or Spin-Polarized Scanning Tunneling Microscopy (ST-STM) [7].

The second imaging approach involves the parallel acquisition scheme well known from conventional light-optical microscopy. The surface is illuminated by a wide beam of electrons or photons and the surface area within the field of view of the microscope is imaged simultaneously, for example, by means of a photo or video camera. In order to combine high spatial resolution with surface sensitivity, the image is formed by the electrons reflected at or emitted from the surface. Technical realizations of this approach, for example, are Low Energy Electron Microscopy (LEEM) [8] and Photoelectron Emission Microscopy (PEEM) [1] for the case of electron and photon beam illumination, respectively. By exploiting specific magnetic contrast mechanisms, both techniques can also be used to study magnetic phenomena at surfaces. This is particularly true for the PEEM technique, which recently became rather popular due to the increasing availability of highly brilliant synchrotron radiation from third generation storage ring facilities. The excitation with polarized soft X-rays (X-PEEM) offers a unique combination of surface sensitivity, element selectivity, and magnetic contrast, as will be discussed in more detail in the following.

1. More precisely this approach is a *photoexcitation* electron emission microscopy. The majority of PEEM experiments generates images with secondary electrons rather than (direct) photoelectrons.

3. *TECHNICAL ASPECTS OF AN ELECTRON EMISSION MICROSCOPE (EEM)*

An EEM maps the spatial distribution of electrons emitted from a surface onto a two-dimensional image detector by means of a dedicated electron optical column. In the following we will review some particular aspects and properties of this set-up.

3.1. *ELECTRON-OPTICAL CONSIDERATIONS*

The principal layout of an EEM (figure 1) shows strong analogies to a light-optical microscope. In order to obtain high spatial resolution the objective lens of a conventional microscope accepts a large solid angle. This is achieved by *immersion lens* objectives, which are brought very close to the object. The EEM uses a similar approach, whereby the EEM immersion lens is placed 2 - 3 mm away from the object surface. The first instrument of this kind has been proposed and constructed by Brüche already in 1934 [9]. Nowadays, the objective usually comprises a set of 3 to 4 electrostatic or magnetic ring lenses [8,10]. In contrast to the light-optical case, however, the sample in an EEM forms an inherent part of the (electron) optics. In order to transfer electrons into the microscope that leave the surface at a large starting angle relative to the surface normal, a strong electrostatic field (~ 10 kV/mm) is applied between sample and the first electrode (extractor).

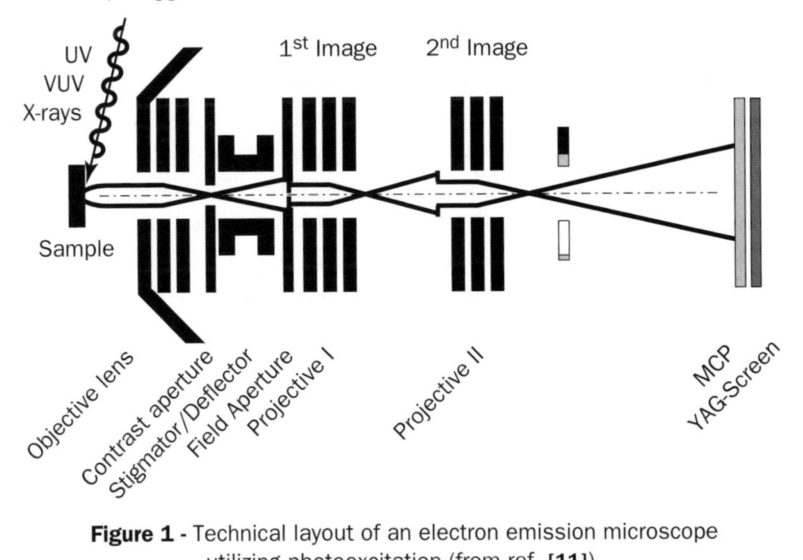

Figure 1 - Technical layout of an electron emission microscope utilizing photoexcitation (from ref. **[11]**).

This set-up has both a drawback and an advantage to it. On the one hand, it geometrically constrains the EEM experiment, as the sample surface normal must be aligned with the electron optical axis in order to ensure cylindrical symmetry of the accelerating field. Non-cylindrical symmetries will cause image distortions. On the other hand, accelerating the electrons significantly reduces their relative energy $\Delta E/E_0$ and angular spreads $\Delta\theta/\theta_0$, before they enter the lens system. The higher the kinetic energy E_0 the lesser the electron trajectories are affected by the electron optical imperfections (e.g., spherical and chromatic aberration) of the cathode lens.

For electrons starting from the same point at the surface with different kinetic energies, the lens imperfections cause the electron trajectories to fan out in the image detector plane, and will thus smear out the image point, thereby impairing the image quality and limiting the spatial resolution. If we want to reduce the effect of the lens errors, we have to select electrons with the proper trajectories, i.e., a defined kinetic energy and direction. This is done by means of a contrast aperture, which may be located at a suitable trajectory crossover, e.g., in the back-focal plane of the objective lens. The aperture must be carefully adjusted with respect to the electron optical axis. The choice of the aperture size is mostly dictated by practical considerations. A small aperture improves the lateral resolution, but impairs the signal-to-noise ratio and increases the image acquisition time. This can be partly compensated by a higher photon flux supplied by dedicated beam lines. The remaining part of the electron optical system mainly consists of projective lenses, which magnify the image onto a multichannel plate/scintillator crystal combination or a phosphor screen. This image converter transforms the "electron" into a "photon" image, which is picked up outside the vacuum chamber by a slow- or dual-scan CCD camera. An improved control of the electron beam and a (partial) compensation of electron-optical imperfections can be achieved by additional elements, such as deflector/stigmator units. The microscope set-up is completed by a sample manipulator, which allows a lateral positioning of the sample in the field of view of the microscope (~ 10 - 500 μm, depending on microscope and lens settings). For the imaging of magnetic domains a sample rotation facilitates optimization of the magnetic contrast. The small distance between immersion lens and sample limits the angle of light incidence with respect to the surface plane to 15 - 25°.

The set-up described above contains the essential elements of a fully electrostatic PEEM with emphasis on the imaging of surfaces. More sophisticated and elaborate electron-optical designs may also include magnetic lenses, active energy filters, and corrective elements, in order to improve the spatial resolution and spectromicroscopic capabilities of the instrument [12].

3.2. TRANSMISSION AND LATERAL RESOLUTION

Photo-excitation of an electronic system with VUV or soft X-ray radiation of energy $h\nu$ generates a rather broad electron spectrum or energy distribution $N(E_{kin})$, ranging from direct photoelectrons with kinetic energy $E_{kin} = h\nu - \Phi$ (Φ denotes the work function) down to low energy secondary electrons with $E_{kin} \leq 1$ eV. Since every electron optical system has a transmission function $T(E_{kin})$, only a part of this electron spectrum will reach the image converter. In fact, the combination of immersion lens and contrast aperture has a pronounced low-pass behaviour (figure 2), i.e., the high energy side of the spectrum is suppressed. The width of the low energy interval transmitted depends – among others – on the diameter of the contrast aperture: the smaller the aperture diameter the smaller the energy width and the lower the image intensity.

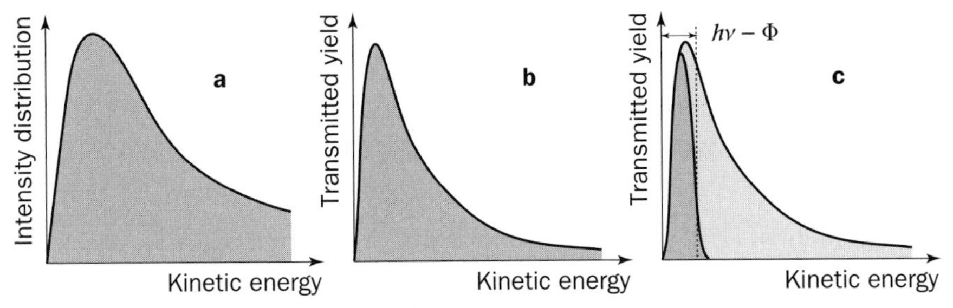

Figure 2 - Schematic secondary electron spectrum before (**a**) and after (**b**) passing through the electron emission microscope optics (**c**) Situation in the case of threshold photoemission (from ref. **[11]**).

This is an important aspect when discussing the issue of spatial resolution. The dominant mechanism determining the spatial resolution in X-PEEM is the chromatic aberration of the acceleration field and the immersion lens elements. An ideally sharp image is obtained only with monoenergetic electrons. The closest approximation to this condition is realized in threshold photoemission, where the width of the electron spectrum may be 1 eV or less. The larger energy spread associated with the high energy excitation in X-PEEM leads to energy-dependent trajectories via the chromatic aberration and results in a blurring of the image. Therefore, even for the same microscope settings, the resolution deteriorates when going from threshold photoemission to excitation with soft X-rays (figure 2). Increasing the acceleration potential and/or reducing the aperture diameter narrows down the trajectory spread at the image detector and improves the lateral resolution. There is an optimum choice of the aperture diameter for a given electron-optical set-up, however, since the role of diffraction effects at the aperture increases with decreasing aperture size. In order to stay away from the diffraction limit, the aperture is typically chosen to have a diameter of 15 - 20 μm. A resolution limit of $\delta x \sim 10 - 20$ nm may be achieved with the present PEEM optics in threshold photoemission [13], and better than $\delta x \leq 50$ nm nm have been demonstrated with synchrotron radiation [14]. A further improvement is predicted with aberration corrected lens systems [12].

It should be kept in mind, however, that the theoretically predicted resolution can be achieved only with an ideal surface. A realistic surface always has a topography determined by defects. These range from microscopic scratches and crystalline facets down to nanoscopic features such as terrace edges and monoatomic steps. Most of these defects are associated with the formation of electrostatic – and in the case of magnetic materials magnetic – microfields, which will affect the electron trajectories. Therefore, the ultimate lateral resolution achievable with a realistic sample will be largely limited by its surface topography [15].

4. NON-MAGNETIC IMAGE CONTRAST IN X-PEEM

The versatility of EEM stems from the variety of physical phenomena available for image contrast formation. Formally, the image contrast may be defined as the variation of intensity or brightness $I(x,y)$ across an image. The contrast disappears, when $I(x,y) = const.$, i.e., the image exhibits a featureless, uniform brightness. The interpretation of an image requires that the image contrast be traced back to the physical mechanism that gave rise to it. For convenience, we may distinguish between *primary* and *secondary* contrast mechanisms in the following.

4.1. PRIMARY CONTRAST MECHANISMS

Primary mechanisms are intimately connected with the local electronic structure of the sample and affect the photocurrent $I(E,\mathbf{k},x,y)$ above the sample surface already during the photo-excitation in the solid. Their origin may be of chemical or structural nature. An example is the work function contrast widely used in threshold PEEM.

4.1.1. WORK FUNCTION CONTRAST

The work function Φ depends sensitively on the chemical state and crystallographic orientation of the surface. Thus, a chemically or crystallographically heterogeneous surface results in a contrast pattern. This way, for example, grains at a polycrystalline surface may show up in a different brightness and the contrast in the image will reflect the grain orientation [8]. This contrast mechanism is dominant at low excitation energies, but may be suppressed or masked in X-PEEM due to the relatively large energy spread of the secondary electrons contributing to the image at high-energy excitation.

4.1.2. CHEMICAL CONTRAST

When working with synchrotron radiation in the soft X-ray regime, a chemical contrast may conveniently be generated by exploiting characteristic absorption edges. For this purpose, the photon energy is tuned such as to excite electrons from a core level into the empty electronic states below the vacuum level, leaving behind core-holes (figure 3a). This process is particularly efficient in elements with only partially filled *d*- or *f*-shells, because the density of empty states is high. If the respective chemical element is inhomogeneously distributed across a surface, the absorption will be high (low) in regions where the element is present (absent). In order to utilize this absorption contrast in X-PEEM the absorption signal has to be translated into electrons emitted from the sample.

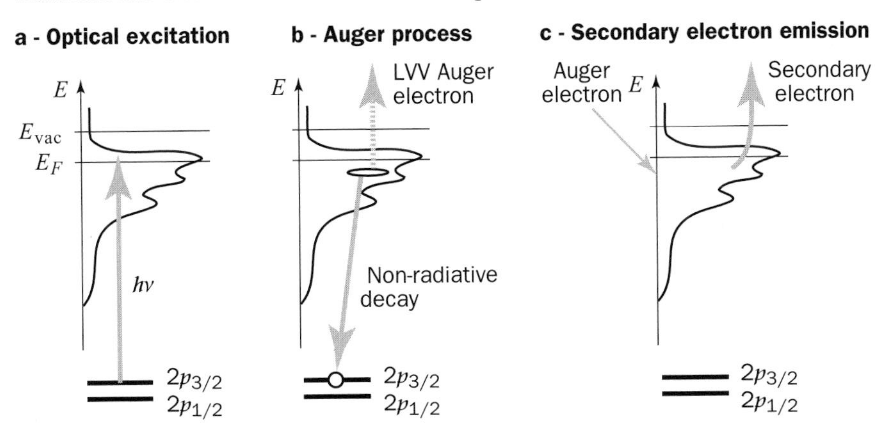

Figure 3 - Chemical contrast mechanism in X-PEEM. **a** - Excitation of a core level. **b** - Decay of the core-hole by an Auger process. **c** - Auger electron induced secondary electron cascade (from ref. **[11]**).

This translation mechanism is provided by the electronic de-excitation of the system. The core-hole created will be filled either by a radiative (X-ray fluorescence) or a non-radiative (Auger) process. In the energy regime in question, the probability for a non-radiative transition is much higher. The core-hole decay leads to the emission of highly energetic Auger electrons, the total number of which

(Auger electron yield) is proportional to the absorption signal (figure 3b). With proper energy discrimination, this signal can already be used to map different elements in EEM.

While passing through the solid the Auger electrons suffer multiple inelastic scattering events, finally leading to a secondary electron cascade. This cascade contains a large number of low energy secondary electrons, the energy distribution of which is cut off by the vacuum level (figure 3c). Since the starting point for this secondary electron distribution is the core-hole decay and the Auger electron generation, the secondary electron yield also contains the chemical information. The low-pass characteristics of an immersion lens EEM is particularly well suited to pick up this signal.

In the simplest case, an X-PEEM image of a chemically heterogeneous surface should directly map the distribution of a particular element selected by the corresponding photon energy. In other words, sample areas containing this element should appear bright in the image. This simple interpretation does not hold for all cases, however, because of the various physical processes involved in the generation of the secondary electron cascade. The secondary electron yield may be strongly affected by the morphology and structure of the specimen, as well as the surface morphology and chemical state. In figure 4 we show an example for a permalloy ($Fe_{20}Ni_{80}$) film on a SiO_2 surface. The film has been patterned into squares of $20\,\mu m \times 20\,\mu m$ size. The image (figure 4b) has been acquired with the photon energy tuned to the Fe L_{III} edge. Naively, one would expect the permalloy areas to appear bright under these circumstances. The opposite is obviously the case.

This observation can be explained on the basis of selected area absorption spectra. For this purpose two alternative approaches may be used. First, a series of images is recorded as a function of photon energy (spectromicroscopy). After the series is completed, the contrast level at the selected area in each image is determined and compiled to result in a spectrum. Second, selected areas are defined by electronic means during the image acquisition, and the contrast level in these areas is measured directly while scanning the photon energy (microspectroscopy). The latter technique has been employed to obtain the spectra shown in figure 4a. In the spectrum taken on the permalloy squares the characteristic absorption lines of Fe and Ni are easily discernible. This permalloy signature is absent in the spectrum taken on the strip between squares, consisting of exposed SiO_2. A very weak signal at the position of the Ni lines suggests that a minute amount of Ni has been incorporated into the SiO_2 either by the ion milling process used for microstructuring or by diffusion of Ni during a short annealing step after the sample was introduced into the X-PEEM chamber. At the energy position of the Fe L_{III} absorption line, however, the average intensity level of the SiO_2 spectrum is significantly higher than the peak intensity at the Fe L_{III} line of the permalloy spectrum. This means that at this photon energy the rate of secondary electron production is higher for SiO_2 than for permalloy. Therefore, the contrast appears "reversed" with respect to intuition. This example emphasizes two important issues. On the one hand, it demonstrates the capability of X-PEEM to obtain local spectroscopic information from areas in the sub-micrometer regime. On the other hand, it illustrates that X-PEEM imaging should always be accompanied by appropriate spectroscopic investigations in order to be able to unambiguously conclude on the physical or chemical origin of the image contrast.

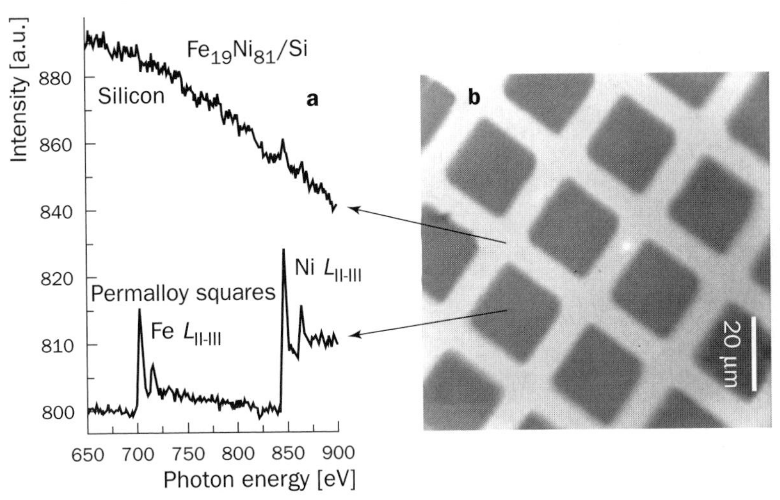

Figure 4 - Microspectroscopy from a patterned permalloy sample. **a** - Local absorption spectra taken from the indicated areas being less than 1 μm × 1 μm in size. **b** - Image acquired at the Fe L_{III} edge (from ref. **[16]**).

The microspectroscopy capabilities of X-PEEM can be widely employed to address materials science related issues. There often the problem arises to discriminate between different modifications of the same chemical element rather than different elements. An example for the analysis of carbon films is given in figure 5 **[17]**. Because of their hardness diamond-like carbon films are used for protecting surfaces against mechanical wear. During preparation of the films, however, also unwanted (because mechanically softer) graphitic phases may be formed. These can be spectroscopically distinguished by their higher amount of sp^2 coordinated bonds which leads to a slightly different energy position of the C–K absorption edge as compared to that observed in sp^3 coordinated diamond. This can be clearly seen by comparing microspot spectra obtained from (001) oriented diamond-like films and highly oriented pyrolithic graphite (HOPG). The predominant spectral features arise due to excitations into specific unoccupied molecular orbitals and are associated with a $1s \longrightarrow \pi^*$ (HOPG) transition and the *C–H** resonance (diamond). The latter is typical for sp^3 coordinated carbon with a high hydrogen content. X-PEEM images recorded at these particular excitation energies reveal graphitic inclusions or contaminations in an otherwise diamond-like film (bright spots in figure 5b) [17].

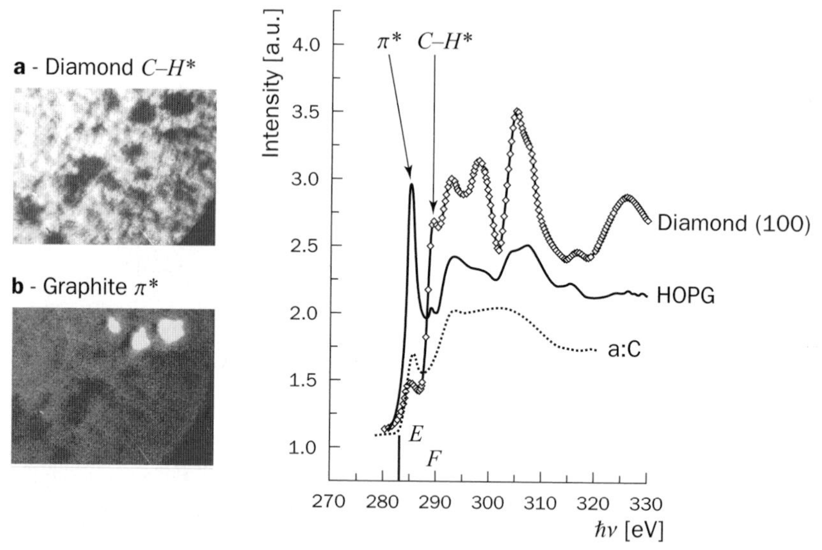

Figure 5 - X-PEEM studies of carbon films. **Right** - X-ray absorption spectra of a diamond (100) film and a graphite film (HOPG), both recorded in $(5 \times 5)\ \mu m^2$ microspots with X-PEEM. An a:C (amorphous carbon) spectrum is shown for comparison. **Left** - X-PEEM images from a DLC (diamond-like carbon) hard coating film recorded at photon energies corresponding to the *C–H** and the π^* electronic excitations. Length of the images 50 μm (taken from ref. **[17]**).

4.2. *SECONDARY CONTRAST MECHANISMS*

After the electrons have left the sample they experience the electrostatic field between sample surface and immersion lens. The field distribution determines the electron trajectories. Secondary contrast mechanisms refer to processes that change the electron trajectories as compared to the ideal situation. This can be achieved by local electrostatic fields (microfields) at the surface. The major source for these microfields is topographical defects, such as scratches, hillocks, or edges of geometrical structures. These create local deviations from the ideal cylindrical electrostatic field distribution in the vicinity of these defects, which in turn leads to topographical image contrast. The way in which this contrast is actually seen in the image depends strongly on the experimental parameters (shape and size of the defect, acceleration voltage, position of contrast aperture, etc.). Some examples are given in figure 6.

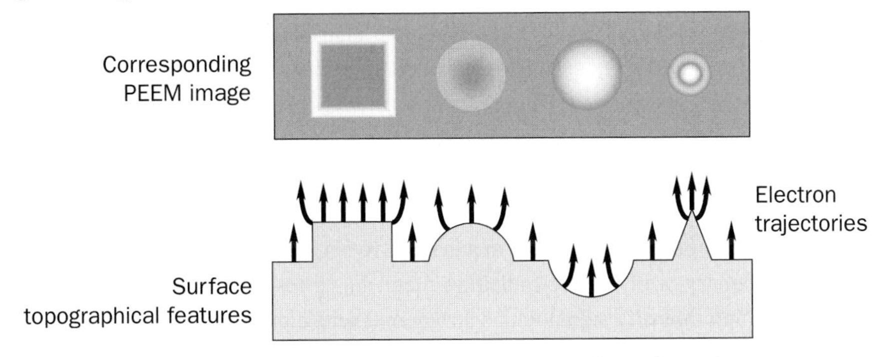

Figure 6 - Topographic contrast in PEEM (after ref. **[18]**).

Regions of different conductivity, which may lead to a partial charging up of surface areas, may also result in an image contrast. As far as magnetic samples are concerned, these are usually associated with a long-ranged magnetic stray field above the sample surface. Electrons moving in this stray field experience a Lorentz force, which will change the electron trajectories and will give rise to a magnetic image contrast. This contrast mechanism was actually employed to obtain the first PEEM images of magnetic domains [19].

It is important to keep in mind that the contrast observed in an arbitrary EEM image usually reflects a complex combination of the above contrast mechanisms. By means of appropriate experimental procedures, for example, taking the difference between two images recorded at and slightly in front of the absorption edge, or comparing images taken with different contrast aperture settings, the individual contrast contributions can be identified and extracted.

5. *MAGNETIC CONTRAST IN X-PEEM*

The full power of the X-PEEM technique becomes available when the physical phenomena giving rise to the image contrast depend on the polarization of the exciting radiation. These may be, for example, spatially oriented electronic orbitals, which are probed by linearly polarized light. The most prominent application of

X-PEEM in polarization dependent studies, however, concerns magnetic materials. In the following, we will therefore concentrate on the *magnetic* contrast mechanisms involving high-energy synchrotron radiation.

Spectroscopic studies of magnetic materials have been greatly facilitated by the discovery of X-ray magnetic dichroism with linearly (XMLD) [20] and circularly polarized light (XMCD) [21]. These effects can also be observed in the total and partial electron yield [22], and are thus well suited as contrast mechanisms in X-PEEM.

5.1. MAGNETIC X-RAY CIRCULAR DICHROISM (XMCD)

XMCD is the appropriate contrast mechanism to image *ferromagnetic* systems. It works particularly well for the transition metal $L_{\text{II-III}}$ absorption edges, because it combines a high intensity signal with a strong magneto-dichroic effect.

5.1.1. PHYSICS OF THE MAGNETIC CONTRAST MECHANISM

The physical principle of the magnetic contrast mechanism is similar to that discussed in section 4.1.2. At the absorption edge, the photo-excitation takes place into the unoccupied density of states (DOS) below the vacuum level. In the $3d$ transition metals, such as Fe, Co, Ni, the incompletely filled d-shell results in a highly unoccupied DOS. As a result, a strong spectral feature ("white lines") is observed. In addition, the ferromagnetic ground state is associated with a *spin-split* DOS (caused by the exchange interaction), which is the first important ingredient in XMCD. The $L_{\text{II-III}}$ absorption process involves the $2p$ core levels, which are spin-orbit split into $2p_{3/2}$ and $2p_{1/2}$ states by about 10 eV. This is the second important ingredient in XMCD, because photo-excitation of these states with circularly polarized (c.p.) light renders the excited electrons *spin-polarized.* The reason for this spin polarization is relativistic dipole selection rules [23]. The spin polarization vector $\boldsymbol{P}$ is aligned with the direction of light incidence $\boldsymbol{q}$. Depending on the light helicity ζ (ζ points parallel/antiparallel to $\boldsymbol{q}$ for left/right handed c.p. light) and the core level involved in the transition, $\boldsymbol{P}$ points either parallel or antiparallel to $\boldsymbol{q}$. The sign of $\boldsymbol{P}$ is opposite for excitation of the $2p_{3/2}$ and $2p_{1/2}$ states.

In a non-magnetic material, reversing ζ changes the sign of $\boldsymbol{P}$, but renders the transition probability (intensity) the same. In a magnetic material, this is no longer true, because of the spin splitting in the empty DOS. As a consequence, there are more empty minority than majority spin states above E_F. Therefore, if the excited core electron has a minority spin character, the transition probability will be higher than for a majority spin character (figure 7a). The result is a magnetic dichroism, i.e., the intensity of the absorption line varies as a function of ζ and the magnetization $\boldsymbol{M}$. The XMCD signal changes sign and magnitude when going from the L_{III} to the L_{II} edge.

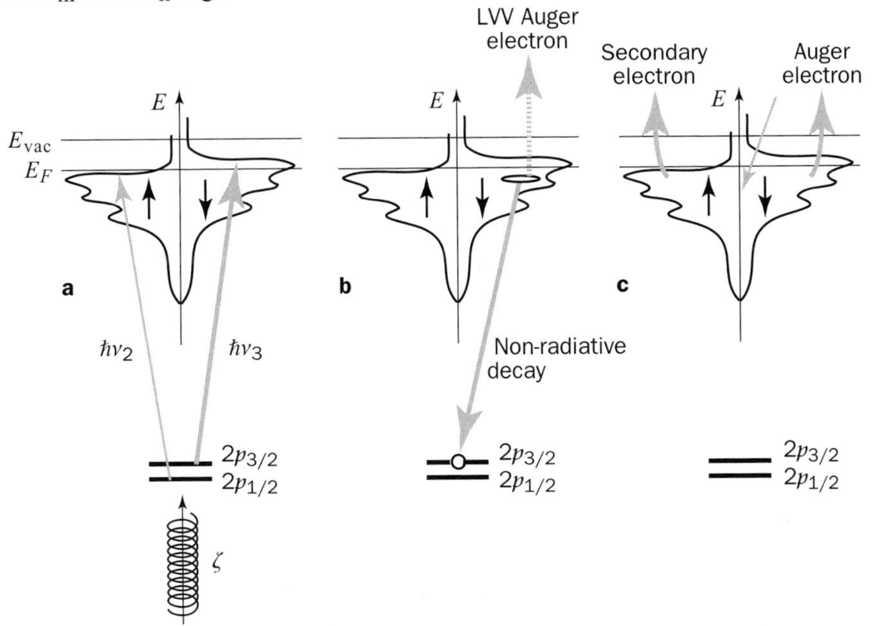

Figure 7 - three-step process leading to magnetic contrast mechanisms based on magnetic circular dichroism. **a** - Photo-excitation and generation of core-holes in an $L_{II\text{-}III}$ absorption process. **b** - Core-hole decay via Auger electron emission. **c** - Auger electron induced secondary electron cascade (from ref. **[11]**).

The translation of the XMCD signal in the photo-absorption process into an electron yield signal involves the same steps already described in section 4.1.2. The first step converts the XMCD into a dichroism in the Auger electron yield (fig. 7b). Given proper energy discrimination, this signal can already be used to obtain a magnetic image in EEM [24]. This approach provides an extreme chemical selectivity, since both the photo-excitation and the electron imaging employ characteristic spectral features. The second step finally translates the Auger XMCD signal into a helicity-dependent difference in the secondary electron yield (fig. 7c), which may be conveniently picked up by X-PEEM [25]. Detailed analyses show that the XMCD in secondary electron yield is proportional to the XMCD absorption signal [26].

5.1.2. CONTRAST ENHANCEMENT

In order to separate magnetic and nonmagnetic contributions to the contrast in the EEM image, one conveniently uses the fact that a reversal of ζ changes the sign of the magneto-dichroic signal C_M, while leaving the nonmagnetic signal C_{NM} essentially unaffected, i.e.,

$$C_M(-\zeta) = -C_M(\zeta) \quad ; \quad C_{NM}(-\zeta) = C_{NM}(\zeta) \qquad \{1\}$$

Therefore, by subtracting two images taken at the same photon energy, but opposite light helicity $I_{\zeta+} - I_{\zeta-}$, the magnetic contrast C_M is enhanced. Summing up the two images extracts the non-magnetic contrast C_{NM}

$$C_M \sim I_{\zeta+} - I_{\zeta-} \quad ; \quad C_{NM} \sim I_{\zeta+} + I_{\zeta-} \qquad \{2\}$$

This way, the images provide both magnetic and nonmagnetic (chemical, topographical) information. In order to describe the magnetic contrast in a more quantitative manner, often the *asymmetry* image A

$$A = \frac{I_{\zeta+} - I_{\zeta-}}{I_{\zeta+} + I_{\zeta-}} \qquad \{3\}$$

rather than the difference image C_M is shown. The quantity asymmetry ranges between +100% and –100%.

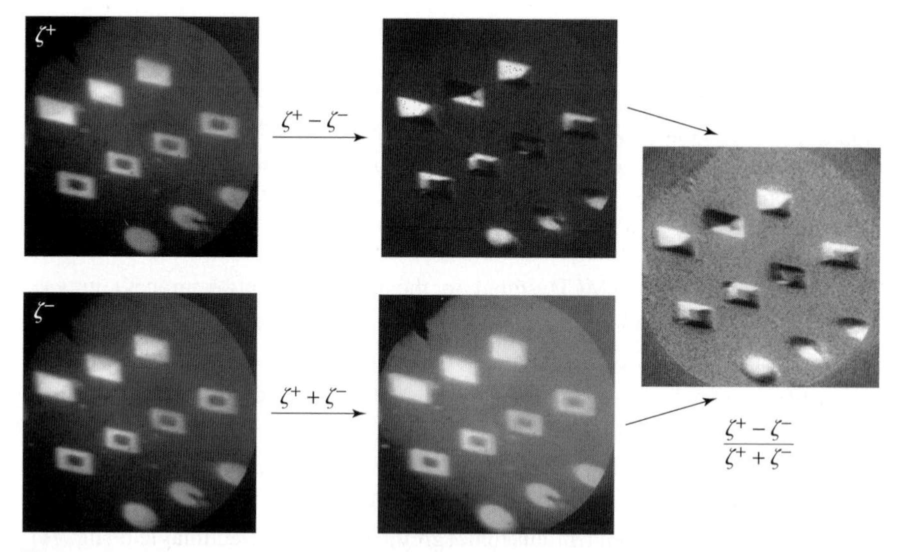

Figure 8 - Magnetic contrast enhancement for the example of permalloy microstructures. Feature size is 12 mm. **Left** - Individual images taken with opposite light helicity $I(\zeta^+)$ and $I(\zeta^-)$ at the Ni L_{III} edge are first subtracted and summed (**center**), respectively. **Right** - The sum and difference images are finally used to calculate an asymmetry image. See also ref. **[11]**.

An example for this contrast enhancement procedure is shown in (figure 8). The sample consisted of permalloy microstructures and the images have been recorded at the Ni L_{III} edge. The magnetic contrast is rather weak and does therefore not show up directly in the images taken at opposite helicity $I(\zeta^+)$ and $I(\zeta^-)$ – the microstructures appear in a uniform grey level. The same is true for the sum image in which the magnetic information should be cancelled. The difference image, however, reveals a clear internal structure in each micropattern. This internal structure reflects the lateral distribution of magnetic domains. The dark and bright areas correspond to magnetic domains with different spatial orientations of the magnetization vector (see below). The asymmetry image shows the same magnetic contrast on a normalized scale. Whether difference or asymmetry images are used for the analysis of problems in surface and thin film magnetism will depend on the actual experimental situation.

5.1.3. ANGULAR DEPENDENCE OF THE IMAGE CONTRAST

The magnetic domain images obtained by X-PEEM contain also quantitative information on the local orientation and distribution of the magnetization vector $\boldsymbol{M}(x, y)$. The geometrical relationship between magnetization vector $\boldsymbol{M}$ and light helicity ζ results in the magnitude of the magnetic contrast scaling as

$$\mathrm{A} \sim \boldsymbol{M} \cdot \zeta \qquad \{4\}$$

This behaviour is nicely illustrated by X-PEEM imaging of magnetic domains at single crystal surfaces. The image in figure 9 has been acquired exploiting XMCD at the Fe $L_{II\text{-}III}$ edges. The magnetic contrast was enhanced using the procedure given in equation {2}.

Figure 9 shows the example of a four-fold symmetric surface a Fe (001) whisker. Iron whiskers are known to exhibit very large regular domains. In the image, we can discern four different contrast levels (black, white, dark grey, light grey). This has been achieved by rotating the direction of light incidence by about $\phi = 15°$ away from the in-plane $\langle 100 \rangle$ direction. This leaves each easy axis of magnetization with a non-zero projection along ζ (projection angles 15°, 75°, 105°, 165°). According to equation {4} this will result in a distinct contrast level for each magnetization direction. This is indeed found in the experiment, when looking at the statistic distribution of contrast (grey) levels in the image. The 8-bit representation of the image results in 256 grey levels, with the histogram revealing four clear broad maxima corresponding to the domain orientations in the image.

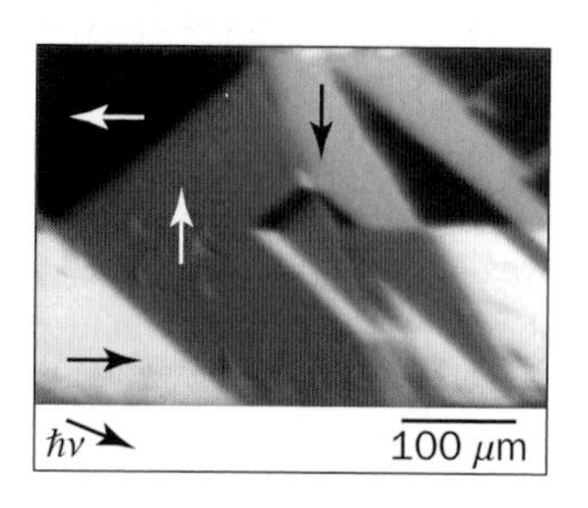

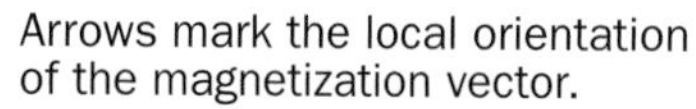
Arrows mark the local orientation of the magnetization vector.

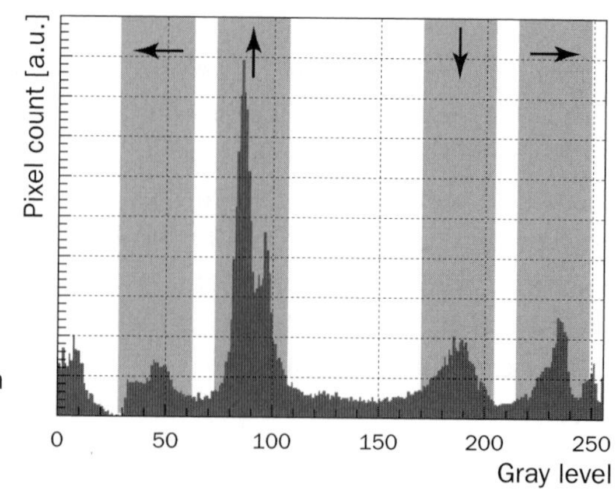

Figure 9 - Left - Magnetic domain pattern on a ferromagnetic Fe(001) whisker surface obtained at the Fe $L_{\text{II-III}}$ edge. **Right** - Histogram of the grey levels in the image revealing four broad maxima (shaded regions) corresponding to distinct magnetization directions (from ref. **[11]**).

With respect to details of the magnetic microstructure, in the left-hand side of the picture a classical flux closure pattern has formed, whereby neighbouring domains are bounded by 90° domain walls. A more complicated situation is found on the right-hand side of the image, where smaller domains appear. This complex domain pattern indicates that the whisker is actually not in an ideal state, and the magnetic microstructure is largely determined by the mechanical strain in the system. An annealing procedure brings the whisker into a strain-free state, which is characterized by a very simple domain pattern (see section 5.1.5).

5.1.4. MAGNETIC DOMAIN WALLS

Mesoscopic spin structures which pose a considerable challenge to magnetic domain imaging experiments are the boundaries of magnetic domains, the domain walls. Given two domains with local orientation of the magnetization $\boldsymbol{m}_1$ and $\boldsymbol{m}_2$ separated by a domain wall, the magnetization $\boldsymbol{M}$ must continuously rotate from $\boldsymbol{m}_1$ and $\boldsymbol{m}_2$ within the width of the domain wall. This can be done in two principal ways [27]. In a Bloch type domain wall $\boldsymbol{M}$ rotates within the plane of the wall, i.e., the component of $\boldsymbol{M}$ normal to the wall plane is always zero. By contrast, in a Néel type domain wall $\boldsymbol{M}$ rotates in a plane perpendicular to the wall plane, and thus always has a perpendicular component perpendicular to the wall. The angular dependence of XMCD equation {4} can be utilized to selectively image magnetic domain walls under certain circumstances [28]. The example shown in figure 10 has been obtained from a thick Fe(001) whisker, which had developed a peculiar domain structure. Since the light was incident along an easy axis of magnetization, the pattern consisting of large domains (figure 10a) is again characterized by three distinct contrast levels. These are associated with $\boldsymbol{M}$ parallel, antiparallel, and orthogonal to the incoming light (ζ). A closer inspection of the image, however, reveals two narrow straight lines in the centre of the pattern bounding a diamond shaped area. These lines start at the boundaries of the left-hand (black) and right-hand (white) domain. Each line has a distinct contrast level, which changes only at the point where they meet. These findings suggest that the lines in the image are caused by domain walls.

The origin of such a domain wall contrast can be understood, if the actual surface termination of a domain wall is taken into account. It is known for Fe (001) that Bloch walls in the bulk form a Néel-like surface termination **[29]**. This is due to an energy (stray field) minimization argument. A Bloch like rotation of $\boldsymbol{M}$ would lead to a magnetization component perpendicular to the surface. In order to avoid this energetically unfavourable situation, the Bloch wall continuously transforms into a Néel-like wall when it reaches the near-surface region **[30]**. As a result, the Néel-type rotation of $\boldsymbol{M}$ takes place within the surface plane. If we take the example of two domains with opposite direction of $\boldsymbol{M}$ (180° domain wall), the magnetization vector must rotate within the surface by 180°. Recalling the angular variation of the XMCD signal (eq. {4}) somewhere during this rotation the $\boldsymbol{M}$ will have a sizable component along ζ, giving rise to distinct contrast. It should also be noted that the rotation of $\boldsymbol{M}$ across the 180° domain wall can take place with either a right- or a left-hand turn, since these two cases are energetically equivalent. Consequently, one should expect *two* distinct contrast levels for a domain wall, just as is observed in the experiment (figure 10).

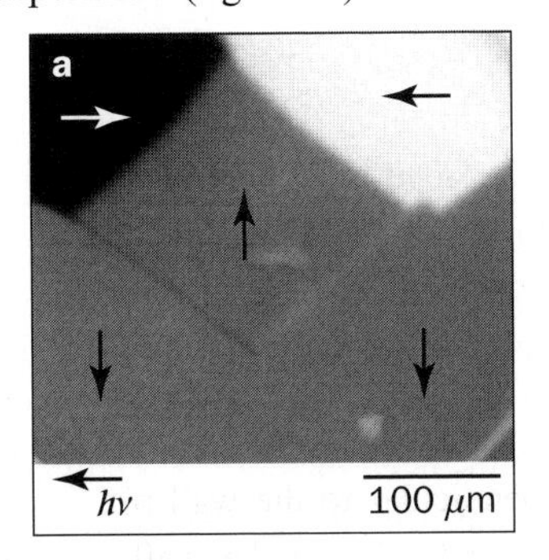

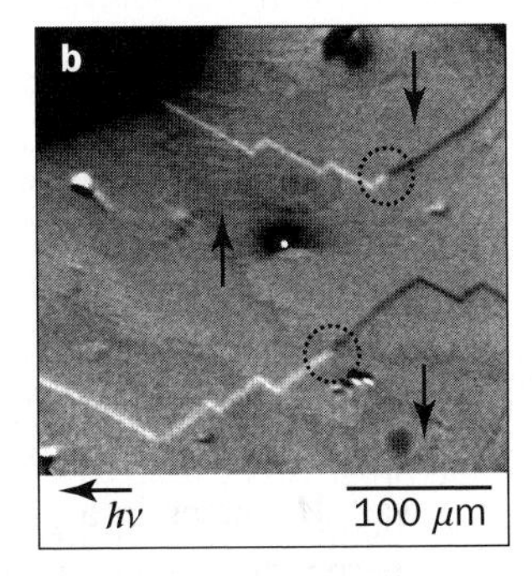

Figure 10 - Magnetic domain pattern at the surface of a strained Fe (001) whisker. **a** - Domain and domain wall contrast. **b** - Selective imaging of domain walls (white and dark lines) (from ref. **[28]**).

The above interpretation can qualitatively explain the experimental findings. The actual situation, however, is more complex than a simple 180° wall. This can be easily seen by reconstructing the magnetization distribution in figure 10a. The domain pattern is energetically very unfavourable, because the head-on orientation of the local magnetization directions generates a magnetic stray field. Such a domain structure cannot be explained by surface effects solely. In fact, the pattern in figure 10a must be interpreted on the basis of closure domains stabilized by the magnetic microstructure in the bulk, with the magnetization direction of the bulk domains pointing perpendicular to the surface [31]. In this case, the domain walls can be identified as so-called "V"-lines [32], being a result of two 90° bulk domain walls meeting at the surface. The formation of V-lines is caused by mechanical strain in the crystal.

A characteristic property of V-lines is a "zig-zag" course of the domain walls, which can be seen in figure 10b. In this image the main magnetic contrast arises from the domain walls only. The angle between two neighbouring segments of the zig-zag line has previously been determined to about 109° [32], which is compatible with the result in figure 10b. Finally, we also observe the predicted jump in the contrast level along the course of a domain wall (encircled regions), associated with a change of the rotation sense of ***M*** at the surface.

5.1.5. INFORMATION DEPTH

An important aspect of the contrast mechanism is the magnetic depth of information. In the case of MXD effects, it is determined by two factors:

- inelastic mean free path of the electrons λ_{in}, and
- penetration depth of the incident light λ_p.

Although the low energy secondary electrons are the same imaged in a SEMPA experiment, there is a fundamental difference. SEMPA determines the actual *spin polarization* of these electrons, which is determined by spin dependent scattering processes in the magnetic material and limits the information depth to about 5 Å [33]. In X-PEEM, however, the *intensity* of the low energy electron cascade (yield) is measured. The information depth is thus determined by the escape depth of the Auger electrons and the physics of the cascade formation process. It reaches values of the order of $\lambda_{in} \approx 15 - 25$ Å in ferromagnetic metals [26]. The penetration depth of the incident light is usually significantly larger than λ_{in}. In grazing incidence geometries, however, it may become relevant and leads to saturation effects in the MXD signals [34,26]. As in a typical X-PEEM geometry the light impinges at an angle of 15 - 25° with respect to the surface, saturation effects may have to be considered in quantitative measurements.

The following example illustrates the combination of chemical and magnetic information that can be extracted from X-PEEM investigations. The sample consisted of an epitaxially grown Cr wedge [ranging from 0 - 4 monolayers (ML)] on a Fe(001) whisker surface. The Cr wedge has then been covered by a 5 monolayer Co film. The magnetic domain pattern in this sample has then been imaged in the "light" of the Fe, Co, and Cr L_{III} absorption lines. Due to the information depth of the XMCD approach discussed above, the domain patterns of the substrate and the Cr layer can be imaged through the respective overlayers. The separation into magnetic and chemical information follows from equation {2}.

The results are compiled in figure 11. The whisker surface has a particularly simple domain pattern, consisting of only two oppositely magnetized domains. This corresponds to the equilibrium domain structure of a strain-free iron whisker. The chemical information about the Cr wedge confirms the onset (marked by the broken line, due to the graphical reproduction the first part of the wedge may appear dark) and a subsequent linear increase of the Cr signal along the wedge direction. Because of technical circumstances the thickness gradient of the wedge is inclined to the whisker main axis. The Co overlayer also shows a clear magnetic signal and a well-defined domain structure. In contrast to the Fe domain pattern, however, the Co domain structure reveals a characteristic change at a critical Cr thickness of about 2 ML. Below this critical thickness, the magnetization direction is the same in both Fe substrate and Co overlayer, i.e., the Co overlayer couples parallel ("ferromagnetically") to the substrate magnetization. Above 2 ML Cr the Co layer reverses its magnetization with respect to the substrate, i.e. it couples antiparallel ("antiferromagnetically"). This behaviour is caused by the interlayer exchange coupling through the Cr wedge [35]. Depending on the thickness of the Cr film, the coupling changes its character from parallel to antiparallel and vice versa.

A surprising result is found in the Cr magnetic signal, demonstrating the high element selectivity and magnetic sensitivity of the X-PEEM approach. First, also on the Cr L_{III} edge a magnetic domain pattern is observed, even in the Cr submonolayer regime. Second, the Cr domain pattern follows closely the Co one, as the change in magnetization direction at about 2 ML Cr is easily discernible. Compared to the magnetic contrast of Fe and Co, the Cr magnetic signal is rather weak. This is consistent with earlier findings, and can be explained by a small magnetic moment of the Cr (either due to magnetic frustration or partially antiferromagnetic order in the Cr patches) [36]. The Cr signal arises due to the exchange coupling to the neighbouring Fe and Co layers, causing a partial polarization of the Cr. The results suggest, however, that the coupling Co–Cr seems to be stronger than the respective Fe–Cr coupling. Below 2 ML Cr we cannot distinguish between these two coupling contributions, as the magnetization directions in Fe and Co are the same. Above 2 ML Cr the Cr signal clearly follows the Co magnetic orientation with a gradual reduction of the Cr XMCD contrast levels. The latter is related to a dilution effect caused by the unpolarized Cr atoms in the bulk of the Cr interlayer.

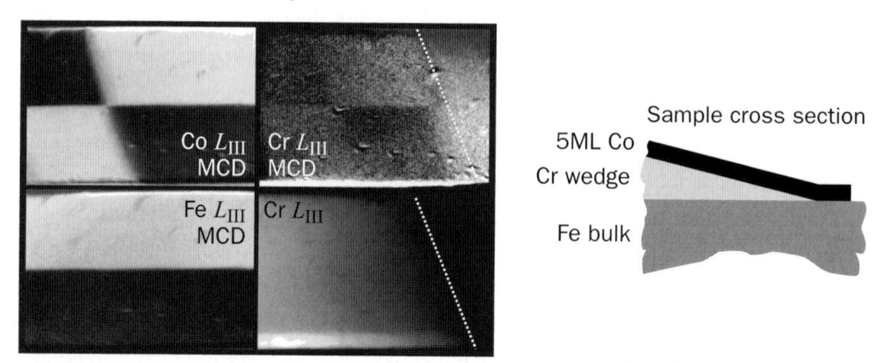

Figure 11 - Exchange coupled thin film system Co/Cr/Fe(001) whisker (broken line marks onset of Cr wedge). **Left** - Compilation of magnetic domain images acquired at the Co, Fe, and Cr L_{III} edges. **Right** - Sample cross section (taken from ref. [37]).

5.2. X-RAY MAGNETIC LINEAR DICHROISM (XMLD)

Another important class of materials for magneto-electronic applications are *antiferromagnets*. However, the expectation value of the local magnetization in antiferromagnets vanishes, i.e. $\langle M \rangle = 0$. Therefore, XMCD cannot be used to image the magnetic domain structure in these materials. Magnetic linear dichroism poses a solution to this problem, since the XMLD signal depends on $\langle M^2 \rangle$ [38].

5.2.1. PROPERTIES OF THE CONTRAST MECHANISM

Antiferromagnets may have very complex spin arrangement. Simple cases are found in Cr or NiO, in which neighbouring lattice planes have opposite spin alignment. This "topological" antiferromagnetism leads to so-called uncompensated planes [(001) in Cr, (111) in NiO]. The orientation of the magnetic moments along a spatial direction defines a quantization axis $\boldsymbol{m}$, which is used as reference for the optical excitation with linearly polarized light (π). XMLD appears between absorption spectra taken with linear polarization parallel $\pi_{//}$ and perpendicular $\pi_{\perp}$ to $\boldsymbol{m}$.

Therefore, the antiferromagnetic contrast C_{AFM} is extracted in analogy to equation {2} as

$$C_{AFM} \sim I_{\pi//} - I_{\pi\perp} \qquad \{5\}$$

The magnetic contrast is quantified by normalizing C_{AFM} to the sum of $I_{\pi//}$ and $I_{\pi\perp}$. The angular dependence between the direction of light polarization and the quantization axis ***m*** is slightly more complex than in the XMCD case, namely

$$C_{AFM} \sim (3\cos^2\theta - 1). \qquad \{6\}$$

Using this approach, domains at the surface of antiferromagnetic materials have been successfully imaged [39, 40].

5.2.2. IMAGING DOMAINS IN ANTIFERROMAGNETS

The last issue addresses the imaging of domains in an antiferromagnetic material on the basis of magnetic linear dichroism. Pioneering experiments in this field have been performed by Spanke et *al.* [39] and Stöhr et *al.* [40] on NiO films. Subsequent studies have concentrated on the microscopic mechanisms of the exchange anisotropy between ferro- and antiferromagnets [41]. The example reproduced in figure 12 shows the antiferromagnetic domain pattern on the (001) surface of a NiO single crystal [42].

The domain image is the result of a specific contrast enhancement procedure. The data have been recorded with *s* polarized light at the Ni L_{II} edge. In NiO the absorption at the L_{II} edge involves a characteristic doublet structure which causes the absorption line to consist of two spectral features ($h\nu_1$, $h\nu_2$) separated by about 1 eV [43]. These features exhibit a pronounced magnetic linear dichroism. The magnetic domain contrast is thus enhanced by calculating the asymmetry distribution equation {3} from two images acquired at $h\nu_1$ and $h\nu_2$, respectively.

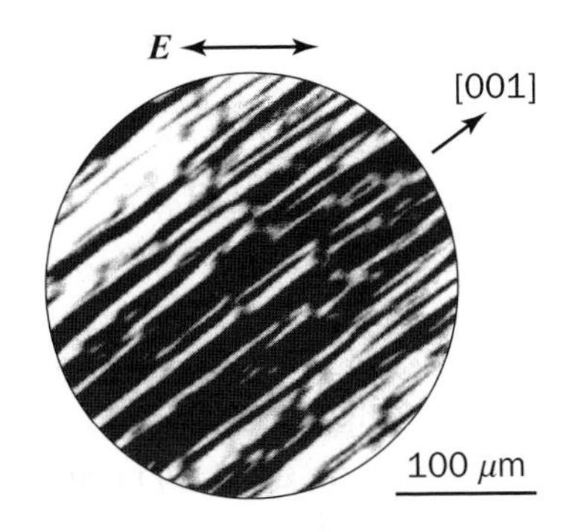

Figure 12 - Antiferromagnetic domains in NiO observed using XMLD [42]. Direction of the light polarization (***E***) with respect to the in-plane crystalline orientation is indicated.

The domain pattern itself is quite complex. The reason for this is the antiferromagnetic spin arrangement in crystalline NiO. The main "easy" axes ***m*** are given by the four equivalent [111] directions, along which lattice planes with in-plane ferromagnetic order are stacked in an antiferromagnetic arrangement. Within a (111) plane, there are three equivalent [211] directions into which the spins can point. As a consequence, in the bulk there are a total of 12 different types of antiferromagnetic domains [44]. These must be projected onto the (001) surface plane in order to obtain the possible surface configurations. Due to the angular dependence of the XMLD contrast equation {6} the X-PEEM picks up the component of the local antiferromagnetic orientation vector $\boldsymbol{m}$ along the electric field vector $\boldsymbol{E}$ of the linearly polarized light. In view of the complexity of the NiO (001) situation, a series of images as a function of the angle of light incidence are needed to unambiguously reconstruct the details of the surface domain pattern.

6. CONCLUDING REMARKS

These lecture notes can give only a brief introduction into the X-PEEM technique and contain only a very limited selection of physical phenomena that can be investigated with this approach. For more concise information on X-PEEM the reader is referred to a number of recent review articles [15, 45, 46, 47, 48]. In future studies, three major issues will be of importance:

- improvement of the lateral resolution by corrected electron optics,
- magnetic micro- and nanostructures in antiferromagnets, and
- time-resolved studies of magnetic phenomena [49, 50].

The investigations associated with the last issue will make use of the intrinsic time structure of the synchrotron radiation generated in storage rings or free-electron lasers. The ultimate goal will be the laterally resolved study of magnetic switching processes on a sub-nanosecond scale.

ACKNOWLEDGEMENTS

The author is indebted to G.H. Fecher, R. Frömter, J. Kirschner, M. Klais, W. Kuch, G. Schönhense, W. Swiech, and C. Ziethen for fruitful collaborations. Financial support by the *Bundesminister f. Bildung und Forschung* (grants N° 05 SC8BDA-2 and 05KS1BDA-9) is gratefully acknowledged.

REFERENCES

[1] *International Technology Roadmap for Semiconductors* Edition 2002 (accessible through website public.itrs.net)

[2] see, for example, the website *www.almaden.ibm.com/sst*

[3] K. KOIKE & K. HAYAKAWA - *Jap. J. Appl. Phys.* **23**, L187 (1984)

[4] L. REIMER - *Scanning Electron Microscopy,* Springer Verlag, Berlin (1985)

[5] H.J. GÜNTHERODT & R. WIESENDANGER eds - *Scanning Tunneling Microscopy,* Springer Verlag, Berlin (1992)

[6] Y. MARTIN & H.K. WICKRAMASINGHE - *Phys. Rev. Lett.* **50**, 1455 (1987)

[7] M. BODE, M. GETZLAFF & R. WIESENDANGER - *Phys. Rev. Lett.* **81**, 4256 (1998)

[8] E. BAUER & W. TELIEPS - *In "Emission and low energy reflection electron microscopy"*, A. HOWIE & U. VALDRÉ eds, Plenum Press, New York (1988)

[9] M. BRÜCHE - *Z. f. Naturforschung* **11**, 287 (1934)

[10] L.H. VENEKLASEN - *Rev. Sci. Instr.* **63**, 5513 (1992)

[11] C.M. SCHNEIDER & G. SCHÖNHENSE - *Rep. Prog. Phys.* **65** (2002) 1785

[12] R. FINK, M.R. WEISS, E. UMBACH, D. PREIKSZAS, H. ROSE, R. SPEHR, P. HARTEL, W. ENGEL, R. DEGENHARDT, R. WICHTENDAHL, H. KUHLENBECK, W. ERLEBACH, K. IHMANN, R. SCHLÖGL, H.J. FREUND, A.M. BRADSHAW, G. LILIENKAMP, T. SCHMIDT, E. BAUER & G. BENNER - *J. Elec. Spec.* **84**, 231 (1997)

[13] G.F. REMPFER, W.P. SKOCZYLAS & O.H. GRIFFITH - *Ultramicroscopy* **36**, 196 (1991)

[14] S. ANDERS, H.A. PADMORE, R.M. DUARTE, T. RENNER, T. STAMMLER, A. SCHOLL, M.R SCHEINFEIN, J. STÖHR, L. SÉVE & B. SINKOVIC - *Rev. Sci. Instr.* **70**, 3973 (1999)

[15] S.A. NEPIJKO, N.N. SEDOV & G. SCHÖNHENSE - *Adv. Imag. Electr. Phys.* **113**, 205 (2000)

[16] W. SWIECH, G. H. FECHER, C. ZIETHEN, O. SCHMIDT, G. SCHÖNHENSE, K. GRZELAKOWSKI, C. M. SCHNEIDER, R. FRÖMTER & J. KIRSCHNER - *J. Elec. Spec.* **84**, 171 (1997)

[17] C. ZIETHEN, O. SCHMIDT, G.K.L. MARX, G. SCHÖNHENSE, R. FRÖMTER, J. GILLES, J. KIRSCHNER, C.M. SCHNEIDER, & O. GRÖNING - *J. Elec. Spec.* **107**, 261 (2000)

[18] J. STÖHR & S. ANDERS - *IBM J. Res. Develop.* **44**, 535 (2000)

[19] G.V. SPIVAK, T.N. DOMBROVSKAIA & N.N. SEDOV - *Sov. Phys. Dokl.* **2**, 120 (1957)

[20] G.V.D. LAAN, B.T. THOLE, G.A. SAWATZKY, J.B. GOEDKOOP, J.C. FUGGLE, J.M. ESTEVA, R. KARNATAK, J.P. REMEIKA & H.A. DABKOWSKA - *Phys. Rev. B* **34**, 6529 (1986)

[21] G. SCHÜTZ, W. WAGNER, W. WILHELM, P. KIENLE, R. ZELLER, R. FRAHM & G. MATERLIK *Phys. Rev. Lett.* **58**, 737 (1987)

[22] C.T. CHEN, F. SETTE, Y. MA & S. MODESTI - *Phys. Rev. B* **42**, 7262 (1990)

[23] J. KESSLER - *Polarized Electrons*, Springer Verlag, Berlin (1985)

[24] C.M. SCHNEIDER, K. HOLLDACK, M. KINZLER, M. GRUNZE, H.P. OEPEN, F. SCHÄFERS, H. PETERSEN, K. MEINEL & J. KIRSCHNER - *Appl. Phys. Lett.* **63**, 2432 (1993)

[25] J. STÖHR, Y. WU, M.G. SAMANT, B.D. HERMSMEIER, G. HARP, S. KORANDA, D. DUNHAM & B.P. TONNER - *Science* **259**, 658 (1993)

[26] R. NAKAJIMA, J. STÖHR & Y.U. IDZERDA - *Phys. Rev. B* **59**, 6421 (1999)

[27] A. HUBERT & R. SCHÄFER - *Magnetic Domains*, Springer Verlag, Berlin (1998)

[28 C. M. SCHNEIDER, R. FRÖMTER, C. ZIETHEN, W. SWIECH, N. B. BROOKES, G. SCHÖNHENSE & J. KIRSCHNER - *Mat. Res. Soc. Symp. Proc.* **475**, 381 (1997)

[29] H.P. OEPEN & J. KIRSCHNER - *Phys. Rev. Lett.* **62**, 819 (1989)

[30] M.R. SCHEINFEIN, J. UNGURIS, R.J. CELOTTA & D.T. PIERCE - *Phys. Rev. Lett.* **63**, 668 (1989)

[31] S. CHIKAZUMI - *Physics of Magnetism*, Wiley & Sons, New York (1964)

[32] R.W. DEBLOIS & J.C.D. GRAHAM - *J. Appl. Phys.* **29**, 931 (1958)

[33] H.P. OEPEN & J. KIRSCHNER - *Scanning Microscopy* **5**, 1 (1991)

[34] J. HUNTER DUNN, D. ARVANITIS, N. MAARTENSSON, M. TISCHER, F. MAY, M. RUSSO & K. BABERSCHKE - *J. Phys.: Condens. Matter* **7**, 1111 (1995)

[35] J. UNGURIS, R.J. CELOTTA & D.T. PIERCE - *Phys. Rev. Lett.* **67**, 140 (1991)

[36] C.M. SCHNEIDER, K. MEINEL, J. KIRSCHNER, M. NEUBER, C. WILDE, M. GRUNZE, K. HOLLDACK, Z. CELINSKI, & F. BAUDELET - *J. Magn. Magn. Mater.* **162**, 7 (1996)

[37] W. KUCH, R. FRÖMTER, J. GILLES, D. HARTMANN, C. ZIETHEN, C.M. SCHNEIDER, G. SCHÖNHENSE, W. SWIECH, & J. KIRSCHNER - *Surf. Rev. Lett.* **5**, 1241 (1998)

[38] D. ALDERS, L.H. TJENG, F.C. VOOGT, T. HIBMA, G.A. SAWATZKY, C.T. CHEN, J. VOGEL, M. SACCHI & S. IACOBUCCI - *Phys. Rev. B* **57**, 11623 (1998)

[39] D. SPANKE, V. SOLINUS, D. KNABBEN, F.U. HILLEBRECHT, F. CICCACCI, L. GREGORATTI & M. MARSI - *Phys. Rev. B* **58,** 5201 (1998)

[40] J. STÖHR, A. SCHOLL, T.J. REGAN, S. ANDERS, J. LÜNING, M.R. SCHEINFEIN, H.A. PADMORE & R.L. WHITE - *Phys. Rev. Lett.* **83**, 1862 (1999)

[41] F. NOLTING, A. SCHOLL, J. STHR, J. FOMPEYRINE, H. SIEGWART, J.-P. LOCQUET, S. ANDERS, J. LÜNING, E.E. FULLERTON, M.F. TONEY, M.R. SCHEINFEIN & H.A. PADMORE - *Nature* **405**, 767 (2000)

[42] F.U. HILLEBRECHT, H. OHLDAG, N.B. WEBER, C. BETHKE, U. MICK, M. WEISS, & J. BAHRDT - *Phys. Rev. Lett.* **86**, 3419 (2001)

[43] D. ALDERS, J. VOGEL, C. LEVELUT, S.D. PEACOR, T. HIBMA, M. SACCHI, L.H. TJENG, C. T. CHEN, G. VAN DER LAAN, B.T. THOLE & G.A. SAWATZKY - *Eur. Phys. Lett.* **32**, 259 (1995)

[44] H. KOMATSU & M. ISHIGAME - *J. Mat. Sci.* **20**, 4027 (1985)

[45] G. SCHÖNHENSE - *J. Phys.: Condens. Matter* **11**, 9517 (1999)

[46] E. BAUER - *J. Phys.: Condens. Matter* **13**, 11391 (2001)

[47] TH. SCHMIDT, U. GROH, R. FINK, E. UMBACH, O. SCHAFF, W. ENGEL, B. RICHTER, H. KUHLENBECK, R. SCHLÖGL, H. J. FREUND, A.M. BRADSHAW, D. PREIKSZAS, P. HARTEL, R. SPEHR, H. ROSE, G. LILIENKAMP & E. BAUER - *Surf. Rev. Lett.* **9**, 223 (2002)

[48] S. GÜNTHER, B. KAULICH, L. GREGORATTI & M. KISKINOVA - *Prog. Surf. Sci.* **70**, 187 (2002)

[49] A. KRASYUK, A. OELSNER, S. NEPIJKO, A. KUKSOV, C.M. SCHNEIDER & G. SCHÖNHENSE - *Appl. Phys. A* **76**, 836 (2003)

[50] J. VOGEL, W. KUCH, M. BONFIM, J. CAMARERO, Y. PENNEC, F. OFFI, K. FUKUMOTO, J. KIRSCHNER, A. FONTAINE & S. PIZZINI - *Appl. Phys. Lett.* **82**, 2299 (2003)

9

X-RAY INTENSITY FLUCTUATION SPECTROSCOPY

M. SUTTON
Centre for the Physics of Materials, McGill University, Montreal, Quebec, Canada

1. INTRODUCTION

At the outset let me stress that X-ray intensity fluctuation spectroscopy (XIFS) is a diffraction technique. As such, the intuition and expertise that you have developed for diffraction carries over to this new technique. The tendency in this chapter is to emphasize the aspects that are different from conventional diffraction but results learned from conventional diffraction will be called upon as needed.

Wave phenomena are very prevalent in nature. Essential features of wave behaviour are the effects of interference and diffraction. For example, most textbooks on X-rays derive the X-ray diffraction by considering X-rays as planes waves and calculating the constructive and destructive interferences of these waves diffracting from atoms. In spite of the predominance of interference phenomena fundamental to wave phenomena, we are accustomed to the lack of such phenomena when dealing with conventional light. In order to describe whether or not one needs to consider these effects one should discuss the coherence of wave sources. Just the mention of coherence almost universally brings to mind lasers. Although lasers are intrinsically coherent sources of light, coherence effects can be seen from any source of waves.

As an introduction to intensity fluctuation spectroscopy (IFS), let me first describe "speckle". Speckle is an effect, often seen in laser light, that results when coherent light reflects or scatters diffusely off disordered material. The intensity at each spot in the "image" is the result of light from many different points in the disordered materials. The essentially random path lengths of the light from these points to the given spot in the image leads to the light being the sum of rays with a random set

F. Hippert et al. (eds.), Neutron and X-ray Spectroscopy, 297–318.

of phases. However, since the light is coherent, the phase of the resulting light, even though it is randomly distributed from point to point, has a definite value at each point. Where the phases add destructively, it results in a dark spot and where they add constructively a bright spot. This is the origin of the speckled appearance of the image. It is a good thing for us that conventional light sources are incoherent, since speckle could be quite annoying in everyday life.

As an introduction to the mathematics needed to deal with speckle and coherence, I would like to give an alternative derivation [1]. Assume we have a set of N particles, randomly distributed throughout some volume. The scattering $I(\boldsymbol{q})$ at wavevector $\boldsymbol{q}$ from this system is (ignoring constants)

$$\begin{aligned} I(\boldsymbol{q}) &= \left|\sum_{j=1}^{N} f_j \, \mathrm{e}^{i\boldsymbol{q}\cdot\boldsymbol{r}_j}\right|^2 \\ &= \sum_{j=1}^{N} \left|f_j\right|^2 + \sum_{k\neq l}^{N} f_k^{\dagger} f_l \, \mathrm{e}^{i\boldsymbol{q}\cdot(\boldsymbol{r}_l-\boldsymbol{r}_k)} \\ &= \sum_{j=1}^{N} \left|f_0\right|^2 + 2\sum_{k<l}^{N} \left|f_0\right|^2 \cos\left(\boldsymbol{q}\cdot(\boldsymbol{r}_l-\boldsymbol{r}_k)\right) \end{aligned} \qquad \{1\}$$

where f_j is the scattering factor for each particle. The last equation has been specialized to identical particles, f_0. From this equation it is easy to see that the average intensity (averaged over particle positions) is N times the scattering of each particle

$$\left\langle I(\boldsymbol{q})\right\rangle = N\left|f_0\right|^2 \qquad \{2\}$$

To see why there is "speckle", one looks at the standard deviation of the intensity. First, calculate the variance. For identically distributed values, this will be the number of objects times the variance of each object. Since the phase factor above varies between plus and minus one we can estimate the variance of each object as

$$\begin{aligned} \sigma_{I(\boldsymbol{q})} &= \left\langle I(\boldsymbol{q})^2\right\rangle - \left\langle I(\boldsymbol{q})\right\rangle^2 \\ &= N(N-1)\left|f_0\right|^4 \\ &\approx \left(N\left|f_0\right|^2\right)^2 \end{aligned} \qquad \{3\}$$

We can now conclude that the scattering from N randomly placed identical particles is $\left\langle I(\boldsymbol{q})\right\rangle \pm \left\langle I(\boldsymbol{q})\right\rangle$ and this varies from zero to about twice the average. A detailed calculation of the distribution of intensities shows, for perfectly coherent

1. This is a variation of an argument I heard in a colloquium by Paul CHAIKIN.

light, that the most probable intensity is actually zero! It is easy to numerically simulate the above ideas. Figure 1a shows a randomly generated set of diamond shaped particles[2] distributed in a plane. Figure 1b shows the 2D fast Fourier transform (FFT) of these particles. One of the striking features in the image is that the speckles have a characteristic size. Since speckle results from the sum of cosines in equation {1}, the sharpest features are due to particles that are the furthest apart. Hence the speckle size is given by one over the size of the scattering region. For comparison, the FFT of a single particle is shown in figure 1c and the FFT of points replacing the particles in figure 1d.

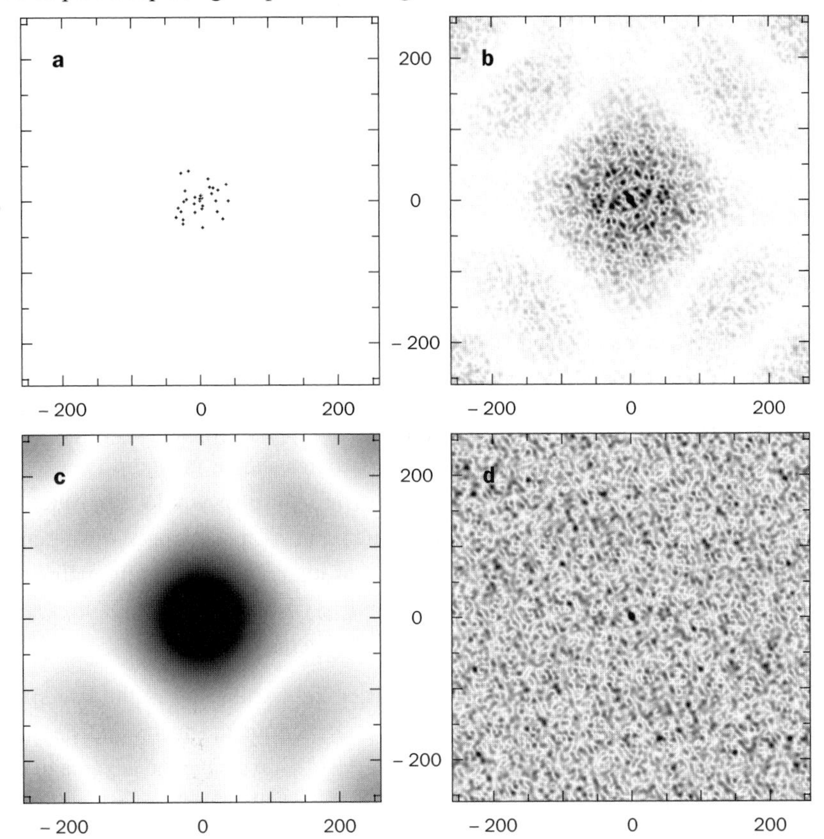

Figure 1 - **a** - Thirty randomly placed diamond shaped particles. **b** - Speckle (diffraction) pattern of the particles from (**a**). **c** - Diffraction pattern of a single diamond shaped particle. **d** - Diffraction pattern found by replacing the particles in (**a**) by point particles.

2. Figure is a 512 by 512 bitmap with 30 "particles" each defined on a 5 by 5 matrix with a 1,3,5,3,1 pattern.

As helped by figure 2, close inspection shows that to a high degree of approximation the speckle pattern is the product of the single pinhole and point position scattering.

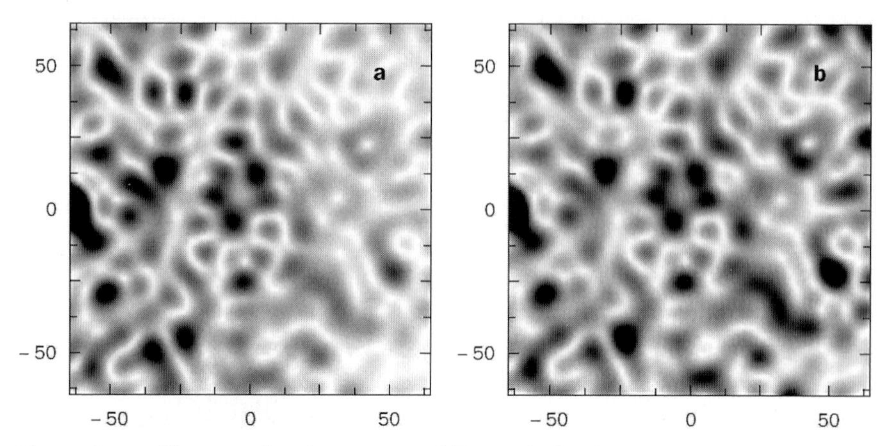

Figure 2 - a - Close up view from a part of figure 1b. **b -** Close up view from figure 1d of the same region.

Reviewing the above arguments it is difficult to imagine why we don't see speckle more often in everyday life. We will leave this as a puzzle until the next section.

With this relatively detailed introduction to speckle it is now easy to see the origin of intensity correlation spectroscopy. If the particles shown in figure 1a where to fluctuate in position, the interference pattern in figure 1b would also fluctuate. Intensity fluctuation spectroscopy is the technique of measuring these intensity fluctuations and relating them to the kinetics of materials undergoing diffraction. One of the powerful features of IFS is that it gives direct access to measuring thermodynamic fluctuations even in equilibrium systems.

This is a technique that was first developed for light scattering in the late sixties and is often called dynamic light scattering. Recently, the technique has been extended to X-rays (which is abbreviated as XIFS). X-rays have the advantage over light that most things are transparent to X-rays, the problems of multiple scattering are not as pronounced or non-existent, and the shorter wavelengths probe smaller distances. Their prime disadvantage is that X-ray sources are not as intense as laser light sources.

This chapter will first describe how to characterize the coherence of a beam and then how coherence effects diffraction. This is followed by a general discussion of kinetics in materials and some examples of the technique.

2. *MUTUAL COHERENCE FUNCTIONS*

An intuitive feeling for coherence can be gained by considering the region over which a wave can be considered as well defined or coherent (its coherence volume). The distance over which the coherence extends along the direction of the wave is called the longitudinal coherence length. It is easy to see that this is related to how monochromatic the wave is. To quantify this, one imagines the different wavelengths λ and $\lambda + \Delta\lambda$ are all in phase at some time. Travelling along the wave the spread in wavelengths leads to a point at which the waves are out of phase. We can define the longitudinal coherence length $\xi_l = \lambda^2/(2\,\Delta\lambda) = c/(2\,\Delta\nu)$, where $\Delta\nu$ is the corresponding frequency spread. Note that this naturally leads to the definition of a coherence time as $\tau_c = \xi_l/c = 1/(2\,\Delta\nu)$. There are also two transverse coherence lengths that determine how far one must travel parallel to the wave fronts before the wave gets out of phase. This is related to the angular spread of the beam. Most often, the primary source of this divergence is determined by the finite size of the source σ_s a distance R away. This gives an equation for the transverse coherence length ξ_t as $\lambda/2 = \xi_l\,\sigma_s/R$, or $\xi_l = \lambda\,R/(2\,\sigma_s)$.

A beam that is monochromatic enough and has a transverse size comparable to its transverse coherence lengths is said to be coherent. When the dimensions of the beam are much larger than the coherence lengths, it is said to be incoherent. Beams that are not quite coherent are called partially coherent. Coherent effects of the interaction of light with matter happen inside the light's coherence volume and incoherent effects occur when light interacts with matter over much longer length scales than its coherence volume. Intuitively, one can consider an incoherent beam as composed of an independent sum of regions each of order the size its coherence volume. Each coherent region will lead to a speckle pattern, and an image or diffraction pattern is then made up of the sum over all these speckle patterns. This incoherent sum leads to the smooth intensity patterns we are used to observing.

To be more quantitative we need to describe the electric field correlations. See Mandel & Wolf [1] for more details. The mutual coherence function is defined as [3]

$$\Gamma(r_1, r_2, t_1, t_2) = \left\langle E^*(r_1, t_1) E(r_2, t_2) \right\rangle \qquad \{4\}$$

where $\boldsymbol{E}(\boldsymbol{r}, t)$ is the electric field.

---

3. We will use the notation of [1] in this chapter. Also, as is conventional we use the *complex analytic signal* representation. An elegant and clear description of its use is given in section 3.1 of ref. [1] .

Here the averages can be considered either over the different coherence regions or over the distribution of random amplitudes and phases of the electromagnetic (em) wave. Just as in statistical mechanics, by dealing directly with correlation functions, the averaging needed to handle the intrinsic randomness of the wave no longer needs to be explicitly taken care of. Usually the incident em wave can be considered as having constant intensity and it makes sense to consider the randomness as stationary as this means the correlation function only depends on the time difference $\tau = t_2 - t_1$. A normalized form of the correlation function, called the degree of coherence, is also useful and it is defined as

$$\gamma(r_1, r_2, \tau) = \frac{\Gamma(r_1, r_2, \tau)}{\Gamma(r_1, r_1, 0)^{1/2}\,\Gamma(r_2, r_2, 0)^{1/2}} \qquad \{5\}$$

Regions with $|\gamma| = 1$ can be considered perfectly coherent and those with $|\gamma| = 0$ are incoherent. As easily seen from the definition, $\Gamma(\boldsymbol{r}, \boldsymbol{r}, t, t)$ is just the intensity of the light at point $(\boldsymbol{r}, t)$.

In optics, problems in diffraction and interference are often treated by Huygens principle. This states that wave propagation problems can be treated by imagining the wave fronts as composed of point sources, each of which generates spherical waves. Propagating these spherical waves and re-summing them can explain most diffraction effects. A similar approach allows one to propagate coherence functions from one surface to another along the wave fronts.

We will not pursue these ideas in much detail here but will give two examples of coherence calculations. First we argue the form of the coherence function for an incoherent light source. The naive assumption that $\Gamma(r_1, r_2, \tau) = 0$ does not make sense because $\Gamma(\boldsymbol{r}, \boldsymbol{r}, 0)$ must equal the intensity distribution of the source. A natural form for the coherence is thus

$$\Gamma(r_1, r_2, \tau) = I(r_1) F(\tau) \delta(r_2 - r_1) \qquad \{6\}$$

where $F(\tau)$ describes the coherence in time (longitudinal coherence) and from the definitions $F(0) = 1$. It is related to the Fourier transform of the frequency spectrum of the source. For a Lorentzian frequency spectrum, centred at frequency ν_0 with a full width at half maximum (FWHM) of $\Delta\nu$, on finds

$$F(\tau) = \mathrm{e}^{-|\tau|/\tau_{\mathrm{cor}}}\,\mathrm{e}^{-2\pi i \nu_0 t} \qquad \{7\}$$

and $\tau_{\mathrm{cor}} = 1/2\pi\Delta\nu$. It follows that for a Lorentzian spectrum, the longitudinal coherence length can be defined as $\xi_l = c\tau_{\mathrm{cor}} = (k\Delta\lambda / 2\lambda)^{-1} = (\lambda/\pi)(E/\Delta E)$, where ΔE is the FWHM of the energy spectrum.

Using Huygens construction, one can easily develop formulae to propagate the mutual coherence from one surface to another a distance z away. For an incoherent source this turns out to be simply the spatial Fourier transform of the source intensity distribution[4]. For synchrotron radiation, the intensity distribution is often described as the product of a Gaussian distribution in the vertical position y with width σ_v and one in the horizontal position x, with width σ_h,

$$I(x,y) = \frac{I_0}{2\pi\sigma_h\sigma_v} e^{-\left(\frac{x^2}{2\sigma_h^2}+\frac{y^2}{2\sigma_v^2}\right)} \qquad \{8\}$$

With this definition I_0 is just the integrated number of photons per second. The resulting mutual coherence is

$$\Gamma(x_1,y_1,x_2,y_2,\tau) = \frac{I_0 F(\tau)}{\pi} \frac{e^{-\left(\frac{(x_2-x_1)^2}{2\xi_h^2}+\frac{(y_2-y_1)^2}{2\xi_v^2}\right)}}{z^2} \qquad \{9\}$$

for which transverse coherence lengths have been defined as

$$\xi_i = \frac{z}{k\sigma_i} = \frac{\lambda z}{2\pi\sigma_i} \qquad \{10\}$$

where i is one of v or h and an inessential phase factor is ignored. Note that by specifying the properties of the source we have reached a specific relation between coherence length and angular size of the source σ_i/z It is informative to write this as $\left(\frac{\sigma_i}{z}\right)\xi_i = \lambda/2\pi$, which relates the angular size of the source σ_i/z to the coherence length ξ_i and is an example of the angle-position uncertainty relation. Angle and position are the appropriate phase space variables for propagating waves and the phase space density of a coherent wave is $\lambda/2\pi$. As in all uncertainty relations, the form of the equation is straightforward to derive by hand waving or dimensional arguments but the numerical factor depends on the details of the specific functions describing the source and the mutual coherence functions. This form for the mutual coherence is quite useful and shows the essential features affecting coherence. It shows how transverse coherence lengths propagates and increase along the beam direction and how they are related to the angular size of the source. Reference **[2]** gives a more complete discussion of mutual coherence functions for synchrotron sources. This includes how to calculate the extra coherence arising from the high collimation of undulators as well as how coherence is connected to brilliance.

---

4. Such a calculation proves the van Cittert-Zernicke theorem.

3. DIFFRACTION BY PARTIALLY COHERENT SOURCES

Using the Huygens-Fresnel construction as above makes it straight forward to calculate the effects of coherence on diffraction. This section is heavily based on section 2.4 of a review article by Pusey [3]. As with conventional X-ray diffraction calculating the scattering of X-rays, it is convenient to use the following form of an incident wave with wavevector $\boldsymbol{k}_i$ and frequency $\omega_0 = 2\pi\nu_0$

$$E(\boldsymbol{r},t) = E_i(\boldsymbol{r},t)\mathrm{e}^{i(\boldsymbol{k}_i \cdot \boldsymbol{r} - \omega_0 t)} \qquad \{11\}$$

For a (quasi-)monochromatic wave, the amplitude function $E_i(\boldsymbol{r}, t)$ must be slowly varying. With this incident wave the far field scattering at a point $(\boldsymbol{R}, t)$ is seen to be

$$E_s(\boldsymbol{r},t) = \frac{k^2}{4\pi R}\mathrm{e}^{i(\boldsymbol{k}_s \cdot \boldsymbol{R} - \omega_0 t)} \int \mathrm{e}^{i(\boldsymbol{Q}\cdot\boldsymbol{r})}\, \delta\rho(\boldsymbol{r},t) E_i\left(\boldsymbol{r}, t + \frac{\boldsymbol{k}_s \cdot \boldsymbol{r}}{\omega_0}\right) d\boldsymbol{r} \qquad \{12\}$$

where k_s is the scattered wave vector and the momentum transfer is $\boldsymbol{Q} = \boldsymbol{k}_s - \boldsymbol{k}_i$. Since the incident wave vector is fixed, a given position $\boldsymbol{R}$ in the scattered beam corresponds to a unique value of $\boldsymbol{Q}$. To arrive at the above form it is assumed that the transit time of the X-rays across the scattering volume V is much less then the time for changes in the density $\rho\,(r, t)$ (static approximation). To emphasize that scattering arises from density fluctuations, $\delta\rho = \rho - \langle\rho\rangle$ is used in the equations. With this form of the scattering, it is easy to discuss the effects of coherence on normal diffraction. First consider the electric field-field correlation function $G^{(1)}(\boldsymbol{R}, \tau)$ for the scattered waves.

$$G^{(1)}(\boldsymbol{R},\tau) = \left\langle E_s^*(\boldsymbol{R},t) E_s(\boldsymbol{R},t+\tau)\right\rangle \qquad \{13\}$$

Several assumptions make calculating $G^{(1)}$ easier. One can assume that fluctuations in the source and those in the sample are independent and at least, for this section, stationary and translationally invariant. For disordered scatterers, the coherence lengths and the sample size will be much bigger then the correlation lengths of the sample (that is many speckles occur in the diffraction pattern). Also used is that the electric field fluctuations are much faster then the sample's correlation times. This allows factoring of $E - E$ and $\rho - \rho$ correlations and leads to

$$G^{(1)}(\boldsymbol{R},\tau) = \frac{k^4}{(4\pi R)^2}\mathrm{e}^{-i\omega_0\tau} \int_V \int_V \mathrm{e}^{i(\boldsymbol{Q}_i \cdot(\boldsymbol{r}_1 - \boldsymbol{r}_2))}$$
$$\left\langle \delta\rho(\boldsymbol{r}_1,t)\delta\rho(\boldsymbol{r}_2,t+\tau) E_i^*\left(\boldsymbol{r}_1, t + \frac{\boldsymbol{k}_s\cdot\boldsymbol{r}_1}{\omega_0}\right) E_i\left(\boldsymbol{r}_2, t+\tau+\frac{\boldsymbol{k}_s\cdot\boldsymbol{r}_2}{\omega_0}\right)\right\rangle d\boldsymbol{r}_1 d\boldsymbol{r}_2$$

$$G^{(1)}(\boldsymbol{R},\tau) = \frac{k^4 V}{(4\pi R)^2}\left\langle E_i^*(0,0)E_i(0,\tau)\right\rangle e^{-i\omega_0\tau}\int_V e^{i\boldsymbol{Q}_i\cdot\boldsymbol{r}}\left\langle\delta\rho(\boldsymbol{r}_1,t)\delta\rho(\boldsymbol{r}_2,t+\tau)d\boldsymbol{r}\right.$$
$$= \frac{k^4 V}{(4\pi R)^2}\left\langle E_i^*(0,0)E_i(0,\tau)\right\rangle e^{-i\omega_0\tau} S(\boldsymbol{Q},t) \qquad \{14\}$$

Since $G^{(1)}(\boldsymbol{R},0) = \langle I(\boldsymbol{R})\rangle$ this equation shows that, just as for incoherent X-ray scattering, the average intensity is proportional to the Fourier transform of the equal time density-density correlation function $S(\boldsymbol{Q},t)$. One must be a little careful as this result depends on the correlation lengths in the sample being short. See Sinha *et al.* [4] for what can be done when correlation lengths are comparable to the scattering volume. As we expect from the introduction there should be a speckle structure superimposed on the scattering. To quantify this we calculate the intensity-intensity correlations $G^{(2)}$, between two points on the detector $\boldsymbol{R}$ and $\boldsymbol{R}+\delta\boldsymbol{R}$. These two detector positions correspond to momentum transfers $\boldsymbol{Q}$ and $\boldsymbol{Q}+\boldsymbol{\kappa}$.

$$G^{(2)}(\boldsymbol{R},\boldsymbol{R}+\delta\boldsymbol{R},\tau) = \left\langle I(\boldsymbol{R},t)I(\boldsymbol{R}+\delta\boldsymbol{R},t+\tau)\right\rangle$$

$$=\frac{k^8}{(4\pi R)^4}\int_V\int_V\int_V\int_V e^{i\boldsymbol{Q}(\boldsymbol{r}_1-\boldsymbol{r}_2+\boldsymbol{r}_3-\boldsymbol{r}_4)}e^{i\boldsymbol{\kappa}\cdot(\boldsymbol{r}_3-\boldsymbol{r}_4)}\left\langle\delta\rho(\boldsymbol{r}_1,0)\delta\rho(\boldsymbol{r}_2,0)\delta\rho(\boldsymbol{r}_3,\tau)\delta\rho(\boldsymbol{r}_4,\tau)\right\rangle$$

$$=\left\langle E_i^*\left(\boldsymbol{r}_1,t+\frac{\boldsymbol{k}_s\cdot\boldsymbol{r}_1}{\omega_0}\right)E_i\left(\boldsymbol{r}_2,t+\frac{\boldsymbol{k}_s\cdot\boldsymbol{r}_2}{\omega_0}\right)E_i^*\left(\boldsymbol{r}_3,t+\tau+\frac{(\boldsymbol{k}_s+\boldsymbol{k})\cdot\boldsymbol{r}_3}{\omega_0}\right)E_i\left(\boldsymbol{r}_2,t+\tau+\frac{(\boldsymbol{k}_s+\boldsymbol{k})\cdot\boldsymbol{r}_4}{\omega_0}\right)\right\rangle d\boldsymbol{r}_1 d\boldsymbol{r}_2 d\boldsymbol{r}_3 d\boldsymbol{r}_4$$

$$=\left\langle I(R)\right\rangle^2+\beta(\boldsymbol{\kappa})\frac{k^8}{(4\pi R)^4}V^2\left\langle|E_i|^2\right\rangle^2\left|\int_V e^{i\boldsymbol{Q}\cdot\boldsymbol{r}}\left\langle\delta\rho(0,0)\delta\rho(\boldsymbol{r},\tau)\right\rangle d\boldsymbol{r}\right|^2$$

$$=\left\langle I(R)\right\rangle^2+\beta(\boldsymbol{\kappa})\frac{k^8}{(4\pi R)^4}V^2\left\langle|E_i|^2\right\rangle^2\left|S(\boldsymbol{Q},t)\right|^2$$

$$=\left\langle I(R)\right\rangle^2+\beta(\boldsymbol{\kappa})\left|G^{(1)}(\boldsymbol{R},\tau)\right|^2 \qquad \{15\}$$

where

$$\beta(\boldsymbol{\kappa})=\frac{1}{V^2\left\langle|E_i|^2\right\rangle^2}\int_V\int_V e^{i\boldsymbol{\kappa}\cdot(\boldsymbol{r}_2-\boldsymbol{r}_1)}\left|\Gamma\left(\boldsymbol{0},\boldsymbol{r}_2^{\perp}-\boldsymbol{r}_1^{\perp},\frac{\boldsymbol{Q}\cdot(\boldsymbol{r}_2-\boldsymbol{r}_1)}{\omega_0}\right)\right|^2 d\boldsymbol{r}_1 d\boldsymbol{r}_2 \qquad \{16\}$$

The key approximation in this chain of equations is to use the short range nature of the correlations to factor the averages over products of four density terms into the products of averages over pairs of density terms. For instance, if any $\boldsymbol{r}_i$ is far removed from the rest, that density term factors by independence and since $\langle\delta\rho\rangle = 0$ such terms drop out of the sum. Therefore significant contributions can only arise if $\boldsymbol{r}_1\approx\boldsymbol{r}_2\neq\boldsymbol{r}_3\approx\boldsymbol{r}_4$ or $\boldsymbol{r}_1\approx\boldsymbol{r}_4\neq\boldsymbol{r}_2\approx\boldsymbol{r}_3$ and the fourth density moments required by the formula become the sum of two products of second moments. Another contribution should occur when $r_1 \approx r_2 \approx r_3 \approx r_4$ but is ignored as it has negligible volume (it is smaller by the ratio of the correlation volume to the scattering volume). These arguments actually show that all higher order density moments become products of second moments meaning that the central limit theorem applies. We stress that this argument only depends on the properties of the material from which the scattering is taking place.

Equation {16} is the central equation of coherent X-ray diffraction and it is worth spending some time discussing its implications. First it shows how the measured diffraction intensity $I(\boldsymbol{R},t)$ factors into terms arising from the sample $S(\boldsymbol{Q},t)$ and coherence properties of the beam $\beta(\boldsymbol{\kappa})$ for which it gives an explicit formula in terms of the mutual coherence function. Thus $\beta(\boldsymbol{\kappa})$ is the speckle pattern modulating the scattering as the discussion about pinholes in tinfoil argued. Several limits are worth exploring.

First with $\tau = 0$ and $\delta\boldsymbol{R} = \boldsymbol{0}$, one gets

$$\left\langle I^2(\boldsymbol{R},t)\right\rangle - \left\langle I(\boldsymbol{R},t)\right\rangle^2 = \beta(0)\left\langle I(\boldsymbol{R},t)\right\rangle^2 \qquad \{17\}$$

which shows that the variance of the scattering intensity is proportional to the square of the scattering intensity. When β is one, this reproduces equation {4} from the introduction. Now we have shown its statistical mechanics basis. Figure 3 shows three small angle diffraction patterns from a slowly phase separating system of borosilicate glass.

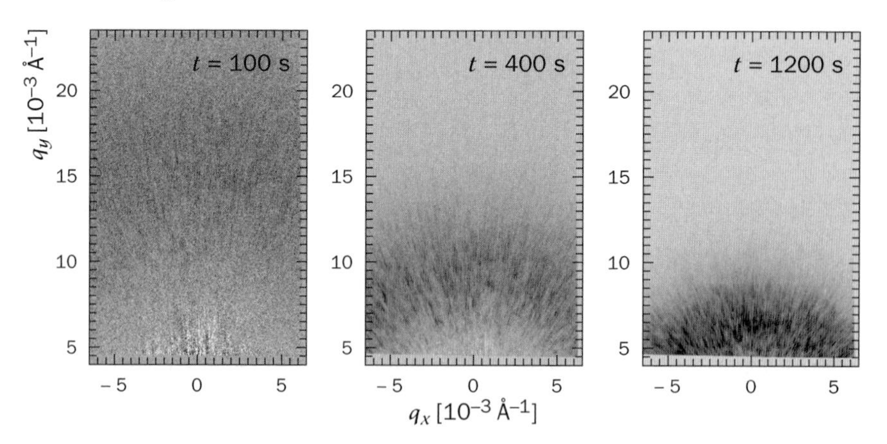

Figure 3 - Speckle patterns produced by a coherent X-ray beam scattering from a sodium borosilicate glass sample undergoing phase separation at 963 K. One-second exposures at 100, 400, and 1200 s after a quench from 1033 K are shown. Higher intensity corresponds to darker greys.

The scattering from this system consists of diffuse rings of scattering and as the system phase separates into sodium-rich and boron-rich silicate glasses the ring of scattering grows in intensity and decreases in angle. During each short exposure the system may be considered as static and so produces a nice speckle pattern when coherent light is used. From these images one can easily see the speckle patterns that modulate the conventional diffraction patterns. These images also serve to show that we should be able to use speckle to study non-equilibrium systems. For a more complete description of the physics in this system see Malik *et al.* **[5]**.

In optics, the term $(I_{max} - I_{min}) / (I_{max} + I_{min})$ is called the visibility or contrast and one can see that this is proportional to $\sqrt{\beta}$. It makes sense to call β the coherence factor although it is sometimes called the contrast.

To estimate $\beta(\mathbf{0})$, one can set $\mathbf{\Gamma}(\boldsymbol{r}_1, \boldsymbol{r}_2, 0)$ to $\langle|E_i|\rangle^2$ $(\gamma = 1)$ inside the coherence volume and zero outside. Thus equation (16) estimates $\beta(\mathbf{0})$ as the ratio of the coherence volume to the scattering volume and for perfectly coherent light $\beta = 1$. This same approximation estimates that $\beta(\boldsymbol{\kappa})$ is the Fourier transform of the scattering volume V. Thus one can estimate the speckle size as the inverse of the appropriate projection of the beam size. This is approximately $R_{det}\lambda/d_{beam}$ in terms of the size of the beam on the sample d_{beam} and the distance to the detector R_{det}. It is also obvious that this projection will slowly change with angle and it should not surprise the reader that speckle size will slowly change with $|\boldsymbol{Q}|$. More details on how to evaluate this integral with a Lorentzian frequency spectrum and comparison to measured speckle patterns are given in references **[6,7]** and so will not be reproduced here.

Another limit of interest is for $\delta\boldsymbol{R} = 0$ (measuring a single speckle) then equation {16} can be rewritten as

$$G^{(2)}(\boldsymbol{R},\boldsymbol{R},\tau) = \langle I(\boldsymbol{R})\rangle^2 + \beta(0)\left|G^{(1)}(\boldsymbol{R},\tau)\right|^2 \qquad \{18\}$$

or by defining $g^{(1)} = G^{(1)}(\boldsymbol{R},\tau)/\langle I(\boldsymbol{R})\rangle$ and $g^{(2)} = G^{(2)}(\boldsymbol{R},\boldsymbol{R},\tau)/\langle I(\boldsymbol{R})\rangle^2$, as

$$g^{(2)} = (\boldsymbol{R},\tau) = 1 + \beta(0)\left|g^{(1)}(\boldsymbol{R},\tau)\right|^2 \qquad \{19\}$$

This equation relates the normalized autocorrelations of the intensity measurements to the normalized autocorrelations of the density. It shows that by measuring $G^{(2)}(\boldsymbol{R},\tau)$ (the speckle intensity as a function time) one can measure the dynamics of the system under study. This is the basic formula of intensity fluctuation spectroscopy showing the explicit relationship between XIFS measurements and the statistical mechanics of the materials under study.

4. KINETICS OF MATERIALS

In order to describe non-equilibrium processes in a material, one needs two things. The first is a description of the state of the material that characterizes the system at each stage of its evolution and the second is an equation of motion that describes how this description evolves in time. The description is often given by a phase field or order parameter field, here labelled $\Psi(\boldsymbol{x},t)$, which describes the local order. It is imagined that Ψ arises from a microscopic description of the system and in equilibrium, this field is considered to have an average that varies slowly on atomic length scales but exhibits thermodynamic fluctuations on short length scales. Depending on the problem being considered this thermodynamic field could be concentration, density, molecular order or one of many other possibilities. Because of fluctuations, the field's values are often different from their thermodynamic equilibrium values and there is a driving force tending to force them back to their appropriate equilibrium values. This is of course, provided there are fluctuation mechanisms available to allow it to happen. An appropriate free energy $F[\Psi(\boldsymbol{x},t)]$ describes how much thermodynamic energy such a fluctuation costs and thus also how many fluctuations exist under given external conditions. (For convenience, we define $\Psi = 0$ as the equilibrium value and this value is defined to have a free energy that is zero.)

With these definitions, the equation of motion for most systems can be written in the deceptively simple form

$$\frac{\partial \Psi(\boldsymbol{x},t)}{\partial t} = -M\frac{\partial F}{\partial \Psi} + \eta(\boldsymbol{x},t) \qquad \{20\}$$

where M is a parameter, called the mobility[5], that characterizes transport mechanisms. The functional F is the free energy or thermodynamic potential needed to describe the thermodynamics of the system. The effect of thermodynamic noise on the evolution is represented by $\eta(\boldsymbol{x},t)$ and to a high degree of approximation it need only be characterized by the strength of its correlations, $\langle \eta(\boldsymbol{x},t)\,\eta(\boldsymbol{x}',t')\rangle = 2\,Mk_bT\delta(\boldsymbol{x}-\boldsymbol{x}')\,\delta(t-t')$. Here k_b is Boltzmann's constant and T is temperature. This equation closely follows the Langevin description of Brownian motion so it makes sense to describe such equations generically as Langevin equations.

---

5. If Ψ corresponds to a conserved quantity then one must replace M by $-M'\nabla^2$ to ensure the equation of motion preserves the conservation law.

Equation {20} is highly appealing and intuitive. It says that there is a thermodynamic driving force pushing the system towards equilibrium and that this force is proportional to how far the system is from equilibrium. Furthermore, it stresses how the thermodynamic fluctuations of the system are the underlying mechanism by that the system evolves in time and which tend to keep the system in equilibrium. On a deeper level, this equation incorporates, equipartition of energy, the fluctuation-dissipation theorem, and the generalized Einstein-Stokes relation, which relates the transport coefficients to the thermal noise strengths.

Not only are Langevin equations useful for describing non-equilibrium processes, but they give useful insight into kinetics of equilibrium systems. Except in special cases, like very near their phase transition temperatures, equilibrium systems can be described by phase fields that have only short-ranged and short-lived thermodynamic fluctuations. Thus these fluctuations are intrinsically small and the free energy can be expanded in powers of deviations of Ψ from its equilibrium value. Any constant term does not contribute to the equation of motion and there can be no linear term because the free energy is minimized at equilibrium. Thus in this limit, the equation of motion for equilibrium systems is

$$\frac{\partial \Psi(\boldsymbol{x},t)}{\partial t} = -M\left(\frac{\partial^2 F}{\partial \Psi^2}\right)_0 \Psi(\boldsymbol{x},t)+\eta(\boldsymbol{x},t) \qquad \{21\}$$

We choose to use a version in terms of $\Psi(\boldsymbol{q},t)$, the spatially Fourier transform of Ψ as this directly compares to the X-rays scattering results below. By defining $S(\boldsymbol{Q}) = \langle \Psi^\dagger(\boldsymbol{Q},t)\,\Psi(\boldsymbol{Q},t)\rangle$ where the brackets reflect thermal averages and using thermodynamic identities to relate the second derivative of a free energy to the inverse of this susceptibility [8] the equation of motion simplifies to

$$\frac{\partial \Psi(\boldsymbol{Q},t)}{\partial t} = \frac{-Mk_bT}{S(Q)}\Psi(\boldsymbol{Q},t)+\eta(\boldsymbol{Q},t) \qquad \{22\}$$

In other words, $\Psi(\boldsymbol{Q},t)$ varies exponentially in time and has Q-dependent time constants that are proportional to $S(\boldsymbol{Q})$. This result is exact in linear response and surprisingly, is not widely known.

One further approximation is often made. The free energy can be written as (remember Ψ is defined such that $\langle\Psi\rangle = 0$ in equilibrium)

$$F = F_0 + \int\left(\frac{r\,\Psi(\boldsymbol{x},t)^2}{2}+\frac{\kappa\left|\nabla\Psi(\boldsymbol{x},t)\right|^2}{2}\right)d\boldsymbol{x} \qquad \{23\}$$

which accounts for the most important thermal fluctuations. Coefficient r is the bulk energy (free energy) cost of microscopic deviations in Ψ and coefficient κ for the energy cost of slow spatial variations in $\Psi(\boldsymbol{x},t)$. Fourier transforming and taking second derivatives easily shows that

$$S(\boldsymbol{Q}) = \frac{k_b T}{r + \kappa Q^2} = \tau(Q) M k_b T \tag{24}$$

In practice, several complications arise. Terms in the free energy of higher order in Ψ may become important. This must happen very near a continuous phase transition where r is small, or below a phase transition where $r < 0$ and $\Psi = 0$ is no longer the equilibrium value. Both these conditions require at least a fourth order term in Ψ. The resulting Langevin equations are then highly non-linear. Another common occurrence is that the system under study (and thus F) depends on more then one thermodynamic-like parameter. Thus an equation of motion for each of the other parameters that couple to the one under study needs to be included. Generally, these equations are not only coupled but non-linear. Such an approach has been used to study dendritic growth [9], eutectic crystallization [10], polymorphic crystallization [11], and also been used to analyze quenches in order-disorder systems [12]. Langevin equations are instructive from a pedagogic perspective. In particular, they highlight one of the advantages of IFS. Since it measures time constants as a function of wavevector, it can be said to directly measure the equation of motion of a system. If the equations are known, they can be compared to the data and if they are not, insight into their form is directly obtained. Since structure obviously influences the time constants of the system, measuring the time constants through the variations of $S(Q)$ should give the most insight. An advantage of X-ray measurements of IFS over light scattering is that it is easier to reach wavevectors corresponding to interesting features in the structure factor. The direct relation between diffraction and density fluctuations (Born approximation) is another advantage of X-rays over light scattering where avoiding multiple scattering can be difficult.

5. *X-RAY INTENSITY FLUCTUATION SPECTROSCOPY*

To demonstrate XIFS results, small angle X-ray scattering from latex spheres (R = 670 Å) in glycerol (volume fraction 28%) at T = –5 C will be presented. Figure 4 shows one quadrant of the small angle X-ray diffraction from the spheres [6].

---

6. This data was take at Beamline 8-ID of the Advanced Photon Source. For an incident beam size of 20 μm (horizontal) by 50 μm (vertical) and $\Delta E/E$ (FWHM) of 3×10^{-4} (Ge(111)) the beamline has an incident intensity of ~ 10^{-10} counts per second (cps). This setup gives $\beta \approx 0.12$.

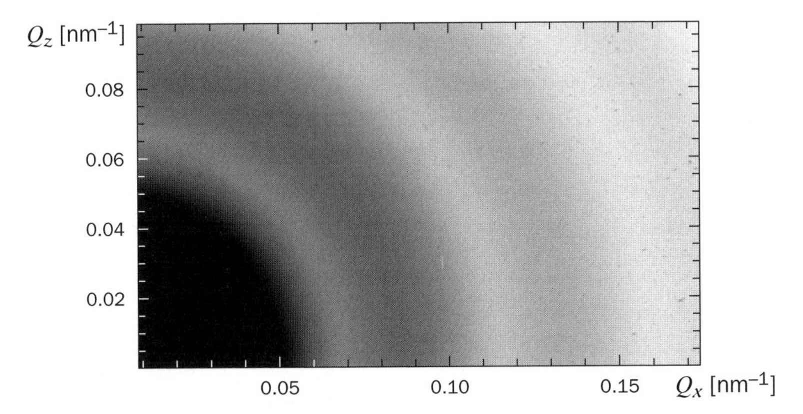

Figure 4 - Small-angle scattering pattern for latex spheres (R = 67 nm) in glycerol (volume fraction 28%) at T = –5°C. The isolated spots result from a low background of high energy X-rays.

The scattered intensity in figure 4 falls from about 100 cps per pixel at the peak (bottom left) to about 1 cps at the high edge Q of the detector (top right). Although a coherent beam was used to measure this pattern, no speckle structure is observed and the diffraction pattern looks like one from conventional incoherent source. This X-ray pattern corresponds to an exposure time of over 10 minutes and the Brownian motion of the particles during the exposure has lead to a temporal averaging of the scattering. In equilibrium $\langle S(Q,t) = S(Q)\rangle$ and which is the same as would be measured by incoherent scattering.

Typically, IFS consists in measuring the intensity in a single speckle as a function of time. To analyze such data one calculates the autocorrelation of this intensity and this is used to find the time scales. Using equation {19}, a normalized correlation function can be calculated. Using $g^{(2)}$ has the advantage of correcting for slow fluctuations in the incident beam. In figure 5 the normalized correlation functions for three wavevectors are shown. The correlation functions cover time scales from 30 ms to 300 s. As discussed below, the correlations have been averaged over pixels with constant $|\boldsymbol{Q}|$ to improve statistics. One sees that the correlation functions are exponential in time and that higher wave vectors give smaller time constants. Figure 6 shows the time constants *versus* Q. For this system, the approximations leading to equation {22} are valid in the limit of low wavevectors or for low volume fractions. The low Q limit of the figure gives a diffusion constant $D \approx 2\times10^{-16}$ m^2/s. The Einstein-Stokes relation gives

$$D_0 = k_b T / 6\pi\eta R \qquad \{25\}$$

T is the absolute temperature, R is the radius and η the viscosity.

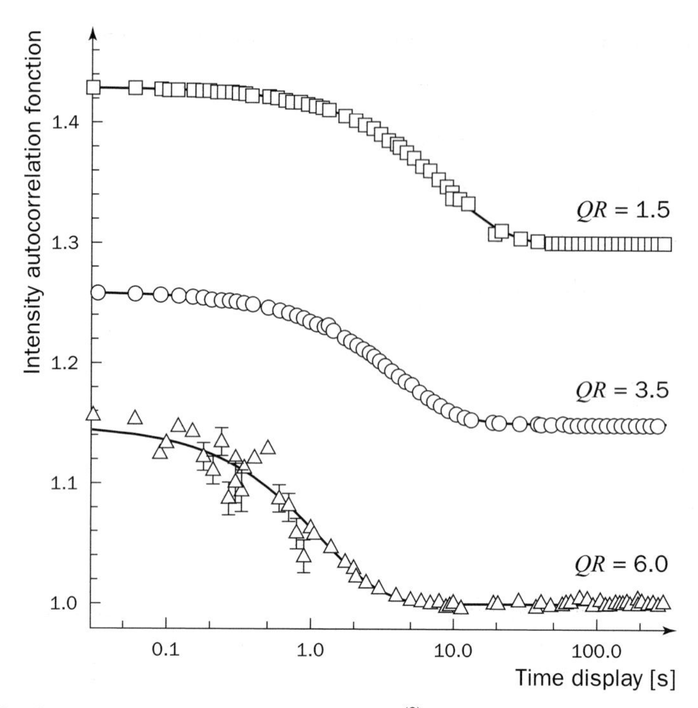

Figure 5 - Intensity autocorrelation function $g^{(2)}$ for latex spheres (R = 67 nm) in glycerol (volume fraction 28%) for QR = 1.5 (offset by 0.2), QR = 3.5 (offset by 0.1) and QR = 6.0. Lines show single exponential fits.

For this system $D_0 \approx 1.72 \times 10^{-16}$ m^2/s, which can be considered good agreement with D. Deviations from the Einstein-Stokes relation are expected for interacting systems and the effective diffusion constant becomes wavevector dependent reflecting the correlations seen in the structure factor. For example the bump in the This data of figure 6 occurs at a Q near that of the peak in the structure factor. This interesting physics is discussed in more detail in references **[13-15]**. This example clearly demonstrates what is involved in measuring XIFS and also that it can be used to obtain valuable information in many systems.

Since count rates are often low in XIFS experiments, it is worth taking some time to discuss how best to optimize experiments. A more complete error analysis is given in references **[16,17,7]**. It can be seen that larger values of β are preferred for measuring $g^{(2)}$ so the first choice is to use as brilliant an X-ray source as possible. From above, it can be seen that β can be controlled by a combination of the incident beam slits, the sample shape and the detection area. Although smaller dimensions of the incident beam lead to higher values of β they also give lower count rates and any benefits primarily cancel. Typically other noise sources exist and so one has to compromise between their effects at high contrast and too low count rates versus systematic errors at high count rates with low contrast. Using values of β in the range of 0.1 to 0.3 is what is typically used.

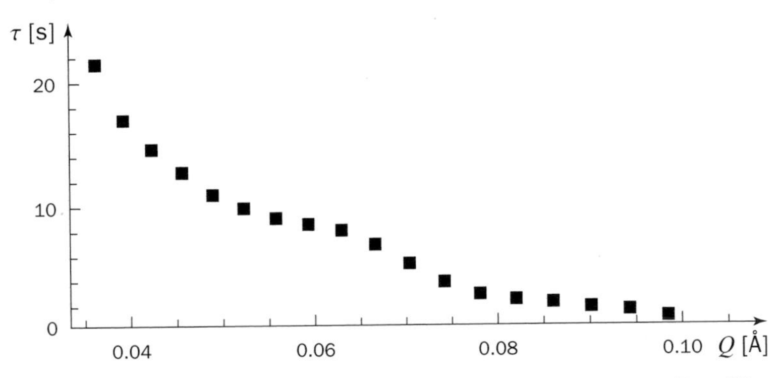

Figure 6 - Time constants *versus* wavevector for latex spheres (R = 67 nm) in glycerol (volume fraction 28%) at T = –5°C.

The accuracy of $g^{(2)}$ can be estimated as $1/\sqrt{N}$ where N is the number of correlation times used. As long as the signal to noise is at least one, this result is independent of how accurately one measures the intensity. Thus, the smallest correlation time measurable is roughly one count per correlation time and one needs about 1000 correlation times to get 3% accuracy. Measuring a single speckle as a function of time is the common method of measurement using light scattering. When time constants are nanoseconds to microseconds, as for typical liquids, it is easy to measure for millions of correlation times and obtain highly accurate correlation functions. When correlations persist for long times (seconds and longer) it becomes more difficult to ensure the stability needed to measure enough correlation times. For systems with isotropic scattering, the requirement for a large number of correlation times can be obtained by simultaneously measuring different angles of the same $|\boldsymbol{Q}|$ and therefore the same correlation function. The use of high resolution direct illumination charge coupled device (CCD) arrays makes this relatively straight forward and in such a setup thousands of detectors have the same $|\boldsymbol{Q}|$. This means that much shorter times can be measured by each detector. Typically, for XIFS the counts rates are low and systems are chosen to have longer time constants. Thus the use of array detectors has been emphasized from the start. It is interesting that CCD are now being used with light scattering as a means to study systems with much longer correlation times [18-19].

The error analysis [16] that leads to errors in $g^{(2)}$ going like $1/\sqrt{N}$ irrespective of the accuracy of the intensity measurements, can also be used to show that using exposure times much smaller than the correlation time leads to less accurate estimates. It can also be shown, that for equilibrium systems, averaging over times much longer than fast time constants does not distort the slow time constants. This has led to the development of the multiple tau algorithm, which averages over varying time windows giving improved estimates and providing a convenient method for covering a large range of times [16].

Just as the optimum counting time is around one correlation time, the optimum detector area is about one speckle size. For low count rates it is important to optimize both these parameters. This is particularly true for detectors, such as CCD, which have an electronic noise over and above the intrinsic noise statistics of the light.

It is clear that XIFS puts very stringent conditions on the detector. If the detector resolution is larger then the intrinsic speckle size, the speckle will be smeared and the apparent contrast will be reduced. Although this is analytically taken into account in the references showing details on how to calculate contrast, it is better not to have this as a limiting aspect. Matching the detection area to the speckle widths gives desired detector sizes of 10 μm to 50 μm. Measuring a large number of pixels with exposure times of milliseconds pushes the limits of commercial CCD electronics, which typically read a million pixels per second. The short exposure times implied by figure (5) where obtained by a "kinetics mode" in which X-rays illuminate a small region of the detector and the fast shift capabilities of a CCD are used to store several (say 16) of these regions on the chip. However, once the CCD is "full" a readout takes 1.6 s. To read 16 regions each exposed for 25 ms has a duty cycle of 4/1.6 = 25% and the number of pixels is only 1/16 of the full array. Faster area detectors would lead to an immediate improvement in XIFS capabilities.

Having made this diversion into details about performing XIFS measurements, this section will end with a review of some of the results reported in the literature. The relationship between kinetics and structure, as emphasized above, gives insight into the type of measurements where IFS can play an important role. Obvious examples are systems that have a microstructure, heterogeneous materials, polymers, non-Newtonian fluids, colloids and other systems where one expects kinetics to be different on different length scales. This bias is reinforced by this review. From the first, XIFS measurements have emphasized colloidal systems. This is in part because in the low concentration limit, these systems provide an ideal model system to demonstrate and test IFS techniques and in part because the interesting and challenging physics of what happens as the concentration is increased is difficult to access by light scattering [13,15, 21-23].

The similarity and contrast between X-ray and light IFS has been nicely demonstrated using a colloidal system in reference [20]. By measurements on index matched particles, a direct comparison between light scattering and X-ray scattering correlation functions shows the agreement between the two techniques. In contrast, by slightly changing the liquid, a very similar but non-indexed suspension is obtained. As stated in the introduction, this mismatched system shows strong multiple scattering effects for light scattering and but none for X-ray scattering. Polymers are another example of a class of systems in which equilibrium motions are not well understood. Although the scattering from a homogeneous sample of a polymers is probably too weak to allow XIFS measurements with current sources, the kinetics in a homopolymer blend can be seen by XIFS [24]. These measurements show non-diffusive relaxation at high wave vectors, consistent with compositional fluctuations of the conventional reptation model. Equilibrium fluctuations of other aspects of polymers systems are also being pursued. One study shows the dynamics in block co-polymer micellar systems [25]. Another demonstrates aging effects in filled polymer systems [19]. In this work, two systems are compared. One has an untreated filler of fumed silica well dispersed in poly(dimethyl siloxane) (PDMS) giving a viscous fluid suspension and a second has a similiar but treated silica to make it hydrophobic, resulting in a thixotropic liquid suspension. Both systems show an aging effect, which leads to a novel wavevector dependence in the time constants measured by IFS and an explanation in terms of flocculation of the silica aggregates is proposed. This article also compares both light and X-ray IFS. Measurement of the long time constants needed to help understand this phenomenon was made possible by the use area detectors for both techniques. Thermally excited fluctuations in interfaces were measured in references [26,27]. The strong Bragg scattering from the smectic layer spacing allowed the authors to measure the very short time constants (1 to 50 μs) associated with short wavelength modes in freely suspended thin films (.3 to 50 μs). A related, but slightly different effect, thermally excited capillary waves at the surface of glycerol surfaces is presented in reference [28]. By using glancing incident X-ray diffraction, the wavevector dependence of the time constants for overdamped capillary waves are shown to agree with theoretical predictions. -

Equilibrium measurements present only one class of experiments where IFS can provide useful information, extending XIFS to non-equilibrium systems is also under way. If a non-equilibrium system is in steady state, stationarity may still apply but typically the intensity-intensity correlations are not stationary and must depend on both time arguments. With the use of area detectors, correlation times can be estimated with many fewer time slices. One suggestion for two-time correlation functions is used to theoretically study the kinetics of phase transitions in references [29, 30] and experimentally demonstrated in measurements of phase separation [5,31]. These ideas should be useful in other types of non-equilibrium systems. In general non-equilibrium problems are much more difficult to deal with than equilibrium problems. However equation {20} is most often a reasonable starting point and similarly IFS should become a powerful technique for their study.

6. CONCLUSIONS

XIFS is in the process of becoming a routine tool in the growing arsenal of diffraction techniques. Synchrotrons around the world are designing beam lines dedicated to this technique. This chapter has tried to emphasize the underlying features that are needed in comparison to conventional diffraction techniques. XIFS only uses one aspect of coherence, the fluctuation in time of the intensity. Mathematically, it is known that using coherent light allows one to invert the diffraction pattern and get a direct real space image. For instance, it has recently been experimentally demonstrated using soft X-rays, that a speckle pattern similar to that seen in figure 1b can be inverted to recover the relative positions of the pinholes [32]. Another example, inverts coherent diffraction patterns to get the shape of small particles [33] and the use of these ideas for solving the large molecular structures is explored in reference [34].

To conclude, the exploitation of coherence is just beginning, obviously much work remains to be done.

ACKNOWLEDGMENTS

Use of the APS was supported by the U.S. DOE (BES, OER) under Contract No. W-31-109-Eng-38.

REFERENCES

[1] L. MANDEL & E. WOLF - *Optical Coherence and Quantum Optics*, Cambridge University Press, Cambridge UK (1995)

[2] M. SUTTON - *In "Third Generation Hard X-ray Synchrotron Radiation Sources: Source Properties, Optics and Experimental Techniques"*, D. MILLS ed., John Wiley and Sons, Inc, New York (2002)

[3] P.N. PUSEY - *In "Photon Correlation Spectroscopy and Velocimetry"*, H.Z. CUMMING & E.R. PIKE eds, Plenum, New York, 45 (1974)

[4] S.K. SINHA, M. TOLAN & A. GIBAUD - *Phys. Rev. B* **57,** 2740 (1998)

[5] A. MALIK, A.R. SANDY, L.B. LURIO, G.B. STEPHENSON, S.G.J. MOCHRIE, I. MCNULTY & M. SUTTON - *Phys. Rev. Lett.* **81**, 5832 (1998)

[6] A.R. SANDY, L.B. LURIO, S.G.J. MOCHRIE, A.MALIK, G.B. STEPHENSON, J.F. PELLETIER & M. SUTTON - *J. Synchrotron Radiation* **6**, 1174 (1999)

[7] D. LUMMA, L.B. LURIO, S.G.J. MOCHRIE & M. SUTTON - *Rev. Sci. Instr.* **71**, 3274 (2000)

[8] One of the simplest descriptions of this is the discussion leading up to Eq. 3.5.22 (page 132) *in* P.M. CHAIKIN & T.C. LUBENSKY - *Principles of condensed matter physics*, Cambridge University Press, Cambridge, UK (1995)

[9] R. KOBAYASHI - *Physica D* **63**, 410 (1993)

[10] K.R. ELDER, F. DROLET, J.M. KOSTERLITZ & M. GRANT - *Phys. Rev. Lett.* **72**, 677 (1994)

[11] B. MORIN, K.R. ELDER, M. SUTTON & M. GRANT - *Phys. Rev. Lett.* **75**, 2156 (1995)

[12] B. MORIN, K.R. ELDER, & M. GRANT - *Phys. Rev. B* **47**, 2487 (1993)

[13] L.B. LURIO, D. LUMMA, P. FALUS, M.A. BORTHWICK, S.G.J. MOCHRIE, J.F. PELLETIER, M. SUTTON, L. REGAN, A. MALIK & G.B. STEPHENSON - *Phys. Rev. Lett.* **84,** 785 (2000)

[14] L.B.LURIO, D. LUMMA, M.A. BORTHWICK, P. FALUS, S.G.J. MOCHRIE, J.F. PELLETIER & M. SUTTON - *Synchrotron Radiation News* **13**, 28 (2000)

[15] D. LUMMA, L.B. LURIO, M.A. BORTHWICK, P. FALUS & S.G.J. MOCHRIE - *Phys. Rev. E* **62**, 8258 (2000)

[16] K. SCHATZEL - *In "Dynamic Light Scattering: The method and some applications"*, W. BROWN ed., Clarendon Press, Oxford (1993)

[17] R. RAINER - *In "Dynamic Light Scattering: The method and some applications"*, W. BROWN ed., Clarendon Press, Oxford (1993)

[18] L. CIPELLETTI & D.A. WEITZ - *Rev. Sci Instr.* **70**, 3214 (1999)

[19] E. GEISSLER, A.M. HECHT, C. ROCHAS, F. BLEY, F. LIVET & M. SUTTON - *Phys. Rev. E* **62**, 8308 (2000)

[20] D.O. RIESE, W.L. VOS, G.H. WEGDAM, F.J. POELWIJK, D.A. ABERNATHY & G. GRÜBEL - *Phys. Rev. E* **61**, 1676 (2000)

[21] S.B. DIERKER, R. PINDAK, R.M. FLEMING, I.K. ROBINSON & L. BERMAN - *Phys. Rev. Lett.* **75**, 449 (1995)

[22] T. THURN-ALBRECHT, W. STEFFEN, A. PATKOWSKI, G. MEIER, E.W. FISCHER, G. GRÜBEL & D.L. ABERNATHY - *Phys. Rev. Lett.* **77**, 5437 (1996)

[23] D.O. RIESE, G.H. WEGDAM, W.L. VOS, R. SPRIK, D. FENISTEIN, J.H.H. BONGAERTS & G. GRÜBEL - *Phys. Rev. Lett.* **85**, 5460 (2000)

[24] D. LUMMA, M.A. BORTHWICK, P. FALUS, L.B. LURIO & S.G.J. MOCHRIE - *Phys. Rev. Lett.* **86**, 2042 (2001)

[25] S.G.J. MOCHRIE, A.M. MAYES, A.R. SANDY, M. SUTTON, S. BRAUER, G.B. STEPHENSON, D.L. ABERNATHY & G. GRÜBEL - *Phys. Rev. Lett.* **78**, 1275 (1997)

[26] A.C. PRICE, L.B. SORENSON, S.D. KEVAN, J. TONER, A. PONIEWIERSKI & R. HOYLE - *Phys. Rev. Lett.* **82**, 755 (1999)

[27] A. FERA, I.P. DOLBNYA, G. GRÜBEL, H.G. MULLER, B.I. OSTROVSKI, A.N. SHALAGINOV & W.H. DE JEU - *Phys. Rev. Lett.* **85**, 2316 (2000)

[28] T. SEYDEL, A. MADSEN, M. TOLAN, G. GRÜBEL & W. PRESS - *Phys. Rev. B* **63**, 073409 (2001)

[29] G. BROWN, P.A. RIKVOLD, M. SUTTON & M. GRANT - *Phys. Rev. E* **56**, 6601 (1997)

[30] G. BROWN, P.A. RIKVOLD, M. SUTTON & M. GRANT - *Phys. Rev. E* **60**, 5151 (1999)

[31] F. LIVET, F. BLEY, R. CAUDRON, D. ABERNATHY, C. DETLEFS, G. GRÜBEL & M. SUTTON - *Phys. Rev. E* **63**, 036108 (2000)

[32] J. MIAO, P. CHARALAMBOUS, J. KIRZ & D. SAYRE - *Nature* **400,** 342 (1999)

[33] I.K. ROBINSON, I.A. VARTANYANTS, G.J. WILLIAMS, M.A. PFEIFER & J.A. PITNEY - *Phys. Rev. Lett.* **87**, 195505 (2001)

[34] J. MIAO, K.O. HODGSON, & D. SAYRE - *Proc. Nat. Acad. Sci.* **98**, 6641 (2002)

10

VIBRATIONAL SPECTROSCOPY AT SURFACES AND INTERFACES USING SYNCHROTRON SOURCES AND FREE ELECTRON LASERS

A. TADJEDDINE
LURE, Université Paris-Sud, Orsay, France

P. DUMAS
Synchrotron SOLEIL, Gif-sur-Yvette, France

1. INTRODUCTION

A detailed description of the inter-atomic interaction determining the structure and dynamics at clean and adsorbate-covered surfaces is one of the most fundamental topics in surface science. It is of paramount importance in any attempt to understand the many microscopic scale phenomena involving interfacial processes such as adsorption, desorption, diffusion, friction, lubrication, epitaxial growth, catalysis and electronic properties. Among key understanding are the bonding characteristics between molecules and surfaces, the structure of the adsorbed layer, and the dynamical aspect of the interaction between the adsorbate and the substrate. Such information is now attainable by using a variety of different experimental tools with sensitivities on the order of a fraction of a monolayer (~ 10^{14} atoms/cm^2).

Vibrational spectroscopy has been widely employed for the last 15 years for the determination of surface interaction and reactivity. It relies, primarily, on the fact that any molecule can be easily identified through its vibrational signature (often referred as its "fingerprint"). Indeed, surface chemical analysis has benefited greatly from the determination of the internal vibrational structure of adsorbate species. Most recently, these studies have been extended, using non-linear vibrational techniques, to encompass high pressure environments and even liquid

F. Hippert et al. (eds.), Neutron and X-ray Spectroscopy, 319–358.

interfacial systems, allowing direct observation of chemical reaction pathways in environments of practical interest.

Notably, the majority of vibrational studies to date have been confined to the mid-infrared region ($\geq 600\ cm^{-1}$), due to limitations in source intensity, detector sensitivity or the optical response of the substrate and other optical components (e.g. salt windows). Consequently, the low frequency substrate-adsorbate modes that contain a wealth of information about the potential energy surface of the molecule bound to the surface have not been studied in detail. The effective mass involved in the motion of the so-called "external" modes is usually significantly larger than for the intra-molecular vibrations, and the binding energy is usually less than the chemical bonds within the molecule, so that the frequency of these modes is generally much smaller than that of the internal modes, and typically spans the 20-1000 cm^{-1} spectral region.

Optical methods are, furthermore, of non-destructive character, and are capable of *in situ* remote sensing with high temporal and spatial resolution. One of the particularly unique aspects of optical techniques is the possibility of probing buried interfaces.

Recent advances in laser technology, and detectors [in particular the development of detector arrays of Charge-Coupled Devices (CCD) for Sum-Frequency Generation (SFG)] have allowed such techniques to reach the required sensitivity for surface studies. Similarly, for InfraRed Absorption Spectroscopy (IRAS), the use of synchrotron infrared sources and improvements in Fourier transform infrared (IR) instrumentation, have also afforded increased potentialities. As a result of these experimental advances, complex surfaces and interfaces (often close to those found in technologically relevant systems), can now be investigated over a wide spectral range, yielding unprecedented insight into the formation and thermal evolution of these systems.

This article is intended to focus on one hand on the description of laser-based and synchrotron vibrational-based spectroscopy techniques applied to surface and interface analysis, and on the other hand, on selected examples demonstrating the new scientific opportunities offered by these new sources.

2. *INFRARED SYNCHROTRON SOURCE AND FREE ELECTRON LASERS*

2.1. *INFRARED SYNCHROTRON SOURCES*

2.1.1. *PRINCIPLES*

Electron storage rings (or synchrotrons) use magnetic fields to bend electrons into a closed orbit [1]. Synchrotron radiation is produced at each of these "bending" magnets. Infrared radiation is generated by electrons travelling at relativistic velocities, either inside a curved path through a constant magnetic field (bending magnet radiation [2]) or by longitudinal acceleration or deceleration when leaving or entering a magnetic section (edge radiation) [3,4]. The emitted radiation spans an extremely broad spectral domain, extending from the X-ray to the very far infrared region.

The effective source size is quite small, and can be on the order of ~ 100 μm. Furthermore, the light is emitted into a narrow range of angles, as illustrated in figure 1. This figure shows the intensity distribution (fig. 1a and 1b) at a wavelength of 10 μm (1000 cm^{-1}), for a storage ring with an energy of 2.75 GeV, a current of 400 mA, and an extraction geometry of 20 mrad (vertical) × 30 mrad (horizontal). It can be seen from figures 1a and 1b that the edge radiation infrared photons are emitted in a narrower angular distribution compared to that from a bending magnet. In figure 1c, the calculated power (in watts per 0.1% bandwidth) for the two types of source is shown. They are quite close, but figure 1b shows that the total power extracted in the case of bending magnet radiation can be slightly increased by enlarging the horizontal and vertical apertures (which often requires considerable modification of the dipole vessel on existing facilities). Calculations reported in figure 1 have been performed using the SRW computer code from Chubar and Elleaume [5]. However, in the case of bending magnet radiation, a simplified formula can be used to calculate the power emitted by a synchrotron source as well as the brightness [6]

$$P(\lambda) = 4.38\times10^4\, I\, \theta\, bw\, (\rho/\lambda)^{1/3} \qquad \{1\}$$

where I is the stored current (in amperes), θ is the horizontal collection angle, bw the bandwidth (in %), λ the wavelength in microns, and ρ the radius of the ring considered (in meters). $P(\lambda)$ is expressed in photons per second, but can be converted to watts by dividing by $5.04\times10^{18}\,\lambda$.

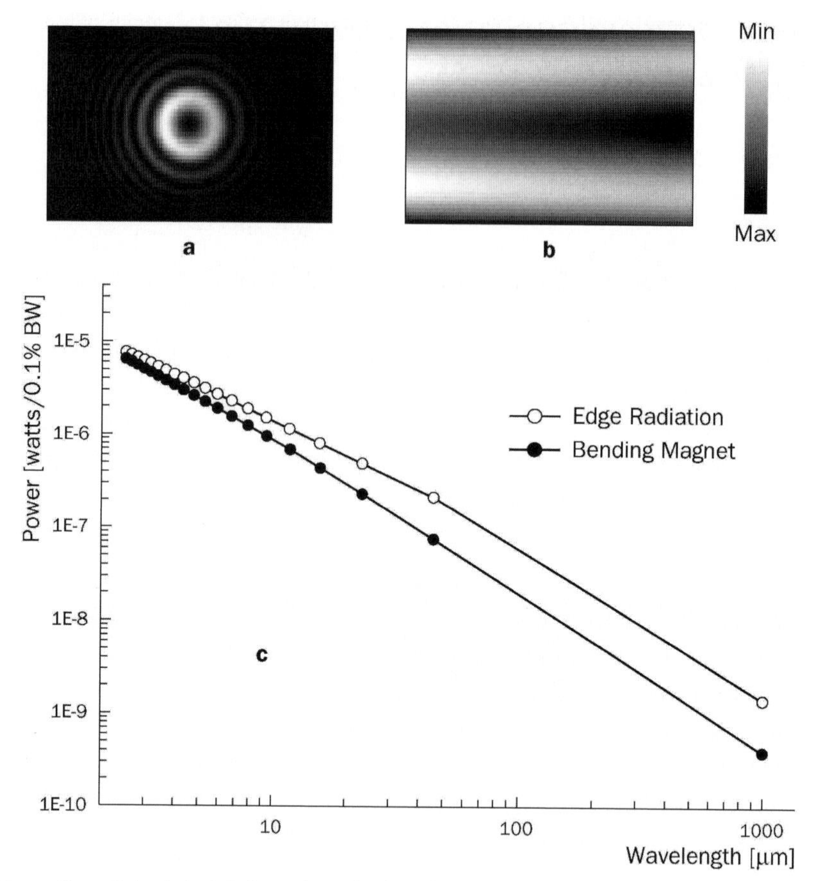

Figure 1 - Calculated infrared emission from a synchrotron source, using the SRW code [5] for the following parameters: electron energy = 2.75 GeV, electron current = 400 mA, magnetic field = 1.56 T, straight section length = 7 meters, distance to point source = 2.3 meters, collection angles: 30 mrad (horizontal) × 20 mrad (vertical). **a** - Intensity profile for an edge radiation. **b** - intensity profile for the bending magnet, collected at 4° deviation. **c** - power emitted by the two sources at a function of wavelength (expressed in watts per 0.1% bandwidth). Open circles for edge radiation, black dots for bending magnet radiation.

Accordingly, the approximate infrared brightness of a storage ring is given by

$$B(\lambda) = 75\, I\, bw/\lambda^3 \qquad \text{in W/mm}^2\text{/sr} \tag{2}$$

Thus, the light has a very high brightness, which is defined as the photon flux or power emitted per source area and solid angle. When compared with a conventional black body IR source, the brightness advantage is two to three orders of magnitude greater [2].

Figure 2 compares the calculated brightness of a storage ring (identical parameters as for figure 1) and a thermal source at 2000 K. As can be seen, a brightness advantage of roughly three orders of magnitude is achieved with the synchrotron source.

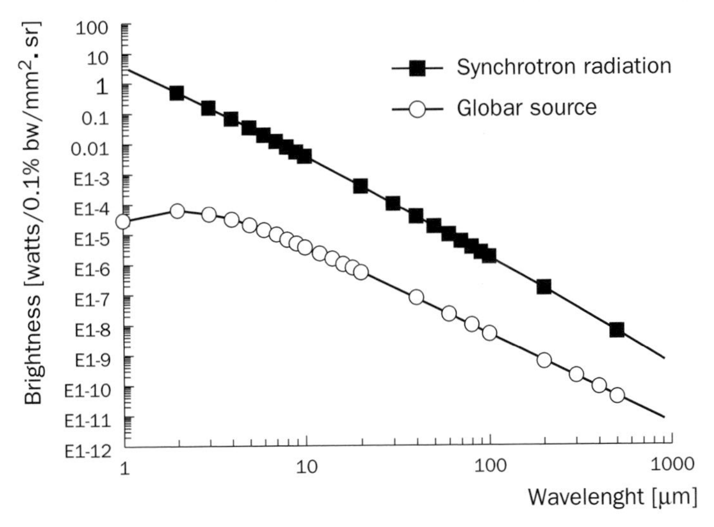

Figure 2 - Calculated brightness for a black body at 2000 K and bending magnet infrared emission. Electron energy = 2.75 GeV, electron current = 400 mA, bending magnet radius = 5.36 meters, collection angles: 50 mrad (horizontal) × 17 mrad (vertical).

High brightness is a key issue for any measurement with a limited "throughput", such as surface studies (and also in microspectroscopy [7]).

One characteristic feature of synchrotron light is that it is pulsed, i.e. the electrons do not form a continuous distribution around the orbit, but instead they travel in "bunches". Depending upon the storage ring, these bunches have lengths on the order of 1-10 cm, and this results in pulses of light on a sub-nanosecond time scale. As a matter of fact, time-resolved studies have been undertaken with a pulsed synchrotron IR beam [8,9]. In addition, synchrotron IR light is highly polarized when bending magnet is used as the infrared emission source. One remarkable feature is the production of coherent radiation in the far-infrared region [10]. The intensity of the emitted radiation scales with the square of the number of electrons, producing a highly intense beam. This coherent emission has promising applications, especially in Surface Science.

2.1.2. EXTRACTION OF THE IR BEAM FROM THE SYNCHROTRON RING

The most important experimental details are of an engineering nature and involve dealing with the large opening angles at long wavelengths. Practical issues make it difficult to extract the infrared, particularly from storage rings of large radius. The "natural opening angle", while less than a milliradian at the critical wavelength $\lambda_C = (5.59\,.\,\rho)/E^3$ (half of the power in the synchrotron radiation spectrum is radiated above λ_C, and half below, ρ is the electron bending radius in meters and E has units of GeV), and γ is related to the electron energy $\gamma = 1957\,E$ [GeV], increases to several tens of milliradians in the infrared depending on the radius of the ring as follows

$$\theta_{nat}\,[\text{radians}] = 1.66\left(\frac{\lambda}{\rho}\right)^{1/3} \qquad \{3\}$$

where λ is the wavelength. θ_{nat} is the angle required to transmit 90% of the emitted light at a given wavelength. Note that although this formula shows that the radiation opening angles are smaller for larger rings, engineering, thermal problems at the first optical element, and radiation shielding difficulties make it more practical to use smaller rings.

Infrared beam lines differ from VUV (Vacuum Ultra Violet) lines in that ultrahigh vacuum is not required except for beam extraction. Pressures of 100 millitorr are sufficient to eliminate problems due to water vapour and carbon dioxide. Ideally the infrared beam lines are separated from the storage ring vacuum chambers by wedged diamond windows which are usually placed near a mirror focus. Only one beam line, SA5 at SUPERACO in Orsay, France, uses thick ZnSe windows for extraction in the mid-IR range [11]. The emerging beams are collimated or re-focussed into a Michelson interferometer.

2.1.3. FOURIER TRANSFORM INTERFEROMETER

Most interferometers used for IR spectroscopy are based on the two-beam interferometer, originally developed by Michelson in 1891 [12]. The principle of such interferometers is based on the division of the incoming beam into two equivalent beams, using a beamsplitter. Each beam trajectory suffers a reflection on either a fixed mirror, or a movable mirror, and is recombined at the beamsplitter level. The moving mirror introduces a phase difference between the two recombined beams (fig. 3).

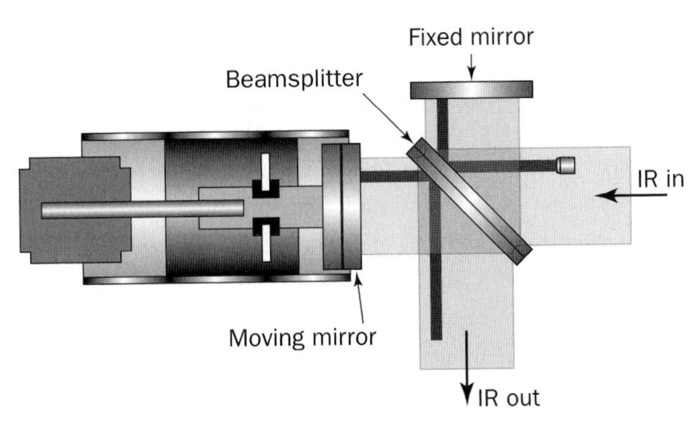

Figure 3 - Schematic of a Michelson interferometer, as mostly used in Fourier Transform infrared spectroscopy.

Interferometry exhibits several advantages over dispersive spectroscopy, namely the simultaneous detection of the complete spectral distribution, the use of a large solid angle (precluded when using entrance and exit slits in dispersive instruments). Based on the simultaneous use of a wavelength reference (typically a He-Ne laser) the position of the moving mirror is known precisely. Finally, the only signal that is contributing to the spectrum is the modulated signal, meaning that any background signal (parasitic IR signal) will not contribute to the Fourier transform signal. This property is important, especially in the far infrared region, where there is a marked contribution from the 300 K emission of the spectrometer itself!

2.2. *NON LINEAR SURFACE SPECTROSCOPY WITH CONVENTIONAL LASERS AND FREE ELECTRON LASERS*

Optical techniques are the only probes which can be utilized in the electrochemical environment. However, linear optical techniques suffer from the undesired contribution from the bulk material, the interface contribution to the overall signal being several orders of magnitude weaker than that of the bulk.

On the other hand, second-order non-linear optical processes, such as second harmonic (SHG) and sum (SFG) and difference (DFG) frequency generation, are forbidden, under the electric-dipole approximation, in the bulk of a medium with inversion symmetry. SHG and SFG are then generated only at the surface of such media where the symmetry is broken and appear to be highly surface-specific for interfaces between centro-symmetric media [13,14]. As a result, submonolayers of species adsorbed on surfaces can be readily detected, for all interfaces accessible by light, including the electrochemical interface. They have inherently high spectral, temporal and spatial resolution and constitute a non-destructive remote sensor of surfaces and interfaces.

The non-linear optical technique of IR-visible sum (difference) frequency generation (SFG) has been established as an efficient probe of the vibrational properties of interfaces, as demonstrated first in the gas environment [15,16] and later, in 1990, in the electrochemical one. The development of high power infrared free electron lasers [FELIX (Free Electron Laser for Infrared eXperiments, Netherlands), CLIO (Centre Laser Infrarouge d'Orsay, France)], tuneable in a wide spectral range, is stimulating an intense research activity in this field.

2.2.1. PRINCIPLE OF SFG

SFG is a second order non-linear optical process in which two incident laser beams, one tuneable infrared of frequency ω_1 and the other visible at fixed frequency ω_2, interact and generate a reflected beam at the Sum Frequency $\omega = \omega_1 + \omega_2$ in the visible spectrum [17]. In the dipolar approximation and for media with inversion symmetry where there is no bulk contribution, the second order polarizability $P_i(\omega_s)$ originates from the interface where the symmetry is broken

$$P_i(\omega_s = \omega_1 + \omega_2) = \chi_{ijk}^{s(2)} E_j(\omega_1) E_k(\omega_2) \qquad \{4\}$$

$\chi^{s(2)}$ is a third rank second order surface susceptibility tensor, E_j and E_k are the components along the j and k axis of the IR and visible electric field amplitudes, respectively. SFG photons are emitted in a coherent direction defined by

$$\omega_s\, n(\omega_s) \sin\theta_s = \omega_1\, n(\omega_1) \sin\theta_1 + \omega_2\, n(\omega_2) \sin\theta_2 \qquad \{5\}$$

θ_1, θ_2 the angle of incidence of the IR and visible laser beams, respectively and θ_s the angle of emission of the SFG beam (fig. 4) and n the refractive index of the electrolyte. The number of detected SFG photons is expressed as

$$S \propto \frac{\left| L_s L_1 L_2 \chi^{s(2)} \right|^2 U_1 U_2}{AT} \qquad \{6\}$$

where L_s, L_1 and L_2 account for the Fresnel factors and beam polarizations of the SFG, the IR and visible beams, respectively. U_1, U_2, A and T are the overlapping IR and visible pulse energy, the irradiated area and the overlapping pulse duration, respectively.

Equation {3} shows that for a given pulse energy, the higher is the number of detected SFG photons, the smaller are the overlapping area and the pulse duration.

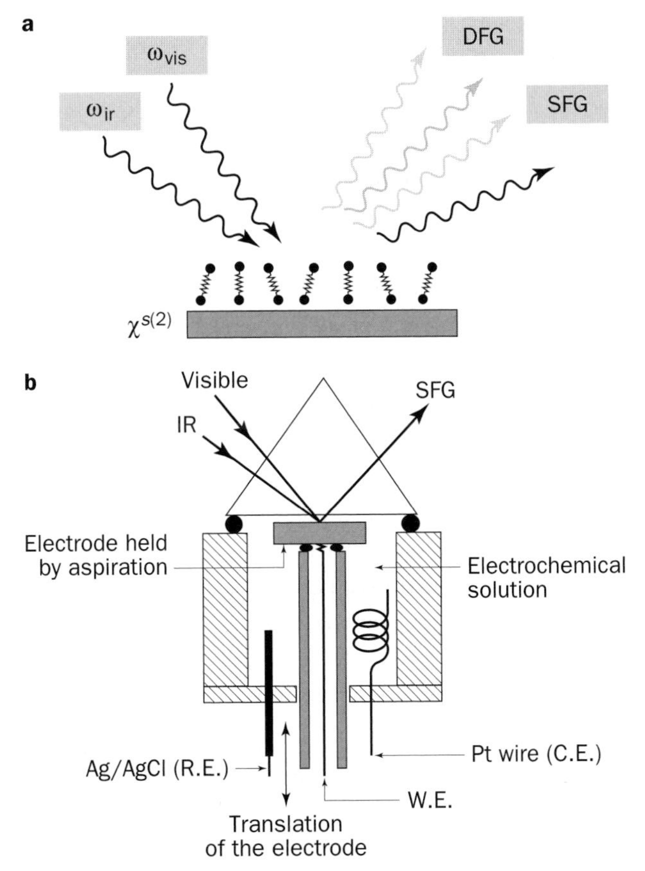

Figure 4 - Principle of Sum Frequency Generation (SFG). **a** - Visible and infrared beam impinge the surface, and a SFG beam is generated, with a precisely defined orientation which is a function of the angle of incidence of the two incoming beams. **b** - Typical spectro-electrochemical cell for SFG experiment.

$\chi^{s(2)}$ is a 27 component tensor (only a few of them are non-zero and independent) containing the information on the ordering and the structure of the adsorbate. Indeed $\chi^{s(2)}$ is enhanced if one of the three waves (IR at frequency ω_1, visible at ω_2 and SFG at ω_s) is in resonance with either an adsorbate vibrational or electronic level. Close to a resonance, $\chi^{s(2)}$ can be modelled as

$$\chi^{s(2)}(\omega) = \chi_{NR}^{s(2)}(\omega) + \chi_{R}^{s(2)}(\omega) \qquad \{7\}$$

where $\chi_{NR}^{s(2)}(\omega)$ is a non-resonant term accounting for all slowly varying (with the IR wavenumber) contributions to $\chi^{s(2)}$.

The resonant term originates from an adsorbate vibration, that must be both IR and Raman active and can be modelled by

$$\chi_R^{s(2)}(\omega) = \sum_v \frac{A_v}{\omega_v - \omega_1 + i\Gamma_v} \qquad \{8\}$$

where A_v, ω_v and Γ_v are the strength, the resonance frequency and the damping constant of the vth mode, respectively. Far from any electronic interband transition in the substrate, the resonant term is dominant and carries information on the vibrational properties of the adsorbate. Finally, it must be noted that unlike IR absorption spectroscopy, equation {8} shows that $\chi_{NR}^{s(2)}$ and adjacent adsorbate vibrational bands, characterized by complex values of A_v, will interfere. Such phase effects carry information on the relative orientation of the bands [18].

2.2.2. *EXPERIMENTAL SFG SET-UP*

An SFG experiment requires two synchronized visible and IR pulsed beams. Our set-up (fig. 5) [19,20] makes use of an actively mode-locked frequency doubled Nd-YAG laser synchronized with the Free Electron Laser (FEL) CLIO. The 2xYAG and IR CLIO-FEL micropulse repetition rate is 62 MHz. They are emitted in 10 μs bunches at 25 Hz. The doubled-YAG and IR CLIO-FEL micropulse duration are 70 ps and 3 ps, respectively. Both laser beam powers arriving on the sample are adjusted at 150 mW. During SFG spectroscopy, the spectral power of the CLIO-FEL is monitored in real-time. The green and infrared beams are focussed on the sample with an angle of 55° and 65° to the surface normal, respectively. Both spot sizes are about $500 \times 1000\ \mu m^2$. The SFG beam is geometrically and spectrally filtered from the green before its detection by a photomultiplier. Normalization to a simultaneously acquired SFG reference signal gives the absolute SFG yield.

2.2.3. *THE CLIO-FEL INFRARED LASER [21,22]*

The FEL offers the advantage over conventional lasers that it is broadly tunable and powerful in a wide IR spectral range (5 to 100 μm).

The gain medium of an FEL is a high energy electron beam (20 to 50 MeV) passing through a transverse magnetic field of an undulator. The relative degree of coherence of the light produced is due to the periodicity of the undulator providing a higher spectral brilliance than ordinary synchrotron radiation. In addition, under certain conditions, the interaction in the undulator of the emitted radiation with the electron beam leads to light amplification [22]. This effect, when operating within an optical cavity, is the principle of the FEL.

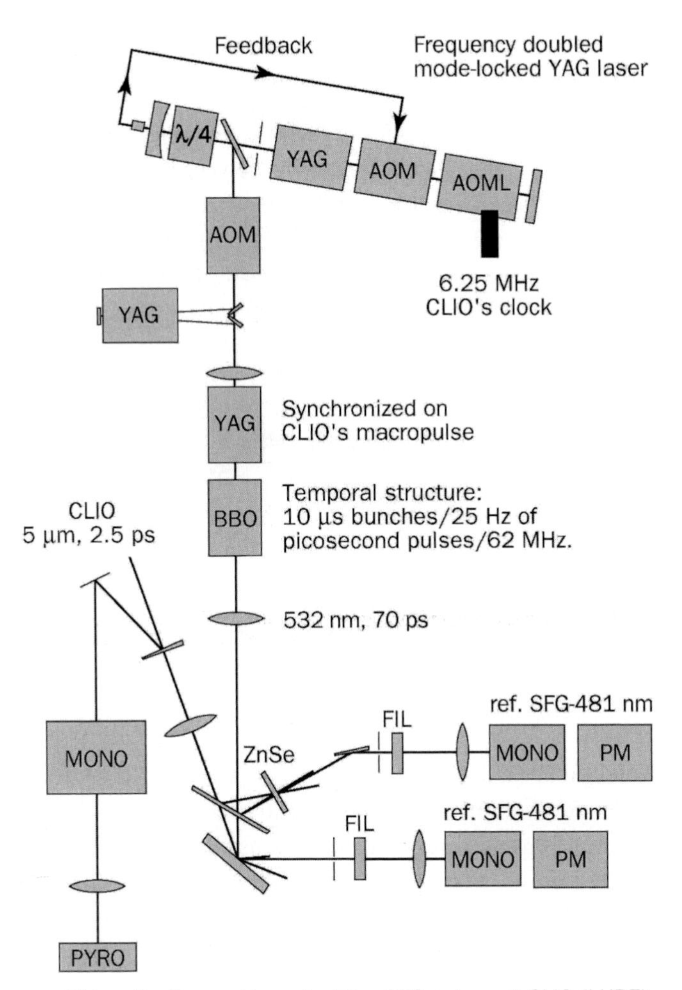

Figure 5 - General layout of the SFG set up at CLIO (LURE).

The laser radiation is emitted by the FEL at the "resonance wavelength" λ_R

$$\lambda_R = \lambda_0 / 2\gamma^2 (1 + K^2/2) \qquad \{9\}$$

where λ_0 is the magnetic period of the undulator, γmc^2 is the electron beam energy, and $K = eB\lambda_0 / 2\pi mc$ is the deflection parameter proportional to the magnetic field B of the undulator. In the case of CLIO, the tuning of the laser wavelength is controlled continuously by the user, by varying the undulator gap, allowing easy use of the laser for spectroscopic measurements. The main characteristic of this kind of laser is that the gain medium is intrinsically broadband. The lasing range is limited only by the energies that can be reached by the accelerator, while attaining sufficiently good beam properties, and by the diffraction losses due to the finite size of the apertures in the optical cavity.

The temporal and spectral characteristics of the FEL are determined by the electron beam properties. The peak and average power can be very large as there is no limit, in principle, to the number of electrons that can be used in the process. CLIO is a user facility operating in the mid-infrared (3-100 μm). It has been designed to produce a high performance and tuneable laser beam. It uses a linear accelerator operated between 20 and 60 MeV. Figure 6 displays a general layout of the FEL.

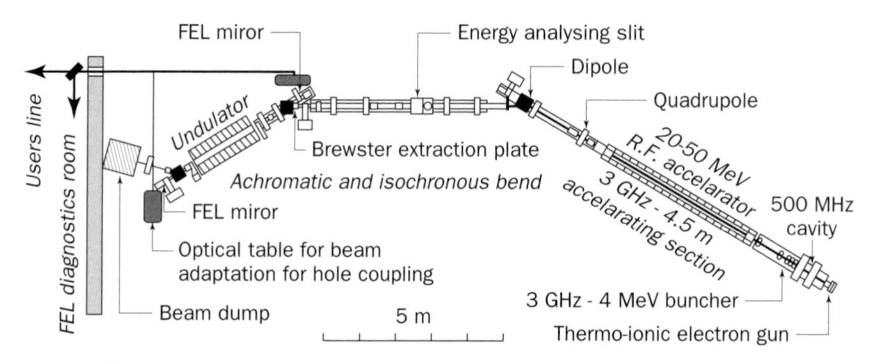

Figure 6 - General layout of the Free Electron Laser CLIO, at LURE.

The electron gun has the unique property to deliver pulses whose separation can be adjusted between 32 ns (optical cavity roundtrip time) and 4 ns within the macropulse. The macropulse repetition rate is adjustable between 1 and 50 Hz. This feature is very useful to synchronize an external laser with the FEL since typical optical cavities of mode-locked lasers are a few meters long. Another interesting feature of CLIO is the versatility of the duration of the laser micropulses within the range 0.5 to 6 ps. This allows users to choose the laser peak power and the relevant spectral width. The repetition rate of the machine is usually the combination 16 ns / 25 Hz, which matches the Nd/YAG laser used for SFG measurements.

The laser performance with respect to wavelength is displayed on figure 7. The lowest wavelength is determined by the highest energy attainable. The production of radiation of wavelength shorter than 3 μm appears quite uninteresting, as this spectral range is easily covered by OPO (Optical Paramagnetic Oscillator) based benchtop lasers [23]. The longest wavelength is determined by the diffraction losses in the vacuum chamber, the minimum accelerator energy and the transmission of the optical cavity output window. With a diamond window, the upper wavelength, limited by diffraction, was found to be approximately 75 μm. The typical average power (at the repetition rate of 62.5 MHz for the micropulses and 25 Hz for the macropulses) is 1 W, while users have typically several hundred mW depending on experimental conditions: repetition rates, user station, working wavelength range (residual absorption of H_2O and CO_2).

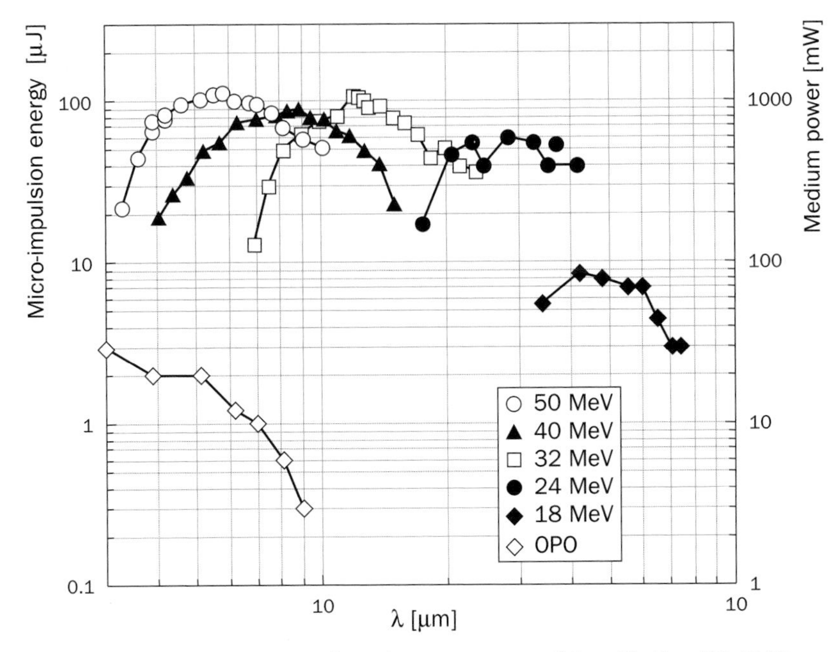

Figure 7 - Frequency range and peak power accessible with the FEL-CLIO, as a function of electron energy.

3. *EXAMPLES OF APPLICATIONS IN SURFACE SCIENCE*

3.1. SYNCHROTRON INFRARED SPECTROSCOPY AT SURFACES

All the experiments reported below have been achieved at the U4IR beam line, National Synchrotron Light Source (NSLS), Brookhaven National laboratory, in collaboration with G.P. Williams [17]. For far-infrared studies in surface science, one is concerned with signal to noise and a lot of attention has been paid to this issue. For the infrared region, the limiting noise is given by the expression

$$\%N = \frac{100\mathrm{NEP}}{\phi(\nu)\Delta\nu\varepsilon t^{1/2}\xi} \qquad \{10\}$$

where NEP is the detector noise equivalent power, $\phi(\nu)$ the brightness,

$\Delta\nu$ the bandwidth, ε the experiment's throughput, t the measuring time interval and ξ the optical efficiency. Clearly there is an upper bound on ε, and for surface science (as well as for microscopy) it can be quite small. Therefore, even with the best detectors, superior signal to noise is possible only if the source *brightness* (sometimes called spectral brilliance = power into a given spectral bandwidth, per unit area-angle product, and thus is in units of watts/mm^2 . steradian). Since the synchrotron source brightness is 1000 times greater than for thermal sources, eq. {10} tells us that, all other things being equal, to attain similar quality spectra with the latter, the data collection time would need to be a million times longer, i.e. a 1 second measurement would take 11 days! To illustrate the effects of a brighter source, figure 8 shows the calculated signal for a sample with a throughput of 5×10^{-2} mm^2 sr. This is used as a typical example of a brightness limited experiment, and is in fact equivalent to a 250 microns diameter sample illuminated at $f/1$, or the grazing incidence surface science experiments discussed below.

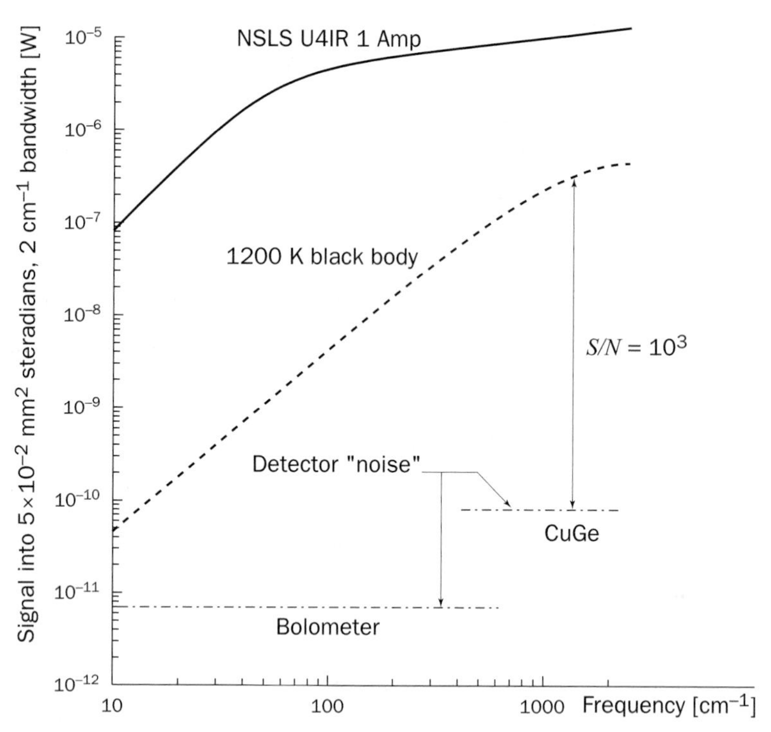

Figure 8 - Calculated signals available at a detector for an infrared experiment, if the sample limits the throughput. This is equivalent to a 20 mm $\times$ 5 mm sample illuminated at an angle of incidence of 87° with $f/10$ optics. Typical values of detector noise are also presented so that the theoretical signal to noise can be estimated.

Also shown on the plot are the actual measured noise levels of two of the U4IR detectors. The theoretical signal to noise advantage is immediately evident and from this one would expect to obtain S/N ratios of 10^5 with the synchrotron, compared to 10-100 with the globar. In practice it is found that this is not fully achieved due to instabilities in the electron orbits in the synchrotron. These instabilities are of several kinds – long term drifts probably due to changes in temperature as the storage ring current decays, short term fluctuations in the 10-100 Hz region due to mechanical vibrations and power supplies, and radio frequency (rf) side-bands in the kHz region from the accelerating cavity. Standard rapid-scan FTIR (Fourier Transform InfraRed) spectrometers and detectors are sensitive to noise in the audio range of frequencies, so reducing the beam instabilities is crucial for realizing the brightness advantage of the source.

For a typical surface science experiment, the beam emerging from the Michelson interferometer was refocussed (*f*/10) into an ultra high vacuum (uhv) chamber with a base pressure of 1×10^{-10} torr, and reflected at a grazing incidence angle of ~ 87° off a sample of typical dimensions 25 mm × 5 mm, the throughput being 6.5×10^{-2} mm^2 sr^{-1}. The UHV (Ultra High Vacuum) was separated from the 100 milliTorr beam line vacuum by either 6 mm CsI or by polyethylene windows. A boron doped silicon bolometer at 4.2 K was used to detect light in the 50-600 cm^{-1} range, while a Cu doped Ge photoconductor was used for frequencies of 350 cm^{-1} and above. This arrangement gave reproducibilities of 1×10^{-4} at a resolution of 1 cm^{-1} over the 50-2500 cm^{-1} range in 3 minutes of measuring time. The CsI windows allowed high resolution spectra to be taken in the range above 180 cm^{-1}. For frequencies below this, diamond had to be used as the beam line window and this restricted the resolution to 6 cm^{-1} due to interference fringes.

Most of the work described here was done using Cu single crystals. These were first prepared with several cycles of mechanical and electro-polishing. Once in vacuum, the crystals were cleaned by sputtering at 300 K with a 0.08 μA/mm^2 beam of 500 eV Ne^+ ions for 20 minutes.

3.1.1. LOW FREQUENCY MODES OF ADSORBED MOLECULES AND ATOMS

Oxygen adsorption on copper is little studied by infrared spectroscopy. It's a very important system, since the adsorption process is the first step of the copper corrosion by oxygen. As oxygen dissociates on copper surfaces, the only expected adsorbed species is atomic oxygen.

Figure 9 shows the spectrum recorded for oxygen exposed to Cu (100) at 300 K, up to saturation coverage. This spectrum results from 256 accumulated scans, requiring 90 seconds of acquisition time. Due to the very weak dynamic dipole moment of the Cu-O stretch mode [24] ($e^* < 0.12\,e$), this mode is hardly observable, but is discernible after removing the baseline, as shown in the inset.

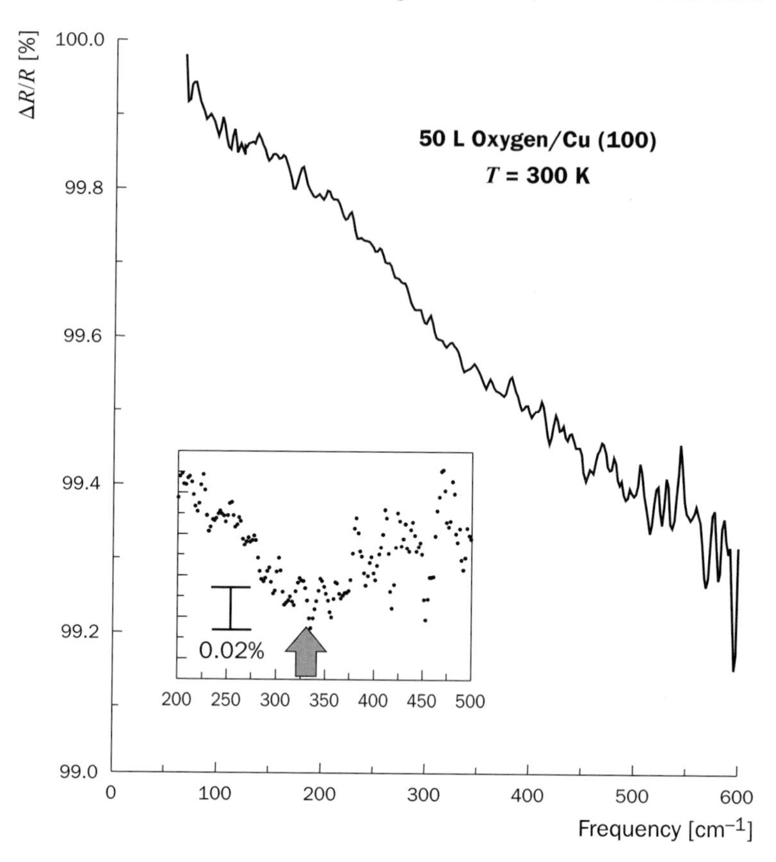

Figure 9 - Measured absolute reflectance changes induced by a 50 Langmuir coverage of oxygen on Cu (100) at 300 K. The broad dip, shown with the background slope removed in the inset, is assigned to the oxygen-Cu vibrational mode. Note that the changes are only ~ 0.03%.

The continuous decrease in the reflectance, starting from unity at zero frequency is commonly observed for metallic substrates, and will be discussed in the following section in terms of diffuse scattering of the conduction electrons of the substrate by the adsorbate overlayer. Adsorbate-subtrate modes of adsorbed molecules have been also detected, as illustrated on figure 10 for NO adsorbed on Cu (111) [25]. In this figure, we have displayed the low frequency region observed after exposing a Cu (111) surface to high exposures of $^{14}N^{16}O$, $^{15}N^{16}O$ and $^{14}N^{18}O$.

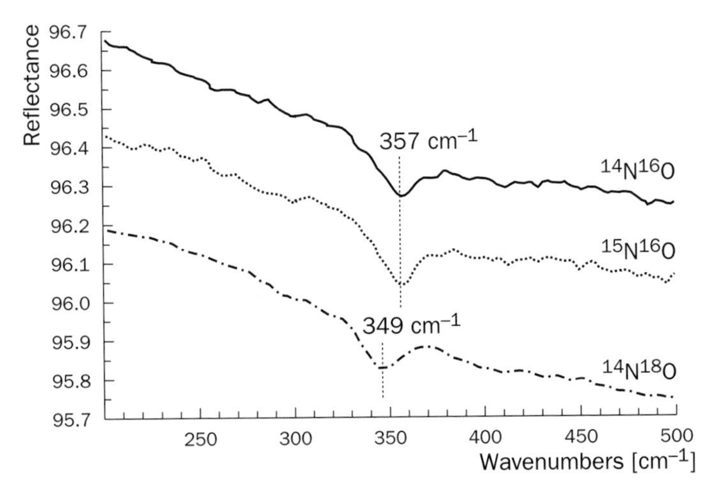

Figure 10 - Frequency dependence of the low frequency mode, for 5 L exposure of NO, on Cu (111), at T = 90K. The low frequency mode is downward shifted upon oxygen labelling (3%). This has been interpreted as a Cu-ON_2 peak, as the Cu-O peak would have shifted by 6%.

The isotopic shift is clearly seen, and the accuracy in determining the frequency position of the band allows definitive assignment of this feature to a Cu-O-N-N mode, rather than the alternative Cu-O stretching mode.

The adsorption of CO molecules on copper surfaces has been studied thoroughly in the mid-infrared region (using an internal source). The intramolecular motion of the molecules (namely the stretching mode) has been investigated in great detail. Using the synchrotron source, the Cu-C stretch mode (the adsorbate-substrate mode) has been clearly identified [26], as shown on figure 11a, at about 340 cm^{-1}. In this figure, a remarkable feature is observed as an anti-absorption band located at around 285 cm^{-1}. This surprising feature is always accompanied by a broadband reflectance change, as it is illustrated on figure 11b.

This observation has led to several studies, trying to explain the origin. As explained below, interactions between adsorbate motion and conduction electrons of the metallic substrate are responsible for such observations.

3.1.2. *VIBRATIONAL DYNAMICS OF LOW FREQUENCY MODES*

Low frequency vibrational modes of adsorbates are precursors to motion, about or away from the surface, as their energy is directly related to the potential energy surface of the molecule or atom bonded. Due to their low energies, they are easily thermally populated and contribute predominantly to the adsorbate entropy.

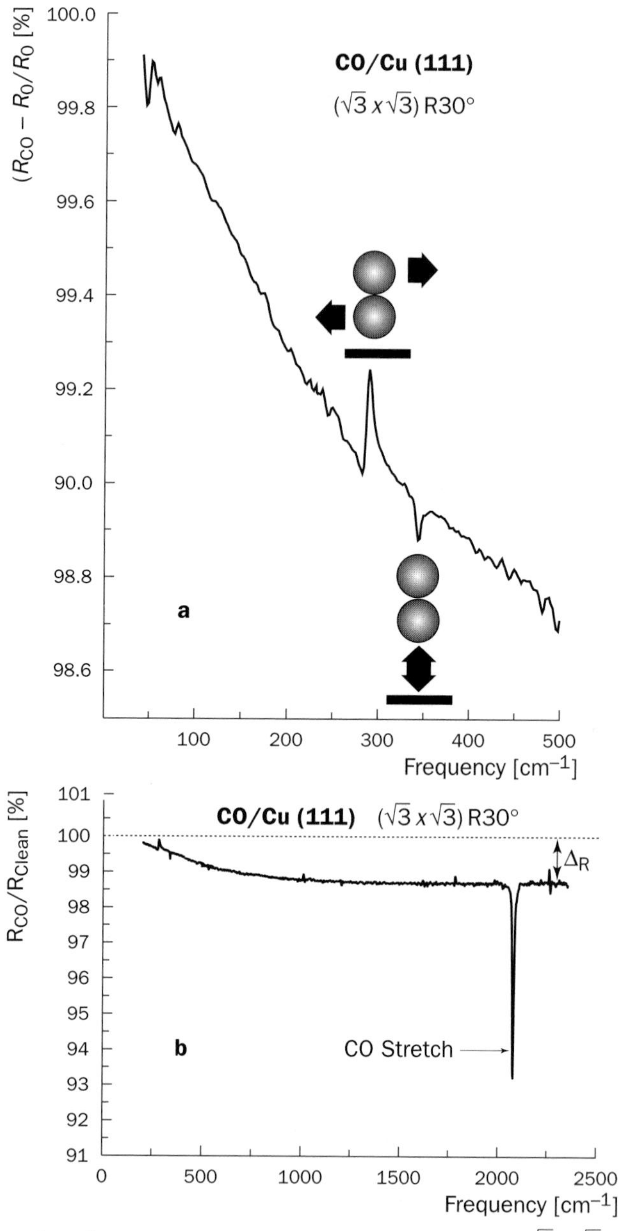

Figure 11 - a - Low frequency Synchrotron infrared spectra of $(\sqrt{3} \times \sqrt{3})$R30° CO/Cu (111). Two features are observed. The peak at 355 cm^{-1} arises from the Cu-CO stretch mode. In (**b**) a continuous broadband reflectance change can be noticed, which is related to the diffuse scattering of the conduction electrons of the substrate by the adsorbate in the anomalous skin depth region.

Therefore, energy transfer through these modes is important in various processes, such as surface diffusion and reactions. Since the beginning of the 90's, interest in the study of the low energy modes has been renewed when Persson suggested that frustrated translation of molecules or atoms is the key motion in the friction force induced by the adsorbate on the metallic surface [27]. A direct relation between the lifetime of the frustrated translation τ and the friction coefficient η, responsible for the diffuse scattering of the conduction electron of the substrate by the adsorbate (resistivity change) has been suggested.

$$\eta = \frac{1}{\tau} \qquad \{11\}$$

The above model relies upon the assumption that the lifetime of the frustrated translation is purely of electronic origin. The CO experiments, using synchrotron infrared spectroscopy, have been important in determining these contributions.

The theoretical details of the elastic scattering of the conduction electrons are reported in [28]. The validation of the theoretical model has required to perform simultaneous d.c. resistivity and IR reflectivity changes, upon adsorption of atoms or molecules. For such a purpose, thin Cu(111) films were grown epitaxially on TiO_2(110). After preparation of the 15 mm×10 mm films, four 200 nm thick copper pads were evaporated onto the edges to allow measurement of the film resistance W. The absolute resistance values were determined with a four terminal method. In order to register small changes ΔW of film resistance, the films were made part of a Wheatstone bridge, which was supplied with alternating current. A lock-in amplifier was used to improve the signal to noise ratio. The film resistance at room temperature was also used to determine the thickness of the clean films by comparison to a calibration curve. This was not trivial because at first there is no film to measure. Thus, in the initial growth process, the film thickness was determined with a quartz microbalance. Then the samples were removed, contacts were put on, and the samples were re-mounted in UHV and cleaned by ion bombardment. During the subsequent epitaxial re-growth process (see below), the resistance and the additional film thickness, again using the quartz microbalance, were recorded continuously. If W_{start} and W_{end}, are the start and end values of the sheet resistance (at T = 293 K) in this process and if the known thickness change is Δd, the film thickness d, corresponding to the resistance W_{start} can be determined by

$$d = \frac{\Delta d}{(W_{start} / W_{end}) - 1} \qquad \{12\}$$

neglecting diffuse scattering at the interface itself.

Thus two points $d(W_{end})$ and $d(W_{start})$ for the calibration curve were obtained. The samples were then mounted in the U4IR chamber on a manipulator, which allows cooling to liquid nitrogen temperature (around 90 K). One hour of sputtering with Ne at 500 V and a current of 5 μA was needed in order to clean each sample for the first time after baking. After the first 20 minutes of this process, the film resistance reached a minimum, but Auger spectra showed carbon contamination of the surface. After each of the experiments that followed, typically 10 minutes of ion bombardment (Ne^+, 500 eV, 5 μA) were needed to clean the sample. The loss in film thickness due to sputtering was compensated by evaporating additional copper at room temperature until the initial resistance of the film was retrieved. The re-growth process was followed by a short annealing at 400 K. The LEED (Low Energy Electron Diffraction) pattern confirmed that the films were again of (111)-orientation. The subsequent cooling down of the sample to 80 K took approximately 20 minutes until the resistance stabilized. The experimental lifetime of the films was limited, though, as the removal of material by sputtering was not completely homogeneous. Repeated recycling of the sample could lead to variations in the film thickness too large to be neglected.

Apart from the *reproducibilities* (noise) of .01% in 1 minute of measuring time at the NSLS, *absolute* reflectance changes can be measured. This is because the brightness of synchrotron radiation is dependent only on the magnitude of the stored beam current which can be measured to one part in 10^6. Variables normally present in IR spectroscopy with a thermal source, such as temperature fluctuations and emissivity changes do not exist. In practice the absolute accuracy of the present experiments was found to be $\pm 0.2\%$ across a broad spectral band from 300 cm^{-1} to 3000 cm^{-1} using a silicon beamsplitter and a liquid helium cooled Cu doped Ge photoconductive detector, provided the data were all taken within 10 minutes or so. Over longer periods of time, as the current decays in the storage ring, there are slight orbit shifts due to changing thermal loads on the synchrotron, which lead to higher discrepancies.

Infrared spectra obtained for different coverage and for a given film thickness are displayed on figure 12, together with the fit according to Persson's model. Linear dependencies, like the one displayed on figure 13 between the asymptotic limit of the IR reflectance change and the resistance change, have been found for different film thicknesses and for different adsorbates (such as ethylene) [28].The electronic lifetime extracted from the IR reflectance change and from the initial slope of the resistivity change, for the case of CO on Cu (111) films are $\approx$ 58 ps and 49 ps respectively, in relatively good agreement [29].

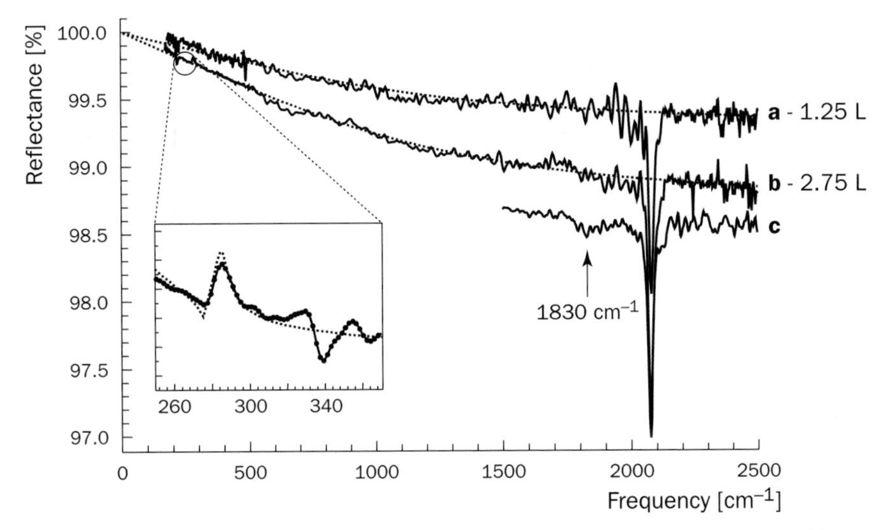

Figure 12 - Infrared spectra of two different exposures of CO on a thin (67 nm) Cu (111) film, at 90 K. Each spectrum has been obtained using two separate runs, since none of the detectors used covers the entire frequency range. The solid lines are fits of the reflectance change, according to **[27]**.

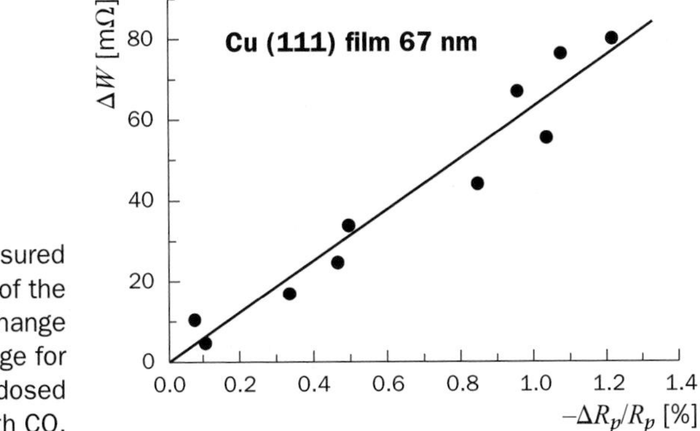

Figure 13 - The measured linear dependence of the DC resistance change on the reflectance change for a 67 nm Cu (111) film dosed with CO.

3.2. SUM FREQUENCY AND DIFFERENCE FREQUENCY GENERATION AT INTERFACES

The investigation of the vibrational properties of interfaces by SFG, using a FEL laser as infrared source is new, and most experiments have been performed at LURE-Orsay, first in the electrochemical environment (the scientific field of the first user group!), then in the ambient and the UHV. We present some most relevant results, which are aimed to be an illustration of the capability of this technique to probe selectively the spectroscopic properties of buried interfaces.

3.2.1. IDENTIFICATION OF ADSORBED INTERMEDIATE OF ELECTROCHEMICAL REACTIONS BY SFG

One of the goals of spectroscopic investigation of the electrochemical interface is the identification of adsorbed species produced by dissociative chemisorption of the reactant, which act as a poison of the reaction. The structure and the chemical composition of the interface is dependent on the applied potential (referred to the Nernst potential of a given system, in well-defined thermodynamic conditions: H/H^+, Hg/Hg^+, $AgCl/Ag^+$...). Absolute SFG spectra, recorded in a wide IR spectral range at each electrode potential allow identification of the adsorbed species and their adsorption configuration. Figure 14 shows the potential-evolved SFG spectra of the platinum electrode in (0.1 M $HClO_4$ and 0.5 M CH_3OH) aqueous electrolyte [30].

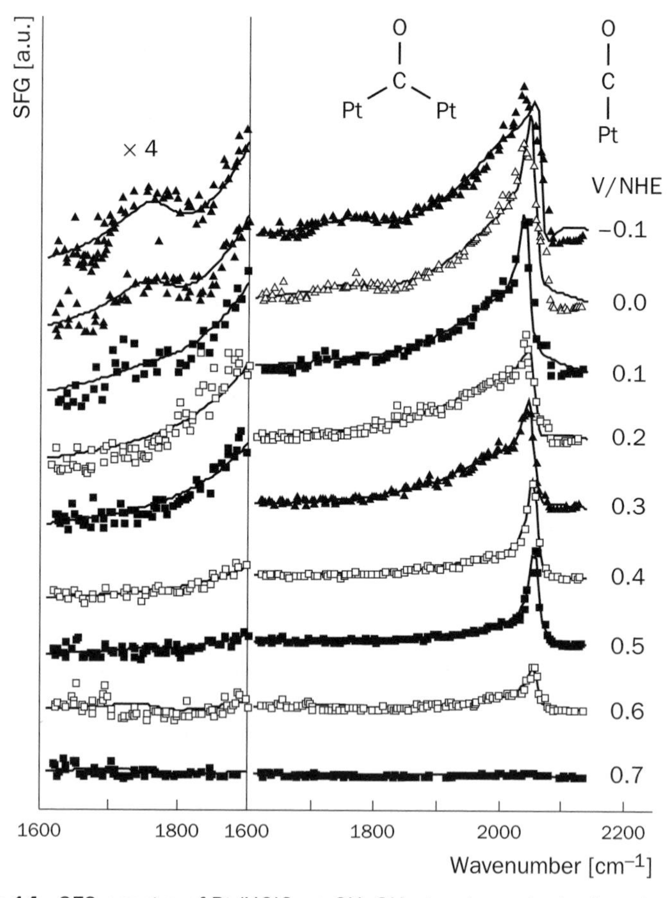

Figure 14 - SFG spectra of Pt/$HClO_4$ + CH_3OH at various electrode potentials.

The strong SFG resonance peaks at approximately 2050 cm^{-1}, corresponding to the stretching vibration of the CO poison adsorbed in a one-fold coordinated site (on top). The broad shoulder at the low frequency edge of the sharp SFG resonance is assigned to CO in a multi-fold coordinated adsorption site, since the infrared bands of CO in two-fold (bridged-bound) and three-fold (hollow site) adsorption site on Pt are located at 1977 and 1820 cm^{-1}, respectively. The relative magnitude of the shoulder versus the sharp resonance increases as the potential is lowered. The CO SFG signal disappears at a positive potential of 0.6 V/NHE (Normal Hydrogen Electrode) where the CO oxidation reaction is effective

$$CO_{ad} + 2\,OH^{-}_{in\ solution} \longrightarrow CO_{2\ in\ solution} + H_2O_{in\ solution} + 2\,e^{-} \qquad \{13\}$$

A third weak resonance, peaked around 1770 cm^{-1}, appears at negative potential, where the hydrogen evolution reaction processes. This absorption at 1770 cm^{-1} has been assigned to the adsorbed intermediate of the hydrogen evolution, identified for the first time, thanks to the combination of SFG and CLIO. Actually, the IR cross-section of the H-Metal bond is very weak and makes difficult its study by IR technique, even in the UHV environment. The relative contribution of linear and multi-bound species depends strongly on the methanol concentration, as shown in figure 15, for c = 0.1 M, 0.02 M, 0.004 M.

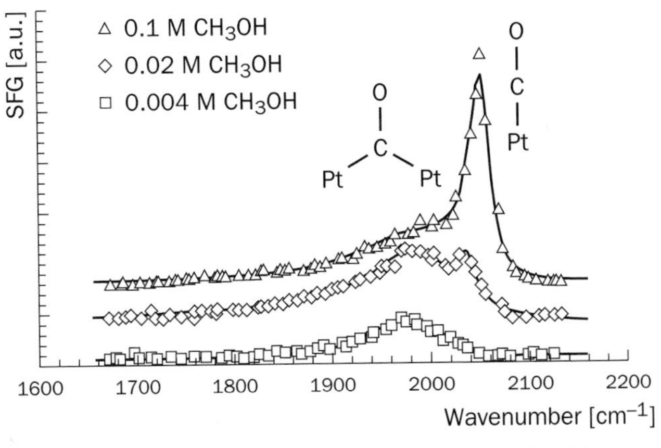

Figure 15 - SFG spectra of CO adsorbed on Pt in $HClO_4$ 0.1 M as a function of CH_3OH concentration.

This behaviour should be connected to a dynamic equilibrium between species in solution and in the adsorbed phase. At low CH_3OH concentration of the electrolyte, the CO poison occupies the higher adsorption energy multi-bound sites of high steric hindrance. At higher CH_3OH concentration, CO is displaced to the lower adsorption energy linearly bound sites of lower steric hindrance. At the lowest concentration, the linearly bound species band disappears totally, leaving the broad peak corresponding to the multi-bound species. The linearly bound CO band appears at c = 0.02 M and becomes dominant at c = 0.1 M.

3.2.2. VIBRATIONAL SPECTROSCOPY OF CYANIDE AT METAL-ELECTROLYTE INTERFACE

The first application of SFG in electrochemistry was performed on the adsorption of pseudo-halide ions (CN^- and SCN^-) on polycrystalline platinum, in neutral solution. Unlike CO, which has been widely studied because of its strong IR cross-section, the adsorption of pseudo-halides on metals has been much less investigated by vibrational spectroscopy. The smaller cross-section and the contribution of species in solution makes IR spectroscopy less straightforward and difference spectra between several potentials are usually taken to ensure the surface nature of the spectra.

However, it is difficult to analyze unambiguously such spectra that are obtained as a difference between two applied potentials, the spectrum taken at the negative one acting as a reference where the electrode is assumed free of adsorbate. In this condition, the method may not detect species present at constant coverage and retaining the same IR absorption properties at both potentials. It is then essential to call for SFG as a surface sensitive technique which gives the absolute vibrational spectra at each state of the interface.

The vibrational signature of adsorbed CN^- and its potential evolution are shown in figure 16 in the case of Pt(110) in contact with a solution of $NaClO_4$ (0.1 M) and KCN (0.025 M). Two distinct resonances, assigned to N and C bound CN^- species are detected, depending on the electrode potential. The same features were observed for OCN^- adsorbed on silver, supporting the presence of two metal-adsorbate IR absorption detected in the far infrared spectrum of this system [31]. Absolute DFG vibrational spectra recorded over a large potential range for Au(111), Au(100) and Au(110) have allowed us to study the adsorption behaviour of CN^- on these surfaces and to bring basic insights into the interfacial properties and the surface-adsorbate interactions [32].

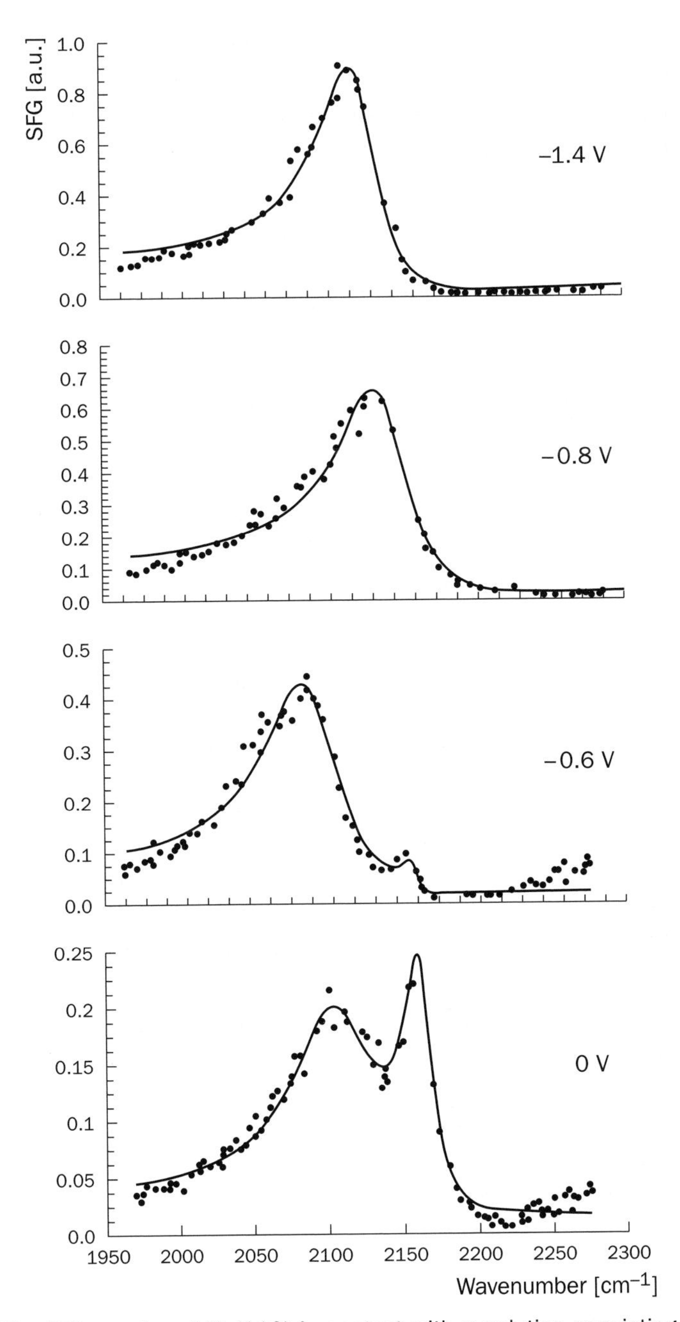

Figure 16 - SFG spectra of Pt (110) in contact with a solution consisting of 0.1 M $NaClO_4$ + 0.025 M KCN, showing the emergence of a second band at –0.6 V; experimental values (points) and fit (continuous lines).

Figure 17 shows the potential-evolved DFG spectra of Au (110) (**a**), Au (100) (**b**) and Au (111) (**c**) electrodes, in contact with an aqueous electrolyte consisting of $NaClO_4$ 0.1 M and KCN 0.025 M, along with their fits. Each spectrum is adjusted with only one potential-position dependent resonance, indicating the presence of only one absorption band localized within the CN^- stretching spectral range. While the disappearance of the resonance upon oxidation of the adsorbed CN^- layer is completed at the same potential for the three investigated orientations, the potential at which the adsorption starts depends on the electrode surface orientation: CN^- adsorption starts at –1.35 V on Au (110), –1.05 V on Au (100) and –0.85 V on Au (111). These results reveal some marked differences in the process of CN^- adsorption on the different surface orientations of gold single crystal electrodes.

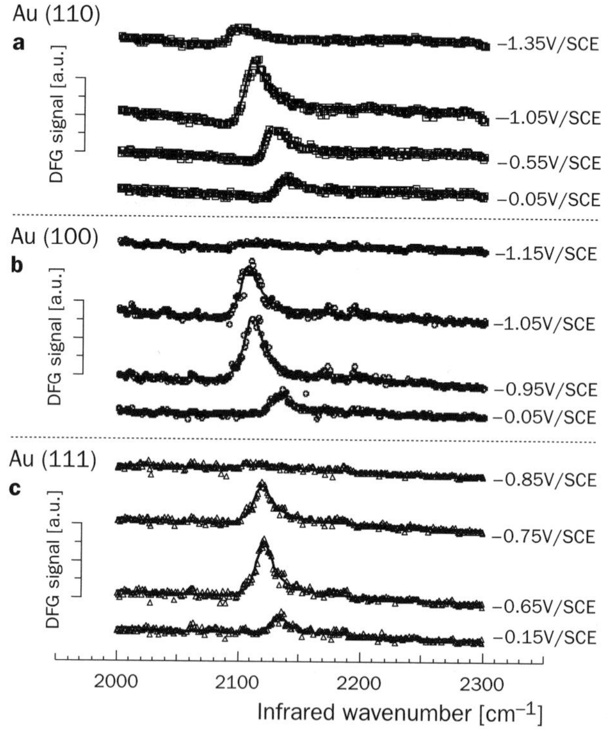

Figure 17 - DFG spectra of gold single crystals in contact with $NaClO_4$ 0.1 M + KCN 0.01 M. **a** - Au (110). **b** - Au (100). **c** - Au (111).

The full width at half maximum (FWHM) of the DFG resonance is plotted in curves a of figure 18. The most striking and unexpected result is the invariability of the FWHM resonance width with either the electrode potential or the surface orientation. Its constant and small value of 17 cm^{-1} can be explained by a homogeneous broadening mechanism, indicative of a same single adsorption site of CN^-.

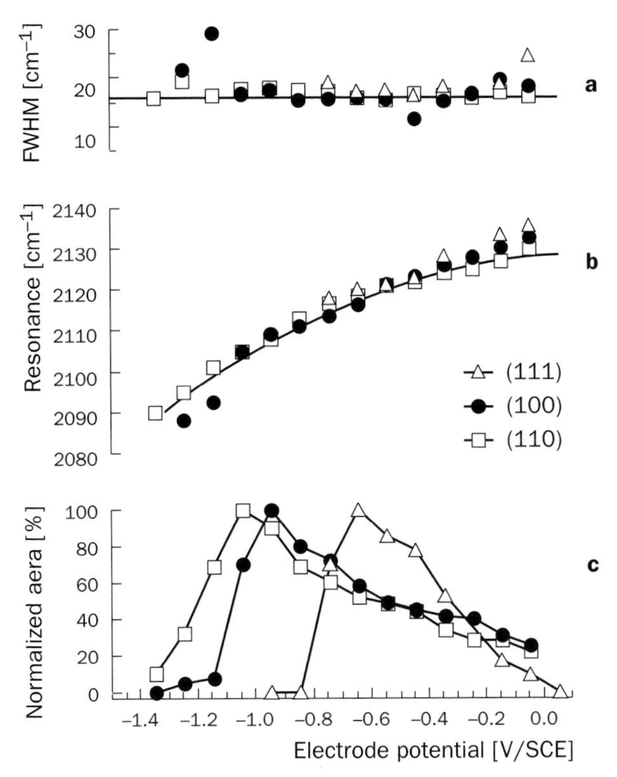

Figure 18 - Potential dependence of (**a**) the Full Width at Half Maximum, (**b**) peak centre and (**c**) integrated intensity of DFG spectra for Au(111), Au(110) and Au(100), in contact with $NaClO_4$ 0.1 M + KCN 0.01 M solution.

The DFG resonance peak frequency (curve **b** of figure 18) obeys a quadratic potential dependence, increasing from 2090 cm^{-1} at –1.3 V to 2130 cm^{-1} at 0 V. The potential dependence of the DFG resonance frequency (Stark effect) is the same for all the investigated surface orientations, which implies that CN^- is tightly bound to the gold surface and experiences only the electric field of the surface atom on which it is adsorbed. Owing to the very different surface symmetry of the investigated orientations considered here, only an adsorption on an atop-site can account for this behaviour.

We have already noted that the adsorption onset depends on the electrode surface orientation. These observations are clearly evidenced on curve **c** of figure 18, where we have plotted the potential dependence of the integrated DFG resonance intensity normalized to its maximum value, for each crystal orientation. CN^- adsorption starts at –1.35 V on Au(110), –1.05 V on Au(100) and –0.85 V on Au(111). This observation indicates that the electrode-adsorbate interactions are stronger on the more open (110) surface, and decrease on the denser surface. This follows the trend of the pzc in CN^--free solution, where the most positive pzc is found for the (111) and the most negative for the (110) surface orientation.

Following the theoretical analysis by Trasatti [33], one can get some insights into the mechanism and the energetic of CN^- adsorption from the position of the onset adsorption potential relative to the pzc of each surface orientation, in CN^- free solution, [ΔE(hkl)]. The larger (more negative) ΔE(hkl), the stronger the metal-ion interaction. They demonstrate also that the CN^--Au interaction is stronger on the less dense (110) face. This is probably due to the weaker adsorbate-adsorbate repulsion attendant on the more open face. It is noteworthy that the DFG resonance is dramatically reduced around 0 V for all three orientations. As a result, the kinetics and the dynamics of the CN^- adsorption processes are faster on Au (111) and slower on Au (100) and Au (110). The rate of CN^- adsorption as measured by the positive slope of the increasing segment is roughly the same for all surface orientations, while the rate of CN^- desorption is much higher for the (111) orientation. This behaviour should be connected to the thermodynamics of the system. The upper stability potential limit of CN^- on the surface electrode is the same for all the orientations while the lower limit is more negative for (110) than (100) and (111) orientations.

3.2.3. VIBRATIONAL SPECTROSCOPY OF SELF-ASSEMBLED MONOLAYERS ON METAL SUBSTRATE

Organic coating on metals is an important field of materials science aimed at tailoring surface properties such as corrosion-inhibition, biocompatibility or chemical reactivity. Anchoring of pure or functionalized *n*-alkanethiols from liquid solutions onto metallic substrates is a convenient method for achieving compact and well ordered organic thin films. SFG has been used to monitor *in situ* the evolution of the interface vibrational fingerprint of *p*-nitroanilino-dodecane thiol (NAT) during film formation on gold. The polycrystalline gold substrate was obtained by evaporating a 100 nm gold film on a Si (100) wafer with a 5 nm chromium interlayer. The sample was moved into a cell filled with spectroscopic grade ethanol.

To prevent the IR laser beam absorption by the solution, the sample was pressed slightly against the cell window during the acquisition of SFG spectra (5 min per spectrum). Figure 19 shows the time evolution of the SFG signature of the gold substrate immersed in the NAT/ ethanol solution. The SFG spectrum of the sample in pure ethanol prior to the adsorption experiment is also shown at the bottom of figure 19. This curve shows the strong non-resonant SFG signal of the gold surface. Four resonances are identified in the SFG signatures. The prominent feature near 1325 cm^{-1} must be decomposed into two resonances, labelled α and β. Two weaker resonances, labelled γ and δ are further detected around 1280 and 1180 cm^{-1}, respectively. Band assignments indicate that the SFG signature in the 1400-1150 cm^{-1} spectral range probes predominantly the NAT end-group. This was expected since the dominant alkene chain vibrations lie out of this investigated spectral range and their detection is further hindered by the centrosymmetric character of the chain skeleton.

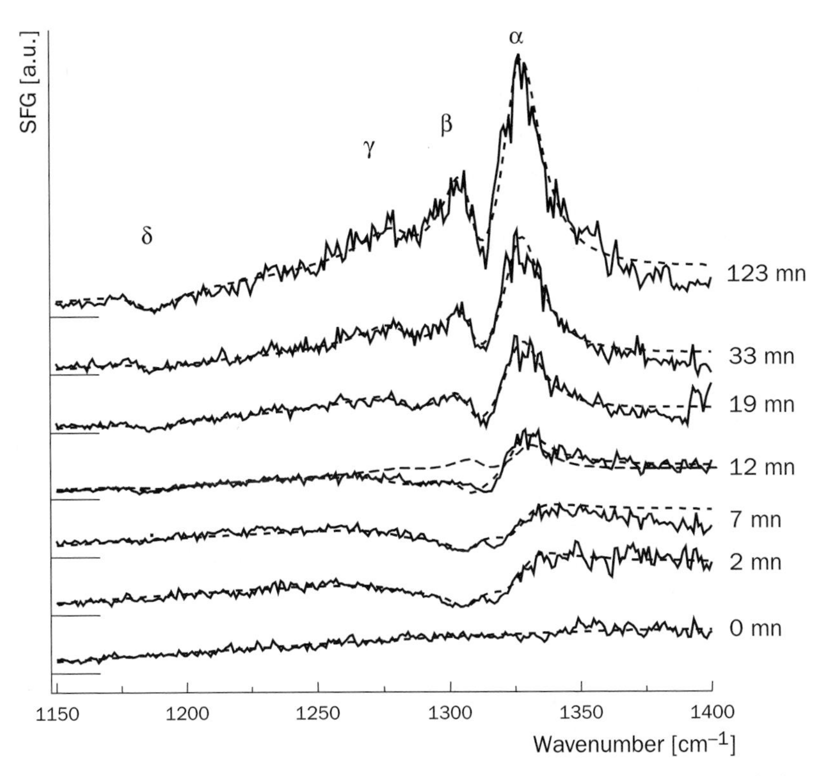

Figure 19 - Time evolved SFG spectra of NAT adsorbed on gold in ethanol solution.

In order to gain more insight into the film anchoring process, we compare the time-evolved spectra of figure 19 to the theoretical SFG response of a set of 4 Lorentz oscillators

$$\mathrm{SFG}(\omega_{ir}) = \left| g(N)\chi_{nr}(\omega_{ir}) + N\sum_{v=1}^{4}\frac{A_v}{\omega_v - \omega_{ir} + i\Gamma_v}\right|^2 \qquad \{14\}$$

where A_v, ω_v and Γ_v are the oscillator strength, the resonance frequency and the damping constant of the vth vibrational mode. N is the film coverage and ω_{ir} the infrared beam pulsation. $[g(N)\, \chi_{nr}(\omega_{ir})]$ is the non-resonant term which includes all other slowly varying contributions to the second order surface susceptibility. The factor $g(N)$ accounts for the possible amplitude and phase variation of the substrate non-resonant response as a function of the film coverage. The mode frequencies and bandwidths, ω_v and Γ_v, are expected to show little or no dependence with respect to the film conformation and film coverage since local field variation and interaction between adjacent NAT end-groups can be neglected in a first approximation. However, the respective amplitudes and phases of the factors A_v depend on the NAT end-group orientation.

The simplest hypothesis of film growth mode assumes a stable molecular orientation. This implies that all oscillator strengths, A_v, keep the same phase and same amplitude with respect to each other but their amplitudes are proportional to the coverage. Although this assumption gives appropriate fits to the lowest and the upper curve of figure 19, no satisfactory fit could be obtained for the SFG spectra at intermediate coverage (dashed line in figure 19). This result points to the necessity to go one step beyond in the complexity of the film formation process by considering molecular reorientation. In this case, the respective amplitudes and phases of the oscillator strengths can vary. Good fits to the experimental data can be obtained by allowing the oscillator amplitudes and phases to vary independently. The evolution of these parameters is shown in figure 20. Important variations of the phases during the first 30 min of exposure indicate the occurrence of strong molecular reorientation. After this first adsorption stage, the film conformation stabilizes and the oscillator strength amplitudes become good indicators of the film coverage. This allows us to ascertain that the film conformation stabilizes when the coverage reaches 70 ±20 % of its saturation value.

3.2.4. *VIBRATIONAL SPECTROSCOPY OF FULLERENES C_{60}, ADSORBED ON AG(111), IN UHV ENVIRONMENT*

In the present example, we have exploited the complementary SFG and infrared absorption technique, for performing vibrational spectroscopy of K-doped fullerene thin-films. SFG enables sensitive and high resolution vibrational spectroscopy of C_{60} to be done in the adsorbed phase. Combining SFG with conventional infrared for studying undoped and K-doped monolayers of C_{60} deposited on Ag(111) allowed us to pinpoint the occurrence of dynamical charge transfer at the interface [34].

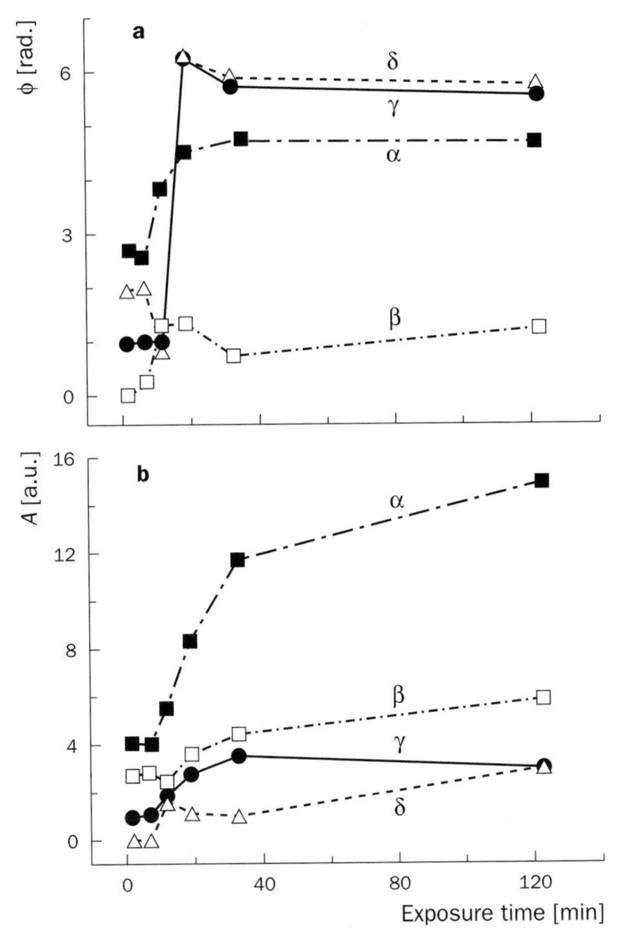

Figure 20 - Time evolution of (**a**) the phase and (**b**) the amplitude of the 4SFG resonances of NAT.

The C_γ epitaxial layer is obtained by saturating the Ag (111) surface with C_{60} sublimated from a tantalum crucible while the substrate temperature is maintained at 570 K to prevent multilayer growth. Doping with potassium evaporated from a commercial getter source is performed at 390 K. Spectroscopic investigations were performed at the doping temperature of 390 K. Infrared reflection-absorption spectroscopy is performed at grazing incidence using a Fourier transform spectrometer (4 cm^{-1} resolution) equipped with a liquid nitrogen cooled MCT (Mercury Cadmium Telluride) detector. The clean Ag (111) sample at 570 K, prior to C_{60} deposition, is taken as reference for the infrared spectra. We present the SFG spectra normalized to the peak height of the ubiquitous high frequency resonance labelled α.

The evolution of the infrared and SFG spectra of C_{60}/Ag(111), as a function of K-doping, is compared in figures 21 and 22. Both spectroscopies reveal two vibrational modes.

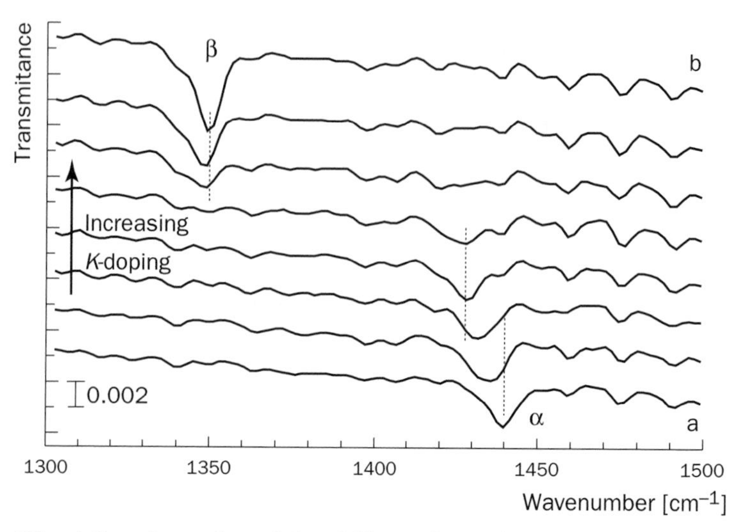

Figure 21 - Infrared spectra of the C60 overlayer deposited on Ag(111), as a function of increasing doping level of potassium, in the frequency region of the pentagonal pitch mode.

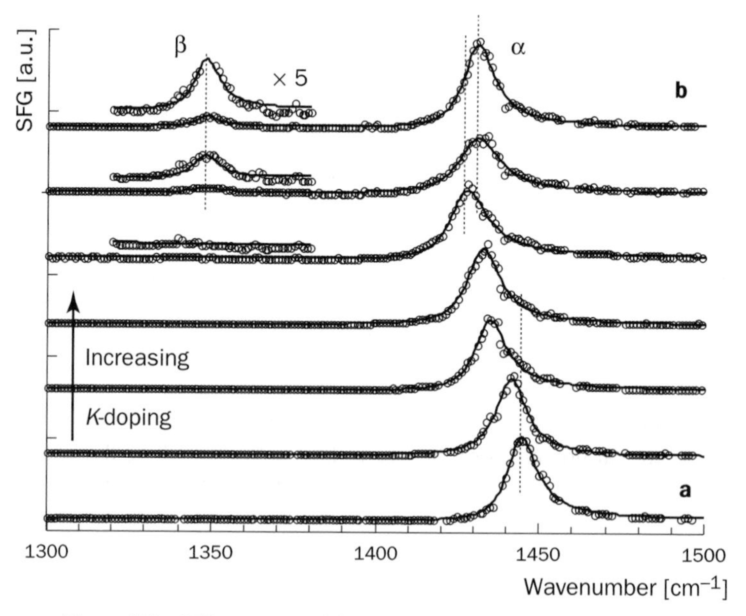

Figure 22 - SFG spectra of the same system as for figure 21.

For the undoped or weakly doped C_{60} layer, both spectroscopies show one peak, labelled α, which softens upon K-doping. When the K-doping level reaches saturation, this α peak vanishes from the infrared spectra but remains detected by SFG. Meanwhile, a second peak labelled β appears in both the infrared and SFG spectra. The centro-symmetry of the free C_{60} molecule implies that its vibrational modes are exclusively infrared or Raman active. This should preclude any SFG signal which is proportional to the product of the infrared and Raman activities of the vibrational modes. However, the perturbation of the molecule by its asymmetric environment, the vacuum on one side and the metallic surface on the other, induces a weak mixing of the vibrational characters, e.g. provides the infrared modes with weak Raman activity and the Raman modes with weak infrared activity, to yield non-vanishing SFG signals. Comparing the peak intensities in the infrared and SFG spectra enables one however to infer the dominant character of the detected vibrations.

In the K-saturated layer, the β peak is strongly infrared active while the α peak Raman activity is at least two orders of magnitude stronger than that of the β peak. These vibrational properties perfectly mimic those of bulk C_{60} with the β and α features assigned to the $T_{1u}(4)$ infrared mode and the $A_g(2)$ Raman mode, respectively. A charge state of 6 electrons per molecule is inferred from the softening of the $T_{1u}(4)$ infrared mode to 1349 cm^{-1} (β peak of curve b in figures 21 and 22 and of the $A_g(2)$ Raman mode to 1430 cm^{-1} (α peak of curve **b** in figure 22). This result agrees with the electronic spectroscopic data on C_{60} adsorbed on metals which showed that, upon K-doping with potassium, C_{60} can accept up to 6 electrons per molecule, corresponding to the complete filling of the six-fold degenerate t_{1u} molecular orbital, a behaviour which parallels that of the bulk fullerene compounds.

The absence of the β peak in both the infrared and SFG spectra of the undoped and weakly doped C_{60}/Ag (111) interface is to be assigned to the drastic two orders of magnitude decrease of the $T_{1u}(4)$ mode infrared activity when the C_{60} charge vanishes. The evolution of the vibrational spectra from the K-saturated state to the weakly doped or neat interface enables us to assign this α infrared peak of curve a in figures 21 and 22 to the $A_g(2)$ mode. This infrared feature contrasts dramatically with the vibrational properties of the free or bulk state C_{60} where the $A_g(2)$ mode is Raman active and has no infrared activity. This observation reveals a strong influence of the adsorbate/substrate interaction on the adlayer vibrational properties. This dramatic infrared enhancement is also highly selective since it only significantly affects the $A_g(2)$ mode while the neat C_{60} presents scores of modes in the scanned spectral range. We have shown that the process of interfacial dynamical charge transfer can fully account for all the observed vibrational features (fig. 23).

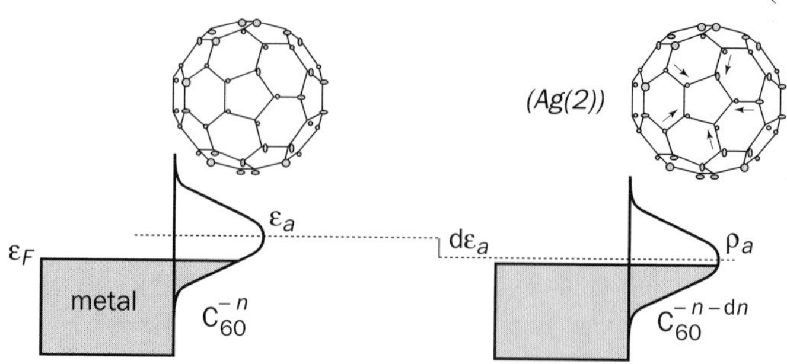

Figure 23 - Schematic drawing of the dynamical charge transfer occurring between the pentagonal pith mode of C_{60} and the metallic substrate.

3.2.5. ADSORPTION OF 4-CYANOPYRIDINE ON AU(111) MONITORED BY SFG

The 4-cyanopyridine (4-CP) molecule belongs to the C_{2v} point group and is made of an aromatic ring of pyridine to which the nitrile group is bound. This molecule presents interesting properties due to its planar geometry, with a strong permanent dipole of 1.4 debye oriented towards the nitrogen of the pyridine ring. It is then worthwhile to investigate the adsorption geometry on an electrode metal substrate and its dependence upon the electrode potential [35].

The electrochemical behaviour of the system is monitored using linear cyclic voltammetry in the double layer potential range, before each SFG experiment. SFG spectra were recorded using the following procedure. A first spectrum is recorded after polarization of the electrode at –1.0 V. Then the potential is increased stepwise to 0.7 V and a SFG spectrum acquired every 0.1 V. The spectrum was acquired in about fifteen minutes, the infrared beam being provided by the Free Electron Laser CLIO tuned within the 830-1250 cm^{-1} range, with a spectral resolution of 5 cm^{-1}. Figure 24 displays the first of these spectra recorded at –1.0 V and its fit to equation (8). Figure 25 presents a three-dimensional plot of the SFG spectra as a function of the IR wavenumber and the electrode potential.

The following observations can be extracted from these spectra:

- At all potentials between –1.0 V and +0.7 V strong vibrational modes of the pyridine ring are detected, indicating that 4-CP is bound to the electrode throughout this range of potential. Therefore, the features identified on the cyclic voltammogram are related to transitions that occur in the 4-CP adsorbed on the surface. In particular, the peak at –0.5 V is not due to the beginning of the adsorption process of the molecule, as suggested by Chen *et al.* [36], but to a change of its adsorption configuration on the electrode.

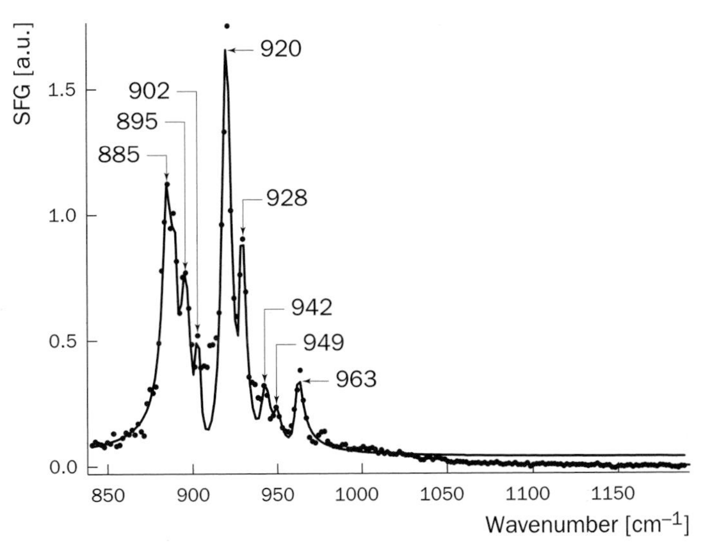

Figure 24 - SFG spectra of Au(111) in 1 mM 4-cyanopyridine and 100 mM $NaClO_4$ solution, at E = –1.0 V, in the ring mode spectral range (dots = experimental data points ; continuous line = fit).

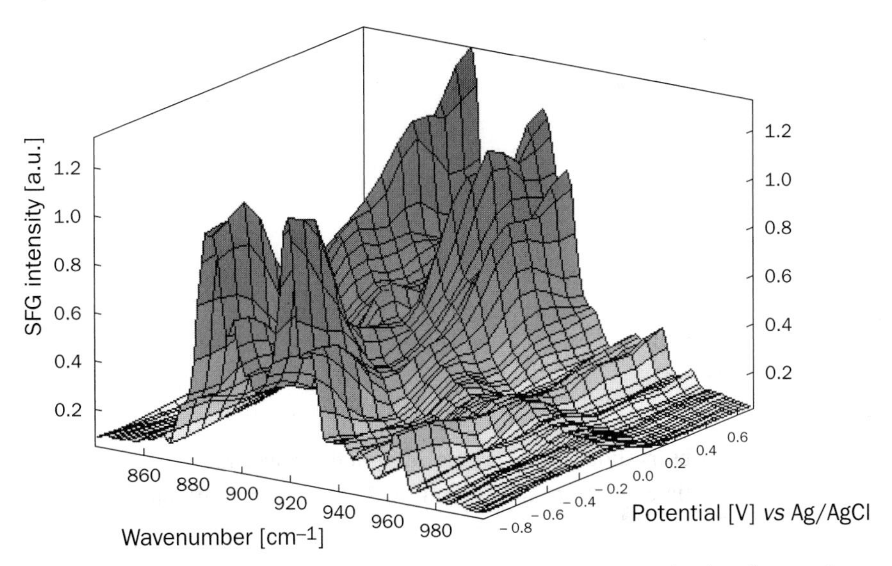

Figure 25 - 3D-plot of the SFG spectra of the same system in the ring mode spectral range.

- Several resonances are detected, each spectrum being adjusted to equation {1} with 9 resonances centred at 885, 890, 895, 902, 920, 928, 942, 949 and 963 cm^{-1}. The position of these resonances does not depend on the electrode potential. Figure 25 shows two kinds of peaks. Most of them are dominant at the lower and the higher and become very weak at intermediate potentials. On the contrary, a few of them have the opposite behaviour (peak *d*). This behaviour is evidence of a change of the adsorption geometry of the 4-CP from a perpendicular to a flat, then again to a perpendicular configuration, when the potential is increased. For example, at intermediate potentials, around 0 V, the vibrational mode *e* has a dipole moment parallel to the interface and becomes SFG-inactive, resulting in the vanishing of peak *e*. A similar explanation can be applied to the potential behaviour of mode *d* which vanishes at low and high potentials. The *e*-type modes are assigned to *in-plane* vibration modes of the pyridine ring. In this way, these modes are dominant at extreme potential limits and reveal a vertical adsorption geometry of 4-CP on the electrode surface. The *d*-type peak at 902 cm^{-1} corresponds to an *out-of-plane* vibrational mode and is SFG-active only when the molecule is in a flat configuration on the surface, which is the case at intermediate potentials. Therefore, by analyzing carefully the behaviour of the modes detected by SFG, it is possible to split the vibrations into two different types: *in-plane* and *out-of-plane*.
- Figure 26 displays the potential evolution of the SFG resonances along with the voltammogram recorded in the same conditions.

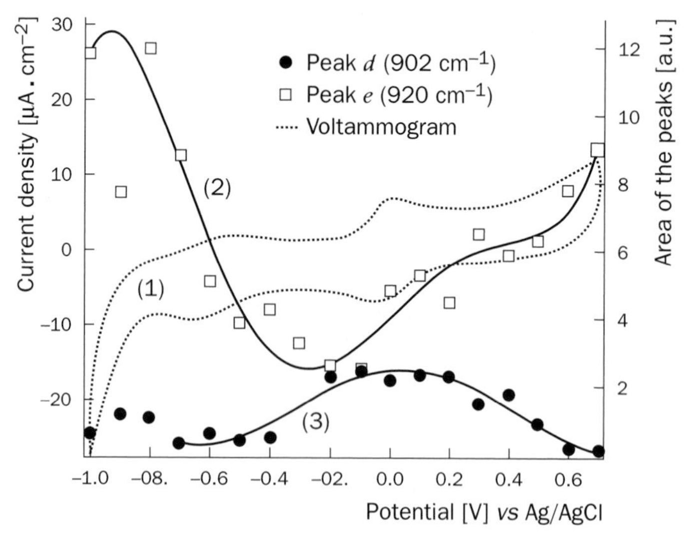

Figure 26 - Intensity of peaks *d* and *e* as a function of the potential. The voltammogram has been recorded simultaneously in a thin electrolyte gap configuration.

The features observed on the voltammogram appear to be in good agreement with the changes of the peak intensity observed in SFG. This allows us to assign the two structures at –0.5 V and +0.3 V to the reorientation of adsorbed 4-CP from vertical to flat configuration and then from flat to vertical again, when the electrode potential is increased from –1.0 V to +0.7 V.

The question of the orientation of the vertically adsorbed molecule on the surface remains, since it can adsorb via the nitrogen of either the pyridine or the cyanide end. To address this issue we have performed SFG measurements of 4-CP adsorbed on gold in the CN stretching spectral range. Results indicate a flipping from a configuration where the 4-CP is bound to the gold surface via the nitrogen of the pyridine ring at negative potential to the one where it is bound via the nitrogen of the cyanide end at positive potential (fig. 27).

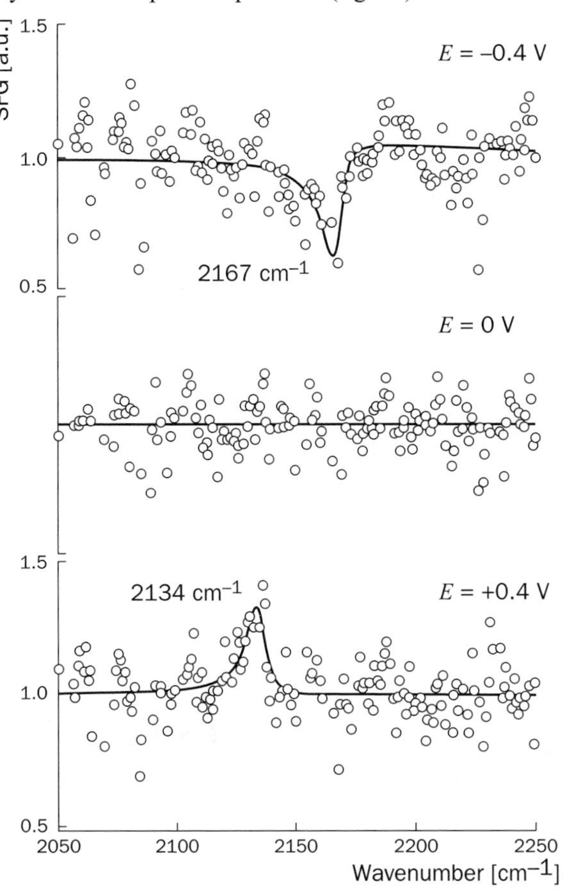

Figure 27 - SFG spectra of the same system at three different increasing potentials in the CN stretching mode spectral range.

4. CONCLUSION AND OUTLOOK

In this chapter, we have addressed the issues of an IR beam extracted from a synchrotron ring, and reported on its potentiality for the investigation of the vibrational properties of interfaces, especially in the far IR spectral range, where conventional sources are too low in brightness. Several FTIR instruments are operating or being developed on several synchrotron rings and are expected to bring valuable quantitative information on the adsorbate-substrate bonding, as well as their vibrational dynamics. Accessing this information should be a critical step towards a better understanding of the energetics and the kinetics of interfacial processes, especially in electrochemical interfaces, where adsorption is always a competitive process between the ionic species and the solvent dipoles, which is controlled by the electrode potential.

The development of optical non-linear techniques benefits directly from the progress of the laser technology. In this respect, the Free electron laser offers a great advantage of producing a high power pulsed beam, tunable within a wide IR spectral range, allowing the simultaneous probe of several vibrational modes of the adsorbate. Combining the unique properties of the FEL with the high sensitivity and interface selectivity of SFG between centro-symmetric media allows the investigation of the vibrational properties of adsorbate species, with no interference from a signal of bulk species, which is forbidden in the electric dipole approximation. The recent experimental examples discussed in this lecture show clearly the advantage of this technique. The use of SFG, as well as various related second order optical processes, is increasing rapidly in various surface and interface research fields such as catalysis, corrosion, electrocatalysis, thanks to its versatility, its capability to probe any interface.

REFERENCES

[1] See for example: H. WINNICK - Properties of Synchrotron Radiation *In "Synchrotron Radiation Research"*, H. WINNICK & S. DONIACH eds., Plenum Press (1980)

[2] W.D. DUNCAN & G.P. WILLIAMS - *Appl. Opt.* **22**, 2914 (1983)

[3] R.A. BOSCH - *Nucl. Instr. Meth. A* **386**, 525 (1997)

[4] Y.L. MATTHIS, P. ROY, B. TREMBLAY, A. NUCARA, S. LUPI, P. CALVANI & A. GERSCHEL - *Phys. Rev. Lett.* **80**, 1220 (1998)

[5] O. CHUBAR & P. ELLEAUME - www.esrf.fr/machine/support/ids/Public/Codes/SRW/srwindex.html

[6] G.P. WILLIAMS - *SPIE*, G.L.CARR & P. DUMAS eds, **3775**, 2 (1999)

[7] P. DUMAS & G.P. WILLIAMS - *In "Chemical Applications of Synchrotron Radiation"*, T.K. SHAM ed., World Scientific (2002)

[8] G.L. CARR, R.P.S.M. LOBO, J. LAVEIGNE, D.H. REITZE & D.B. TANNER - *Phys. Rev. Lett.* **8**, 3001 (2000)

[9] L. NAHON, E. RENAULT, M.E. COUPRIE, D. NUTARELLI, D. GARZELLA, M. BILLARDON, G.L. CARR, G.P. WILLIAMS & P. DUMAS - *SPIE*, G.L.CARR & P. DUMAS eds, **3775**, 145 (1999)

[10] M. ABO-KAHR, J. FEIKES, K. HOLLDACK, G. WÜSTEFLED & H.W. HÜBERS - *Phys. Rev. Lett.* **88**, 1 (2002)

[11] F. POLACK, R. MERCIER, L. NAHON, C. ARMELLIN, J.P. MARX, M. TANGUY, M.E. COUPRIE & P. DUMAS - *SPIE*, G.L.CARR & P. DUMAS eds, **3775**, 13 (1999)

[12] A.A. MICHELSON - *Phil. Mag.* **31**, 256 (1891)

[13] T.H. DUNNING Jr - *J. Chem. Phys.* **90**, 1007 (1989)

[14] D. ANDRAE, U. HAEUSSERMANN, M. DOLG, H. STOLL & H. PREUSS - *Theoret. Chim. Acta* **77**, 123 (1990)

[15] P. FUENTEALBA, H. STOLL, L.V. SZENTPALY, P. SCHWERDTFEGER & H. PREUSS - *J. Phys. B* **16**, L323 (1983)

[16] M. DUPUIS, A. FARAZDEL, S.P. KARNA & S.A. MALUENDES - *Modern Techniques in Computational Chemistry*, E. Clementi, Leiden (1990)

[17] Y.R. SHEN - *Nature* **337**, 519 (1989)

[18] J.H. WOOD & A.M. BORING - *Phys. Rev. B* **18**, 2701 (1978)

[19] A. PEREMANS, A. TADJEDDINE, P. GUYOT-SIMONET, R. PRAZERES, F. GLOTIN, J.M. JAROZINSKI, J.M. BERSET & J.M. ORTEGA - *Nucl. Instr. Meth. A* **341**, 146 (1994)

[20] A. TADJEDDINE & A. PEREMANS - *Surf. Sci.* **368**, 377 (1996)

[21] J.M. ORTEGA - *Synchr. Rad. News* **9**, 20 (1998)

[22] R. PRAZERES, J.M. BERSET, F. GLOTIN, J.M. JAROZINSKI & J.M. ORTEGA - *Nucl. Instr. Meth. A* **331**, 15 (1993)

[23] A. PEREMANS, A. TADJEDDINE, W.Q. ZHENG & A. LE RILLE - *Ann. Physique* **20**, 527 (1996)

[24] K.C. LIN, R.G. TOBIN & P. DUMAS - *Phys. Rev. B* **49**, 17273 (1994)

[25] P. DUMAS, M. SUHREN, Y.J. CHABAL, C.J. HIRSCHMUGL & G.P. WILLIAMS - *Surf. Sci.* **371**, 200 (1997)

[26] C.J. HIRSCHMUGL, G.P. WILLIAMS, F.M. HOFFMANN & Y.J. CHABAL - *Phys. Rev. Lett.* **65**, 408 (1990)

[27] B.N.J. PERSSON - *Phys. Rev. B* **44**, 3277 (1991)

[28] M. HEIN, P. DUMAS, A. OTTO & G.P. WILLIAMS - *Surf. Sci.* **419**, 308 (1999)

[29] M. HEIN, P. DUMAS, A. OTTO & G.P. WILLIAMS - *Surf. Sci.* **465**, 249 (2000)

[30] A. PEREMANS & A. TADJEDDINE - *J. Electroanal. Chem.* **395**, 313 (1995)

[31] G.A. BOWMAKER, J.R. LÉGER, A. LE RILLE, C.A. MELENDRES & A. TADJEDDINE - *Trans. Far. Soc.* **94**, 1309 (1998)

[32] A. LE RILLE, A. TADJEDDINE, W.Q. ZHENG & A. PEREMANS - *Chem. Phys. Lett.* **271**, 95 (1997)

[33] S. TRASATTI - *Surf. Sci.* **335**, 1 (1995)

[34] A. PEREMANS, Y. CAUDANO, P. THIRY, P. DUMAS, A. LE RILLE, W.Q. ZHENG & A. TADJEDDINE - *Phys. Rev. Lett.* **78**, 2999 (1997)

[35] O. PLUCHERY & A. TADJEDDINE - *J. Electroanal. Chem.* **500**, 379 (2000)

[36] A. CHEN, D. YANG & J. LIPKOWSKI - *J. Electroanal. Chem.* **475**, 130 (1999)

NEUTRON SPECTROSCOPY

11

INELASTIC NEUTRON SCATTERING: INTRODUCTION

R. SCHERM
Physikalische Technische Bundesanstalt, Braunschweig, Germany

B. FÅK
Département de recherche fondamentale sur la matière condensée, CEA, Grenoble, France

1. INTERACTION OF NEUTRONS WITH MATTER

Thermal neutrons are often praised as the (nearly) ideal probe to inform us about structure and motion on an atomic scale. Indeed, their wavelengths, similar to X-rays, fit perfectly interatomic distances and their energies in the range 1 meV to ~1 eV match excitation energies typical for atomic or molecular oscillations. However, neutrons are only nearly perfect because there are so few of them. Starting from a high vacuum corresponding to 10^{-7} mbar inside a reactor (or spallation target), a beam of say 10^8 n cm^{-2} s^{-1} is scattered by a cm^3-sized sample. Often, only a few neutrons arrive in the detector after angular and energy analysis. This is why instrument design is so important. Use X-rays first! Formulate precise questions and optimize the experiment accordingly. Then inelastic neutron scattering becomes a powerful tool to supplement what you already know. For studies of condensed matter with neutrons, it is useful to distinguish between three cases, which provide different information on the sample properties.

- **Nuclear scattering** - Neutrons interact with nuclei *via* the very short-range nuclear force. With its wavelength λ of a few Å (10^{-10} m), a thermal neutron cannot "see" the internal structure of the nucleus, hence the scattering is isotropic (independent of angle). The fancy name for this is *s*-wave scattering. Therefore, the interaction is characterized by a scattering length b, which is of the order of the size of the nucleus, i.e. fm (10^{-15} m). X-ray scattering, on the other hand, is not isotropic (it depends on angle) and is characterized by a form factor, $f(Q)$, since the photons are scattered from all the electrons of an atom. Coherent nuclear scattering probes density correlations.

F. Hippert et al. (eds.), Neutron and X-ray Spectroscopy, 361–381.

- **Spin-dependent nuclear scattering** - The neutron-nucleus interaction depends actually on the total spin of the compound nucleus formed during the scattering event between the neutron (spin 1/2) and the nucleus (spin I). To the two possible states, ($I + 1/2$ and $I - 1/2$), correspond two different scattering lengths (b_+ and b_-) respectively. This spin dependence gives rise to incoherent scattering, which probes single-particle motion (see section 5). It also allows to study the ordering of nuclear spins at ultra-low temperatures, $\sim \mu K$.
- **The magnetic dipole moment** of the neutron senses the dipolar field from unpaired electrons. It is thus the total magnetic moment of the sample, i.e. spin S plus orbital L, which is probed. The magnetic scattering length from one electron is $b_m = \gamma r_0 = 1.348$ fm, i.e. of the same order of magnitude as the nuclear scattering length. Because of the dipole-dipole character of the interaction, the scattering is not isotropic, and only magnetic moments $\boldsymbol{M}_\perp$(or moment fluctuations) perpendicular to the wave vector transfer $\boldsymbol{Q}$ are observed. As the extension of the wave function of the unpaired electrons is of the same order as the neutron wavelength, there is a magnetic form factor $f(Q)$, similar to the case of X-ray scattering.

In addition to the useful interactions discussed above, there are so-called "nuisance" interactions, which provide no useful information about the system [1]. They are small, but might need to be corrected for. Corrections to the neutron dipole interaction include spin-orbit effects (Schwinger term), relativistic quantum effects (Foldy interaction), and nuclear dipole moments. The internal structure of the neutron gives also rise to electrostatic and electric polarizability terms. One should also keep in mind that the weak interaction limits the life time of a free neutron to some odd 10 minutes while gravity causes cold neutrons to fall as fast as Newton's apple.

2. *KINEMATICS*

2.1. *ENERGY AND MOMENTUM CONSERVATION*

A neutron is characterized by its wave vector $\boldsymbol{k}$ and its spin state σ. From $\boldsymbol{k}$ we can calculate its momentum $\boldsymbol{p}$ and velocity $\boldsymbol{v}$ *via* $\boldsymbol{p} = \hbar \boldsymbol{k} = m\boldsymbol{v}$ and also the kinetic energy, $E = p^2/2m = \hbar^2 k^2/2m = mv^2/2$, where m is the neutron mass. In cases where one deals with unpolarized neutrons, one can forget the neutron spin state σ, and the neutron state is thus fully characterized by $\boldsymbol{k}$.

In an inelastic neutron scattering experiment, one measures the number of neutrons scattered from a sample as a function of their final wave vector $\boldsymbol{k}_f$ given their initial wave vector $\boldsymbol{k}_i$. The scattering event is thus characterized by 6 variables: three components of $\boldsymbol{k}_i$ and three components of $\boldsymbol{k}_f$. However, what we are really interested in is the energy transfer $\hbar\omega$ and momentum (or wave vector) transfer $\boldsymbol{Q}$ from the neutron to the sample, given by the energy and momentum conservation rules,

$$\hbar\omega = E_i - E_f = \frac{\hbar^2}{2m}(k_i^2 - k_f^2) \qquad \{1\}$$

and

$$\boldsymbol{Q} = \boldsymbol{k}_i - \boldsymbol{k}_f \qquad \{2\}$$

respectively. This corresponds to only four variables, i.e. there are two redundant variables, which should be carefully chosen to optimize the experiment. For non-crystalline samples, only $Q = |\boldsymbol{Q}|$ and $\hbar\omega$ are meaningful, and there are four redundant variables.

2.2. SCATTERING TRIANGLE

It is useful to study the diagram in real (fig. 1a) and reciprocal (fig. 1b) space of a scattering event. The latter is called the scattering triangle, and shows the kinematical conditions $\boldsymbol{k}_i$ and $\boldsymbol{k}_f$ must fulfil to obtain a particular energy and momentum transfer; one talks about "closing the scattering triangle". From figure 1b, it is easily seen that for a given $\hbar\omega$ and $\boldsymbol{Q}$, any combination of $\boldsymbol{k}_i$ and $\boldsymbol{k}_f$ that lies on the dotted line in the figure can be chosen. This corresponds to one of the two redundant variables in the scattering event mentioned above. It is used to optimize the intensity and resolution of the experiment, as discussed in the following chapters.

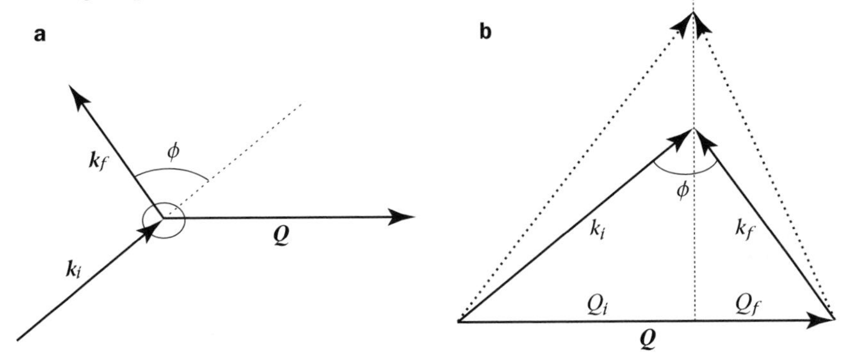

Figure 1 - Scattering event in (**a**) real space and (**b**) reciprocal space Q_i and Q_f are the projections of $\boldsymbol{k}_i$ and $\boldsymbol{k}_f$ onto $\boldsymbol{Q}$ and ϕ is the scattering angle.

Also, $\boldsymbol{k}_i$ and $\boldsymbol{k}_f$ can be rotated together around $\boldsymbol{Q}$, which corresponds to the second redundant variable.

For elastic scattering, the lengths of Q_i and Q_f in figure 1b are equal. For inelastic scattering, the dotted line moves to the right for positive energy transfers and to the left for negative energy transfers. Note that the energy transfer is linear in the displacement $\Delta Q = Q_i - Q_f$ of this line, since

$$\hbar\omega \propto (k_i^2 - k_f^2) = (Q_i - Q_f)(Q_i + Q_f) = \Delta Q\, Q$$

2.3. PARABOLAS

Combining the equations for energy and momentum transfer, equations {1-2}, gives

$$Q^2 = k_i^2 + k_f^2 - 2k_i k_f \cos\phi \qquad \{3\}$$

where ϕ is the scattering angle. Scattering at a constant angle corresponds to a parabolic trajectory in $(Q, \hbar\omega)$ space. Most neutron spectrometers work at either constant k_i or constant k_f, and the corresponding trajectories are drawn in figure 2. However, one can in principle change both of them, keeping some other quantity (like ϕ) constant.

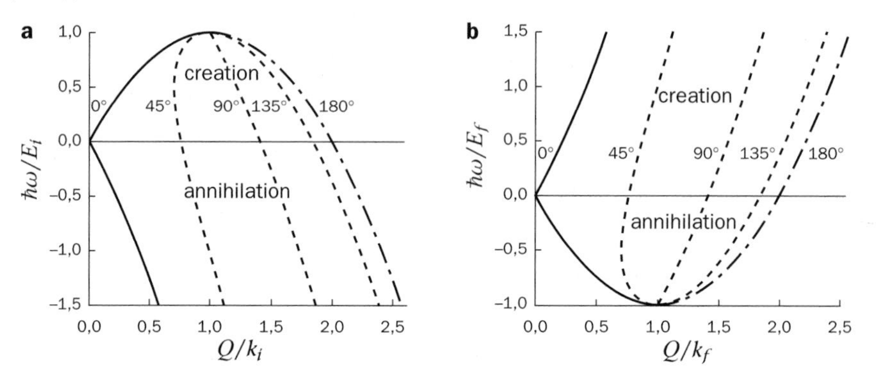

Figure 2 - Scattering parabolas for different scattering angles of scans with (**a**) k_i constant and (**b**) k_f constant.

3. MASTER EQUATION AND $S(Q, \omega)$

As neutrons interact rather weakly with matter, one may use a result from perturbation theory, Fermi's golden rule, to calculate the double differential cross-section. The result is the so-called Master Equation [2-5],

$$\frac{d^2\sigma}{d\Omega dE_f} = \left(\frac{m}{2\pi\hbar^2}\right)^2 \frac{k_f}{k_i} \sum_{n_o} p(n_0) \sum_{n_1} \left| \langle k_f \sigma_f n_1 | V | k_i \sigma_i n_0 \rangle \right|^2 \delta(\varepsilon_1 - \varepsilon_0 - \hbar\omega) \qquad \{4\}$$

We use the indices i and f for the incident and final neutron, respectively, and the indices 0 and 1 for the initial and final states of the sample, whereas ε is the energy of the sample and E the energy of the neutron. The factor k_f / k_i is related to the fact that the cross-section is defined as the ratio of the outgoing and incoming neutron currents. The first sum runs over all initial states n_0 of the system, each occurring with a probability $p(n_0) = \mathrm{e}^{-\varepsilon n_0 / k_B T} \Big/ \sum_{n_0} \mathrm{e}^{-\varepsilon n_0 / k_B T}$ at a given temperature T.

The second sum is over all final states n_1 of the system. The delta function singles out only those states (n_1, ε_1) for which energy is conserved. The matrix element $\langle \ldots \rangle$ is between the initial state of the total system, i.e. sample (n_0) plus neutron ($\boldsymbol{k}_i, \sigma_i$), and the final state of the total system, given by n_1, $\boldsymbol{k}_f$, and σ_f.

For nuclear scattering (magnetic scattering will be treated in section 7), the neutron-nucleus interaction can be written in terms of the Fermi pseudo potential, where each nucleus is "decorated" by a δ-function,

$$V(\boldsymbol{r}) = \frac{2\pi\hbar^2}{m} \sum_j b_j \delta(\boldsymbol{r} - \boldsymbol{R}_j) \qquad \{5\}$$

$\boldsymbol{R}_j$ is the position of nucleus j and b_j its scattering length. Inserting equation {5} into {4} and assuming (for simplicity) unpolarized neutrons, we obtain (after some algebra involving $|\boldsymbol{k}_i\rangle = \mathrm{e}^{i\boldsymbol{k}_i \cdot \boldsymbol{r}}$ and δ-functions)

$$\frac{d^2\sigma}{d\Omega\, dE_f} = \frac{k_f}{k_i} \sum_{n_o} p(n_0) \sum_{n_1} \left| \langle n_1 | \sum_j b_j \mathrm{e}^{i\boldsymbol{Q}\cdot\boldsymbol{R}_j} | n_0 \rangle \right|^2 \delta(\varepsilon_1 - \varepsilon_0 - \hbar\omega) \qquad \{6\}$$

This is an intermediate result, which will be used in section 5 to discuss coherent and incoherent scattering.

To proceed, one needs to introduce quantum mechanical Heisenberg operators, a treatment that can be found in standard text books, see for instance Squires [2] and Lovesey [3]. The final result, still for unpolarized neutrons, is

$$\frac{d^2\sigma}{d\Omega\, d\omega} = \frac{k_f}{k_i} \overline{b^2} S(\boldsymbol{Q},\omega) \qquad \{7\}$$

The beauty of this equation is that the scattering function or the dynamic structure factor, $S(\boldsymbol{Q},\omega)$ which describes the properties of the sample, is separated from how neutrons interact with the sample, given by $(k_f / k_i)\overline{b^2}$. Unfortunately, it becomes a bit more complicated for samples containing several types of atoms or isotopes, where the b for the different atoms have to be included in $S(\boldsymbol{Q},\omega)$. The expressions for polarized neutrons are considerably more complicated, and we refer the interested reader to references **[6-9]**.

4. CORRELATION FUNCTION

As the dynamic structure factor $S(\boldsymbol{Q},\omega)$ is a measurable quantity, it has to be a real function. Normal life happens in space $\boldsymbol{r}$ and time t, and it is therefore illustrative to inspect the double Fourier transform of $S(\boldsymbol{Q},\omega)$,

$$G(r,t) = \frac{1}{(2\pi)^3}\int d\boldsymbol{Q}\, \mathrm{e}^{-i\boldsymbol{Q}\cdot\boldsymbol{r}} \int_{-\infty}^{\infty} d\omega\, \mathrm{e}^{i\omega t} S(\boldsymbol{Q},\omega) \qquad \{8\}$$

which is called the time-dependent pair-correlation function. A rigorous derivation of $G(r,t)$ takes several pages and we refer to references **[2-5]** for details. The important point is that $G(r,t)$ is a density-density correlation function,

$$G(r,t) = \int dr' \langle \rho(r',0)\, \rho(r'+r,t)\rangle_T \equiv \langle \rho(\mathbf{0},0)\, \rho(r,t)\rangle_T \qquad \{9\}$$

Here, ρ is the density operator for all the nuclei

$$\rho(r,t) = \sum_{j=1}^{10^{23}} \delta[\boldsymbol{r} - \boldsymbol{R}_j(t)] \qquad \{10\}$$

where each nucleus is marked by a δ-function. The notation $\langle\ldots\rangle_T$ means that one needs to average over all initial states at a temperature T; it also implies that one needs to shift around the arbitrary chosen "starting point" $\boldsymbol{r} = 0$ and $t = 0$. The correlation function $G(r,t)$ describes the probability to find an atom at a position $\boldsymbol{r}$ at time t, provided one found an atom (the same or another) at $\boldsymbol{r} = 0$ and $t = 0$. This picture is correct for $T >> \hbar\omega / k_B$ (the classical limit), where $G(r,t)$ becomes a purely real function.

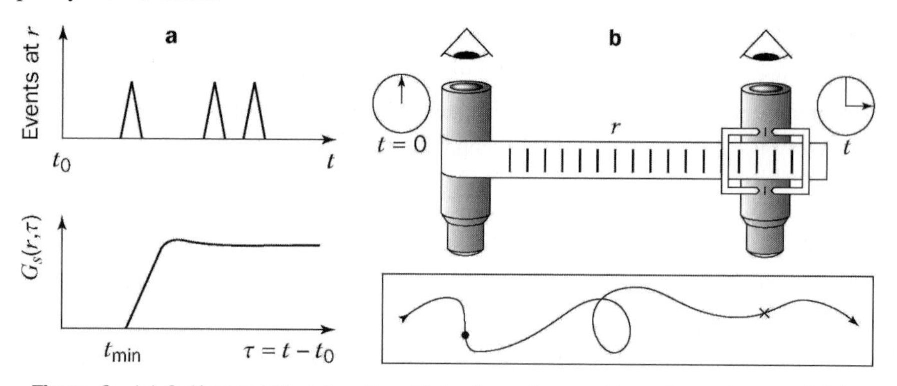

Figure 3 - (**a**) Self correlation function $G_s(r,t)$ as observed by a "correloscope" (**b**).

To visualize what $G(r,t)$ means, imagine an instrument with two looking devices separated by a distance r (see fig. 3b). Look through this "correloscope" into a box with one fly in it. Press the stop-watch once the fly has passed the left hole and note the time t when it passes the right hole. Do this repeatedly and plot the self-correlation function $G_s(r,t)$ (of one fly). The gap in $G_s(r,t)$ (see fig. 3a) reflects the minimum time $t_{\min} = r/v$ the fly needs to fly a distance r. However, if there are several flies in the box, the correlation function $G(r,t)$ will look quite different. At large r, $G(r,t)$ is constant since the flies wont know of each other then – they are uncorrelated. Only at small r and t there will be a gap in $G(r,t)$, since two flies cannot be at the same position simultaneously.

For the initiated reader, we point out the important fact that $\boldsymbol{R}(t)$ in equation $\{10\}$ is a Heisenberg operator, and hence does not commute with $\boldsymbol{R}(t')$. This implies that $\rho(t)$ does not commute with $\rho(t')$, which makes $G(r,t)$ a complex function. However, $S(\boldsymbol{Q},\omega)$ is of course still real.

$S(\boldsymbol{Q},\omega)$ or its Fourier transform $G(r,t)$ contain all information neutrons can gather about condensed matter:

- The $\boldsymbol{Q}$ dependence of $S(\boldsymbol{Q},\omega)$ or the r dependence of $G(r,t)$ tells us about structure: *where do atoms sit.*
- The ω dependence of $S(\boldsymbol{Q},\omega)$ or the t dependence of $G(r,t)$ tells us about dynamics: *how do atoms move.*
- Coherent inelastic scattering measures both the $\boldsymbol{Q}$ and ω dependence of $S(\boldsymbol{Q},\omega)$, and tells us *where and how atoms move.*

Similarly, magnetic scattering tells about structure and motion of electronic spins.

Why is $G(r,t)$ called a correlation function? Why can neutrons (or X-rays) not directly see an image of atoms? To form an image of atoms at their sites, we would need to detect the phase and amplitude of their wave function ψ, or use a neutron (or X-ray) lens. This device does not exist. The only thing accessible to measurement is the intensity $\psi^* \psi$ of the scattered neutrons, but not their phase, because of the $|\,|^2$ factor of the matrix element in the Master equation $\{4\}$. This product in $\boldsymbol{Q}$ and ω transforms into a convolution in r and t.

5. *Coherent and Incoherent Scattering*

A confusing concept for beginners in neutron scattering is that of coherent and incoherent (nuclear) scattering. To explain this, our starting point is equation $\{6\}$. For simplicity, we consider first only the part within $|\,|^2$,

$$| \, |^2 \equiv \left| \sum_j b_j \langle n_1 | \mathrm{e}^{i\boldsymbol{Q}\cdot\boldsymbol{R}_j} | n_0 \rangle \right|^2 \equiv \left| \sum_j b_j \langle j \rangle \right|^2 = \sum_{j,j'} b_{j'} b_j \langle j' \rangle^* \langle j \rangle \qquad \{11\}$$

where $\langle j \rangle$ is just an abbreviated notation and *is the complex conjugate. We have assumed for simplicity that the b are real quantities (the b are complex only for strongly absorbing nuclei). If all nuclei have the same scattering length, b can simply be taken out of the sum. However, if the nuclei have spin, or if there are several isotopes, one needs to average over the b_j. This can be done easily only if the b_j are randomly distributed, a valid assumption except at ultra-low temperature, $T \lesssim 1$ mK, where nuclear spins may be correlated. We have

$$\overline{b_{j'} b_j} = \overline{b_{j'}}\,\overline{b_j} = \bar{b}\,\bar{b} = \bar{b}^2 \qquad \text{if } j' \neq j$$

but

$$\overline{b_{j'} b_j} = \overline{b_j^2} = \overline{b^2} \qquad \text{if } j' = j \qquad \{12\}$$

Inserted in equation {11} we find two terms

$$| \, |^2 \equiv \bar{b}^2 \sum_{j \neq j'} \langle j' \rangle^* \langle j \rangle + \overline{b^2} \sum_j \langle j \rangle^* \langle j \rangle \qquad \{13\}$$

The second term describes the motion of *one* particle and is called a self-correlation function. The first term, called "distinct", would describe correlations between different particles ($j \neq j'$). However, this function cannot be measured, as quantum mechanics forbids the labelling of identical particles. In other words, one can measure $\Sigma_{jj'}$ but not $\Sigma_{j \neq j'}$. To obtain a more useful function, we can add and subtract a term with $j = j'$ and obtain

$$| \, |^2 \equiv \bar{b}^2 \sum_{j j'} \langle j' \rangle^* \langle j \rangle + \left(\overline{b^2} - \bar{b}^2 \right) \sum_j |\langle j \rangle|^2 \qquad \{14\}$$

Here, the first term is now the coherent cross-section that describes correlations between all atoms in the sample, which in full is written as

$$\left. \frac{d^2\sigma}{d\Omega\, d\omega} \right|_{\text{coh}} = \frac{k_f}{k_i} b_{\text{coh}}^2 S_{\text{coh}}(\boldsymbol{Q}, \omega) \qquad \{15\}$$

where $b_{\text{coh}} \equiv \bar{b}$. The second term is the incoherent cross-section,

$$\left. \frac{d^2\sigma}{d\Omega\, d\omega} \right|_{\text{inc}} = \frac{k_f}{k_i} b_{\text{inc}}^2 \, S_{\text{inc}}(\boldsymbol{Q}, \omega) \qquad \{16\}$$

which as before is a self-correlation function, but now has another pre-factor, $b_{\text{inc}}^2 \equiv \overline{b^2} - \bar{b}^2$. We note that if all the b are equal, then the incoherent cross-section vanishes.

6. *GENERAL PROPERTIES OF* $S(Q, \omega)$

Luckily enough, $S(\boldsymbol{Q},\omega)$ is not a completely arbitrary function, but has to follow certain rules. Some of these are most useful for nuclear scattering from monatomic systems. Our first observation is that $S(\boldsymbol{Q},\omega)$ is a distribution function in ω, i.e. it has dimension $1/\omega$ (or time). Many people use $S(\boldsymbol{Q},\hbar\omega)$, which has dimension 1/energy. On the other hand, $S(\boldsymbol{Q},\omega)$ is simply a function of $\boldsymbol{Q}$ and *not* a distribution function in Q, i.e. it has *not* dimension $1/Q$. This follows essentially from the definition of the cross-section in equation {15}, since the solid angle $d\Omega$ has dimension 1. Also, since $S(\boldsymbol{Q},\omega)$ is essentially an intensity, it is always positive.

6.1. *DETAILED BALANCE*

The second observation is the so-called detailed-balance condition,

$$S(-\boldsymbol{Q},-\omega) = e^{-\hbar\omega/k_B T} S(\boldsymbol{Q},\omega) \qquad \{17\}$$

which relates the intensity of up-scattering ($\hbar\omega < 0$) to that of down-scattering ($\hbar\omega > 0$), i.e. the annihilation (a) of an excitation to the creation ($a^\dagger$) of the excitation (see fig. 4). Clearly, at $T = 0$, no energy levels above the ground state are populated, and up-scattering is not possible. The intensity at $\omega < 0$ goes to zero: you cannot steal from a beggar. At high temperatures, the intensity is nearly the same on both sides, since the transition rate between two nearly equally populated levels is nearly the same.

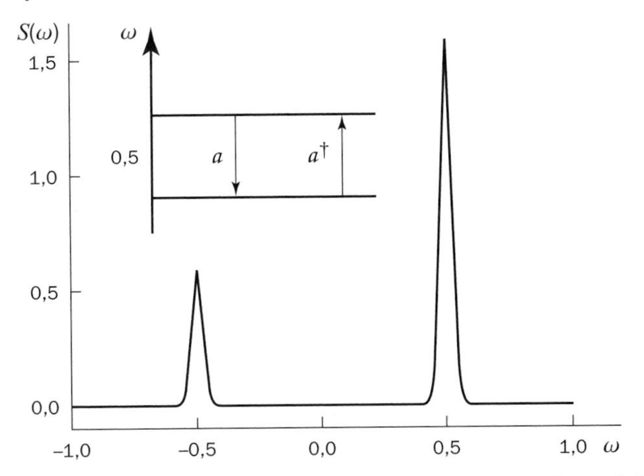

Figure 4 - Example of $S(\omega)$ for a two-level system with $\chi''(\omega = 0.5) = 1$ at a temperature corresponding to $\omega = 0.5$.

The detailed balance condition can be traced back to the sum $\sum_{n_0} p(n_0)$ in the Master equation {4} and is applicable to systems in thermal equilibrium. Sometimes a particular degree of freedom, like e.g. the rotation of H_2 molecules, is not in equilibrium with the rest of the system. Then measuring $S(\boldsymbol{Q},\omega)$ for both positive and negative ω tells about the non-equilibrium population of this subsystem. For systems with inversion symmetry (they are very common), one can replace $-\boldsymbol{Q}$ by $\boldsymbol{Q}$ in equation {17}.

As a consequence of the detailed balance condition, one can define another (complex) function, the dynamic susceptibility $\chi(\boldsymbol{Q},\omega)$. Its imaginary part χ'' is an odd function of ω, and is related to $S(\boldsymbol{Q},\omega)$ through

$$S(\boldsymbol{Q},\omega) = \frac{1}{\pi}\frac{1}{1-\mathrm{e}^{-\hbar\omega/k_BT}}\chi''(\boldsymbol{Q},\omega) \qquad \{18\}$$

The advantage with $\chi''(\boldsymbol{Q},\omega)$ is that the trivial temperature dependence of $S(\boldsymbol{Q},\omega)$ related to the population of excitations has been removed. The remaining T dependence of $\chi''(\boldsymbol{Q},\omega)$ is often more useful for understanding the underlying physics.

6.2. *MOMENTS*

Moments of the scattering function $S(\boldsymbol{Q},\omega)$ can be quite useful, in particular for liquids and gases. Generally, the nth moment is defined as

$$S_n(\boldsymbol{Q}) = \int_{-\infty}^{\infty} d\omega\, \omega^n S(\boldsymbol{Q},\omega) \qquad \{19\}$$

The zeroth moment of the coherent scattering function,

$$\int_{-\infty}^{\infty} d\omega\, S_{\text{coh}}(\boldsymbol{Q},\omega) = S_{\text{coh}}(\boldsymbol{Q}) \qquad \{20\}$$

For an isotropic system, $S(Q) \equiv S_{\text{coh}}(Q)$ is simply the static structure factor. This quantity can be measured directly with X-rays from the angular intensity distribution without energy analysis (but after correction for the atomic form factor), because of the high incoming energy of the X-rays. This is not the case with thermal neutrons, since a detector at a fixed angle traces a trajectory in Q-energy space (see fig. 2), i.e. the energy integration is not performed at constant Q. To correct for this, quite elaborate corrections need to performed to extract $S(Q)$ from neutron diffraction data.

The zeroth moment of the incoherent scattering function, on the other hand, is unity

$$\int_{-\infty}^{\infty} d\omega\, S_{\text{inc}}(\boldsymbol{Q},\omega) = 1 \qquad \{21\}$$

For solids, where most of the scattering is elastic, this relation is useful, as the intensity of the incoherent *elastic* scattering is then nearly constant, only modulated by the Debye-Waller factor $e^{-2W_Q(T)}$, which is unity for small Q's and T's.

The first moment, often called the f-sum rule, is

$$\int_{-\infty}^{\infty} d\omega\, \omega\, S(\boldsymbol{Q},\omega) = \frac{Q^2}{2M} \tag{22}$$

The first moment simply reflects the recoil energy $\hbar^2 Q^2/2M$, which is the energy a free atom, initially at rest, would take up if struck by a neutron. The f-sum rule is valid for both coherent and incoherent scattering (separately), provided that the Hamiltonian is intrinsically spin and velocity independent. This is true for most systems. As an example, $S(\boldsymbol{Q},\omega)$ of a gas is a Gaussian centred at the recoil energy, the width of which is proportional to Q and $\sqrt{T}$, due to the Doppler effect. Note that all these moments are meaningful only if the integration is between $-\infty$ and $+\infty$ *and* taken at constant Q and not at a fixed scattering angle. A question that often arises is: What is M in the f-sum rule? The answer: It depends! At low energy, M is in general the mass of the recoiling molecule (H_2, e.g.), while at higher energies it is the quasi-free atom (H, e.g.). At very high incoming energies, ~1 eV, there is no true elastic scattering, even for very heavy atoms.

6.3. TOTAL VERSUS ELASTIC SCATTERING

The dynamic structure factor $S(\boldsymbol{Q},\omega)$ and the time-dependent pair correlation function $G(\boldsymbol{r},t)$ are related *via* the double Fourier transforms

$$S(\boldsymbol{Q},\omega) = \frac{1}{2\pi}\int d\boldsymbol{r}\, e^{i\boldsymbol{Q}\cdot\boldsymbol{r}} \int_{-\infty}^{\infty} dt\, e^{-i\omega t} G(\boldsymbol{r},t) \tag{23}$$

and

$$G(\boldsymbol{r},t) = \frac{1}{(2\pi)^3}\int d\boldsymbol{Q}\, e^{-i\boldsymbol{Q}\cdot\boldsymbol{r}} \int_{-\infty}^{\infty} d\omega\, e^{i\omega t} S(\boldsymbol{Q},\omega) \tag{24}$$

Inserting the limit $t=0$ in equation {24} and letting FT denote a space Fourier transform,

$$G(\boldsymbol{r},t=0) = \Phi\mathrm{T}\int d\omega\, S(\boldsymbol{Q},\omega) = \Phi\mathrm{T}\, S(\boldsymbol{Q}) \tag{25}$$

shows that *total scattering is a snap shot*, while inserting $\omega \longrightarrow 0$ in equation {23},

$$S(\boldsymbol{Q},\omega=0) = \Phi\mathrm{T}\int dt\, G(\boldsymbol{r},t) \tag{26}$$

shows that *elastic scattering is a long exposure*. In a solid, total and elastic scattering will be quite similar, as the atoms only vibrate about their equilibrium positions. Therefore, neutron diffraction, where no energy analysis is done and which hence corresponds to a snap shot, is a good approximation to the long-time structure sought after. In a liquid or gas, on the other hand, the elastic (long exposure) scattering will have hardly any relevant Q structure, as the atoms have lost memory where they were, while the total (snap shot) scattering will show quite some structure of the instantaneous "order" around every atom. In real life, the finite energy resolution $\Delta\omega > 0$ of a neutron instrument implies that the neutron observes the system during a finite coherence time, typically $1/\Delta\omega \sim 10^{-12}$ s. Fluctuations on time scales of 10^{-9} s appear then as quasi-static in neutron scattering but can be observed to be dynamic in other (longer-time) techniques, such as NMR, μSR, and neutron spin-echo.

7. MAGNETIC SCATTERING

For magnetic scattering, i.e. the scattering of neutrons by unpaired electrons, the double differential cross-section can be written

$$\frac{d^2\sigma}{d\Omega dE_f} = \frac{k_f}{k_i}(\gamma r_0)^2 S_{\text{mag}}(\boldsymbol{Q},\omega) \qquad \{27\}$$

where $\gamma r_0 = 1.348$ fm is the strength of the dipolar neutron-electron interaction and plays the same role as the scattering length b_c for neutron-nucleus scattering (see also the chapter by R. Currat in this volume for a detailed discussion of scattering from magnetic excitations). The dynamic structure factor is given by

$$S_{\text{mag}}(\boldsymbol{Q},\omega) = \frac{1}{2\pi}\int_{-\infty}^{\infty} dt\, \mathrm{e}^{-i\omega t}\langle \boldsymbol{M}_{\perp}^{\dagger}(-\boldsymbol{Q},0)\cdot \boldsymbol{M}_{\perp}(\boldsymbol{Q},t)\rangle \qquad \{28\}$$

The magnetic interaction operator $\boldsymbol{M}(\boldsymbol{Q},t)$ is the Fourier transform of the *total* magnetization density $\boldsymbol{M}(\boldsymbol{r},t)$, i.e. neutrons observe the sum of spin and orbital contributions. Only magnetic moments (or fluctuations) that are perpendicular to the wave vector transfer $\boldsymbol{Q}$ are observed, which is expressed by that $\boldsymbol{M}_{\perp} = \hat{\boldsymbol{Q}}\times(\boldsymbol{M}\times\hat{\boldsymbol{Q}})$ rather than $\boldsymbol{M}$ occurs in the above equation. As for nuclear scattering, $S_{\text{mag}}(\boldsymbol{Q},\omega)$ describes *correlations* in space and time of the system, in this case of the magnetic moments. Similarly to equation {18}, $S_{\text{mag}}(\boldsymbol{Q},\omega)$ can be written in terms of the dynamic magnetic susceptibility $\chi(\boldsymbol{Q},\omega)$,

$$S_{\text{mag}}(\boldsymbol{Q},\omega) = \left(\frac{1}{2}gf(\boldsymbol{Q})\right)^2 \mathrm{e}^{-2W_Q}\frac{1}{\pi}\frac{1}{1-\mathrm{e}^{-\hbar\omega/k_BT}}\times\sum_{\alpha\beta}(\delta_{\alpha\beta}-\hat{Q}_\alpha\hat{Q}_\beta)\chi''_{\alpha\beta}(\boldsymbol{Q},\omega) \qquad \{29\}$$

The Debye-Waller factor e^{-2W_Q} reflects the movements of the nuclei, which drag the electron clouds with them. The sum over the Cartesian coordinates α and β expresses that only components perpendicular to $\boldsymbol{Q}$ are observed, and $\hat{Q}_\alpha$ is the component $\alpha = x, y, z$ of the unit vector along $\boldsymbol{Q}$. In contrast to nuclear scattering, the extension of the electron wave function in real space gives rise to a magnetic form factor $f(\boldsymbol{Q})$, which in general decreases with increasing Q. Hence, magnetic scattering is best observed at small Q values. In neutron scattering, the form factor is due to unpaired electrons only, while in X-ray scattering it is due to all electrons. The magnetic dynamic susceptibility can be related to the static susceptibility, χ, measured in a Squid magnetometer, for instance, through

$$\chi = \lim_{Q\to 0} \chi'(Q) = \lim_{Q\to 0} \frac{1}{\pi}\int_{-\infty}^{\infty} d\omega \frac{\chi''(\boldsymbol{Q},\omega)}{\omega} \qquad \{30\}$$

The static structure factor can also be useful in isotropic paramagnetic systems, as the total moment of the electron of an atom is obtained,

$$S_{\text{mag}}(\boldsymbol{Q},\omega) = \left(\frac{1}{2} g f(\boldsymbol{Q})\right)^2 \frac{2}{3} J(J+1) \qquad \{31\}$$

where $\boldsymbol{J} = \boldsymbol{L} + \boldsymbol{S}$ is the total moment and g is the Landé factor.

8. *RESPONSE FROM SIMPLE SYSTEMS*

8.1. *EXAMPLES*

There is a huge variety of excitations that can be observed by neutron scattering, and it is of course impossible to include them all here. It is even difficult to make a useful classification, but we will nevertheless try to do so. In table 1, we classify a few typical excitations according to their approximate energy ranges, separating nuclear from magnetic (there are even excitations of mixed character).

Another useful classification is to distinguish ***local single-particle excitations*** from (Q dependent) collective excitations. In the first case, an object rattles where it sits (rotating CH_3, H in cage, vibrating molecule, single magnetic ion). Its frequency does not depend on Q. However, the Q dependence of the intensity describes the "path" of the oscillating entity. Usually, energy spectroscopy with rather crude Q resolution in a big solid angle is the adapted method.

Table 1 - Typical excitations in condensed matter, as seen by nuclear and magnetic scattering. Single-particle excitations are denoted by S.

Energy (meV)	Nuclear	Magnetic
10^{-3}	rotational tunnelling of molecular groups (S)	
0.1-1	diffusion	spin fluctuations
1-80	phonons (acoustic, optical)	magnons
1-300		crystal field excitations (S)
~ 200	molecular vibrations (S) multiphonons recoil of molecules	
10^3	recoil of single atoms (S)	intermultiplet splitting (S) Stoner excitations

A collective excitation, on the contrary, is supported by simultaneous motion of many atoms or spins. Moving in concert, they create a propagating mode with a Q-dependent frequency. The width of this mode reflects damping, i.e. the life time of the corresponding excitation. In crystalline materials, frequencies are periodic in $\boldsymbol{Q}$ from one Brillouin zone to the next, and depend only on the reduced wave vector $\boldsymbol{q}$ from the Brillouin zone centre $\boldsymbol{\tau}$. Hence, the eigenfrequency is $\omega(\boldsymbol{q})$ with $\boldsymbol{q} = \boldsymbol{Q} - \boldsymbol{\tau}$. However, the intensity of the mode does depend on $\boldsymbol{Q}$, and measuring the same excitation $\omega(\boldsymbol{q})$ at different $\boldsymbol{Q}$ values can be extremely helpful in distinguishing different theoretical models. Technically, an instrument with good energy *and* Q resolution is here required. When dispersive excitations are measured *via* the incoherent scattering cross-section, the Q information is lost, and the density of states of the excitation spectrum is observed.

Diffusion of a single particle or spin in space, on the other hand, leads to a broadened so-called quasielastic peak around $\omega = 0$ with a Q-dependent width. This is an intermediate case between localized and collective excitations.

Multi-excitations - If $S(\boldsymbol{Q},\omega)$ contains an excitation at say ω_0, it will in general also contain multiples of this excitation. This is analogous to that a violinist playing a' always generates overtones a'', e''... If the interaction is harmonic, the frequencies of the multiexcitations are $n\,\omega_0$. If the interaction is anharmonic, the single excitations interact, and the energy levels will not be equidistant. Similarly, neutrons always generate multi-phonon excitations which includes all possible combinations $\boldsymbol{Q} = \boldsymbol{q}_1 + \boldsymbol{q}_2 + \ldots$ with $\omega = \omega_1(\boldsymbol{q}_1) + \omega_2(\boldsymbol{q}_2) + \ldots + \Delta$, where Δ is a correction for interacting phonons. While the one-phonon cross-section varies with Q^2, that for two-phonon scattering is $\propto Q^4$. Multiphonon scattering is thus important only at large Q, where it forms a broad continuum centred around the recoil energy $E = \hbar^2 Q^2 / 2M$. At very large Q (beyond 10 Å^{-1}), the image is billiard ball type collisions: a neutron kicks a single atom, which recoils. The width of this so-called "deep inelastic" peak reflects the initial momentum distribution of atoms, $n(p_0)$.

It is important to distinguish multi-phonons (or multi-excitations) from multiple scattering. In the first case, several excitations are created simultaneously in a single scattering event. In the second case, single excitations are created in (at least two) distinct scattering events, separated in time and space. Reducing the sample size reduces the ratio of multiple to single scattered neutrons. As a rule of thumb, a sample should not scatter more than 10% of all incoming neutrons to reduce the multiple-scattering contribution.

In order to distinguish magnetic from nuclear excitations with neutrons, there are two methods: Magnetic scattering decreases in general with increasing Q due to the magnetic form factor, while nuclear scattering goes approximately as Q^2. Polarization analysis is a more precise method (coherent phonons cannot flip a neutron spin while atomic spins can), but this technique is severely intensity limited.

Other experimental probes are also used to directly measure excitations. Inelastic X-ray scattering can nowadays be used to measure phonons with an energy resolution that in the best cases is comparable to neutron scattering. However, inelastic magnetic scattering is not (yet) possible. Raman scattering measures optically active modes in transparent materials, but is limited to $Q \longrightarrow 0$. NMR is a very sensitive technique measuring on long time scales (high resolution). However, it is difficult to extract any Q information, as a weighted sum over Q is measured.

8.2. *RESPONSE FUNCTIONS*

An excitation at a given frequency ω_q and with an infinite life time can be represented by a δ-function, i.e.

$$S(Q,\omega) = \left\{n(\omega)+1\right\} Z_Q \left[\delta(\omega-\omega_q)-\delta(\omega+\omega_q)\right] \tag{32}$$

where $n(\omega)+1=\left(1-e^{-\hbar\omega/k_B T}\right)^{-1}$ and Z_Q is a dimensionless structure factor. Note that one has to introduce a second δ-function at $-\omega_q$ corresponding to the annihilation of the excitation (the so-called anti-Stokes term), and that there is a minus sign in front of this second δ-function, since $\chi''(Q,\omega) \propto Z_Q\,[\ldots]$, has to be an odd function of ω. $G(r,t)$ corresponds here to a sine wave.

If the excitation has a finite life time, but still remains well-defined, one can replace the two δ-functions in equation {32} by Lorentzians, $\delta(\omega \pm \omega_q) \longrightarrow \pi^{-1}\Gamma_q \big/ \left[(\omega \pm \omega_q)^2 + \Gamma_q^2\right]$, to obtain

$$S(Q,\omega) = \frac{1}{\pi}\{n(\omega)+1\} Z_Q \left[\frac{\Gamma_q}{(\omega-\omega_q)^2+\Gamma_q^2} - \frac{\Gamma_q}{(\omega+\omega_q)+\Gamma_q^2}\right] \qquad \{33\}$$

Here, Γ_q is a q-dependent linewidth, corresponding to the *half* width at half maximum of the peak. This Lorentzian form corresponds to an exponential decay of the excitation in time, i.e. the total time response is a damped sine wave. Equation {33} is often rewritten in terms of a damped harmonic oscillator,

$$S(Q,\omega) = \frac{1}{\pi}\{n(\omega)+1\} Z_Q \frac{4\omega\omega_q\Gamma_q}{(\omega^2-\Omega_q^2)^2+4\omega_q^2\Gamma_q^2} \qquad \{34\}$$

where $\Omega_q^2 = \omega_q^2 + \Gamma_q^2$. Overdamped excitations, which are not eigenmodes of the system, can be obtained from the above formula taking $\Omega_q / \Gamma_q \longrightarrow 0$ in the limit of small ω. One then obtains a quasielastic Lorentzian,

$$S(Q,\omega) = \frac{1}{\pi}\{n(\omega)+1\} \frac{Z_Q}{\Omega_q} \frac{\omega\Gamma_Q}{\omega^2+\Gamma_Q^2} \qquad \{35\}$$

(after having redefined Ω_q and Γ_q). A quasielastic Lorentzian describes a relaxation behaviour, and the Fourier transform describes an exponential decay, $e^{-t/\tau}$, where $\tau^{-1} = \Gamma_q$ is the half-width at half maximum. Note that equation {35} is valid only for small ω, and violates e.g. the f-sum rule if integrated to $\pm\infty$. For magnetic scattering, Z_Q should in the above formulae simply be replaced by $\chi'(Q)$.

9. *INSTRUMENTATION*

Inelastic neutron scattering covers more than six orders of magnitude in energy transfer (see tab. 1), and there is no way a single instrument can cover, and even less be optimized for, such a large range. We have also seen that some inelastic experiments on single crystals require information about both $\hbar\omega$ and $\boldsymbol{Q}$, in which case both the vectors $\boldsymbol{k}_i$ and $\boldsymbol{k}_f$ need to be determined. Other experiments requiring only $\hbar\omega$ and Q from an isotropic sample, e.g. a powder or a liquid, need k_i, k_f, and the scattering angle. Finally, experiments recording only $\hbar\omega$ need only k_i and k_f. Most instruments use equation {1} to determine the energy transfer, i.e. they determine the incoming and outgoing energies separately. One exception is the neutron spin-echo (NSE) technique, discussed in length in reference **[10]** and in the chapter of R. Cywinsky in this volume, where the energy transfer (or rather the time, its conjugate variable) is determined directly, knowing E_i and E_f within only 10% precision or so.

Two principal methods are used to monochromate neutrons:

- Bragg reflection from a (single) crystal produces a beam of more or less monochromatic neutrons.
- Time-of-flight: the beam is chopped into short pulses of some 20 μs and its velocity is analyzed over a flight path of several meters.

Both monochromating techniques can be used for k_i and k_f, which yields the classification of inelastic instruments given in table 2.

Table 2 - Classification of instruments for inelastic scattering, with typical examples from ILL and ISIS. NSE stands for neutron spin-echo and TOF for time-of-flight.

Instrument	**Monochromator k_i**	**Analyzer k_f**	**Example**
Triple-axis	Crystal	Crystal	IN8, IN14
High-resolution backscattering	Crystal	Crystal	IN10
Direct (hybrid) TOF	Crystal	TOF	IN6
Crystal analyzer or indirect TOF	TOF	Crystal	IRIS
Chopper spectrometer	TOF	TOF	IN5, MARI
Neutron spin-echo	NSE	NSE	IN11, ZETA

For completeness, we mention that filters, which selectively absorb neutrons within some energy range, work similar to crystals, but with reduced flexibility. In addition, there are spin-echo spectrometers as already mentioned. The coverage in energy and Q space is shown in figure 5 for some typical neutron spectrometers. Optimization of an inelastic neutron scattering experiment is a delicate balance between a sufficiently high k_i (or k_f) so that the interesting Q and $\hbar\omega$ range is covered and a sufficiently low k_i (or k_f) to obtain good resolution. One also needs to take into account how the flux varies with k_i as well as the resolution volumes (see the chapter of R. Currat in this volume for a detailed discussion). In cases where the neutron absorption ($\sigma_a \propto 1/k_i$ in most cases) is strong, a higher k_i might be necessary.

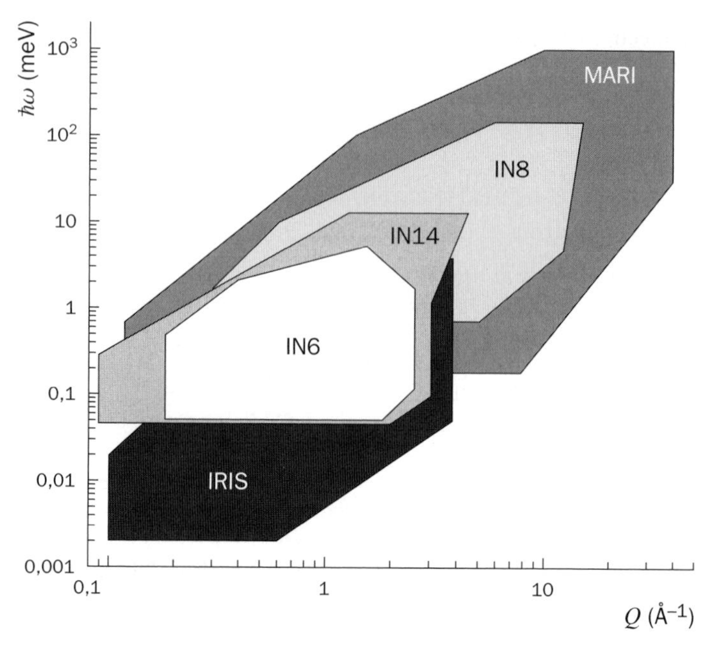

Figure 5 - Approximate coverage in energy and wave-vector transfer for some world-leading neutron spectrometers. Only positive energies (neutron energy loss) are considered, and the low-energy limit is given by the energy resolution.

9.1. TAS

Triple-axis spectrometers (TAS), discussed in length in the chapter of R. Currat in this volume, use crystals as both monochromator and analyzer. They are ideal to study dispersive excitations in crystalline samples, as both $\boldsymbol{Q}$ and $\hbar\omega$ are measured. The sample scatters of course into all directions, but the instrument picks up only a very small solid angle. However, the scan is done in a highly controlled fashion to answer precise questions at carefully selected $\boldsymbol{Q}$ values, where the most important information about the system can be obtained. One of the huge advantages with TAS is the possibility to polarize the incoming neutron beam and also to analyze the polarization of the outgoing beam. The triple-axis technique is used exclusively on continuous neutron sources, such as the ILL high-flux reactor.

A special case of triple-axis spectrometer is the backscattering (BS) instrument, which uses the monochromator and analyzer in backscattering ($\theta \longrightarrow 90°$) to improve the resolution. In this case, the flexibility of a TAS is lost and mostly non-dispersive systems are studied (see chapter by B. Frick in this volume). A recent development is the use of the resonant (zero-field) spin-echo technique on top of a triple-axis spectrometer, which extends the classical quasi-elastic applications of NSE to measurements of linewidths at finite energy transfers, even for dispersive excitations.

9.2. TOF

Time-of-flight spectrometers (TOF), discussed in length in the chapter of R. Eccleston in this volume, can be classified according to two schemes:

- If the incoming beam is monochromatic (fixed k_i), they are said to be "direct" geometry, while if a fixed energy (i.e. k_f) of the outgoing beam is selected, one talks about "indirect" or "inverted" geometry.
- If a TOF spectrometer uses a mixture of crystals and time-of-flight, the key words are hybrid or crystal-analyzer spectrometers; if TOF is used for both incoming and outgoing beam, one talks about chopper spectrometers.

Pulsed neutron sources, such as the ISIS spallation source, are ideal for instruments where the incoming beam is analyzed with TOF. TOF spectrometers simultaneously measure neutrons scattered into a large solid angle, subdivided into many angular channels.

Time-of-flight spectrometers are very good for isotropic samples, non-dispersive excitations, or to obtain an overview of dispersive excitations (so-called mapping), but are intrinsically less suited to measure dispersive excitations in three-dimensional crystals. However, progress has been made for low-dimensional systems, or even certain three-dimensional systems with well-defined and well-understood excitations. While TOF spectrometers are increasingly "invading" TAS territory, new techniques on the TAS makes it more useful for mapping excitations and thus "invading" TOF territory. Recent progress is discussed in the chapters of R. Currat and R. Eccleston in this volume. Polarized neutrons are used mostly on TAS, with the exception of D7 at the ILL, where polarization is combined with time-of-flight.

10. HOW TO BEAT STATISTICS

Consider the case where a young brilliant scientist wants to determine the area and peak position of a peak measured on a triple-axis spectrometer to some given precision. Let us assume for simplicity that the peak is Gaussian, there is no background, each point is counted for the same amount of time, and the scanning variable x is scanned with equidistant steps. At each point i (or x_i) of the scan, N_i

counts are observed. Since the count rate in a detector follows Poisson statistics, this means that the one-sigma uncertainty in N_i is $\delta N_i = \pm\sqrt{N_i}$ for each point i. The total number of counts in the whole peak, assuming there is no background, is $N = \sum_i N_i$, with uncertainty $\pm\sqrt{N}$. Counting for a longer time decreases the relative uncertainty, $\delta N/N = 1/\sqrt{N}$.

Less trivial is the uncertainty in the peak position. The first moment, $\bar{x} = \sum_i x_i N_i / N$, is a good measure of the peak position. It is easy to show that its uncertainty is $\delta\bar{x} = \sigma/\sqrt{N} = \mathrm{FWHM}/(2.355\sqrt{N})$, where σ is the standard deviation, FWHM the full width at half maximum of the peak, and the factor $2.355 = (8 \ln 2)^{1/2}$ comes from the assumption of a Gaussian line shape. From this equation, the young scientist realizes that there are two ways to improve the precision in the determination of the peak position. One is to decrease the peak width by improving the resolution (assuming that the peak is a δ-function and the Gaussian shape comes from the instrumental resolution), the other is to improve the counting statistics by counting for longer times, hence increasing N. While the first one seems natural (the narrower the peak, the better one can determine its position), this is often not the best way in practice. The reason is that improving the resolution comes at some expense: the intensity goes in general as W^n of the resolution width W, with $n \approx 2$-4. If $n > 2$, it is more efficient to keep the resolution fix and just count longer. Note also that some points are worth more than others: counting on the maximum gives no information on the peak position, counting near the half height is most sensitive.

There are a few more important aspects here, related to systematic errors:

- Positioning errors in x_i can give a bigger uncertainty in $\bar{x}$ than the statistics.
- It is in general better to repeat a scan many times than just counting longer at each point.
- Be sure there is only one peak. You cannot know this if you only count longer and do not improve the resolution. This is a safe way to miss a Nobel prize!

REFERENCES

[1] V.F. SEARS - *Physics Reports* **141**, 281 (1986)

[2] G.L. SQUIRES - *Introduction to the theory of thermal neutron scattering,* Cambridge University Press (1978)

[3] S.W. LOVESEY - *Theory of neutron scattering from condensed matter,* Clarendon Press, Oxford (1984)

[4] D.L. PRICE & K. SKÖLD - *In "Neutron scattering"*, K. SKÖLD & D.L. PRICE eds, *Methods of Experimental Physics* **23A**, Academic Press, New York (1986)

[5] R. SCHERM - *Ann. Physique* **7**, 349 (1972)

[6] W.G. WILLIAMS - *Polarized Neutrons*, Clarendon Press, Oxford (1988)

[7] L.P. REGNAULT, H.M. RØNNOW, C. BOULLIER, J.E. LORENZO & C. MARIN - *Physica B* **345**, 111 (2004)

[8] M. BLUME, *Phys. Rev.* **130**, 1670 (1963)

[9] S.V. MALEYEV - *Physica B* **297**, 67 (2001);
S.V. MALEYEV - *Physica B* **267-268**, 236 (1999);
S.V. MALEYEV, G. BAR'YAKHTAR & R.A. SURIS - *Sov. Phys. Solid State* **4**, 2533 (1963)

[10] F. MEZEI ed. - Neutron spin-echo, *Lecture Notes in Physics* **128**, Springer, Berlin (1980)

12

THREE-AXIS INELASTIC NEUTRON SCATTERING

R. CURRAT
Institut Laue-Langevin, Grenoble, France

1. PRINCIPLE OF THE TECHNIQUE

In the field of inelastic neutron scattering, *three-axis* (TAS) and *time-of-flight* (TOF) spectroscopy are the two most general and most widely used techniques. Their respective merits, with reference to specific experiments, will be discussed in some detail further on (see section 3). In general terms, the TAS technique is the technique of choice whenever it is desirable to have a precise control of the positions in $(\boldsymbol{Q},\omega)$ space where data are to be collected. With the TAS method $(\boldsymbol{Q},\omega)$ space is explored in a step-by-step manner, each spectrometer configuration corresponding to a well-defined value of $\boldsymbol{k}_i$ and $\boldsymbol{k}_f$, the incident and scattered neutron wavevectors, and hence of $\boldsymbol{Q}$ and $\hbar\omega$ [1]

$$\boldsymbol{k}_f - \boldsymbol{k}_i = \boldsymbol{Q} \qquad \text{(a)}$$

$$\frac{\hbar^2 k_i^{\,2}}{2M_n} - \frac{\hbar^2 k_f^{\,2}}{2M_n} = \hbar\omega \qquad \text{(b)} \qquad \{1\}$$

A typical three-axis spectrometer (TAS) set-up is shown schematically in figure 1. The incident neutron wavevector $\boldsymbol{k}_i$ is selected by Bragg reflection on a crystal monochromator (angles labelled $\mathrm{A1} \equiv \theta_m$ and $\mathrm{A2} \equiv 2\theta_m$ in the figure). The orientation of the vector $\boldsymbol{k}_i$ in the specimen's reciprocal space is controlled by orienting the specimen with respect to $\boldsymbol{k}_i$ (rotation around a vertical axis (A3) + double-goniometer or eulerian cradle). The scattered neutron wavevector $\boldsymbol{k}_f$ is

1. Different sign conventions are possible for the momentum and energy transfer variables. The convention in eq. (1.1a) ensures that in Ewald-type constructions (cf. fig. 2 and 3) the vector $\boldsymbol{Q}$ has its origin at the origin of reciprocal space, while the incident and scattered neutron wavevectors, $\boldsymbol{k}_i$ and $\boldsymbol{k}_f$, have a common origin at the centre of the Ewald sphere. With the sign convention in eq. (1.1b) for $\hbar\omega$, positive energy transfer values correspond to neutron energy loss processes (Stokes processes), which are dominant at low temperature.

F. Hippert et al. (eds.), Neutron and X-ray Spectroscopy, 383–425.

selected by Bragg reflection on a crystal analyzer (A5 ≡ θ_a, A6 ≡ $2\theta_a$). The orientation of the vector $\boldsymbol{k}_f$ in the specimen's reciprocal space is determined by the value of the scattering angle at the sample position (A4).

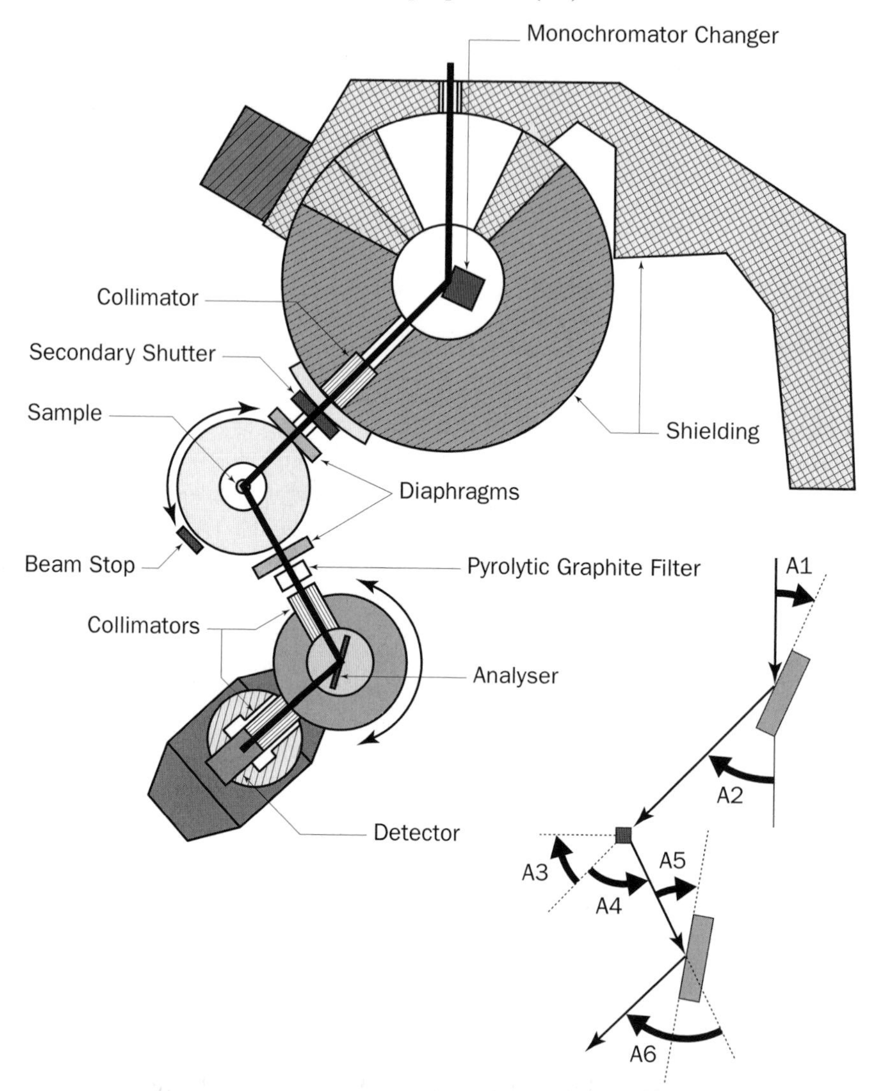

Figure 1 - A typical three-axis spectrometer set-up at a reactor thermal beam-port (IN20 at ILL).

Figure 2 shows the reciprocal space diagram (solid lines) corresponding to the spectrometer configuration in figure 1. We have assumed $k_i \neq k_f$ (finite energy transfer $\hbar\omega$) and we have decomposed the total momentum transfer $\boldsymbol{Q}$ into a reciprocal lattice vector τ_{hkl} and a wavevector q. In such a configuration, the spectrometer is set to measure a collective excitation with a dispersion $\omega(q)$, and the measurement is said to "take place in the [hkl] Brillouin zone" in the specimen's reciprocal lattice. In principle the same excitation can be measured in any Brillouin zone (i.e. near any τ_{hkl}), albeit with varying intensities.

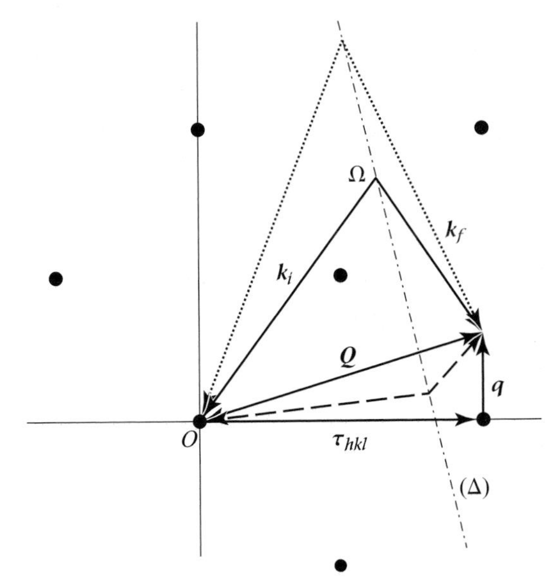

Figure 2 - Reciprocal space representation of an inelastic measurement in a single-crystal specimen. The solid lines correspond to the TAS configuration in figure 1. Dashed and dotted lines show alternate configurations leading to the same momentum and energy transfers (Q, ω).

The same measurement, at the same (Q,ω) point, can be performed using an infinity of different instrument configurations: there are 4 degrees of freedom associated with a particular (k_i, k_f) choice in the horizontal scattering plane, and only 3 constraints associated with the (Q,ω) requirement. This is illustrated by the broken and dotted lines in figure 2, which show two alternate (k_i, k_f) pairs corresponding to the same (Q,ω) point. All these configurations can be obtained by changing both k_i and k_f in such a way that the common origins (Ω) of the vectors k_i and k_f lie on the same line (Δ) normal to Q. In practice, however, all these configurations are *not equivalent*, as they lead to different intensity/resolution characteristics, and the choice of the value of k_i or k_f is an important part of the spectrometer optimization procedure, prior to any inelastic measurement (see section 11).

One characteristic feature of the TAS technique is the possibility of performing arbitrary scans in ($\boldsymbol{Q}$,ω) space, by step-by-step modifications of the spectrometer configuration. The *constant-$\boldsymbol{Q}$ scan* is the most commonly used mode of operation, since information collected in that mode can be directly related to the specimen's *dynamic susceptibility*. Model predictions, to which the experimental results are to be compared, often refer to the form of the dynamic susceptibility $\chi(\boldsymbol{Q},\omega)$ at specific high-symmetry points in $\boldsymbol{Q}$-space. The relevant susceptibility may be either the magnetic susceptibility or a generalized mode-susceptibility, in the lattice-dynamical case. In all cases, the direct experimental determination of $\mathrm{Im}\,\chi(\boldsymbol{Q},\omega)$ at constant-$\boldsymbol{Q}$ [and, by energy integration, of $\chi(\boldsymbol{Q},\omega=0)$] is of major interest and TAS constant-$\boldsymbol{Q}$ spectra have provided unique and often decisive experimental evidence in many physical problems of past and current interest.

For the same reason that a particular ($\boldsymbol{Q}$,ω) point can be obtained from an infinite set of different TAS configurations, a particular scan can be performed in an infinite number of different ways. Figures 3a and 3b show two different ways of performing the same constant-$\boldsymbol{Q}$ scan, keeping $k_i = |\boldsymbol{k}_i|$ constant in the first case and $k_f = |\boldsymbol{k}_f|$ constant in the second case. These two modes are the most commonly used, for reason of simplicity, but other modes (constant sample orientation A3, constant scattering angle A4) are in principle also possible.

In the constant-$\boldsymbol{Q}$ constant-k_i mode, angles A3, A4, A5 and A6 are readjusted at each scan-point (A1, A2, A3 and A4 in the constant-$\boldsymbol{Q}$ constant-k_f mode), while only A3 and A4 move in constant-energy scans.

2. *THE THREE-AXIS SPECTROMETER*

TAS instruments can be found on all the steady-state neutron sources where the in-pile neutron flux is high enough to permit inelastic scattering experiments. Normally, each TAS spectrometer is optimized for a specific incident neutron energy range. As an example, the ILL operates:

- ♦ one cold-source TAS (IN14; $2 < E_i < 15$ meV) installed on a cold neutron guide;
- ♦ one thermal-beam TAS (IN8; $15 < E_i < 100$ meV);
- ♦ one thermal-beam TAS optimized for polarized neutron work (IN20; same energy range);
- ♦ one thermal-guide TAS for instrumentation developments (IN3; $15 < E_i < 50$ meV);
- ♦ one hot-source TAS (IN1; $80 < E_i < 500$ meV).

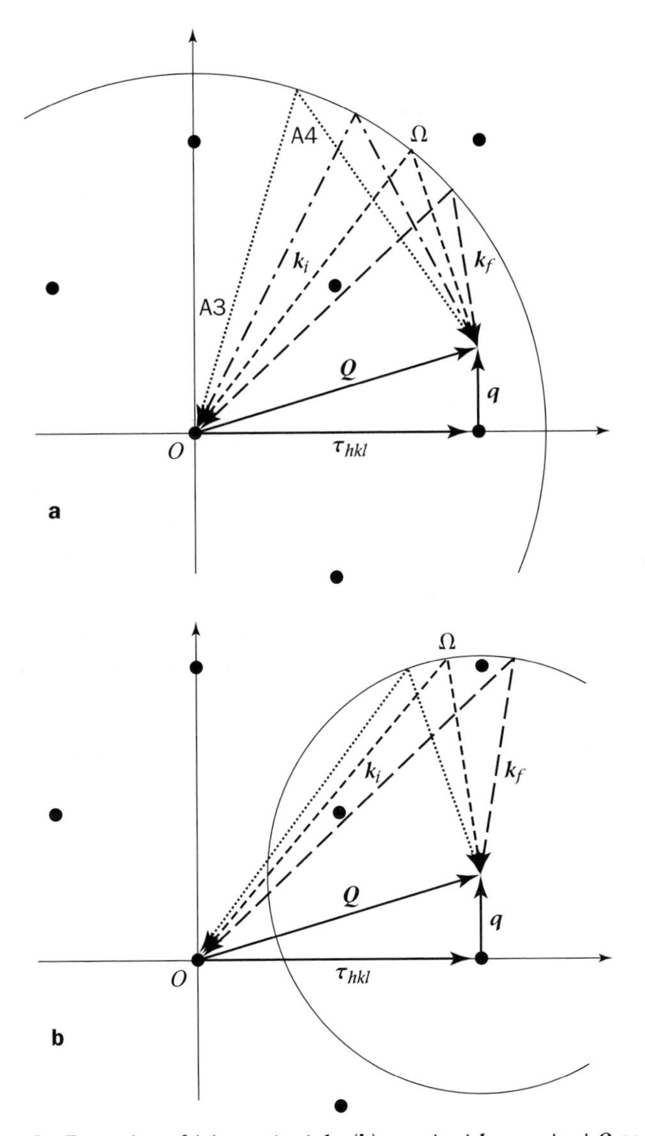

Figure 3 - Examples of (**a**) constant-k_i, (**b**) constant-k_f constant-Q scans.

Two other TAS instruments (IN12, on a cold guide, and IN22, on a thermal-supermirror guide) are operated by Collaborative Research Groups (CRGs) and are partially accessible to ILL users. Except for IN1 and IN8, all instruments can be used for either unpolarized- or polarized-beam experiments.

The neutron beam *monochromation* is achieved by Bragg reflection of a large single crystal or, more frequently, an assembly of smaller single-crystal pieces oriented in such a way as to focus the monochromatic beam onto the specimen. Figure 4 shows a state-of-the-art focussing monochromator set-up on a thermal-beam TAS (IN8 at ILL). In that case, 3 monochromator crystals are permanently installed inside the primary protection and are remotely interchanged and oriented.

Figure 4 - 3-face doubly-focussing monochromator on a thermal-beam TAS (IN8 at ILL). The two identical faces shown in the left picture are to be equipped with 99 (9×11) pyrolytic graphite (PG002; d_m = 3.355 Å) and copper (Cu200; d_m = 1.8074 Å) pieces, respectively. Each copper piece (25×17 mm^2) is itself composed of several copper single-crystal plates, purposely miscut in order to simulate an anisotropic mosaic distribution. The third face (right picture) is composed of 11 elastically bent perfect silicon strips (Si111; d_m = 3.1354 Å).

The choice of the monochromator *d*-spacing d_m affects the energy width ΔE_i of the "monochromatic" beam, through

$$\Delta E_i / E_i = 2\, \Delta k_i / k_i = 2 \cot \theta_m\, \Delta\theta_m \qquad \{2\}$$

with $\lambda_i = 2\,\pi / k_i = 2\, d_m \sin \theta_m$.

For a given incident energy, a large d_m entails a small monochromator Bragg angle θ_m and hence a large $\Delta E_i / E_i$ (poor energy resolution). The uncertainty on the Bragg angle $\Delta\theta_m$ in equation {2}, is a function of the beam *collimations* (before and after the monochromator crystal) and of the *mosaic* width of the monochromator crystal itself.

The important monochromator characteristics are therefore the *peak reflectivity* and *mosaic widths*. The horizontal mosaic width η_m should be consistent with horizontal collimations, typically 20' to 40', while the vertical mosaic width η'_m should be as narrow as possible. The techniques used to prepare single-crystal pieces with anisotropic mosaic distributions are discussed in **[1,2]**. Typical monochromator materials are pyrolytic graphite, copper, germanium and silicon. Odd-index reflections for diamond structure materials, such as Si111 or Ge111, are free from second-harmonic contamination (no $\lambda_i/2$ neutrons in the reflected beam) since the (Si,Ge)222 reflection is, in principle, extinct and nearly so in practice. We shall see in section 4 how the mosaic width and peak reflectivity of the monochromator (and analyzer) influence the luminosity (i.e. the transmission function) of the spectrometer.

For polarized work, the monochromation and spin-polarization of the incident beam can be achieved *in one step* using a ferromagnetic monochromator crystal, such as Cu_2MnAl (Heusler alloy; see fig. 5) or Fe_8Co_{92}. In these two materials, there exists a strong, low-index Bragg reflection [Heusler(111) and Fe_8Co_{92}(200)] for which the magnetic and nuclear scattering amplitudes are of equal magnitudes and thus add up or cancel each other depending on the neutron spin state.

Figure 5 - Horizontally-focussing Heusler analyzer. Focussing is achieved by rotating the 13 Cu_2MnAl blades about vertical axes. The magnetic circuit consists of NdFeB magnets and polar pieces (here covered with an absorbing cadmium foil), placed above and below the crystals and connected by a soft-iron C-shaped yoke. The magnetic saturation of the crystals is complete when the applied vertical field is everywhere larger than ≈ 0.12 tesla.

In principle, the reflected beam then contains only one spin state, parallel or antiparallel to the monochromator magnetization direction (antiparallel, in the Heusler case). In practice polarization efficiencies are in the range 0.90-0.95 depending upon neutron wavelength and field homogeneity. The monochromation and spin-polarization can also be performed as two separate steps, using a non-magnetic monochromator and a polarizing supermirror bender [3,4] or a ^{3}He spin-filter [5,6].

The nature and characteristics of the crystal used as *analyzer* are similar to those of the monochromator. The analyzer *d*-spacing d_a must be adapted to the range of scattered neutron energies E_f to be analyzed and to the energy resolution ΔE_f required. The detector is generally a simple ^{3}He-gas proportional counter. The thickness of the counter and the gas pressure are chosen in order to achieve a counting efficiency of $\approx$ 80%, in the relevant neutron energy range.

The monochromator stage is embedded in a heavy biological shielding, 60 to 80 cm thick, designed to absorb thermal, epithermal and fast neutrons, as well as γ-rays present in the primary beam or produced by neutron capture. The analyzer and detector are protected against stray radiation by a $\approx$ 30 cm thick protection, with a hydrogenous material such as polyethylene or paraffin wax, to thermalize epithermal and fast neutrons, and an absorbing element such as cadmium, gadolinium or boron. The importance of this protection is easily understood if one compares the value of the detector count rate for a typical inelastic signal (5-20 neutrons.min^{-1}) to the neutron flux incident on the specimen (10^7-10^8 neutrons.cm^{-2}.s^{-1}). The elimination of parasitic signal and background sources is therefore essential.

As seen in figure 1, a variety of devices and optical elements may be inserted anywhere along the neutron beam path: parallel blade collimators (Soller collimators) and adjustable diaphragms to limit, respectively, the horizontal divergence of the beam and its lateral or vertical dimensions; low efficiency (10^{-2}-10^{-5}) counters to monitor the intensity of the beam incident on the specimen or the intensity of the beam incident on the analyzer; filters (oriented pyrolytic graphite, polycrystalline Be or BeO, resonance filters) to eliminate harmonic contamination (n k_i or n k_f with n = 2, 3...). Pyrolytic graphite (PG) is used mainly in the range $k_{i,f}$ = 2.55-2.67 Å^{-1}, to eliminate 2 $k_{i,f}$ neutrons (transmission: 2×10^{-3}) and, to a lesser extent 3 $k_{i,f}$ and 4 $k_{i,f}$ neutrons. It can also be used to reduce 2 $k_{i,f}$ contamination at a number of other $k_{i,f}$ values: 1.48, 1.64, 1.97, 3.85 and 4.1 Å^{-1}. Cooled polycrystalline Be or BeO filters are well suited to the cold neutron range since they allow to eliminate all neutrons with energies above the Bragg cut-off (respectively 1.58 $Å^{-1}$ and 1.38 $Å^{-1}$). In all cases (PG, Be, BeO) the elimination takes place through a Bragg scattering process, which implies that, unless carefully shielded, the filter may act as an efficient source of background!

For polarized neutron experiments, the neutron beam path must be equipped with *guide fields* (20-100 Gauss) in order to prevent depolarization of the neutron beam, between the spin-polarizer and the spin-analyzer (Heusler monochromator and analyzer crystals, polarizing supermirror benders or ^{3}He spin-filters). To be able to measure spin-dependent scattering cross-sections, the spectrometer must be equipped with at least one *spin-flipper* that reverses the beam polarization when energized. Different types of flipper devices are available: radio-frequency coils [7], Mezei-type flippers [8], cryogenic flippers [5,9]. Because of its simplicity, the Mezei-type flipper is the most widely used on TAS instruments. It consists of two flat solenoidal coils generating magnetic fields in two perpendicular directions normal to the beam axis. The currents in the coils are adjusted to obtain a 180° precession of the polarization. Although compact and inexpensive, the Mezei-type flipper has two disadvantages: the current settings vary with neutron velocities and they are sensitive to stray magnetic fields, which can lead to severe problems when the stray fields vary during the measurements. The cryoflipper, which makes use of the diamagnetic properties of a superconducting Nb foil, is free from such problems. However it does require permanent cryogenic maintenance. A similar technique is used in the CRYOPAD (Cryogenic Polarization Analysis Device) device (see fig. 6) which allows the polarization of the incident (scattered) beam to be directed (analyzed) along any direction (*spherical neutron polarimetry*).

3. THE TOF VERSUS TAS CHOICE

A simple way of performing an inelastic neutron scattering experiment consists in directing a beam of monochromatic neutrons onto the specimen to be studied, while monitoring the energy and angular distribution of the scattered neutrons. This technique is used in direct geometry *time-of-flight* (TOF) spectroscopy (see chapter by R. Eccleston in this volume) where the incident monochromatic beam is chopped into short pulses and the energy of the scattered neutrons is measured by monitoring the time of arrival in the detector. When used in conjunction with large arrays of position-sensitive detectors (PSDs), the TOF technique enables one to probe the specimen's response function *simultaneously* over a wide angular range of scattered neutrons (i.e. over a wide portion of reciprocal space).

Figure 6 - The CRYOPAD device for spherical polarization analysis mounted on the IN14 cold-source TAS at ILL. With this set-up the incident beam polarization can be directed along three orthogonal directions and, in each case, the scattered beam polarization is analyzed along the same three directions. The technique, routinely used in magnetic structure analysis [9], has been recently applied to inelastic scattering studies [10].

One disadvantage of the TOF method, however, can be seen directly from the form of the momentum and energy conservation equations (eq. {1a} and {1b}) which shows that at fixed incident neutron wavevector $\boldsymbol{k}_i$, a variation in $k_f = |\boldsymbol{k}_f|$, entails a variation in both the magnitude and direction of the momentum transfer $\boldsymbol{Q}$, in addition to a variation in the energy transfer $\hbar\omega$. Hence in the TOF method $\boldsymbol{Q}$ and ω are *coupled* in a way which depends only on $\boldsymbol{k}_i$ and on the scattering angle. For a *fixed orientation* of a single-crystal specimen, $\hbar\omega$-scans at constant $\boldsymbol{Q}$ (constant-$\boldsymbol{Q}$ scans) and reciprocal space surveys at constant-$\hbar\omega$ (constant-energy scans) are not possible. Of course, if necessary, these scans can be reconstructed from series of data sets taken with different specimen orientations or different $\boldsymbol{k}_i$, but the data collection time is correspondingly increased.

As pointed out in section 1, TAS is a flexible and selective technique which allows data collection along arbitrary paths in $(\boldsymbol{Q},\omega)$ space, with the possibility to *optimize* the experimental conditions for each scan. With a single-crystal sample, this implies reorienting the sample at each scan point. As a consequence, TAS *multiplexing* (i.e. operating several analyzer-detector arms in parallel) cannot be achieved without sacrificing part of the instrument's selectivity (see section 7). TOF on the other hand, is inherently non-selective ($\boldsymbol{Q}$ and ω are coupled) but very easy to multiplex using large PSD arrays (at least for direct geometry spectrometers).

Hence the choice between TAS and TOF, for a given experiment, corresponds to a trade-off between selectivity and efficiency:

- ♦ TAS is ideal for experiments where information is needed on specific, well-localized features in $(\boldsymbol{Q},\omega)$ space, e.g. repetitive measurements as a function of one or more external parameters (temperature, pressure, dopant concentration…).
- ♦ TOF is ideal for dynamical studies on isotropic (liquids, glasses, powders) or low-dimensional systems (when selectivity in $\boldsymbol{Q}$ is less important) and, more generally, for studies where large sections of $(\boldsymbol{Q},\omega)$ space need to be explored for each set of external parameter values.

In terms of monochromatic flux at the sample, today's best chopper spectrometers, such as MARI, HET or MAPS at ISIS [2], achieve time-averaged flux values in the range 10^4-10^5 neutrons cm^{-2} s^{-1} at thermal energies (50 meV), to be compared with 10^8-10^9 on a thermal TAS instrument at ILL (IN8 [11]). This large difference in flux is due partly to the characteristics of the two sources and partly to the ability of modern crystal spectrometers to use focussing Bragg optics to concentrate large diverging beams onto small samples.

On the other hand, in the TOF technique, the pulsed nature of the beam enables one to measure simultaneously a wide range of energy transfers. More importantly, the use of wide solid-angle PSD arrays allow information to be collected simultaneously in a large number of independent pixels (typically 10^3 to 10^5). Hence, if even a limited fraction of this information is of actual physical interest, the chopper technique can readily overcome its disadvantage in terms of lower incident flux.

In principle, when large sections of $(\boldsymbol{Q},\omega)$ space need to be explored, the TAS data acquisition rate can be increased by multiplexing the secondary spectrometer, in the same way, albeit not to the same extent, as on a TOF instrument. In section 7 we shall review a few of the ideas which have emerged in this field.

---

2. http://www.isis.rl.ac.uk/

4. WHAT DETERMINES THE TAS COUNT RATE ?

The intensity in the detector, for a given spectrometer configuration (corresponding to nominal values $(\boldsymbol{Q}_0, \omega_0)$ of momentum and energy transfer) is written as [12]

$$I(\boldsymbol{Q}_0, \omega_0) = N \int \boldsymbol{J}(\boldsymbol{k}_i, \boldsymbol{k}_f)\, \mathrm{d}\boldsymbol{k}_i\, \mathrm{d}\boldsymbol{k}_f \qquad \{3\}$$

with

$$\boldsymbol{J}(\boldsymbol{k}_i, \boldsymbol{k}_f) = A(\boldsymbol{k}_i)\, p_i(\boldsymbol{k}_i)\, S(\boldsymbol{Q}, \omega)\, p_f(\boldsymbol{k}_f) \qquad \{4\}$$

where $A(\boldsymbol{k}_i)$ describes the spectrum of the source, and $p_i(\boldsymbol{k}_i)$ $[p_f(\boldsymbol{k}_f)]$ refers to the transmission of the monochromator (analyzer) for each incident (scattered) neutron wave-vector. N is the number of scattering particles in the sample volume. The variables $(\boldsymbol{k}_i, \boldsymbol{k}_f)$ and $(\boldsymbol{Q}, \omega)$ are related through the energy and momentum conservation equations {1}.

The integration {3} can be performed over the variables $\boldsymbol{k}_i$ and $\boldsymbol{k}_f$ *at fixed* $\boldsymbol{Q}$ *and* ω, and, subsequently, over the variables $\boldsymbol{Q}$ and ω

$$I(\boldsymbol{Q}_0, \omega_0) = N \int S(\boldsymbol{Q}, \omega)\, \mathrm{d}\boldsymbol{Q}\, \mathrm{d}\omega$$

$$\times \int A(\boldsymbol{k}_i)\, p_i(\boldsymbol{k}_i)\, p_f(\boldsymbol{k}_f)\, \delta[\boldsymbol{Q} + \boldsymbol{k}_i - \boldsymbol{k}_f]\, \delta[\hbar\omega - \hbar^2(k_i^2 - k_f^2)/2M_n]\, \mathrm{d}\boldsymbol{k}_i\, \mathrm{d}\boldsymbol{k}_f \qquad \{5\}$$

or

$$I(\boldsymbol{Q}_0, \omega_0) = N \langle A(\boldsymbol{k}_i) \rangle \int R(\boldsymbol{Q} - \boldsymbol{Q}_0, \omega - \omega_0)\, S(\boldsymbol{Q}, \omega)\, \mathrm{d}\boldsymbol{Q}\, \mathrm{d}\omega$$

which defines the instrumental resolution function $R(\boldsymbol{Q} - \boldsymbol{Q}_0, \omega - \omega_0)$ as

$$R(\boldsymbol{Q} - \boldsymbol{Q}_0, \omega - \omega_0) = \int p_i(\boldsymbol{k}_i)\, p_f(\boldsymbol{k}_f)\, \delta[\boldsymbol{Q} + \boldsymbol{k}_i - \boldsymbol{k}_f]\, \delta[\hbar\omega - \hbar^2(k_i^2 - k_f^2)/2M_n]\, \mathrm{d}\boldsymbol{k}_i\, \mathrm{d}\boldsymbol{k}_f \qquad \{6\}$$

Equation {6} may be used to obtain the norm of the resolution function

$$\int R(\boldsymbol{Q} - \boldsymbol{Q}_0, \omega - \omega_0)\, \mathrm{d}\boldsymbol{Q}\, \mathrm{d}\omega = \int p_i(\boldsymbol{k}_i)\, p_f(\boldsymbol{k}_f)\, \mathrm{d}\boldsymbol{k}_i\, \mathrm{d}\boldsymbol{k}_f = V_I\, V_F \qquad \{7\}$$

with

$$V_I = \int p_i(\boldsymbol{k}_i)\, \mathrm{d}\boldsymbol{k}_i \text{ and } V_F = \int p_f(\boldsymbol{k}_f)\, \mathrm{d}\boldsymbol{k}_f \qquad \{8\}$$

- The resolution function {6} is the *convolution product* of the two reciprocal space distributions $p_i(\boldsymbol{k}_i)$ and $p_f(\boldsymbol{k}_f)$.
- The norm of the resolution function is the *product* of the two integrated distributions $V_I\, V_F$ and is independent of the scattering angle.

Equations {3} to {8} are completely *general* and apply equally to TOF and TAS instruments. For a TAS spectrometer the "*volumes*" V_I and V_F can be evaluated in terms of the characteristics of the monochromator and analyzer crystals and in terms of the angular divergences of the neutron beam.

For a flat (i.e. *non-focussing*) mosaic crystal in the Gaussian approximation, one gets [12]

$$V_I = P_m(k_I)k_I^{\,3}\cot\theta_m(2\pi)^{3/2}\frac{\beta_0\,\beta_1}{\sqrt{4\sin^2(\theta_m)\,\eta'^{\,2}_m+\beta_0^2+\beta_1^2}}\frac{\eta_m\,\alpha_0\,\alpha_1}{\sqrt{4\,\eta_m^2+\alpha_0^2+\alpha_1^2}} \qquad \{9\}$$

where $\boldsymbol{k}_I$ is the average incident neutron wavevector

$$\boldsymbol{k}_I = V_I^{-1}\int \boldsymbol{k}_i\, p_i(\boldsymbol{k}_i)\,\mathrm{d}\boldsymbol{k}_i \qquad \{10\}$$

and α_0, β_0, α_1 and β_1 are respectively, the horizontal and vertical beam collimation before and after the monochromator; η_m and η'_m are the horizontal and vertical mosaic widths of the monochromating crystal and $P_m(k_I)$ its peak reflectivity. The expression for V_F is completely analogous.

The 4-dimensional resolution function $R(\boldsymbol{Q}-\boldsymbol{Q}_0,\omega-\omega_0)$ depends *parametrically* on the spectrometer configuration, i.e. on $\boldsymbol{Q}_0$ and ω_0.

In the Gaussian approximation [13]

$$R(\boldsymbol{Q}-\boldsymbol{Q}_0,\omega-\omega_0) = R_0(\boldsymbol{Q}_0,\omega_0)\exp\left[-\frac{1}{2}\sum_{k=1}^{4}\sum_{l=1}^{4}M_{kl}(\boldsymbol{Q}_0,\omega_0)X_kX_l\right] \qquad \{11\}$$

where the matrix M_{kl} defines the resolution ellipsoid and the 4 coordinates X_k are *linear combinations* of the components of $\boldsymbol{Q}-\boldsymbol{Q}_0$ and of $\omega-\omega_0$.

During a constant-$\boldsymbol{Q}$ scan across a dispersion curve, as sketched in figure 7 below, the nominal value of the energy transfer ω_0 varies at each point and hence the quantities $R_0(\boldsymbol{Q}_0,\omega_0)$ and $M_{kl}(\boldsymbol{Q}_0,\omega_0)$ vary as well and these variations affect the experimental lineshapes.

However, for the purpose of estimating the measured *integrated intensity* of the mode, one may, to a good approximation, neglect these changes and assume that the ellipsoid is translated along the ω-axis without deformation during the constant-$\boldsymbol{Q}$ scan, in which case the mode integrated intensity is simply proportional to the norm of the resolution function

$$\int I(\boldsymbol{Q}_0,\omega_0)\,\mathrm{d}\omega_0 \sim NA(\boldsymbol{k}_I^*)\int R(\boldsymbol{Q}-\boldsymbol{Q}_0,\omega-\omega_0^*)\,\mathrm{d}\boldsymbol{Q}\,\mathrm{d}\omega \sim NA(\boldsymbol{k}_I^*)\,V_I^*\,V_F^* \qquad \{12\}$$

where the (*) refers to the point of maximum overlap between the ellipsoid and the dispersion surface.

Another instance when the norm of the resolution function enters directly into the measured intensity occurs when the scattering function varies slowly on the scale of the resolution ellipsoid (*diffuse scattering* in both $\boldsymbol{Q}$ and ω)

$$
\begin{aligned}
I(\boldsymbol{Q}_0,\omega_0) &= N A(\boldsymbol{k}_I) \int R(\boldsymbol{Q}-\boldsymbol{Q}_0,\omega-\omega_0)\, S(\boldsymbol{Q},\omega)\, \mathrm{d}\boldsymbol{Q}\, \mathrm{d}\omega \\
&\approx N A(\boldsymbol{k}_I) \int R(\boldsymbol{Q}-\boldsymbol{Q}_0,\omega-\omega_0)\, S(\boldsymbol{Q}_0,\omega_0)\, \mathrm{d}\boldsymbol{Q}\, \mathrm{d}\omega \qquad \{13\} \\
&= N A(\boldsymbol{k}_I)\, V_I\, V_F\, S(\boldsymbol{Q}_0,\omega_0)
\end{aligned}
$$

As a rule, the intensities measured in a constant-$\boldsymbol{Q}$ scan should be corrected *point by point* for the variation of the norm of the resolution function. Such a correction may be looked upon as a *zeroth-order resolution correction*. It is a useful first step, particularly when one is interested in integrated intensities (mode strengths) rather than lineshapes. This applies also to more general scans when both $\boldsymbol{Q}$ and ω are varied simultaneously (for a constant-ω scan the norm of the resolution is constant).

For a scan obtained in the *constant-k_f mode,* V_F is constant across the scan and the measured intensities should be corrected for the variation of $A(\boldsymbol{k}_I)\, V_I$. Since the quantity $A(\boldsymbol{k}_I)\, k_I\, V_I$ measures the neutron flux incident on the sample [12], it is sufficient therefore to use a monitor with a $1/k_I$ characteristic in the incident beam to normalize the counting time per point [3].

For a scan obtained in the *constant-k_i mode* the measured intensities must be corrected for the variation of $P_a(k_f)\, k_f^3 \cot\theta_a$, as seen from the analogue of equation {9} for V_F. Note that $P_a(k_f)$ may vary *rapidly* as a function of k_f due to parasitic multiple Bragg reflections in the analyzer crystal.

5. *WHAT DETERMINES THE SIZE AND SHAPE OF THE RESOLUTION FUNCTION ?*

Since the resolution function is the convolution product of the distributions $p_i(\boldsymbol{k}_i)$ and $p_f(\boldsymbol{k}_f)$ [see eq. {6}], its shape is determined by the shape of each distribution and by the value of the scattering angle φ(A4) which controls the way in which the two distributions are combined.

The size and shape of the distribution $p_i(\boldsymbol{k}_i)$ depends on a number of independent parameters: the Bragg angle at the monochromator θ_m; the horizontal (α_0, α_1) and vertical (β_0, β_1) beam collimation before and after the monochromator; the horizontal and vertical monochromator mosaic widths (η_m, η'_m). A similar set of independent parameters determines the size and shape of the distribution $p_f(\boldsymbol{k}_f)$.

---

3. In practice this procedure must be modified to take into account the proportion of higher-order harmonics ($2k_i$, $3k_i$, ... neutrons) which may be present in the incident beam and which do not contribute to the measured signal.

With so many independent input parameters it is not easy to trace the influence of each input parameter on the final result. To get a feeling for the complexity of the problem, let us define the deviation of $\boldsymbol{k}_i$ from its average value $\boldsymbol{k}_I$ as $\delta\boldsymbol{k}_i = (\delta k_i^{||}, \delta k_i^{\perp}, \delta k_i^{z})$ where the superscripts $||$, $\perp$, and z, refer to the directions, respectively, parallel to $\boldsymbol{k}_I$, normal to $\boldsymbol{k}_I$ in the horizontal scattering plane and vertical. Using a similar convention for $\delta\boldsymbol{k}_f = \boldsymbol{k}_f - \boldsymbol{k}_F$ and $\delta\boldsymbol{Q} = \boldsymbol{Q} - \boldsymbol{Q}_0$, we see from any of the reciprocal space diagrams in section 1, that a finite value of $\delta k_i^{||}$ $(\delta k_f^{||})$ induces a finite value of $\delta Q^{||}$, $\delta Q^{\perp}$, and $\delta\omega$, while $\delta k_i^{\perp}(\delta k_f^{\perp})$ couples to $\delta Q^{||}$, $\delta Q^{\perp}$, and not to $\delta\omega$ and δk_i^{z} (δk_f^{z}) couples to δQ^{z} only.

On the other hand, when dealing with *dispersive* excitations, it is of practical importance to optimize the *orientation* of the resolution ellipsoid with respect to the slope of the dispersion curve. *Reciprocal space focussing*, as shown in figure 7, allows one to minimize instrumental broadening without reducing the overall size of the ellipsoid, i.e. at no cost in integrated intensity (and hence with a gain in peak intensity).

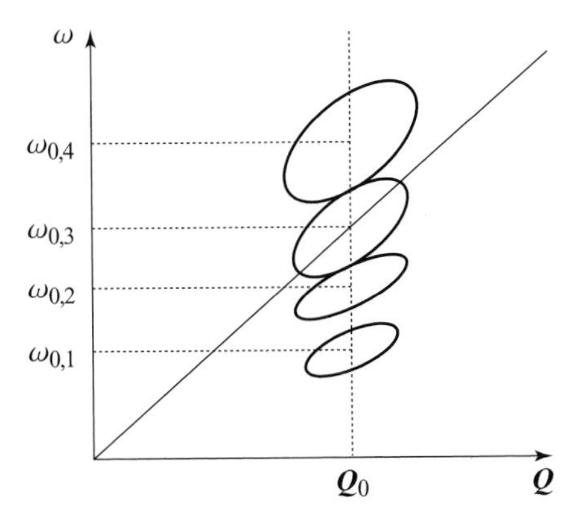

Figure 7 - Constant-$\boldsymbol{Q}$ scan across a dispersion curve in a focussed configuration. The ellipsoids correspond to the $(\boldsymbol{Q},\omega)$ projection of the instrumental resolution function at each scan point $(\boldsymbol{Q}_0,\omega_{0n})$, n =1,4.

The first step is to find the shape of the distributions $p_i(\boldsymbol{k}_i)$ and $p_f(\boldsymbol{k}_f)$. To this end one makes use of scattering diagrams drawn in the reciprocal space of the monochromator (analyzer) crystal. Figure 8 shows a graphical illustration of Bragg scattering on a perfect $(\eta_m = 0)$ monochromator crystal: a neutron wavevector $\boldsymbol{k}_0$ incident on the monochromator satisfies Bragg law if its origin lies on the bisector (Δ) of the Bragg scattering vector $\boldsymbol{\tau}_m$:

$$\boldsymbol{k}_0 - \boldsymbol{k}_i = \boldsymbol{\tau}_m \;;\; |\boldsymbol{\tau}_m| = 2\pi/\mathrm{d}_m = 2k_i \sin\theta_m$$

The direction of the incident ($\boldsymbol{k}_0$) and scattered ($\boldsymbol{k}_i$) wavevectors is limited by the collimations α_0 and α_1, respectively.

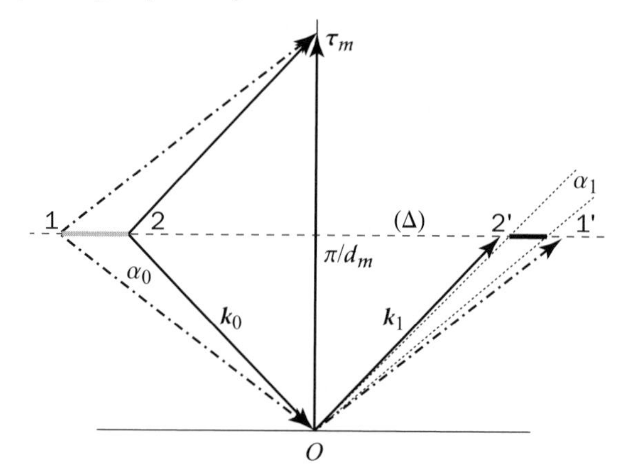

Figure 8 - *k*-space distribution delivered by a perfect monochromator. The segment 1-2 is transformed into 1'-2'. The length of the segment 1'-2' is further reduced by the outgoing collimation α_1 ($\alpha_1 < \alpha_0$).

Introducing a finite mosaic width η_m amounts to rotating the above diagram by $\pm 1/2\eta_m$ as shown in figure 9a for the case $\eta_m < \alpha_0$, α_1 and figure 9b for the case $\eta_m > \alpha_0, \alpha_1$. The resulting outgoing **k**-space volume V_I is elongated along a direction inclined by an angle θ from the direction of $\boldsymbol{k}_I$, with $\theta = \theta_m$ in case (a) and $\theta_m < \theta = \tan^{-1}(2 \tan \theta_m) < 2\,\theta_m$ in case (b). Its thickness is controlled by the mosaic width in case (a) and its length by the collimations. As expected, in case (b), the dimensions of the volume V_I are entirely determined by the collimations.

Figure 10a shows the volumes V_I and V_F drawn in the reciprocal space of the sample. The scattering geometry is of the "W-type" with scattering to the right on the monochromator, to the left on the sample and to the right again on the analyzer. Two projections of the resolution ellipsoid obtained by convoluting the two volumes are shown in figures 10b and 10c. Figure 10b shows that the largest positive (negative) values of $\delta Q^{||}$ are obtained by coupling incident wavevectors near point 1 in V_I with scattered wavevectors near point 3 in V_F (resp. points 2 and 4). The corresponding wavevectors $\boldsymbol{k}_i$ and $\boldsymbol{k}_f$ are shorter (longer) than $\boldsymbol{k}_I$ and $\boldsymbol{k}_F$ by about the same amount so that $\delta\omega \sim 0$. Conversely the largest positive (negative) values of $\delta\omega$ are obtained by coupling 2 and 3 (resp. 1 and 4) which corresponds to $\delta Q^{||} \sim 0$. Altogether this leads to a low degree of correlation between the variables $\delta Q^{||}$ and $\delta\omega$, and to a projection of the ellipsoid onto the ($\delta Q^{||}$, $\delta\omega$) plane as shown in figure 10b.

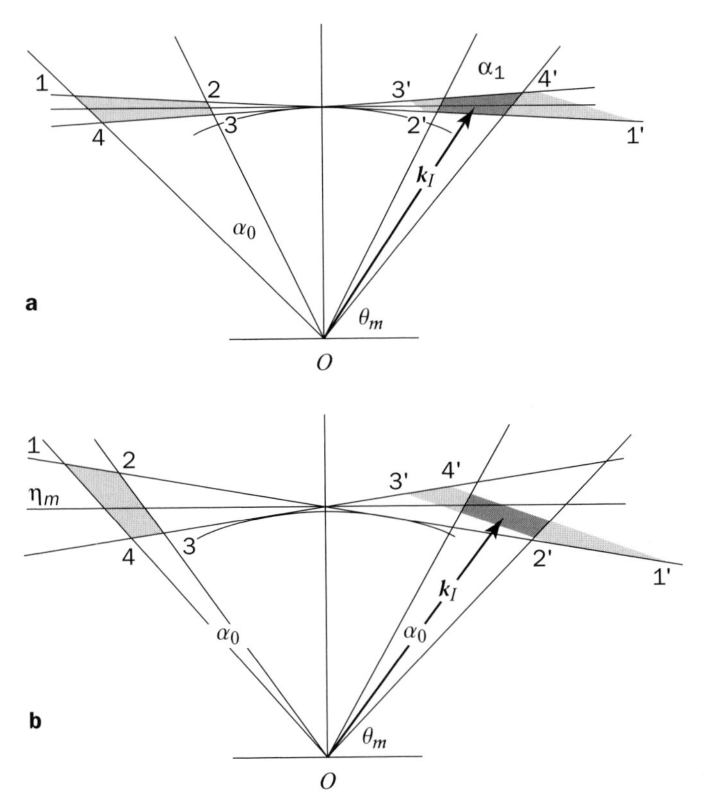

Figure 9 - k-space volume delivered by a mosaic monochromator. **a** - Case $\eta_m < \alpha_0, \alpha_1$. **b** - Case $\eta_m > \alpha_0, \alpha_1$. The acceptance volume 1-2-3-4 is transformed into 1'-2'-3'-4', out of which the dark grey volume V_I is selected by the outgoing collimation α_1.

The situation is entirely different for the $(\delta Q^{\perp}, \delta\omega)$ projection where the strong correlation between the two variables is evident by inspection of figure 10a, leading to a characteristic "*cigar*" shape of the $(\delta Q^{\perp}, \delta\omega)$ projection. One may also convince oneself that the thickness of the cigar vanishes in the limit of small scattering angles, as the two conditions $\delta Q^{\perp} = 0$ and $\delta\omega = 0$ become identical.

- The strong correlation between $\delta Q^{\perp}$ and $\delta\omega$ is a very general feature. Changing the input parameters or the spectrometer configuration can change the value of the slope $(\delta\omega/\delta Q^{\perp})$ *but not its sign*, as long as $\delta Q^{\perp}$ is always defined opposite to $\boldsymbol{k}_F$ as in figure 10 above. This strong correlation gives the possibility of *focussing* on dispersive excitations measured in *transverse* geometry $(\boldsymbol{Q} \perp \boldsymbol{q})$, as, e.g., for transverse acoustic phonons.

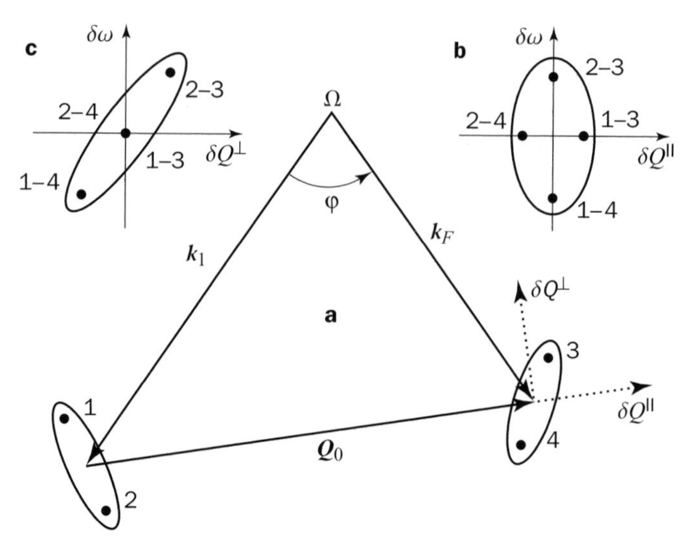

Figure 10 - Convolution of the distributions $p_i(k_i)$ and $p_f(k_f)$ showing the strong $(\delta Q^{\perp}, \delta\omega)$ correlation.

- Nothing as general can be said about the correlation between $\delta Q^{||}$ and $\delta\omega$, which is weak in the case of figure 10 but could become significant for different configurations, different scattering angles and different energy transfers.
- The fourth dimension of the ellipsoid δQ^{z} is decoupled from the other three, in first order. Hence the vertical axis is always a principal direction of the ellipsoid.

A quantitative discussion of the TAS resolution function, relevant to the case of flat mosaic monochromator and analyzer crystals, can be found in the original paper of Cooper & Nathans or in the recent review by Shirane *et al.* [13].

6. DECOUPLING ENERGY AND MOMENTUM RESOLUTIONS: DIRECT SPACE FOCUSING

The coupling between the uncertainties on the energy transfer and on the in-plane components of momentum transfer is an *unwanted* characteristic of the 3-axis technique. Ideally one would like to be able to tune the momentum and energy resolutions *independently* of one another to best match the experimental requirements.

In many cases energy resolution is the more important factor and $\boldsymbol{Q}$-resolution can be readily sacrificed in favour of luminosity. The use of bent focusing crystals allows the neutron beam to be concentrated onto small samples at the expense of $\boldsymbol{Q}$-resolution, but without degrading the energy resolution (*monochromatic direct space focusing*). This is *by far the most effective way*of trading $\boldsymbol{Q}$-resolution for improved luminosity.

Focusing in the vertical plane is used systematically on monochromator crystals. The principle is shown in figure 11. In first order, vertical focusing has no effect on the Bragg angle at the monochromator and hence on the energy definition of the neutron beam incident on the sample. It only affects the uncertainty on k_i^z and hence on Q_z.

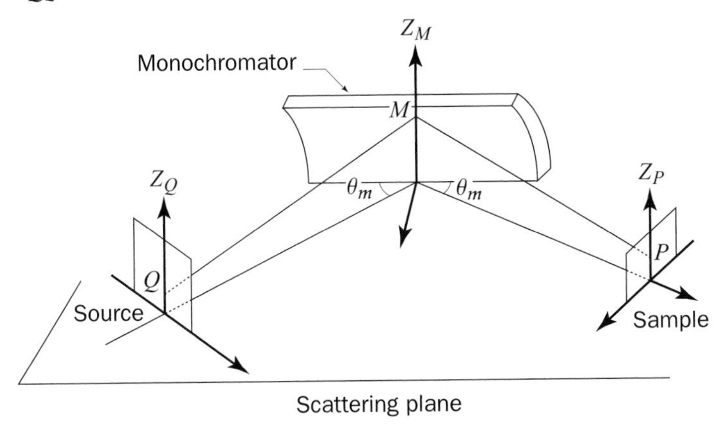

Figure 11 - Vertically-focusing bent monochromator.

Figure 12 shows the ray-tracing diagram corresponding to figure 11. The horizontal axis is taken along the beam path and the vertical curvature $1/R_V$ of the monochromator is assumed to satisfy the "lens" equation

$$2 \sin\theta_m/R_V = 1/L_0 + 1/L_1$$

Each vertical source element dZ_Q emits rays which are focussed by the monochromator onto a sample element $dZ_P = dZ_Q(L_1/L_0)$. The source of height h_S has an image in the sample plane of height $h_{im} = h_S(L_1/L_0)$. For a sample of height $h < h_{im}$, the intensity gain G_V over a non-focusing geometry can be estimated by comparing the vertical beam divergences in both cases, viz.: $h_S/(L_0 + L_1)$ in the non-focusing case vs. $h\ _m/L_1$ in the focusing case. Hence

$$G_V = \frac{h_m}{h_S}\frac{L_0 + L_1}{L_1}$$

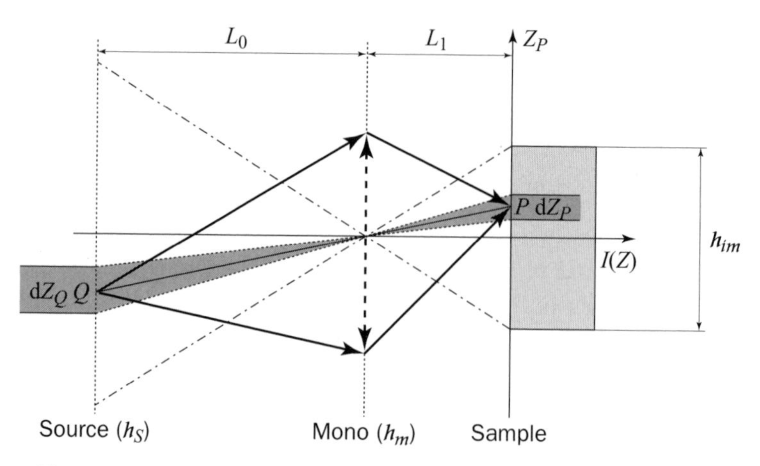

Figure 12 - Vertically-focusing monochromator: ray-tracing diagram.

In practice, measured gains are reduced by *blurring* effects connected with the monochromator vertical mosaic spread and by the use of composite assemblies of small flat crystal pieces instead of bent crystals. Nevertheless measured values are typically in the range: $2 < G_V < 4$.

For an instrument installed on a guide the corresponding expressions are obtained by setting $1/L_0 = 0$ and $h_S/L_0 = \alpha_g$ (the characteristic divergence of the guide).

Similar gains are also possible on the analyzer side although one often finds it more expedient to use a cylindrical detector of suitable size mounted with its axis vertical.

Horizontal focusing is more delicate. Figure 13 shows the general principle.

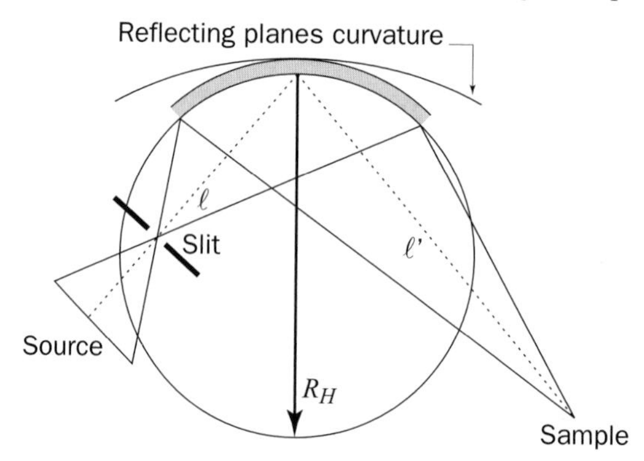

Figure 13 - Horizontally focusing monochromator. The focussed beam is monochromatic when both slit and sample lie on the circle of diameter R_H (Rowland circle).

The entrance slit plays the role of a virtual source, which is imaged onto the sample by an appropriate choice of the crystal plane curvature $1/R_H$

$$(2 \sin\theta_m R_H)^{-1} = 1/\ell + 1/\ell'$$

Furthermore when $\ell = \ell' = R_H \sin\theta$, the Bragg angle is constant all along the surface of the monochromator and the beam incident on the sample is *monochromatic*.

The intensity gain G_H over the non-focusing geometry is again given by the ratio of the acceptance angles. From figure 14 one gets

$$G_H = \alpha_{\text{bent}} / \alpha_{\text{flat}} = (L + \ell) / (L - \ell)$$

for the monochromatic focusing case.

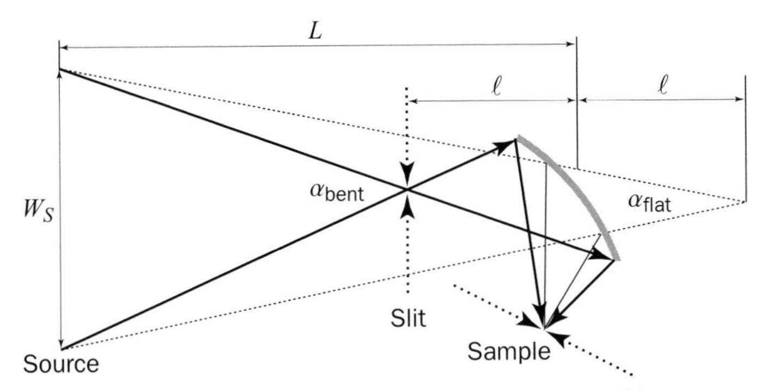

Figure 14 - Horizontally bent monochromator in monochromatic focusing geometry.

7. TAS MULTIPLEXING

When momentum resolution can no longer be relaxed the next natural step, in order to increase the TAS data acquisition rate, is to *multiplex* the secondary spectrometer, i.e. to operate several analyzer-detector arms in parallel. For a single-crystal experiment, this means that one accepts to give up part of the instrument's selectivity. To be specific, this means that most of the collected information will no longer refer to constant-$\boldsymbol{Q}$ scans at preselected high-symmetry points or high-symmetry directions in reciprocal space. But recent experience has shown that, in many cases, there is an interest in having a *continuous* coverage of $(\boldsymbol{Q},\omega)$ space and that crystal spectrometers can do a reasonably good job at it.

There are several levels at which multiplexing can be implemented on a TAS instrument, depending upon the type of coverage one wishes to achieve.

The first level consists in using a multi-blade analyzer in conjunction with a 2-dimensional position-sensitive detector (PSD). The so-called RITA (Re-Invented Three-Axis) spectrometer [14], now at PSI-Switzerland, as well as the SPINS spectrometer [4] at NIST-USA, fall in that class of instruments. The $(\boldsymbol{Q},\omega)$ space coverage which they provide is more or less continuous but limited to the neighbourhood of a specific preselected point $(\boldsymbol{Q}_0,\omega_0)$ (*local reciprocal space imaging*).

Figure 15 below illustrates some of the possible modes of operation of such a device.

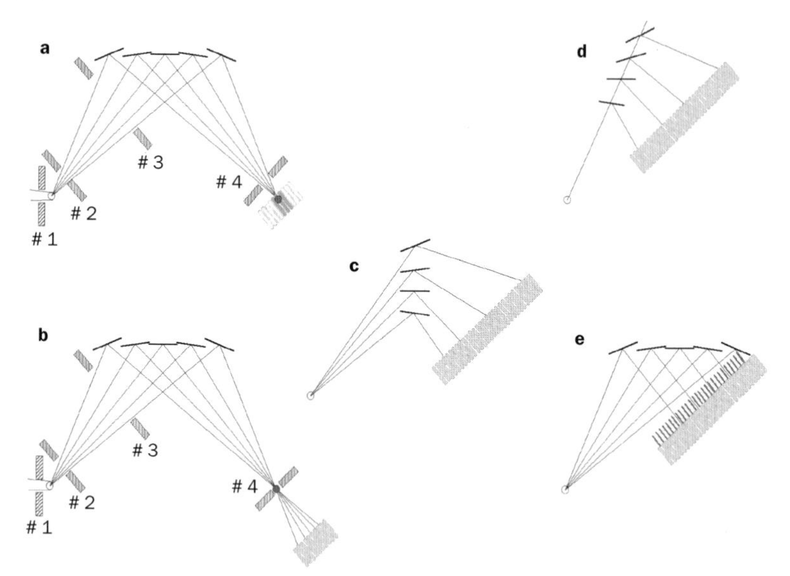

Figure 15 - A selection of configurations of a multi-analyzer secondary spectrometer

a -Horizontal focusing: the PSD is used in conjunction with mechanical slit # 4, as a fine-tuned software slit; the neutrons propagate from left to right in the figure; the sample is located between slit # 1 and slit # 2; the detector is located behind slit # 4.

b - Monochromatic $\boldsymbol{Q}$-space imaging in reflection geometry: the PSD is placed at a suitable distance behind slit # 4.

c - Monochromatic $\boldsymbol{Q}$-space imaging using a pseudo-transmission geometry: the accessible range in $Q_{\perp f}$ can be adjusted by changing the distance between blades.

d - Energy-dispersive imaging at constant $Q_{\perp f}$.

e - Energy-dispersive imaging with a parallel beam impinging on the PSD: a Soller collimator in front of the PSD can be used for background reduction.

4. SPINS : http://rrdjazz.nist.gov/instrument.html

The three parameters which may or may not vary across the PSD's horizontal dimension are the neutron final energy E_f (or equivalently the energy transfer ω), and the components of $\boldsymbol{Q}$ ($Q_{//f}$, $Q_{\perp f}$) parallel and perpendicular to $\langle \boldsymbol{k}_f \rangle_{\text{av.}}$. The third component of $\boldsymbol{Q}$, Q_z, varies along the vertical dimension of the PSD. It is beyond the scope of this chapter to discuss in detail the possible applications of each configuration, some of which have not yet been fully explored. A comprehensive review of the work done so far on the RITA-type instruments can be found in ref. **[14]**.

The main difficulty in designing and operating such devices is the open beam geometry associated with the use of a PSD, which can lead to high background levels and low signal to noise values. In particular it is important that both the multi-analyzer and PSD are enclosed in a common evacuated protection and that the analyzer mount is designed in such a way as to minimize parasitic scattering processes.

The next multiplexing level consists in having a set of independent analyzer / detector arms, tuned to transmit the same final neutron energy, and covering a range of scattering angles as wide as possible (60-90°). This is the principle of the Multi-arm Analyzer-Detector (MAD) spectrometer [15] shown in figure 16a. The crystal analyzers are located on an arc (~ 1 m radius) centred on the sample position. The vertical counters are placed on a concentric circle (~ 1.5 m radius). Absorbing wedges prevent cross-talk between channels. The crystals and counters are enclosed in a common shielding box.

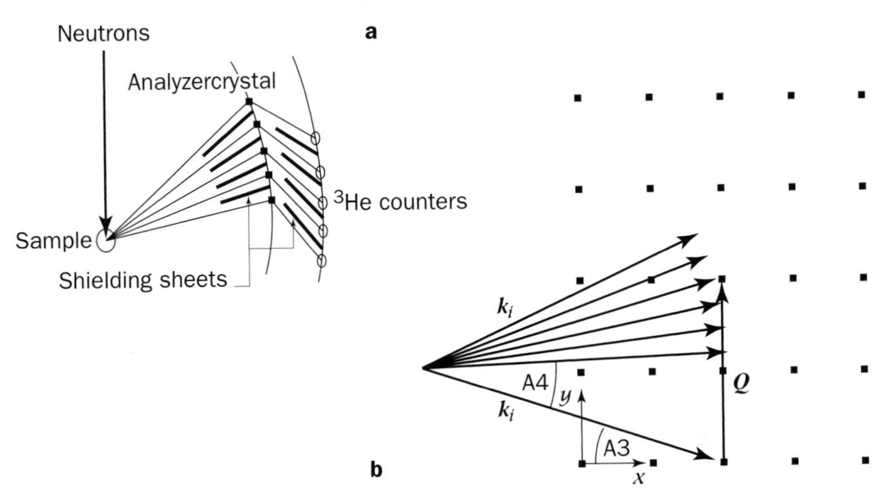

Figure 16 - a - Sketch of the Multi-arm Analyzer-Detector (MAD) set-up. **b** - The MAD set-up in reciprocal space.

The MAD set-up can be used for constant-energy *mapping* of a plane in reciprocal space. For each sample orientation (A3 angle) and for each position of the analyzer-detector bank (A4 angle), constant-energy data are collected along a circular arc (arrowheads in figure 16b). The efficiency gain with respect to the conventional TAS set-up is of the order of the number of arms (20-60).

By stepping the incident neutron energy it is possible to perform a constant-$\boldsymbol{Q}$ scan with *one of the arms only*, while the other arms describe mixed trajectories in $(\boldsymbol{Q},\omega)$ space. Nevertheless the MAD spectrometer can be used quite effectively for excitation mapping in *low-dimensional systems*. Figure 17 shows the kinematics in the case of a 1D dispersion. The dispersive direction $\boldsymbol{Q}_{//}$ is normal to $\boldsymbol{k}_i$ in the horizontal scattering plane. By stepping E_i for a fixed scattering geometry (A3, A4 fixed) each analyzer-detector arm performs a *constant-*$Q_{//}$ *scan*. Hence in that case the complete dispersion can be mapped in a single E_i-scan. A similar argument applies to the case of an isotropic planar dispersion. As before, the sample orientation should be such that $\boldsymbol{k}_i$ is normal to the dispersive plane.

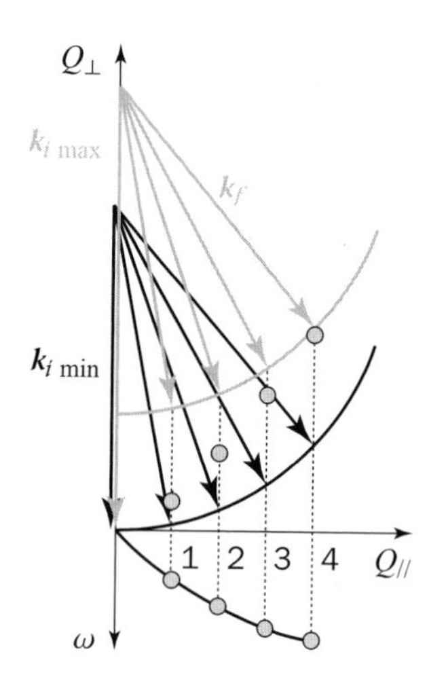

Figure 17 - Mapping a 1D dispersion with MAD. During a k_i scan each arrow head describes a straight line parallel to k_i and perpendicular to the dispersion direction (constant-$Q_{//}$ scan).

Again the efficiency gain factor over a conventional TAS instrument is of the order of the number of analyzer-detector arms. This gain comes however at a cost in selectivity. In the constant-energy mapping mode the positions of the mesh points cannot be independently selected. Furthermore these positions will not be evenly spaced in reciprocal space and they will not coincide for different maps at different energy transfers. Similarly, in the constant-$Q_{//}$ mode, the $Q_{//}$ values at which the 1D dispersion is probed cannot be independently selected and they are not evenly spaced. They do not change with energy transfer as long as $\boldsymbol{k}_i \perp \boldsymbol{Q}_{//}$.

Figure 18 shows an alternative way of multiplexing a TAS secondary spectrometer. In this so-called *flat-cone* geometry, each analyzer-detector arm operates in a vertical plane. The method has been successfully implemented at HMI-Berlin

(E2 spectrometer [16]) and a similar set-up, is under construction at ILL. The main advantage over the MAD design lies in the fact that the detectors are located above the horizontal scattering plane, and thus can be more efficiently shielded against stray radiation.

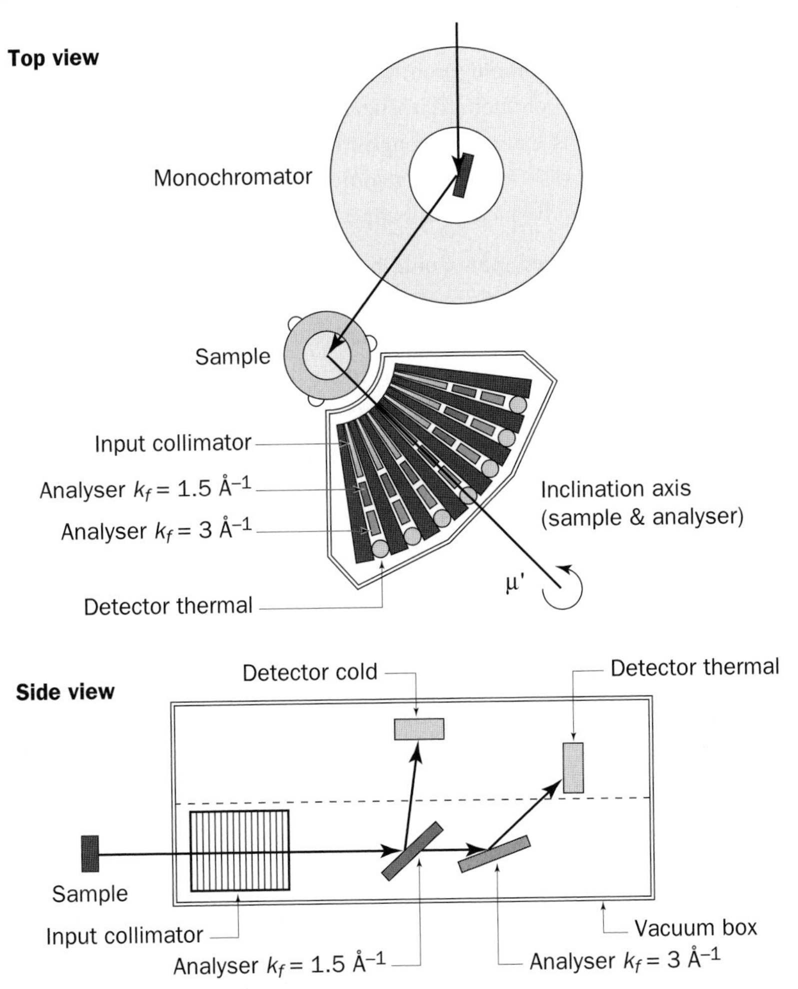

Figure 18 - The *flat cone* set up for excitation mapping on a three-axis spectrometer (only seven analyzer channels are drawn for clarity). Each channel contains 2 pairs of crystal analyzers and detectors in vertical scattering geometry. One pair, aligned for $k_f = 1.5$ Å^{-1}, is used for *cold* neutron work and a second one ($k_f = 3$ Å^{-1}) for *thermal* neutron work, without necessity of any remote control positioning units. By tilting the sample and the analyzer-detector bank, as shown in the figure, reciprocal layers (h, k, $l \neq 0$) can be accessed (after ref. **[17]**).

8. *PHONON STUDIES WITH TAS*

Lattice vibrations (*phonons*) are labelled by a wavevector $\boldsymbol{q}$ and a branch index j ($j = 1$ to $3r$; $r =$ number of atoms in the primitive unit cell of the crystal). In principle, a three-axis spectrometer performing a constant-$\boldsymbol{Q}$ scan at $\boldsymbol{Q} = \boldsymbol{\tau}_{hkl} + \boldsymbol{q}$ (as in figure 3a or 3b), should record $3r$ intensity maxima (phonon peaks), corresponding to the $3r$ phonon frequencies $\omega_j(\boldsymbol{q})$ with $j = 1 - 3r$ (distinct or degenerate). If the scan is extended enough to cover both the energy gain ($k_f > k_i$; $\omega < 0$) and energy loss ($k_f < k_i$; $\omega > 0$) regimes, and if the sample temperature is high enough, each phonon frequency will appear twice, at $\omega = \pm\omega_j(\boldsymbol{q})$.

In practice, such an extended scan would be highly unusual and ineffective, for at least two reasons. First, a TAS configuration can only be optimized within a limited range of energy transfer values and hence one would prefer to partition the extended scan into a set of shorter scans, each performed under different experimental conditions. Second, one does not expect in general all phonon branches to have measurable intensities in the same Brillouin zone, i.e. for the same $\boldsymbol{\tau}_{hkl}$. Apart from instrumental factors, as discussed in section 4, the measured phonon intensities are given by the coherent one-phonon scattering function

$$S_j(\boldsymbol{Q},\omega) = \pi^{-1}(1 - \mathrm{e}^{-\hbar\omega/k_B T})^{-1} |F_j(\boldsymbol{Q},\boldsymbol{q})|^2 \,\mathrm{Im}\,\chi_j(\boldsymbol{q},\omega) \qquad \{14\}$$

where the variable ω may assume positive and negative values, corresponding to, respectively, phonon creation (neutron energy loss) and phonon annihilation (neutron energy gain) processes [5].

Apart from the mode susceptibility $\chi_j(\boldsymbol{q},\omega)$, which will be discussed later, equation {14} contains the one-phonon *inelastic structure factor*, $F_j(\boldsymbol{Q},\boldsymbol{q})$, which may be expressed in terms of the phonon polarization vectors $\{\boldsymbol{e}_{jd}(\boldsymbol{q})\}$

$$F_j(\boldsymbol{Q},\boldsymbol{q}) = \sum_{d=1}^{r} \frac{b_d}{\sqrt{M_d}} \left[\boldsymbol{Q}\cdot\boldsymbol{e}_{jd}(\boldsymbol{q})\right] \mathrm{e}^{i\boldsymbol{Q}\cdot\boldsymbol{r}_d}\, \mathrm{e}^{-W_d(\boldsymbol{Q})} \qquad \{15\}$$

where M_d, $\boldsymbol{r}_d$, b_d, and $W_d(\boldsymbol{Q})$ are, respectively, the mass, position vector, coherent neutron scattering length and Debye-Waller factor for atom d ($d = 1, .., r$) in the unit cell.

Equation {15} is the expression of the *selection rules* which control the observability of a given phonon frequency $\omega_j(\boldsymbol{q})$ in the $\boldsymbol{\tau}_{hkl}$ Brillouin zone. When

5. One may easily convince oneself that the first bracket on the righy hand side of equation {14} is equivalent to $n(\omega) + 1$ for $\omega > 0$ and $-n(|\omega|)$ for $\omega < 0$, where $n(\omega) \equiv [\exp(\hbar\omega/k_B T) - 1]^{-1}$ stands for the Bose-Einstein population factor. Note also that the imaginary part of the mode susceptibility $\chi_j(\boldsymbol{q},\omega)$ is an odd function of ω, so that $S_j(\boldsymbol{Q},\omega)$ remains positive for $\omega < 0$.

the phonon wavevector $\boldsymbol{q}$ lies in a symmetry direction, crystallographic symmetry may impose restrictions on the form of the polarization vectors $\{\boldsymbol{e}_{jd}(\boldsymbol{q})\}$, which, when inserted into equation {15}, lead to systematic *extinctions* [18,19]. The situation is analogous to systematic extinctions for Bragg reflections, associated with crystallographic symmetry operations.

Independent of lattice symmetry, equation {15} shows that only lattice vibrations polarized along the momentum transfer $\boldsymbol{Q}$ are visible. For long-wavelength acoustic modes, atoms within a given unit cell vibrate with the same amplitude in the same direction. This implies that the quantities $\dfrac{\boldsymbol{e}_{jd}(\boldsymbol{q})}{\sqrt{M_d}}$ can be factored out in equation {15}, whence

$$F_j(\boldsymbol{Q},\boldsymbol{q}) \sim (\boldsymbol{Q}\cdot\boldsymbol{\sigma}_j) \sum_{d=1}^{r} b_d \, e^{i\boldsymbol{Q}\cdot\boldsymbol{r}_d} \, e^{-W_d(\boldsymbol{Q})} \sim (\boldsymbol{Q}\cdot\boldsymbol{\sigma}_j) \, F_{\text{Bragg}}(\boldsymbol{\tau}_{hkl}) \qquad \{16\}$$

where $\boldsymbol{\sigma}_j$ is a unit vector in the polarization direction and $F_{\text{Bragg}}(\boldsymbol{\tau}_{hkl})$ is the elastic structure factor for the (hkl) Bragg peak. For transverse acoustic modes $\boldsymbol{\sigma}_T \perp \boldsymbol{q}$ and therefore one must choose a Brillouin zone $\boldsymbol{\tau}_{hkl}$ such that $\boldsymbol{\tau}_{hkl}$ ($\approx \boldsymbol{Q}$) $\perp \boldsymbol{q}$, while for a longitudinal mode, with $\boldsymbol{\sigma}_L \,//\, \boldsymbol{q}$, one must take $\boldsymbol{\tau}_{hkl} \,//\, \boldsymbol{q}$. The corresponding scattering geometries are shown in figure 19.

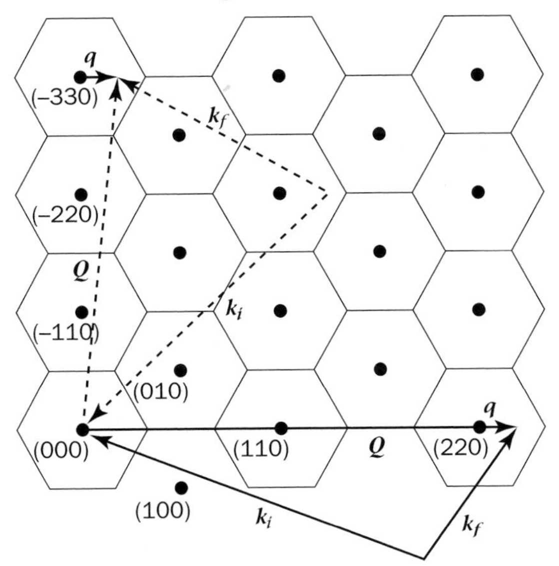

Figure 19 - Scattering diagrams for transverse (top) and longitudinal modes (bottom). The scattering plane is the $(hk0)$ reciprocal plane (hexagonal symmetry). Both modes propagate along the hexagonal [110] direction. The third acoustic mode propagating along [110] is polarized along [001] and is not visible in the $(hk0)$ scattering plane.

Equation {16} shows that long-wavelength *acoustic* modes should be measured near strong Bragg peaks. The converse statement for long-wavelength *optic* modes is only valid for simple diatomic structures such as GaAs (see below). For more complicated crystal structures with many atoms in the unit cell (and many optic branches) no general statement can be made. In fact, it is possible to determine *experimentally* the polarization vectors (or *eigenvectors*) associated with a particular phonon mode $\{\boldsymbol{q};j\}$ by measuring the integrated intensity of the phonon response near $\omega = \omega_j(\boldsymbol{q})$ and $\boldsymbol{Q} = \boldsymbol{\tau}_{hkl} + \boldsymbol{q}$, for as many values of $\boldsymbol{\tau}_{hkl}$ as possible. This yields a set of experimental values for the phonon inelastic structure factor $|F_j(\boldsymbol{Q} = \boldsymbol{\tau}_{hkl} + \boldsymbol{q})|$ which can be used to obtain the form of $\{\boldsymbol{e}_{jd}(\boldsymbol{q})\}$ by inverting equation {15}. This procedure is the inelastic analogue of a structural determination based on measured Bragg intensities. Details and examples can be found in **[20]**.

Figure 20 shows a comparison between measured [21] and calculated [22] phonon dispersion curves in GaAs at T = 12 K. The dispersion curves consist of 6 branches, 3 acoustic and 3 optic branches. Near the zone centre (Γ-point) the two sets are well separated in frequency and each branch is characterized by a polarization direction, either parallel (LA, LO) or perpendicular ($TA_{1,2}$,$TO_{1,2}$) to $\boldsymbol{q}$. For $\boldsymbol{q}$ lying in the [100] (Γ-*X*) and [111] (Γ-*L*) directions, the two TA (TO) branches are degenerate. More details about the measurements and the comparison of the data with various phenomenological model predictions can be found in reference **[21]**. Because of the high degree of accuracy of the experimental dispersion curves, none of the models available at the time turned out to be fully satisfactory. By far the best agreement was subsequently obtained by Giannozzi *et al.* [22], using a parameter-free *ab initio* technique (solid line in figure 20).

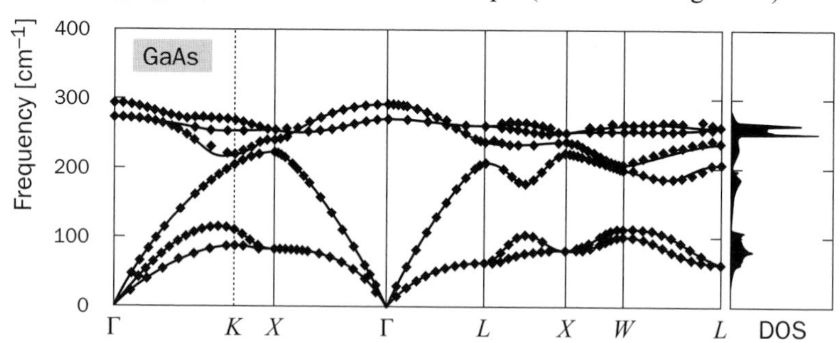

Figure 20 - Phonon dispersion curves for GaAs. The experimental points are taken from the TAS measurements of Strauch & Dorner [21]. The experimental uncertainties are less than the size of the symbols. The solid lines show the result of an *ab initio* model calculation by Giannozzi *et al.* [22]. The letters refer to the zone centre (Γ) and various high symmetry points (*K*, *X*, *L*, *W*) on the fcc Brillouin zone boundary. The calculated phonon density of states (DOS) is shown on the right (1 meV = 8.065 cm^{-1}).

Such an accurate comparison between theory and experiment is only possible for simple structures (i.e. with few phonon branches to be measured and when *ab initio* calculations are possible). The availability of large single-crystal specimens (several cm^3, in the present case) is also essential in order to perform accurate and complete dispersion curve measurements

This does not mean, however, that all attempts to observe phonons in single-crystal specimens less than a golf ball in size are necessarily bound to fail! It is important to realise that high frequency phonons are most time-consuming to measure, because phonon integrated intensities fall off at least as $\omega_i(\boldsymbol{q})^{-1}$ and because the TAS resolution deteriorates at high energies and spreads the phonon response over a broad ω-range. At low frequencies ($\hbar\omega \leq 8$ meV), volume requirements are much less stringent: typically, a sample volume of 10-20 mm^3 is sufficient to determine acoustic slopes with a 3-5% accuracy.

At finite temperature, phonon-phonon interactions (and electron-phonon interactions in metals) are responsible for *anharmonic* effects. The theoretical description of these effects [23, 24] is embodied in the concept of phonon *self-energy* $\Pi_j(\boldsymbol{q},T,\omega)$

$$\Omega_j^2(\boldsymbol{q},T,\omega) = \omega_{0j}^2(\boldsymbol{q}) + \Pi_j(\boldsymbol{q},T,\omega) = \omega_{0j}^2(\boldsymbol{q}) + \Delta_j(\boldsymbol{q},T,\omega) - i\omega\Gamma_j(\boldsymbol{q},T,\omega) \qquad \{17\}$$

where $\Delta_j(\boldsymbol{q},T,\omega)$ and $\omega\Gamma_j(\boldsymbol{q},T,\omega)$ are the real and imaginary parts of the complex self-energy $\Pi_j(\boldsymbol{q},T,\omega)$ and $\Omega_j(\boldsymbol{q},T,\omega)$ is a complex renormalized frequency associated with the harmonic ($T \approx 0$) phonon frequency $\omega_{0j}(\boldsymbol{q})$. Since the phonon self-energy is a function of frequency ω the renormalized (real) phonon frequency ω_{Tj} observed in a scattering experiment is obtained self-consistently as a solution of

$$\omega_{Tj}^2(\boldsymbol{q}) = \omega_{0j}^2(\boldsymbol{q}) + \Delta_j[\boldsymbol{q},T,\omega = \omega_{Tj}(\boldsymbol{q})] \qquad \{18\}$$

For weak anharmonicity, the self-energy can also be evaluated at the harmonic phonon frequency, setting $\omega = \omega_{0j}$ in equations {17} and {18}.

Quite generally, the mode susceptibility associated with the anharmonic phonon is given as

$$\chi_j(\boldsymbol{q},T,\omega) = [\Omega_j^2(\boldsymbol{q},T,\omega) - \omega^2]^{-1} \qquad \{19\}$$

Inserting expressions {19} for $\chi_j(\boldsymbol{q},T,\omega)$ and {17} for $\Omega_j(\boldsymbol{q},T,\omega)$, into expression {14}, the one-phonon scattering function reads

$$S_j(\boldsymbol{Q},\omega) = \pi^{-1}\left(1-e^{-\hbar\omega/k_BT}\right)^{-1} |F_j(\boldsymbol{Q},\boldsymbol{q})|^2 \times \frac{\omega\Gamma_j(\boldsymbol{q},T,\omega)}{[\omega_{0j}^2(\boldsymbol{q}) + \Delta_j(\boldsymbol{q},T,\omega) - \omega^2]^2 + [\omega\Gamma_j(\boldsymbol{q},T,\omega)]^2} \qquad \{20\}$$

As long as the ω-dependence of Δ_j and Γ_j can be ignored, equation {20} is equivalent to a *damped harmonic oscillator* (DHO) response function multiplied by the detailed balance factor. In this approximation, widely used in practice, equation {20} reads

$$S_j(\boldsymbol{Q},\omega) = \pi^{-1} \frac{\omega}{1-\mathrm{e}^{-\hbar\omega/k_BT}} |F_j(\boldsymbol{Q},\boldsymbol{q})|^2 \frac{\Gamma_{Tj}}{\left(\omega_{Tj}^2-\omega^2\right)^2+\left(\omega\Gamma_{Tj}\right)^2} \qquad \{21\}$$

with ω_{Tj} given by equation {18} and $\Gamma_{Tj} \equiv \Gamma_j(\boldsymbol{q},T,\omega=\omega_{Tj})$. In a standard data analysis procedure, expression {21} is convoluted with the TAS resolution function (as discussed in section 4), and the resulting line profile is compared to the observed phonon response. The oscillator parameters ω_{Tj} and Γ_{Tj} can then be obtained via an iterative fitting procedure. Figure 21 (top frame) shows an example of a DHO fit for a constant-$\boldsymbol{Q}$ scan across a zone-boundary phonon in orthorhombic α-uranium [25]. The same data are shown in the bottom frame together with the deconvoluted (i.e. intrinsic) DHO lineshape.

In the *classical* limit ($\hbar\omega \ll k_BT$) the first factor in equation {21} reduces to $k_BT/\hbar$ and the energy-integrated phonon response is simply

$$S_j^{\mathrm{cl}}(\boldsymbol{Q}) = \int S_j^{\mathrm{cl}}(\boldsymbol{Q},\omega)\mathrm{d}\omega = (k_BT/\hbar)\frac{|F_j(\boldsymbol{Q},\boldsymbol{q})|^2}{\omega_{Tj}^2} \qquad \{22\}$$

a very useful relation that does not depend on the value of the phonon damping Γ_{Tj}.

In the limit of vanishing damping, the phonon response reduces to two δ-function peaks centred at $\omega = \pm\omega_{Tj}$ and broadened by the instrumental resolution. In the opposite limit of overdamping ($\omega_{Tj} \ll \Gamma_{Tj}$), the phonon lineshape becomes Lorentzian-like

$$S_j(\boldsymbol{Q},\omega) = \pi^{-1} \frac{\omega}{1-\mathrm{e}^{-\hbar\omega/k_BT}} |F_j(\boldsymbol{Q},\boldsymbol{q})|^2 \frac{\gamma_{Tj}}{\omega_{Tj}^2\left(\omega^2+\gamma_{Tj}^2\right)} \qquad \{23\}$$

centred at $\omega = 0$ and with a width $\gamma_{Tj} = \omega_{Tj}^2/\Gamma_{Tj} \ll \omega_{Tj}$.

9. THE INS VERSUS IXS CHOICE

There are a number of instances when inelastic X-ray scattering (IXS), using the backscattering technique, can compete with inelastic neutron scattering (INS) as a tool to study collective *structural* excitations. For technical reasons the X-ray backscattering technique cannot take advantage of the magnetic X-ray scattering enhancement near resonance. Consequently, the study of *magnetic* excitations by IXS is presently not possible.

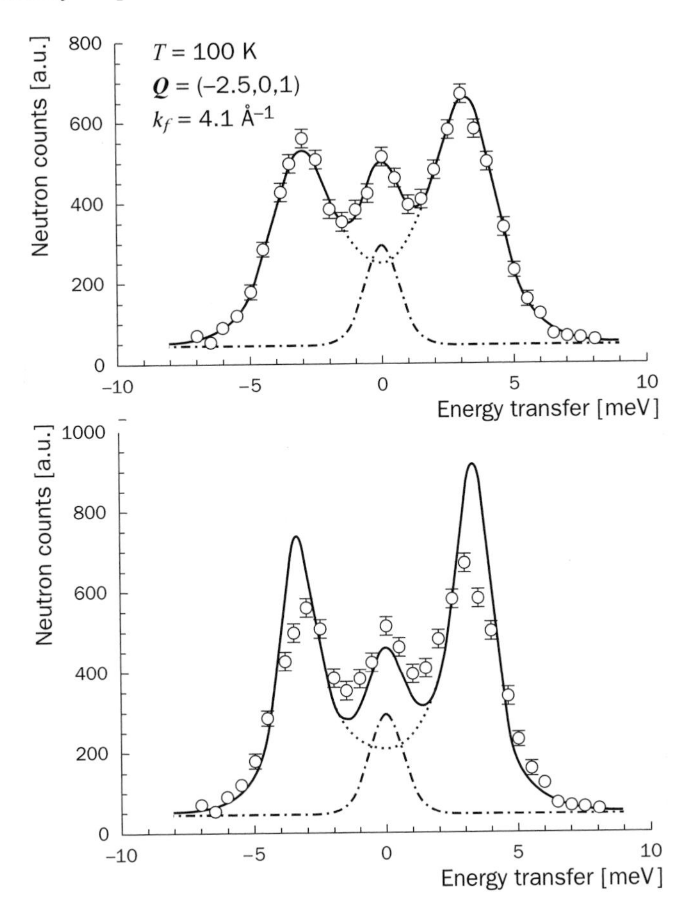

Figure 21 - Constant-$\boldsymbol{Q}$ scan across a low-energy zone boundary phonon in orthorhombic α-uranium (T = 150 K). The top frame shows the fit to the raw data in terms of a DHO function (as in equation {21}), convoluted with the instrumental resolution, and a Gaussian component centred at $\omega = 0$, sitting on a flat background. The Gaussian component corresponds to the elastic incoherent scattering cross-section from the sample and sample mount; its energy width (≈ 1.65 meV) is instrumental. The bottom frame shows the same Gaussian component together with the intrinsic best-fit DHO lineshape. The best-fit oscillator parameters are: ω_{Tj} = 3.57 ±0.03 meV; Γ_{Tj} = 1.68 ±0.12 meV. The asymmetry between the energy gain (ω < 0) and energy loss (ω > 0) sides is a consequence of the detailed balance factor in equation {21}; after **[25]**.

The IXS technique is presented in detail in the chapter by F. Sette and M. Krisch in this volume. Its characteristic features can be summarized as follows:

- In the energy transfer range corresponding to collective excitations in solids ($\hbar\omega < 1$ eV), $\boldsymbol{Q}$ and ω are decoupled, i.e., at fixed $\boldsymbol{k}_i$, the variation of $k_f = |\boldsymbol{k}_f|$ with energy transfer can be ignored. This means that constant-$\boldsymbol{Q}$ scans can be performed by simply scanning k_f, without having to modify the scattering triangle as in the neutron TAS technique (see figure 3). More importantly, *finite energy transfers* become possible *in the limit of low* Q, a regime which neutrons cannot readily access due to the so-called *kinematical limitations*, as seen by inspection from figures 2 and 3 and from the structure of equation {1}. This regime is important for the study of long-wavelength collective excitations (sound waves) in liquid and amorphous systems (X-ray and neutron *Brillouin scattering* [26-28]). In the opposite limit, the X-ray technique suffers more and more from the decay of the atomic form factors at large Q;
- The effective sample volume is always very small in the IXS case, because beam cross-sections are in the sub-mm range and absorption is high. As a result, inelastic signals are generally of comparable intensity with the two techniques, in spite of the higher incident photon flux. This assumes of course that single-crystal specimens of appropriate size (~ 0.1 – 1.0 cm^3) are available and can be used for the INS measurements. When this is not the case, as e.g. for high-pressure work, IXS is the technique of choice;
- The IXS energy resolution is of the order of a few meV (~ 1 meV in the best cases), which is comparable with the energy resolution of a thermal-beam TAS, but does not compete with that available on a cold-neutron spectrometer (typically 0.1 meV). As a result the IXS technique is better suited for *high energy transfer work*. Comparative studies of optic phonon branches in quartz [29] and diamond [30, 31] have shown that IXS is generally preferable for $\hbar\omega > 100$ meV. The INS-TAS resolution function is approximately Gaussian while the IXS one is Lorentzian with long tails. This can be a serious handicap for *low energy work* ($\hbar\omega \leq 20$ meV) in the presence of defect- or disorder-induced elastic diffuse scattering, as is often found to be the case in real specimens. The IXS momemtum resolution is significantly better than for the INS-TAS technique, but, as discussed in section 6, this is generally not a decisive factor of choice.

In summary, the IXS technique should be considered as a potential alternative to INS for the study of structural excitations, whenever at least one of the following criteria is fulfilled:

- low Z material (for which the ratio $\sigma_{scatt.}/\sigma_{abs.}$ is most favourable in the X-ray case),
- small sample volume,
- finite energy transfer at low Q
- good energy resolution at high energy transfer.

10. MAGNETIC EXCITATION STUDIES WITH TAS

Magnetic scattering arises from the interaction between the neutron magnetic moment $\boldsymbol{\mu}_n$ and the magnetic field created by the unpaired electrons in the specimen. This interaction is long range and non-central; therefore the calculation of the scattering cross-sections is more complicated than for the nuclear interaction [32, 33]. In standard notations, the double differential scattering cross-section is given by Fermi's golden rule

$$\left(\frac{d^2\sigma}{d\Omega\, dE_f}\right) \propto (\gamma r_0)^2 \frac{k_f}{k_i} \sum_{\sigma\lambda\sigma'\lambda'} p_\sigma p_\lambda \left|\left\langle \sigma'\lambda' \middle| \boldsymbol{\sigma}\cdot\boldsymbol{M}_\perp \middle| \sigma\lambda \right\rangle\right|^2 \delta\left(E_\lambda - E_{\lambda'} + \hbar\omega\right) \qquad \{24\}$$

where $\sigma\,(\sigma')$ refer to the initial (final) polarization state of the neutron and $\lambda(\lambda')$ to the initial (final) state of the scattering system, with energy $E_\lambda\,(E_{\lambda'})$. The probabilities p_λ and p_σ are obtained by thermal or statistical beam averaging; $\boldsymbol{\sigma}$ is the Pauli spin operator for the neutron and $\boldsymbol{M}(\boldsymbol{Q})$ is the quantum mechanical operator associated with the Fourier transform of the *total* magnetization density $\boldsymbol{M}(\boldsymbol{r})$ (including both spin and orbital contributions), expressed in Bohr magnetons ($\mu_B = e\hbar/2m_e$). The subscript $\perp$ refers to the component of $\boldsymbol{M}$ perpendicular to the momentum transfer $\boldsymbol{Q}$

$$\boldsymbol{M}_\perp = \hat{\boldsymbol{Q}}\times(\boldsymbol{M}\times\hat{\boldsymbol{Q}}) = \boldsymbol{M} - (\boldsymbol{M}\cdot\hat{\boldsymbol{Q}})\hat{\boldsymbol{Q}} \qquad \{25\}$$

with $\hat{\boldsymbol{Q}}$ a unit vector along $\boldsymbol{Q}$.

For an *unpolarized* neutron beam, equation {24} reduces to

$$\left(\frac{d^2\sigma}{d\Omega\, dE_f}\right) = \left(\gamma r_0\right)^2 \frac{k_f}{k_i} S\left(\boldsymbol{Q},\omega\right) \qquad \{26\}$$

with

$$S\left(\boldsymbol{Q},\omega\right) = \frac{1}{2\pi\hbar}\int_{-\infty}^{+\infty} \left\langle \boldsymbol{M}_\perp(-\boldsymbol{Q},0)\boldsymbol{M}_\perp(\boldsymbol{Q},t)\right\rangle e^{-i\omega t}\, dt \qquad \{27\}$$

where $\langle\ \rangle$ stands for thermal averaging.

The magnetic scattering cross-section can also be expressed in terms of the magnetic susceptibility tensor $\{\chi_{\alpha\beta}(\boldsymbol{Q},\omega)\}$ defined by

$$\boldsymbol{M}_\alpha(\boldsymbol{Q},\omega) = \sum_\beta \chi_{\alpha\beta}(\boldsymbol{Q},\omega)\, \boldsymbol{H}_\beta(\boldsymbol{Q},\omega) \qquad \{28\}$$

where α, β are Cartesian coordinates. Using fluctuation-dissipation theory, expression {27} becomes

$$S(\boldsymbol{Q},\omega) = \left[\frac{1}{2} g\, f(\boldsymbol{Q})\right]^2 \left(\frac{\mathrm{N}}{\pi}\right) \mathrm{e}^{-2W(\boldsymbol{Q})} \frac{1}{1-\mathrm{e}^{-\hbar\omega/k_B T}} \times \sum_{\alpha\beta} (\delta_{\alpha\beta} - \hat{Q}_\alpha \hat{Q}_\beta) Im\, \chi_{\alpha\beta}(\boldsymbol{Q},\omega) \qquad \{29\}$$

where we have assumed a Bravais lattice of magnetic ions with localized moments. Equation {29} is written in the *dipole approximation*, valid at low Q, where the orbital motion of the unpaired electrons is described by the Landé g factor. $f(\boldsymbol{Q})$ is the magnetic form factor that describes the spatial extent of the unpaired electrons wavefunction [$f(0) = 1$; $f(Q) \longrightarrow 0$ for $Q \longrightarrow \infty$]. The term $(\delta_{\alpha\beta} - \hat{Q}_\alpha \hat{Q}_\beta)$ ensures that only magnetic fluctuations normal to $\boldsymbol{Q}$ are sampled.

In ordered magnetic systems with localized moments the elementary excitations are called *spin waves* (even though the magnetic moments may include orbital components). For a simple ferromagnet in its low temperature ground state, all magnetic moments are aligned along a common direction, within each ferromagnetic domain. If we consider a domain polarized in the z-direction, a spin wave corresponds to a small *transverse* deviation of the magnetic moments, analogous to a small-angle precession around the z-axis, as sketched in figure 22a. These excitations propagate across the lattice with a wavevector $\boldsymbol{q}$ and the amplitude of the circularly-polarized transverse spin-component (see fig. 22b) is quantized. Spin waves (or *magnons*) are quantized bosons, which play the same role in the description of the dynamics of magnetic systems as do *phonons*, in the dynamics of atomic lattices.

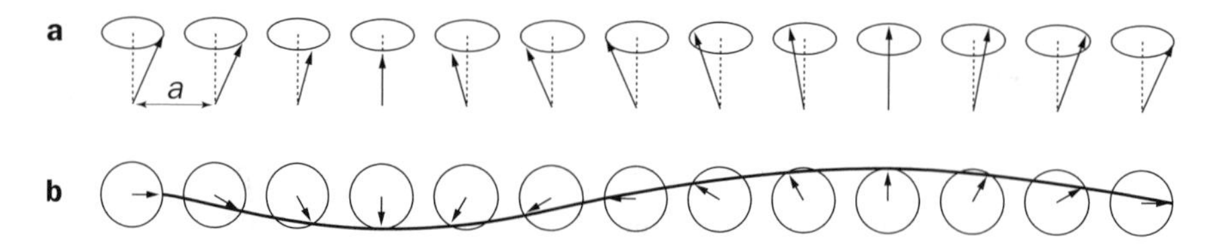

Figure 22 - Spin wave in a ferromagnet.

Formally, spin waves are obtained from the Heisenberg exchange Hamiltonian

$$H = -\sum_{i,j} J(r_i - r_j)\, \boldsymbol{S}_i \cdot \boldsymbol{S}_j \qquad \{30\}$$

where J is the exchange integral coupling the spins (or other magnetic moments) at neighbouring magnetic sites. The nearest neighbour exchange integral is positive for ferromagnets and negative for antiferromagnets. For a Heisenberg ferromagnet, the spin wave dispersion relation is given by

$$\hbar\omega(\boldsymbol{q}) = 2S\left[\mathcal{J}(0) - \mathcal{J}(\boldsymbol{q})\right] \qquad \{31\}$$

where $\mathcal{J}(\boldsymbol{q})$ is the Fourier transform of the exchange integral $J(\boldsymbol{r})$: $\mathcal{J}(\boldsymbol{q}) = \sum_r J(\boldsymbol{r})\, \mathrm{e}^{-i\boldsymbol{q}\cdot\boldsymbol{r}}$. In the limit $\boldsymbol{q} \longrightarrow 0$, the spin wave dispersion is quadratic: $\hbar\omega(\boldsymbol{q}) \approx D\boldsymbol{q}^2$ and the coefficient D is called the spin-wave *stiffness constant.* In the presence of a magnetic field, or any other source of magnetic anisotropy, the spin wave dispersion develops a gap at $q = 0$ (the $q = 0$ spin wave corresponds to an homogeneous rotation of all spins which costs energy in the presence of a magnetic anisotropy). The spin-wave dispersion of an antiferromagnet is linear at low $\boldsymbol{q}$ (as for an acoustic phonon) but becomes quadratic with a gap in the presence of magnetic anisotropy. At low temperature, the spin wave structure factor is written as

$$S_j(\boldsymbol{Q},\omega) = f^2(\boldsymbol{Q})(1+\hat{Q}_z^{\,2})\frac{1}{2}S\,\mathrm{e}^{-2W(\boldsymbol{Q})} \qquad \{32\}$$

$$\times \sum_{\boldsymbol{q},\boldsymbol{\tau}} \left[\left\langle n_q + 1\right\rangle \delta(\boldsymbol{Q}-\boldsymbol{\tau}+\boldsymbol{q})\,\delta(\hbar\omega - \hbar\omega_q) + \left\langle n_q\right\rangle \delta(\boldsymbol{Q}-\boldsymbol{\tau}+\boldsymbol{q})\,\delta(\hbar\omega + \hbar\omega_q)\right]$$

where the first (second) term in the summation corresponds to magnon creation (annihilation) and $\langle n_q \rangle = (\mathrm{e}^{\hbar\omega_q/k_B T} - 1)^{-1}$ is the Bose thermal population factor. The orientation factor $(1+\hat{Q}_z^{\,2})$ is equal to 2 when $\boldsymbol{Q}$ lies along the z-direction ($\hat{Q}_z = 1$): in that case both components of the fluctuating moments, say $\delta M_x(t)$ and $\delta M_y(t)$, are normal to $\boldsymbol{Q}$ and hence both contribute to the spin-wave structure factor. If $\boldsymbol{Q}$ lies in the (x,y) plane ($\hat{Q}_z = 0$), one can decompose the circularly polarized spin wave into two components with linear polarizations parallel and normal to $\boldsymbol{Q}$ and only the latter will contribute to $S(\boldsymbol{Q},\omega)$.

Spin waves renormalize with temperature due to magnon-magnon interactions. The situation is analogous to anharmonic frequency shifts and anharmonic broadening for phonons. As T approaches T_c, the magnetic ordering temperature, long-wavelength spin waves become overdamped and the magnetic fluctuation spectrum becomes diffusive. At short wavelengths, the situation is less clear. In principle "sloppy" spin waves can still propagate above T_c, due to magnetic short-range order, as long as the wavelength of the spin-wave is short on the scale of the magnetic correlation length. Intense neutron scattering activity has been devoted to establishing this point experimentally, particularly in the case of 3d itinerant ferromagnets such as metallic iron or nickel [34-36].

Figure 24 shows an example of magnon dispersions obtained from thermal and cold TAS data [37]. The compound is a Sr-doped bilayer manganite ($La_{1.2}Sr_{1.8}Mn_2O_7$) whose structure is sketched in figure 23 and which orders ferromagnetically below $T_c \approx 120$ K. Like most other doped manganites, the compound shows metallic behaviour near T_c, with a resistivity that drops dramatically in an applied magnetic field (the so-called colossal magneto-resistance "CMR" effect). With two Mn ions per primitive unit cell, the magnon spectrum consists in two branches: an *acoustic* branch for which the two Mn moments precess in phase and an optical branch where they precess in anti-phase.

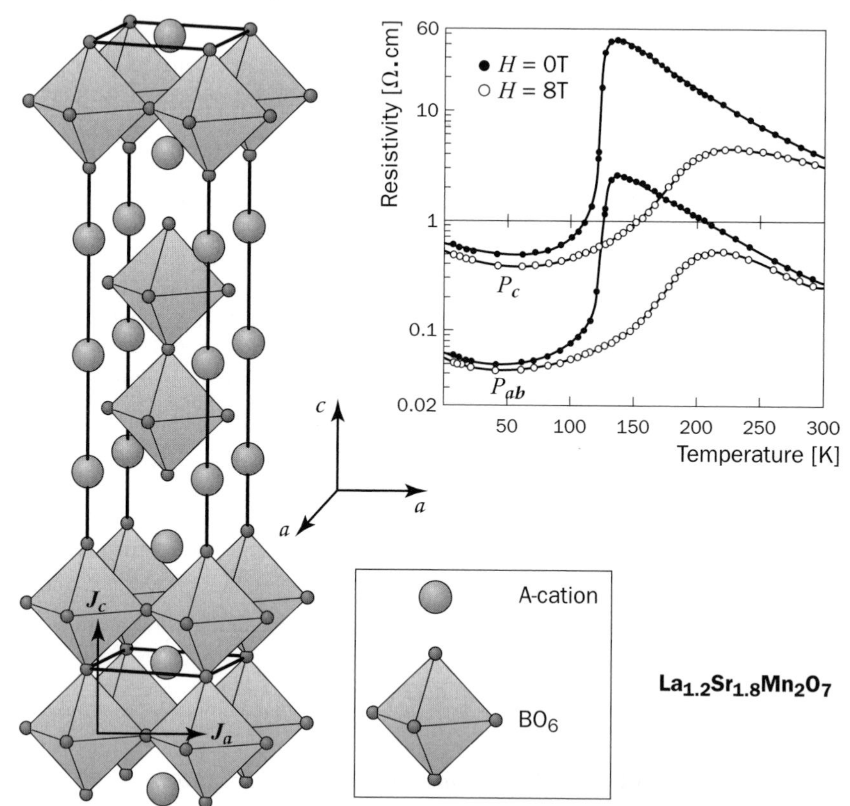

Figure 23 - The body-centred tetragonal unit cell of the bilayer manganite (after ref. **[38]**). The Mn ions are located at the centre of the oxygen octahedra; each primitive unit cell contains two octahedra; the inset shows the CMR effect for currents applied along the [001] (top) and [100] (bottom) directions.

Figure 24a shows the dispersion in the basal plane [100] direction, which is the direction of strong near-neighbour magnetic coupling (J_a in figure 23). Figure 24b shows the acoustic magnon dispersion along ***c*** which reflects the much weaker *inter-bilayer* Mn-Mn coupling (note the change in vertical scale between figures 24a and 24b). The *intra-bilayer* coupling along ***c*** (J_c in figure 23) gives rise to the splitting between the optic and acoustic branches. The $q = 0$ gap of the acoustic branch is smaller than the experimental resolution, indicating a very low degree of anisotropy with respect to the orientation of the Mn moments.

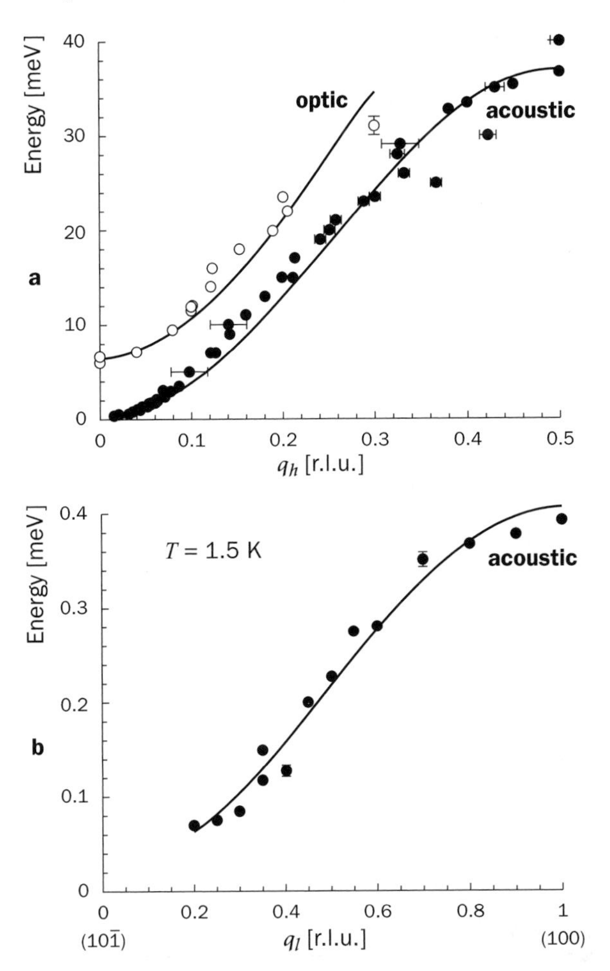

Figure 24 - Dispersion of acoustic and optic magnon branches in $La_{1.2}Sr_{1.8}Mn_2O_7$ along the (**a**) [100], (**b**) [001] propagation directions (after ref. **[37]**). In (**b**) the optical mode at ~ 6 meV is off-scale and shows no measurable dispersion along [001].

The solid lines in figure 24 show a fit of the experimental dispersions in terms of an effective quasi-2D Heisenberg Hamiltonian (cf. eq. {30}), with near-neighbour exchange interactions between localized magnetic moments. As discussed in **[37]** the fit is not fully satisfactory, due to the dual localized-itinerant character of the 3d electron system in doped metallic manganites.

More detailed information on magnetic structures and magnetic excitations can be obtained from the use of spin-polarized neutron beams and *polarization analysis* **[39,40]**. As discussed in section 2, the implementation of polarization techniques on TAS instruments is straightforward using Heusler crystals, polarizing supermirror benders or neutron spin filters. In the simplest case, a guide field at the sample position defines the direction of the incident beam polarization $\boldsymbol{P}_0$. One measures separately the scattering cross-sections corresponding to a change (*spin-flip*) and no change (*non-spin-flip*) in the neutron spin state. This allows magnetic and nuclear $S(\boldsymbol{Q},\omega)$ components to be distinguished, whenever they may overlap. Polarization techniques are also useful to determine the direction of the (static or fluctuating) magnetic moments, This is possible because the spin-flip and non-spin-flip magnetic cross-sections depend on the orientation of these moments with respect to the polarization direction $\boldsymbol{P}_0$. Even in non-magnetic materials polarization techniques are sometimes used to distinguish between coherent (non-spin-flip) and incoherent (partly spin-flip) contributions.

11. PRACTICAL ASPECTS

The first choice to be made in preparing a TAS experiment is to select the incident neutron energy range. This choice determines in turn the type of neutron source (cold, thermal or hot) and the instrument best suited for the experiment. Thermal-beam TAS are typically used for incident neutron energies in the range from 15 to 100 meV, although the precise crossover values depend on the detailed characteristics of each instrument.

The choice of the incident neutron energy should be made in relation with a number of often conflicting requirements:

- to be able to "close the triangle" i.e. to satisfy the kinematical constraints over the required $(\boldsymbol{Q},\omega)$ range. These constraints are particularly severe in the low-Q large-ω limit.

- to satisfy the existing angular limitations on the selected instrument, notably the maximum value of the scattering angle (A4) which determines the maximum value of Q that may be reached for a given set of (k_i, k_f) values.
- to be able to perform the experiment under acceptable resolution/intensity conditions. One may recall from the discussion in sections 2 and 4 (eq. {2} and {9}) that resolution and integrated intensities vary in opposite ways as a function of k_i.

Optimizing the experimental conditions also involves making choices and compromises with respect to: the *d*-spacing of the monochromator and analyzer crystals; the collimation; the use of flat or focusing bent crystals; the scattering configurations. In carrying out this, often complex, optimization process, the availability of reliable *simulation* programs is essential [41,42]. For example the RESTRAX computer package [41] is able to simulate scan profiles and resolution corrections on several levels of approximation: Gaussian approximations for flat or bent [43] crystals as well as general Monte-Carlo ray-tracing simulations.

The same programs, embedded in appropriate fitting routines, are used at the data treatment stage, to correct the experimental spectra from instrumental effects. As discussed in section 7, the design of modern TAS spectrometers emphasizes luminosity at the expense of momentum resolution, with the aim to allow a broader range of small and/or weakly scattering specimens to be investigated. The price to pay is the reliance on more sophisticated data treatment programs involving the modelling of the scattering function and its convolution with the full 4D instrumental resolution function.

An element that may often perturb the measurements is the presence of harmonic wavelengths $2\,k_i$, $3\,k_i$, ... in the primary beam or $2\,k_f$, $3\,k_f$, ... in the detected beam due to higher-order diffraction at the monochromator or analyzer position. For instance, the contribution of unfiltered $2\,k_i$ and $3\,k_i$ harmonics to the monitor count rate can be quite significant. Since the monitor count rate is used to normalize the acquisition times in the constant-k_f mode (cf. section 4), this effect may lead to large scan profile distorsions, unless corrected for.

Harmonics also lead to spurious signals for commensurate values of k_i and k_f

$$\mathrm{n}\,k_i = \mathrm{m}\,k_f \qquad (\mathrm{n, m} = 1,2,3,4\ldots)$$

due to elastic (diffuse or incoherent) scattering in the sample (or sample environment).

To minimize the probability of such processes one must limit the magnitude of the energy transfer with respect to the incident neutron energy. A safe rule of thumb is given by $2/3 < k_i/k_f < 3/2$.

As discussed in section 2, harmonic contamination may also be suppressed by using appropriate filters or by using elastically bent perfect Si111 monochromator or analyzer crystals that have reflecting properties comparable to PG002 and are free from 2nd order contamination.

12. SUMMARY AND OUTLOOK

The three-axis technique continues to be a basic neutron scattering technique for inelastic work on single-crystal specimens. As discussed in section 3, there is, at the moment, a fair degree of complementarity between TAS instruments on steady-state sources and TOF instruments on steady-state and pulsed sources. For the study of structural excitations, there is also complementarity with the synchrotron-based IXS technique whose capabilities have dramatically improved in recent years. In the long term, the availability of more powerful pulsed neutron sources will certainly tip the balance more in favour of the TOF technique and the same probably applies to synchrotron sources and the IXS technique. On the other hand, the capabilities of the TAS technique in certain fields, such as the study of low energy magnetic excitations, with or without polarization analysis, are likely to remain unchallenged for some time in the future.

As seen in section 7, there are a number of possible extensions of the TAS technique, which are presently under consideration or under development and which should lead to major gains in data acquisition rate in the future. There are other developments which aim at improving the quality of the information accessible from TAS experiments. One such development is the spherical polarization analysis (CRYOPAD) technique, [44] which was briefly mentioned in section 2. Another important development, which could not be covered within the framework of this chapter, consists in combining the TAS and NSE (neutron spin-echo) techniques. NSE is widely used to study slow relaxation phenomena in solids (see chapter by R. Cywinski in this volume). It was suggested long ago [45,46] that it could also be applied to the study of collective excitations and that it would then permit inelastic work to be done, such as phonon linewidth studies, with the equivalent of a μeV energy resolution. The feasibility of such inelastic NSE experiments was demonstrated by the pioneering work of Mezei [47] on the roton linewidth in superfluid ^{4}He and by several more recent experiments on ^{4}He [48] and Ge [49]. The combined use of a polarized-beam TAS instrument, acting as background spectrometer, together with a NSE set-up optimized for inelastic work, may open new fields of research in the near future. Several TAS spectrometers at HMI, ILL and FRM-II are presently being equipped with NSE coils (using the precession [50] or resonance [51] method) in order to explore the potential applications of the technique.

ACKNOWLEDGMENTS

The author is indebted to B. Dorner, B. Fåk, A. Hiess, J. Kulda and A. Stunault for critical reading of the manuscript and to E. Lelievre-Berna and T. Chatterji for assistance in writing parts of section 10.

REFERENCES

[1] T. VOGT, L. PASSELL, S. CHEUNG & J.D. AXE - *Nucl. Instr. Meth. A* **338**, 71 (1994); J. SCHEFER, M. MEDARDE, S. FISCHER, R. THUT, M. KOCH, P. FISCHER, U. STAUB, M. HORRISBERGER, G. BÖTTGER & A. DOENNI - *Nucl. Instr. Meth. A* **372**, 229 (1996)

[2] B. HAMELIN - *ILL Annual Report*, Institut Laue-Langevin, Grenoble, 80 (1996)

[3] T. KRIST, R. GÖTTEL, P. SCHUBERT-BISCHOFF & F. MEZEI - *J. Phys. Soc. Japan* **65**-Suppl. A, 226 (1996)

[4] P. BÖNI - *Physica B* **267-268**, 320 (1999)

[5] F. TASSET - *Physica B* **174**, 506 (1991)

[6] W. HEIL, J. DREYER, O. HOFMANN, H. HUMBLOT, E. LELIÈVRE-BERNA & F. TASSET - *Physica* B **267-268**, 328 (1999)

[7] R. NATHANS, C.G. SHULL, G. SHIRANE & A. ANDRESEN - *J. Phys. Chem. Sol.* **10**, 138 (1959)

[8] F. MEZEI - Neutron Spin Echo: A New Concept in Polarized Thermal Neutron Techniques, *Z. Physik* **255**, 146 (1972)

[9] F. TASSET - *Physica B* **156-157**, 627 (1989); V. NUNEZ, P.J. BROWN, J.B. FORSYTH & F. TASSET - *Physica B* **174**, 60 (1991); P.J. BROWN, J.B. FORSYTH & F. TASSET - *Proc. R. Soc. London A* **442**, 147 (1993);

[10] L.P. REGNAULT, H.M. RØNNOW, J.E. LORENZO, R. BELLISSENT & F. TASSET - *Physica B* **335**, 19 (2003); L.P. REGNAULT, B. GEFFRAY, P. FOUILLOUX, B. LONGUET, F. MANTEGEZZA, F. TASSET, E. LELIÈVRE-BERNA, E. BOURGEAT-LAMI, M. THOMAS & Y. GIBERT -

Physica B **335**, 255 (2003);
L.P. REGNAULT, B. GEFFRAY, P. FOUILLOUX, B. LONGUET, F. MANTEGEZZA, F. TASSET, E. LELIÈVRE-BERNA, S. PUJOL, E. BOURGEAT-LAMI, N. KERNAVANOIS, M. THOMAS & Y. GIBERT - *Physica B* **350**, E811 (2004)

[11] A. HIESS, R. CURRAT, J. SAROUN & F.J. BERMEJO - *Physica B* **276-278**, 91 (2000)

[12] B. DORNER - *Acta Cryst. A* **28**, 319 (1972)

[13] M.J. COOPER & B. NATHANS - *Acta Cryst.* **23**, 357 (1967);
G. SHIRANE, S.M. SHAPIRO & J.M. TRANQUADA - *Neutron scattering with a triple-axis spectrometer*, Cambridge University Press, 256 (2002)

[14] K. LEFMANN, D.F. MCMORROW, H.M. RØNNOW, K. NIELSEN, K.N.CLAUSEN, B. LAKE & G. AEPPLI - *Physica B* **283**, 343 (2000)

[15] F. DEMMEL, A. FLEISCHMANN & W. GLÄSER - *Nucl. Instr. Meth. A* **416**, 115 (1998)

[16] R. BORN & D. HOHLWEIN - *Z. Phys. B: Condens. Matter* **74**, 547 (1989)

[17] J. KULDA, H. SCHOBER & S. HAYDEN - *Flat cone geometry for excitation mapping on TAS*, The ILL Millennium Programme-Second Phase Proposals, ILL Report 00.CA02T, Institut Laue-Langevin, Grenoble (2000)

[18] A.A. MARADUDIN & S.H. VOSKO - *Rev. Mod. Phys.* **40**, 1 (1968)

[19] J.M. PEREZ-MATO, M. AROYO, J. HLINKA, M. QUILICHINI & R. CURRAT - *Phys. Rev. Lett.* **81**, 2462 (1998)

[20] H. BOYSEN, B. DORNER, F. FREY & H. GRIMM - *J. Phys. C: Sol. St. Phys.* **13**, 6127 (1980)

[21] D. STRAUCH & B. DORNER - *J. Phys.: Condens. Matter* **2**, 1457 (1990)

[22] P. GIANNOZZI, S. DE GIRONCOLI, P. PAVONE & S. BARONI - *Phys. Rev. B* **43**, 7231 (1991)

[23] A.A. MARADUDIN & A.E. FEIN - *Phys. Rev.* **128**, 2589 (1962)

[24] R.A. COWLEY - *Adv. Phys.* **12**, 421 (1963); *Phil. Mag.* **11**, 673 (1965)

[25] J.C. MARMEGGI, R. CURRAT, A. BOUVET & G.H. LANDER - *Physica B* **276-278**, 272 (2000) and unpublished results

[26] B. DORNER - *In "Frontiers of Neutron Scattering"*, A. FURRER ed., World Scientific, Singapore, 27 (2000)

[27] F. SETTE, G. RUOCCO, M. KRISCH, C. MASCIOVECCHIO, R. VERBENI & U. BERGMANN - *Phys. Rev. Lett.* **77**, 83 (1996)

[28] C. PETRILLO, F. SACCHETTI, B. DORNER & J.B. SUCK - *Phys. Rev. E* **62**, 3611 (2000)

[29] E. BURKEL, C. SEYFERT, CH. HALCOUSSIS, H. SINN & R.O. SIMMONS - *Physica B* **263-264**, 412 (1999)

[30] J. KULDA, B. DORNER, B. ROESSLI, H. STERNER, R. BAUER, T. MAY, K. KARCH, P. PAVONE & D. STRAUCH - *Solid State Comm.* **99**, 799 (1996)

[31] J. KULDA, H. KAINSMAIER, D. STRAUCH, B. DORNER, M. LORENSEN & M. KRISCH - *Phys. Rev. B* **66**, 241202 (2002)

[32] G.L. SQUIRES - *Introduction to the theory of thermal neutron scattering*, Cambridge University Press, Chapter 7 (1978)

[33] S.W. LOVESEY - *Theory of neutron scattering from condensed matter*, Oxford University Press, Vol. 2 (1988)

[34] O. STEINSVOLL, C.F. MAJKRZAK, G. SHIRANE & J. WICKSTED - *Phys. Rev. Lett.* **51**, 300 (1983)

[35] J.D. WICKSTED, G. SHIRANE & O. STEINSVOLL - *Phys. Rev. B* **29**, 488 (1984)

[36] J.W. LYNN - *Phys. Rev. Lett.* **52**, 775 (1984); H.A. MOOK & J.W. LYNN - *J. Appl. Phys.* **57**, 306 (1985)

[37] T. CHATTERJI, L.P. REGNAULT, P. THALMEIER, R. VAN DE KAMP, W. SCHMIDT, A. HIESS, P. VORDERWISCH, R. SURYANARAYANAN, G. DHALENNE & A. REVCOLEVSCHI - *J. of Alloys and Compounds* **326**, 15 (2001)

[38] T. CHATTERJI, R. SCHNEIDER, J.-U. HOFFMANN, D. HOHLWEIN, R. SURYANARAYANAN, G. DHALENNE, A. REVCOLEVSCHI - *Phys. Rev. B* **65**, 134440 (2002)

[39] R.M. MOON, T. RISTE & W.C. KOEHLER - *Phys. Rev.* **181**, 920 (1969)

[40] W.G. WILLIAMS - *Polarized Neutrons*, Clarendon Press-Oxford (1988)

[41] J. SAROUN & J. KULDA - *Physica B* **234-236**, 1102 (1997)

[42] K. LEFMAN, K. NIELSEN, A. TENNANT & B. LAKE - *Physica B* **276-278**, 152 (2000)

[43] M. POPOVICI, A.D. STOICA & M. IONITA - *J. Appl. Cryst.* **20**, 90 (1987)

[44] F. TASSET - *Physica B* **297**, 1 (2001); P.J. BROWN - *Physica B* **297**, 198 (2001)

[45] F. Mezei - *In "Neutron Inelastic Scattering"*, IAEA, Vienna, 125 (1978)

[46] R. PYNN - *J. Phys. E* **11**, 1133 (1978)

[47] F. MEZEI - *Phys. Rev. Lett.* **44**, 1601 (1980)

[48] E. FARHI, B. FÅK, C.M.E. ZEYEN & J. KULDA - *Physica B* **297**, 32 (2001)

[49] J. KULDA, E. FARHI & C.M.E. ZEYEN - *Physica B* **297**, 37 (2001)

[50] R. GÄHLER, R. GOLUB & T. KELLER - *Physica B* **156-157**, 653 (1989)

[51] C.M.E. ZEYEN - *J. Phys. Chem. Sol.* **60**, 1573 (1999)

13

NEUTRON SPIN ECHO SPECTROSCOPY

R. CYWINSKI
School of Physics and Astronomy, University of Leeds, UK

1. INTRODUCTION

Quasi-elastic neutron scattering has long been an invaluable tool for studying the complexities of very slow dynamical processes associated with stochastic phenomena as diverse as molecular rotation and tunnelling, diffusion, polymer reptation, magnetic relaxation and glassy dynamics. The characteristic time constants of such processes are typically in the range 10^{-12} s < τ < 10^{-6} s, and hence quasi-elastic neutron scattering studies must always be made at very small energy transfers with a correspondingly high energy resolution. However, whilst time-of-flight and triple-axis-based backscattering spectrometers provide the highest resolution currently available with conventional neutron instrumentation, this resolution, at best $\Delta E \sim 0.3$ μeV with an incident neutron energy of ~ 2 meV, is unfortunately inadequate for many studies of dynamical processes for which characteristic time scales are longer than ~ 10^{-10} s. This significant constraint on the dynamic range of quasi-elastic neutron scattering, and the associated implications for the investigation of slow dynamics in matter, is simply a consequence of the so-called Liouville theorem which, through conservation of the phase space density of neutron trajectories, strongly couples the effective neutron count rate of a conventional neutron spectrometer directly to its resolution. For an ideally optimized generic neutron spectrometer, the Liouville theorem shows that the count rate is at very best proportional to the square of the instrumental resolution. Therefore to extend the time window of a conventional backscattering spectrometer by the necessary three orders of magnitude would correspondingly reduce the effective count rate by six orders of magnitude. This is clearly an unacceptable sacrifice in what is an already severely intensity-limited experimental technique.

F. Hippert et al. (eds.), Neutron and X-ray Spectroscopy, 427–455.

Fortunately, in 1972, Mezei successfully demonstrated that it was possible to circumvent the restrictions imposed by the Liouville theorem by using an extremely ingenious and elegant technique known as neutron spin echo (NSE) [1]. By effectively decoupling instrumental energy resolution from incident neutron monochromatization, and hence from beam intensity, neutron spin echo offers gains of at least two orders of magnitude in resolution over conventional quasi-elastic scattering spectrometers. This gain is achieved not simply by producing a neutron beam with better defined incident energy and by analyzing more accurately the energy of the scattered neutrons, but instead by determining the relative number of Larmor precessions executed by spin polarized neutrons travelling through a well defined magnetic field before and after the scattering sample. By utilizing Larmor precession in this way each neutron effectively carries its own microscopic stop watch through which its change of velocity, and hence its change of energy, in the scattering process is uniquely and accurately defined.

In this chapter we shall consider the basic principles of the neutron spin echo technique, its practical realisation, and some of its recent applications in the study of slow dynamics and relaxation phenomena in physics and chemistry.

2. *POLARIZED NEUTRON BEAMS, LARMOR PRECESSION AND SPIN FLIPPERS*

As the Larmor precession of polarized neutrons lies at the very heart of the spin echo technique, it is perhaps worth reviewing some of the more fundamental properties of polarized neutron beams and the techniques employed in their manipulation.

The neutron is a fermion. It carries a spin 1/2 and an associated angular momentum of $\pm 1/2\hbar$ The spin angular momentum component operators of the neutron expressed in matrix form are:

$$s_x = \frac{\hbar}{2}\begin{bmatrix} 0 & 1 \\ 1 & 0 \end{bmatrix} \quad s_y = \frac{\hbar}{2}\begin{bmatrix} 0 & -i \\ i & 0 \end{bmatrix} \quad s_z = \frac{\hbar}{2}\begin{bmatrix} 1 & 0 \\ 0 & -1 \end{bmatrix} \qquad \{1\}$$

whilst the operators

$$\sigma \equiv \frac{2}{\hbar} s \qquad \{2\}$$

are known as the Pauli spin operators.

By virtue of its spin the neutron also has a magnetic moment, μ_n with

$$\mu_n = \gamma_n \mu_N \sigma = 2\gamma_n \mu_N s/\hbar = -9.66 \times 10^{-27}\,\mathrm{J.T^{-1}} \qquad \{3\}$$

where $\gamma_n = -1.91304275(45)$ is the neutron moment in units of the nuclear magneton, μ_N. Additionally, the gyromagnetic ratio of the neutron is defined by $\boldsymbol{\mu}_n = \gamma_L \boldsymbol{s}$ with $\gamma_L = -1.832 \times 10^8 \, \text{rad.s}^{-1}.\text{T}^{-1}$.

The neutron spin wave function χ, can be written as a coherent superposition of the two possible states of its spin quantum number, $m_s = \pm 1/2$, i.e.

$$|\chi\rangle = a|+\rangle + b|-\rangle \qquad \{4\}$$

where $|a|^2$ and $|b|^2$ are the probabilities that a measurement of the spin will show it to be in the $m_s = +1/2$ or $m_s = -1/2$ state. Clearly $|a|^2 + |b|^2 = 1$.

The polarization, p, of an individual neutron, relative to a magnetic field (i.e. the z-direction) is given by the expectation value of the Pauli spin operator,

$$\boldsymbol{p}_i = \langle \chi | \boldsymbol{\sigma} | \chi \rangle \qquad \{5\}$$

and correspondingly the net polarization $\boldsymbol{P}$ of a beam of neutrons is simply the mean polarization of the N neutrons in the beam, so

$$\boldsymbol{P} = \frac{1}{N} \sum_i^N \boldsymbol{p}_i \qquad \{6\}$$

It is important to note that whilst $\boldsymbol{s}$ and $\boldsymbol{\sigma}$ are quantum operators or quantum vectors, the Ehrenfest theorem tells us that because $\boldsymbol{p}_i$ and $\boldsymbol{P}$ are defined by an expectation value, i.e. by $\boldsymbol{p}_i = \langle \chi | \boldsymbol{\sigma} | \chi \rangle$, they can thus be treated as classical vectors. It is then straightforward to write down the classical equation of motion of the beam polarization vector, $\boldsymbol{P}$, in a magnetic flux density, $\boldsymbol{B}$. We therefore have

$$\frac{d\boldsymbol{P}}{dt} = -\gamma_L (\boldsymbol{P} \times \boldsymbol{B}) \qquad \{7\}$$

So, with $\boldsymbol{B}$ aligned along the z-axis, we find

$$\frac{d\boldsymbol{P}_x}{dt} = -\omega_L \boldsymbol{P}_y, \quad \frac{d\boldsymbol{P}_y}{dt} = \omega_L \boldsymbol{P}_x, \quad \frac{d\boldsymbol{P}_z}{dt} = 0 \qquad \{8\}$$

the solutions of which are

$$\begin{aligned} P_x(t) &= \cos(\omega_L t) P_x(0) - \sin(\omega_L t) P_y(0) \\ P_y(t) &= \sin(\omega_L t) P_x(0) + \cos(\omega_L t) P_y(0) \\ P_z(t) &= P_z(0) \end{aligned} \qquad \{9\}$$

These coupled equations of motion show that those components of the neutron beam polarization transverse to an applied magnetic flux density precess at an angular frequency of ω_L ($= |\gamma_L| B$), i.e. the Larmor precession frequency, whilst the component parallel to the flux density remains constant (fig. 1).

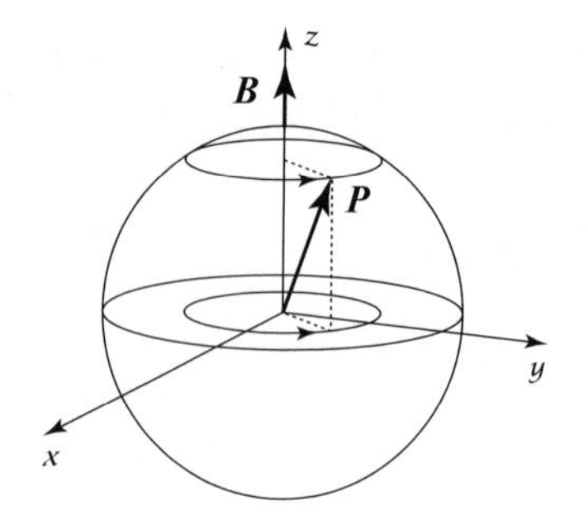

Figure 1 - Larmor precession of the neutron beam polarization vector in a magnetic flux density $\boldsymbol{B}$ applied along the z-axis. Note that the z-component of the polarization remains constant, whilst only the x- and y-components vary with time.

If the direction of the flux density B changes sufficiently slowly in the rest frame of the neutrons, then the component of polarization parallel to $\boldsymbol{B}$ is conserved and there is a corresponding adiabatic rotation of the polarization. An adiabaticity parameter, E, can be defined in terms of the Larmor precession frequency of the polarization, $\boldsymbol{P}$, relative to the rate of angular rotation, ω_B, of the magnetic flux density away from the z-axis along the neutron beam direction (also chosen to be the z-axis), in the rest frame of the neutron, i.e.

$$E = \frac{\omega_L}{\omega_B} \qquad \{10\}$$

For a neutron travelling with velocity v along the z-direction in a region in which the magnetic flux density is rotated at a constant angular rate of $d\theta_B/dz$ we may write

$$\omega_B = \frac{d\theta_B}{dt} = \frac{d\theta_B}{dz}\frac{dz}{dt} = \frac{d\theta_B}{dz}v \qquad \{11\}$$

Hence

$$E = \frac{\omega_L}{\omega_B} = |\gamma_L| B \bigg/ \left(\frac{d\theta_B}{dz} v \right) \qquad \{12\}$$

For an adiabatic rotation without loss of polarization an adiabaticity parameter, E, greater than 10 is required. This corresponds to an angular rotation of the flux density associated with a magnetic guide field ($H_G = B/\mu_o$) along the z-direction of $d\theta_B/dz < 26.5\, B\lambda$ degrees cm^{-1} with B in mT, θ in degrees, distance z in cm and neutron wavelength, λ in Å.

Although an adiabatic rotation of the guide field direction can be used to reorient adiabatically the beam polarization in the laboratory frame, the component of polarization along the guide field direction under this condition will, of course, remain constant. However, for an extreme non-adiabatic or sudden reorientation of the guide field (equivalent to $E \sim 0$) the polarization cannot follow the guide field and instead the beam polarization will preserve its initial direction through the region of sudden field change, and afterwards begin to process about the new guide field direction. On the one hand this implies that any sudden changes in direction of the guide field H_G must be avoided if beam polarization is to be preserved, on the other hand it also provides a mechanism for changing the direction of the polarization with respect to the guide field (for example by $\pi/2$ or π) in a controlled fashion.

The Mezei spin flipper [2], an essential component of all neutron spin echo spectrometers, relies upon such a controlled non-adiabatic process to rotate, or flip, the direction of the polarization of the neutron beam with respect to the guide field. In its simplest form the Mezei flipper consists of a solenoidal coil of rectangular cross-section with its axis, and therefore internal flux density $\boldsymbol{B}_C$ perpendicular to the beam direction, and at a specified angle to the external guide field, $\boldsymbol{B}_G$ (fig. 2). The neutrons entering and leaving the energized coil through the current-carrying coil windings experience an abrupt non-adiabatic flux density reversal, but whilst inside the coil their polarization precesses through an angle determined by the direction and magnitude of the internal resultant flux density, $\boldsymbol{B}_R = \boldsymbol{B}_G + \boldsymbol{B}_C$, the thickness of the coil, d, and the neutron velocity, v.

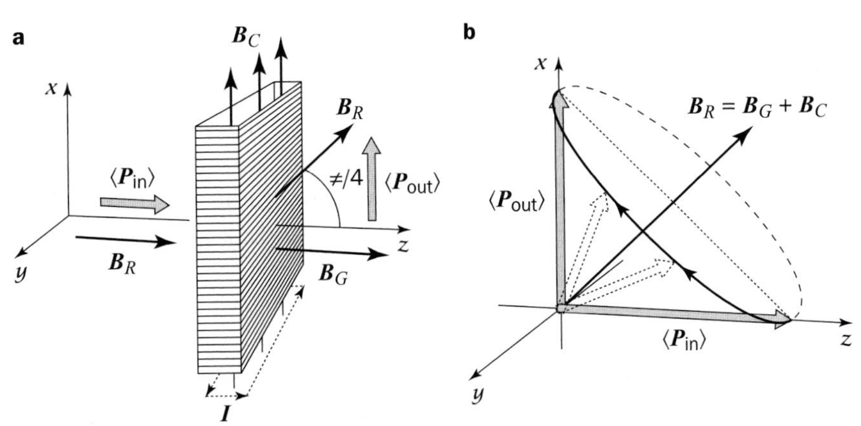

Figure 2 - a - A schematic layout of the Mezei $\pi/2$ spin flipper coil assembly in the neutron beam, showing the relative orientations of the guide field $\boldsymbol{B}_G$, the coil field $\boldsymbol{B}_C$ and the resultant field within the spin flipper, $\boldsymbol{B}_R$. **b -** The π-precession of the neutron polarization around the resultant field, $\boldsymbol{B}_R$, on traversing the $\pi/2$-flipper coil, resulting in a $\pi/2$ rotation of the polarization vector from the z- to the x-direction. Note that a similar geometry, but with $\boldsymbol{B}_R \approx \boldsymbol{B}_C >> \boldsymbol{B}_G$, enables a π-flip of the neutron polarization around the x-axis thereby reversing its y-component.

So, for example, to effect a $\pi/2$-rotation of the polarization with respect to the external guide field, $\boldsymbol{B}_R$ should be at an angle of 45° with respect to the direction of the guide field, whilst the magnitude of $\boldsymbol{B}_R$ should be such that the neutron polarization undergoes a precession of precisely π whilst within the coil. Similarly a π rotation, or spin flip, of the polarization direction with respect to the guide field is achieved if the resultant flux density $\boldsymbol{B}_R$ is now perpendicular to the guide field, with $|\boldsymbol{B}_R|$ again adjusted such that a precession of the polarization by π radians takes place within the thickness of the coil. In both cases the necessary π-precession within the coil requires

$$|\boldsymbol{B}_R| = \frac{v\pi}{\gamma_L d} \qquad \{13\}$$

Typically, neutrons of wavelength 2 Å will undergo a precession of π radians in a rectangular solenoid of 1cm thickness if the resultant flux density is $|\boldsymbol{B}_R| = 3.5$ mT.

3. *GENERALIZED NEUTRON SPIN ECHO*

We now have the all the elements necessary to discuss the underlying principles of the neutron spin echo technique. A schematic diagram of a generic spin echo spectrometer is illustrated in figure 3a. We shall assume that a broadly monochromatic beam of neutrons is longitudinally polarized by a polarizing supermirror assembly [3], and then rotated by $\pi/2$ into the x-direction just prior travelling along the z-axis of a solenoid of length L_1 in which the longitudinal flux density is B_1. As the neutrons enter the solenoid with a polarization perpendicular to the field direction the polarization will precess in the (x,y) plane orthogonal to the axis of the solenoid. The total precession angle φ_1 (in radians) over the length of the solenoid L_1 for those neutrons with velocity v is

$$\varphi_1 = \frac{\gamma_L B_1 L_1}{v} \qquad \{14\}$$

The total number, N, of 2π precessions expressed in terms of the neutron wavelength λ is therefore

$$N = \frac{\varphi_1}{2\pi} = 7361\, B_1 L_1 \lambda \qquad \{15\}$$

where λ is in Å, L_1 in metres and B_1 in T. For a typical flux density line integral, $B_1 L_1$ (or more strictly $\int_0^{L_1} \boldsymbol{B}\cdot d\boldsymbol{l}$) of 0.3 T.m, the polarization of 10 Å neutrons will undergo 22083 precessions. It is apparent that if the initially polarized neutron beam is broadly monochromatic, the precessing spins associated with neutrons of differing velocities within the beam will rapidly dephase and the final net polarization of the neutron beam at the exit of the first solenoid will be zero.

If the neutron beam is now spin-flipped through π radians at the exit of the first solenoid, before passing through the second solenoid of length L_2 and longitudinal flux density B_2 (directed in the same sense as that in the first solenoid) the neutron polarization will again precess in the (x,y) plane orthogonal to the axis of the solenoid, but in the reverse sense to that in the first solenoid. The total precession angle, φ_T, accumulated along the flight path through the two solenoids is therefore

$$\varphi_T = \varphi_1 - \varphi_2 = \gamma_L \left(\frac{B_1 L_1 - B_2 L_2}{v} \right) \quad \{16\}$$

In otherwords, providing the line integrals $B_1 L_1$ and $B_2 L_2$ are identical the total precession angle φ_T is zero, and the polarization of the beam is fully recovered on exit from the second solenoid, *irrespective of the spread of neutron velocities*. This final polarization can then be measured by flipping the neutron spins back into the longitudinal z-direction using a second $\pi/2$-flipper and finally analyzing the beam with a second supermirror polarizer (fig. 3b and 3c).

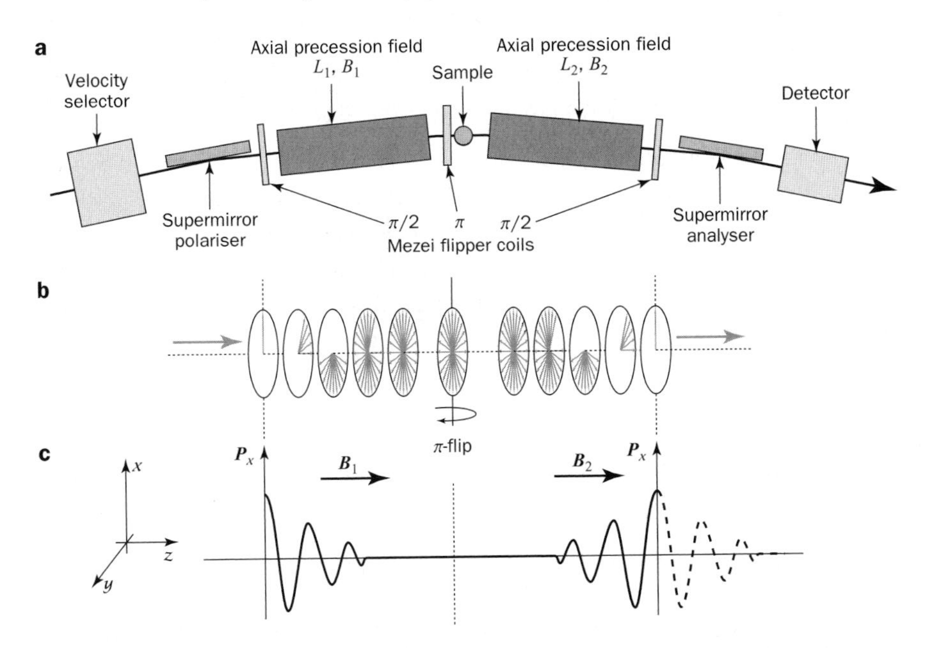

Figure 3 - (**a**) The layout of a generic neutron spin echo spectrometer showing the principal components and (**b**) a schematic representation of the dephasing of the precessing neutron spins along the first arm of the spectrometer, followed by a π-spin flip and subsequent rephasing along the second arm, with (**c**) the corresponding x-component of the net neutron polarization.

It should be noted that in this instrumental geometry the first $\pi/2$ spin flipper provides an effective "start time" of the neutron beam's precession "clock", whilst the second $\pi/2$ spin flipper provides the "stop time". The π-flipper between the two precession solenoids acts essentially as "time reversal" device in terms of the sense of the neutron precession.

It is clear that if the line integral of the flux density along the path through the second solenoid $B_2 L_2$ is varied from the optimal "echo" condition and the measured polarization P_z will also vary. For a perfectly monochromatic neutron beam this variation is a simple cosine function. However because of the distribution of neutron velocities $f(v)$ in the beam, the cosine response of P_z is heavily damped:

$$P_z = \langle \cos(\varphi_T) \rangle = \int f(v) \cos\left(\gamma_L (B_1 L_1 - B_2 L_2)/v\right) dv \qquad \{17\}$$

This expression, which defines the shape of the spin echo envelope or group, is simply the Fourier transform of the distribution function for $1/v$, which, in turn, is equivalent to the Fourier transform of the wavelength spread of the beam. Correspondingly, the period of the oscillations within the spin echo group is determined by the inverse of the mean neutron wavelength (fig. 4).

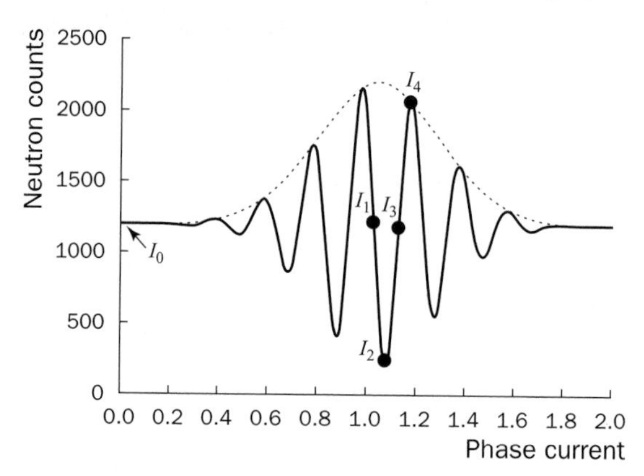

Figure 4 - Representation of the spin echo group measured by varying the phase current to the correction coil used to change the longitudinal guide field in one of the spectrometer's precession coils. I_0 is the average count rate, whilst I_1, I_2, I_3 and I_4 are the count rates at points on the spin echo group separated by $\pi/4$ in phase which are used to determine the echo amplitude and phase. Note the (Gaussian) envelope of the echo group, which represents the Fourier transform of the (Gaussian) distribution of incident neutron wavelengths.

The wavelength spread in the incident beam of a typical neutron spin echo spectrometer is of the order of 15%. However, given that the final polarization of the beam, and hence the total precession angle, can be measured to an accuracy of at least 1%, and remembering that typically the neutron spins will precess 10^4 times along the precession solenoids, it is possible to achieve a neutron energy resolution of better than 1 in 10^6 despite this broad wavelength distribution. It is through this decoupling of incident monochromation and energy resolution that Mezei's spin echo technique is able to circumvent the restrictions of the Liouville theorem.

So far we have assumed that the neutrons travelling down the length of the spectrometer have precisely the same velocity in the two precession solenoids. If instead we introduce a scattering sample at the position of the π-flipper, any quasi-elastic scattering processes will change the energy and hence velocity of the neutrons. The accumulated precession angle will now be

$$\varphi_T = \gamma_L \left(\frac{B_1 L_1}{v_1} - \frac{B_2 L_2}{v_2} \right) \qquad \{18\}$$

where v_1 and v_2 are the velocities of the incident and scattered neutrons respectively.

The (kinetic) energy transfer in the scattering process is

$$E_2 - E_1 = \hbar\omega = \frac{m}{2}\left(v_2^2 - v_1^2\right) \qquad \{19\}$$

For small energy transfers (i.e. in the quasielastic limit) we may use the approximations

$$\delta E_1 = \hbar\omega = m v_1 \delta v_1 \qquad \{21\}$$

$$v_2 = v_1 + \delta v_1 = v_1 + \hbar\omega / m v_1 \qquad \{22\}$$

and the total accumulated precession angle, φ_T, becomes

$$\varphi_T = \frac{\gamma_L B_1 L_1}{v_1} - \frac{\gamma_L B_2 L_2}{v_1 + \hbar\omega / m v_1} \qquad \{23\}$$

which, when expanded to first order in ω, becomes

$$\varphi_T = \frac{\gamma_L (B_1 L_1 - B_2 L_2)}{v_1} + \frac{\gamma_L B_2 L_2 \hbar\omega}{m v_1^3} \qquad \{24\}$$

The accumulated precession angle with respect to the echo condition for purely elastic scattering (for which $\omega = 0$ and $B_1 L_1 = B_2 L_2$) is therefore

$$\varphi'_T = \frac{\gamma_L B_2 L_2 \hbar\omega}{m v_1^3} = t_f \omega \quad \left(= \frac{\gamma_L B_1 L_1 \hbar\omega}{m v_1^3} \right) \qquad \{25\}$$

t_f is a constant of proportionality which has the units of time. It should be noted that not only is t_f directly proportional to the line integrals $B_{1(2)}L_{1(2)}$, it is also directly proportional to λ^3, where λ is the incident wavelength.

Therefore, in the presence of quasi-elastic scattering, the final polarization close to the elastic echo condition will decrease to a value of $\cos(\varphi'_T)$, or $\cos(\omega t_f)$. Noting that the scattering law, $S(\boldsymbol{Q},\omega)$, defines the probability of any scattering process at the sample giving rise to an energy transfer of $\hbar\omega$ at a scattering vector $\boldsymbol{Q}$, the final polarization of the beam can be averaged over all quasi-elastic scattering processes to give

$$P_z(\boldsymbol{Q},t_f) = \langle\cos(\omega t_f)\rangle = \frac{\int S(\boldsymbol{Q},\omega)\cos(\omega t_f)d\omega}{\int S(\boldsymbol{Q},\omega)d\omega} \qquad \{25\}$$

Whilst the numerator represents the Fourier transform of $S(\boldsymbol{Q},\omega)$ with respect to ω, the denominator is the integral of $S(\boldsymbol{Q},\omega)$ over all energy, in other words the static structure factor $S(\boldsymbol{Q})$. We therefore have

$$P_z(\boldsymbol{Q},t_f) = \frac{S(\boldsymbol{Q},t_f)}{S(\boldsymbol{Q})} \qquad \{26\}$$

So, a measurement of the final polarization of the beam for a given t_f is simply a measurement of the normalized intermediate scattering law, $S(\boldsymbol{Q},t_f)/S(\boldsymbol{Q})$ where, in the quasi-elastic limit, the Fourier time, t_f, is equivalent to real time. It is important to note that not only has neutron spin echo reduced the process of measuring an energy transfer to one of measuring a final polarization of the neutron beam, but that spin echo also provides the intermediate scattering law directly without any need for a numerical Fourier transformation of the scattering data. These features are yet further illustrations that neutron spin echo is unlike any other neutron scattering technique.

4. POLARIZATION DEPENDENT SCATTERING PROCESSES

The above analysis has assumed that the neutron spin direction does not change in the scattering process. This assumption is valid only in situations where coherent nuclear scattering processes dominate. In general, a further term, P_S, must be introduced to account for any polarization dependence in the scattering process. The final, or neutron spin echo polarization, P_{NSE}, is therefore

$$P_{NSE}(\boldsymbol{Q},t_f) = P_S\frac{S(\boldsymbol{Q},t_f)}{S(\boldsymbol{Q})} \qquad \{27\}$$

For the case of non-spin flip nuclear scattering P_S takes the value 1, and it is implicit that a π-flipper is needed at the sample position in order to realise the echo condition. More generally, there will be additional scattering contributions from, for example, nuclear spin incoherent scattering, from isotopic incoherent scattering and from disordered, fluctuating paramagnetic spins. P_S must therefore be evaluated for each of these cases. This is best achieved by considering Blume's generalized equation for the polarization dependent double differential cross-section reduced in this case to the six partial cross-sections associated with spin flip (SF) and non spin flip (NSF) scattering for the spins directed along each of the three orthogonal x, y and z-axes [4,5,6].

Assuming that the scattering is measured in the (y,z) plane, and that at the sample position the beam direction and magnetic field define the z-axis we have

$$\left(\frac{\partial^2\sigma}{\partial\Omega\partial\omega}\right)_x^{SF} = \frac{1}{2}\left(\frac{\partial^2\sigma}{\partial\Omega\partial\omega}\right)_{MAG} + \frac{2}{3}\left(\frac{\partial^2\sigma}{\partial\Omega\partial\omega}\right)_{SI}$$

$$\left(\frac{\partial^2\sigma}{\partial\Omega\partial\omega}\right)_x^{NSF} = \frac{1}{2}\left(\frac{\partial^2\sigma}{\partial\Omega\partial\omega}\right)_{MAG} + \frac{1}{3}\left(\frac{\partial^2\sigma}{\partial\Omega\partial\omega}\right)_{SI} + \left(\frac{\partial^2\sigma}{\partial\Omega\partial\omega}\right)_{NC} + \left(\frac{\partial^2\sigma}{\partial\Omega\partial\omega}\right)_{II}$$

$$\left(\frac{\partial^2\sigma}{\partial\Omega\partial\omega}\right)_y^{SF} = \frac{1}{2}(1+\cos^2\alpha)\left(\frac{\partial^2\sigma}{\partial\Omega\partial\omega}\right)_{MAG} + \frac{2}{3}\left(\frac{\partial^2\sigma}{\partial\Omega\partial\omega}\right)_{SI} \qquad \{28\}$$

$$\left(\frac{\partial^2\sigma}{\partial\Omega\partial\omega}\right)_y^{NSF} = \frac{1}{2}(\sin^2\alpha)\left(\frac{\partial^2\sigma}{\partial\Omega\partial\omega}\right)_{MAG} + \frac{1}{3}\left(\frac{\partial^2\sigma}{\partial\Omega\partial\omega}\right)_{SI} + \left(\frac{\partial^2\sigma}{\partial\Omega\partial\omega}\right)_{NC} + \left(\frac{\partial^2\sigma}{\partial\Omega\partial\omega}\right)_{II}$$

$$\left(\frac{\partial^2\sigma}{\partial\Omega\partial\omega}\right)_z^{SF} = \frac{1}{2}(1+\sin^2\alpha)\left(\frac{\partial^2\sigma}{\partial\Omega\partial\omega}\right)_{MAG} + \frac{2}{3}\left(\frac{\partial^2\sigma}{\partial\Omega\partial\omega}\right)_{SI}$$

$$\left(\frac{\partial^2\sigma}{\partial\Omega\partial\omega}\right)_z^{NSF} = \frac{1}{2}(\cos^2\alpha)\left(\frac{\partial^2\sigma}{\partial\Omega\partial\omega}\right)_{MAG} + \frac{1}{3}\left(\frac{\partial^2\sigma}{\partial\Omega\partial\omega}\right)_{SI} + \left(\frac{\partial^2\sigma}{\partial\Omega\partial\omega}\right)_{NC} + \left(\frac{\partial^2\sigma}{\partial\Omega\partial\omega}\right)_{II}$$

where the subscripts MAG, SI, NC and II refer to paramagnetic, nuclear spin incoherent, nuclear coherent and isotopic incoherent scattering respectively, and α is the angle between the scattering vector and the direction of incident polarization in the horizontal (y,z) plane.

Inspection of these six equations reveals that nuclear coherent scattering and isotopic incoherent scattering occurs entirely with non spin flip, irrespective of the scattering geometry. Recalling that just before the sample the neutron spins are dephased and precessing in the (x,y) plane, the beam polarization can be described by the vector $(P_x, P_y, 0)$. Assuming the π-flipper is oriented to operate on the y-component of the neutron spins, i.e., flipping the precession plane of the spins around the x-axis, the spin flipped polarization becomes $(P_x, -P_y, 0)$, thereby reversing the direction of precession. As the scattering process is non spin flip, the precessing spins of the scattered neutrons are therefore brought back into phase by the echo condition, confirming our assumption that $P_S = 1$.

Spin incoherent scattering, which arises from the interaction of the neutron with the (randomly oriented) nuclear spins, is distributed between spin flip and non spin flip scattering in the ratio 2:1, irrespective of the precise scattering geometry. With the π-flipper active the polarization just before the sample is again $(P_x, -P_y, 0)$. However, as two thirds of spins are spin flipped along each of the x- and y-directions, the scattered beam polarization is now $1/3(-P_x, P_y, 0) = -1/3(P_x, -P_y, 0)$. Whilst this represents a reversal of the direction of precession, and therefore allows rephasing of the spins by the echo condition, this polarization is not only of opposite sign to that obtained from nuclear coherent scattering, it is also reduced to one third of its initial value. Hence $P_S = -1/3$ for nuclear spin incoherent scattering. However, it should be noted that this significantly diminished echo is observed together with the echo associated with the purely nuclear coherent scattering.

Paramagnetic scattering, on the other hand, is sensitive to the scattering geometry, and occurs only with spin flip when the polarization is directed along the scattering vector, $\boldsymbol{Q}$, (in this case the y-direction) and is divided equally between spin flip and non spin flip when the polarization is perpendicular to the scattering vector (i.e. the x- and z-directions). So, assuming again that the π-flipper is activated and the polarization before the sample is $(P_x, -P_y, 0)$, the scattered beam polarization must be considered as a superposition of two components, namely $1/2(-P_x, P_y, 0)$ and $1/2(P_x, P_y, 0)$. The latter component represents a precession in the same direction as that in the primary arm of the spectrometer, and the associated precessing neutron polarization will therefore not be re-phased by the echo condition. However, the first component does represent a reversal of precession, although the polarization is of opposite sign to that obtained from nuclear coherent scattering. Hence, with the π-flipper activated, $P_S = -1/2$ for paramagnetic scattering, although as for nuclear spin incoherent scattering, this echo will be superimposed upon that associated with the nuclear coherent scattering. Fortunately this problem can easily be overcome simply by de-activating the π-flipper: With the π-flipper switched off the polarization before the sample is $(P_x, P_y, 0)$ and the two components of the scattered beam polarization are $1/2(-P_x, -P_y, 0)$ and $1/2\ (P_x, -P_y, 0)$. In this case it is the precessing spins described by the first component of the polarization, together with those associated with the nuclear coherent scattering, that are not re-phased by the echo condition, whilst those described by the second component are re-phased. Correspondingly $P_S = 1/2$ and the spin echo signal associated with the paramagnetic scattering is seen in isolation.

The methodology for separating the spin echo signal from paramagnetic and nuclear coherent scattering can be applied to any disordered magnetic systems, such as spin glasses, random anisotropy magnets and superparamagnets. Magnetically ordered antiferromagnets can also be studied, and for polycrystalline multi-domain samples P_S is also 1/2. However, for single domain antiferromagnetic single crystal samples the orientation of the atomic spins with respect to the scattering vector becomes important, and whilst for some orientations a full echo signal can be observed without activating the π-flipper, for others a π-flipper must be used in order to see any echo. In general $1/2 \leq |P_S| \leq 1$. Ferromagnets introduce an additional level of complexity as the random orientation of ferromagnetic domains in the relatively low fields at the sample position leads to strong depolarization of the neutron beam, and a complete loss of the echo signal. This can be overcome in some circumstances by applying a sufficiently high field at the sample position, parallel to the z-axis, in order to saturate the ferromagnet and correspondingly, in the absence of a π-flipper, flipping the spins by $\pi/2$ into the z-direction, and then back to the x-direction before and after the sample respectively. In this geometry $P_S = 1/2$.

A summary of these conditions can be found in table 1.

Table 1 - Spin flipper configuration, sample field and resulting polarization factor, P_S, for a range of scattering samples

Type of scatterer	Sample region		Polarization factor P_S
	Spin flip coils	Sample field	
Coherent nuclear	π	small	1
Nuclear spin incoherent	π	small	1/3
Paramagnets	none	small	1/2
Ferromagnets	$\pi/2$ - sample - $\pi/2$	high (saturating)	1/2
Antiferromagnets	none	small	$1/2 \leq P_S \leq 1$

5. PRACTICALITIES: MEASUREMENT OF THE SPIN ECHO SIGNAL

We have seen that in the quasi-elastic limit a measurement of the final, or neutron spin echo polarization, P_{NSE}, provides a direct measurement of the normalized intermediate scattering law, i.e.

$$P_{NSE}(\boldsymbol{Q},t_f) = P_S \frac{S(\boldsymbol{Q},t_f)}{S(\boldsymbol{Q})} \qquad \{27\}$$

As an example, many diffusion and relaxation phenomena are characterized by a Lorentzian scattering law, namely

$$S(\boldsymbol{Q},\omega) \propto \frac{\Gamma}{\Gamma^2+\omega^2} \qquad \{29\}$$

so the Fourier time-dependent component of the neutron spin echo amplitude can be evaluated as

$$P_{NSE}(t_f) = \frac{P_S \int_{-\infty}^{+\infty}\left[\Gamma^2+\omega^2\right]^{-1}\cos(\omega t_f)d\omega}{\int_{-\infty}^{+\infty}\left[\Gamma^2+\omega^2\right]^{-1}d\omega} = P_S\, e^{-\Gamma t_f} \qquad \{30\}$$

Spin echo therefore measures directly the exponential decay of the correlation function $S(\boldsymbol{Q})$ in Fourier time, t_f, as shown in figure 5. But how is this measurement made in practice?

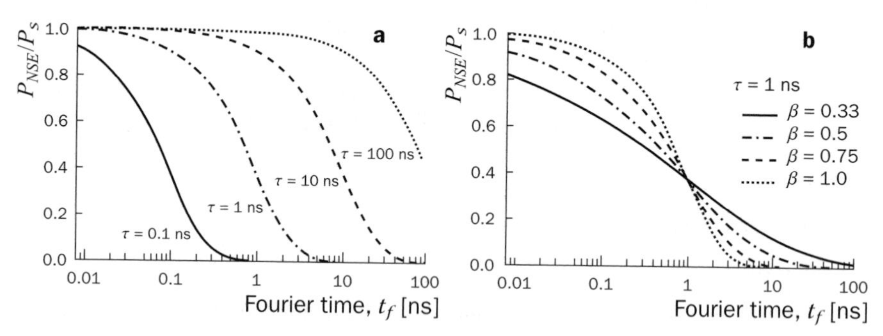

Figure 5 - **a** - Simulated spin echo spectrum associated with a simple exponential form for the intermediate scattering law $S(Q,t)$ for a range of characteristic relaxation times, τ. **b** - Another commonly observed relaxation function, the stretched exponential or Kohlrausch form **[26,27]**, $S(Q,t) \sim e^{-(t/\tau)^\beta}$ evaluated for a number of values of β.

In a typical spin echo measurement an appropriate value of the field in the first precession arm of the spectrometer, B_1, is chosen, and the field in the second precession arm, B_2, (or, more precisely, the current in the second solenoid) is scanned in a so-called *asymmetric scan* in order to find the centre of the spin echo group for which $B_1 L_1 = B_2 L_2$. At this setting the difference between the average number of precessions in the two arms of the spectrometer, ΔN, is therefore zero. A *symmetric scan* in Fourier time, t_f, is then performed by varying B_1 and B_2 simultaneously, but keeping the ratio B_1/B_2 fixed, such that all appropriate values of t_f

$$t_f = \frac{\gamma_L B_2 L_2 \hbar}{m v_1^3} = \frac{\gamma_L B_1 L_1 \hbar}{m v_1^3} \tag{31}$$

are sampled, as shown in figure 6.

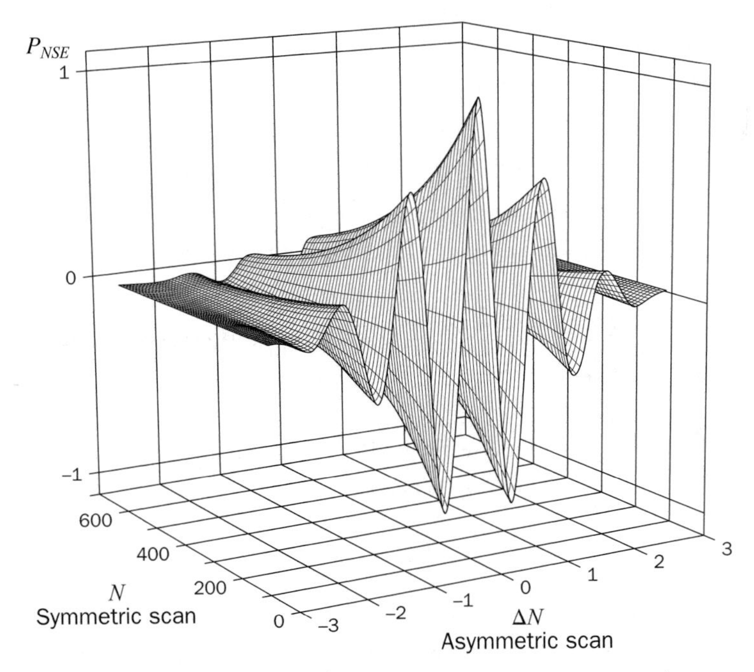

Figure 6 - The spin echo group showing the effects of an asymmetric scan, where the line integral along only one of the precession arms of the spectrometer is varied, thereby allowing the difference, ΔN, between the average number of precessions in the two arms to be minimized, and of a symmetric scan where the line integrals along both precession arms are varied, but their ratio is kept constant. The latter measurement is equivalent to varying the Fourier time, t_f.

In an ideal situation the amplitude of the spin echo signal, P_{NSE}, could be defined simply in terms of the maximum and minimum count rates in the spin echo group.

$$P_{NSE} = \frac{(I_{max} - I_{min})}{2I_0} \quad \{32\}$$

where I_0 is the average count rate of the group. However this would assume that the total precession phase difference φ_τ is precisely zero for this particular spin echo group. Alternatively, a measurement of the shape of the whole echo group would ensure that both the spin echo amplitude and φ_τ were measured, but the process would be extremely time consuming and inefficient. As a compromise the count rates, I_n, at only four points spaced at phase intervals of $\pi/2$ around the centre of the group are measured for each spin echo group [7]. With reference to figure 4 the four count rates are

$$I_1 = I_0 + E\sin\varphi_\tau \quad I_2 = I_0 - E\cos\varphi_\tau \quad I_3 = I_0 - E\sin\varphi_\tau \quad I_4 = I_0 + E\cos\varphi_\tau \quad \{33\}$$

and E is the amplitude of the echo. Clearly I_0, E and φ_τ can each be determined from the four count rates, I_n.

The average count rate, defined by

$$I_0 = \frac{1}{4}\sum_{n=1,4} I_n \quad \{34\}$$

represents the ideal maximum amplitude of the spin echo group, a quantity that can also be determined by setting a guide field to maintain the beam polarization of the scattered beam, and measuring the total scattered count rates, I_{on} and I_{off}, with the π-flipper on and off respectively, i.e.

$$I_0 = \frac{1}{2}(I_{on} - I_{off}) \quad \{35\}$$

Using either (or both) of these methods for determining I_0 leads to a final, normalized, spin echo amplitude for a particular group of

$$P_{NSE} = \frac{E}{I_0} \quad \{36\}$$

It should be noted that the count rates I_n do not all contribute equally to the statistical accuracy of the calculation of the final spin echo amplitude. In practice it is desirable to spend additional time counting I_2 (and I_{off}).

In the discussions so far we have assumed an ideal spin echo spectrometer in which the line integrals of the precession fields along the incident and scattered neutron beam paths are both precisely known and uniform for all scattering trajectories. In practice this is rarely the case as inhomogeneities in the line integral of the precession fields occur for any beam of finite size or of finite divergence. Some of these inhomogeneities can be minimized to the level of 1 part in 10^5 or 10^6 by using "Fresnel"-type field coils [8]. Those field integral inhomogenities that cannot be corrected for in this way, such as those arising from the variation in neutron flight paths caused by scattering from a finite sample, lead to a total precession angle φ_T

$$\varphi_T = \varphi_1 - \varphi_2 = \gamma_L \left(\frac{B_1 L_1}{v_1} - \frac{B_2 L_2}{v_2} \right) \tag{37}$$

that is different for each neutron trajectory. The final polarization, even for a purely elastic scatterer, will correspondingly be reduced to $\langle \cos\varphi_T \rangle$.

Such field integral inhomogeneities correspond to an effective instrumental resolution function. However, providing that the resulting $\langle \cos\varphi_T \rangle$ is not too small, it is relatively straightforward to correct the spin echo signal, P_{NSE}, for the related resolution broadening. Indeed, $\langle \cos\varphi_T \rangle$ can be determined directly by measuring the spin echo amplitude, P^E, as a function of Fourier time, t_f, from a purely elastic scatterer of similar dimensions to the sample, providing that this calibration sample reproduces the spread of trajectories (and hence angular distribution) of those neutrons scattered by the sample. The resolution corrected spin echo signal, P'_{NSE}, is then obtained by the division

$$P'_{NSE}(\boldsymbol{Q}, t_f) = \frac{P_{NSE}(\boldsymbol{Q}, t_f)}{P^E(\boldsymbol{Q}, t_f)} \tag{38}$$

This equation highlights yet another advantage of the spin echo technique over conventional quasi-elastic neutron scattering methods: the often complex and model-dependent deconvolution of the instrumental resolution that is necessary in conventional neutron spectroscopy is, in the case of spin echo, replaced by a simple division of the observed spectrum by an experimentally measured model-independent resolution function. This is possible precisely because the spin echo measurement is made directly in Fourier time.

Finally, it should be noted that in this and the preceding section we have focussed entirely upon the underlying principles of the archetypal spin echo spectrometer as first introduced by Mezei in 1972 [1,8]. Although the details of neutron spin echo instrumentation have improved considerably over the past three decades, for example providing longer Fourier times and wide angle detectors, e.g. **[9]**, most spin echo spectrometers used today largely conform to the same basic design and mode of operation. However several new and innovative variations on the spin echo technique have also been introduced, or are currently being developed. Unfortunately it is beyond the scope of this chapter to consider these in detail. Nevertheless, it should be mentioned that the coupling of spin echo with conventional triple-axis spectrometry enables precise measurement of the line widths of coherent, dispersive excitations [10]; in neutron resonance spin echo (NRSE) the large precession coils of the basic spin echo spectrometer are replaced with small radio-frequency spin flippers [11]; and of particular importance for the high power third generation spallation neutron sources currently under construction, the time-of-flight neutron spin echo technique, in which the incident beam is pulsed, has been successfully demonstrated [7]. It is also interesting to note that the spin echo technique has even been adapted for high intensity, high resolution, *elastic* small angle neutron scattering studies (SESANS) of large-scale structural inhomogeneities, avoiding the necessity for tight angular collimation [12,13]. A description of many of these novel techniques can be found in [14].

6. APPLICATIONS OF NEUTRON SPIN ECHO SPECTROSCOPY

We have seen that neutron spin echo spectroscopy is ideally suited to measuring slow relaxation or diffusion phenomena on mesoscopic time scales intermediate between those associated with the motion of individual atoms and those associated with macroscopic dynamics. In most cases relaxation on such time scales is related to the collective motion of large molecular units or correlated assemblies of atoms, molecules or magnetic spins. Indeed it is the time dependence of these correlations that is often the focus of interest. Fortunately, as neutron spin echo is generally a long wavelength technique, it provides access to a Q-range that is well suited to exploring such extended correlations in space as well as in time. Consequently, neutron spin echo has been most successfully exploited in studies of polymer and soft matter dynamics, glassy dynamics in molecular glasses, and dynamical spin correlations in spin glasses and similar disordered magnets. In this section we shall briefly review just a few of the recent applications of neutron spin echo to such studies.

6.1. SOFT CONDENSED MATTER

Neutron spin echo has made many significant contributions to our understanding of a wide range of slow dynamical processes in "soft condensed matter" including polymers, polymer melts, polymer solutions, e.g. [15], complex fluids [16], micro-emulsions [17] and materials of biological relevance such as crowded protein solutions [18].

A notable example is the first experimental confirmation, at a molecular level, of the phenomenon of reptation in dense polymeric systems [19]. Reptation is the snake-like motion of an individual polymer chain that is constrained laterally by the continually changing topological conformations of its entangled neighbours to move within and along an effective tube defined by its own chain contour (fig. 7).

Figure 7 - A schematic representation of reptation where a polymer chain is constrained laterally by the continually changing topological conformations of its entangled neighbours to move within and along an effective tube defined by its own chain contour.

The reptation model, introduced by de Gennes [20], provides a successful phenomenological theory of polymer dynamics that is appealing because it is both simple and readily visualized. However many other theoretical models have also been proposed to describe the relaxation and entanglement in polymer melts. Although the concept of reptation is entirely absent from these theories they can, nevertheless, reproduce many of the experimentally observed features of polymeric interdiffusion. Indeed, on Fourier time scales shorter than 50 ns, there is little to choose between the reptation and non-reptation models, and neutron spin echo measurements of polymer relaxation are equally well described by the theories of de Gennes, and the generalized Rouse models of Ronca [21] and Chaterjee [22], and the rubber-like model of des Cloizeaux [23]. However, each of these models predict a markedly different dynamical response at longer times, and with the recent development of very high resolution spin echo instrumentation capable of measuring to extended Fourier times of 200 ns (i.e. IN15 at the Institut Laue-Langevin, Grenoble) an effective and unambiguous discrimination between these various models is possible.

The single-chain normalized dynamic structure factor, $S(Q,t)/S(Q)$, was measured using the IN15 spin echo spectrometer on a polyethylene sample (PEB-2, molecular weight 36000) prepared as a dilute mixture of fully protonated chains within a background of deuterated, but otherwise identical, polymer chains [19]. Such deuteration provides a scattering length contrast between the protonated (i.e., labelled) polymer and the deuterated background thereby allowing single-chain correlations to be measured. It should be noted that a further advantage of such deuteration is that the intense nuclear spin incoherent scattering, arising from protons within the sample, is significantly reduced. Coherent nuclear scattering therefore dominates and the spin echo polarization factor $P_s = 1$, maximizing the amplitude of the spin echo group.

The neutron spin echo measurements of the normalized intermediate scattering law, $S(Q,t)/S(Q)$, shown in figure 8, clearly follow de Gennes' analytical expression for reptation [24] as Fourier time extends to 200 ns. Indeed, the reptation model provides an excellent quantitative description of the experimental data over all Q and t. Remarkably, this description is achieved with the entanglement distance (i.e. the diameter of the tube, d (= 4.6 nm) within which the polymer chain reptates) as the only free parameter in the fitting procedure.

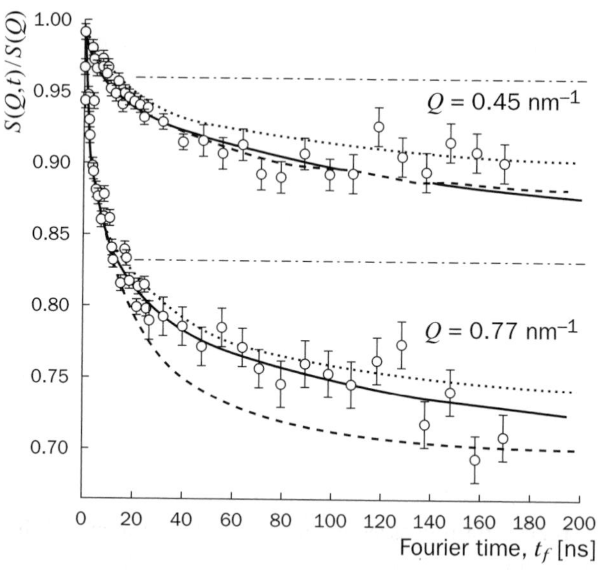

Figure 8 - A plot of the experimentally measured $S(Q,t_f)$ for polyethylene at $Q = 0.45\ \text{nm}^{-1}$ and $0.77\ \text{nm}^{-1}$, together with theoretical predictions for reptation (solid lines) and local reptation (dotted lines). It can be seen that the non-reptation models of des Cloizeaux [22] and Ronca [20] (dashed and dashed-dotted lines respectively) do not adequately describe the data at long Fourier times (after [18]).

6.2. GLASSY DYNAMICS

It has been said that "the deepest and most interesting unsolved problem in solid state theory is probably the theory of the nature of glass and the glass transition" [25], and therefore any experimental technique which can provide new and direct insights into the physics of the underlying dynamical processes is particularly important. Consequently neutron spin echo has been used extensively to probe glassy dynamics, and their relation to structure, particularly in glass forming polymeric and molecular liquids, with much of the experimental attention focusing upon the non-exponential, or more precisely stretched exponential, relaxation that is observed above the glass transition e.g. references [17,26].

In general, at temperatures well above any glass transition, the correlation function of a liquid, $\Phi(t)$, decays rapidly, on a picosecond time scale, as a simple exponential function, i.e.

$$\Phi(t) \propto e^{-t/\tau} \qquad \{39\}$$

in which the relaxation time τ follows a simple Arrhenius temperature dependence. However, as the system cools to the glass transition the $\Phi(t)$ develops into a two-stage process. The initial, and very rapid, first or β-relaxation stage remains exponential, but the increasingly dominant second or α-relaxation stage of $\Phi(t)$ is associated instead with a slow, stretched exponential relaxation of the form

$$\Phi(t) \propto e^{-(t/\tau)^{\beta}} \qquad \{40\}$$

This functional form, with $0 < \beta < 1$, is known as Kohlrausch [27], or Kohlrausch-Williams-Watts (or KWW) [28] relaxation. First observed 150 years ago, and for many years treated as a simple empirical convenience, KWW relaxation is now considered as a fundamental physical phenomenon which occurs across a wide range of complex, strongly relaxing electronic, structural and magnetic systems [29]. There remains some debate as to whether the characteristic KWW form arises from a broad distribution of relaxation rates, $g(\ln \tau)$, within a sample, such that

$$\Phi(t) = \int_{-\infty}^{\infty} g(\ln \tau)\, e^{-t/\tau} d\tau \propto e^{-(t/\tau)^{\beta}} \qquad \{41\}$$

or whether all relaxation processes within a sample are identical, but intrinsically non-exponential. Neutron scattering measurements have suggested, at least for polymeric glasses, that the stretching of the α-relaxation arises from the latter "homogeneous" relaxation model, rather than from any dynamical heterogeneities [30].

For α-relaxation the intrinsic relaxation time, τ, often follows closely a Vogel-Fulcher temperature dependence

$$\tau(T) \approx \tau_\infty \, e^{-A/(T-T_0)} \qquad \{42\}$$

where, T_0, the temperature at which τ is expected to diverge, is generally below the glass temperature, T_g. The exponent β may also be temperature dependent. Indeed, Ngai [31] has attempted to classify structural glasses according to their relaxation dynamics above T_g, defining Type A or "Fragile" glasses as those which display stretched exponential relaxation with a temperature dependent β decreasing from 1 to 1/3 as T_g is approached, and Type B or "Strong" glasses as those showing stretched exponential relaxation but with $\beta < 1$ and independent of temperature.

Neutron spin echo measurements of the KWW α-relaxation in many polymeric glasses yield $0.4 < \beta < 0.5$, apparently with relatively little temperature dependence [26]. However it should be noted that for these polymeric systems β and its temperature dependence are difficult to extract from the neutron spin echo measurements, unless the characteristic relaxation rate, τ, is well centred in the spin echo experimental time window: at low temperatures, close to the glass temperature, β is difficult to determine, as there is very little observable decay, whilst at high temperatures only the tail of the relaxation is seen. Nevertheless, significant information can be obtained from spin echo measurements of the dynamic structure factor of polymer glasses, as illustrated by a study of polybutadiene [32], where the temperature dependence of the intermediate scattering law was determined at scattering vectors corresponding to the first and second peaks of the static structure factor (fig. 9). The first peak corresponds to *interchain* correlations, and as the polymer chains bond only through weak Van der Waals forces, the position of this peak is strongly temperature dependent. The second peak in the structure factor, on the other hand, is dominated by *intrachain* correlations, and as the corresponding bonding is covalent, it is relatively insensitive to temperature. Each spin echo spectrum was rescaled by the experimentally determined characteristic time, τ_η, for bulk viscous relaxation for the corresponding temperature. The suitably rescaled time-dependent correlation functions measured at the first peak of the structure factor are found to follow a single, stretched exponential ($\beta = 0.4$) universal curve, indicating that the interchain dynamics very closely follow the α-relaxation behaviour associated with macroscopic flow of polybutadiene. Those correlation functions measured at the second peak of the structure factor, however, do not rescale in the same way. Instead they are characterized by a simple exponential function, with an associated temperature dependent relaxation rate which follows an Arrhenius dependence with the same activation energy as the dielectric

β-process in polybutadiene, and which is unaffected by the glass transition. This is clear evidence that the α- and β-relaxation processes in polybutadiene are statistically independent, with the former related to interchain relaxation processes, and the latter to intrachain relaxation.

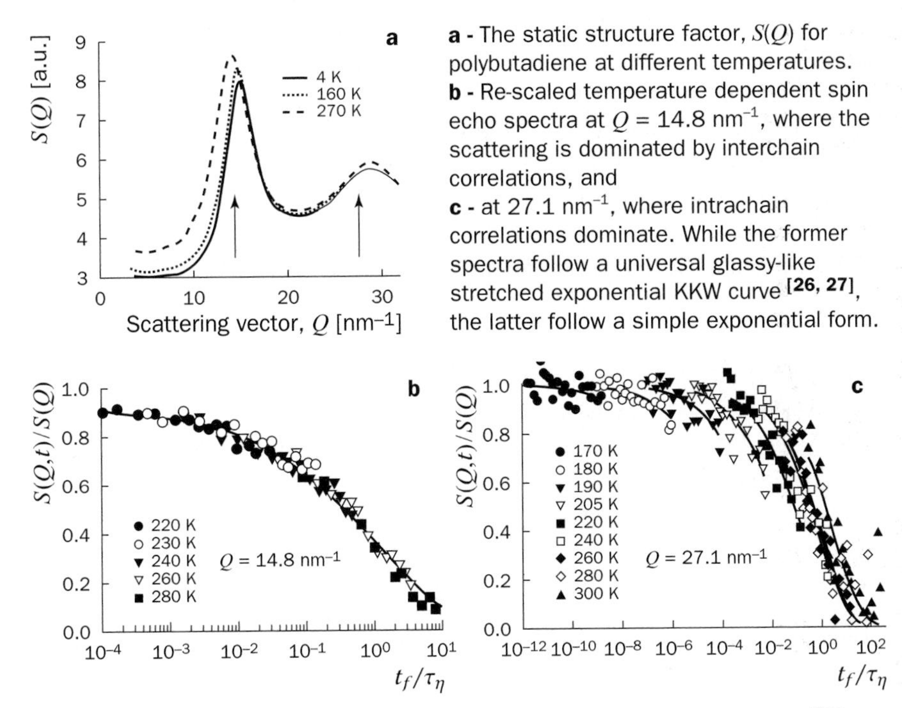

a - The static structure factor, $S(Q)$ for polybutadiene at different temperatures.
b - Re-scaled temperature dependent spin echo spectra at $Q = 14.8\ \text{nm}^{-1}$, where the scattering is dominated by interchain correlations, and
c - at $27.1\ \text{nm}^{-1}$, where intrachain correlations dominate. While the former spectra follow a universal glassy-like stretched exponential KKW curve [26, 27], the latter follow a simple exponential form.

Figure 9 - Neutron spin echo studies of glassy dynamics in polybutadiene [31].

6.3. SPIN RELAXATION IN MAGNETIC SYSTEMS

Polarized neutrons are virtually synonymous with the study of magnetic systems. Not only do polarization techniques greatly enhance the sensitivity of neutron scattering to magnetic phenomena, they often allow an unambiguous separation of magnetic scattering from nuclear scattering. However such sensitivity comes a price, and in this case, even with ideally efficient neutron polarizers, the cost is at least one half of the available incident beam intensity. Moreover, we have already seen that in spin echo studies of paramagnetic relaxation the polarization factor of the spin echo group is reduced to $P_S = 1/2$. It is therefore clear that spin echo studies of magnetic systems are severely intensity limited. This is particularly true in the case of *dilute* magnetic alloys such as spin glass systems where frustrated exchange carried by the long range but oscillatory Rudermann-Kittel-Kasuya-Yoshida (RKKY) interaction leads to a freezing of the dilute spins in random orientations at low temperatures. Nevertheless the dynamics of magnetic spins in spin glass systems have been the focus of considerable experimental and theoretical interest for almost four decades, not least because spin glasses provide a relatively simple magnetic analogue of real structural glasses. It is not surprising that spin glasses were the subject of some of the very first spin echo studies [33]. More recently neutron spin echo has been used to compare the dynamic autocorrelation function of real spin glasses with detailed predictions extracted by Ogielski from large scale Monte Carlo simulations of the $3d \pm J$ Ising spin glass model [34]. Ogielski's simulations suggest that the relaxation process above the glass temperature, T_g, should take the strongly non-exponential form

$$\langle S_i(0)\, S_i(t)\rangle \propto t^{-x} e^{-(t/\tau)^{\beta}} \qquad \{43\}$$

τ, β and x are all temperature dependent with the relaxation time, $\tau(T)$, diverging at T_g, whilst the stretch exponent $\beta(T)$ decreases from unity at 4 T_g to ~1/3 at T_g. Correspondingly $x(T)$ decreases from 0.5 to 0.1 at T_g. Within the framework of dynamic scaling, $x(T)$ is particularly important. Ogielski relates $x(T)$ at the glass temperature to the static and dynamic universal critical exponents,

$$x = \frac{(d-2+\eta)}{2z} \qquad \{44\}$$

where d is the dimensionality of the system, η is the static Fisher exponent and z is the dynamic exponent.

However, more detailed studies, with improved spin echo instrumentation, for example on the concentrated spin glass $Au_{86}Fe_{14}$ with $T_g = 41K$ [35], have recently shown that the normalized intermediate scattering law above T_g for short Fourier times and the below T_g over the entire time range is fully consistent with Ogielski's power law dependence (fig. 10). The results yield a value of $x(T = T_g)$ of 0.12, in precise agreement with estimates obtained from complementary techniques such as ac susceptometry [35]. It is also very interesting to note that a re-evaluation of the early neutron spin echo measurements on $Cu_{95}Mn_5$ [33] yields values for $x(T)$ which are in excellent agreement with those obtained for $Au_{86}Fe_{14}$ [36], as can be seen in figure 10b.

Unfortunately Ogielski's predictions for the temperature dependence of β above T_g could not be confirmed. For the $Au_{86}Fe_{14}$ alloy the stretched exponential relaxation influences only the tail of the relaxation. $\beta(T)$ therefore remains rather ill-defined, with a value of approximately *1* at temperatures greater than 50 K and ~0.65

at 46 K. Indeed, most spin glass systems studied by neutron spin echo have shown stretched exponential relaxation above T_g, with typical exponents of $\beta \sim 0.5$ [33,37,38] similar to those found for the KWW-like α-relaxation in glassy polymeric systems. However, as with the structural glassy systems, it has not generally been possible to establish within statistical accuracy whether β varies significantly with temperature for spin glasses. A notable exception is the amorphous random anisotropy magnet a-Er_3Fe_3 [39].

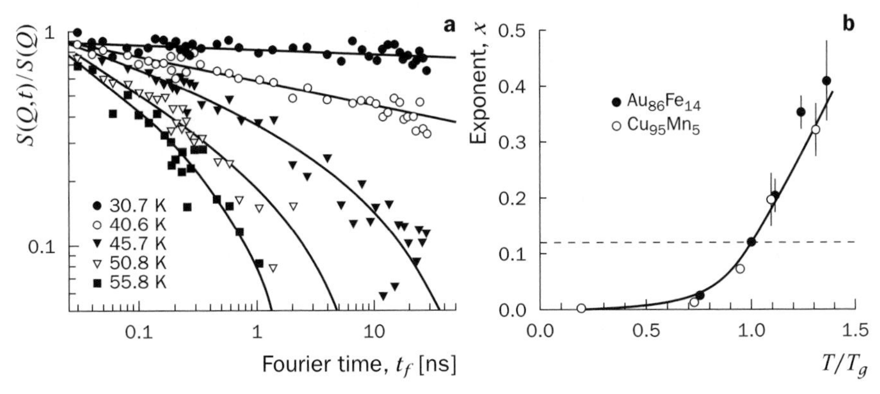

Figure 10 - a - Neutron spin echo spectra obtained from $Au_{86}Fe_{14}$ [34]. The solid lines represent least square fits to a relaxation function of the form predicted by Ogielski [33] for spin glass systems. **b -** The temperature dependence of the Ogielski exponent, x, for $Au_{86}Fe_{14}$ determined from the fits in (a) and for $Cu_{95}Mn_5$ obtained from re-evaluation of the data in ref. [32]. The dashed line corresponds to $x = 0.12$.

Random anisotropy magnets are distinct from conventional spin glass systems. Spin "freezing" at low temperatures in systems where the single ion anisotropy is significantly stronger than the exchange interactions is driven by a site-to-site randomness of the local anisotropy axes, rather than by random exchange interactions or topological frustration. In amorphous alloys random anisotropy is a result of the absence of any translational structural order, and the consequent lack of significant correlations between the single ion anisotropy axes [40].

For a-Er_3Fe_3 the relaxation function as measured by neutron spin echo is non-exponential and of the KWW form (fig. 11). It is found that β is strongly temperature dependent, approaching 1/3 at T_g, as predicted by Ogielski, but with no evidence of any t^{-x} power law contribution. Moreover, whilst the relaxation time, τ, is essentially Arrhenius-like, τ also continues to evolve below T_g, although with a small increase in the characteristic activation energy at T_g.

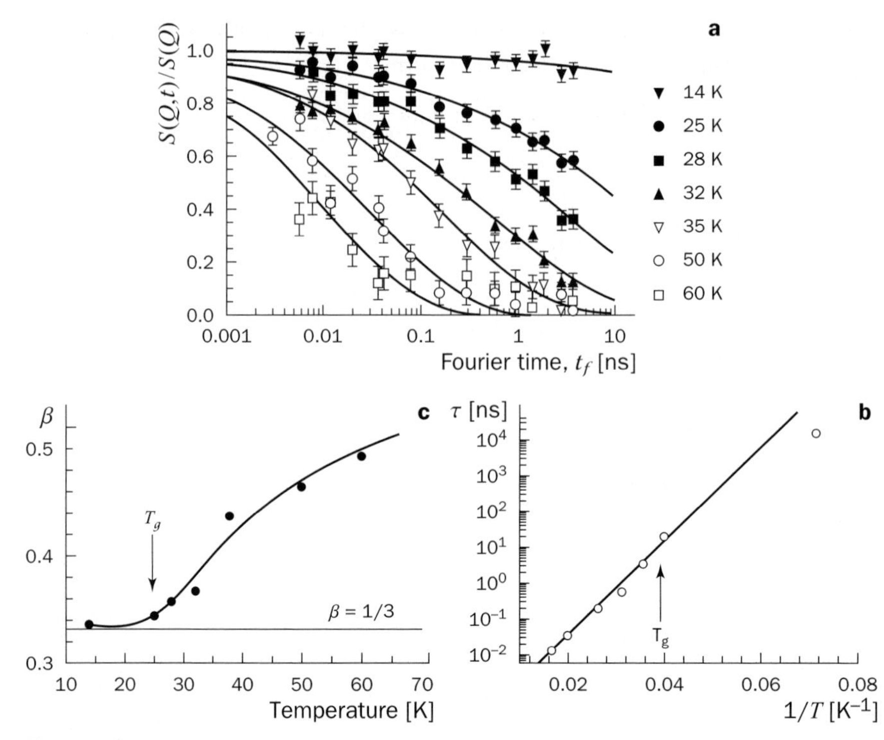

Figure 11 - **a** - Neutron spin echo data from the random anisotropy magnet a-Er_7Fe_3 over the temperature range 14 K to 60 K. The solid lines are fits of the Kohlrausch relaxation function $e^{-(t/\tau)^\beta}$. **b** - The relaxation time τ and (**c**) the exponent β extracted from least squares fits of the Kohlrausch relaxation $e^{-(t/\tau)^\beta}$ to the neutron spin echo data from a-Er_7Fe_3. The solid lines are guides to the eye. Note that β approaches 1/3 at the glass temperature (from ref. **[38]**).

It is tempting to conclude that the spin dynamics above T_g in the glassy random anisotropy magnet a-Er_7Fe_3 are perhaps analogous to the structural relaxation in Ngai's Type A fragile structural glasses [31], whilst more conventional random-exchange driven spin glasses, such as CuMn and AuFe are more analogous to a Type B strong glass. This is question of fundamental interest, but before it can be resolved it must be established whether the relative temperature independence of β for the conventional spin glasses is intrinsic, or simply a consequence of poor statistical accuracy resulting from the characteristic relaxation time not being well centred in the neutron spin echo experimental time window.

Stretched exponential KWW-like relaxation might also be expected in magnetic systems with topologically frustrated spin-spin interactions, such as the so-called spin ice compounds. Surprisingly, neutron spin echo studies indicate that, unlike for spin glass and random anisotropy systems, the relaxation in such topologically frustrated systems is purely exponential in form [41,42].

Of course, neutron spin echo studies of magnetic systems extend well beyond the field of spin glass dynamics. Experimental and theoretical interest in systems as diverse as nanoscale superparamagnetic particles [43], colossal magnetoresistance and spin relaxation in ferromagnetic manganites [44] and vortex (flux line) motion in Type II superconductors [45] has precipitated detailed spin echo studies which not only have helped to elucidate the underlying dynamical physical processes in such model magnetic systems, but also have further demonstrated the power of the neutron spin echo technique.

7. CONCLUSIONS

Over the last thirty years neutron spin echo spectroscopy has evolved from an ingenious concept to a powerful workhorse. Its many unique characteristics, including its wide dynamic range and ultra-high resolution, the ability to measure the intermediate scattering law directly, the simplicity of the resolution corrections, and the added advantage of polarization analysis of the scattering from the sample has ensured its position as an indispensable tool in studies of slow relaxation phenomena in all fields of condensed matter research. There is little doubt that the continually increasing sophistication of spin echo instrumentation will enable the effective implementation of powerful time-of-flight NSE techniques at the new high intensity third generation spallation sources now being built in the US and Japan, and being planned in Europe.

BIBLIOGRAPHY

Neutron Spin Echo, *Lecture Notes in Physics* **128**, F. MEZEI ed., Springer Verlag, Berlin (1980)

Polarized Neutrons, *Oxford Series on Neutron Scattering in Condensed Matter* **1**, W GAVIN WILLIAMS (Oxford University Press, Oxford, 1988)

Neutron Spin Echo Spectroscopy: Basics, Trends and Applications, *Lecture Notes in Physics* **601**, F. MEZEI, C. PAPPAS & T. GUTBERLET eds, Springer Verlag, Berlin, Heidelberg (2003)

REFERENCES

[1] F. MEZEI - *Z. Phys.* **255**, 146 (1972)

[2] J. HAYTER - *Z. Phys. B.* **31**, 117 (1978)

[3] F. MEZEI - *Com. in Physics* **1**, 81 (1976) &
F. MEZEI & P. DAGLEISH - *Com. in Physics* **2**, 41 (1977)

[4] M. BLUME - *Phys. Rev.* **130**, 1670 (1963)
S.V. MALEYEV, V.G. BAR'YAKTHAR & R.A. SURIZ - *Fiz. Tv. Tela* **4**, 3461 (1962)
[*Sov. Phys. Solid State* **4**, 2533 (1963)]

[5] R.M. MOON, T. RISTE & W.C. KOEHLER - *Phys. Rev.* **181**, 920 (1969)

[6] O. SCHÄRPF & H. CAPELLMANN - *Phys. Stat. Sol.* **a135**, 359 (1993)

[7] B. FARAGO - *In* "Neutron Spin Echo Spectroscopy, Basics, Trends and Applications", F. MEZEI, C. PAPPAS & T. GUTBERLET eds, *Lecture Notes in Physics* **601**, 15, Springer Verlag, Berlin, Heidelberg (2003)

[8] F. MEZEI - *In* "Neutron Spin Echo ", F. MEZEI ed., *Lecture Notes in Physics* **128**, 3, Springer Verlag, Berlin (1980)

[9] C. PAPPAS, A. TRIOLO, F. MEZEI, R. KISCHNIK & G. KALI - *In* "Neutron Spin Echo Spectroscopy, Basics, Trends and Applications", F. MEZEI, C. PAPPAS & T. GUTBERLET eds, *Lecture Notes in Physics* **601**, 35, Springer Verlag, Berlin, Heidelberg (2003)

[10] T. KELLER, B. KEIMER, K. HABRICHT, R. GOLUB, & F. MEZEI - *In* "Neutron Spin Echo Spectroscopy, Basics, Trends and Applications", F. MEZEI, C. PAPPAS & T. GUTBERLET eds, *Lecture Notes in Physics* **601**, 74, Springer Verlag, Berlin, Heidelberg (2003), and references therein.

[11] R. GOLUB & R. GÄHLER - *Phys. Lett. A* **123**, 43 (1987)

[12] M.TH. REKVELDT - *Nucl. Instr. Meth. B* **114**, 366 (1996)

[13] M.TH. REKVELDT, W.G. BOUWMAN, W.H. KRAAN, O. UCA, S.V. GRIGORIEV, K. HABICHT & T. KELLER - *In* "Neutron Spin Echo Spectroscopy, Basics, Trends and Applications", F. MEZEI, C. PAPPAS & T. GUTBERLET eds, *Lecture Notes in Physics* **601**, 87, Springer Verlag, Berlin, Heidelberg (2003)

[14] R. PIKE & P. SABATTIER eds - *Scattering,* Academic Press (2002)

[15] B. EWEN, D. RICHTER, T. SHIGA, H.H. WINTER & M. MOURS - Neutron Spin Echo Spectroscopy - Viscoelasticity - Rheology, *Advances in Polymer Science* **134**, Springer-Verlag, New York (1997)

[16] W. HÄUSSLER & B. FARAGO - *J. Phys.: Condens. Matter* **15**, S197 (2003)

[17] M. MONKENBUSCH - *in* "Neutron Spin Echo Spectroscopy, Basics, Trends and Applications", F. MEZEI, C. PAPPAS & T. GUTBERLET eds, *Lecture Notes in Physics* **601**, 246, Springer Verlag, Berlin, Heidelberg (2003)

[18] S. LONGEVILLE, W. DOSTER, M. DIEHL, R. GAHLER & W. PETRY - *in* "Neutron Spin Echo Spectroscopy, Basics, Trends and Applications", F. MEZEI, C. PAPPAS & T. GUTBERLET eds, *Lecture Notes in Physics* **601**, 325, Springer Verlag, Berlin, Heidelberg (2003)

[19] P. SCHLEGER, B. FARAGO, C. LARTIGUE, A. KOLLMAR & D. RICHTER - *Phys. Rev. Lett.* **81**, 124 (1998)

[20] P.G. DE GENNES - *J. Chem. Phys.* **55**, 572 (1971)

[21] G. RONCA - *J. Chem. Phys.* **79**, 1031 (1983)

[22] A. CHATTERJEE & R. LORING - *J. Chem. Phys.* **101**, 1595 (1994)

[23] J. DES CLOIZEAUX - *J. Physique I* **3**, 1523 (1993)

[24] P.G. DE GENNES - *J. Physique* **42**, 735 (1981)

[25] P.W. ANDERSON - *Science* **267**, 1615 (1995)

[26] J. COLMENERO, A. ARBE, D. RICHTER, B. FARAGO & M. MONKENBUSCH - *in* "Neutron Spin Echo Spectroscopy, Basics, Trends and Applications", F. MEZEI, C. PAPPAS & T. GUTBERLET eds, *Lecture Notes in Physics* **601**, 268, Springer Verlag, Berlin, Heidelberg (2003)

[27] R. KOHLRAUSCH - *Ann. Phys. Lpz* **12**, 393 (1847)

[28] G. WILLIAMS & D.C. WATTS - *Trans. Faraday Soc.* **66**, 80 (1970)

[29] J.C. PHILLIPS - *Rep. Prog. Phys.* **59**, 1133 (1996)

[30] A. ARBE, J. COLMENERO, M. MONKENBUSCH & D. RICHTER, *Phys. Rev. Lett.* **81**, 590 (1998)

[31] K.L. NGAI - *J. Appl. Phys.* **55**, 1714 (1984)

[32] A. ARBE, D. RICHTER, J. COLMENERO & B. FARAGO - *Phys. Rev. E* **54**, 3853 (1996)

[33] F. MEZEI & A.P. MURANI - *J. Magn. Magn. Mater.* **14**, 211 (1979)

[34] A.T. OGIELSKI - *Phys. Rev. B* **32,** 7384 (1985)

[35] C. PAPPAS, F. MEZEI, G. EHLERS & I.A. CAMPBELL - *Appl. Phys. A* **74**, S907 (2002)

[36] C. PAPPAS - private communication (2003)

[37] B. SARKISSIAN - *J. Phys.: Condens. Matter* **2**, 7873 (1990)

[38] B. SARKISSIAN & B.D. RAINFORD - *Solid State Comm.* **78**, 185 (1991)

[39] P. MANUEL & R. CYWINSKI - in preparation (2005)

[40] E.M. CHUDNOVSKY - *J. Appl. Phys.* **64**, 5770 (1988)

[41] G. EHLERS, H. CASALTA, R.E. LECHNER & H. MALETTA - *Phys. Rev. B* **63**, 224407 (2001)

[42] G. EHLERS, A.L. CORNELIUS, M. ORENDÁ, M. KAJAKOVÁ, T. FENNEL, S.T. BRAMWELL & J.S. GARDNER - *J Phys.: Condens. Matter* **15**, L9 (2003)

[43] H. CASALTA, P. SCHLEGER, C. BELLOUARD, M. HENNION, G. EHLERS, B. FARAGO, J.L. DORMANN, M. KELSCH, M. LINDE & F. PHILLIPP - *Phys. Rev. Lett.* **82**, 1301 (1999)

[44] R.H. HEFFNER, J.E. SONIER, D.E. MACLAUGHLIN, G.J. NIEUWENHUYS, G. EHLERS, F. MEZEI, S.W. CHEONG, J.S. GARDNER & H. RÖDER - *Phys. Rev. Lett.* **85**, 3285 (2000)

[45] E.M. FORGAN, P.G. KEALEY, S.T. JOHNSON, A. PAUTRAT, CH. SIMON, S.L. LEE, C.M. AEGERTER, R. CUBITT, B. FARAGO & P. SCHLEGER - *Phys. Rev. Lett.* **85**, 3488 (2000)

14

TIME-OF-FLIGHT INELASTIC SCATTERING

R. ECCLESTON
ISIS Facility, Rutherford Appleton Laboratory, Chilton, Didcot, UK

1. INTRODUCTION

Thermal neutrons travel at a velocity of approximately 2 km s^{-1}. As a consequence, the velocity of a neutron and hence its energy can be determined from its time-of-flight over a distance of a few metres. By fixing the energy of the neutrons incident on the sample and measuring their energy after scattering, or by fixing the energy of the scattered neutrons and determining their energy before scattering, the energy transferred from neutron to sample can be determined. This is the basis of time-of-flight (TOF) spectroscopy.

These two scenarios are referred to as direct and indirect (or inverted) geometry spectroscopy respectively. I will discuss the relative merits of both techniques and the different ways they are employed to address scientific problems. I will also draw on some scientific examples to illustrate the advantages and disadvantages of techniques or instrument designs.

At pulsed sources all spectrometers are TOF spectrometers because in this way the high peak flux and inherent time structure are utilized optimally. On steady state sources however, choppers can used to provide the pulse structure, and TOF spectrometers provide additional and complementary capabilities to triple-axis spectrometers (TAS).

The TOF technique has the following advantages:

- measurements are performed simultaneously over a broad range in momentum transfer/energy transfer ($\boldsymbol{Q}$,ω) space, making it an ideal technique for surveying,
- the instruments all tend to be of fixed geometry, which can have advantages for time dependent measurements or when measuring under difficult sample environment conditions.

F. Hippert et al. (eds.), Neutron and X-ray Spectroscopy, 457–482.

To set against the above there are the following disadvantages:

- TOF spectrometers offer less flexibility in the selection of scans and the tuning of resolution,
- it is difficult to collimate the scattered beam before detection in the wide detector arrays usually found on TOF spectrometers,
- wide detector arrays are not compatible with standard polarization analysis techniques. It is only now with the development of hyperpolarized ^{3}He filters and some innovative solid state devices that there are practical solutions for polarizing a scattered beam over a wide angular range.

In general, TOF instruments are well suited to broad surveys of $(\boldsymbol{Q},\omega)$ space while TAS instruments are better suited to detailed investigations at well-defined $(\boldsymbol{Q},\omega)$ points. Recent developments in both TOF and TAS instrumentation however, mean that this distinction is becoming increasingly less valid and the choice of the most appropriate instrument is often based on more subtle factors. TOF instruments do not have the ability a TAS instrument has to perform directly scans along well-defined crystallographic directions at fixed energy transfer or to scan energy transfer at a fixed point in the reciprocal lattice.

2. CLASSES OF TOF SPECTROMETERS

Time-of-flight spectrometers may be divided into two classes:

- ***Direct geometry spectrometers:*** in which the incident energy, E_i, is defined by a device such as a crystal or a chopper, and the final energy, E_f, is determined by time-of-flight or
- ***Indirect (inverted) geometry spectrometers:*** in which the sample is illuminated by a pulsed white incident beam and E_f is defined by a crystal or a filter and E_f is determined by time-of-flight.

The figure 1 [1] shows examples of generic direct geometry and indirect geometry types. In the following paragraphs I will describe the characteristics of instruments in each class and provide some scientific examples.

2.1. DISTANCE-TIME PLOTS

Distance-time plots illustrate the mode of operation of an instrument and are a useful way of understanding contributions to the energy resolution and the optimization of a spectrometer. Figures 2a and b are distance-time plots for a direct and an indirect geometry spectrometer respectively.

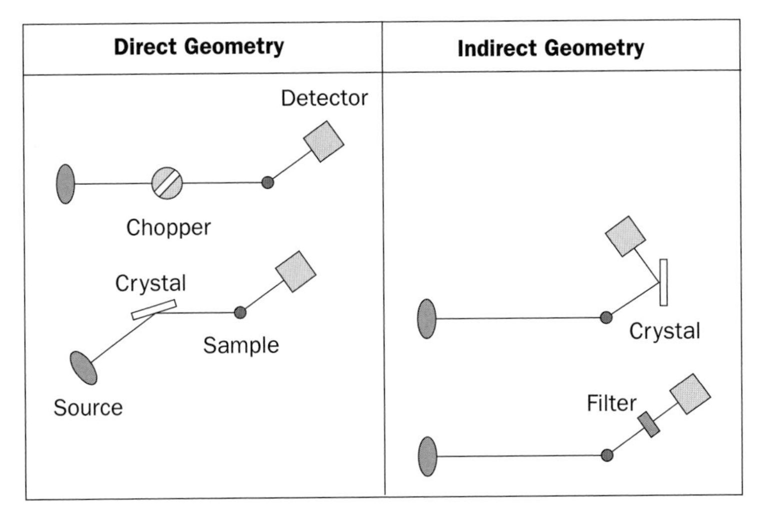

Figure 1 - Examples of generic instrument types of TOF spectrometer.

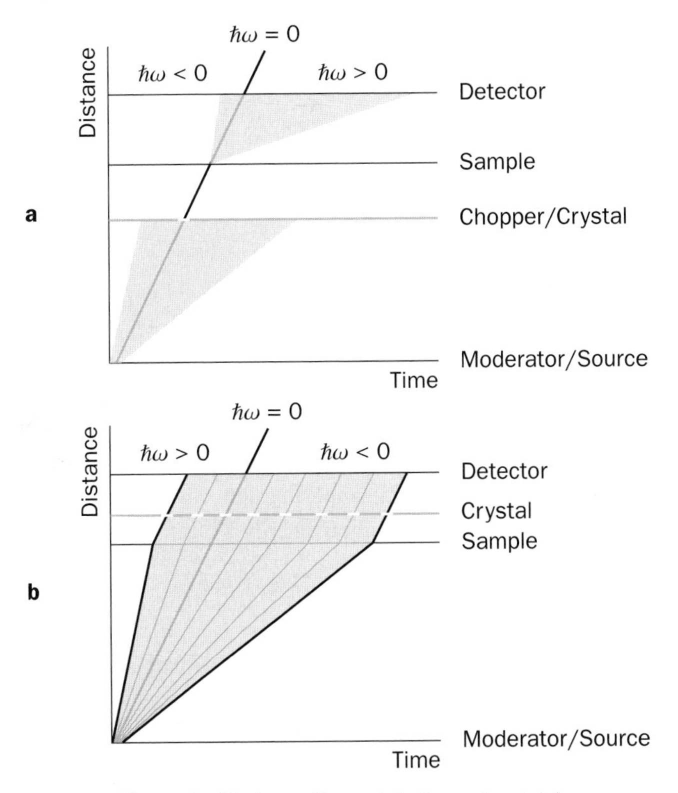

Figure 2 - Distance-time plots for a direct (**a**) and an indirect (**b**) geometry spectrometer respectively.

The abscissa is the time from the initial neutron pulse. On a pulsed source this is the time after the proton beam hits the target, on a steady state source it is the time the pulsing chopper is opened. The ordinate is the distance from the moderator in the case of the pulsed source, or the pulsing chopper in the case of a steady state source. We will return to distance-time plots later in the chapter when instrumental design and resolution is discussed.

2.2. *KINEMATIC RANGE*

To understand the kinematic range, and the trajectories traced by a TOF scan in any given detector through $(\boldsymbol{Q},\omega)$ space consider the scattering triangles. Clearly for direct geometry spectrometers the incident momentum of the neutron, $\boldsymbol{k}_i$, is fixed and the momentum of the scattered neutron $\boldsymbol{k}_f$ varies as a function of time whereas the reverse is true for indirect geometry instruments. The scattering triangles are shown in figure 3. The scattering angle, ϕ, is the angle between the un-scattered incident beam and the detector.

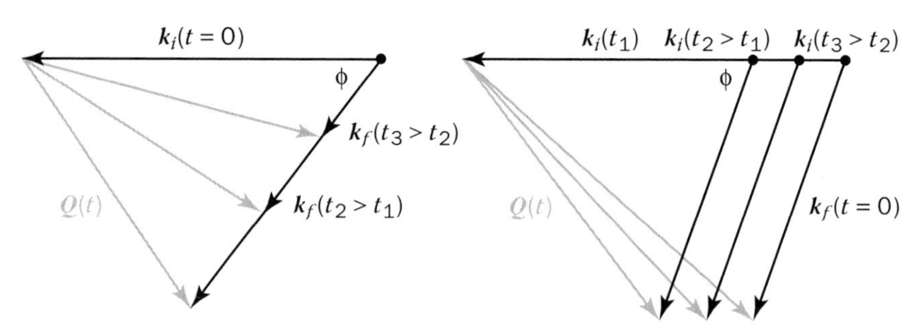

Figure 3 - Scattering triangles for a direct (**a**) and an indirect (**b**) geometry spectrometer respectively.

In order to understand the way $(\boldsymbol{Q},\omega)$ space is probed the relationship between $\boldsymbol{Q}$ and ω needs to be understood. By applying the cosine rule to the scattering triangles and converting from momentum to energy we get

$$\boldsymbol{Q}^2 = k_i^2 + k_f^2 2k_i k_f \cos\phi \qquad \{1\}$$

$$\frac{\hbar^2 \boldsymbol{Q}^2}{2m} = E_i + E_f - 2(E_i E_f)^{1/2} \cos\phi \qquad \{2\}$$

For direct geometry E_f can be eliminated as follows

$$\frac{\hbar^2 \boldsymbol{Q}^2}{2m} = 2E_i - \hbar\omega - 2\left[E_i(E_i - \hbar\omega)\right]^{1/2} \cos\phi \qquad \{3\}$$

which defines the trajectory a detector at a scattering angle ϕ traces through $(\boldsymbol{Q},\omega)$ space.

Similarly for indirect geometry E_i can be eliminated giving

$$\frac{\hbar^2 \boldsymbol{Q}^2}{2m} = 2E_f + \hbar\omega - 2\left[E_f(E_f + \hbar\omega)\right]^{1/2}\cos\phi \qquad \{4\}$$

and the trajectories are inverted. An important feature of the indirect geometry instrument is the access to a wide range of energy transfers for neutron energy loss. The trajectories through $(\boldsymbol{Q},\omega)$ space for direct geometry and indirect geometry instruments with detectors at a range of scattering angles are shown in figure 4.

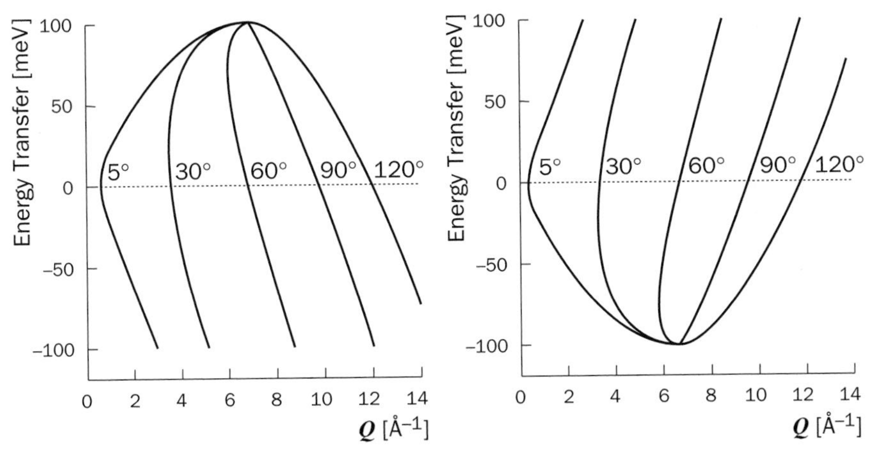

Figure 4 - TOF trajectories for detectors at a range of scattering angles for direct (left) and indirect (right) geometry.

3. *BEAMLINE COMPONENTS*

Before embarking on a more detailed description of the design and operation of spectrometers I will briefly describe the beamline components of a typical TOF spectrometer.

3.1. *CHOPPERS*

As mentioned above, choppers are key components of a direct geometry spectrometer. For instruments using neutrons in the thermal to high energy ranges Fermi choppers are most appropriate, for cold neutron instruments disk choppers offer inherent advantages.

3.1.1. FERMI CHOPPERS

A Fermi chopper is effectively a drum with a hole through the middle. The hole is filled with alternating sheets of neutron absorbing material (slats) and transparent material (slits). The slits and slats are curved, with the radius of curvature and the slit/slat ratio optimized for specific energy ranges. On the ISIS chopper spectrometers for example three slit packages are used to cover the incident energy range from 15 meV to 2 eV. An additional slit package with very broad slits is used to improve intensity at the expense of resolution. The Fermi chopper can be rotated at frequencies of up to 600 Hz at ISIS. The IN4 TOF spectrometer at the ILL uses a Fermi chopper rotating at frequencies of up to 40,000 rpm (666 Hz). The incident energy is defined by the phase of the chopper relative to the incident neutron pulse while the resolution is dictated by the slit/slat ratio of the slit package and the chopper frequency. Fermi choppers allow continuous selection of E_i.

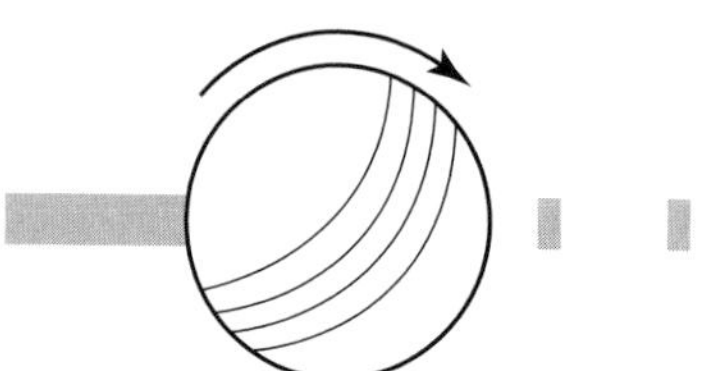

Figure 5 - A Fermi chopper.

3.1.2. DISK CHOPPERS

Disk choppers are simply rotating disks with holes in them. They offer higher transmission than Fermi choppers with the same flexibility in the selection of E_i. Using two disks or a variable aperture chopper overcomes the need to change slit package and they are also compact. However their frequency is limited by engineering constraints, as is the thickness of the disks. Both these factors limit the suitability of disk choppers for neutrons using thermal to high-energy neutrons, but they are very effective for cold neutron chopper spectrometers such as IN5 at ILL, NEAT at HMI and CNCS at NIST. Disk choppers on these instruments are currently operated at frequencies of up to 330 Hz (20,000 rpm).

3.1.4. $T = 0$ CHOPPERS

On a pulsed source, the high-energy neutron pulse and γ flash from the target are potential sources of background. A $T = 0$ chopper is simply a propeller blade structure that operates at the source frequency and is phased in such away that the line of sight from the target to the sample is blocked when the proton beam hits the target, hence $T = 0$. The 'blades' need to be thick and absorbing to be effective and sufficiently strong to survive being rotated. The $T = 0$ choppers have blades that are typically 50 cm in length, made of an alloy known as nimonic. These choppers are often referred to a nimonic choppers.

3.2. *MONOCHROMATING AND ANALYZING CRYSTALS*

The use of crystal monochromators for time-of-flight spectroscopy is limited, because they impose some geometrical constraints, and reflectivity falls as energy rises. However, by focusing a monochromating crystal large flux enhancements are possible, at the expense of beam divergence. This is a technique that is used very effectively on IN4 at the ILL for example. The cold neutron spectrometer IN6 also uses monochromating crystals, combined with a chopper to provide geometrical and time focusing. These instruments will be described in greater detail later.

Analyzing crystals are as ubiquitous for indirect geometry spectrometers as choppers are for direct geometry spectrometers. The choice of material, and crystal plane used, depends upon the analyzing energy and the optimization of reflectivity and mosaic spread, which dictates precision. The table below (from ref. **[1]** p. 342) shows the crystal planes and backscattering neutron energy and precision for silicon and graphite.

Material	**Plane**	E [meV]	ΔE [meV]
Pyrolytic graphite	002	1.828	0.009
	004	7.333	0.027
Silicon	111	2.074	0.0007

Geometrical constraints imposed by using analyzer crystals can often limit detector coverage and flexibility. The geometrical constraints can be lifted by using a double bounce arrangement, but at the expense of additional losses arising from the second reflection.

3.3. *FILTERS*

Be filters effectively remove neutrons with an energy greater than the Bragg cut-off of 5.2 meV. Incident neutrons are scattered in the beryllium and then absorbed in absorbing sheets within the filter. They are used to remove higher order contamination in indirect geometry instruments. Their efficiency can be improved by cooling, which reduces transmission of neutrons with energies above the cut-off that arise from up-scattering by thermally excited phonons.

Nuclear resonant absorption from a foil is used for very high-energy measurements. The eVS spectrometer at ISIS for example uses a filter difference technique using uranium or gold foils to measure atomic momentum distributions in the eV range [2].

3.4. *DETECTORS*

As we shall see later the optimization of a direct geometry spectrometer requires the largest detector array in terms of solid angle consistent with the space available and financial constraints. The detectors of choice for most TOF instruments are ^3He tubes because they offer low γ sensitivity and high efficiency. Their ability to work in magnetic fields is also advantageous. Developments of position sensitive tubes [3] have represented a leap forward in the technology and allow large detector areas to be tiled with pixelated arrays.

Scintillator detectors are employed when high spatial resolution is required, for a low angle bank for example. (Scintillator detectors are available with spatial resolution of the order of mm whereas gas tubes are currently limited to spatial resolution of the order of cm). Also, current scintillator detectors are faster than gas tubes so are preferred in high rate applications.

3.5. *NEUTRON GUIDES*

The use of guides provides instrument designers with the opportunity to move their instrument further away from the source without a large flux penalty in order to improve the resolution. The walls reflect neutrons incident at an angle less than the critical angle making the transport of neutron beams over relatively long distances possible. In the past nickel has been used because of its high scattering length density. Nickel guides have a reflectivity of approximately 99.5%. The critical angle, γ_c, is proportional to the wavelength (for ^{58}Ni $\gamma_c = 0.12° Å^{-1}$); consequently guides are most gainfully employed for cold neutron instruments. An important development over recent years has been the development of supermirror guides [4]. The realization of multi-layer devices with critical angle of up to four times that of nickel have resulted in significant gains, particularly for cold neutron applications.

Supermirrors are also used as benders, which will only transmit long wavelength neutrons, thereby preventing high-energy neutrons, which may produce a back-ground signal, entering the instrument.

3.6. POLARIZERS AND POLARIZATION ANALYSIS

The use of polarization analysis on TOF spectrometers has been limited in the past by the difficulty in both polarizing a broad-band incident beam and analyzing a scattered beam over a large solid angle. Developments in both ^{3}He filter technology [5] and in solid state devices are now opening up new opportunities in this field.

4. DIRECT GEOMETRY SPECTROMETERS

4.1. CHOPPER SPECTROMETER ON A PULSED SOURCE

For the first example consider chopper spectrometers like HET, MARI and MAPS on a pulsed source. Their operation can easily be explained by referring to the schematic (fig. 6).

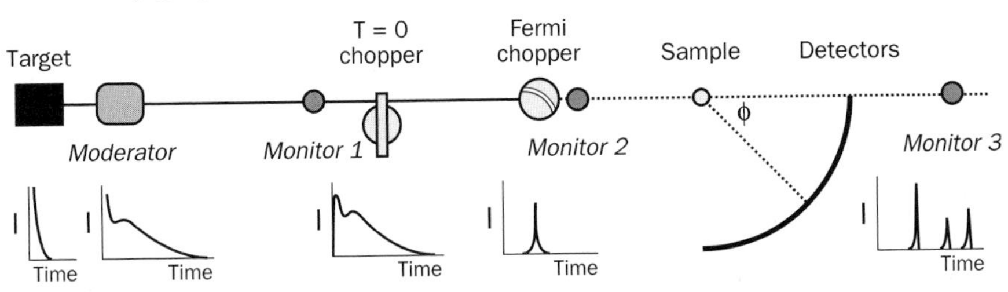

Figure 6 - A schematic diagram of a chopper spectrometer on a pulsed source.

The proton pulse incident on the target produces a sharp pulse of high-energy (MeV) neutrons. A hydrogenous moderator, such as water or methane slows these towards thermal energies. The moderators are small to ensure that the beam is not fully moderated. This preserves the inherent sharpness of the incident pulse and also provides a high flux of hot neutrons in the under-moderated or epithermal region. A T = 0 chopper is used to suppress γs and very high-energy neutrons entering the spectrometer where they would thermalize producing a background signal. The beam is monochromated by a Fermi chopper before being scattered from the sample into a large array of detectors. Monitors are positioned to record the incident neutron flux (monitor 1). The incident energy is calculated from the time-of-flight between monitors 2 and 3. The detector arrays cover as broad a solid angle as physical and financially practical, and in most cases ^{3}He tubes are used. New instruments tend to be equipped with position sensitive detectors. The MAPS spectrometer at ISIS (fig. 7) is currently the most advanced spectrometer of its kind. Examples of experiments performed on instruments like MAPS include studies of quantum fluids [6] and low dimensional magnets [7].

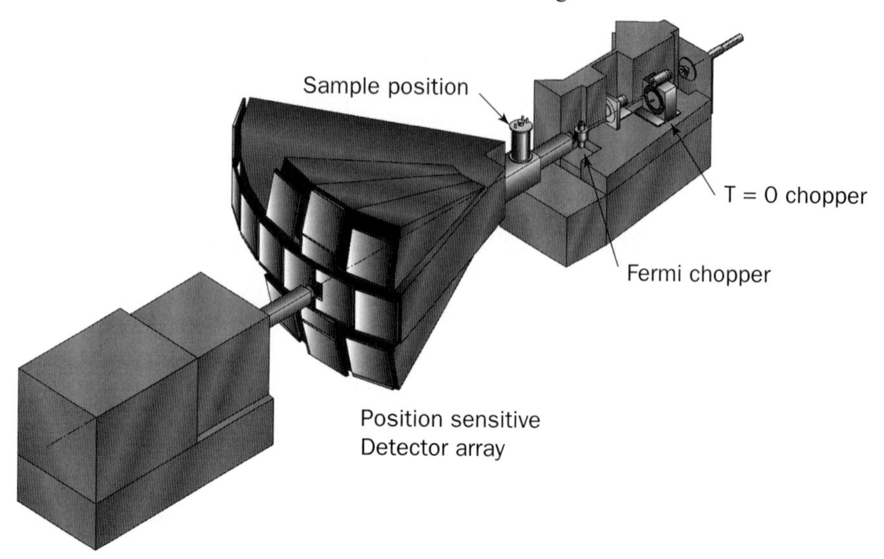

Figure 7 - The MAPS spectrometer at ISIS.

4.2. CHOPPER SPECTROMETERS ON A STEADY STATE SOURCE

In the case of some time-of-flight spectrometers on steady state sources the chopper simply provides the pulsed structure of the beam and a crystal monochromator is used to select the incident energy. IN6 at the ILL is an excellent example of this type of instrument and is ideal for quasielastic and inelastic scattering measurements [8] (fig. 8). The incident energy is selected by three crystal monochromators that focus the beam vertically. The monochromators are aligned in such a way as to extract neutrons of three slightly different energies from the incident beam. The rotation of the Fermi chopper is such that it scans the crystal with the highest angle and thereby scattering the slower neutrons, first, and the crystal scattering the fastest neutrons last. Slight path length differences are also optimized to produce time focusing. A cooled beryllium filter is used to remove second order reflections from the monochromators, the first chopper stops neutrons that would overlap into the next measurement frame and the pulsed structure is provided by a Fermi chopper. The scattered beam is detected in a large detector bank populated by ^{3}He tubes.

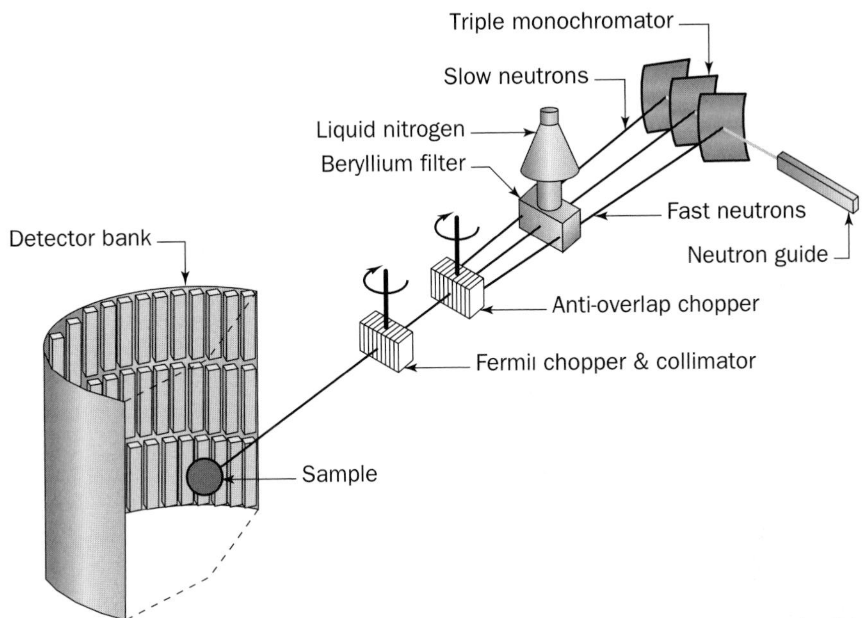

Figure 8 - The IN6 spectrometer at the ILL.

IN4, also at the ILL, uses a large focusing crystal monochomator to select the incident energy with a Fermi chopper to provide the pulse structure [9]. The focusing of the monochromator provides a large flux enhancement for experiments where beam divergence and consequently Q resolution is less important. The curvature of the monochromator can be carefully controlled to provide optimal space and time focusing conditions [10].

Two background choppers are used to eliminate fast neutrons and γs and produce broad pulses that are then incident on the monochromator. The Fermi chopper is located between the monochromator and the sample.

TOF instruments on a steady state source have an advantage over instruments on a pulsed source in that their maximum operating frequency is not dictated by the source. However, innovative instrumentation designs may provide the opportunity to increase the effective repetition rate on a pulsed source by phasing an array of chopper in such a way that the pulse can effectively be used several times.

4.3. MULTI-CHOPPER TOF SPECTROMETERS

I will refer to the last group of chopper spectrometers as multi-chopper TOF instruments. There are several examples on steady state sources, such as NEAT at

HMI (shown in fig. 9) and IN5 at ILL [11], but not yet any on pulsed sources. Design work [12,13] has demonstrated their potential to provide flexible, high flux and high resolution cold neutron spectrometers on the next generation pulsed sources.

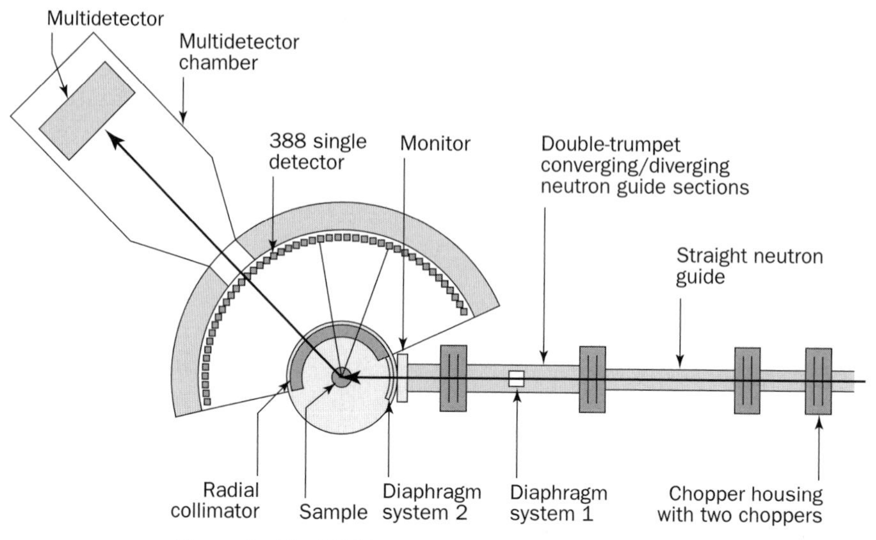

Figure 9 - The NEAT spectrometer at BENSC at HMI.

On a steady state source, the first chopper is used to provide the pulse structure. On a pulsed source of course this chopper is not necessary. A further array of choppers (at least two are required) then combine to select the incident energy by selection of the appropriate phases and the incident energy resolution by the transmission width of the choppers. One of the great advantages of this design of instrument is that it provides a high level of control of the resolution function. The resolution function of the chopper spectrometer comprises two components; one arising from the monochromating chopper, the other from the moderator. In the case of a multi-chopper instrument, choppers dictate both components. Furthermore, there is a matching condition for the two components, which results in an optimization of the flux (see reference **[12]**).

The resolution is limited by the maximum speed of the choppers, which are subject to engineering constraints, but compressing the beam and the use of pairs of counter-rotating disk choppers allow energy resolutions of approximately 2 μeV to be achieved to good scientific affect [14].

5. RESOLUTION AND SPECTROMETER OPTIMIZATION

I will now discuss the resolution of a chopper spectrometer using for the sake of simplicity a chopper spectrometer on a pulsed source. For a more thorough discussion refer to reference [1]. The resolution functions for mutli-chopper spectrometers and spectrometers that use crystals as the primary monochomators are well described elsewhere [10], but much of what is given below applies equally well to those cases except the moderator component of the resolution arises from the initial pulse width.

The energy transfer resolution $\Delta\varepsilon$ at the elastic line of a chopper spectrometer is given by

$$\frac{\Delta\varepsilon}{E_i} = 8.7478\times10^{-4}\frac{\sqrt{E_i}\ [\text{meV}]}{L_2\ [\text{m}]}\Delta t\ [\mu\text{s}] \qquad \{5\}$$

where L_2 is the flight path between the sample and the detector and Δt is the pulse width at the detector. Δt comprises a component (Δt_{md}) arising from the moderator pulse width at the detector and a component from the chopper pulse width at the detector (Δt_{cd}) (see fig. 10).

We can consider these two components as follows, assuming we are restricting the discussion to elastic scattering for the time being for the sake of simplicity.

$$\begin{aligned} \Delta t_{md} &= \frac{L_2 + L_3}{L_1}\Delta t_{mm} \\ \Delta t_{cd} &= \frac{L_2 + L_1 + L_3}{L_1}\Delta t_{cc} \end{aligned} \qquad \{6\}$$

Δt_{md} and Δt_{mm} is the moderator pulse width at the detector and the moderator pulse width at the moderator respectively. Similarly, Δt_{cd} and Δt_{cc} represent the chopper pulse width at the detector and at the chopper. L_1 is the moderator to chopper distance and L_3 the chopper to sample distance.

Clearly, to optimize resolution L_2 is made as long as possible. The only limit being the expense of large detector arrays and the availability of space. For high incident energies large portions of $\boldsymbol{Q}$ space can be accessed at relatively low angles, at thermal to low energies however large detector areas are important to provide reasonable $\boldsymbol{Q}$ coverage.

In considering the optimal configuration for a chopper spectrometer one can calculate the minimum value of L_1, which will give the desired resolution for the given moderator pulse width. Scattering from the Fermi chopper is a source of background, so L_3 should be sufficiently long so as to remove the chopper from the direct line of sight of the detectors.

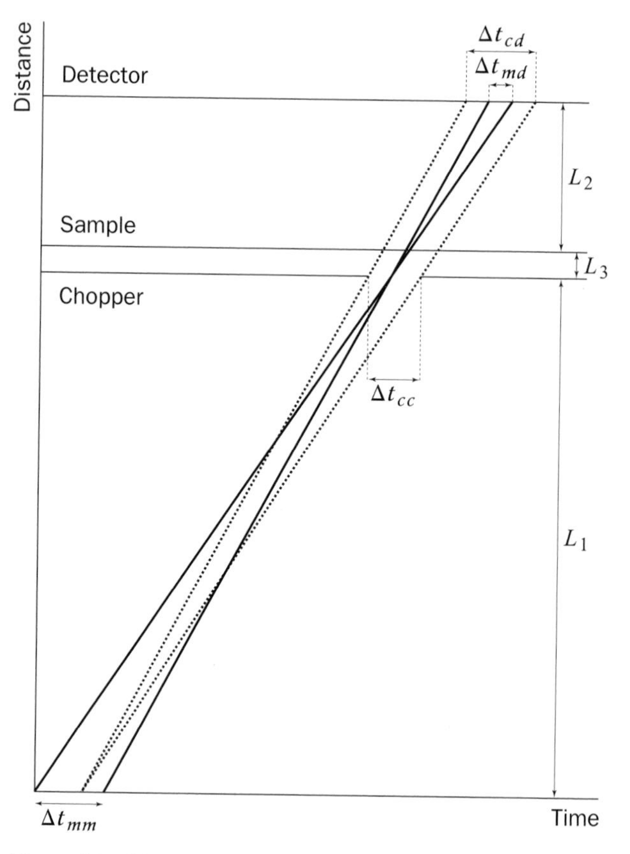

Figure 10 - Distance-time plot for a chopper spectrometer.

6. *FLUX*

The flux at the sample for a given incident energy is proportional to the solid angle of the moderator as seen by the sample and the open time of the chopper as a fraction of the flight-time from moderator to chopper.

$$\Phi = \left(\frac{p}{d}\right)\left(\frac{W_m H_m}{(L_1 + L_3)^2}\right)\left(\frac{\Delta t_{cc}}{L_1}\right) \qquad \{7\}$$

The term (p/d) refers to the ratio of the width of the chopper transmitting slits to the width of the absorbing slats. W_m and H_m are the width and the height of the moderator as viewed by the sample respectively. For high incident energies, supermirrors will not reduce the $(L_1 + L_3)^2$ flux loss. It is clear that to optimize flux, L_1, should be as short as possible (10 m is the practical minimum allowing for choppers…). For thermal and cold neutron energy regimes, supermirror guides become increasingly effective as the energy decreases. The optimization of the multi-chopper spectrometer described above suggest that an instrument of 30 - 40 m would be ideal. It is worth pointing out however that the third term in this equation is not affected by the presence or otherwise of the guide. Consequently there is always a flux penalty to extending the instrument, although this may be offset by other factors, such as the use of a more intense moderator to provide the equivalent resolution.

7. *RESOLUTION AS A FUNCTION OF ENERGY TRANSFER AND EXPERIMENTAL CONSIDERATIONS*

If we now consider the resolution as a function of the energy transfer the moderator and chopper terms are summed in quadrature to give.

$$\frac{\Delta\hbar\omega}{E_i} = \left[\left\{ \frac{2\Delta t_c}{t_c} \left(1 + \frac{L_1 + L_3}{L_2} \left(1 - \frac{\hbar\omega}{E_i} \right)^{3/2} \right) \right\}^2 + \left\{ 2\frac{\Delta t_m}{t_c} \left(1 + \frac{L_3}{L_2} \left(1 - \frac{\hbar\omega}{E_i} \right)^{3/2} \right) \right\}^2 \right]^{1/2} \quad \{8\}$$

where t_c is the time-of-flight at the chopper, Δt_c is the chopper burst time and Δt_m is the moderator pulse width. The important fact to be drawn from this equation is that the fractional energy resolution improves with energy transfer.

On a chopper spectrometer, the experimenter has control over the incident energy and can optimize flux and resolution by selecting the appropriate type and frequency of chopper. In selecting the correct incident energy the experimenter will tend to choose an incident energy close to but above the feature of interest in order to view it with the best resolution. However a rule of thumb that should be borne in mind when selecting experimental parameters for a measurement on a direct geometry spectrometer is to not measure features with more than 75% energy transfer. There are two reasons for this. One is the fact that intensity will decrease with k_f/k_i, from the master formula for the neutron cross-section, the other is that the time bins over which data is being collected are increasing as the square. This has the effect that once the conversion to energy transfer is made a background signal that is constant in time will appear to rise rapidly at the end of the measurement frame. Time-independent background can usually be estimated and subtracted by using empty portions of the frame.

SINGLE CRYSTAL EXPERIMENTS ON A CHOPPER SPECTROMETER

Developments in both instrumentation and experimental technique have demonstrated the opportunities for studying excitations from single crystal sample afforded by chopper spectrometers (for a review see Perring [15]). The use of position-sensitive ^{3}He detectors in particular means that large surveys of reciprocal space can be accomplished simultaneously.

To explain the use of a direct geometry instrument for measuring excitations in single crystals I will start with the scattering triangle (fig. 3) again. For any given detector at a scattering angle ϕ, k_i is constant and k_f decreases as a function of the time-of-flight. If we draw the scattering triangle over the reciprocal lattice and consider and array of detector covering a range of scattering angles, the elastic line ($\hbar\omega = 0$) is a circle of radius $k_i = k_f$, and as k_f changes the array of detectors cover a sector through reciprocal space as illustrated (fig. 11).

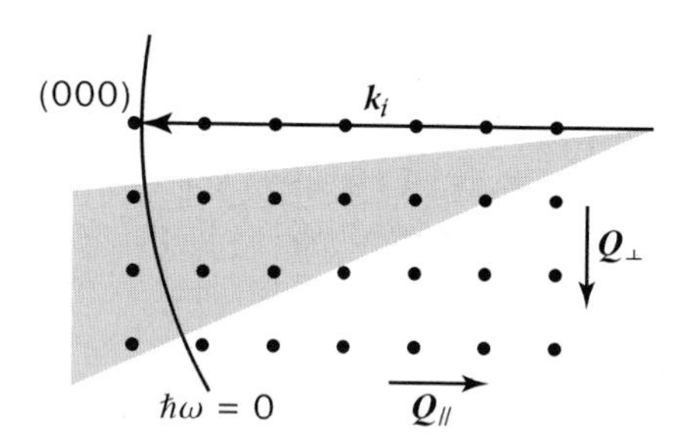

Figure 11 - Trajectories traced by an array of detectors at scattering angles of 5° to 30°.

The trajectory through reciprocal space is parabolic as explained above, so we can view figure 11 in three dimensions (fig. 12).

By way of an example, consider a ferromagnetic spin wave in a 3 dimensional magnetic system. The spin waves emerge from a reciprocal lattice point looking like a cone. As the dispersion intersects the surface traced out by the time-of-flight trajectories for the detectors, scattering will be observed. This results in smoke rings appearing when the scattering intensities are plotted on a contour plot in the same format as figure 12. These data can be treated in software to produce cuts akin to constant energy transfer ($\hbar\omega$) scans on a triple-axis spectrometer, although it is important to remember that they are not strictly speaking constant $\hbar\omega$ scans because $\boldsymbol{Q}$ and ω are coupled. In such a way, a full dispersion relation can be followed using relatively few crystal orientations. Just a few data sets of the type shown in figure 13 were required to produce the full dispersion plotted in figure 13 [16].

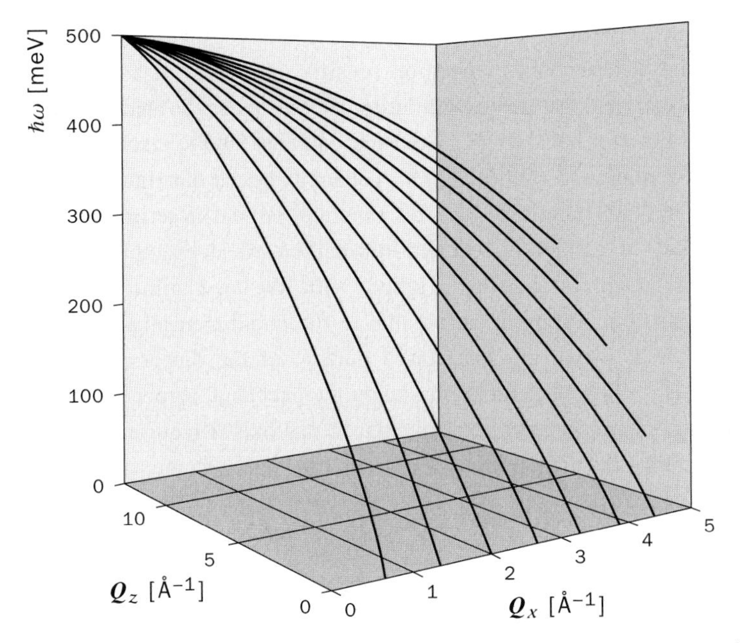

Figure 12 - Trajectories traced through ($\boldsymbol{Q}_z$, $\boldsymbol{Q}_X$, $\hbar\omega$) space by detectors spaced at 5° intervals from a minimum scattering angle of 5° for an incident energy E_i = 500 meV.

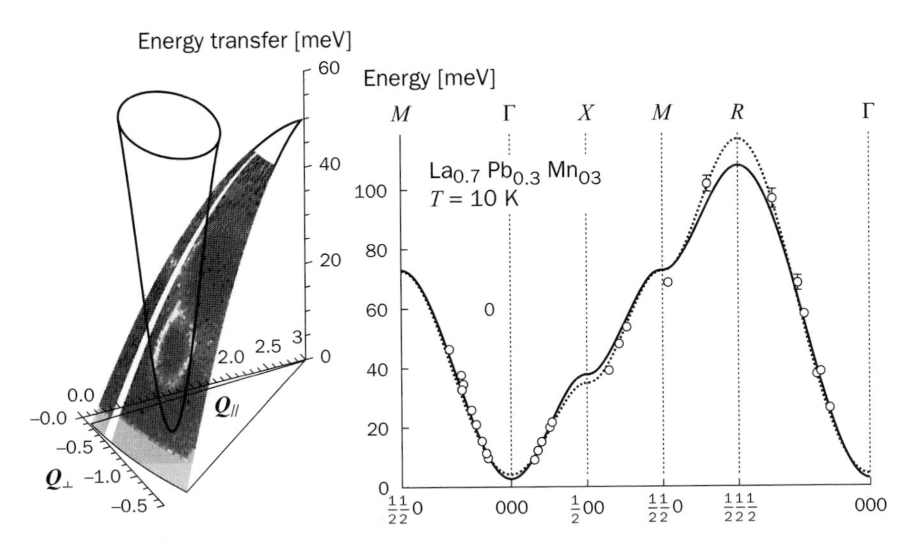

Figure 13 - **Left** - Data collected from a single crystal sample of $La_{0.7}Pb_{0.3}MnO_3$ on the HET spectrometer at ISIS with and incident energy of 50 meV. **Right** - The spin wave dispersion [16].

This simple description is true for an array of detector lying in the scattering plane. However the development of position sensitive detectors mean that instruments like MAPS can be constructed with large detector areas divided into many pixels (the MAPS detector bank cover 17 m^2 and contains 50,000 pixels). Using such an instrument a volume in (Q_x, Q_y, Q_z, $\hbar\omega$) space is traced out simultaneously (Q_z is the component of $\boldsymbol{Q}$ parallel to $\boldsymbol{k}_i$). This generates large data sets that can be cut in a variety of ways to extract the required information. As an example, consider a one-dimensional magnetic system aligned with the one dimensional direction perpendicular to $\boldsymbol{k}_i$. Taking a cut parallel to the one-dimensional direction, a data set is extracted, which allows a large portion of the dispersion to be imaged simultaneously. However, in order to study the exchange perpendicular to the one-dimensional direction a scan perpendicular to that axis is required. This is possible by simply taking the appropriate slice through the data volume perpendicular to the previous dataset, and in this example at the point along Q_x that coincides with the minimum in the dispersion. This is illustrated in figure14.

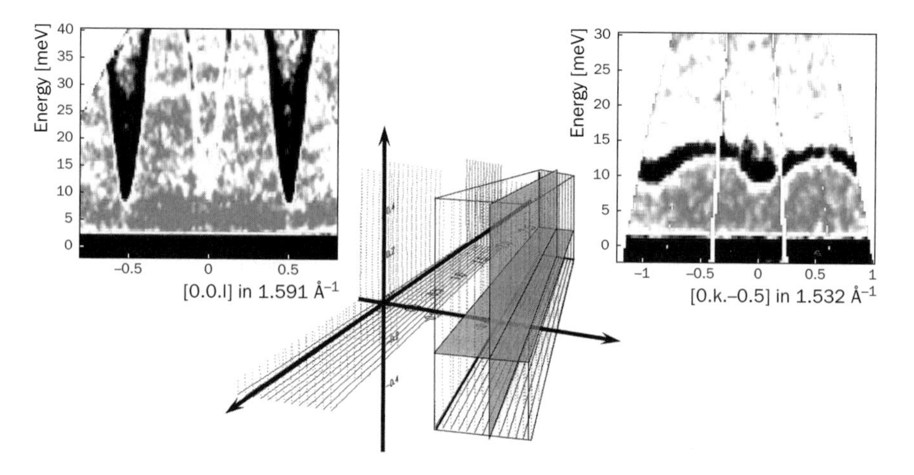

Figure 14 - Datasets extracted from software cuts through the volume swept out by a large detector bank.

8. *INDIRECT GEOMETRY SPECTROMETERS*

Indirect geometry spectrometers offer access to a wide range of neutron energy loss values, they also offer high resolution at the elastic line and tend to offer broader coverage in terms of energy transfer with reasonable resolution.

A variety of spectrometer designs offer different capabilities:

- crystal analyzer spectrometer – molecular spectroscopy,
- back scattering spectrometer – high resolution,
- coherent excitations spectrometer – coherent excitations in single crystals.

Before discussing each of these in term, it is useful to briefly consider the resolution characteristics of an indirect geometry spectrometer.

8.1. THE RESOLUTION OF INDIRECT GEOMETRY SPECTROMETERS

For the indirect geometry spectrometer the energy resolution contains terms pertaining to the uncertainty in the angular spread of neutrons scattered from the analyzer, $\Delta\theta_A$ and timing errors over the incident flight path, Δt. Timing errors arise from several contributions, including moderator thickness, sample size broadening, analyzer thickness broadening, detector thickness and the finite width of data collection time channels. If Δt is expressed in terms of an equivalent distance $\delta = \hbar\, k_i\, \Delta t/m$ the energy resolution can be concisely expressed as

$$\frac{\Delta\hbar\omega}{E_i} = 2\left[\left(\frac{\delta}{L_1}\right)^2 + \left\{\frac{E_f}{E_i}\cot\theta_A\Delta\theta_A\left(1+\frac{L_2}{L_1}\left(\frac{E_i}{E_f}\right)^{3/2}\right)\right\}^2\right]^{1/2} \qquad \{9\}$$

where L_1 is the moderator to sample distance, and L_2 the sample to analyzer distance. Clearly increasing θ_A or L_1 or decreasing L_2 improves resolution but physical and instrumental limitations apply. Also of interest is the fact that fractional energy resolution becomes worse with energy transfer, the opposite of the case for the direct geometry instruments.

8.2. BACKSCATTERING SPECTROMETER

In order to achieve high resolution the $\cot\theta_A$ term can be reduced to almost zero by using a backscattering or near backscattering geometry, thereby optimizing the definition of E_f. In a matched spectrometer, E_i should be determined to the same precision, hence L_1 tends to be long. IRIS (fig. 15) at the ISIS Facility represents a good example of a near back scattering instrument. In the most frequently used mode of operation the 002 reflection from graphite analyzers defines an E_f of 1.845 meV and provides resolution of 17 μeV. The analyzers are cooled to reduce thermal diffuse scattering (TDS), which broadens the elastic peak and gives rise to additional background. To ensure that the spectrometer of the primary spectrometer is matched to that of the secondary spectrometer, the incident flight path is long (35 m). An array of disk choppers prevent frame overlap and allow the possibility of running at 25 Hz rather than the source frequency of 50 Hz to enhance the bandwidth.

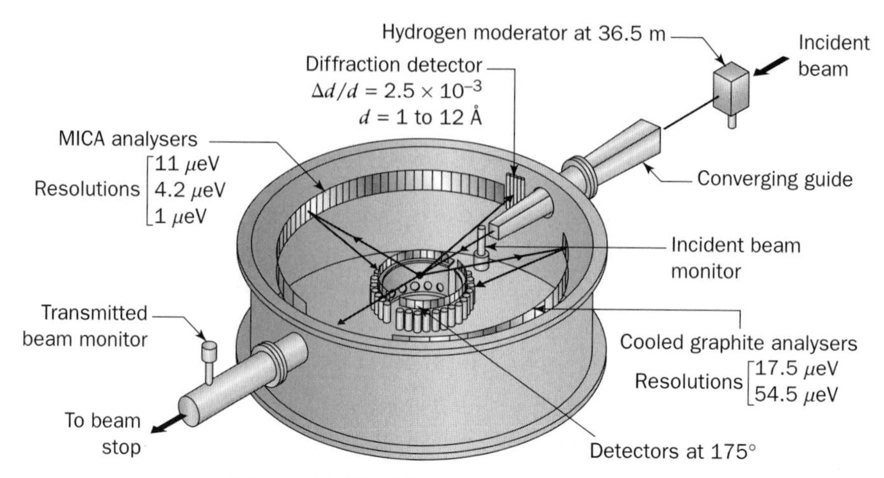

Figure 15 - The IRIS spectrometer at ISIS.

Instruments like IRIS are referred to as near-backscattering instruments because the detectors are positioned below the sample. In the case of true back scattering instruments the analyzed beam is scattered back directly toward the sample and is detected behind the sample. A chopper in front of the sample is required to avoid cross-talk.

Recent design work [17] has demonstrated that backscattering and near back-scattering instruments on next generation pulsed sources could have access to 1 μeV resolution, with broader bandwidths and considerably higher fluxes than existing instruments. In order to achieve such high resolutions, the instruments would be long, of the order of 200 m.

Instruments like IRIS are well suited to a range of scientific applications, such as dispersion, quantum tunnelling, quasi-elastic scattering. The example I have chosen is a study of the spin dynamics of a one dimensional spin system $Cu(NO_3)_{2.2.5}D_2O$ [18], which is a dimerized antiferromagnetic chain. Figure 16 shows the temperature evolution of the spin excitations compared to a theoretical model. The excitations range form 0.35 to 0.5 meV

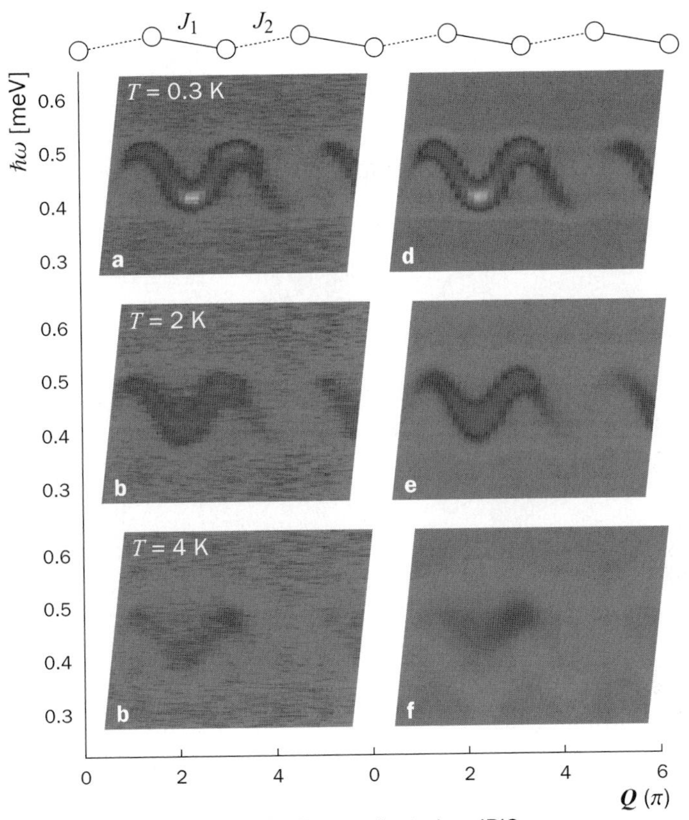

Figure 16 - Data collected on IRIS from the dimer antiferromagnetic chain compound $Cu(NO_3)_2 \cdot 2.5D_2O$.

8.3. CRYSTAL ANALYZER SPECTROMETERS

For molecular spectroscopy, energy information is often more important than $\boldsymbol{Q}$ information. Consequently, measuring energy transfer along a single trajectory provides an efficient method for measuring excitations across a broad spectral range [19].

TOSCA at ISIS (fig. 17) provides a good example. A graphite analyzer defines an E_f of 3 meV, a cooled beryllium filter is used to remove higher order reflections. The geometry of the secondary spectrometer is such that the sample, analyzer and detectors are parallel; this effectively time-focusses the scattered neutrons, reducing uncertainty in the scattered neutron flight time. The simultaneous broad access to a wide energy range has a wide range of scientific applications [20] where INS is often complementary to other spectroscopic techniques.

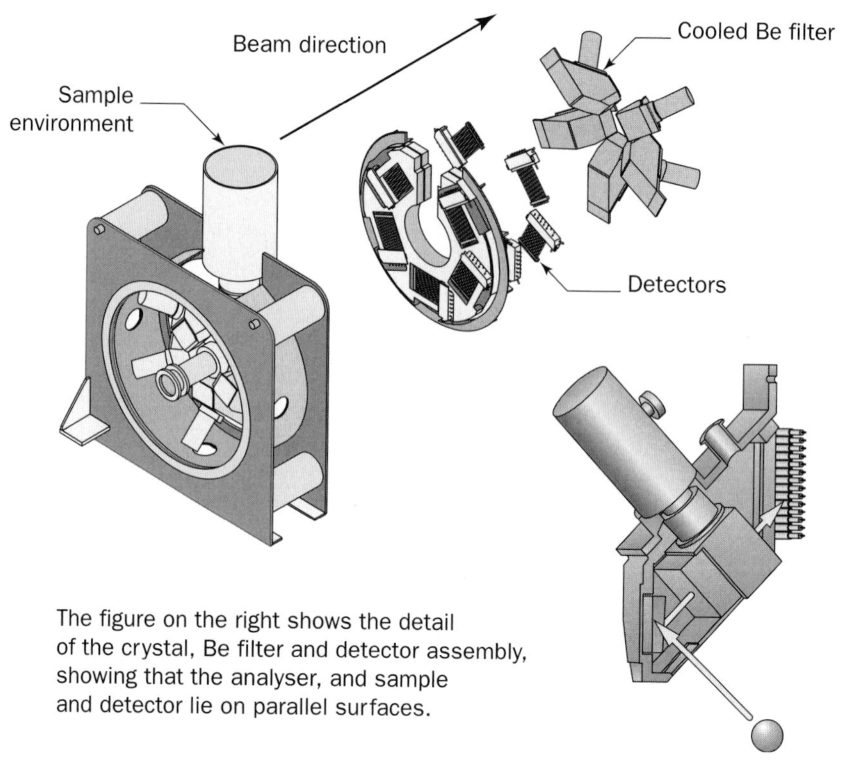

Figure 17 - The TOSCA detector array and sample vessel.

8.4. DEEP INELASTIC NEUTRON SCATTERING

Pulsed sources provide unrivalled access to very high energy transfer regimes where it is possible to measure atomic momentum distributions. This technique is known as deep inelastic neutron scattering [21] (DINS) or neutron Compton scattering by analogy to the X-ray measurement of electron momentum distributions. For a review see Mayers [22].

The interpretation of the data is performed with the framework of the impulse approximation, which states that if the momentum lost by the incident neutron is sufficiently large, scattering occurs from a single atom with conservation of kinetic energy and momentum. If the scattered neutron loses momentum $\hbar\boldsymbol{Q}$ and energy $\hbar\omega$ on being scattered from an atom of mass M, given that the momentum of the atom before the collision is $\boldsymbol{p}$, the momentum after collision will be $\boldsymbol{p} + \hbar\boldsymbol{Q}$ after the collision. To conserve momentum the following equation must be satisfied.

$$\hbar\omega = \frac{(\boldsymbol{p}+\hbar\boldsymbol{Q})^2}{2M} - \frac{\boldsymbol{p}^2}{2M} \qquad \{10\}$$

This equation can be rearranged as follows, where $\hat{\boldsymbol{Q}}$ is the unit vector along the direction of $\boldsymbol{Q}$ and y if the component of atomic momentum along the same direction:

$$y = \boldsymbol{p}\cdot\hat{\boldsymbol{Q}} = \frac{M}{Q}\left(\omega - \frac{\hbar \boldsymbol{Q}^2}{2M}\right) \qquad \{11\}$$

The VESUVIO spectrometer at ISIS is dedicated to the development of DINS and is at present in the process of further upgrades such as an increase in detector coverage. A filter difference technique is used to analyze the energy of the scattered neutrons. The filter is a thin foil of either gold or uranium, which absorbs neutrons strongly over narrow bands of energy (absorption resonances). Two TOF spectra are collected: one with the foil between the sample and the detectors and the other with it removed. The difference between the two spectra is due to the neutrons absorbed in the foil and is effectively the TOF spectrum for neutrons with final energy equal to the foil absorption energy.

Deep inelastic neutron scattering has been employed for the measurement of kinetic energies in quantum fluids, and also to explore the local structure and dynamics of amorphous materials and glasses.

8.5. COHERENT EXCITATIONS

One of the perceived shortcomings of chopper spectrometers and the indirect geometry spectrometers described above is their inability to perform scans uniquely along high symmetry directions. By adopting a scattering geometry whereby the condition $\sin\phi/\sin\theta_A$ = constant is met (fig. 18), a TOF scan corresponds to a scan along a defined direction in $\boldsymbol{Q}$ space.

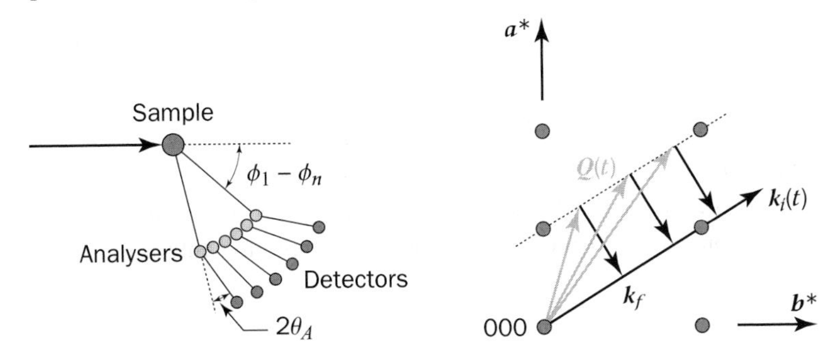

Figure 18 - A schematic representation of a PRISMA type condition and the implications for time-of-flight scans.

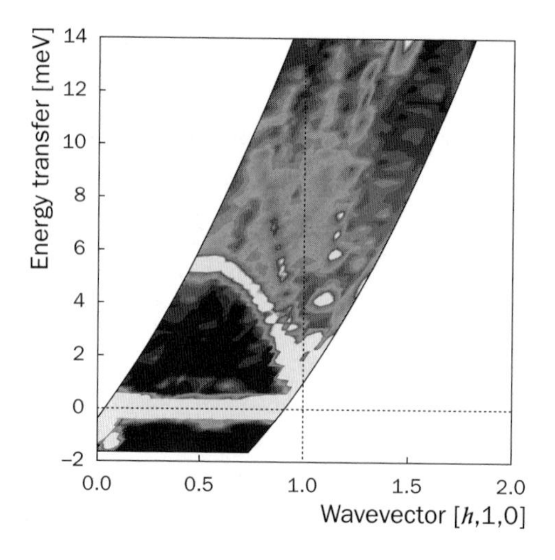

The PRISMA spectrometer can operate in this mode. The data show in figure 19 were collected on PRISMA from an ice crystal [23], a n d s h o w t h e phonon dispersion.

Figure 19 - Phonons in ice 1h measured on PRISMA.

Operating in this mode, is technically challenging because the movement of analyzers is limited by the risk of clashing and consequently coverage of $(\boldsymbol{Q},\omega)$ space can be sparse with a consequent reduction in data rate. The use of analyzers in a double bounce configuration provides more flexibility and also permits access to higher scattering angles and consequently higher resolution. PRISMA has a modular design, which enables the experimenter to use different configurations of the instrument for both spectroscopy and diffraction.

Recent developments on TAS spectrometers such as RITA at Risø may be adapted to improve the flexibility of PRISMA type instruments. For example the use of large position sensitive detectors, combined with an analyzer array that is sufficiently flexible to provide several options for focusing could be adapted for TOF studies of excitations in crystals.

9. *CONCLUSIONS*

TOF spectroscopy is well suited to experiments that involve surveying large ranges in either energy transfer or energy and momentum transfer space. Developments in both instrumentation and experimental techniques have opened up new scientific opportunities and produced some outstanding results to date. These developments show every sign of continuing, fuelled by the opportunities for constructing state-of-the-art instrumentation for the next generation sources currently under construction or in planning.

FURTHER INFORMATION

Further details about the instruments described in this chapter can be found at the websites of the appropriate sources:

- www.ill.fr
- www.isis.rl.ac.uk
- www.hmi.de

REFERENCES

[1] C.G. WINDSOR - *Pulsed Neutron Scattering,* Taylor and Francis Ltd. (1981)

[2] R.J. NEWPORT, P.A. SEEGER & W.G. WILLIAMS - *Nucl. Inst. Meth. A* **238**, 177 (1985)

[3] C.D. FROST - *Proceedings of the 14th Meeting of the International Collaboration on Advanced Neutron Sources,* Starved Rock Lodge, Utica, Illinois, US (1998)

[4] F. MEZEI - *Commun. Phys.* **1**, 81 (1976); J.B. HAYTER & H.A. MOOK - *J. Appl. Cryst.* **22**, 35 (1989)

[5] R. SURKAU, J. BECKER, M. EBERT, T. GROSSMAN, W. HEIL, D. HOFMANN, H. HUMBLOT M. LEDUC, E.W. OTTEN, D. ROHE, K. SIEMENSMEYER, M. STEINER, F. TASSET & N. TRAUTMANN - *Nucl. Inst. Meth. A* **384**, 444 (1997)

[6] H.R. GLYDE, R.T. AZUAH & W.G. STIRLING - *Phys. Rev. B* **62**, 14337 (2000)

[7] S. ITOH, Y. ENDOH, K. KAKURAI, H. TANAKA, S.M. BENNINGTON, T.G. PERRING, K. OHOYAMA, M.J. HARRIS, K. NAKAJIMA & C.D. FROST - *Phys. Rev. B* **59**, 14406 (1999)

[8] For example: M.R. GIBBS, K.H. ANDERSEN, W.G. STIRLING & H. SCHOBER - *J. Phys.: Condens. Matter* **11**, 603 (1999)

[9] G. CICOGNANI, H. MUTKA, D. WEDDLE, B. HAMELIN, P. MALBERT, F. SACCHETTI, C. PETRILLO & E. BABUCCI - *Physica B* **276**, 85 (2000)

[10] H. MUTKA - *Nucl. Instrum. Meth. in Phys. Research A* **338**, 144 (1994)

[11] H. SCHOBER, A.J. DIANOUX, J.C. COOK & F. MEZEI - *Physica B* **276**, 164 (2000)

[12] R.E. LECHNER - *Physica B* **276-278**, 67 (2000)

[13] R. BEWLEY & R. ECCLESTON - *Appl. Phys. A* **74**, S218 (2002)

[14] For example: M. RUSSINA, E. MEZEI, R. LECHNER, S. LONGEVILLE & B. URBAN - *Phys. Rev. Lett.* **84**, 3630 (2000)

[15] T.G. PERRING - *In "Frontiers of Neutron Scattering"*, A. FURRER ed., World Scientific Co. Pte. Ltd., Singapore (2000)

[16] T.G. PERRING, G. AEPPLI, S.M. HAYDEN, S.A. CARTER, J.P. REMEIKA & S.W. CHEONG - *Phys. Rev. Lett.* **77**, 711 (1996)

[17] K. ANDERSEN - *In "The ESS Project Vol. IV - Instruments and User Support"*, ISSN 1443-559X (2002)

[18] G.Y. XU, C. BROHOLM, D.H. REICH & M.A. ADAMS - *Phys. Rev. Lett.* **84**, 4465 (2000)

[19] S.F. PARKER - *Internet Journal of Vibrational Spectroscopy* Vol.2 Ed. 1 (http://www.ijvs.com/volume2/edition1)

[20] For examples: P.W. ALBERS, J. PIETSCH, J. KRAUTER & S.F. PARKER - *Phys. Chem. Chem. Phys.* **5**, 1941 (2003); A.J. RAMIREZ-CUESTA, P.C.H. MITCHELL, S. PARKER & P. M. RODGER - *Phys. Chem. Chem. Phys.* **1**, 5711 (1999)

[21] G.F. REITER, J. MAYERS & J. NORELAND - *Phys. Rev. B* **65**, 104305 (2002)

[22] J. MAYERS - *Neutron News* **6**, 2 (1995)

[23] S.M. BENNINGTON, J.C. LI, M.J. HARRIS & D. KEITH ROSS - *Physica B* **263**, 396 (1999)

15

NEUTRON BACKSCATTERING SPECTROSCOPY

B. FRICK
Institut Laue-Langevin, Grenoble, France

1. INTRODUCTION

Neutron backscattering (BS) spectroscopy is an important scattering technique for achieving high energy resolution and thus for accessing long times. What does backscattering mean? The term "backscattering" signifies, that the neutron energy is determined by Bragg reflection from crystals under a Bragg angle Θ of 90° in order to minimize the energy resolution. To avoid misunderstanding among newcomers, we note that backscattering is not related to the scattering process at the sample (BS measurements are of course possible at small scattering angles). Thus "backscattering" concerns only the neutron optics such as monochromators and analyzers. As we will see later, the principle is to define the incident wave vector $|\boldsymbol{k}_i|$ precisely and to vary its length for energy analysis, whereas the final wave vector $|\boldsymbol{k}_f|$ is always kept constant.

The neighbouring neutron instruments of BS in phase space are neutron spin echo (NSE) and time-of-flight (TOF) instruments. Neutron-BS, like TOF, measures in frequency space, thus determines $S(\boldsymbol{Q},\omega)$, whereas NSE measures in time space and determines $S(\boldsymbol{Q},t)$. Neutron-BS is a unique technique for measuring low lying, in frequency peaked excitations down to energy transfers of 0.2 μeV and it is complementary to NSE with respect to quasielastic scattering. Concerning inelastic X-ray spectrometers, there is no equally versatile spectrometer with similar energy resolution today.

We show in figure 1 the energy-momentum-range (phase space or (Q,ω)-range), which can be covered by neutron backscattering. The different dark grey areas belong to the already available regions on reactor-BS (r-BS) instruments. The light-grey areas indicate possible extensions. The Q-range is normally covered for BS in one run simultaneously (e.g. not so for IN15) and one can potentially access a time

F. Hippert et al. (eds.), Neutron and X-ray Spectroscopy, 483–527.

range between 0.01 microseconds and 100 picoseconds. All regions are limited by the instrumental resolution towards the low energies. On TOF spectrometers, in order to achieve a high energy resolution, one has to work with a long wavelength, which consequently limits the maximum Q strongly. We compare the NSE spectrometer IN15 in its longest time resolution mode and IN11 optimized for short times. These spectrometers cover with an excellent energy resolution a wide time range.

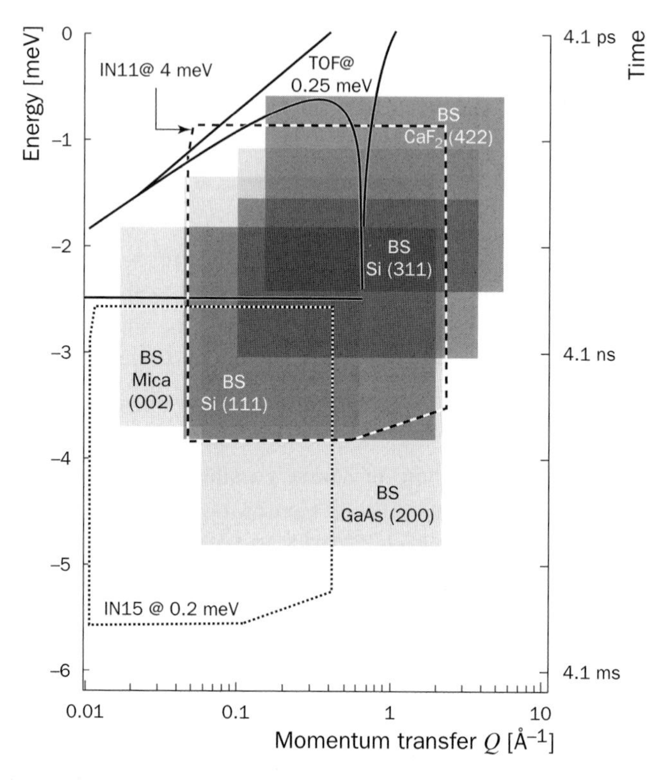

Figure 1 - Region in energy momentum transfer or (Q,ω)-range which is accessible to neutron backscattering instruments. The lowest accessible energy is limited by the instrumental resolution. Regions for existing reactor-BS instruments are presented by the dark areas, their possible extensions by grey areas. NSE and TOF spectrometers are compared at certain incident energies (IN11 at 4 meV, IN15 at 0.2 meV and a typical chopper TOF instrument at 0.25 meV) and the lines show the limits of the accessible range. BS instruments on spallation sources are not shown. Their energy resolution is somewhat better (typically 2-4 μeV) than for TOF spectrometers, but worse than for reactor-BS instruments. The main advantage of BS over TOF is that the excellent energy resolution can also be achieved at high Q values.

2. *REFLECTION FROM PERFECT CRYSTALS AND ITS ENERGY RESOLUTION*

Let us recall that the objective for high-resolution spectroscopy is to know the energy of the neutrons before and after the scattering process as precisely as possible. One way to obtain this information is to use the crystal optics in a favourable way. Therefore it is helpful to get acquainted with the crystal reflection representations in reciprocal space (fig. 2).

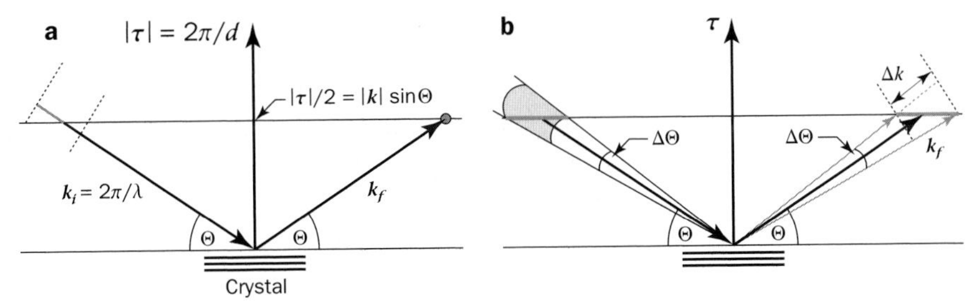

Figure 2 - Reciprocal space representation of the reflection of a neutron beam with divergence $\Delta\Theta$ and a wavelength spread $\Delta\lambda$ at a perfect crystal with reciprocal lattice vector τ. Hypothetical cases: (**a**) a perfectly collimated incident beam leading to a monochromatic beam after reflection and (**b**) a divergent incident beam resulting in a divergent and non-monochromatic outgoing beam.

Let's start with a neutron beam of zero divergence, the simplest hypothetical case. Neutrons with wave vector $\boldsymbol{k}_i$ are incident and reflected under the Bragg angle Θ with respect to the lattice planes (lattice distance d) of a *perfect crystal* (see fig. 2a), to which belongs the modulus of the reciprocal lattice vector $|\boldsymbol{\tau}| = 2\pi/d$. According to Bragg's law, the lattice planes reflect only neutrons if the condition

$$|\boldsymbol{k}_i| \sin\Theta = |\boldsymbol{\tau}/2| \qquad \{1\}$$

is fulfilled (with $|\boldsymbol{k}| = 2\pi/\lambda$ and $|\boldsymbol{\tau}| = 2\pi/d$ we get the more common form of Bragg's law: $2d \sin\Theta = n\lambda$).

As we can see from figure 2a, a perfect crystal would convert a "white" neutron beam into a *perfectly monochromatic beam* – just what we want. Because the intensity is proportional to the phase space volume this also means zero intensity. In reality one has a wider collimation for $\boldsymbol{k}_i$, thus allowing for a finite divergence $\Delta\Theta$ as shown in figure 2b . For a certain wavelength-spread $\Delta\lambda$, a divergent incident beam would then be transformed by a perfect crystal into an outgoing beam with the same divergence $\Delta\Theta$. Unfortunately the reflected beam is no longer monochromatic, which leads to a length difference $\Delta \boldsymbol{k}$ of the outgoing $\boldsymbol{k}_f$ vectors. What can we do about this?

The monochromaticity of the outgoing beam depends on the Bragg angle, thus the spread of Δk becomes small at large Bragg angles, close to backscattering. This is also easily recognizable from the Bragg equation which is plotted in figure 3a for a Si(111) single crystal. At a constant incident divergence the wavelength spread produced after Bragg reflection is considerably smaller close to $\Theta = 90°$ than at small Θ, because the differential of the Bragg equation gives

$$\Delta\lambda/\lambda \sim \Delta k/k \sim \cot\Theta\,\Delta\Theta + \Delta d/d$$

The previous discussion was in one respect too much idealized. We have discussed only a simplified case, the *kinematic* and not the *dynamic scattering theory,* which has to be invoked for BS from perfect crystals. The differentiation of the Bragg equation gives besides the Θ dependence a resolution term which depends on the lattice distance spread $\Delta d/d \sim \Delta\tau/\tau \sim \Delta k/k$. For perfect crystals and within the kinematic theory $\Delta\tau$ would become infinitely small. Compared to the Θ dependent term it might then be neglected for many crystal spectrometers. Not so for close-to-backscattering geometry for which we have to apply the dynamic scattering theory.

Within the *dynamic scattering theory*, $\Delta\tau/\tau = \Delta d/d$ is always finite, even for a perfect crystal. This so-called *primary extinction* term is caused by the interaction of both, the incoming and the reflected waves with the perfect crystal lattice planes (scattering, absorption…) and it can be calculated within the dynamical scattering theory [1]. Its contribution to the energy resolution may be estimated from the width of either the *Ewald* or *Darwin reflection curves* [2], which both have a finite plateau for the reflectivity $R = 1$, and then fall off steeply but also with some wings. We show the Darwin curve for the reflectivity as a function of a variation parameter y [3] in figure 3b (y can represent a variation of $|\boldsymbol{k}|$, of λ or of Θ)

$$R = \begin{Bmatrix} 1 & \text{for } |y| \le 1 \\ \left(|y| - \sqrt{y^2 - 1}\right)^2 & \text{for } |y| > 1 \end{Bmatrix} \qquad \{2\}$$

In our case y arises from a $|\boldsymbol{k}|$ variation only. The plateau width of the Darwin curve presents the primary extinction term $\Delta\tau/\tau$ to the energy resolution and depends on the type of crystal used. For the Darwin reflectivity curve, shown in figure 3b, the plateau width is given by

$$\frac{\Delta\tau}{\tau} = \frac{16\pi F_\tau N}{\tau^2} \qquad \{3\}$$

Here N is the number density of unit cells and F_τ the structure factor corresponding to the lattice vector $\boldsymbol{\tau}$, $F_\tau = \sum_{\text{sites } i \text{ in cell}} b^i_{\text{coh}}\, DWF_i\, e^{-i\boldsymbol{Q}\cdot\boldsymbol{R}_i}$, which depends on the coherent scattering length b_{coh} of atoms in the cell, the symmetry of the unit cell and the Debye-Waller factor (DWF).

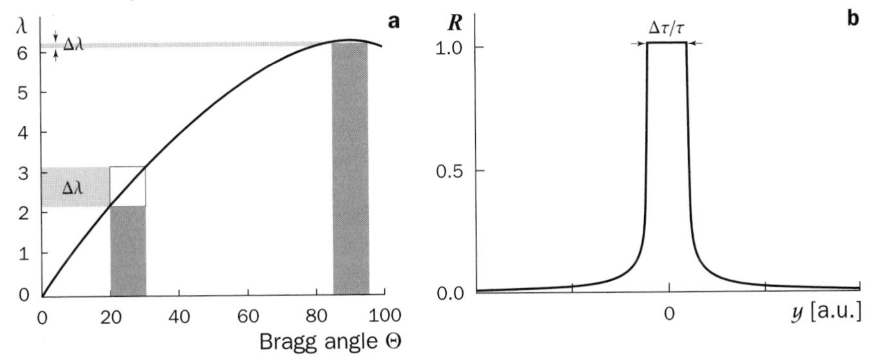

a - Bragg equation: reflected wavelength as a function of Bragg angle for Si(111) crystals, showing the influence of the incident divergence far off and close to backscattering. **b** - Darwin curve, showing the extinction contribution $\Delta\tau/\tau$ to the energy resolution, given by the plateau width of the reflection curve.

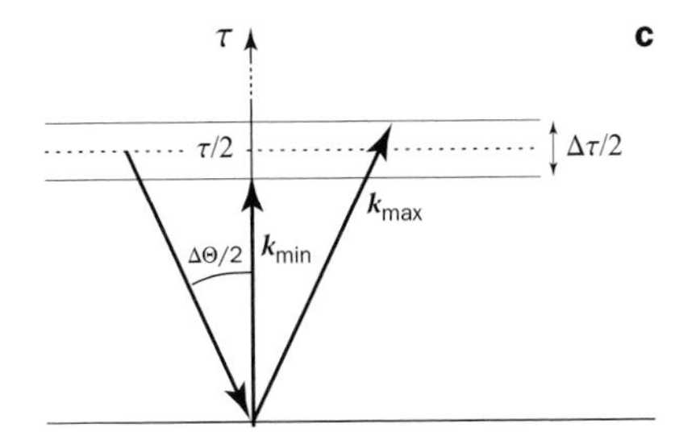

c - Bragg reflection in BS from a perfect crystal and for a divergence $\Delta\Theta$. The divergence contribution Δk_{div} can be estimated from the difference of the shortest and longest wave vectors.

Figure 3

For minimizing the energy resolution one has to find suitable crystals with small enough extinction terms. Table 1 summarizes values of $\Delta\tau/\tau$ for a few perfect crystals which are often applied in neutron-BS; $\Delta\tau/\tau$ is of the order of 10^{-5} (after ref. [4]). The primary extinction is related to the energy resolution by $\Delta E = 2E\dfrac{\Delta\tau}{\tau}$ and from the inverse proportionality of $\Delta\tau/\tau$ to τ^2 in equation {3} one could hope to improve the energy resolution by using higher order reflections for τ. For *neutrons* this is unfortunately not possible! Multiplying $E = \dfrac{\hbar^2(\tau^2/4)}{2m}$ with $\Delta\tau/\tau$ from equation {3} the τ dependence drops out. This is different for *X-rays*, for which $E_X = \hbar k c$, thus one can improve ΔE by using high order reflections on a X-ray-BS spectrometer. For instance Si(11,11,11) monochromators and analyzers are used on ID16 and ID28 at ESRF. In spite of this one achieves today with neutron-BS an energy resolution, which is by three orders of magnitude better, due to the low incoming neutron energy.

Table 1 - Some crystals used in neutron-BS and their relative and absolute extinction contributions, as well as the wavelength for perfect BS [3].

Crystal	$(d\tau/\tau)\times 10^5$	ΔE_{ext} [μeV]	λ [Å] @ $\Theta = 90°$
Si(111)	1.86	0.077	6.2709
Si(311)	0.98	0.077	3.2748
CaF_2(111)	1.52	0.063	6.307
CaF_2(422)	0.54	0.063	2.23
GaAs(400)	0.75	0.153	2.8269
GaAs(200)	0.16	0.008	5.6537
PG(002)	12	0.44	6.70
mica(002)	?	?	19.8
mica(004)	?	?	9.9

A quantitative estimate of the *divergence contribution* Δk_{div} close to backscattering, i.e. for small deviations $\varepsilon = 90° - \Theta \approx 0$, is achieved by calculating the difference between the maximal and minimal length of the k_f vectors (see fig. 3c for $\varepsilon = 0$)

$$\frac{\Delta k_{div}}{k} = \frac{k_{max} - k_{min}}{k} = \frac{1}{\cos(\Delta\Theta/2)} - 1 \approx \frac{(\Delta\Theta)^2}{8} \qquad \{4\}$$

Thus the neutron beam *divergence* contributes in BS only in second order to the energy resolution [5,6]. A similar formalism deals with small *deviations* ε of the incident beam from *backscattering,* which then leads to another contribution $\frac{\Delta k_{ang}}{k} \approx \frac{[\varepsilon + (\Delta\Theta/2)]^2}{2}$ [3,5,6]. The above presented considerations led Maier-Leibnitz in the 60's to propose neutron backscattering for inelastic crystal spectrometers as mean to minimize the energy resolution [7] and to compensate the flux loss by increased divergence.

3. *GENERIC BACKSCATTERING SPECTROMETER CONCEPTS*

For convenience we split the instruments up into a primary and a secondary spectrometer. The *primary spectrometer* defines the incoming neutron beam before the sample. The *secondary spectrometer* analyzes the energy transfer after the scattering process on the sample. As we will see, most *secondary* spectrometers for BS instruments are very similar and differ only in how closely the backscattering condition is fulfilled. However, the *primary* spectrometers look quite different, both, on reactors and on spallation sources. At spallation source-BS instruments (s-BS) one uses in addition to crystal optics TOF techniques for determining the energy transfer. Because TOF spectrometers are discussed in another chapter of this book, we will not go into detail about this.

3.1. NEUTRON OPTICS OF THE PRIMARY SPECTROMETERS OF REACTOR-BS INSTRUMENTS

A *perfect crystal monochromator* is used in the primary spectrometer of reactor-BS instruments to define a highly monochromatic beam, which then is sent to the sample. Mosaic crystals are used either before or after the perfect crystal to select a suitable wavelength band or to deflect the beam and they do in general not affect the energy resolution.

The **first generation BS instruments** (sketch fig. 4a) have a flat perfect crystal monochromator placed in the white beam of a neutron guide, which sends the neutrons slightly out of backscattering back to a mosaic deflector, which then deviates the monochromatic beam towards the sample. The deviation ε from BS results in a degradation of the energy resolution and should be matched with the given primary beam divergence $\Delta\Theta$ of the incident beam. ε is kept small by choosing large distances between the monochromator and the deflector. The IN10 at ILL and the BSS at Jülich (see tab. 2a) are examples for such spectrometers, both characterized by a relatively small incident beam divergence from the straight neutron guides ($\Delta\Theta \sim 1.2°$ for a Ni guide and 6 Å neutrons). At the exit of the primary spectrometer one produces a pulsed neutron beam using a chopper of about 50% duty cycle. The need for this chopper is explained later.

Second generation backscattering instruments use focusing optics to compress the incident beam and thus to increase the neutron flux. Examples are the IN16 at ILL [10] and the HFBS at NIST [11] (tab. 2a). The resulting large divergence would be disastrous in combination with flat monochromators. However, if the focusing optics is combined with a spherical monochromator, then one can maintain perfect backscattering conditions at the latter and only the focal size F and the distance focus-monochromator L should determine the energy resolution (see fig 4b). Technically, the neutron beam has to be coupled somehow into the sample-monochromator line. This is achieved with a mobile deflector crystal, which moves out of the way at the time when the neutrons come back from the monochromator. The moving deflector is built as a chopper, which is needed anyway for maintaining perfect BS in the secondary spectrometer. In the latest generation of BS spectrometers this deflector chopper takes an additional function as *phase space transformation* device (see section 6.3).

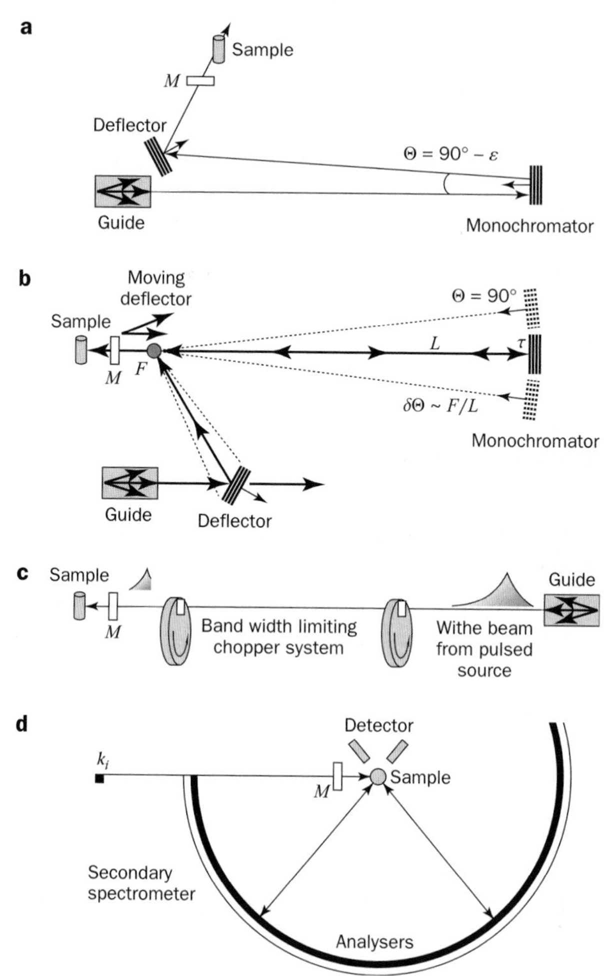

Figure 4 - Schematic view of BS *primary spectrometers* at reactors:
a - first generation spectrometer with monochromator close to backscattering located in the primary beam. **b** - second generation spectrometer in perfect backscattering (after ref. [51]). **c** - spallation source-BS instrument with chopper system to determine TOF. **d** - Schematic view of the *secondary spectrometer* of BS instruments (M = monitor).

Table 2a - Summary of reactor backscattering instruments (updated from ref. **[50]**).

Instruments	Monochromator/ analyser configuration	Wavelength [Å]	E-resolution [μ eV]	E-transfer range [μ eV]	Q-range [$Å^{-1}$]	Comment
IN10, ILL (F)	Si(111)	6.2709	0.8	±16	0.07 - 2	Doppler **[8]**
	Si(311)	3.275	~ 3.5	±28	1.7 - 3.7	Doppler
	$Si_{1-x}Ge_x(111)$	6.3	~ 1	[−2 , +32]	0.07 - 2	offset; x = 10 at-%
	IN10B	variable	~ 1	offset/D ~ 100	0.07 - 2	T-variation ; offset
IN16, ILL (F)	Si(111)	6.2709	0.85/0.4	±14	0.07 - 1.9	Doppler; simultaneous diffraction **[10]**
	Si(311)	3.275	~ 2	±28	1.7 - 3.7	Doppler
IN16B, ILL (F)	*Si(111) & Si(311)*	*see IN16*	*see IN16*	*similar RSSM*	*similar RSSM*	*Doppler/PST project/triple axis option*
BSS, Jülich (D)	Si(111)	6.2709	1	±16	0.2 - 1.9	Doppler **[9]**
	$Si_{1-x}Ge_x(111)$	6.3	~ 1	[−2 , +32]	0.2 - 1.9	offset ; x = 10 at-%
HFBS, NIST (USA)	Si(111)	6.2709	0.8-1.1	±36	0.25 - 1.75	Doppler/PST **[11]**
IN13, ILL (F)	$CaF_2(422)$	2.23	8 - 12	[−100 , +400]	0.7 - 5.2	T-variation **[22]**
RSSM, FRM II (D)	*Si(111)*	*6.2709*	*0.9*	*±32*	*0.12 - 1.87*	*Doppler/PST in construction* **[14]**
	Si(311)	*3.275*	*~ 3.5*	*±60*	*0.23 - 3.41*	*in construction*

Table 2b - Summary of near backscattering instruments at spallation source (updated from ref. **[50]**).

Instruments	Monochromator/ analyser configuration	Wavelength [Å]	E-resolution [μ eV]	E-transfer range [μ eV]	Q-range [$Å^{-1}$]	Comment
IRIS, ISIS (GB)	PG(002)	6.705	15	[–800 , +2200]	0.3 - 1.8	TOF-X; near BS [12]
	PG(004)	3.35	50	[–3000 , +5000]	0.5 - 3.7	TOF-X; near BS
	Mica(002)	19.8	> 1	[–100 , +300]	0.1 - 0.6	TOF-X; near BS
	Mica(004)	9.9	4	[–400 , –500]	0.2 - 1.2	TOF-X; near BS
	Mica(006)	6.6	11	[–500 , –1000]	0.3 - 1.8	TOF-X; near BS
OSIRIS, ISIS (GB)	PG(002)	6.69	25	[–800 , –5000]	0.3 - 1.8	TOF-X ; near BS; polarisation analysis [51]
LAM-80ET, KENS (J)	Mica(002)	19.6	1.5	±30	0.08 - 0.55	TOF-X; near BS [24]
	Mica(004)	9.8	5.6	[–400 , +500]	0.17 - 1.1	TOF-X; near BS
	Mica(006)	6.53	15	[–1000 , +1500]	0.25 - 1.65	TOF-X; near BS
MARS,PSI,project (CH)	*Mica(002)-(008)*	*5 D-ch*	*1 - 100*	*up to 10000*	*0.2 - 3.5*	*TOF-X; 5 disc choppers; in construction* [52]
SNS-BS-project (USA)	*Si(111)*	*6.267*	*2.2*	*±258*	*0.1 - 2.0*	*in construction* [25,53]
	Si(311)	*3.273*	*10*	*±279*	*0.2 - 3.8*	*in construction*
SNS-nsBS-project (USA)	*Mica(002)*	*20*	*0.22*	*±60*	*0.05 - 0.6*	*project* [54]
JSNS (J)	*not yet defined*					*2 BS planned*

3.2. NEUTRON OPTICS OF THE PRIMARY SPECTROMETER OF SPALLATION SOURCE-BS INSTRUMENTS

The s-BS instruments are inverted TOF spectrometers or TOF-X spectrometers (X = crystal) and therefore BS plays only a role in the secondary spectrometer (see sketch fig. 4c or IRIS as an example [12]). On s-BS instruments a pulsed, white beam arrives from the moderator source. A chopper system restricts the pulses to a certain wavelength band. This adds flexibility to these spectrometers, because the incident energy can easily be shifted with respect to the fixed analyzer energy. The price to pay is an additional time-of-flight resolution term, $\Delta t/t$, which usually dominates over the resolution terms from the crystal optics. Therefore one has to choose long flight paths in order to minimize this contribution. Beam focusing is achieved with focusing neutron guides and therefore the incoming neutron beam may have a large divergence.

3.3. SECONDARY SPECTROMETER

We idealize for a point sample and a point beam size with zero divergence. With this simplification most neutron backscattering spectrometers at reactors fulfil the perfect backscattering condition on the secondary side. A schematic view of the secondary spectrometer is shown in figure 4d. The neutrons impinging onto the sample at point S are scattered into all directions. They are analyzed for a fixed final wavelength λ_f using perfect crystals, arranged on a spherical segment (we idealize for infinitely small crystals) with the sample as centre and on only one side of the incoming beam. Neutrons which fulfil the Bragg condition (eq. {1}) for $\Theta = 90°$, i.e. with $k_i = \pi/d_A$, are focussed back to the sample. Usually a detector array, located on the opposite side of the analyzer sphere with respect to the sample, but as close as possible to it, collects the analyzed neutrons. For the case that the monochromator and analyzer crystals are identical, that they are both at rest and at the same temperature, only elastically scattered neutrons can arrive at the detector (see section 9.1).

Finally, those neutrons, which are scattered directly from the sample into the detector, i.e. without energy analysis, have to be eliminated. Now we see the reason, why the incident neutron beam must be pulsed for a perfect BS geometry. On reactors one uses choppers for absorbing about 50% of the time the incoming neutrons, which also sends an electronic signal for each transmitted neutron pulse. This signal allows to close the detector electronically during the passage time of the intense incoming neutron beam at the sample and to open the detector again at the time when the first elastic scattered neutrons are expected back from the analyzers.

In some cases, e.g. at spallation sources, the perfect backscattering condition is voluntarily abandoned. Then the detectors are located out of BS and are screened from the direct view to the sample and the primary beam. "Out-of-BS" one needs at r-BS instruments no longer a pulsed neutron beam, and thus one can increase the flux at the sample by about a factor of two. Yet, already small deviations from BS (eq. {4}) degrade the energy resolution dramatically and lead to at least one asymmetric wing of the resolution function, which means also that the dynamic range, which is for most r-BS instruments already quite narrow, will be even more restricted. This is the reason why the "out-of-BS configuration" is found mainly on s-BS instruments, which offer a much wider energy transfer range (see tab. 2b).

4. TOTAL ENERGY RESOLUTION OF THE SPECTROMETERS

Now we can summarize the different contributions to the energy resolution. The precision at which the neutron energy can be obtained, either in the primary or in the secondary spectrometer, is

$$\frac{1}{2}\frac{\Delta E}{E} = \frac{\Delta\lambda}{\lambda} = \frac{\Delta k}{k} = \sqrt{\left(\frac{\Delta\tau}{\tau}\right)^2 + \left(\cot\theta\cdot\Delta\theta\right)^2 + \left(\frac{\Delta t}{t}\right)^2} \qquad \{5\}$$

The first two terms under the root arise from the crystal optics and can be derived from Bragg's equation. $\Delta\tau$ is defined by the crystal properties and $\Delta\Theta$ by the neutron beam divergence or deviation from BS. The third term presents the neutron flight time contribution to the energy resolution, which we may neglect for reactor-BS spectrometers (r-BS), but which is important for spallation source-BS instruments (s-BS). *The total energy resolution of the instrument is then given by the convolution of the resolution values in the primary and secondary spectrometer.*

5. HOW TO DO SPECTROSCOPY?

5.1. SPECTROSCOPY ON REACTOR BASED INSTRUMENTS

For spectroscopic investigations on r-BS instruments one varies the incoming energy without changing the Bragg angle $\Theta \sim 90°$, i.e. by variation of the modulus of $\boldsymbol{k}_i$, and by changing the lattice constant d of the monochromator. Two different methods are applied: *Doppler motion*, and *monochromator heating*:

- The backscattering monochromator is moved with a velocity $\boldsymbol{v}_D$ parallel to the reciprocal lattice vector $\boldsymbol{\tau}$. Thus the energy of the reflected neutrons is modified via a longitudinal *Doppler effect*: $\frac{\delta E_D}{E_i} \approx 2\frac{v_D}{v_i}$ (in first order and for $v_D \ll v_i$, the neutrons "see" a different lattice constants in case of a moving lattice). One registers the scattered neutrons as a function of the Doppler velocity v_D. The maximum achievable Doppler speed v_D determines the maximum energy transfer that can be reached (fig. 5a and tab. 2). The Doppler velocity is varied periodically (sinusoidal function in $v(t)$ diagram of fig. 5a) and for identical analyzer- and monochromator-crystals (same temperature and orientation) the corresponding energy transfer is centred around E = 0. If we count a neutron at time t_D then we have to know which speed $v(t_M)$ the monochromator had earlier at time $t_M = t_D - T_{MD}$ (T_{MD} = neutron flight time from the monochromator to the detector) when the neutron was at the monochromator. Neutrons are then accumulated into the corresponding equidistant velocity channels $v(t_M)$ (fig. 5a; rhs for detector and lhs for monitor). The neutron flight times between the monochromator and the detector or monitor, T_{MD} and T_{MM} are assumed to be constant, because k_f is fixed and we neglect the variation of the monochromator position and the small change of neutron speed (valid for IN10, IN16 and BSS; HFBS and RSSM have 3 times larger monochromator amplitudes and higher speed, thus T is no longer a constant [13]). Data are acquired with constant time bins of typically several 10 μs and thus, for a non-linear Doppler velocity, a constant rate signal (e.g. background) shows up as the inverted function of the periodic Doppler motion (see monitor spectrum in the left vertical graph of fig. 5a). The monitor spectrum indicates how much time is spent measuring each velocity channel, $g(t_C)$, multiplied by the distribution function $g(v_0)$, the number of neutrons available in the primary beam around velocity v_0. The Doppler velocity profile is often sinusoid-like [8-10], but in order to get a more equal measuring time distribution over the spectrum one might try to build Doppler drives which approach a triangular velocity profile [11,14]. As for the sample scattering, elastically scattered neutrons will show up as a peak in the middle of the spectrum, the width of which corresponds to the instrumental resolution $R(\omega)$. The right hand side of fig. 5a shows that a Lorentzian-like scattering law will be distorted due to the non-linear Doppler motion and thus each spectrum has to be divided by the monitor spectrum (assuming a low background). An offset of the measured energy transfer range from zero can be achieved by choosing Doppler-monochromator-crystals with different d-spacing (e.g. $Si_{1-x}Ge_x$), orientation or temperature from that of the analyzer system (typically Si(111)).

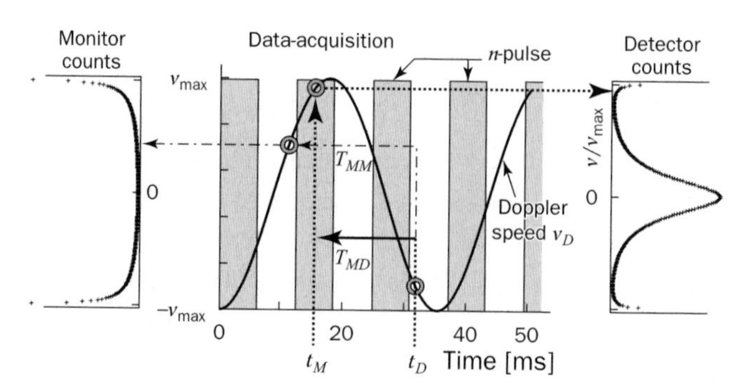

Figure 5 a - Data acquisition on a reactor-BS instrument with Doppler monochromator (Doppler frequency 14 Hz sinusoid; pulse frequency 81 Hz rectangular). Neutrons which are counted at time t_D have been at the moving monochromator at time $t_M = t_D - T_{MD}$. The flight time T_{MD} is assumed to constant.
l.h.s.: monitor spectrum; r.h.s.: sample scattering for a Lorentzian scattering law.

- *The lattice distance d of the backscattering monochromator is changed by temperature* using monochromators with a large thermal expansion coefficients α: $d(T) = d_0(1 + \alpha T + O(2)\ldots)$. Thus the energy of the reflected neutrons changes with temperature: $\frac{\delta E}{E_i} = \frac{\delta d(T)}{d_0}$. One registers the detected neutrons as a function of the monochromator temperature (tab. 2 and fig. 5b). The energy transfer range is limited by the maximum lattice spacing change which can be reached at the monochromator before damaging the crystal at high temperature.

Suitable crystals have, besides a reasonably large thermal expansion coefficient, a good energy resolution and sufficient intensity (eq. {5}). The final choice of crystals is based on the energy range of interest for the physical problem. Figure 5b shows the principle of data acquisition. The monochromator temperature is varied slowly within a cryo-furnace and the neutrons are counted as a function of its temperature. With increasing monochromator temperature the lattice distance increases and the reflected wavelength (energy) becomes longer (lower). Deflector crystals deviate the neutron beam either before or after the monochromator. For a fixed deflector angle, the reflectivity and thus the flux on the sample would decrease and therefore the take-off angles of the deflectors have to be re-aligned continuously. A division of the sample spectrum by the monitor spectrum will eliminate these re-alignment discontinuities (wiggles on the monitor; l.h.s. of fig. 5b). The main practical disadvantage is that the whole spectrum is only known after typically a few hours of scanning time, whereas with the Doppler drive method the spectrum is scanned periodically a few times per second.

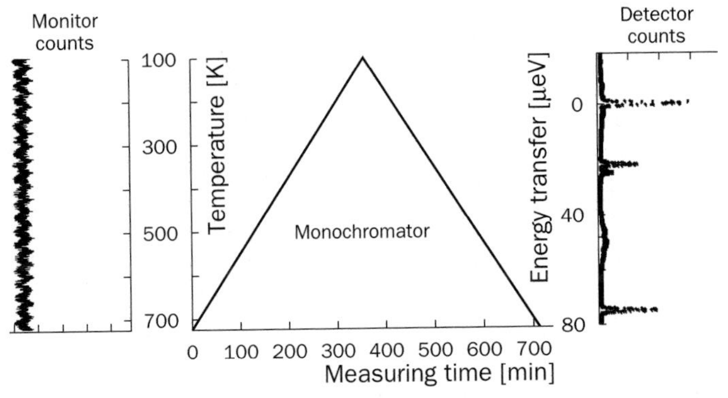

Figure 5 b - Spectroscopy with a temperature controlled monochromator offers a wide energy range with μeV energy resolution. A simulation of a scan with 6 hours heating and 6 hours cooling (about the minimum time needed) of the monochromator is shown. The l.h.s shows the monitor spectrum, the r.h.s. a sample spectrum as a function of the monochromator temperature or energy transfer. Wiggles in the monitor spectrum are due to the reflectivity curve of the mosaic deflector.

5.2. SPECTROSCOPY ON SPALLATION SOURCE-BS INSTRUMENTS

As mentioned, on s-BS instruments the neutron energy is determined by TOF. The analyzers accept only a fixed final neutron energy, which means that the flight time between the last element that could have modified the neutron energy, i.e. the sample, and the detector is known. At the detector the neutrons are then sorted according to their arrival time with respect to the elastically scattered neutrons (see fig. 5c). In order to keep the flight time deviations small, the sample-analyzer-detector distance should be the same for the whole analyzer surface. When operating out-of-BS-like on s-BS, the analyzer surface should therefore not be perfectly spherical. Higher order scattering must eventually be removed with filters in front of the analyzers (see IRIS for PG-analyzers), if the analyzer crystals themselves have not forbidden higher order reflections [like Si(111)].

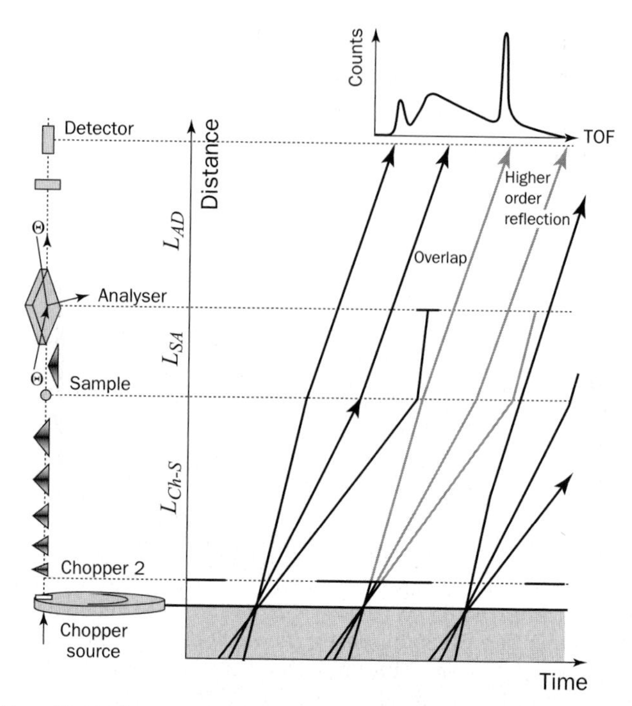

Figure 5 c - Flight distance-time diagram for an inverse TOF spectrometer. s-BS are of that type with the analyzer oriented close to BS (after ref. **[51]**).

6. MORE DETAILS ON OPTICAL COMPONENTS

6.1. BS MONOCHROMATORS AND ANALYZERS

Flat monochromators of relatively small size (see fig. 6, monochromator c) are used in the primary spectrometer at first generation BS instruments, in combination with low divergent beams from straight neutron guides. A small monochromator size simplifies technical issues, like moving or heating of the monochromator. As we have seen in figure 4a, small deviations from backscattering are accepted at the monochromator. On a flat crystal the finite size and the divergence of the beam and the size of the flat crystal itself will contribute to a resolution broadening (see eq. {4}).

One can decouple the divergence and energy resolution with **spherical crystals**. In that idealized case the straight line for $\tau/2$ in figure 3b would be replaced by a circle of radius $\tau/2$, corresponding to a spherical perfect crystal ($\forall\, \Delta\Theta : k_{\max} = k_{\min}$).

Single crystals are of course never spherical but one tries to approach the ideal situation by using sufficiently small flat crystals which cover a spherical support (fig. 6, monochromator a). Deforming crystals spherically is another possibility. Spherical crystal arrays can be found on the secondary side of all BS spectrometers, but on the monochromator side only in second generation r-BS instruments. Deviations from perfect BS are again induced by the finite beam, crystal and sample size. Figure 6 shows three different types of monochromators used on the r-BS instruments IN16 (fig. 6a,b) and IN10 (fig. 6c).

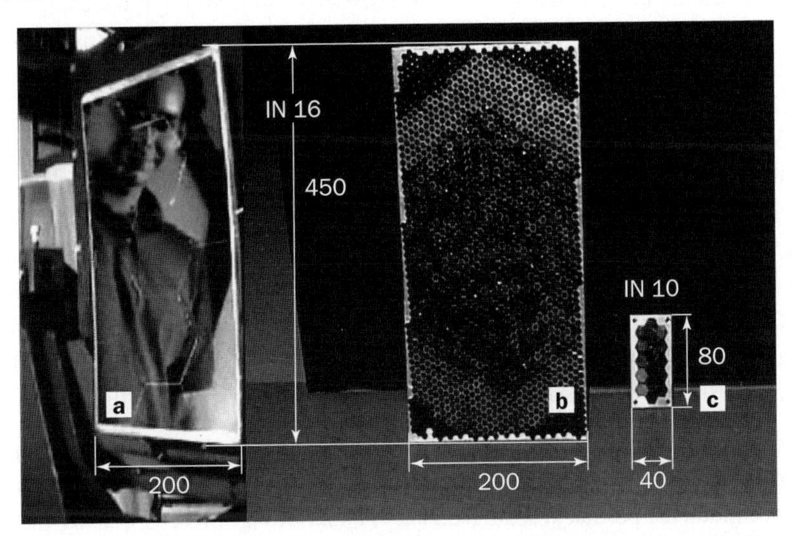

Figure 6 - BS monochromators. **Left** - IN16, **a** - *spherical* and "polished curved monochromator", glued with flat crystals of 4×4 mm^2 each. **b** - chemically etched, 10×10mm^2 crystals. **Right** - **c** - *flat* IN10 monochromator (all measures in mm)

6.2. *HOW TO OBTAIN THE BEST ENERGY RESOLUTION IN BACKSCATTERING?*

For r-BS instruments the TOF resolution term in equation {5} can be neglected, but the crystal term is of outmost importance, especially if one aims to achieve the highest energy resolution, let's say better than HWHM = 0.5μeV. For s-BS instruments the TOF resolution terms dominates.

To achieve the highest possible energy resolution one uses on r-BS polished and non-deformed small crystals (*polished crystal setup*). The *surface quality* of the crystals plays some role as well, because it can be regarded as an imperfection of the crystal lattice and one expects additional reflectivity in the wings of the resolution function. Etching can remove these defects.

Further resolution degradation is due to *misaligned crystals*. Misaligned crystals on the analyzer or monochromator sphere lead to a Bragg angle $\Theta < 90°$, which means that a shorter than the nominal $\lambda = 2d$ wavelength is reflected. Tails, displacement and broadening of the resolution function are the consequence. The elastic line will be shifted to the neutron energy loss (gain) side for a misalignment on the monochromator (analyzer) side. This is also the reason why the size of flat crystals on a spherical support must be small enough to cause an acceptable angular deviation.

Elastic deformation or bending of the crystals will also lead to a broadening of the resolution function, but enhances at the same time the reflectivity (more lattice planes with different d-spacing contribute to the reflection). Because sufficiently high intensity is always a concern with high-energy-resolution spectrometers, one applies crystal deformation by purpose. The additional resolution term from elastic bending of the crystals is proportional to $\Delta d/d \sim \mu t/R$ (μ = Poisson ratio, $\mu \sim 0.41$ for Si, and t = crystal thickness, R = radius of analyzer sphere). Nearly all BS spectrometers operate today with spherically deformed, unpolished crystals (*deformed crystal setup*). Especially successful is the use of large deformed single crystal wafers, which guarantee a rather homogeneous strain pattern over large surfaces. A Gaussian-like resolution function is the consequence and a high intensity [15]. Measured resolution functions from IN16 are shown in figure 7.

We mention only briefly other ways to increase $\Delta\tau/\tau$: by *temperature gradients* within the crystal along the penetration trajectory of the neutrons (IN13), by lattice distance gradients produced with *gradient crystals* of e.g. SiGe [16], or by internal strains, produced by *doping of crystals* (e.g. oxygen doping). Furthermore *ultrasound* can potentially be used to tailor the width and reflectivity of the monochromator crystals. This was shown for a small monochromator at IN10 [17].

6.3. *MOSAIC CRYSTAL DEFLECTORS AND PHASE SPACE TRANSFORMER*

Now let us consider *mosaic crystals*, which have a distribution of small perfect crystallites around the reciprocal lattice vectors $\boldsymbol{\tau}$ and therefore a much higher acceptance for incoming neutrons, thus offer a higher flux. As mosaic crystals one uses mostly graphite (e.g. pyrolytic graphite PG(002) which has a high reflectivity R, $R \sim 90\%$ for $\lambda = 6$ Å). The mosaic η is adapted to accept the primary beam divergence and wavelength band. Some care has to be taken to avoid a too large vertical beam divergence $\Delta\Theta^V$ after deflection ($\Delta\Theta^V \sim 2\eta \sin\Theta$) if a large isotropic mosaic η is used.

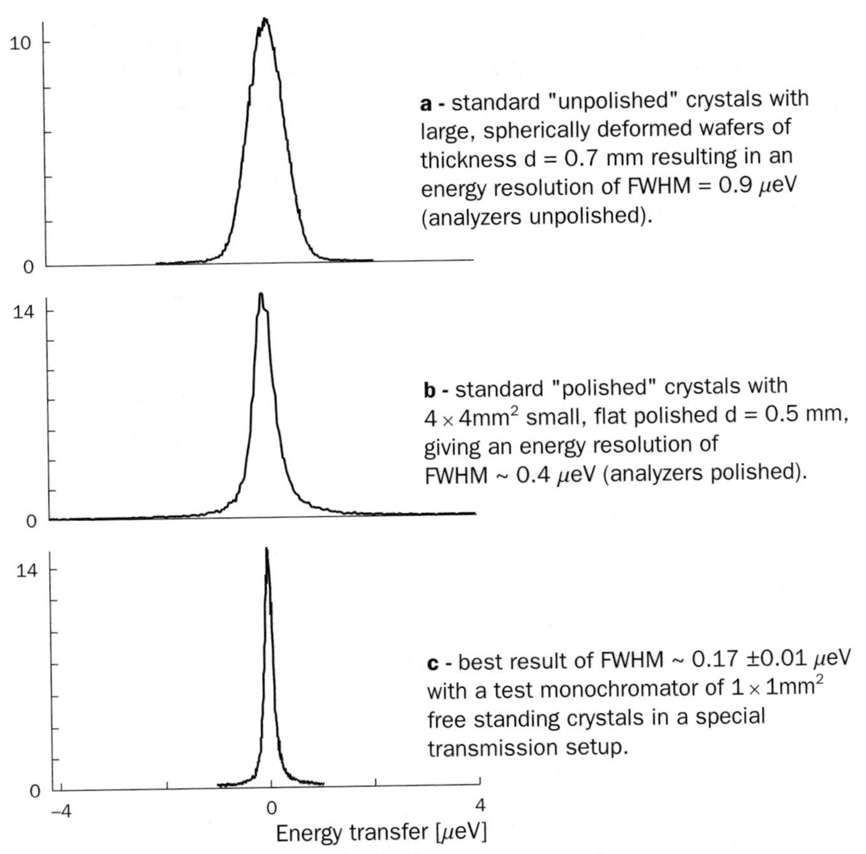

Figure 7 - Energy resolution function of different type of Si(111) BS monochromators at IN16 (after ref. **[15]**).

For this purpose several (n) graphite crystals with a low intrinsic mosaic η are assembled to an artificially wider horizontal mosaic $\eta^h = n\,\eta$, maintaining at the same time a narrow vertical mosaic $\eta^V = \eta$.

On IN16 graphite(002) deflectors are mounted before the monochromator and have the task to offer an optimized wavelength band and divergence to the *moving* perfect crystal monochromator (fig. 3a).

On IN10 (or IN13) the deflector is located in the monochromatic beam (fig. 4a). The deflector reflectivity curve must be broad enough to reflect all Doppler shifted monochromatic neutrons. For the heated monochromator the graphite take-off angle is re-aligned with changing monochromator temperature. The energy resolution is not affected by the deflector. On the IRIS s-BS instrument **[12]** PG(002) and PG(004) are used as analyzers near-BS.

On second generation BS spectrometers the second deflector (fig. 4b) is mounted as segments on the circumference of a chopper disc, alternating with open segments. The main aim of such a *deflector-chopper* is to allow for exact backscattering conditions (see fig. 4b). The rotation speed is matched to the neutron flight time for the distance chopper-monochromator-chopper, the distance sample-analyzer-detector and for the angular width of the reflecting graphite segments on the chopper. The IN16 deflector-chopper has two reflecting segments and a rotation speed of 2800 rpm, corresponding to a crystal speed of 70 m/s.

Additional intensity gain was predicted [18] for a sufficiently fast moving mosaic deflector crystal. This device is called a *Phase Space Transformation* (*PST*) *chopper*. To understand how it works, we have a look on figure 8. A static mosaic crystal transforms a polychromatic incoming beam of zero divergence into a divergent and polychromatic beam. The phase space element after reflection is correspondingly a curved line or in case of a divergent beam a curved reflected phase space element (fig. 8a). We note, that the latter is inclined with respect to the outgoing wave-vector $\boldsymbol{k}_f$.

After deflection from the mosaic crystal, a spherical perfect crystal monochromator shall collect the neutrons. Such perfect crystal monochromator will accept only neutrons with wave vectors located within a thin spherical shell around the origin in phase space. The area, which is common to this shell and the inclined phase space element from the mosaic deflector, is a measure for the monochromatic neutron flux. One can see from figure 8 that the cut is not optimal. A horizontal shift would be needed for a much better alignment of the deflector phase space element with respect to the acceptance of the perfect monochromator. Such a shift can be produced using a deflector crystal which moves with a speed $\boldsymbol{v}_{\text{cryst}}$, perpendicular to the reciprocal lattice vector $\boldsymbol{\tau}$. Now reflection takes place in the moving crystal frame (fig. 8b). Therefore we translate the phase space element before reflection by $\boldsymbol{v}_{\text{cryst}}$. A wider energy band is accepted in the moving crystal system, which transforms into a wide divergence after reflection. We then transform the reflected wave vector $\boldsymbol{k}'_f$ back to the laboratory system, which furthermore results in a more favourable orientation of the reflected phase space element if these neutrons are offered to a spherical BS monochromator and thus it corresponds to a gain of intensity. Of course Liouville's theorem is valid and the phase space volume is conserved during reflection. We have converted a wide energy band before reflection into a large divergence afterwards. Such a PST chopper was first built at the HFBS, NIST [11,19] and a gain factor of 4.2 was measured. The RSSM [14] and the new IN16B will also implement such a PST device.

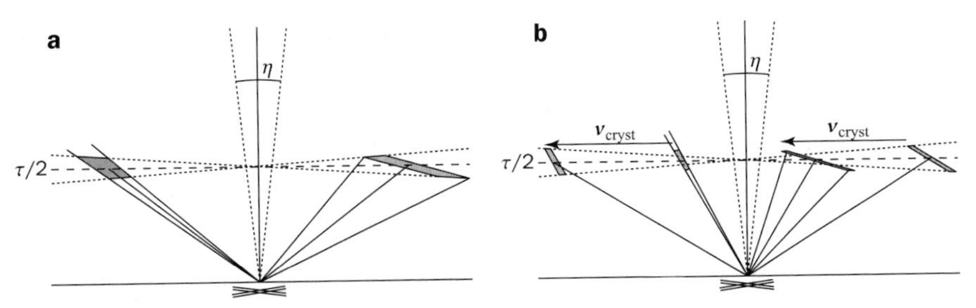

Figure 8 - Schematic reciprocal space representation of the reflection **a** - from a static mosaic crystal, **b** - from a moving mosaic crystal - the principle used with the Phase Space Transformation chopper [18].

6.4. *NEUTRON GUIDES*

A few words on guides are important, because they contribute to the beam divergence and thus to the energy resolution. Neutron guides are used in BS to deliver the beam over some distance at a desired place and to shape (e.g. compress) the beam. Only neutrons with a reflection angle smaller than the critical angle $\Theta_c = \lambda\sqrt{nb/\pi}$ are totally reflected and transported (n = atomic density and b = coherent scattering length; for the natural abundance of nickel isotopes ^{nat}Ni $\Theta_c \sim 0.1° \lambda$, defined as $m = 1$). The critical angle determines the divergence $\Delta\Theta = 2\,\Theta_c$ at the exit of a straight guide. "Supermirror" multilayer coatings are used to increase the critical angle above the value of ^{nat}Ni ($m = 1$) and reach today $m \sim 4$. Usually the reflectivity of such coatings is lower than that of Ni, but can nowadays reach $R \sim 90\%$ for $m = 3$ [20]. High m values are needed for short wavelength or if converging guides are used to compress the beam, in the latter case because the guide inclination angle adds for every reflection to the incident angle.

6.5. *HIGHER ORDER SUPPRESSION*

Undesired higher order reflections must be suppressed, both in the primary beam and in the secondary spectrometer. This can be achieved with filter crystals (e.g. beryllium), chopper or selector systems, crystals with forbidden or weak higher order reflections, selection by the neutron flight time or with combinations of these.

For example, on reactors in first generation BS instruments the monochromator is located in BS in the white primary beam and it would reflect neutrons with $\lambda = 2d/n$, $n = 1,2\ldots$ Choosing crystals with forbidden second order reflections (e.g. for Si, the Si(111) is alloyed, the Si(222) reflection is forbidden), the problem can be reduced, because further higher orders may be weak, given the low flux at the corresponding wavelength in the cold neutron guide. Accepting some loss, a late opening of the detector gate for discriminating 1st and 2nd elastically scattered neutrons, is in some cases another possibility. On IN16, mainly due to background reasons, a beryllium filter is used before the deflector chopper, though the Si(111) monochromator would not reflect the second order. At HFBS, RSSM and future IN16B a velocity selector removes higher orders. On s-BS spectrometers, incident higher harmonics can be avoided by tuning the width of the incident wavelength band with choppers. Additional cooled filters avoid higher harmonics. One can remove slower neutrons by a pulse suppression chopper.

6.6. Q-RESOLUTION

In order to gain intensity, large solid angles are analyzed after scattering, which relaxes the divergence demand for the incoming beam. In fact the analyzer space angle seen by one detector determines mainly the Q-resolution. A 2D-detector, which offers a better Q-resolution, might be used as well. Collimation of the analyzed beam or covering a part of the analyzer sphere with an absorbing screen allows to cut the Q-resolution down to the divergence contribution of the incident beam. The latter can be reduced further as desired by using a smaller monochromator surface and/or a small sample. The low limit of the Q-resolution is in most cases given by the minimal acceptable count rate.

6.7. A SECOND TIME THROUGH THE SAMPLE?

In exact BS a part of the neutron beam which comes back from the analyzers has to pass again through the sample to reach the detector. Thus there is a certain probability that the neutron is scattered in a second scattering process towards a "wrong" detector and energy channel. With some assumptions we can estimate that this introduces only a small correction f_S. For a reasonably high sample transmission T and assuming incoherent, i.e. isotropic, scattering σ_{inc}, $f_S \sim \sigma_{\text{inc}} (1 - T) f^{2\text{nd}} \Omega_D/(4\pi) \sim 10^{-3} - 10^{-4} \sigma_{\text{inc}}$; where Ω_D is the space angle accepted by the detector and $f^{2\text{nd}}$ the fraction of neutrons passing a 2nd time through the sample. Furthermore these neutrons are not correlated with the Doppler velocity (see monitor spectrum in fig. 5a) and therefore contribute as a flat background. One estimates then the signal (from 1st scattering) to noise (from 2nd scattering) to be of the order of 10^4.

7. EXAMPLES FOR BACKSCATTERING INSTRUMENTS

7.1. REACTOR INSTRUMENTS

In this section we present different realizations of backscattering instruments. Based on the previous sections a brief description should be sufficient to understand these instruments.

♦ IN10 at ILL

This is an example for a first generation backscattering instrument (see fig. 9a, the BSS Jülich is similar; tab. 2a). Both instruments were developed following the pioneering work by Maier-Leibnitz, Alefeld and Heidemann at Munich, Jülich and at the ILL, Grenoble (see e.g. ref. **[6]** and references therein). The IN10 monochromator is placed in the white beam at the end of a neutron guide and it is oriented slightly out of backscattering ($\varepsilon = 0.3°$). It reflects monochromatic neutrons to a PG(002) deflector crystal which is located at about 8 m distance above the primary beam and which sends the neutrons to the sample. The perfect monochromator crystals (by standard Si(111) with size 8×4 cm^2) are mounted on a Doppler drive ($v_{max} = 13.5$ Hz) and can be changed easily to the offset monochromators $Si_{0.9}Ge_{0.1}(111)$, $CaF_2(111)$ or Ge(111). The energy resolution for the deformed Si(111) analyzers is about 0.82 μeV and can be improved for a part of the analyzers to 0.45 μeV for polished crystals accepting a reduced flux (tab. 2a).

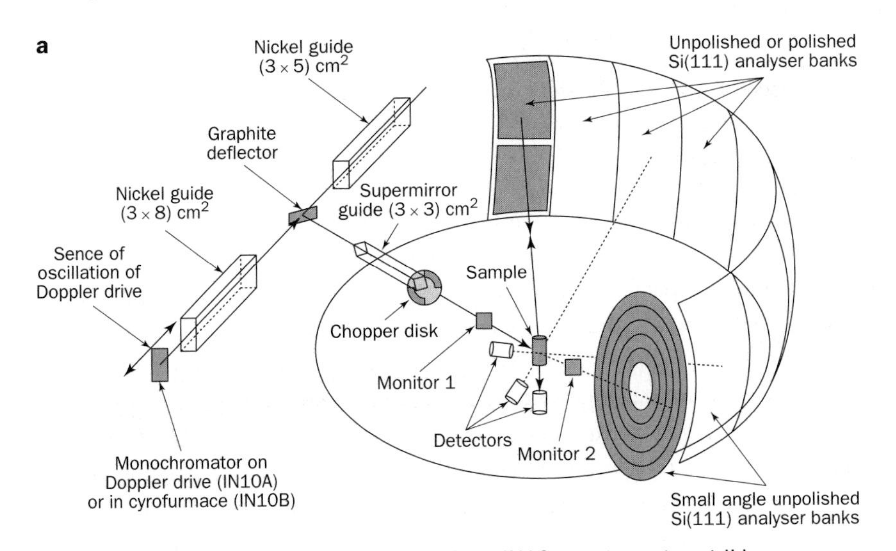

Figure 9 - a - Schematic view of the IN10 spectrometer at ILL.

Both, monochromator and analyzers at large angle can be exchanged for Si(311), which allows an access to higher Q values. Uniquely on IN10 one can replace the Doppler drive with a temperature-controlled monochromator in a cryo-furnace (IN10B) (see section 5.1). Several crystals with large thermal expansion coefficients allow for a large energy offset (e.g.750 μeV; see fig. 9b) at a constant energy resolution of about 1 μeV and a typical dynamic range of about 120 μeV.

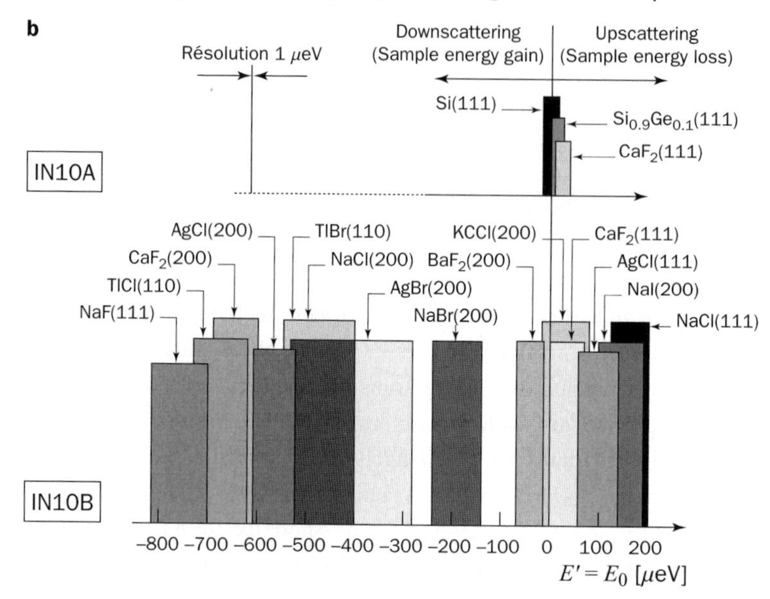

Figure 9 - b - Monochromators and energy transfer-range on IN10 for the Doppler drive (IN10A) or for the heated monochromators (IN10B), assuming a Si(111) analyzer sphere. Energy transfer-ranges of IN10B correspond to a monochromator temperature-range of $80 \leq T\ [K] \leq 700$.

♦ **IN16 at ILL**

This is the first second generation high flux BS instrument; it maintains perfect backscattering conditions at both the monochromator and the analyzer and it is currently the BS spectrometer with the best signal to noise ratio at an energy resolution below 1 μeV (fig. 10). A double deflector arrangement of graphite(002) mosaic crystals offer a suitable wavelength band to the BS monochromator. A large neutron beam is compressed by a vertically focusing PG(002) deflector and the following supermirror neutron guide from $h \times v = 60 \times 120\ mm^2$ down to $27 \times 27\ mm^2$, close to a second PG(002) deflector chopper. This graphite deflector chopper has on its circumference alternating open and reflecting 90º segments and rotates with a speed of 2400 rpm (crystal speed ~ 70 m/s, which is not sufficient for PST; see above). After the second deflector the divergent beam is reflected in BS by a large spherical perfect crystal monochromator (height = 45 mm; width = 25 mm). Polished and deformed Si(111) or Si(311) crystals can be chosen for the monochromator. The monochromatic beam travels back through the deflector chopper towards the sample.

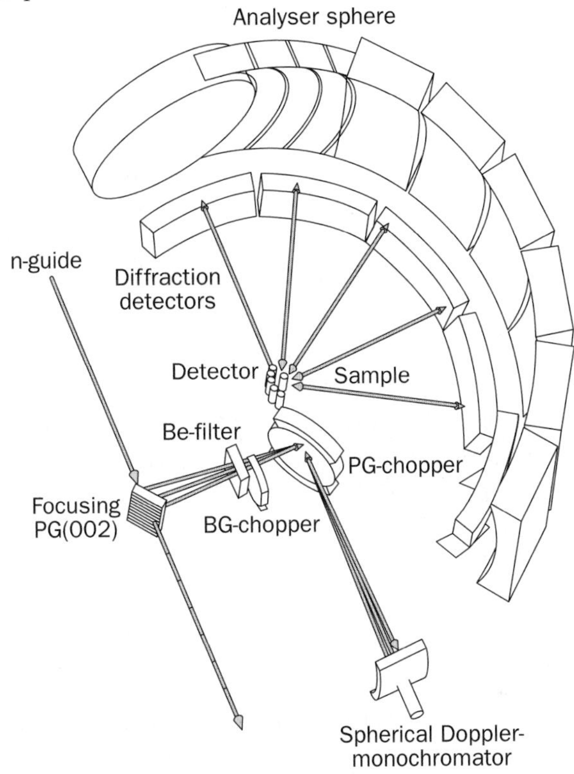

Figure 10 - Schematic view of the IN16 spectrometer at the ILL

The scattered neutrons are analyzed on a sphere of 2 m radius (1.4 m high), covered with Silicon single crystals. For the "polished" Si(111) version more than a quarter million of small crystals of $4 \times 4 \text{mm}^2$ size are covering a part of the analyzer sphere ($\sim 4.5 \text{ m}^2$). For "unpolished" Si(111) or Si(311) large wafers are glued under deformation onto the total analyzer sphere. Neutrons with wavelength $2d/\lambda$ are sent back from the analyzers to a multi-tube detector array with 20 tubes. Additionally, up to four single detectors can be arranged out of the scattering plane.

Three different monochromator and analyzer configurations can be chosen on IN16: perfect Si(111) with the energy resolution of FWHM $< 0.42\ \mu\text{eV}$, deformed Si(111) crystals with a resolution of FWHM $\sim 0.85\ \mu\text{eV}$ and Si(311) crystals with an energy resolution of about FWHM $\sim 2\ \mu\text{eV}$. The Q and energy transfer ranges are given in table 2a. Uniquely, a diffraction detector array is located below the analyzer sphere, which allows for a simultaneous diffraction measurement and it can also be moved into the scattering plane for improved Q resolution.

- **HFBS at NIST [21]**

Further important progress in instrumentation was made with the NIST HFBS spectrometer, where for the first time a PST chopper (see section 6.3) was introduced. Thus a neutron flux gain of about a factor 4 due to PST was achieved [19]. The layout of the spectrometer is similar to the one shown in figure 10 for IN16, if one imagines the instrument to be placed at the end of a focusing neutron guide, with the PG chopper as the first component. The neutron wavelength band, which is offered to the PST chopper, is prepared by a combination of a velocity selector and focusing neutron guides. The dynamic range on this instrument is also the widest which was reached up to now with a Doppler drive: about $\Delta E \approx \pm 36\,\mu eV$ at 1 μeV energy resolution. For lower Doppler speeds corresponding to $\Delta E \approx \pm 11\,\mu eV$ the resolution decreases to about FWHM = 0.8 μeV. The flux on the sample is currently the highest available in n-BS, although with a somewhat worse signal to noise ratio than on IN16.

- **Cold neutron-BS projects**

A new high flux r-BS instrument, RSSM, is under construction at the FRMII, Munich. Similarly to the NIST-HFBS this instrument will use PST to increase the flux and will also use a large dynamic range. The count rate will also increase by the use of a considerably larger analyzer surface. A new linear motor Doppler drive concept was developed, which should allow for exceeding an energy transfer of $\Delta E \approx \pm 32\,\mu eV$ at 6.271 Å, i.e. in the Si(111) configuration. The RSSM is expected to be operational in 2006. Facing the progress at HFBS and RSSM, the ILL has just decided to upgrade IN16. The upgraded instrument (*IN16B*) will be moved to an end position and the PST technique will be applied as well. Together with a new neutron guide focusing optics a flux gain close to a factor 10 is expected with respect to the current IN16.

- **IN13 at ILL**

This is the only reactor BS spectrometer on a thermal source (see ref. [22]). Similar to IN10B it uses a thermally controlled focusing CaF_2(422) monochromator near BS. The monochromatic neutrons are deflected from a graphite deflector to the sample. For different monochromator temperatures, i.e. different energy transfer, the instrument has to be realigned. A chopper, located at the entrance to the secondary spectrometer, creates a pulsed beam and reduces higher order contamination. The secondary spectrometer operates in exact BS with CaF_2 crystals. IN13 is a unique instrument because it can access high momentum transfer ($Q < 5.2$ Å^{-1}) at an excellent energy resolution of about FWHM = 8 μeV, which is very important, e.g. for the determination of molecular spatial structure factors. The energy range can extend over –125 μeV to 300 μeV. Due to the short incoming wavelength on IN13, Bragg reflections from the sample environment (Al, Nb…) or from the sample can be problematic and have to be shielded carefully. The main disadvantage at the moment is the every long counting time in the order of 24 h per inelastic spectrum. The count rate was increased recently by doubling the analyzer sphere (either in the low or the large Q-range) and furthermore a rebuild of the neutron guide is foreseen. For the first time a vertically position sensitive detector tube array (distance 2 cm ; 8 mm vertical resolution) with conical collimator allows exact BS conditions at small angles [22].

7.2. SPALLATION SOURCE INSTRUMENTS

♦ IRIS at ISIS

BS spectrometers on spallation sources are inverted TOF spectrometers with the analyzer crystals in the secondary spectrometer near backscattering. The distance-time diagram for such a spectrometer is shown in figure 5c. The IRIS spectrometer at ISIS is the first successful example of such a spectrometer [23] (fig. 11).

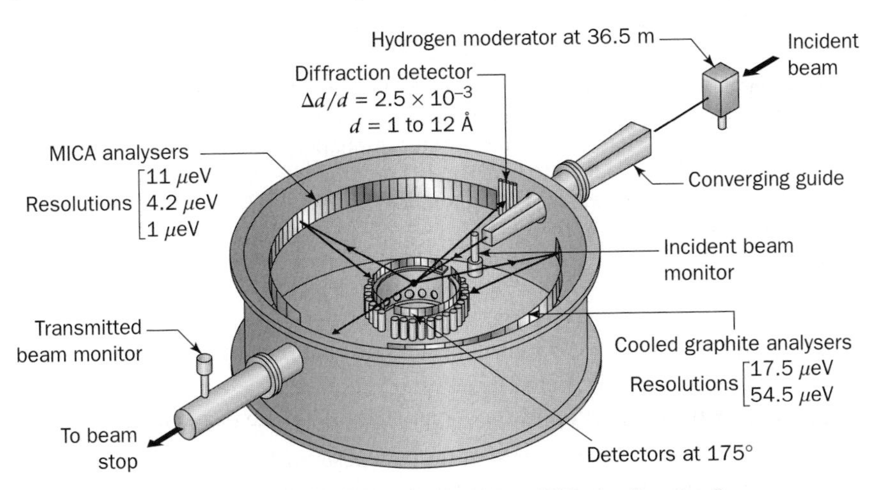

Figure 11 - Schematic view of IRIS at ISIS, the "mother" of the spallation source-BS instruments.

The design of the IRIS spectrometer has inspired many other projects for s-BS (see tab. 2b). Two different types of analyzer banks are used: cooled PG(002) analyzers (cooled to reduce the thermal diffuse scattering) and mica analyzers. In addition Be filters can be installed to remove higher order scattering. As can be seen from figure 11, the detectors are located out of BS. Further similar s-BS instruments (LAM [24]) are operating at the Japanese n-spallation source KEK (see tab. 2b).

♦ BS project at SNS

The most advanced spallation BS project is under construction at the SNS, in Oak Ridge, US (source operational in 2006). The layout of this instrument is presented in figure 12 [25]. The elastic energy resolution is expected to reach about 2.5-3 μeV for the Si(111) analyzers, and about 9 μeV for the Si(311) and the flux is predicted to be higher than on any of today's reactor-BS instruments. Flux comparison with r-BS has to be made with care, because s-BS instruments usually cite "white flux" numbers, whereas r-BS instruments give it for a highly monochromatic beam.

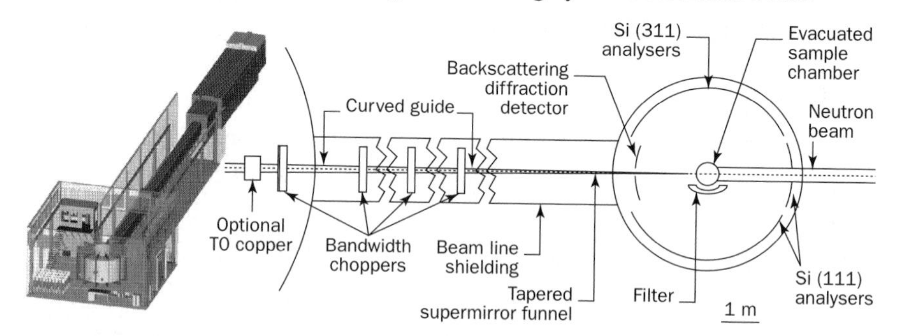

Figure 12 - Schematic view of the SNS s-BS project. The sample is located at the end of a neutron guide, about 85 m downstream from the moderator, within a symmetric analyzer sphere of 2.5 m diameter. A chopper system selects the useful wavelength band. It is planned to use Si(111) on one side, Si(311) analyzers on the other side of the sample, with the detectors located out of the scattering plane near backscattering [25,54].

♦ OSIRIS at ISIS

This is up to now the only s-BS instrument, which plans to use polarized neutrons for high resolution inelastic scattering. With the development of ^{3}He filters and the increased neutron flux on BS instruments, this situation might change in the future. Other s-BS instruments projects exist at the *Japanese spallation source* (planned to be operational around 2006) and at the *European Spallation Source* (no decision yet) (see tab. 2b). In future, the most powerful neutron sources will be spallation sources, and therefore s-BS instrument types will gain on importance.

8. DATA TREATMENT

Here we sketch very briefly the typical data evaluation procedures: The neutrons are measured within a certain time channel bin, which has to be converted to an energy bin, using the above relations between either the Doppler velocity, the monochromator temperature or the neutron flight time and the neutron energy.

The incoming neutron flux is measured in front of the sample using a monitor M, which is a detector with low efficiency ; 10^{-3}-10^{-4}. The division of the detector counts by the monitor counts removes the incoming flux dependence for each channel (see fig. 5).

Before the experimental result from BS can be compared to a theoretical scattering law $S(Q,\omega)$, we have to apply several corrections to the detected spectra. The number of neutrons, N, counted in a detector at the scattering angle 2Θ, is proportional to the double differential cross-section $d^2\sigma/d\Omega\, d\omega$, i.e. the probability that the sample is scattering into the considered space angle Ω and into the frequency bin $d\omega$. It will also be proportional to other factors like the incoming neutron flux, the sample and beam size, the detector efficiency, the analyzer efficiency and analyzer area seen by the detector, etc.

Furthermore the sample is generally held in a container with some scattering probability and there is background scattering from the instrument. If the latter is independent of the sample it can be simply subtracted. However, the "empty can" measurement and sample dependent instrument background must be subtracted from the sample counts by taking the geometry, the scattering and the absorption cross-section dependent correction factors into account. For these corrections one can assume that the measured intensity is related to the double differential scattering cross-section by ref. [26])

$$I(\theta,\omega) \propto \int T_{\mathrm{i}}(x,E_{\mathrm{i}})n(x)\frac{d^2\sigma}{d\Omega\, d\omega}T_{\mathrm{f}}(x,\theta,E_{\mathrm{f}})d^3x \qquad \{6\}$$

where T_{i} and T_{f} are the (i = incoming and f = final) transmission probabilities to reach a point x in the container-sample assembly and to leave it from x under an angle Θ without interaction and n is the local number density of scatterers at point x. The transmission is of the usual form $T = n\sigma l$, where σ is the total scattering and absorption cross-section and l the path length. The main purpose of data evaluation programs, which are specific for different instruments, is to calculate these transmission factors for the given experimental conditions. The scattered intensity is then corrected by $I^s = c_1 * I^{s+c} - c_2 * I^c$, where c_i are the correction factors, calculated from the transmission expressions above for the sample s and the container c [26].

For conversion of the intensities into absolute units, one uses a *scattering standard* with known isotropic, thus incoherent, scattering, normally *Vanadium*. Its scattering probability and geometry should be similar to the sample and the above-mentioned corrections must be applied as well. With the elastic scattering from the corrected standard one can finally obtain the differential cross-section $d^2\sigma/d\Omega\, d\omega$ of the sample (see e.g. ref. [27]).

The measured scattering cross-section is still convoluted with the *instrumental resolution*, which is normally measured with the standard (Vanadium) or with the sample at very low temperature (if one can be sure that there is only elastic scattering). Usually one convolutes a theoretical model function with the measured resolution function for analysing the data. With some care to avoid possible artefacts, a deconvolution of the resolution function is also feasible by applying inverse Fourier transform into time-space. In time-space the convolution converts to a product and one can remove the resolution contribution by a simple division, $S(Q,t) = S''(Q,t)/R(Q,t)$. A Fourier transform back to frequency-space is usually not reliable and therefore not advisable.

The differential cross-section is related to the *scattering law* by

$$\frac{d^2\sigma}{d\Omega\, d\omega} = \frac{k_f}{k_i}\left(\frac{\sigma_{coh}}{4\pi} S_{coh}(Q,\omega) + \frac{\sigma_{inc}}{4\pi} S_{inc}(Q,\omega)\right) \qquad \{7\}$$

Like all other non-polarized neutron scattering instruments, BS measures the sum of both, the coherent and incoherent scattering law. Therefore one must either know one of the additive terms of equation {7}, or one has to measure a sample with dominantly incoherent or coherent scattering. Samples with large hydrogen content possess such a high incoherent scattering cross-section that the coherent part can generally be neglected. If only the *static* coherent part contributes to the first term of equation{7} (Bragg peaks or short range order), then it can be separated by an independent measurement (on a few instruments even simultaneously: IN16, IRIS). Intermediate cases are more difficult to interpret, even though there is increasing tendency for comparing experimental data to *molecular dynamics* (MD) simulations, where both contributions can be calculated.

One correction is still missing: the correction for *multiple scattering* (MS). Of course multiple scattering is present in all experiments. But, in order to reduce the problem, one usually prepares a sample with low scattering probability, typically of the order of 10%. The main difficulty is that one would have to know the scattering law in advance for a proper correction. A way out is to simulate the MS contribution for the theoretical scattering law up to a certain order, by taking exact experimental details into account. Then one fits the MS-corrected law to the data. The resulting best-fit function is again corrected for MS, and so on. After a few iterations the procedure should converge.

9. *TYPICAL MEASURING METHODS AND EXAMPLES*

9.1. *FIXED WINDOW SCANS*

We start with a unique method for r-BS instruments. "Fixed window scans" or "elastic scans" are very powerful for getting a fast overview over the dynamics of a system and are often the starting point for inelastic neutron investigations. How is it done?

The monochromator and the analyzers are chosen to reflect neutrons with fixed wave vectors k_i and k_f (i.e. they are not moving with time t and the lattice spacing does not change with temperature T). If $k_\mathrm{i}(t,T) = \mathrm{const}$ and $k_\mathrm{f}(t,T) = \mathrm{const}$ we analyze scattered neutrons within a *fixed energy window* $\Delta E = \dfrac{\hbar^2}{2m}(k_\mathrm{f}^2 - k_\mathrm{i}^2)$. Now we can *scan* a sample parameter (like temperature, pressure…) which then might influence the dynamics of the sample. For $k_\mathrm{i} = k_\mathrm{f}$ we have the usual "elastic scans" with $\Delta E = 0$ (this is the case if the monochromator is at rest and has the same crystals, orientation and temperature as the analyzers). For $k_\mathrm{i} \neq k_\mathrm{f}$ we deal with inelastic "fixed window" scans [28].

We can associate a frequency or time range with the fixed energy window: e.g. if $\Delta E = 0$ and the instrumental resolution has a typical value for cold neutron r-BS of HWHM ~ 0.4 meV, then this corresponds to a frequency of $\nu_R \sim 10^8$ Hz = 0.1 GHz, thus to a time window of observation of about 10ns or 0.01μs (using the conversion 1 THz = 4.136 meV). Dynamic processes on a time scale slower than the instrumental resolution are not resolved and thus are counted within the "elastic window". Faster motions of scattering particles can be resolved and will induce an energy loss or gain of the scattered neutrons ($k_\mathrm{i} \neq k_\mathrm{f}$), which then are no longer reflected by the analyzers to the detectors. In that case one observes a decrease of the elastic window intensity as function of the recorded variation parameter.

If we analyze for $k_\mathrm{i} \neq k_\mathrm{f}$, (e.g. monochromator and analyzer are at a different temperature, or have a different orientation), then one can find an increase of intensity in the fixed observation window or even a maximum as function of the variation parameter, when e.g. the dominating relaxation frequency of a dynamic process corresponds to the "fixed window" frequency $\Delta E \pm$HWHM (earliest example for such a scan [28]).

The fixed window method is limited to r-BS instruments and has the advantage that the total counting time is spent in one channel, which corresponds to a large intensity gain. Elastic scans can be simulated assuming a model scattering law for e.g. the temperature dependence. It should be done with some caution, because direct spectroscopic information is lost and the number of free fit parameters is usually larger than for spectroscopy. Therefore it is in any case advisable to complete the scans by inelastic spectra at several parameter points.

The parameter dependence of the scattering can give information e.g. on the activation energy of a dynamic process. Furthermore the spatial spatial information is deduced form the angular or Q-dependence ($Q_{el} = 4\pi \sin(\Theta/\lambda)$ of the scattering.

For incoherent scattering the integral over the dynamic scattering law must be one: $\int_{-\infty}^{\infty} S_{inc}(Q,\omega) dQ d\omega = 1$. If the measurement is carried out at very low temperatures, then we can consider the dynamics to be so slow that a BS instrument is not able to resolve it. Furthermore the Q-dependence will be flat (no motion) if we assume that the Debye-Waller factor is unity (we neglect zero-point motion). Thus an elastic window experiment at low temperatures can be used for normalization onto the total scattering, after subtraction of background. This also removes other correction factors, which usually are more difficult to handle. It does not treat correctly multiple scattering effects, however, and artefacts may be produced if the sample scattering is coherent and changes with the scan parameter. For dominantly incoherent scattering and quasielastic processes that are fast on the time scale of the BS resolution, the elastic intensity is a good measure of the Elastic Incoherent Structure Factor (EISF).

An example for such a fixed window scan is given in figure 13. Figure 13a shows the measured elastic intensity as a function of temperature for nylon-12, a polymer with long methylene sequences. With temperature the motion of these $-(CH_2)-$ units becomes increasingly faster (e.g. vibrations, librations, trans-gauche transitions...) and thus leads to a loss of elastic intensity. As explained above, the data are normalized to the lowest temperature. From the Q-dependence one can deduce information on the spatial extend of the motion and from the T-behaviour in simple cases information on the activation energies. Model fits have to be carried out then for the temperature and Q-dependence, as shown here for two simple cases.

In figure 13b a fit with a harmonic vibrational model is attempted ($I = I_0 e^{-1/3Q^2\langle u^2\rangle}$); linear fits of ln I *versus* Q^2 , plus an additional constant contribution y_{frac}, which may be partially due to multiple scattering events). In figure 13b an addition exponential term ~ Q^4 is added: non-Gaussian model).

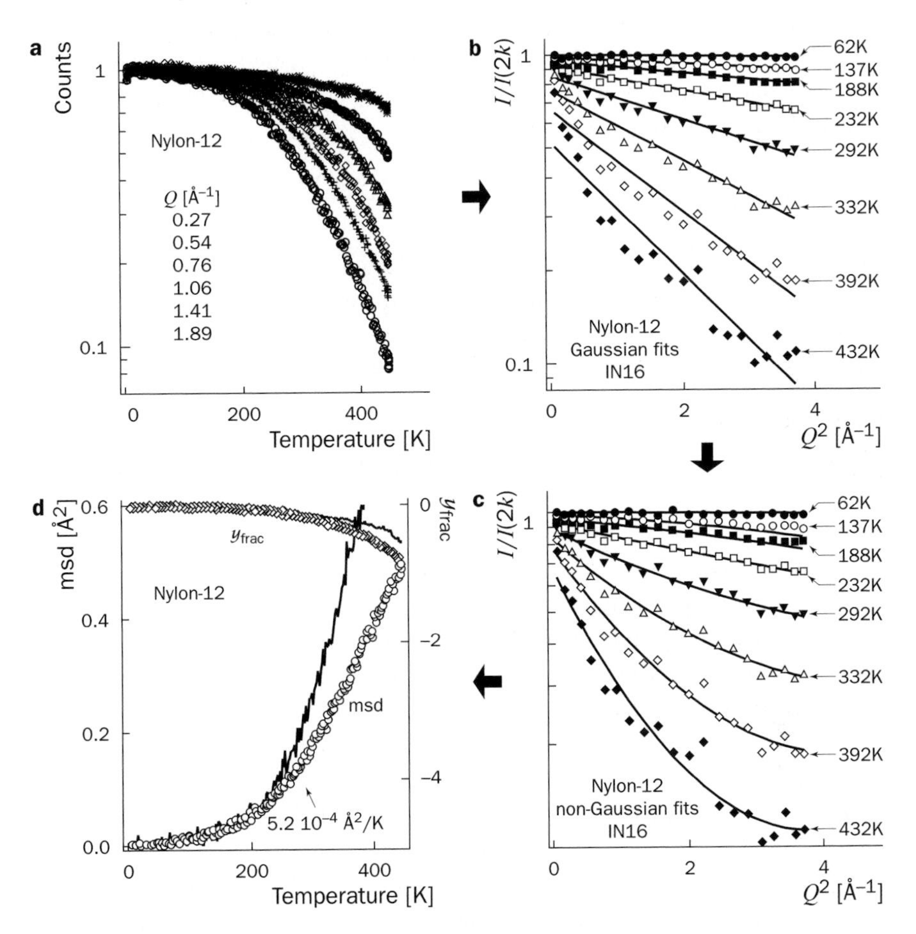

Figure 13 - Elastic temperature scan on nylon-12 (IN16, ILL). **a** - Temperature dependence of the elastic intensity for different Q, normalized to the low temperature values. **b** and **c** - Q dependence of the normalized intensities at different temperatures (symbols) and model fits (lines) for a normal Debye-Waller factor, i.e. Gaussian Q dependence (**a**) or for taking an additional non-Gaussian Q^4 term in the exponential (**b**). From such data an *effective mean squared displacement* can be deduced, as it is shown in figure (**c**) as a function of temperature for both models (symbols: result from Gaussian, lines: from non-Gaussian fits). y_{frac} is the ordinate fraction which result from linear fits of ln I vs Q^2.

An important resulting fit parameter is the T-dependence of the average mean squared displacement (of all protons in the sample), which is shown in figure 13d, together with the constant y_{frac} term. The non-Gaussian term is not shown. As can be seen a harmonic vibrational model is only valid at low temperatures (roughly linear increase of $\langle u^2 \rangle$ with T). Above about 200 K a different model has to be used. This can be a form factor for a specific molecular motion involving also an activation energy term or a more simplified non-Gaussian fit as done here. In our case the latter describes the data somewhat better.

9.2. SPECTROSCOPY

Concerning spectroscopy we can discriminate between *inelastic* (signal located at a finite energy transfer $E \neq 0$) and *quasielastic* experiments (signal centred at $E = 0$). We present here only a few applications to illustrate the instrumental possibilities rather than the scientific background, which the reader can study following the given references. In addition there is no space to treat fundamentals on inelastic and quasielastic neutron spectroscopy for which we refer to standard text books, e.g. to Bée [29].

Two typical applications for **inelastic neutron-BS spectroscopy** are investigations of quantum rotational tunnelling (methyl groups, methane, ammonia...) or the spectroscopy of nuclear hyperfine splitting. Even though spectroscopy on *tunnel splitting* is the most important application for BS, we restrict its discussion here and refer to the chapter by M.R. Johnson in this volume). Most measurements over the last years are compiled in a "tunnelling Atlas" [30]. Here we mention only two cases which illustrate that often the best energy resolution available is needed, either for detecting a very small tunnel splitting like for aspirin (see fig. 4, in the chapter by M.R. Johnson in this volume): the T dependence of a low lying excitation at $\pm 1.22\ \mu$eV was measured on IN16 with an energy resolution of HWHM = 0.16 μeV using polished Si(111) analyzers and a special Si(111) crystal monochromator [31]) or for separating different tunnel excitations like for ammonium perchlorate, NH_4ClO_4, where a few at. % of H atoms in the ammonium group NH_4^+ where exchanged against D (fig. 14, measured with the standard polished monochromator on IN16, HWHM = 0.23 μeV) [32].

Some other systems need excellent energy resolution at large energy transfer. This is mainly the case at low temperatures, because at higher temperatures inelastic lines are often intrinsically broadened. Again we mention a study of rotational tunnelling in γ-picoline, employing IN10B in a wide energy transfer range with 1 μeV resolution as an example (see fig. 3c, in the chapter by M.R. Johnson in this volume). If the resolution can be relaxed, then such experiments may also be suited for s-BS instruments with their wide dynamic range.

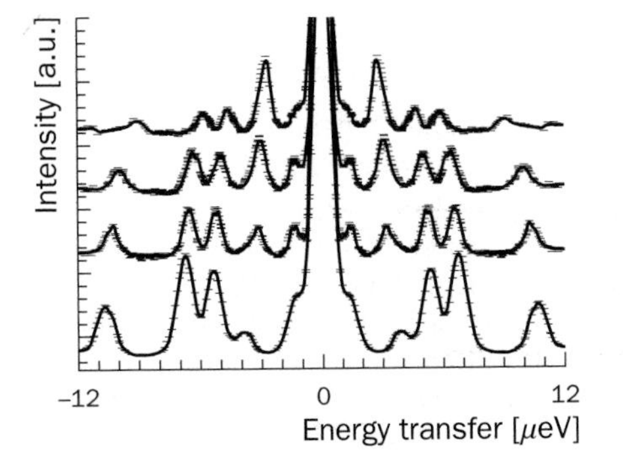

Figure 14 a - Dependence of the NH_4 tunnelling spectrum on the partial deuteration of the ammonia group (0 at. %, 7 at. %, 15 at. %, 30 at. % deuterium; after ref. [32]), measured on IN16.

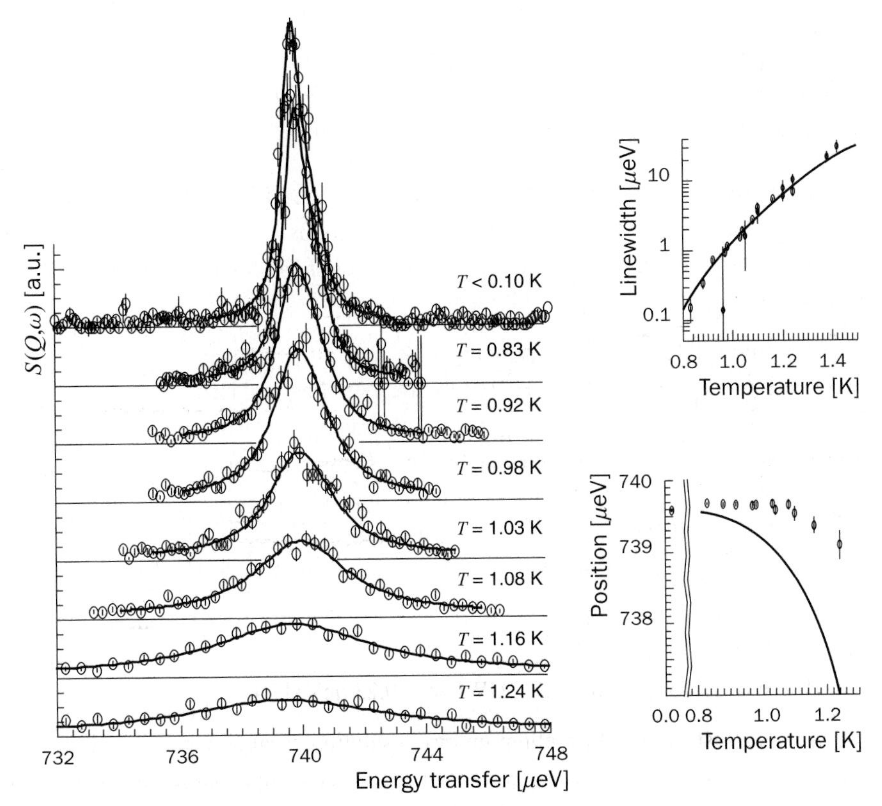

Figure 14 b - Roton peak in ^{4}He measured on IN10B with a NaF(111) monochromator [33]

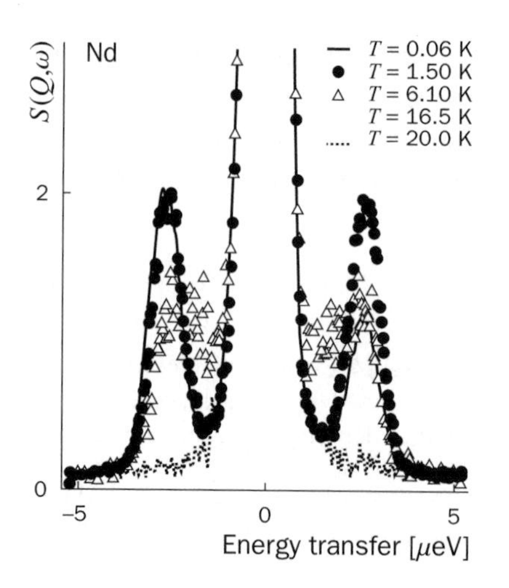

Figure 14 c - A recent example of the nuclear excitation spectrum of Nd measured on IN16 [36]. The inelastic peaks show detailed balance for the lowest temperature (solid line) and move with increasing temperature into the elastic line

Probably the best illustration of the performance of IN10B with 1 μeV energy resolution at 740 μeV energy transfer is a measurement by Andersen *et al.* on quantum excitations, the roton peak, of ^{4}He [33]. At temperatures below 0.7 K this excitation is extremely sharp and limited by the experimental resolution. Its line width changes very fast with increasing temperature (fig. 14c). The T dependence of the line width and the position of the roton peak are of large theoretical interest and the 1 μeV energy resolution on IN10B (additionally restricting severely the Q-range with Cadmium shields) did allow to get the most reliable low temperature values [33].

Another interesting application for inelastic BS is the measurements of *nuclear hyperfine splitting* [34], which is caused by the local field induced at the nuclear site by the electronic magnetism. Neutrons are scattered by inelastic spin-flip-scattering on the nuclei and the energy exchange between the nucleus and the neutron induces a change of the nuclear spin state and thus of the neutron spin, due to momentum conservation. The neutron cross-section, assuming no correlation among nuclear spins or between nuclear spins and the lattice, was deduced by A. Heidemann [34].

$$\frac{d^2\sigma}{d\Omega\, d\omega} = \frac{k_1}{k_0} e^{-2W} \frac{1}{3} I(I+1)\alpha'^2 \left[\delta(\hbar\omega)+\delta(\hbar\omega-\Delta)+\delta(\hbar\omega+\Delta)\right] \qquad \{8\}$$

where α'^2 is the spin incoherent scattering length and Δ the hyperfine splitting. Thus the excitation spectrum consists of a simple three line spectrum was shown e.g. for Cobalt in the early days of IN10 by Heidemann *et al.* (ref. [34] and references therein). Here we show in figure 14c another only recently detected nuclear excitations in Nd and several magnetic Nd compounds [35,36]. Though this excitation is probably again due to nuclear hyperfine interaction, there is also effort to search for a dispersion, which would indicate the presence of nuclear spin waves, i.e. a coupling of the nuclear spins via the electronic magnetism, as was predicted by de Gennes [37]. Figure 14c also shows that only in the mK temperature-range detailed balance can be observed.

Quasielastic BS scattering is applied for investigating the diffusion and eventually the rotational motion of atoms and molecules [3]. Traditionally the main applications were the investigation of diffusion mechanisms in metals, alloys, intercalated compounds, or of hydrogen in metals. Today the centre of research has clearly shifted towards investigations of the dynamics of complex systems, like polymers, biological systems, glass-forming liquids or confined molecular liquids. Still, diffusion studies on single crystals attract our attention, because they make use of the vector properties to determine the microscopic details of the diffusion mechanism.

An example is the *self diffusion* of metal atoms, which move at high temperatures very fast, mainly via vacancies within a crystal lattice. This is the case for the technologically promising alloy B2-Ni_xGa_{1-x} ($x \sim 0.5$) for which the Ni diffusion in single and polycrystals was recently investigated by BS [38]. The diffusion constant of Ni in these alloys is relatively slow and one needs therefore the good energy resolution of backscattering or spin echo instruments to explore the Q and temperature dependence of the line width and intensities. Measurements on IN16 showed a weak concentration dependence of the diffusion constant and by exploiting different single crystal orientations one could proof that the Ni diffusion proceeds by *nearest neighbour jumps*. The *residence time* of Ni on the Ga sublattice sites was found to be only 8-12% as long as on the Ni sublattice sites for temperatures at 1060°C-1130°C. Figure 15a shows a quasielastic spectrum measured for a crystal of $Ni_{51.2}Ga_{48.8}$ at $Q = 0.52$ and 1.6 Å^{-1} and $T = 1130$°C, together with the resolution function (line). More details on different experimental techniques for this application can be found in reference [39] and references therein as well as some comments on spin echo versus backscattering for diffusion studies. Though both methods are to a large extent complementary it is easy to find examples, which are better suited to either one of these techniques. Some discussions conclude pessimistic in what concerns the development potential of n-BS, neglecting possible resolution improvements or the different possibilities to extend the energy transfer range (see fig. 1). On the other hand the extreme usefulness of BS is evidenced by an impressive list of publications in the field of diffusion studies. The ease of BS spectroscopy (most sample environments possible, wide Q-range simultaneously sampled, etc.) may be part of the success of BS in this field, in spite of its narrower dynamic range compared to NSE.

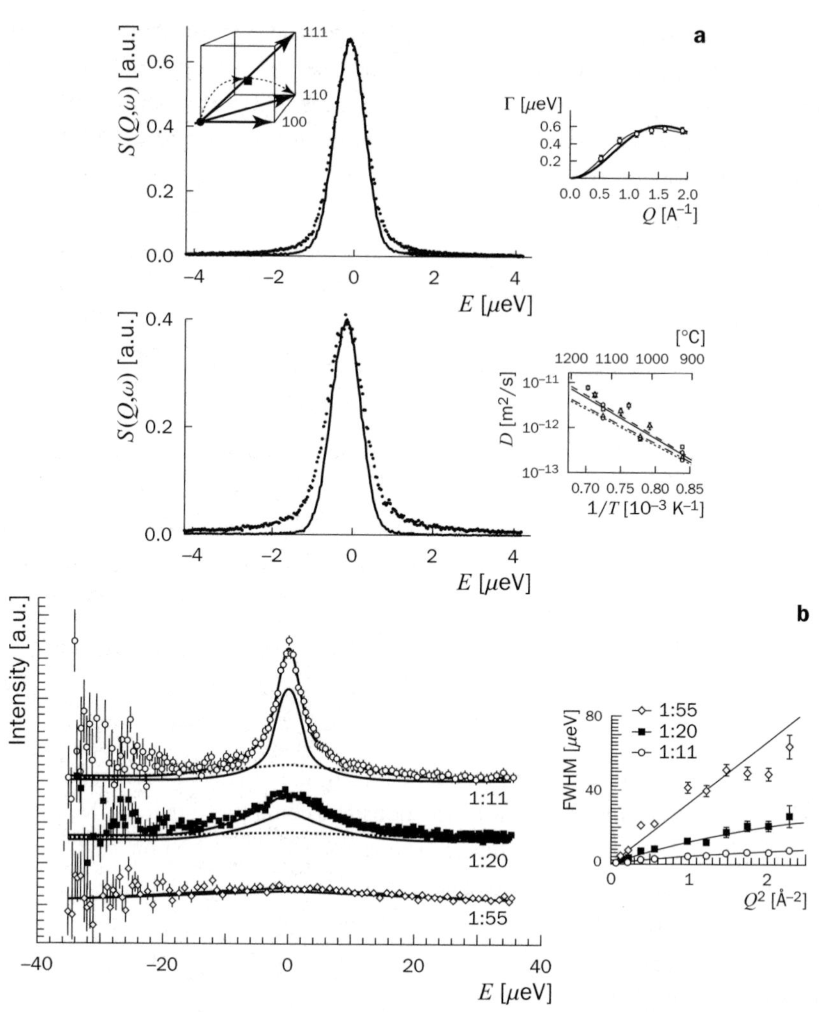

Figure 15 - Two examples for quasielastic scattering. **a** - Atomic diffusion study on a Ni_xGa_{1-x} single crystal [38] measured on IN16, ILL. The orientation of $\boldsymbol{Q}$ with respect to the crystallographic axis can be exploited to determine the jump vectors for atom displacement. **b** - Quasielastic scattering from a Glucose/D_2O solution at concentrations 1:11, 1:20 and 1:55 [40] measured at the NIST, HFBS, with 1 μeV energy resolution and an enlarged dynamic range due to a fast Doppler drive and PST chopper (see text). The inset shows the linewidth as a function of Q^2 and fits for different diffusion models.

The next example for quasielastic scattering illustrates the enlarged energy transfer range, which can be exploited by a r-BS instrument using a fast Doppler drive and a PST device. The energy transfer can be nearly three times wider than on a conventional Doppler based BS spectrometer, still at an energy resolution of 1 μeV.

Figure 15b shows a recent study at the HFBS of the dynamics of glucose (protonated, thus dominating incoherent scattering) in D_2O solutions (coherent scattering and in addition very fast, i.e. its large linewidth contributes as a flat background; complementary measurements with TOF are not shown here) [40]. The aim of this study was to determine the diffusion mechanism of glucose in D_2O and its possible change with D_2O concentration. As one would have intuitively expected, one finds that the dynamics of glucose accelerates drastically with increasing dilution. The inset shows the linewidths which result from fits with a jump diffusion model. The authors interpret a change in diffusion mechanism when going to high dilutions. The bad statistics of the spectra at negative energy transfer origins from the PST reflectivity curve at HFBS and the fact that there are only a few incident neutrons at these energies (monitor spectrum).

Frequently BS data are converted by time-Fourier-transformation (FT) from $S(Q,\omega)$ to $S(Q,t)$ by using a low temperature resolution spectrum for normalization and for division with the resolution function. The reason for converting to time space can be different. In principle there is not more information after FT in time space than one has in energy space (rather less, because of FT artefacts and statistics) and theoretical models can be easily calculated for both, ω and t space. In most cases the motivation comes from the need to combine spectra for which one is obliged to remove the influence of the finite resolution function. The different spectra may be either from different instruments (to enlarge the instrumental dynamic range) or from different temperatures, pressures etc. after rescaling the time axis (again to enlarge the investigated spectral range). The example we have chosen in figure 16 applies FT for the latter purpose [41] and was measured at IRIS [12] with PG(002) analyzers giving "only" 15 μeV energy resolution (see tab. 2b), but a wide dynamic range. The dynamics of disordered materials is usually stretched over a very wide frequency range because due to structural disorder the energy barriers and thus the relaxation times possess a wide distribution. It is often empirically found that the spectral shape of the α relaxation (structural relaxation which freezes close to T_g) is well described by the FT of a stretched exponential function

$$S(Q,\omega) = \mathrm{FT}\left[S(Q,t)\right] = \mathrm{FT}\left[Ae^{-(t/\tau_{KWW})^{\beta}}\right] \qquad \{9\}$$

with exponent $\beta < 1$ and the so-called Kohlrausch-Williams-Watts relaxation time τ_{KWW} (narrow component in fig. 16). The so-called time-temperature-principle can be used to re-scale spectra, which were initially measured at different temperatures, to one and the same reference temperature. For this the resolution effect has to be removed by FT of $S(Q,\omega)$ by dividing with the resolution (fig. 16b). More details can be found in the original work [41] and references therein.

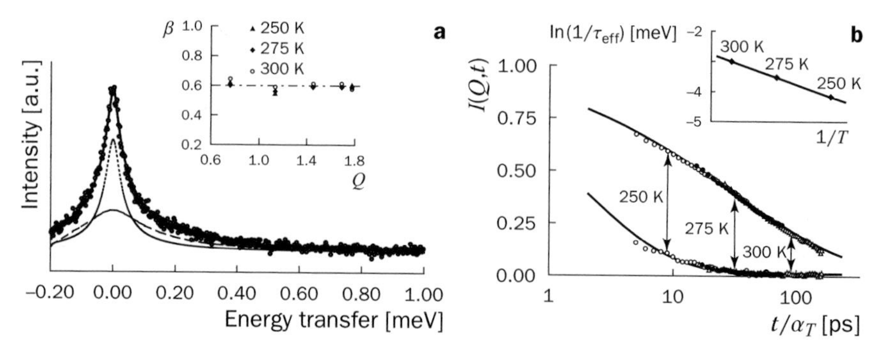

Figure 16 - $S(Q,\omega)$ (**a**) and Fourier transformed $S(Q,t)$ data (**b**) of the polymer PDMS = polymethylsiloxane [–O–Si–$(CH_3)_2$–] measured at the s-BS instrument IRIS with 15 μeV energy resolution [41]. Model fits were made with a fast (methyl groups) and a slow relaxation (α relaxation; stretched exponential with exponent β, see inset) component (lines in both figures are model fits). In (**a**), spectra measured at different temperatures were superimposed to a master curve, using rheological information for re-scaling the time axis (upper and lower curves correspond to two different Q values: Q = 0.55 Å^{-1} and 1.58 Å^{-1}). The inset in (**b**) shows the relaxation time as function of temperature from which an activation energy can be extracted.

Another example is shown in figure 17 and illustrates the unique possibility of IN13 [22] to measure up to Q values of Q = 5.2 Å^{-1} with an energy resolution of 8 μeV. The scientific background behind this example is certainly too complicated to be explained within a few sentences, though it is related to the previous example. We refer again to the literature [42]. The glass transition is one of the less understood physical phenomena and many scattering experiments, triggered by the event of "mode coupling theory (MCT)" [43] have tried over the last 15 years to contribute to its understanding. A recent controversy arose around the question if the self correlation function and its FT $S(Q,t)$ relating to the α relaxation follows a Gaussian behaviour. This can be experimentally verified if in the above mentioned equation for the stretched exponential function the relaxation time τ_{KWW} followed a power law with $\tau_{KWW} \propto Q^{\beta}$. Such a law has been found experimentally for different polymers by n-BS [44] at Q values below the maximum of the static structure factor peak. Molecular dynamic simulations had detected deviations from this law at higher Q values and therefore it was experimentally searched for it as well. This explains the need to measure up to as high as possible Q values (see spectra in figure 17 for polyisoprene in which the methyl group was deuterated in order to measure mainly the polymer backbone). IN13 data in combination with NSE date proof such a deviation (last figure in fig. 17) and suggest a power law close to $\tau_{KWW} \propto Q^{-2}$ and thus a non-Gaussian behaviour [42].

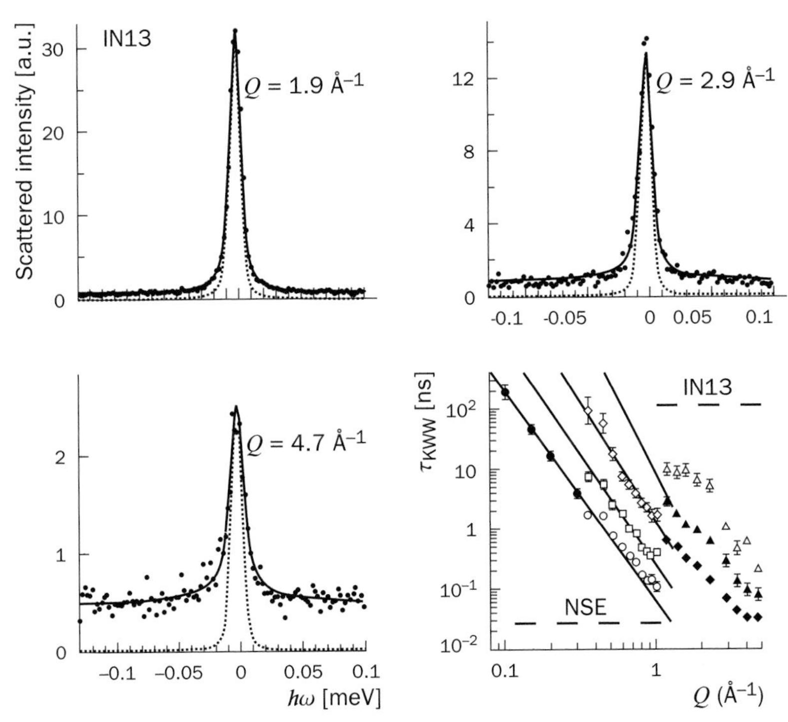

Figure 17 - Combination of IN13 spectra with NSE data on polyisoprene [$-CH_2-CH=C(CD_3)-CH_2-$]. On IN13 one can measure up to very high Q values and extend the Q-range considerably. Three spectra at different Q values are shown. The relaxation times at high Q, resulting from fits to the BS spectra, are combined in a double logarithmic plot with the results from NSE at low Q. The lines show the Gaussian behaviour (see text and ref.[42]) and clear deviations of the IN13 data from this line can be seen.

Many other BS experiments might be cited, such as studies on conducting polymers, the 'Nobel price subject' of the year 2000 (e.g. ref.[45,46]). There is also a rapidly growing number of BS measurements on biological systems, where the objective is often to establish a link between the local microscopic dynamics that is studied by neutron backscattering and the functionality of a protein (see e.g. ref.[40,47]). As a last example we mention a new application, which shows both inelastic and quasielastic features, and which was proposed to measure *shear gradients* or velocity profiles in sheared liquids[48]. The idea is simple and appealing at the same time. In the same way as a moving monochromator adds a longitudinal Doppler shift to the neutron speed, scattering atoms that are moving with some velocity component $v_{\parallel}$ parallel to Q add a Doppler shift to the neutron velocity. If we measure e.g. an empty shear cell (second row in fig. 18), which consists of a rotating and a static disc from elastic scattering materials, then we will observe two lines, one elastic line, originating from the static disc (left column fig. 18), and a second "inelastic" line, originating from the moving disc (right column fig. 18). Now, if we fill the shear apparatus with a liquid, then we will scatter from molecules with a certain velocity distribution. In the laminar flow regime one expects e.g. a constant shear gradient. For shear velocities, sufficiently large with respect to the resolution, one can probe the Doppler shift spectrum, which arises from such a velocity distribution. The last row in figure 18 shows the spectrum, which is found for the transmission geometry ($\boldsymbol{Q} \parallel v$), allowing to measure the shear velocity. In the reflection geometry ($\boldsymbol{Q} \perp v$) the shear velocity becomes invisible and thus only the microscopic dynamics is observed [49].

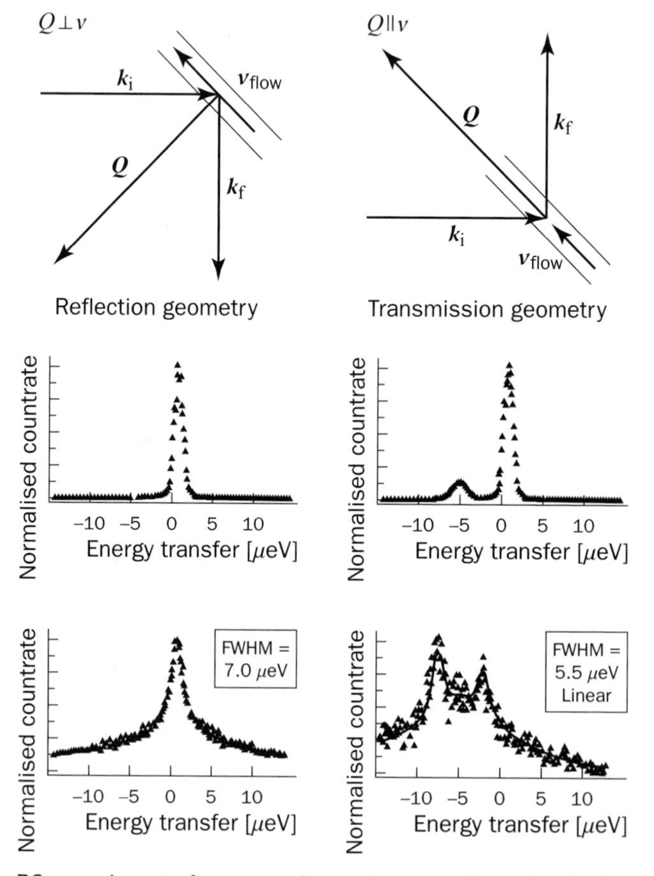

Figure 18 - BS experiments for measuring a shear gradient distribution in a shear cell filled with a liquid. Orienting the $\boldsymbol{Q}$ vector parallel and perpendicular to the shear velocity $\boldsymbol{v}$, respectively, allows to measure both, the shear gradient profile and the microscopic dynamics. For more details see ref. [48,49] .

ACKNOWLEDGMENTS

This article is dedicated to Berthold Alefeld and Anton Heidemann who have developed neutron backscattering during many years of their life; also parts of this article are inspired by their excellent review published as an ILL Report [3]. Further I like to thank K. Andersen, A. Arbe, V. Arrighi, H. Büttner, T. Chatterji, Z. Chowdhuri, K. Herwig, M. Kaisemayr and M. Wolff who have provided support to this article by communicating figures or experimental results.

REFERENCES

[1] W.H. ZACHARIASEN - *Theory of X-ray diffraction in crystals*, J. Wiley & Sons, New York & London (1946)

[2] P.P. EWALD - *Z. Phys.* **30**, 1 (1924);
C.G. DARWIN - *Phil. Mag.* **27**, 315 & 675 (1914)

[3] A. HEIDEMANN & B. ALEFELD - *Review of the Applications of Backscattering Techniques in Neutron & X-ray Scattering,*
ILL Report 91.HE22G, Institut Laue-Langevin, Grenoble (1991)

[4] A. HEIDEMANN - *Physica B* **202**, 207 (1994)

[5] M. BIRR, A. HEIDEMANN & B. ALEFELD - *Nucl. Instr. Meth.* **95**, 435 (1971)

[6] B. ALEFELD, T. SPRINGER, & A. HEIDEMANN -
Nuclear Science and Engineering **110**, 84 (1992)

[7] H. MAIER-LEIBNIZ - *Nukleonik* **8**, 61 (1966)

[8] IN10, http://www.ill.fr/YellowBook/IN10/

[9] BSS, http://www.fz-juelich.de/iff/wns_bss

[10] IN16, http://www.ill.fr/YellowBook/IN16/

[11] HFBS, http://rrdjazz.nist.gov/instruments/hfbs/

[12] IRIS, http://www.isis.rl.ac.uk/molecularspectroscopy/iris/

[13] A. MEYER, R.M. DIMEO, P.M. GEHRING, & D.A. NEUMANN -
Review of Scientific Instruments **74**, 2759 (2003)

[14] RSSM, http://iffwww.iff.kfa-juelich.de/rssm/index_en.html

[15] B. FRICK, A. MAGERL, Y. BLANC & R. REBESCO - *Physica B* **234-236**, 1177 (1997)

[16] A. MAGERL, K.D LISS, C. DOLL, R. MADAR, & E. STEICHELE -
Nucl. Instr. Meth. A **338**, 83 (1994)

[17] R. HOCK, T. VOGT, J. KULDA, Z. MURSIC, H. FUESS, & A. MAGERL - *Z. Phys. B,* **90**, 143 (1993)

[18] J. SCHELTEN & B. ALEFELD - presented at the *Proceedings of Workshop on Neutron Scattering Instrumentation for SNQ,* 1984 (unpublished)

[19] P.M. GEHRING & D.A. NEUMAN - *Physica B* **241-243**, 64 (1998)

[20] I. ANDERSON - *In "Neutron Data Booklet",* A.J. DIANOUX & G. LANDER eds, Institut Laue-Langevin, Grenoble (2002)

[21] http://rrdjazz.nist.gov/instruments/hfbs/

[22] IN13, http://www.ill.fr/YellowBook/IN13/ & http://www.ill.fr/News/40/INSTR_P6.HTM

[23] C.J. CARLILE & M.A. ADAMS - P*hysica B* **182**, 431 (1992)

[24] LAM http://neutron-www.kek.jp/Neutron-Spectrometers/LAM/LAM(E).html

[25] K.W. HERWIG & W.S. KEENER - *Appl. Phys. A* **74**, S1592 (2002)

[26] H.H. PAALMANN & C.J. PINGS - *J. Appl. Phys.* **33**, 2635 (1962)

[27] O. RANDL - ILL Report 96.RA07T, Institut Laue-Langevin, Grenoble (1996)

[28] H.H. GRAPENGETER - B. ALEFELD & R. KOSFELD - *Colloid & Polymer Sci.* **265**, 226 (1987)

[29] M. BÉE - *Quasielastic Neutron scattering,* Adam Hilger, Bristol (1988)

[30] M. PRAGER & A. HEIDEMANN - *Chem. Rev.* **97**, 2933 (1997)

[31] M. JOHNSON, B. FRICK, & H.P. TROMMSDORFF - *Chem. Phys. Lett.* **258**, 187 (1996)

[32] H G BÜTTNER, G J KEARLEY, & B FRICK - *Chem. Phys.* **214**, 425 (1997)

[33] K.H. ANDERSEN, J. BOSSY, J.C. COOK, O.G. RANDL & J.L. RAGAZZONI - *Phys. Rev. Lett.* **77**, 4043 (1996)

[34] A. HEIDEMANN - *Z. Phys.* **238**, 208 (1970)

[35] T. CHATTERJI & B. FRICK - *Physica B* **276-278**, 252 (2000); T. CHATTERJI & B. FRICK - *Solid State Comm.* **131**, 453 (2004)

[36] T. CHATTERJI & B. FRICK - *Applied Physics a-Materials Science & Processing* **74**, S652 (2002)

[37] P.G DE GENNES, P.A. PINCUS, F. HARTMANN-BOUTRON & J.M. WINTER - *Phys. Rev.* **129**, 1105 (1963)

[38] M. KAISERMAYR, J. COMBET, H. IPSER, H. SCHICKETANZ, B. SEPIOL, & G. VOGL - *Phys. Rev. B* **61**, 12038 (2000)

[39] M. KAISERMAYR, B. SEPIOL, J. COMBET, R. RÜFFER, C. PAPPAS, & G. VOGL - *J. Synchrotron Radiation* **9**, 210 (2002)

[40] L.J. SMITH, D.L. PRICE, Z. CHOWDHURI, J.W. BRADY & M.L. SABOUNGI - *J. Chem. Phys.* **120**, 3527 (2004)

[41] V. ARRIGHI, F. GANAZZOLI, C.H. ZHANG, & S. GAGLIARDI - *Phys. Rev. Lett.* **90**, 058301 (2003)

[42] A. ARBE, J. COLMENERO, F. ALVAREZ, M. MONKENBUSCH, D. RICHTER, B. FARAGO & B. FRICK - *Phys. Rev. E* **67**, 51802 (2003)

[43] W. GÖTZE - *in "Liquids, Freezing and the Glass Transition"*, J.P. HANSEN, D. LEVESQUE, & J. ZINN-JUSTIN eds, North-Holland, Amsterdam, 287 (1991)

[44] J. COLMENERO, A. ALEGRIA, A. ARBE & B. FRICK - *Phys. Rev. Lett.* **69**, 478 (1992); A. ARBE, J. COLMENERO, M. MONKENBUSCH & D. RICHTER - *Phys. Rev. Lett.* **81**, 590 (1998)

[45] M. BÉE, D. DJURADO, J. COMBET, M. TELLING, P. RANNOU, A. PRON, & J.P. TRAVERS - *Physica B* **301**, 49 (2001)

[46] D. DJURADO, M. BÉE, J. COMBET, B. DUFOUR, P. RANNOU, A. PRON, & J.P. TRAVERS - ILL Report 96.RA07T, Institut Laue-Langevin, Grenoble, 62 (2000); D. DJURADO, M. BÉE, M. SNIECHOWSKI, S. HOWELLS, P. RANNOU, A. PRON, J.P. TRAVERS & W. LUZNY - *Phys. Chem. Chem. Phys.*, **7**, 1235 (2005)

[47] G. ZACCAI - *Science* **288**, 1604 (2000); T. BECKER & J.C. SMITH - *Phys. Rev. E* **67**, 021904 (2003)

[48] A. MAGERL, H. ZABEL, B. FRICK & P. LINDNER - *Appl. Phys. Lett.* **74**, 3474 (1999)

[49] M. WOLFF, A. MAGERL, B. FRICK & H. ZABEL - *Chem. Phys.* **288**, 89 (2003); M. WOLFF, A. MAGERL, R. HOCK, B. FRICK & H. ZABEL - *Appl. Phys. A* **74**, 374 (2002); M. WOLFF, A. MAGERL, R. HOCK, B. FRICK & H. ZABEL - *J. Phys.: Condens. Matter* **15**, 337 (2003)

[50] B. FRICK & B. FARAGO - *In "Scattering"*, P.S.R. PIKE ed., Academic Press, San Diego, Vol. 2, 1198 (2001)

[51] OSIRIS, http://www.isis.rl.ac.uk/molecularspectroscopy/osiris/

[52] MARS, http://sinq.web.psi.ch/sinq/instr/mars.html

[53] K. HERWIG (private communication)

[54] H.N. BORDALLO, K.W. HERWIG, & G. ZSIGMOND - *Nucl. Inst. Meth. A* **491**, 216 (2002)

16

NEUTRON INELASTIC SCATTERING AND MOLECULAR MODELLING

M.R. JOHNSON
Institut Laue-Langevin, Grenoble, France
G.J. KEARLEY
Interfacultair Reactor Instituut, Technische Universiteit Delft, Netherlands
H.P. TROMMSDORFF
Laboratoire de Spectrométrie Physique, Université J. Fourier, Grenoble, France

1. INTRODUCTION

From cold neutrons to hot, the range of energies offered by neutron sources covers the whole spectrum of molecular excitations in the electronic ground state. Since spectrometer resolution is a small fraction of the incident neutron energy, excitations down in to the sub μeV range can be measured. The quantum tunnelling of molecular rotors provides the most widespread example of low frequency (sub meV) inelastic excitations [1]. Rotational tunnelling is a wide-amplitude motion connecting rotational configurations of the molecular group, protons typically being exchanged between positions separated by several Å. The shape and amplitude of the potential energy barrier separating stable configurations is probed by the tunnelling motion and the tunnelling probability, which determines the energy of the inelastic excitation, depends exponentially on the barrier. Quantum tunnelling is an extremely sensitive probe of the Van der Waals (VDW), Coulomb and torsional interactions that determine the rotational potential. The value of the probe is seen in the importance of these weak inter and intra-molecular interactions, which determine details of molecular packing in the solid state and conformational flexibility [2] and physical processes which can be thermally activated under ambient conditions, notably for biological molecules [3].

F. Hippert et al. (eds.), Neutron and X-ray Spectroscopy, 529–556.

At higher energies, up to ~ 400 meV, molecular vibrations are measured by inelastic neutron scattering (INS). In contrast to tunnelling dynamics, vibrations are small displacements (~ 0.05 Å) of atoms about equilibrium positions. The lowest energy molecular vibrations tend to be torsional in character and, in the case of molecular rotors, they can often be determined from the same rotational potential that governs tunnelling dynamics. With increasing energy, molecular vibrations become less sensitive to inter-molecular VDW and Coulomb interactions, vibrational modes above ~ 50 meV generally being less dependent on the solid state. An important exception to this rule occurs for hydrogen-bonded solids, in which case the frequencies of the bending and stretching modes of hydrogen-bonded protons can change by up to a factor of two compared to the free complex. Tunnelling excitations are similarly sensitive to molecular and solid state structure. Spectroscopy can be regarded as microscopy where the length scale probed is determined by the interaction range (~ 6 Å) and is therefore complementary to diffraction techniques which report average structures over longer length scales.

Hundreds, if not thousands, of INS measurements of quantum tunnelling and molecular vibrations have been accumulated [4]. Spectral peaks can be likened to diffraction peaks in that they are measurements in a reciprocal space (ω and $\boldsymbol{Q}$) which makes their interpretation indirect. In-depth data analysis is required to extract the time-dependent (t) evolution of a system about its average structure ($\boldsymbol{r}$). However, with such a wealth of spectroscopic data, empirical rules for understanding spectra have been established. But extracting a maximum of information from individual spectra, revealing the subtle effects of the solid state, requires modelling those inter and intra-molecular interactions that determine the potential energy surface (PES) of the molecular system [5].

While experimental techniques have been improving at central neutron facilities, giving better quality data, significant developments have also been made in the field of scientific computing. For computers that have become so powerful that bio-molecules can be simulated on desktop machines, a range of software has evolved allowing scientists to perform different kinds of total energy calculations on molecular systems. These calculations fall into two categories, namely force field and quantum chemistry methods. The former are rapid, parametric methods, which can be applied to large systems (> 1000 atoms) and long timescales (> 1 ns), while the latter determine electronic wavefunctions (or densities) giving higher accuracy for smaller systems.

Total energy calculations can be used in different ways to calculate spectra. Molecular dynamics simulations allow classical particles to evolve in time according to Newton's equations of motion. Since, in this case the neutron scattering function $S(\boldsymbol{Q},\omega)$ is just the Fourier transform of the pair correlation function $G(\boldsymbol{r},t)$, the INS spectra can be calculated directly from the atomic trajectories. In complex systems, the disadvantage of real space becomes apparent in such calculations since the evolution of any atom depends on a number of lattice and molecular vibrations. Any atomic trajectory cannot be associated uniquely with a spectral peak. Molecular mechanics (MM) methods give a more direct understanding of spectra and also allow quantum dynamics to be calculated. The PES is determined by calculating the total energy as atoms are displaced from their equilibrium positions. Diagonalizing the Hamiltonian or dynamical matrix, constructed from these calculations, gives eigenfrequencies corresponding to spectral frequencies and eigenmodes from which spectral intensities can be determined. Each eigenmode, describing the relative displacements of all atoms, can be associated with a spectral peak. The eigenmode can even be visualized in an animation giving an unequivocal, real space representation of the spectral mode.

Extracting precise information from spectra depends on them being accurately reproduced by the numerical modelling simulation. In a molecular mechanics approach, the dynamical model, which determines how the atoms are displaced, must be appropriate. For example, a rigid-rotor, one-dimensional model is normally adequate for analysing the rotational tunnelling of methyl groups, but in certain cases, for coupled rotors, a higher-dimensional PES has to be constructed [6]. In the case of molecular vibrations, all Cartesian co-ordinates of atoms are handled on an equal footing. The calculated PES also depends critically on the total energy calculation. All calculations involve approximations and their effect has to be evaluated. Force fields allow solids to be treated but are they accurate enough for determining potential energy profiles precisely [7]? Accurate ab initio, Hartree-Fock-based calculations have been performed on single molecules for many years [8], but to what extent is the single molecule approximation valid for the solid state?

Testing the simulation ultimately depends on reproducing measured spectra and this requires the spectrum to be calculated from the molecular model. In comparison to optical methods, neutron scattering has the advantage that neutrons with zero charge are not scattered by electrons but simply by nuclei. Eigenmodes of atomic displacements or atomic trajectories are sufficient to calculate the INS spectral profile.

Traditionally, force constants, obtained from empirical force fields or ab initio calculations, have been refined to reproduce spectral profiles, forcing the dynamical model to agree with the measurement. As a consequence, the force constants obtained this way have limited transferability to other systems. Reversing the situation, spectra can be calculated and the following question asked, "is the dynamical model appropriate and is the total energy calculation sufficiently accurate?". In this way INS data can be used to test the relevance of dynamical models and the accuracy of total energy calculations.

Finally, a molecular model is required in any simulation. When attempting to determine a reliable method for calculating INS spectra, the most accurate solid state structure information must be included in the initial structure. For model systems, INS measurements must be performed in parallel with elastic scattering (diffraction) experiments. Numerical modelling of dynamics forges the link with structure that ultimately allows spectroscopy to be used for microscopy when the local solid state structure is unknown.

The remainder of this chapter is presented as follows. Section 2 contains the theoretical framework for developing the relevant dynamical models for quantum tunnelling systems and molecular vibrations and the panoply of total energy calculations, required to calculate PES and force constants, are introduced. In section 3, experimental techniques and instruments for INS from reactor and spallation sources are presented. Instruments most familiar to the authors, at ILL and ISIS, are discussed. Information about equivalent instruments at other sources can be found on web-sites, accessible from the ILL web-site (http://www.ill.fr). The results of numerical modelling are shown in section 4 and the chapter concludes with a discussion (section 5). Note that all measurements and modelling methods discussed here have been applied to molecular crystals at low temperatures, close to 5 K. Since this chapter is a rather concise account, interested readers should seek further details in the cited articles, or contact the authors.

2. THEORY FRAMEWORK FOR TUNNELLING, VIBRATIONS AND TOTAL ENERGY CALCULATIONS

2.1. HAMILTONIANS FOR QUANTUM TUNNELLING

Tunnelling is a purely quantum phenomenon which is related to the description of particles by delocalized wavefunctions. A classical particle of energy E is reflected by a potential energy barrier of amplitude V if $V > E$. In the same conditions a wavefunction leaks through the barrier.

For a plane wave of energy E, incident on a barrier with a width, a, and height, V, the probability of transmission, T is given by [9]

$$T \approx \frac{E}{V} e^{-2(a/\hbar)\sqrt{2m(V-E)}} \qquad \{1\}$$

There are two important points to note in this equation:

1. The probability depends exponentially on the mass of the particle, so tunnelling effects are most commonly observed for light particles, like protons.
2. The probability decreases exponentially with the height and width of the barrier. This is the origin of the practical sensitivity of tunnelling in the context of characterizing the PES.

For a small ensemble of tunnelling protons (< 4) a can be as large as ~ 2 Å. Tunnelling is a large amplitude motion that probes the whole profile of a potential energy barrier, in contrast to molecular vibrations which only probe within a small radius (< 0.1 Å) of the minima of a potential energy surface.

When the tunnelling motion connects equivalent molecular configurations, the potential energy minima corresponding to each configuration are equal and the tunnel effect is coherent; tunnelling persists at zero Kelvin, that is in the absence of external (lattice) excitations. The spectroscopic signature is the splitting, Δ, of vibrational energy levels. Notable examples are the inversion of ammonia [10] and rotational tunnelling of molecular groups (CH_3, NH_3, CH_4, NH_4^+). Otherwise molecular configurations are inequivalent, either because particles are transferred to new, unoccupied sites, for example in hydrogen-bonds [11], or because the reorienting particles are inequivalent, for example in the rotor CH_2D [12]. Potential energy minima are no longer degenerate, the tunnel effect is dissipative, or incoherent, with a rate proportional to Δ^2, and it allows a tunnelling particle to escape from a metastable well. In INS we are concerned with the coherent tunnelling of molecular rotors, and in particular methyl groups in this work.

The simplest model of rotation is one-dimensional: a single rotational co-ordinate, ϕ, describes the orientation of a symmetric triangle of protons and the axis of rotation passes through the central carbon atom, usually along a covalent-bond. The Hamiltonian for the single particle motion (SPM) is

$$H = \frac{\hbar^2}{2I}\frac{\partial^2}{\partial\phi^2} + \sum_{n\geq 1}\frac{V_{3n}}{2}\{1+\cos[3n(\phi+\alpha_{3n})]\} = \frac{\hbar^2}{2I}\frac{\partial^2}{\partial\phi^2} + \sum_n b_n e^{i3n\phi} \qquad \{2\}$$

where $I = 3\,mr^2$ is the moment of inertia for three protons of mass m and radius of rotation r. The potential energy is expressed as a Fourier series, generally limited to $n \leq 2$ in which case experimental data can be required to determine three parameters, V_3, V_6 and the relative phase, $\alpha = \alpha_6 - \alpha_3$. H can be expressed in matrix form, typically using a basis set of free rotor wavefunctions. Diagonalization then leads to a set a of tunnelling and vibrational (torsional) energy levels and wavefunctions. The cos 3ϕ form of the rotational potential is shown in figure 1a along with schematic energy levels for a CH_3 rotor. Each torsional level is split into a singlet and a doublet, with symmetry species A and E, respectively. The exponential decay of the tunnel splitting and the approximately linear increase in torsion frequency as a function of barrier height are shown in figure 1b.

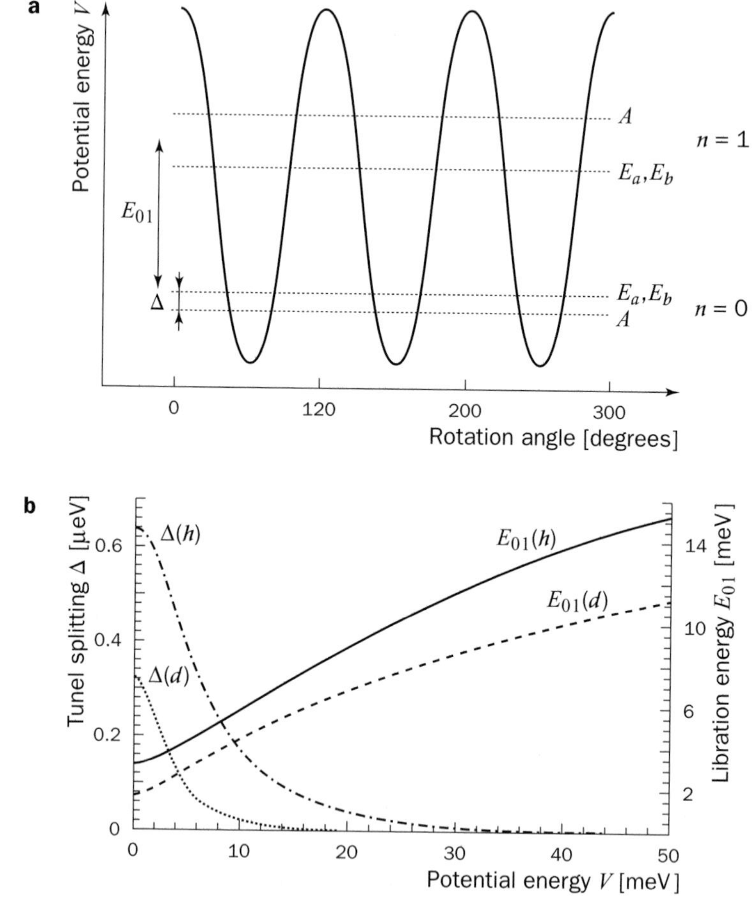

Figure 1 - **a** - Schematic representation of a simple rotational potential and energy levels for a methyl group. **b** - Tunnel splittings and torsional energies for protonated and deuterated rotors as a function of barrier height V.

Since methyl group rotation connects indistinguishable molecular configurations, the total rotor wavefunction, the product of a spatial and a nuclear spin part, must be symmetric. The total wavefunction has A symmetry that imposes the following correlation between spatial (s) and nuclear spin (n) terms: A_sA_n, $E^a{}_sE^b{}_n$, $E^b{}_sE^a{}_n$. The allowed combinations are called the spin symmetry species. They have important consequences for tunnelling spectroscopy since a transition between A and E rotational states entails a "matching" change in the nuclear spin state. Neutrons with nuclear spin therefore constitute the only radiation that can directly induce transitions between tunnelling levels, NMR (Nuclear Magnetic Resonance) depending, for example, on the small dipolar mixing of rotational energy levels. Intensities are calculated according to the spin-dependent double-differential cross-section, elaborated specifically in terms of free rotor basis functions by Matuschek and Hüller [13].

When the rotational potential of a methyl group depends on the orientation of a neighbouring rotor, a coupling potential exists. A network of coupled rotors may extend from a pair to a large cluster. In all cases a set of n one-dimensional rotors and pairwise coupling potentials Vp is considered according to the following Hamiltonian,

$$H = -\sum_{i=1}^{n} \frac{\hbar^2}{2I} \frac{\partial^2}{\partial \phi^2{}_i} + \sum_{i=1}^{n} V_i^s(\phi_i) + \sum_{i,j=1,i<j}^{n} V_i^p(\phi_i, \phi_j) \qquad \{3\}$$

V^s is referred to as the static potential and is independent of the orientation of neighbouring rotors. Resolving the coupled Hamiltonian is tractable by matrix diagonalization for two or three coupled rotors using free rotor basis functions [14,15] and for larger clusters using more appropriate Gaussian pocket states [16]. An eigenstate-by-eigenstate search using the variational principle has been used for the highest dimensional systems [17]. Due to spin symmetry species, a selection rule for INS tunnelling spectroscopy exists by which the spin of the total rotor wavefunction can only change by ±1.

A second type of coupled dynamics has been elaborated for a number of rotational tunnelling systems, namely rotation/translation coupling [18]. Rotation of the CH_3/NH_3 group is accompanied by a small amplitude displacement of the rotor centre-of-mass, effectively the rotation axis (and the C/N atom) precesses during the rotation (figure 2). The Hamiltonian to be resolved then contains kinetic terms for the centre-of-mass and the rotor and the potential term is three dimensional, two additional co-ordinates being required to describe the amplitude (r_c) and the precession (α_c) of the centre-of-mass [19].

$$H = \frac{P_c^2}{2M} + \frac{\hbar^2}{2I} \frac{\partial^2}{\partial \phi^2} + V(r_c, \alpha_c, \phi) \qquad \{4\}$$

The Hamiltonian is solved in terms of products of two-dimensional oscillator functions and one-dimensional rotator functions.

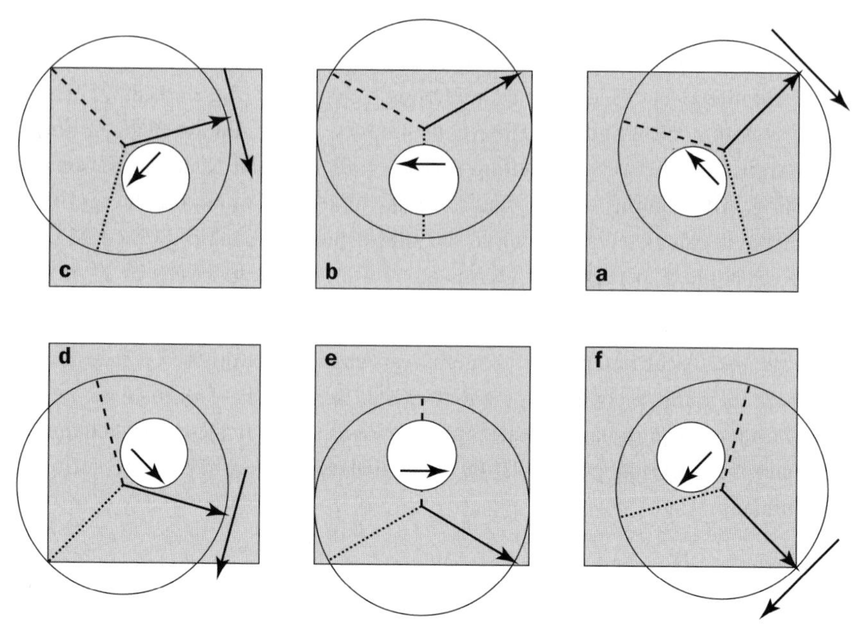

Figure 2 - Schematic representation of rotation-translation coupled dynamics. The centre of mass traces out a complete period in the anti-clockwise direction (**a-f-a**), while the protons undergo a reorientation of $2\pi/3$ in the opposite sense. Quantum mechanically, the central atom of the rotor occupies a larger volume than in SPM and the proton distribution is approximately square.

The above Hamiltonians refer to molecular systems for which the dynamics can be described within a limited number of degrees of freedom. They are only valid at liquid helium temperatures when the lattice motion is quenched. Accordingly, as will be illustrated in measured spectra, tunnelling spectra are rather simple (see figures 2 and 3).

2.2. THE DYNAMICAL MATRIX FOR MOLECULAR VIBRATIONS

In contrast to tunnelling spectroscopy, vibrational spectra for most molecular systems are rather complex, there being $3n - 6$ non-zero modes for a molecule of n atoms. In a crystal with N atoms per unit cell, Bloch waves describe the modes, and $3N$ modes with wavevector $\boldsymbol{k} = 0$, of which 3 translational (acoustic) modes have zero frequency, have to be calculated.

The ground state energy of the solid (at $T = 0$ K) can be expanded as a Taylor series for small amplitude displacements about the minimum energy configuration ($\partial E/\partial r_i = 0$),

$$E = E_0 + \frac{1}{2}\sum_{i,j=1}^{n}\left(\frac{\partial^2 E}{\partial r_i \partial r_j}\right) r_i r_j + \ldots \qquad \{5\}$$

Where higher order terms are negligible for small amplitude oscillations. The dynamical matrix is defined as

$$D_{i,j} = \frac{1}{\sqrt{m_i m_j}}\left(\frac{\partial^2 E}{\partial r_i \partial r_j}\right) = \frac{1}{\sqrt{m_i m_j}} K_{i,j} \qquad \{6\}$$

where $K_{i,j}$ is the force constant matrix. Diagonalizing the dynamical matrix, which amounts to solving Newton's equation of motion for the ensemble of atoms, gives the vibration frequencies (eigenvalues) and normal modes.

For isotropic harmonic oscillators, the spectral intensity of the normal mode u is given by the simple scattering function

$$S(\boldsymbol{Q},n) = \sum_i \left(\frac{\boldsymbol{Q}^2 \cdot \boldsymbol{u}_i^2}{n!}\right)^n e^{-\boldsymbol{Q}^2 \cdot \boldsymbol{u}_i^2} \sigma_i \qquad \{7\}$$

where σ_i is the scattering cross-section of the i'th atom, Q is the momentum transfer given by the spectrometer geometry, $n = 1$ for fundamental excitations and $n > 1$ for overtones. There are no selection rules in INS vibrational spectroscopy and a single neutron can excite overtones and combinations of modes at sum and difference frequencies of the component modes. Calculation of INS spectral profiles are performed with the program CLIMAX [20], originally developed to allow refinement of force constants against measured spectra, or a simpler version of the same program which simply calculates the spectrum for a given set of normal modes [21].

Equations {5-7} allow a description of the molecular vibrations in the harmonic approximation in which the restoring forces acting on atoms is proportional to atomic displacements. A potential energy variation of the type shown in figure 1 is clearly not harmonic, the finite height of the rotational potential causing a progressively more significant reduction in the splitting between successive torsional levels. Due to this effect, the harmonic approximation also overestimates the energy of the fundamental splitting E_{01}. As in low-dimensional tunnelling Hamiltonians, the limitations of the dynamical model, here the harmonic approximation, must be considered when evaluating the agreement between measured and calculated spectra.

2.3. TOTAL ENERGY CALCULATIONS FOR DETERMINING PES AND FORCE CONSTANTS

In principle the dynamical models presented above require the molecular model of the solid state to be optimized so that all forces acting on the atoms are effectively zero. Thereafter the crystal energy is calculated as a function of different perturbations of the stable structure, in which particular atoms are displaced from their equilibrium positions, in order to map out the relevant PES. Dynamical matrices and PES are assumed to be independent of isotopic composition so that calculating spectra for isotopomers amounts to solving equations 1 to 6 for different particle masses.

For the SPM model of methyl group rotation, one rotor in a unit cell is symmetrized (all bond lengths and angles averaged to common values) and then rotated in discrete steps. Fitting the potential energy profile, that is the crystal energy variation with rotor orientation, enables the Fourier coefficients in equation {2} to be obtained and thus the spectroscopic observables. For unit cells with short axes (< 6 Å), a number of unit cells (a supercell) may have to be used so that images of the probe rotor in the periodic model are spatially isolated from the probe itself. In this way the variation in crystal energy is due to a single, decoupled rotor. Methyl-methyl coupling potentials (eq. {3}) are calculated as in the SPM case for a probe rotor as a function of the orientation of a neighbouring rotor and then repeating for different neighbours. Finally for rotation-translation coupling, the crystal energy has to be calculated as a function of the three dynamic variables, centre-of-mass co-ordinates (r, α) and rotor orientation (ϕ).

For molecular vibrations, force constants can be calculated from the energy changes due to the displacements of pairs of atomic Cartesian co-ordinates $r_i r_j$ and K_{ij} is constructed element-by-element. However if the ground state energy is a known function of the atomic co-ordinates the second derivatives can be calculated analytically. In certain cases, like periodic density functional theory (DFT) calculations, only first derivatives are available analytically, so the forces on all atoms are obtained as a function of the displacement of one of the atoms and K_{ij} is constructed line by line.

Empirical force fields provide the computationally quickest way to calculate these energies and forces in the solid state. A force field is a collection of parameters and functional forms that describe the result of the interactions within an ensemble of nuclei and electrons. Bonds are treated as harmonic springs, as are the angle variations between bonds and torsions. In more sophisticated force fields a number of additional intra-molecular terms are included to describe the coupling between different degrees of freedom [22]. Intermolecular interactions are separated in to VDW, Coulomb and hydrogen-bond terms, the latter being treated in a way similar to covalent-bonds. VDW interactions can have several forms, most commonly the Lennard-Jones 12-6 form, which is also the most computationally efficient,

$$E_{LJ12\text{-}6} = D_0\left[\left(\frac{R_0}{R_{ij}}\right)^{12} - 2\left(\frac{R_0}{R_{ij}}\right)^{6}\right] \qquad \{8\}$$

and the exp-6 form which is physically more realistic due to the exponential description of the repulsive term.

$$E_{\text{exp-6}} = D_0\left(\left[\left(\frac{6}{\gamma-6}\right)\mathrm{e}^{\gamma\left[1-(R_0/R_{ij})\right]}\right] - \left[\left(\frac{\gamma}{\gamma-6}\right)\left(\frac{R_0}{R_{ij}}\right)^{6}\right]\right) \qquad \{9\}$$

These two expressions use the same parameters R_0 and D_0. Equation {9} is equivalent to the well-known Buckingham potential $E = a\,\mathrm{e}^{b/R} + cR^{-6}$. Coulomb interactions depend on partial atomic charges q_i, q_j according to the following expression

$$V = \frac{q_i\, q_j}{4\pi\varepsilon_0 R_{ij}} \qquad \{10\}$$

Partial charges can be determined empirically, for example from ionization potentials [23], from quantum chemistry calculations or they can be optimized as part of the set of force field parameters [22]. Examples of force fields are the UNIVERSAL force field [24], which is a simple force field allowing all atoms to be treated , COMPASS [22] which is a sophisticated force field, parameterized from quantum chemistry calculations on molecular fragments, and CHARMM [25], which has been optimized for macromolecules.

Quantum chemistry calculations determine the electronic structure for a given configuration of nuclei by solving the electron-nuclei Schrödinger equation self-consistently. Such calculations are generally based on localized atomic orbitals that are combined in Hartree-Fock wavefunctions that respect electron exchange requirements. Electron correlation effects are treated post Hartree-Fock, typically by perturbation theory. Well-known examples of such quantum chemistry codes are GAUSSIAN [26] and GAMESS [27], which are based on atomic orbitals constructed from Gaussian basis functions. Semi-empirical methods, as embodied in MOPAC [28] for example, offer reasonable accuracy at greater computational speed by ignoring some of the integrals when solving the Schrödinger equation and parameterizing others. GAUSSIAN, GAMESS and MOPAC only allow single molecules or small clusters to be treated, although recent extensions of these methods (CRYSTAL [29] and MOPAC2000 [30]) do allow solids to be treated.

Whereas the aforementioned quantum chemistry codes determine the electronic wavefunction, a computationally efficient way of solving the electronic structure of relatively large systems of molecules is provided by DFT (see ref. [31]). DFT is based on the theorem of Hohenberg and Kohn [32] which demonstrates that the energy of a system of electrons and nuclei is determined by a unique functional of the electron density, and that the minimum of this functional corresponds to the ground state. Kohn and Sham (KS) [33] showed that the problem of a strongly interacting electron gas in the "external" field of a set of nuclei maps onto a single particle moving in an effective non-local potential, so the many electron problem is replaced by a self-consistent, mean-field, one-electron problem. However, the exact form of the exchange-correlation potential as a function of the electron density is unknown for a system of electrons and nuclei and local density (LDA) or gradient corrected (GGA) approximations are used. While DFT describes accurately strong interactions, dispersive interactions are underestimated, causing non-hydrogen-bonded molecular crystal structures to expand significantly from their measured volume. A number of periodic DFT codes have been used and evaluated in the analysis of INS spectra [34], namely VASP [35-38] and CASTEP [39], which use plane-wave basis functions, and SIESTA [40-42] and DMOL3 [43], which use numerical, localized basis sets.

3. EXPERIMENTAL TECHNIQUES FOR MEASURING ROTATIONAL TUNNELLING AND MOLECULAR VIBRATIONS

Tunnel splittings greater than about 0.5 μeV, that is methyl groups with rotational potentials of amplitude less than 50 meV (500 K), are best measured with INS techniques. NMR is used to measure smaller tunnel splittings. This lower limit corresponds to the resolution of the backscattering technique [44]. Larger tunnel splittings (> 20 μeV) are measured using variations on the time-of-flight technique or even triple-axis methods [44]. At a reactor source (ILL) the cold neutron spectrometers typically used to cover this energy range are IN10, IN16 and IN5. At a spallation source, backscattering spectrometers such as IRIS [45] have a slightly lower resolution but a bigger dynamic range and therefore cover most of the energy range of interest.

γ-picoline was one of the first examples of rotational tunnelling measured with neutrons [46] (see figure 3). In fact, the spectrum is almost that of a free rotor, and it was originally interpreted in terms of a weak cos 6ϕ rotational potential. As an illustration of the improvement in instrumentation, the resolution FWHM (Full Width at Half Maximum) of the original triple-axis measurement was about 100 μeV (as in figure 3a). This is to be compared with a number of time-of-flight measurements with resolution as good as ~ 10 μeV [47] (figure 3b), and the most recent spectrum of this compound which was measured on IN10 with a resolution of 2 μeV [48] (figure 3c). The increase in resolution has revealed that the free-rotor peak is split into a number of lines, and as will be shown, spectra of this complexity can only be interpreted with the help of modern computational methods.

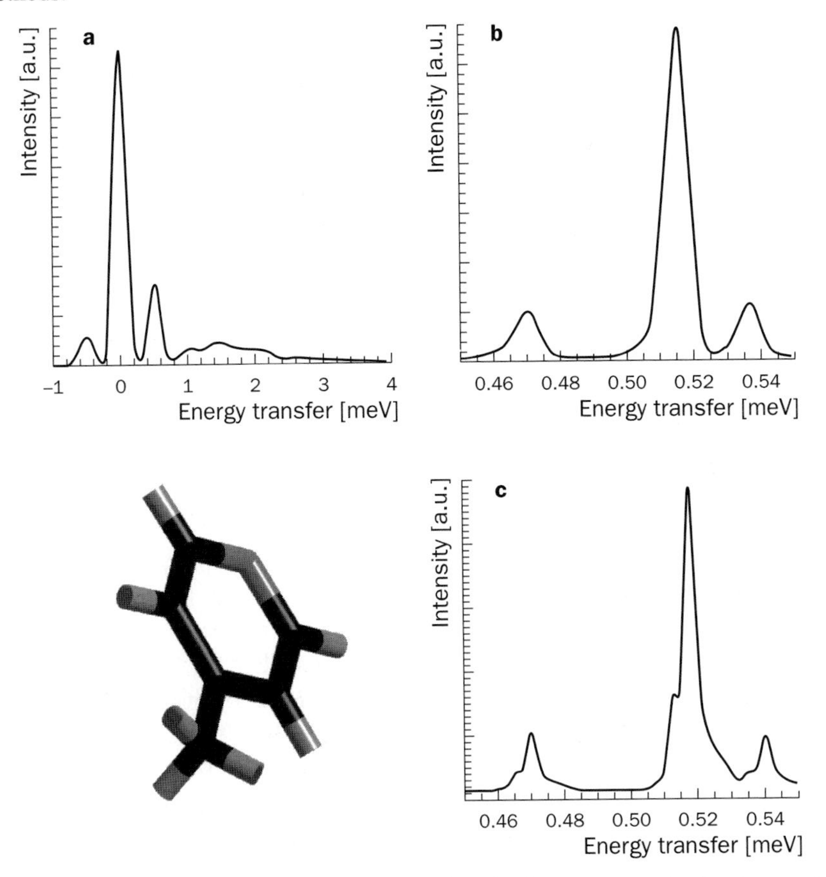

Figure 3 - Rotational tunnelling spectra of -picoline measured with increasing resolution (FWHM) (from ref. **[58]**). **a** - ~ 100 μeV, **b** - ~ 10 μeV on IN5 (ILL) and **c** - ~ 2 μeV on IN10 (ILL).

A simpler, more typical tunnelling spectrum for the methyl group in aspirin (acetyl salicylic acid) [49], is illustrated in figure 4, the symmetric sidebands arising from neutrons which are scattered by upward or downward transitions between the tunnel-split torsional ground-state. Assuming a dominant cos 3ϕ term in the SPM rotational potential, the barrier height V can be determined directly from the tunnel splitting. The importance of additional Fourier terms can be assessed from complementary data, although this must be done with caution. The CD_3 tunnel splitting of 11.2 neV was measured using quadrupole NMR techniques [50] and based on the pair of tunnelling measurements, a set of possible rotational potentials was derived. However, while deuteration changes the moment of inertia by a factor of 2, it can also lead to lattice compression and reduce the quantum delocalization of rotors, which are more difficult effects to evaluate. Torsional excitations (fig. 4) can also be used to characterize the rotational potential, but these are often coupled to other vibrational modes, which goes beyond the scope of the SPM model.

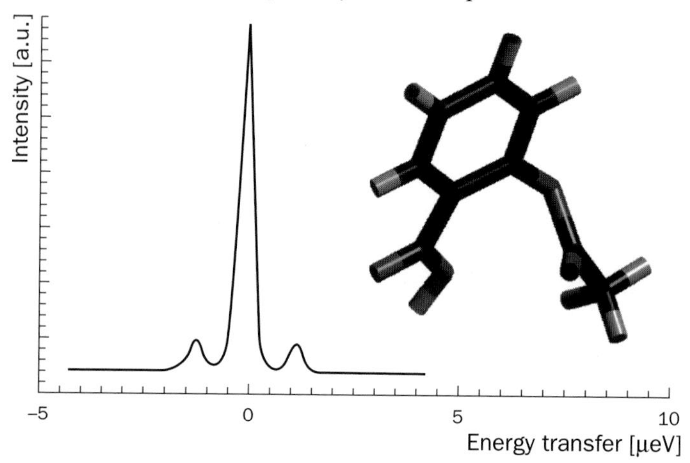

Figure 4 - The simple tunnel spectrum of the crystallographically unique methyl group in aspirin, measured on IN16 (ILL) (from ref. [49]).

For vibrational spectroscopy and larger energy transfers (< 400 meV) higher energy neutrons are required if the spectrum is to be measured in the preferred set-up of neutron energy loss, which allows the sample to be maintained at low temperature and thermal broadening to be minimized. At a reactor source like ILL, the energy range up to ~ 100 meV can be measured with the time-of-flight spectrometer IN4, or the whole energy range can be measured at lower resolution on IN1BeF, a fixed-final-energy, fixed-scattering-geometry variation of the triple-axis spectrometer. At a spallation source, again there is a single spectrometer TOSCA [51] which covers the whole energy range with an energy resolution of about 1% of the incident energy. While TOSCA gives an excellent single-shot overview of the whole spectrum, IN1BeF with its higher flux and monochromator allow measurements to be focussed on modes in particular spectral regions and for smaller samples.

Figure 5 shows a typical vibrational spectrum, measured on TOSCA on a sample of aspirin. Although the molecule is not particularly big (21 atoms), the spectrum is rather congested and initial spectral analysis is limited. As in tunnelling spectroscopy, isotopic substitution is a useful tool to gain insight, changing the effective mass and therefore frequency of modes, but also intensity since isotopes like hydrogen and deuterium have very different scattering cross-sections. Partial deuteration of molecules can be used to hide most of the vibrational modes, highlighting the modes of the remaining protonated part of the molecule. In figure 5, the strongest mode at ~ 20 meV in the lower spectrum (fully protonated molecule) is the methyl group torsion. On deuteration of the rotor (upper spectrum), this band essentially disappears.

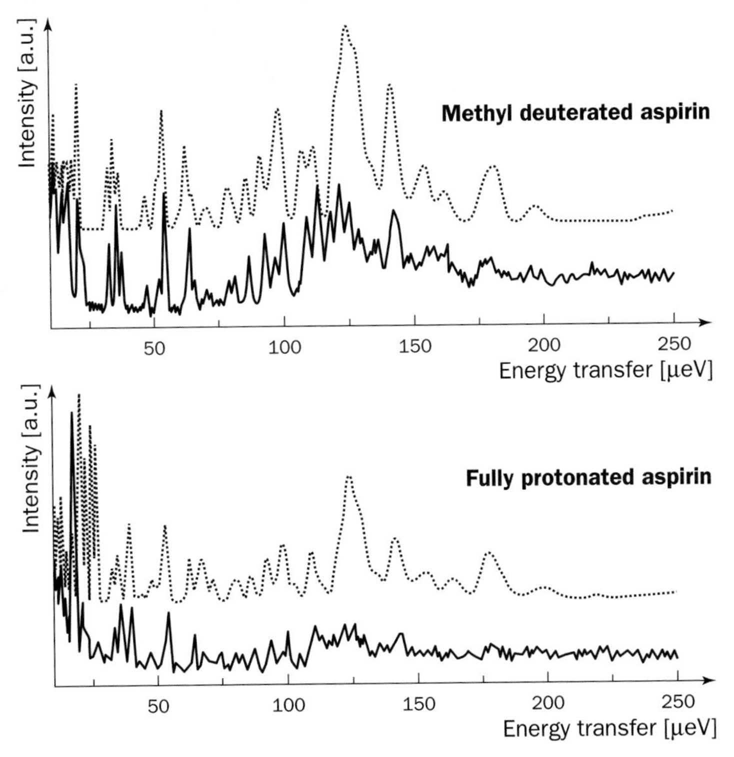

Figure 5 - INS vibrational spectrum of aspirin, measured (lower curve in each frame) on TOSCA (ISIS). The upper curves are calculated.

4. NUMERICAL SIMULATIONS FOR UNDERSTANDING INS SPECTRA

4.1. SPM METHYL GROUP TUNNELLING

Quantum molecular dynamics at low temperature are modelled numerically using the molecular mechanics method outlined in section 2. Since the object of this work is to probe the weak inter-atomic interactions that determine the rotational potential, the crystal structures upon which the calculations are based must be reliable (preferably measured at 5 K), and the rotational model must be appropriate. In addition, if the rotational model is simple, then the information about inter-atomic interactions will be more clearly interpreted. The SPM model fulfils these criteria and numerical modelling should therefore confirm its validity. Finally, a large database of structural and spectroscopic data should be available in order to eliminate accidental agreement between calculation and measurement. To this end, many neutron diffraction experiments, on powder (mainly) and single crystal samples, have been performed in the course of this work, so that about 30 molecular crystals are now characterized, corresponding to about 50 independent tunnelling measurements (for details of samples see ref. [6]).

Initial calculations were performed with a non-optimized, simple force field (UNIVERSAL [24]) and empirical method for assigning partial charges [23]. The results of the numerical modelling are quantified by the correlation (*corr*) between the two data sets (1 for identical data sets) and the root-mean-square difference of the logarithms of the measured and calculated tunnel frequencies ($rms = 0$ for identical data sets). For these force field-based calculations, $corr = 0.57$ and $rms = 2.5$. The correlation between measured and calculated values is poor and there is a systematic overestimate of the rotational barrier height V and therefore an underestimate of the tunnel frequency by up to several orders of magnitude. The precision available in such a basic force field is not adequate for accurately modelling rotational potentials.

Improvements to the total energy calculation were then sought [52-54]. These include:

1. replacing the Lennard-Jones 12-6 form for the VDW interactions with the more realistic exp-6 form,
2. calculating the intra-molecular rotational potential from semi-empirical quantum chemistry methods (MOPAC – AM1 Hamiltonian) and using the charges from these calculations to calculate inter-molecular Coulomb interactions, and

3. repeating 2 using ab initio methods, namely GAMESS with Hartree-Foc6-31G* basis functions and Moller-Plesset second order perturbation corrections for electron correlation effects.

At the expense of increased computational time, a systematic improvement in accuracy is obtained; *corr* = 0.68 and *rms* = 1.1 for 2, while *corr* = 0.94 and *rms* = 0.82 for 3. With method 3 calculated tunnel frequencies are typically within a factor of ~ 3 of the measured values. The corresponding rotational potentials have an accuracy of the order of 90%. In all cases, crystal structures were not optimized prior to calculating the rotational potentials since the pure force field calculations tend to distort significantly the measured structures and geometry optimization is rather difficult to implement in such a simple hybrid approach.

The computational technique of choice for this type of calculations is solid state quantum chemistry using DFT since all interactions, intra and inter-molecular are treated on an equal footing. Geometry optimization is now possible and performed accurately, the only constraint being that the unit cell dimensions are fixed, to compensate for the lack of dispersive interactions in DFT. While dispersive interactions are important for determining the unit cell shape and size, force field and semi-empirical cluster calculations show that the interaction range for determining rotational potentials is of the order of 6 Å. The correlation curve obtained with the plane-wave DFT code CASTEP is shown in figure 6 (*corr* = 0.96 and *rms* = 0.52). The accuracy is better than the hybrid method 3 above, for about the same computational cost, and the method is more complete since it includes optimization of the atomic co-ordinates in the crystal structure.

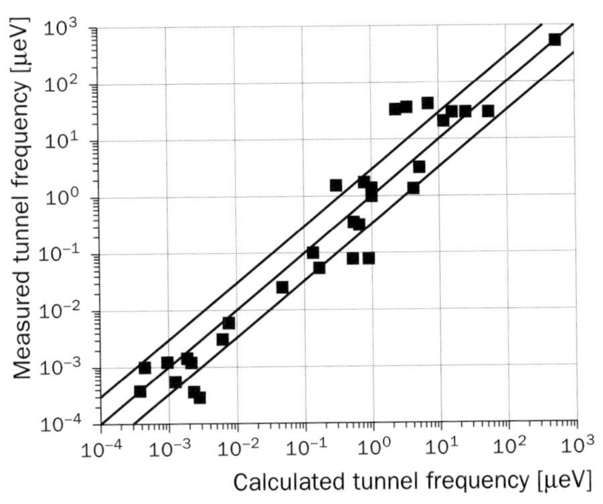

Figure 6 - Comparison of measured tunnel splittings and those calculated using periodic DFT methods (CASTEP).

The solid state quantum chemistry method is particularly important for methyl groups in hydrogen-bonded solids like paracetamol [55]. The quantum tunnelling of the rotors are sensitive to the electronic structure of the hydrogen-bond, but the hybrid methods as used here don't allow accurate determination of the perturbation of the molecular electronic structure by neighbouring molecules.

4.2. MULTI-DIMENSIONAL TUNNELLING DYNAMICS OF METHYL GROUPS

While the SPM-based analysis, described above, generally gives an accurate value for the amplitude of the rotational potential, it does not allow the structure in the tunnelling spectra to be reproduced if the rotational model is too simple [56,57]. In the case of γ-picoline, a multiplet of tunnelling peaks is observed for one crystallographically unique methyl group (fig. 3). A model of four coupled methyl groups was developed which reproduces spectral frequencies *and* intensities [58] (fig. 3c), the latter as a function of the populations of rotational levels which can be controlled by quenching from different temperatures to 2 K, due to slow nuclear spin conversion at this temperature [59,60].

Lithium acetate has also been considered as an example of coupled methyl groups [14,15]. In the low temperature crystal structure, acetate groups are coaxial with methyl groups face-to-face and they are surrounded by molecules of water of hydration [61] (see figure 7). There is however, only one crystallographically unique methyl group. In the tunnelling spectrum, three strong peaks are observed and the model of coupled pairs was elaborated, with values of 4 meV for the amplitude of the SPM potential and 10 meV for the coupling potential, obtained from the spectroscopic data [14]. Numerical modelling however reveals values of 7.2 and 3.9 meV respectively [19], that is the model of coupled pairs is not consistent with known inter-atomic interactions and the measured crystal structure.

In lithium acetate, water molecules make the strongest contribution to the rotational potential of the methyl group and modelling shows that the major cause of the structure in the tunnel spectrum is the departure of the methyl dynamics from uniaxial-rotation [19]. Mapping-out the PES as described in section II for rotation/translation coupling, allows Schrödinger's equation {4} to be constructed and solved. Excellent agreement is obtained between measured and calculated spectra. Indeed a new weak peak was predicted and then measured at 500 μeV. However, the clearest confirmation of the dynamical model for lithium acetate comes from a single crystal diffraction experiment at 15 K since the rotation-translation motion gives rise to a square of nuclear density rather than three crescents on a circle. Figure 7 shows the agreement between nuclear densities calculated from the eigenfunctions of the Hamiltonian and measured nuclear densities, extracted by maximum entropy methods [62]. The calculation has been performed for the ground state, whereas, at 15 K, excited rotational and vibrational states begin to be populated.

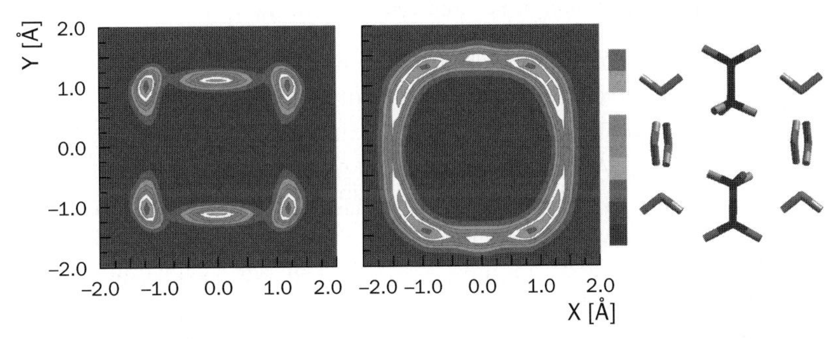

Figure 7 - Contour plots of calculated (left) and measured (right) nuclear density for the methyl group in lithium acetate dihydrate (from ref. **[19]**). Part of the crystal structure, co-axial acetate groups and water molecules, is shown on the far right.

4.3. *Vibrational Spectroscopy of Molecular Crystals*

Numerical modelling of tunnelling data has shown that the most appropriate total energy calculations for the solid state are those based on periodic DFT methods. This technique has to be evaluated against traditional approaches based on optimized force fields [7] and single molecule [8], or cluster [63], Hartree-Fock ab initio methods. While both of these methods require force constants to be refined, either because they are explicitly parameters or because they fail to take into account certain (inter-molecular) interactions, the periodic DFT should allow accurate spectra to be calculated from measured crystal structures, without any refinement whatsoever. To what extent this is true is illustrated in figure 4, for protonated and methyl deuterated aspirin. The upper curve in each pair is calculated using VASP and both calculated curves are derived from the same force constant matrix. Generally, the agreement between calculated and measured spectra is excellent, small changes between protonated and partially deuterated spectra being well reproduced. However two vibrational bands are worthy of comment. For the fully protonated molecule, the measured band at 20 meV is the methyl group torsion. The frequency of this mode can be derived from the tunnelling measurement (fig. 4) and the assignment is confirmed by the absence of the peak in methyl deuterated spectrum. The frequency of the mode is however over-estimated in the calculation. VASP correctly reproduces the SPM rotational potential for aspirin so the error is due to the harmonic approximation. In the absence of the libration mode in the methyl deuterated spectrum, the external mode spectrum (21 modes calculated at the Brillouin zone centre) is clearly seen and is similar to the measured phonon density of states. The second vibrational band of note is discussed in the context of benzoic acid (BA) below.

BA is a molecule closely related to aspirin, the acetyl group on the latter molecule being replaced by a hydrogen atom. In the solid state, the structural similarity persists in that both molecules form dimers, pairs of hydrogen-bonds connecting carboxylic acid groups. Accordingly the INS vibrational spectra of BA (fig. 8) bear a certain resemblance to those of aspirin. BA has been studied as a model system for proton transfer along hydrogen-bonds, these quantum proton dynamics having been measured directly by quasi-elastic neutron scattering and NMR [64]. Proton transfer in this system is an example of a chemical reaction, as indicated by the change in hybridization of the oxygen atoms. The modification by proton transfer of the electronic structure of the molecular skeleton results in coupling between small amplitude vibrational modes of the skeleton and large amplitude proton transfer. Since no normal modes follow precisely the tunnelling trajectory, a number of modes promote proton transfer while others hinder it. A description of the coupling of small and large amplitude motions goes beyond the harmonic approximation and semi-classical theory is used to characterize the promoting or hindering role of different modes [65]. However the vibrational modes themselves must be characterized and this has been done with VASP [66], as for aspirin, for several isotopomers (H6, D6, D5H) including a fully protonated oxygen-18 sample. Again the general agreement between calculation and measurement is excellent, even the small changes due to the ^{16}O-^{18}O substitution are quantitatively reproduced. Whereas partial deuteration in aspirin effectively enabled badly calculated modes to be hidden, the ring-deuterated, acid protonated molecule (D5H) has been measured to highlight those modes associated with the inter-molecular hydrogen-bonds. The doublet in the measured spectrum at about 115 meV is the in-phase and out-of-phase, out-of-plane wagging modes of the pair of O-H-bonds. By mapping out the potential energy variation for these modes, the over-estimated spectral frequency is again seen to be due to the harmonic approximation. Fitting a fourth order polynomial to the potential energy profile and applying the corresponding quartic perturbation to the harmonic oscillator accounts for about one-third of the over-estimate. The remaining error is due to the optimized hydrogen-bonds being shorter than the measured bonds. This error is clearly seen in the spectra of fully protonated benzoic acid and aspirin, but it is more difficult to analyze due to the spectral congestion. According to theoretical models based on formic acid [65], the principal modes that promote proton transfer are the intra-dimer stretch and the COOH rocking mode. These modes have calculated frequencies of 16.0 meV and 16.9 meV, which are in excellent agreement with the low temperature activation energy of 15.5 meV, determined from QENS (Quasi-Elastic Neutron Scattering) measurements.

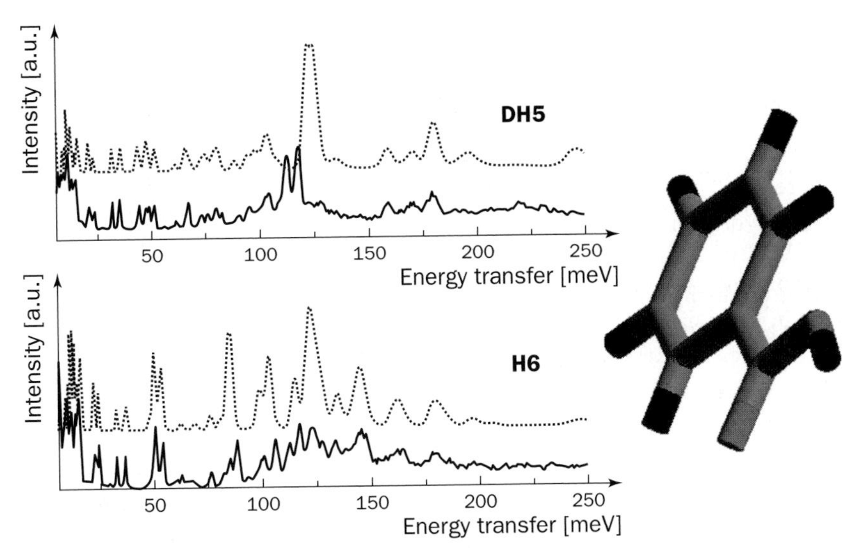

Figure 8 - INS vibrational spectrum of benzoic acid, measured (lower curve in each frame) on TOSCA (ISIS) (from ref. **[66]**). The upper curves are calculated. The benzoic acid molecule is shown on the right.

Possibly one of the most important applications of periodic DFT calculations in vibrational spectroscopy is extended hydrogen-bond networks, which cannot easily be modelled by small clusters. N-methyl acetamide (NMA) is one of the smallest molecules containing the peptide linkage (CONH) and is therefore a model system to be fully characterized before going on to study polypeptides. Hydrogen-bonding between peptide groups gives rise to linear chains of molecules in the solid state. Optical spectroscopy has been used to investigate NMA as a function of temperature and modes around ~ 200 meV were found to show anomalous temperature dependence [67]. INS measurements were then performed at low temperature on a complete set of isotopomers and the data was analyzed by refining force constants for a molecule in a simplified molecular environment. Controversially the N-H stretching mode was assigned precisely to the spectral region at ~200 meV on the basis that the hydrogen-bonds are strong and significant electron density is transferred from the covalent-bond to the hydrogen bond [68]. Subsequently accurate crystal structure determinations have shown the N-H-bond to have a typical length of 0.95 Å [69]. Vibrational spectra calculated from these structures using DMOL3 are shown in figure 9 [70]. Without any parameter refinement, agreement between measured and calculated spectra is good. The most prominent modes in the methyl-deuterated spectrum are associated with the peptide proton. In particular the mode at ~ 100 meV is the out-of-plane wagging mode and it is slightly overestimated as in the case of BA. This mode is particularly sensitive to the hydrogen-bond network, occurring at 55 meV in an isolated molecule calculation. The controversial N-H stretch mode is calculated at ~ 400 meV in this analysis. Very recently, similar calculations have been performed on measured structures showing short, strong N–H···O hydrogen-bonds (about 0.2 Å shorter than the hydrogen-bond in NMA). N-H-bond stretching modes are predicted at 250 meV in the harmonic approximation and as low as 175 meV when the precise shape of the hydrogen-bond potential is taken in to account, although these have yet to be confirmed by INS.

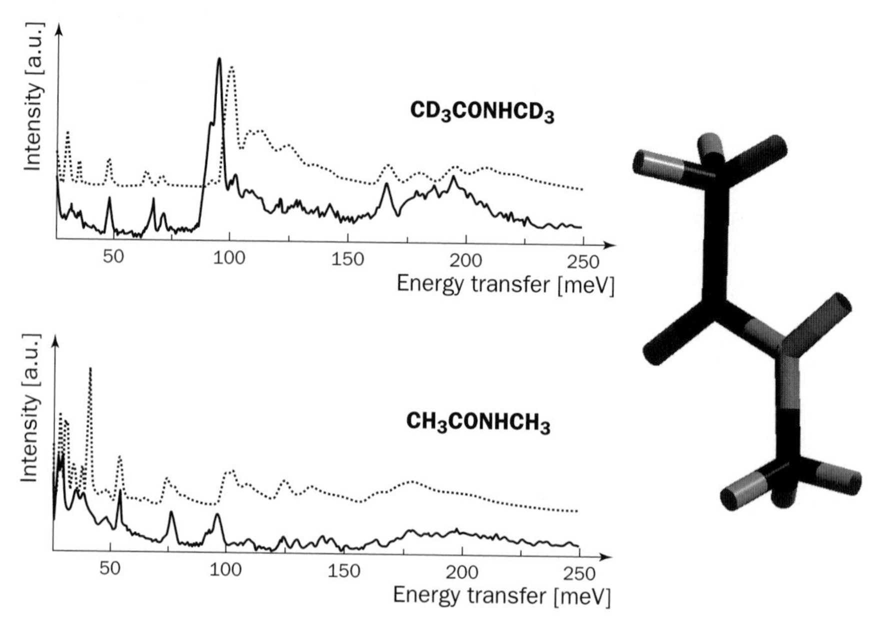

Figure 9 - INS vibrational spectrum of n-methyl acetamide, measured (lower curve in each frame) on TOSCA (ISIS) (from ref. **[70]**). The upper curves are calculated. The n-methyl acetamide molecule is shown on the right.

5. *DISCUSSION*

For quantitative understanding of INS spectra of proton tunnelling and molecular vibrations, periodic DFT gives the most appropriate total energy calculation for deriving the relevant PES and force constants. Analysis of a set of simple tunnelling spectra shows that the lack of dispersive interactions in DFT method is not a significant limitation since the interaction range determining the rotational potential is of the order of 6 Å. Applying periodic DFT to more complex tunnelling problems and molecular vibrations enables quantitative insight to be obtained from eigenfrequencies and eigenmodes, derived from the multidimensional Hamiltonian or dynamical matrix. Since no parameters have been refined when applying DFT calculations, the method of applying simple dynamical models and total energy calculations of this kind to molecular dynamics problems is generally applicable. In addition to the large body of analyzed tunnelling data presented here, we have successfully calculated INS spectra for about 20 molecular crystals [71]].

There are however limitations to the approach employed here. The harmonic approximation is a problem for torsional modes and for the PES for hydrogen-bonds. The most appropriate solution, alluded to here, entails mapping-out more of the PES and correcting the harmonic force constant prior to obtaining the normal modes by matrix diagonalization.

In order to calculate phonon modes, or simply the phonon density of states, the dynamical matrix has to be generalized for all points in reciprocal space. Although this has been done for non-molecular systems [72], molecular crystals tend to show a number of unphysical negative frequency modes. While phonons are propagated by the same interactions that govern rotational tunnelling dynamics, with the same interaction range, the delocalized nature of the excitations makes them sensitive to the shape and size of the unit cell. In molecular crystals, these parameters are determined in part by dispersive interactions that are absent from DFT methods. To what extent this problem is alleviated in hydrogen-bonded solids is currently being investigated. Finally accurate phonon calculations require DFT calculations to be performed on supercells that cover the full, spatial interaction range and are therefore of the order of 2000 Å^3 in volume. Cells of this size are tractable for DFT calculations, especially with a linear scaling code like SIESTA [40], but the difficulty also arises in constructing supercells this small from unit cell building blocks. Phonon calculations also challenge the precision of the numerical methods since the potential energy variations of interest are smaller than those determining torsional dynamics and can be as small as a few meV.

Parameterized force fields give analytical expressions for total energy calculations and consequently the numerical precision of these methods is much higher. In addition they can be applied to very large systems of atoms and molecules such as proteins in aqueous solutions. However force fields need to be optimized for specific applications. The analysis of SPM tunnelling data shows the accuracy that should be obtained, for a given dynamical model, while higher accuracy may be indicative of parameter refinement concealing other errors.

Hybrid methods also have an important role to play. In this work, applied in a simplistic way, they have given respectable results in the analysis of a body of tunnelling data. The essence of the approach as used here is to treat the most sensitive interactions with a quantum chemistry calculation and the less critical, longer-range interactions with a force field-based calculation. This approach is precisely that required to describe accurately active sites in large bio-molecules, in which the overall structure and its modulation also play an important role in regulating physical processes. For this reason, codes that combine quantum chemistry and force field calculations, so as to allow geometry optimizations and molecular dynamics, have recently become available.

Equipped with these numerical modelling tools, it is now possible to accurately interpret INS spectra from known solid state structures. The methods should now be applied in earnest to real systems of interest. Tunnelling studies are often made for probe molecules in porous media [73], with a view to characterizing free volumes and internal surfaces. By constructing plausible molecular models and comparing calculated tunnel frequencies or distributions with measured spectra, quantitative structural information can be extracted. The same approach should be applied in vibrational spectroscopy, where modes with specific frequencies can determine molecular conformation and the presence of hydrogen-bonding in structural units. For example Raman (optical vibrational) spectroscopy is already used to locate cancerous cells in situ [74]. A quantitative understanding of these spectra may allow more precise cell identification.

ACKNOWLEDGEMENTS

We are grateful to a number of thesis students, post-docs and collaborators for their contributions to this work, namely Beatrice Nicolai, Marcus Neumann, Marie Plazanet, Petra Schiebel and Nobuhiro Fukushima. The authors are also grateful to the Centre Grenoblois de Calcul Vectoriel of the Commissariat à l'Energie Atomique (CEA-Grenoble) for the use of computing resources.

REFERENCES

[1] W. PRESS - *"Single-Particle Rotations in Molecular Crystals", Springer Tracts in Modern Physics*, vol. 92 (1981)

[2] P. VERWER & F.J.J. LEUSEN - *Rev. Comp. Chem.*, K.B. LIPKOWITZ & D.B. BOYD Eds, Wiley, 327 (1998)

[3] P.G. WOLYNES & W.A. EATON - *Physics World* **39** (1999)

[4] M. PRAGER & A. HEIDEMANN - *Chem. Rev.* **97**, 2933 (1997)

[5] M.R. JOHNSON, G.J. KEARLEY & J. ECKERT Eds – *"Condensed phase structure and dynamics: a combined neutron scattering and numerical modelling approach", Chem. Phys.* **261**, 1 (2000)

[6] M.R. JOHNSON & G.J. KEARLEY - *Ann. Rev. Phys. Chem.* **51**, 297 (2000)

[7] R.L. HAYWARD, H.D. MIDDENDORF, U. WANDERLINGH & J.C. SMITH - *J. Chem. Phys.* **102**, 5525 (1995)

[8] G.J. KEARLEY, J. TOMKINSON, A. NAVARRO, J.J. LÓPEZ GONZÁLEZ & M. FERNÁNDEZ GÓMEZ - *Chem. Phys.* **216**, 323 (1997)

[9] R.M. EISBERG - *Fundamentals of Modern Physics*, John Wiley and Sons, 232 (1961)

[10] F. HUND - *Z. Physik* **43**, 805 (1927)

[11] J.L. SKINNER & H.P. TROMMSDORFF - *J. Chem. Phys.* **89**, 897 (1988)

[12] M.R. JOHNSON, K. ORTH, J. FRIEDRICH & H.P. TROMMSDORFF - *J. Chem. Phys.* **105**, 9762 (1996)

[13] M. MATUSCHEK & A. HÜLLER - *Can. J. Chem.* **66**, 495 (1988)

[14] S. CLOUGH, A. HEIDEMAN, A.J. HORSEWILL & M.N.J. PALEY - *Z. Phys. B: Condens. Matter* **55**, 1 (1984)

[15] W. HÄUSLER & A. HÜLLER - *Z. Phys. B: Condens. Matter* **59**, 177 (1985)

[16] M. TIMANN, G. VOLL & W. HÄUSLER - *J. Chem. Phys.* **100**, 8307 (1994)

[17] A. HULLER & D. KROLL - *J. Chem. Phys.* **63**, 4495 (1975)

[18] P. SCHIEBEL, A. HOSER, W. PRANDL, G. HEGER & P. SCHWEIß - *J. Physique I* **3**, 982 (1993)

[19] P. SCHIEBEL, G.J. KEARLEY & M.R. JOHNSON - *J. Chem. Phys.* **108**, 2375 (1998)

[20] G.J. KEARLEY - *Nucl. Instr. Meth. A* **354**, 53 (1995)

[21] http://www.isis.rl.ac.uk/molecularSpectroscopy/aclimax/

[22] H. SUN - *J. Chem. Phys.* **102**, 7338 (1998)

[23] A.K. RAPPE & W.A. GODDARD - *J. Chem. Phys.* **95**, 3358 (1991)

[24] A.K. Rappe, C.J. Casewit, K.S. Colwell, W.A. Goddard & W.M. Skiff - *J. Am. Chem. Soc.* **114**, 10024 (1992)

[25] http://mir.harvard.edu/

[26] M.J. Frisch, G.W. Trucks & H.B. Schlegel - *GAUSSIAN 98*, Revision A.5, Gaussian Inc., Pittsburgh, PA (1998)

[27] http://www.dl.ac.uk/TCSC/Software/GAMESS/main.html

[28] J.J.P. Stewart - *MOPAC: A Generalised Molecular Orbital Package*, Frank J. Seiler Research Lab., U.S. Air Force Academy, Colorado

[29] http://www.cse.clrc.ac.uk/Activity/CRYSTAL

[30] http://www.fqspl.com.pl/mopac2000/overview2.htm

[31] M.C. Payne, M.P. Teter, D.C. Allan, T.A. Arias & J.D. Joannopoulos - *Rev. Mod. Phys.* **64**, 1045 (1992)

[32] P. Hohenberg & W. Kohn - *Phys. Rev. B* **868**, 136 (1964)

[33] W. Kohn & L.J. Sham - *Phys. Rev.* **140**, 1133 (1965)

[34] M. Plazanet, M.R. Johnson, J.D. Gale, T. Yildirim, G.J. Kearley, M.T. Fernández-Díaz, D. Sánchez-Portal, E. Artacho, J.M. Soler, P. Ordejón , A. Garcia & H.P. Trommsdorff - *Chem. Phys.* **261**, 189 (2000)

[35] http://cms.mpi.univie.ac.at/vasp

[36] G. Kresse & J. Hafner - *Phys. Rev. B* **47**, 558 (1993) ; *Phys. Rev. B* **49**, 14251 (1994)

[37] G. Kresse & J. Furthmüller - *Comput. Math. Sci.* **6**, 15 (1996)

[38] G. Kresse & J. Furthmüller - *Phys. Rev. B* **54**, 11169 (1996)

[39] http://www.accelrys.com/cerius2/castep.html

[40] http://www.uam.es/departamentos/ciencias/fismateriac/siesta

[41] D. Sanchez-Portal, P. Ordejon, E. Artacho & J.M. Soler - *Int. J. Quant. Chem.* **65**, 453 (1997)

[42] E. Artacho, D. Sanchez-Portal, P. Ordejon, A. Garcia & J.M. Soler - *Phys. Stat. Sol. B* **215**, 809 (1999)

[43] http://www.accelrys.com/cerius2/dmol3.html

[44] http://www.ill.fr/YellowBook/

[45] http://www.isis.rl.ac.uk/molecularSpectroscopy/iris/index.htm

[46] B. Alefel, A. Kollmar & B.A. Dasannacharya - *J. Chem. Phys.* **63**, 4495 (1975)

[47] F. Fillaux, C.J. Carlile & G.J. Kearley - *Phys. Rev. B* **44**, 12280 (1991)

[48] M.A. Neumann, M. Plazanet & M.R. Johnson - *A.I.P. Conference Proceedings: "Neutrons and Numerical Methods" workshop*, M.R. Johnson, G.J. Kearley & H.G. Buttner Eds, **479**, 212 (1999)

[49] M.R. JOHNSON, B. FRICK & H.P. TROMMSDORFF - *Chem. Phys. Lett.* **258**, 187 (1996)

[50] A. DETKEN, P. FOCKE, H. ZIMMERMAN, U. HAEBERLIN, Z. OLEJNICZAK & Z.T. LALOWICZ - *Zeitschrift Naturforschung* **50 a**, 95 (1995)

[51] http://www.isis.rl.ac.uk/molecularSpectroscopy/tosca/index.htm

[52] M.R. JOHNSON, M. NEUMANN & B. NICOLAI - *Chem. Phys.* **215**, 343 (1997)

[53] M. NEUMANN & M.R. JOHNSON - *J. Chem. Phys.* **107**, 1725 (1997)

[54] B. NICOLAI, G.J. KEARLEY & M.R. JOHNSON - *J. Chem. Phys.* **109**, 9062 (1998)

[55] M.R. JOHNSON, M. PRAGER, H. GRIMM, M.A. NEUMANN, G.J. KEARLEY & C.C. WILSON - *Chem. Phys.* **244**, 49 (1999)

[56] M.A. NEUMANN, M.R. JOHNSON, A. AIBOUT & A.J. HORSEWILL - *Chem. Phys.* **229**, 245 (1998)

[57] M.A. NEUMANN, M.R. JOHNSON, P.G. RADAELLI, H.P. TROMMSDORFF & S.F. PARKER - *J. Chem. Phys.* **110**, 516 (1999)

[58] M.A. NEUMANN, M. PLAZANET, M.R. JOHNSON & H.P. TROMMSDORFF - *J. Chem. Phys.* **120**, 85 (2004)

[59] W. HÄUSLER - *Z. Phys. B* **81**, 265 (1990)

[60] A. WÜRGER - *Z. Phys. B* **81**, 273 (1990)

[61] G.J. KEARLEY, B. NICOLAI, P.G. RADAELLI & F. FILLAUX - *J. Solid State Chem.* **126**, 184 (1996)

[62] R. PAPOULAR, W. PRANDL & P. SCHIEBEL - *Maximum Entropy and Bayesian Methods*, Seattle 1991, C.R. SMITH, G.J. ERICKSON & P. NEUDORFER Eds, Kluwer, Dordrecht, 359 (1991)

[63] B.S. HUDSON - *J. Phys. Chem. A* **105**, 3949 (2001)

[64] M.A. NEUMANN, D.F. BROUGHAM, C.J. MCGLOIN, M.R. JOHNSON, A.J. HORSEWILL & H.P. TROMMSDORFF - *J. Chem. Phys.* **109**, 7300 (1998)

[65] V.A. BENDERSKII, E.V. VETOSHIN, L. VON LAUE & H.P. TROMMSDORFF - *Chem. Phys.* **219**, 143 (1997)

[66] M. PLAZANET, N. FUKUSHIMA, M.R. JOHNSON, A.J. HORSEWILL & H.P. TROMMSDORFF - *J. Chem. Phys.* **115**, 3241 (2001)

[67] G. ARAKI, K. SUZUKI, H. NAKAYAMA & K. ISHII - *Phys. Rev. B* **43**, 12662 (1991)

[68] F. FILLAUX, J.P. FONTAINE, M.H. BARON, G.J. KEARLEY & J. TOMKINSON - *Chem. Phys.* **176**, 249 (1993)

[69] J. ECKERT, M. BARTHES, W.T. KLOOSTER, A. ALBINATI, R. AZNAR & T.F. KOETZLE - *J. Phys. Chem. B* **105**, 19 (2001)

[70] G.J. KEARLEY, M.R. JOHNSON, M. PLAZANET & E. SUARD - *J. Chem. Phys.* **115**, 2614 (2001)

[71] M. PLAZANET, N. FUKUSHIMA & M.R. JOHNSON - *Chem. Phys.* **280**, 53 (2002)

[72] http://wolf.ifj.edu.pl/phonon/

[73] D. BALSZUNAT, B. ASMUSSEN, M. MÜLLER, W. PRESS, W. LANGEL, G. CODDENS, M. FERRAND & H. BÜTTNER - *Physica B* **226**, 185 (1996)

[74] T.C. BAKKER SCHUT, M.J.H. WITJES, J.C.M. STERENBORG, O.C. SPEELMANN, J.L.N. ROODENBURG, E.T. MARPLE, H.A. BRUINING & G.J. PUPPELS - *Analytical Chemistry* **72**, 6010 (2000)

INDEX

Numbers in boldface type are the first page of the relevant chapter.

A AAS..see anomalous scattering
absorbing atom..3,67,116,133
absorption..3,67,103,240,377,511
  edge..3,90,108-125,132,149,229,240,277
  self..46,144
activation energy..448,514,549
adsorption..319,333
Ag..156,191,207,348
ammonia..517,533
amorphous systems..173,249,414,479
analyzer
  crystal..176,377,389,463,480,488
  electron..190,198
Anderson model..47,232
angle resolved photoelectron spectroscopy (ARPES)..189
angular
  acceptance..176,190,403
  divergence..176,301,394
anharmonic effects..374,411
anti-absorption band..335
antiferromagnetic..108,291,417,439,476
anomalous
  dispersion..245
  scattering..239
  scattering anisotropy (AAS)..260
  differences..246
  site analysis..252
ARPES..see angle resolved photoelectron spectroscopy
Auger decay..9,13,44,144,277,338
autocorrelation..307,450

B backscattering
  process 136,159
  spectrosocopy 377,475,483
band mapping 193,234
BCS .... see superconductor
Bijvoet differences 246
Born approximation 310
Bragg reflection 244,387,485
brownian motion 308
brightness 276,321,331
Brillouin zone (BZ) 170,196,374,408

C C60 .... see fullerene
capillary waves 315
catalysis 163
centro-symmetry 253,347
charge density waves (CDW) 220
charge transfer 45-58,348
chopper 379,393,461,467,489-508
chromatic aberration 275
cluster 52,78-96
coherence 139,208,301,328
collective dynamics 169
collective excitation 179,206,373,412
collimation 176,388,395,485
commensurate 224
1D compound 220
2D compound 209
Compton scattering 478
conduction band 48,104,191,232
constant-Q scan 386,413
contrast 11
  chemical 247,277
  magnetic 281
  method 247
  secondary 280
  work function 276
coordination number 138,162
core
  hole 5-15,87,104,119,144,277
  level 3,71,105,132,151,193,249,277

correlated electrons 14,189,219,230
correlation function 172,302,366,440,522,531
Coster-Kronig process 22,229
Coulomb interaction 18,69,83,113,529
count rate 180,312,380,394
cross-section
  absorption 3-10,57,71-95,103,119,229,244
  differential 67,170-174,365,372,415,512,518
crystal field 31-63,69,115,374
Cu 127,148,163,230,244,333
cuprate 216,228

D DAFS .... see Diffraction Anomalous Fine Structure
camped harmonic oscillator 180,376,412
DANES .... see Diffraction Anomalous Near-Edge Structure
Darwin reflectivity curve 486
Debye-Waller factor 138,253,371,408,487,515
deep inelastic neutron scattering 478
deflection parameter 329,478
deflector 489,500
density
  correlation 169,305,366
  fluctuation 169,304
  functional theory (DFT) 11,48,538
density of states (DOS)
  vibrational 410,548
  electronic 3,49,71,151,191,215,282
desorption 346
dichroism 3,103,260,282,290
difference frequency generation (DFG) 327,339
diffraction anomalous fine structure (DAFS) 239
diffraction anomalous near-edge structure (DANES) 258
diffusion 278,311,440,519
dilute alloys 148
dipole approximation 6,68,105,133,151,194,325,416
Dirac-Hara potential 85
disordered systems 170,297,439,521
dispersive excitations 182,374,397,444
domain walls 286
DOS .... see density of states
dynamic structure factor 169,365,371,408,446,483,531
dynamic susceptibility 370,386,408

E EDAFS .... see extended DAFS
EDC .... see energy distribution curve
EELS .... see electron energy loss spectroscopy
EEM .... see electron emission microscope
elastic constants 186
electrochemistry 162,325,340
electron
  emission microscope (EEM) 273
  energy loss spectroscopy (EELS) 10
  electron interaction 14,83,212
  phonon interaction 212,221
  secondary 10,272
  yield 10,278
element-selectivity 3,244,272
energy
  conservation 170,362,392
  distribution curve (EDC) 190
  gap 215-229,417
  resolution 175,193,262,372,401,414,435,475,485-522
  transfer 171,363,383,400,413,435,461,471,483,522,541
EPR spectrum 32
escape depth (electron) 232,288
EXAFS .... see extended X-ray absorption spectroscopy
exchange interaction 18,87,121,289,417,450,474
excitation annihilation/creation 179,364,408
extended X-ray absorption spectroscopy (EXAFS) 3,125,131,250
extended-DAFS (EDAFS) 251

F Fano profile 31,229
fast sound 179
Fe 93,113,182,246,256,286
FEFF method 97,142,257
FEL .... see free electron laser
Fermi
  golden rule 6,71,105,133,194
  level 3,85,145,191-234
  liquid 85,205-215
ferromagnetic 63,118,282,416,439,453
fluctuation-dissipation theorem 309,416
fluorescence 9,144,255,277

flux
  neutron 390,468,500-511
  photon 174,322
force field 530
form factor
  atomic 172,250,414
  magnetic 362,416
Fourier transform interferometer 324
free electron laser (FEL) 319
free energy 308
friction 337
Friedel pairs 245
fullerene 220,348

G giant magnetoresistance (GMR) 126,155
grazing incidence 288,315,333
glassy dynamics 447,522
graphite pyrolytic 280,388,500
group theory 31,33

H harmonic contamination 143,389,421,503
Hedin-Lundqvist potential 87
Heusler analyzer/monochromator 389,391
high pressure 182
Hund's rules 19,28
Huygens 303
hydrogen bond 179,550
hydrodynamics 180
hyperfine splitting 518

I imaging techniques 63,271
inelastic neutron scattering (INS) 186,361,383,457,483,529
inelastic X-ray scattering (IXS) 169,412
infrared
  source 321
  spectroscopy 331
interface 315,319
inverse photoelectron spectroscopy (IPES) 191
IXS .... see inelastic X-ray scattering

J Jahn-Teller distortion 38,92,261
j-j coupling 20

K Kondo effect 219,230
Kramers-Kronig relation 241

L Langevin equations 308
lattice distortion 154
Larmor precession 428
LEED .... see low energy electron diffraction
LEEM .... see low energy electron microscopy
Lennard-Jones potential 539
lens objectives 273
lifetime 8,31,86,152,205
ligand field multiplet theory (LFMT) 53,96
liquid 179,313,370,414,447,523
local structure 136,240
low energy electron diffraction (LEED) 338
low energy electron microscopy (LEEM) 272
Luttinger liquid 227

M MAD .... see multiwavelength anomalous diffraction
Magnetic
- domains 281,291,439
- EXAFS 124
- scattering 239,372,415,438,449

magnon 374,416,472,419
manganite 210,453
MDC .... see momentum distribution curve
metal-insulator transition 224
methyl group 516,531-547
microspectroscopy 278,323
molecular
- crystal 540,547
- dynamics simulation 179,512,530
- interactions 529,539
- modelling 529
- rotor 529
- vibration 540

momentum
- conservation 136,195,362,394,518
- distribution curve (MDC) 208
- photoelectron 85,134,197
- resolution 176,189,400
- transfer 169,304,363,378,384,392,457,484

monochromator 143,176,388-403,421,463,487-504
monolayer........ 220,289,325,346
mosaicity 395,500
Mott insulator........ 227
muffin-tin potential 78
multi-arm analyzer/detector........ 405
multilayer 126,249
multiple
    scattering 142,300,375,512
    scattering theory (MST) 67
multiplet structure 3,118
multiwavelength anomalous diffraction (MAD) 245

N near-edge structure .... see X-ray absorption near edge structure
Nernst potential........ 340
neutron
    filter 390,458,463,503
    guides........ 386,464,503
    reactor........ 378,489,494,540
    spin echo (NSE) 376,422,427,484
    spin polarization .... see polarized neutrons
Non-linear surface spectroscopy........ 325
NSE ....see neutron spin echo

O orbital momentum 14,67,115

P pair correlation function 172,366,371
Peierls transition........ 220,249
permalloy 278
PES........ see photoelectron spectroscopy
phase shift (EXAFS) 137
phonon........ 179,374,408,480
photoelectron........ 67,87,108,132,142,190
    emission microscopy (PEEM)........ 11,271
    spectroscopy (PES) 58,189
photoemission 190,275
pinhole........ 178,300
plasmon 87
polarized
    light........ 59,106,230,272
    neutrons 366,389,420,428,510

polymers ... 186,314,444,515,522
potential energy surface ... 320,530

Q quadrupole transition ... 4,34,68,126
quasielastic scattering ... 179,374,427,435,466,483,514,549
quasiparticle ... 206

R Racah parameter ... 19,36
Raman scattering ... 10,182
rare earths ... 28,60,108,121
refraction ... 240
resolution
  ellipsoid ... 395
  function ... 180,394,468,499
  spatial ... 272
resonant
  photoemission (RESPES) ... 229
resonant
  X-ray scattering ... 6,53,247
rotational potential ... 529
roton ... 422,517
Rowland circle ... 176,402
Russell-Saunders coupling (RS-coupling) ... 20

S scanning electron microscopy (SEM) ... 272
  with spin polarization analysis (SEMPA) ... 272,288
scattering
  coherent ... 367,420,437,512
  incoherent ... 311,367,420,437,512
  length ... 173,240,365,408,446,464,487
  Thomson ... 173,240
selection rules ... 8,22,78,106,229,282
SEXAFS ... see surface EXAFS
SFG ... see sum frequency generation
signal to noise ... 243,274,313,331,380,405,504
Slater-Condon parameter ... 18-52
small angle neutron scattering (SANS) ... 444
small angle X-ray scattering (SAXS) ... 157,249,306
soft X-rays ... 9,104,190,240,271
sound velocity ... 182
speckle ... 292

spectral function 191,203
spectromicroscopy 278
spherical-neutron polarimetry 391
spin
  density waves 216
  flipper 428
  glass 439
  orbit coupling 11-57,68,104-121
  polarized electrons 104,282
  transition 95
  wave .... see magnon
stretching mode 334
sum frequency generation (SFG) 326,339
sum rules 114
superconductor 40,48,149,217-221,247,258,453
surface 319
surface EXAFS (SEXAFS) 145

T Tanabe-Sugano diagram 36
TAS .... see three-axis specroscopy
TAS multiplexing 403
thermodynamic fluctuations 309
three-axis spectroscopy (TAS) 178,378,383,457
time-of-flight spectroscopy (TOF) 377,391,427,457,509
transition metals 21,108,282
triple-axis spectroscopy .... see three-axis spectroscopy
tunnelling 516,532,540

V vacuum ultra violet (VUV) 12,273,324
valence 10,160,249
valence band 10,48,228
Van der Waals interaction 448,529
vibrational spectroscopy 319,547
voltammetry 161,352

X X – α potential 85
XAFS .... see X-ray absorption fine structure
XANES .... see X-ray absorption near edge structure
XAS .... see X-ray absorption spectroscopy
XIFS .... see X-ray intensity fluctuation spectroscopy
XMCD .... see X-ray magnetic circular dichroism

XMLD .... see X-ray magnetic circular dichroism
XNLD .... see X-ray natural linear dichroism
X-PEEM .... see photoelectron emission microscopy
X-ray
  absorption fine structure (XAFS) 6,136,240
  absorption near-edge structure (XANES) 3,67,153,258
  absorption spectroscopy (XAS) 3,67,103,131
  intensity fluctuation spectroscopy (XIFS) 297
  magnetic circular dichroism (XMCD) 4,59,70,103,282
  magnetic linear dichroism (XMLD) 59,290
  natural linear dichroism (XNLD) 94